During the 1970s and 1980s, the technique of mutation breeding made a substantial, although often poorly appreciated, contribution to crop improvement, leading to an increased understanding of the genetics of plant breeding as well as to the development of new crop varieties. Now, as the techniques of molecular biology become more widely adopted by plant breeders, this comprehensive summary sets mutation breeding in a contemporary context and relates it to other breeding methods, including those most recently developed. The book opens with a general introduction to plant breeding and a review of the development of mutation breeding, including consideration of the strengths and weaknesses of the technique. Chapters covering the underlying theory of the technique are followed by sections which consider more practical aspects such as *in vitro* methods, techniques used for seed propagated crops and techniques used for vegetatively propagated crops. Case studies and examples are included throughout and a list of 1400 references provides a valuable resource for those new to the field.

Mutation Breeding
Theory and Practical Applications

Mutation

Breeding

Theory and Practical Applications

A.M. van Harten
Principal Lecturer, Department of Plant Breeding,
Wageningen Agricultural University, The Netherlands

CAMBRIDGE UNIVERSITY PRESS
Cambridge, New York, Melbourne, Madrid, Cape Town, Singapore, São Paulo

Cambridge University Press
The Edinburgh Building, Cambridge CB2 8RU, UK

Published in the United States of America by Cambridge University Press, New York

www.cambridge.org
Information on this title: www.cambridge.org/9780521470742

First published 1998
This digitally printed version 2007

A catalogue record for this publication is available from the British Library

Library of Congress Cataloguing in Publication data

Harten, A. M. van.
    Mutation breeding : theory and practical applications / A.M. van
    Harten
          p.   cm.
    Includes bibliographical references (p.).
    ISBN 0 521 47074 9 (hardbound)
    1. Plant mutation breeding.   I. Title.
    SB123.H336   1998
    631.5'23–dc21   97-28369 CIP

ISBN 978-0-521-47074-2 hardback
ISBN 978-0-521-03682-5 paperback

To Alex Micke, former head Plant Breeding and
Genetics Section of the Joint FAO/IAEA Division
of Nuclear Techniques in Food and Agriculture,
International Atomic Energy Agency, Vienna, Austria.

Contents

Preface

In 1990 the Joint FAO/IAEA Division of Nuclear Techniques in Food and Agriculture in Vienna, Austria, organized a symposium to assess the results obtained so far with induced mutations for plant breeding and to discuss persisting problems. During the concluding panel meeting, Prof. Dr Alexander Micke, at the time Head of the Plant Breeding and Genetics Section of the Joint Division, stated that, during the past 25 years, the practical application of plant mutation breeding had made good progress, as was shown, for instance, by 1300 mutant cultivars that had been derived from induced mutations (Micke, 1991a). But Micke also remarked that our understanding of the mutation process in plant breeding had not improved much and that, in recent years, research on mutagenesis had dwindled considerably. In addition, some years later, Cassells & Jones (1995), who overviewed a conference on genetic manipulation in plant breeding, organized by EUCARPIA – the European organization of professional plant breeders – in 1994 in Cork, Ireland, concluded that induced mutagenesis was widely used in developing countries, but is 'currently under-represented' in European breeding programmes.

An explanation for the decreasing interest in mutation breeding, at least in most 'developed' countries, may be that during the past two decades attention has become more and more directed towards studying the possibilities offered to plant breeding by various new molecular technologies. It appears that for the next few years the interest of plant breeders – predominantly – will continue to be focussed on these new techniques, despite the fact that it has become clear that the anticipated 'major break-through' in procedures for breeding new cultivars is not as easy to achieve and not as fast as has been anticipated by many.

As a result of these developments mutation breeding seems to have lost part of its previous attraction for young researchers. It is even not inconceivable that mutation breeding, as a discrete branch of plant breeding, may sink into oblivion and that, as a consequence,

much valuable knowledge on this topic built up throughout the years, will be lost.

This book is an attempt to relate mutation breeding to contemporary plant breeding, to discuss present applications and practical achievements of the method and to point out its virtues and weaknesses.

The initial set-up was to start from my own lecture notes that have been used – with constant updating – during a period of more than 25 years in annual short courses on mutation breeding for advanced students, majoring in plant breeding at Wageningen Agricultural University. With these notes as a backbone, it was the intention that each section of the book, as much as possible, should be illustrated by examples covering various breeding goals for a wide range of crops and many references should be added.

During the previous years of work on this job, I discovered that the previously described approach was rather ambitious for a single person. In retrospect, it appears that a certain amount of 'Dutch courage' was required to finish this (self-imposed) task. Questions that kept coming up all the time were: who are the potential readers; what would they hope to find in the book; and what knowledge do they have already of genetics, agronomy and plant breeding? Gradually, it became clear to me that it would be impossible to suit all possible groups of readers equally well. In any case, I decided that the book had to give an outline of the history and the development of mutation breeding during the past century, that methods and techniques should be described without becoming too 'technical' and that the contents should be presented in such a way that practical breeders are stimulated to compare the merits of various breeding methods, including mutation breeding, between which they can choose for specific cases.

The result of many hours of writing and much deliberation lies before the reader. I have to leave it to them to judge whether the efforts have been worthwhile.

In Chapter 1 a general outline is given of the field of plant breeding, followed by a short introduction to mutations in higher plants and to mutation breeding. Frequently, reference is made to subjects that are treated in later chapters. Chapter 2 covers the history of mutation breeding. In Chapter 3, different systems of classifying mutations and their effects are discussed, illustrated with many examples and references. Chapter 4 contains a survey of the most important types of irradiation and chemical mutagens and, briefly, their mode of action. Various methods of mutagenic treatment are presented and, in addition, a short description is given of the most important terms used in this context. The occurrence of spontaneous mutations *in vitro* (nowadays often called 'somaclonal variation') and the induction of mutations *in vitro* by means of purposeful mutagenic treatments are the main subjects treated in Chapter 5. Results obtained so far and further prospects of both approaches are further discussed and compared. Finally, in Chapters 6 and 7 procedures for mutation breeding for seed propagated and vegetatively propagated crops, respectively, are outlined and specific problems which may be met are discussed in detail. The effectiveness of various methods is assessed and different situations are described. Both chapters are completed with a number of case studies, representing for instance mutation breeding for particular traits or for a specific group of crops.

Many colleagues have contributed to this project. My thanks are due to all of them. Some of them have to be mentioned in particular. First of all, a – if not **the** – most important and indispensable support came from Alex Micke, who spent much time going critically through all chapters more than once and who made many valuable suggestions for improvement. This book is dedicated to him. The final choice of subjects to be discussed and the omission of other topics, the mistakes and incorrect interpretations that, of course, still will be found in this book – sometimes against the advice of this experienced specialist on 'mutation management' – are all my own responsibility.

Prof. Dr Katarina Borojevic from Novi Sad in the former Yugoslavia, accompanied by her husband, Prof. Dr Slavko Borojevic, stayed on a senior fellowship from Wageningen Agricultural University for slightly more than six months at our department in 1993–4. She helped in particular with collecting material for the chapter on the history of mutation breeding, prepared drafts for several 'boxes' and commented on many other sections of the draft chapters. Katarina, thank you for all your help. Dr Udda Lundqvist, previously from – what is called now – the Svalöf Weibull Breeding Company in Sweden, also very much helped me by providing useful details about the Scandinavian mutation breeding work and about the great Prof. Åke Gustafsson, and by correcting the concerning sections of this book. My colleague Dr Gijsbert van Marrewijk suggested many text improvements about plant breeding topics, in particular in the first two chapters, and also advised on many linguistic improvements. The help of Dr Klaas Puite of the Centre for Plant Breeding and Reproduction Research (CPRO-DLO) of the Dutch Ministry of Agriculture, Nature Conservation and Fisheries, who made useful comments on the first part of Chapter 4 concerning sources of radiation and radiation treatments for mutation induction, and of Dr Jan de Jong, also from the CPRO-DLO, on the section about chrysanthemum (Chapter 7) is also greatfully acknowledged here. Prof. Dr J. Grunewaldt, director of the Bundesanstalt für Züchtingsforschung, Ahrensburg, Germany, and Mrs Ir. Tineke Schavemaker of our Department suggested some improvements for Chapter 5. My colleague Dr Johan Dourleijn advised on the formulas used in Chapter 6, and how to explain them in the text.

Mrs Letty Dijker-Lefers of the Department of Plant Breeding of Wageningen University, my home base, in fact taught me how to use the computer and corrected without any complaint the many silly mistakes of an inexperienced lap-top user. Mrs Annie Schouten, representing the Central University Library at our department, was of enormous help in my hunt after hundreds of (often difficult to access) publications and reprints from everywhere. I am grateful to Mr P. Kostense of Wageningen University, who helped with the 'basic design' of the jacket of this book.

The reassuring presence, competence and interest shown throughout the years that I have been working on this book by Dr Maria Murphy, commissioning editor of Cambridge University Press, should be gratefully acknowledged as well. Many thanks also go to the copy-editor Mrs Beverley Lawrence, whose kind and patient help with editing and improving on the English has been indispensable. During the time that we exchanged e-mails and faxes, almost on a daily basis, she became something of a 'friend of the family'. Thank you Beverley.

Of course, I have to apologize (again) to my wife Ineke for the many lonely hours and I wish to thank her for her understanding.

It is my sincere hope that this book will help to establish, for the years to come, the – relatively modest but nevertheless very valuable – role of mutation breeding and to stimulate plant breeders to seriously consider the various options of this relatively simple but sometimes quite effective approach.

A.M. van Harten, Wageningen, The Netherlands

1   General introduction

1.1.   The beginning of mutation breeding

A **mutation** can be briefly defined as a sudden heritable change in the DNA of a living cell, not caused by the common phenomena of genetic segregation or genetic recombination. Reliable reports about the occurrence of mutations in higher plants trace back to as early as the year 1590 but **mutation breeding**, the purposeful application of mutations in plant breeding, has only a history of about one century.

Early contributions to our present understanding of mutations and a number of useful suggestions towards a further exploration of mutation breeding around 1900, make the work of the Dutch scientist Hugo de Vries (1848–1935) a good starting point for a study on what, so far, has been achieved in this field. De Vries, an already internationally renowned professor of botany in Amsterdam, was the first of three scientists (Correns, von Tschermak and de Vries) who, independently, in 1900, rediscovered Mendel's laws of inheritance (see for instance de Vries, 1900). Based on the evolution theory of Charles Darwin (*The Origin of Species*, 1859), de Vries had published in 1889 his intracellular 'pangenesis' theory (in 1909 abbreviated to 'gene' theory by the Danish scientist W. Johannsen). 'Pangenes' were considered the physical carriers of the different traits of an adult individual; these carriers (or hereditary units), according to de Vries, must be present in the nuclei of all cells of an organism. It is clear that this view represented an early statement about present-day **genes**.

'Mutations' in the sense of de Vries
In *Die Mutationstheorie*, published in two volumes (in German; see Fig. 1.1), de Vries (1901; 1903) presented an integrated concept concerning the occurrence of sudden, shock-like changes ('leaps') of existing traits, which lead to the origin of new species and varieties. In this concept, which was based also on the aforementioned 'pangenesis' theory, de Vries partly used results from his own experimental work with the genus *Oenothera*, in particular *Oenothera lamarckiana* (= evening primrose), in which species de Vries, following self-pollination, observed many aberrant types which he called 'mutants'. According to de Vries's theory, the parent type itself remained unchanged and repeatedly might give rise to such new forms, an opinion which was opposite to the prevailing idea that species and varieties slowly and continuously change into new types. The notion 'mutation', used by de Vries to indicate the phenomenon of sudden genetic changes as a major cause for evolution, quickly became firmly established and, despite the fact that his own observations did not add essentially to the proposed theory, de Vries became one of the most celebrated scientists of his time.

The decision to work on *Oenothera* was, only to a certain extent, a fortunate one. The aberrations observed by him, for instance, in practically all cases did not produce reliable Mendelian segregation ratios. De Vries indeed was able to prove that sudden changes did occur but, later on, it was shown that most of de Vries's extensive experimental work on aberrant types of *Oenothera lamarckiana* did not refer to **gene mutations in the strict sense** (i.e. to mutations in single genes), but – as has been described on many occasions and, recently, for instance by Harte (1994) – to other effects that, nowadays, most often would be defined as **mutations in a wide sense** (e.g. including polyploidy, aneuploidy, etc.). Moreover, several of de Vries's ideas on mutations and evolution are not accepted anymore these days. Early 'Mendelians', including de Vries, for instance, rejected continuous variation as a possible source of evolution because this variation, apparently, did not segregate according to Mendel's laws and, therefore, in their opinion could not be 'real' genetic variation on which to base an evolutionary theory. In later years, the existence of spontaneous mutations became generally accepted, but it was shown that most mutations did not produce

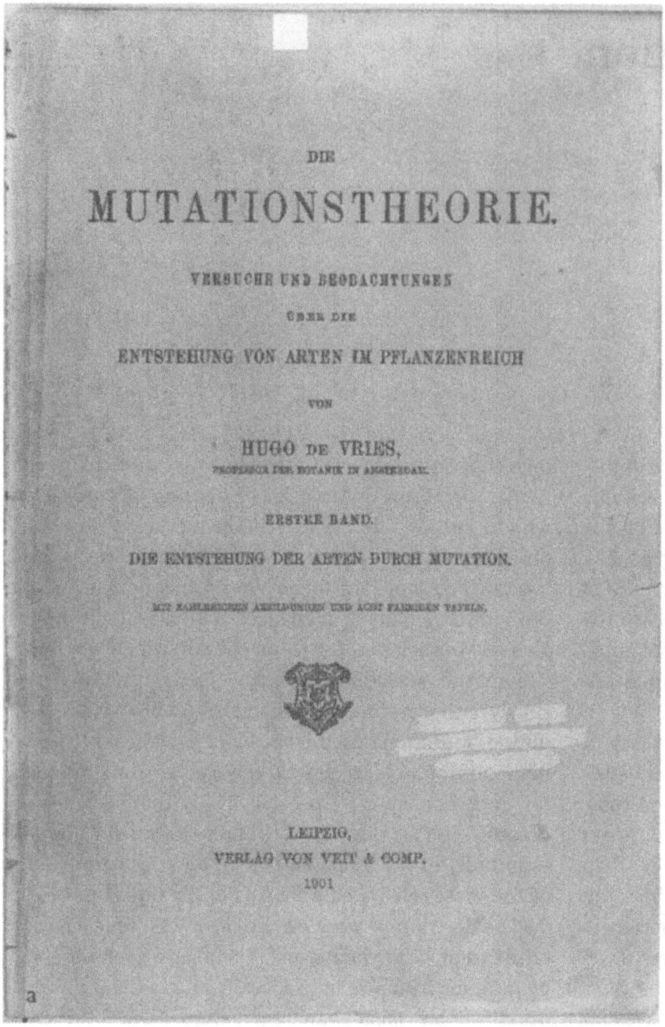

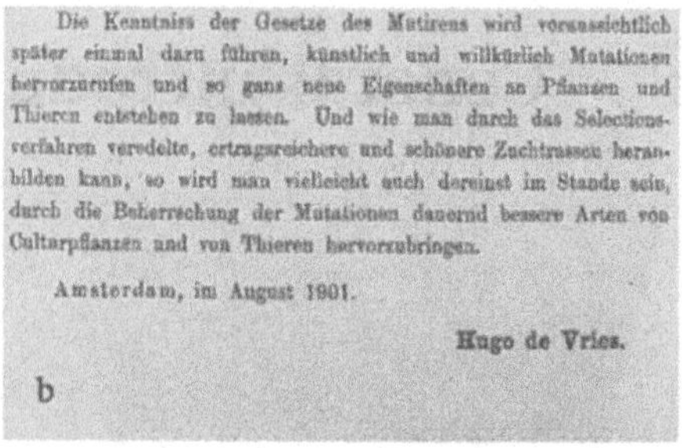

Die Kenntniss der Gesetze des Mutirens wird voraussichtlich später einmal dazu führen, künstlich und willkürlich Mutationen hervorzurufen und so ganz neue Eigenschaften an Pflanzen und Thieren entstehen zu lassen. Und wie man durch das Selections-verfahren veredelte, ertragsreichere und schönere Zuchtrassen heran-bilden kann, so wird man vielleicht auch dereinst im Stande sein, durch die Beherrschung der Mutationen dauernd bessere Arten von Culturpflanzen und von Thieren hervorzubringen.

Amsterdam, im August 1901.

Hugo de Vries.

b

c

Figure 1.1a. Frontispiece of *Die Mutationstheorie* by Hugo de Vries.

Figure 1.1b. Part of the introduction to *Die Mutationstheorie*.

Figure 1.1c. Picture of Hugo de Vries with signature (around 1920). (Picture of H. de Vries: courtesy Hugo de Vries-Laboratory, University of Amsterdam.)

significant phenotypic effects. Moreover, the **major mutations** or sudden, discontinuous 'jumps' that were hypothesized by de Vries and others to lead suddenly and completely to new species did not occur. Evolution was found to go in small steps and, moreover, gradually the insight grew that selection referred more to whole phenotypes than to single genes and that genetic recom-bination is the direct source of genetic variation. Nevertheless, despite the – to a certain extent – wrong interpretation of his own results by de Vries, the notion of 'mutation' to indicate the phenomenon of sudden genetic changes as a major cause for evolution remains firmly established.

Stubbe (1963, in German; or 1972) has mentioned the essential points of the Mutation Theory by de Vries, which are summarized here:

1. New elementary species appear suddenly without intermediate steps.
2. New elementary species usually are constant, right from their origin.
3. Most new types, in their traits, correspond exactly to elementary species and not to real (botanical) varieties.
4. Elementary species usually appear at the same time (or at least during the same period) in a large number of individuals.

5. New characters (or traits) exhibit no obvious relationship to the individual variability.
6. Mutations which lead to new elementary species occur undirectionally. Alterations affect all organs and occur in all directions.
7. Mutations occur periodically.

Stubbe (loc.cit.) added that de Vries studied mutations in nature, both by collecting mutants from the locality of the original stock, as well as by collecting seeds from that spot. Sowing these seeds under favourable conditions enabled him to investigate also mutants that, under natural conditions, would have been lost because of their lower fitness.

De Vries's ideas about mutation breeding and artificially induced mutations

In the introduction to the first volume of *Die Mutationstheorie*, de Vries (1901) expressed the opinion that by understanding the mutation process, it probably might become possible to artificially induce new mutations 'on call' and to obtain new traits in plants and animals. He further predicted at various occasions, for instance in his book *Species and Varieties: Their Origin by Mutation* (de Vries, 1905, p. 688) that, in this way, it probably might become possible in the future to produce new varieties of plants and animals on a permanent base. De Vries stated literally;

We may search for mutable plants in nature, or we may hope for species to become mutable by artifical methods. The first promises to yield results most quickly, but the scope of the second is much greater and it may yield results of far more importance. Indeed, if it once should become possible to bring plants to mutate at our will and perhaps even in arbitrarily chosen directions, there is no limit to the power we may finally hope to gain over nature.

It turned out that in 1904, at the Station for Experimental Evolution of the Carnegie Institution of Washington at Cold Spring Harbor, USA, de Vries had already referred to the new types of radiation (X-rays, γ-rays) discovered by W.K. von Röntgen in 1895, H. Becquerel in 1896 and by P. and M. Curie in 1897, and suggested that they might be applied to artificially induce mutations (Blakeslee, 1935; 1936). On the occasion of the death of Hugo de Vries an obituary was written in *Science* by Blakeslee (1935) in which de Vries is quoted as follows: '... If the same holds good for our dormant representatives in the egg, we may hope some day to apply the physiological activity of the rays of Röntgen and Curie to experimental morphology'.

Because of de Vries's concept of mutations as the source of genetic variation and his early ideas about their potential value for plant breeding, his work around the turn of the century may be marked as the starting point of what is indicated as the discipline of **mutation breeding**, by which designation is meant the use of **spontaneous** as well as **induced mutations** for the genetic improvement of crop plants. After this start it almost took 25 years, before it was proved in 1927, that mutations could be artificially induced and still much longer before mutation breeding became an accepted method in plant breeding.

The word 'mutation' had already been used in a different sense in the seventeenth century to indicate changes in the life cycle of insects, and again in the nineteenth century by palaeontologists, to mark anomalies in fossils (Mayr, 1963). Instead of the word 'mutation', other expressions have also been suggested to describe sudden variations, for instance the word '**heterogenesis**', as proposed by Korschinsky (1901), but these alternatives have never found general acceptance.

Many examples referring to topics like heterogenesis, sudden (genetic) changes in wild and cultivated plants, bud variations (or sports) and related subjects were described before 1900 by a number of scientists like Darwin, Carrière, Meehan, Korschinsky and Bateson.

1.2. Plant breeding

In 1935 Vavilov referred to plant breeding as 'evolution in human hands'; others call it an art rather than a science. In any case, plant breeding involves human activities directed towards the production of crop plants which are **genetically improved** for certain **traits** (or **characters**) and which, therefore, are better suited to human needs. (N.B. The expression **trait** or **character** is used in a rather general sense to indicate any recognizable feature of an individual arbitrarily distinguished by man.) The aforementioned activities include practical breeding work for the production of new **cultivars** (or **varieties**), as well as the underlying research aiming at acquiring increased knowledge of, for instance, breeding strategies, germ plasm, crossing methods, selection procedures, etc.

The expression 'better suited to human needs' is a rather subjective term. It deserves some attention because various categories of traits can be discerned, such as <u>yield</u>, <u>quality</u>, <u>resistances</u> (to <u>diseases</u> and <u>pests</u>) or <u>tolerance</u> (<u>abiotic stress factors</u>), <u>developmental pattern of flowering</u>, <u>maturity</u> and <u>harvestability</u> (<u>earliness</u> or <u>lateness</u>), and <u>costs of cultivation</u>. More details are presented in Box 1.1.

The genetic control of a (plant) trait may be exercised by one major gene (**monogenic inheritance**), by a few

Box 1.1 The most important breeding goals

Plant breeding activities are directed towards a very broad range of consumer's products including for instance: food for men and fodder for animals, building materials (wood, bamboo, reed), raw material for clothing (cotton, flax), fire wood, luxury goods (alcoholic drinks, tobacco), drugs for medicinal use, insecticides, ornamentals, etc.

The most important groups of traits for which breeding work is performed are:

1. Yield. In most cases yield is considered in relation to food and fodder crops, but yield could also refer to the production of latex, fibres, wood, or number of flowers per unit of area. In affluent countries the yield level is often considered to be more important for the grower than for the consumer because the grower is usually paid on the basis of number or weight of produce. If consumers should prefer cultivars with improved quality, for instance higher protein content or better taste, growers often are willing to grow such cultivars only if their yields are not lower than those of the current cultivars, or if higher prices are paid to compensate for the extra effort or the loss of yield.

Yield is a very complex trait in which various **yield components** are involved. Increased yield in cereals, for instance, could be the result of an increased number of ears or kernels per ear, of a more efficient root system, or of better resistances against diseases and pests. Higher yields also may be obtained when a more efficient system of photosynthesis would be available or by a higher **harvest index**, which refers to the distribution of the dry matter within the plant (e.g. a higher total seed

weight instead of straw or leaves). Many yield components are polygenically determined and, as a consequence, such traits are difficult to handle from a breeding point of view. Environmental factors strongly affect the final result, as is indicated by the low degree of heritability of the yield trait.

2. Quality. From the plant breeder's point of view the concept 'quality' covers a wide range of traits concerning the final product which, often, are the 'translation' of various wishes from consumers. Some traits can be determined in an 'objective' way like: baking quality of flour; colour and size of cut flowers; fibre length and strength in flax (*Linum usitatissimum*) and sisal (*Agave sisalana*) and seedlessness in citrus fruits or cucurbits. For other traits no objective criteria are available whereas the appreciation of quality may be very personal and change throughout the years. Examples are, for instance, taste of fruits and colour and shape of flowers.

3. Earliness. An early harvest may allow the production of a second or even a third crop during one season; early crops may escape from pests or drought; better prices may be paid for 'primeurs' of vegetables; a shorter juvenile phase of fruit trees, forest trees, oilpalms, etc. may result in considerable economic profit.

4. Resistance to pathogens and pests and tolerance to abiotic stress factors. Perhaps two-thirds or more of all breeding work refers to breeding for resistance. There is an increased interest in breeding for so-called durable resistance, which often is polygenically determined. Durable resistance is

much more difficult to obtain than the traditional monogenic resistances which often are broken down rather quickly. Resistance breeding becomes increasingly important because of more stringent regulations for the application of fungicides and pesticides.

Tolerance against abiotic stress includes salt tolerance, pH-tolerance, winter hardiness, drought tolerance, Al-tolerance, etc.

5. Costs of cultivation. Reduced plant height often relates to a better **harvest index**. Short straw types in cereals, moreover, do not lodge and allow a higher level of nitrogen fertilization. Monogerm cultivars of sugarbeet (*Beta vulgaris*) are easier to sow and seedlings do not have to be thinned out. Synchronous maturation of Brussels sprouts (*Brassica oleracea* var. *gemmifera*) allows one single harvest period. Short stolon length in potato (*Solanum tuberosum*) facilitates mechanical harvesting.

Different traits may be negatively correlated – high grain yield of maize (*Zea mays*), for instance, is difficult to combine with high protein quality of the kernels. Another complication of many breeding programmes is that the wishes of various parties involved may oppose each other. To mention one example: lawn owners prefer slow-growing, non-flowering and non-seed bearing grass cultivars, whereas farmers who produce on contract the seed for those cultivars, like to grow profusely flowering and heavily seed bearing plants, because of the common system of paying the farmer on the base of kilograms of seed delivered to the breeder.

genes (oligogenic inheritance) or by many genes (polygenic inheritance) with various systems of gene interaction. Before starting a breeding programme it is, in most cases, essential to have at hand information about the mode of inheritance of the traits concerned, the prevailing system of propagation of the crops concerned – self-fertilizing (autogamous), cross-fertilizing (allogamous), vegetatively propagated or hybrid – the flowering system, the level of ploidy, etc. In later chapters these aspects will be considered in relation to handling different groups of crops in mutation breeding programmes.

As a consequence of the enormous variation in crop characteristics (including ways of propagation), breeding goals, system and purpose of cultivation, etc., breeding programmes differ very much; in fact two completely identical breeding programmes may not exist. However, in all breeding procedures a number of major steps – sometimes repeated in a cyclic way – can be discerned.

1. Choice of the starting material.
2. Combination of desired traits.
3. Selection of the genotypes desired.
4. Continued breeding work (perfection) towards a new cultivar.
5. Maintenance and propagation of the new cultivar.

In Box 1.2 more details about these steps are presented.

Despite the fact that plant breeding, as a science, became established only about a century ago, a wide range of methods and techniques for almost every specific aspect of breeding work has been developed by practical breeders, starting from the early beginning of plant domestication and, more recently, by research workers. Handbooks and a number of journals on plant breeding and related topics provide detailed information on all important aspects of breeding. Methods and techniques will be discussed only in so far as they are directly related to mutation breeding, for instance when the relative efficiency of various breeding methods is compared or when use is made of special techniques in relation to mutation breeding.

Some economic figures about plant breeding are given in Box 1.3.

1.3. Mutations and mutation breeding

1.3.1. Introductory remarks

Mutations are the ultimate source of genetic variation (Stebbins, 1950). They provide the raw material upon which other factors of evolution act, and therefore all new species ultimately arise from mutations. The process of occurrence of mutations, also called mutagenesis, followed by recombination of genes and chromosomes, and by natural selection, is the fundamental force in evolution. In natural populations, other factors like migration, geographic isolation and genetic drift contribute as well to changes in populations.

In Section 1.1 mutations were defined as heritable changes in the genetic material not caused by recombination or segregation. Some additional comments should be made here. Recombination results in new gene combinations within the genetic material, following the occurrence of crossing over between two homologous DNA molecules, but this process does not alter the structure of the genes or the number of the chromosomes. For higher plants an addition to be made is that mutations may occur in the chromosomes, which are situated inside the cell nucleus, as well as in the extranuclear DNA that is present in the chloroplasts and mitochondria. In plant breeding sometimes the expression idiotype is used to indicate the sum total of all nuclear and extranuclear DNA present in a plant. Accordingly, the expression genotype refers to the sum total of the genetic information (genes) in the chromosomes, whereas the expression phenotype connotes the particular expression of a gene or, more generally, refers to the observed properties (for given traits: yellow seed coat, short internode) of an organism as a result of the interaction between the organism's genotype and its environment. (For definitions, consult for instance Rieger, Michaelis & Green, 1991, or general text books on genetics.)

For mutation breeding it is crucial whether mutations arise in the sporophytic plant tissue (i.e. in the somatic cells), or in the gametophytic tissue (also called the germ cell line), from which the gametes arise. Until recently in seed propagated crops only those mutations which arose in gametophytic tissue were of value for breeding work. Recently, this situation has changed, as will be discussed later.

This may be the place to give a more suitable working definition for a mutation for higher plants, namely: any heritable change in the idiotypic constitution of sporophytic or gametophytic plant tissue, not caused by normal genetic recombination or segregation. The process of mutation formation at the molecular level is called mutagenesis. Mutagenesis results in mutants, which could be applied to either mutated cells, mutated organelles or mutated plants.

In the heading of this section the expression mutation breeding has been used with the understanding

Box 1.2 Different steps in breeding programmes

For annual, seed propagated crops like bread wheat (*Triticum aestivum*) or barley (*Hordeum vulgare*) it may take 10–15 years to conduct a complete breeding programme. The phase of repeated testing under various conditions and the official trials for registration and application for breeders rights may require another 5–6 years. Shortening the breeding cycle, for instance through increasing the number of generations that are tested annually or by applying 'early testing' for specific traits, wherever possible, are important subjects of research.

A plant **genome** (i.e. the basic chromosome set in eukaryotes and the sum of all nuclear genes of that plant) may contain 10 000–100 000 genes in all of which, basically, spontaneous mutations may occur. As a consequence and theoretically speaking, the amount of genetic variation that could be piled up, even within one species, may be enormous. However, not all changes in the DNA last. At the same time, not all permanent changes within genes result in recognizable changes and not all recognizable (or **phenotypic**) variation is heritable because the **idiotype** (i.e. the sum of all nuclear and extranuclear genetic information within a plant) as well as the environment contribute to the phenotype. (N.B. **genotype** refers to nuclear genes only).

Many important 'traits' like yield require rather sophisticated breeding techniques because of their polygenic inheritance. The **heritability** in narrow sense (h_n^2) expresses the extent to which phenotypes are determined by genes transmitted from the parents. This h_n^2 determines the frequency of resemblance between relatives and, therefore, is of great importance for plant breeding. The

heritability in broad sense (h_w^2) expresses the extent to which individual phenotypes are determined by their genotypes. This heritability is mainly of theoretical interest.

1. Sources of genetic variation. Basically, the choice of parent material is the most important decision in a breeding programme. In most cases use is made of genetic variation from existing gene pools. Another approach may be to combine various sources of genetic variation as a new gene pool from which by selection and repeated crosses, basic material is derived for further cultivar breeding. This could be called 'pre-breeding'.

The breeder can choose between modern cultivars (either local ones or from abroad), old varieties or land races, which are still grown in some places and which may contain some useful traits, primitive varieties, cultivated ancestors and even wild relatives. New sources of genetic variation become available as the result of various techniques of **plant biotechnology**.

Crossing between a cultivar and a wild relative often appears to be difficult or even impossible. If crosses are successful, in addition to the desired gene(s) from the donor parent, many undesired genes may be introduced as well. The same may happen after applying newer methods like somatic hybridization or protoplast fusion. In such cases a backcrossing programme is necessary to get rid again of those unwanted genes.

A single trait can be changed by mutation breeding or a single gene can be added to an existing cultivar by more recent techniques like genetic transformation. In this way also man-made gene constructs

can be brought into existing cultivars.

2. Crossing. Knowledge of flower biology as well as considerable technical skill are often required when performing crossings. The number of crossings required for a specific combination depends on the system of propagation of the crop and on the mode of inheritance of the trait that has to be incorporated. When crosses are made between parents with different ploidy levels, it may be advantageous to make use of spontaneous or artificially induced doubling of the genome of the parent with the lower ploidy level.

Modern techniques like fusion of protoplasts give another option for combining genetic variation. In this way it sometimes may be possible to combine the genomes of unrelated species. By fragmentation of the donor DNA by irradiation it is possible to transfer randomly only a part of the genome to the recipient protoplast cells.

3. Selection. The selection phase is a most essential (and often very laborious and time-consuming) part of the whole breeding process. Even a thorough knowledge of all relevant crop particulars, of the genetics of the particular traits of interest, as well as of the prevailing breeding strategies for the specific group of crops, does not guarantee that efforts will result in new, successful cultivars. The method applied for fixation of selected, improved genotypes largely depends on the propagation system of the crop (vegetatively propagated, self- or cross-fertilizing) and on its multiplication rate. Various selection strategies – such as clonal selection, pedigree selection, line selection, mass selection, family selection, progeny testing or more complicated schemes like the remnant (spare, reserve) seed method – have been developed for special traits or specific situations.

In breeding programmes for hybrid cultivars, the selection of suitable inbred lines, which show a good combining ability, is an essential part of the procedure.

Two examples of breeding schemes are given in Fig. 1.2.

The efficiency of the selection procedure in cross-fertilizing crops is strongly determined by the fact of whether a specific trait can be selected before or after flowering. The length of the selection procedure depends on the genetic complexity of the trait and the heritability.

4. From selection towards a new cultivar. Although the potential of selection, of course, is restricted by the genetic variation present in the unselected population, it is the selection method (criteria, techniques, intensity of screening) that determines which part of the population will be advanced to the next generation. In vegetatively propagated crops, after the selection phase, the resulting superior idiotype can be used directly as a new cultivar. A population of related plants with identical genotypes in vegetatively propagated crops is called a **clone**; in seed propagated crops, a **line** or a **family**. In seed propagated crops further selection for homozygosity (in self-fertilizing crops) and homogeneity is required.

For self-fertilizing crops the desired genetic fixation can be reached; for cross-fertilizing crops it is impossible to obtain a progeny that is completely 'true to type' and homogenous. In the last case continued selection throughout the years remains necessary.

Procedures followed during the stage of final perfecting of the cultivars are identical to those during the selection stage.

Most selection failures are the result of inaccurate definition of breeding objectives and inadequate identification of the traits related to these objectives and the interrelationships of these traits.

5. Maintenance and multiplication. When, in a vegetatively propagated crop, in an advanced stage of selection, a certain clone performs well, a limited multiplication of this clone is started in a special nursery. When the clone continues to be outstanding, the breeder has already built up enough plant material for official trials as well as for distribution to growers. During maintenance and multiplication the breeder must check very carefully for aberrant types and for virus infections, and avoid any risk of mixing up different stocks of starting material.

For seed propagated crops often detailed schemes are followed when producing breeders seed and commercial seed (N.B. commercial seed is the seed sold to the farmer). For economically important crops this work is often done under supervision of government-controlled organizations. For sowing, for instance, 250 000 ha of bread wheat (*Triticum aestivum*), about 30 million kg of seed is needed. Starting from 1000 kg of breeders seed, four multiplication cycles are necessary to produce the required amount of seed, because of the relatively low multiplication rate of wheat.

Breeders of cross-fertilizing crops in this stage must continually select to maintain a certain degree of homogeneity in their commercial seed in order to maintain the identity of the cultivar.

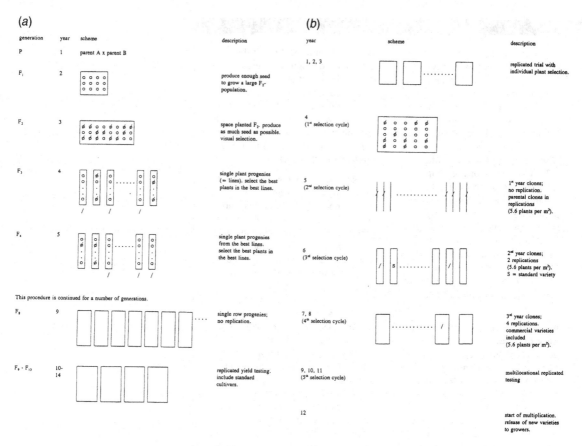

Figure 1.2. Schematic presentation of two selection schemes: a) pedigree selection for seed propagated crops; b) selection of vegetatively propagated crops (example *Pyrethrum* sp.)

that in most cases additional breeding work has to be done before a crop plant in which an agronomically useful mutation has occurred, can be released as a new cultivar. Many failures of mutation breeding programmes are caused by unfounded expectations and inappropriate procedures due to inadequate knowledge of plant breeding science by the performers. This subject will be illustrated later.

In the previous lines mention has been made a few times of DNA, which is the abbreviation for deoxyribonucleic acid. DNA molecules contain, in code, the instructions for the production of proteins. The code consists of four letters (nucleotides, organic bases) which are indicated as A, C, G and T. Combinations of three letters (triplets or codons) code for one of the amino acids which are the building blocks of proteins. DNA has the structure of a double-stranded helix and the DNA molecules are composed of sequences of the aforementioned four nucleotides. The bases of the nucleotides in the two different strings of the helix form so-called **base** pairs (abbreviated: bp) by their hydrogen bonds. The most common bonds are A–T and G–C.

A single gene may consist of 10^3 or more base pairs, whereas a single cell, depending on the species or organism, may contain from several million (10^6 or 1 Mbp) to even a few billion (10^9) base pairs. One genome of wheat (*Triticum aestivum*) with seven chromosomes, for instance, consists of about 5.3 billion base pairs. In other words, a (plant) cell may contain 10^4–10^5 genes, each of which, in principle, could mutate.

In terms of molecular genetics, a **mutation**, in general, is often defined as <u>any heritable change in the nucleotide sequence of a genome</u>. Plant breeders are in particular interested in **gene mutations** which occur primarily <u>within</u> the genes themselves. The smallest change possible within the DNA is a mutation affecting one single base pair. Gene mutations and several other kinds of mutations, including mutations in the extranuclear DNA (chloroplast, mitochondria) will be discussed in Chapter 3.

Box 1.3 Some economic figures about plant breeding*

The world market for seed as sowing material for plant production, very roughly, may be worth about 45 billion US dollars at an annual base. About one-third refers to commercial seed, one-third to seed provided by governments and one-third to so-called farmers-saved seeds, i.e. seed that is kept by the farmer for next year's sowing (de Kleijn, Boers & Heijbroek, 1992; de Kleijn & Heijbroek, 1992; van Gaasbeek et al., 1994). It is estimated that, on a global scale, 1500 breeding companies (or other companies with a large breeding division) are active, of which 600 operate from the USA and 400 from Europe. About 20 big companies control some 50% of the total market.

The largest breeding company of the world is Pioneer Hi-Bred International in the USA with a retail value of about 600 million US dollars. More than 40% of the total acreage of maize (*Zea mays*) in the USA is sown with cultivars of this company, whereas the share of the second largest company is less than 10%. Other large companies are Limagrain in France and the breeding division of Sandoz (now

combined with Ciba-Geigy), with its homebase in Switzerland. Each of them employs thousands of workers and has branches all over the world.

For barley (*Hordeum vulgare*) in the USA, the total annual production value is estimated at one billion US dollars. About 3–5% of this sum, or 30–50 million US dollars, refers to sales of commercial seed from breeders to farmers (assuming that all farmers each year would buy fresh seed to start their new crop).

The world market for vegetable seed – which operates almost exclusively in affluent countries – is estimated at 1-2 billion US dollars. For the European Common Market the costs of seeds and young plants for vegetable production in 1990 were 0.5-1 billion US dollars. Seed costs corespond to 5-10 % of the total production value for vegetables.

The world cut flower production amounts to at least 6 billion US dollars. The production value of pot plants at a world scale in 1990 can be estimated at 4 billion US dollars. If the grower's costs for starting material (seeds, rooted cuttings, *in vitro* material, etc.) would be 10% of this figure – a

very low estimation – breeders of starting material would earn about 400 million US dollars. Very few (less than 20) big breeding companies operate on a global scale in the field of horticultural crops (vegetables and ornamentals) and they together have a production value of more than 100 million US dollars.

Another way to assess the economic value of breeding activities is to look at the contribution of breeding work to increasing yields, for instance in specific groups of crops. For cereals (except maize) a reliable estimation for Europe is that during the last decades an average annual increase in yield of 1–1.5% has been achieved on a total area of almost 800 000 ha. More than 50% of this annual increase is contributed by genetic improvement. (N.B. This figure does not include other genetic improvements like for quality.)

*Detailed, reliable information on economic activities in plant breeding is often very hard to obtain, and in many cases it is unclear how figures are put together (e.g. with or without deducting costs, etc.).

1.3.2. Mutations and mutants in plant breeding

For plant breeding, mutations are of interest in two different ways. They provide us with new starting material (or building blocks) for the production of new cultivars and they give us the tools for identifying new genes, for studying the nature of genes and their way of controlling biochemical pathways (Konzak, 1984; Konzak et al., 1984; Rick, 1986; Old & Primrose, 1989; Micke et al., 1990). In this chapter and the following ones most attention will be paid to the use of mutations in practical plant breeding programmes, the most suitable starting material, mutagenic treatments, selection procedures, etc. As the starting material for applied mutation breeding programmes is often identical to the material that is used for the various categories of crops (self- or cross-fertilizers; vegetatively propagated) in conventional plant breeding, this subject will be discussed more extensively in the specific chapters (Chapters 5, 6 and 7) in which is dealt with applied mutation breeding for those different categories of crops.

The value of mutations for the better understanding of the genetic regulation of specific traits was realized soon after mutations became known as heritable changes, and many new investigations were initiated. But only in 1981 a symposium called '*Induced Mutations – A Tool in Plant Research*' (Anon., 1981a), held in Vienna, was organized by the International Atomic Energy Agency (IAEA) and the Food and Agriculture Organization (FAO) of the United Nations. In the opening lecture of this conference MacKey (1981) discussed the value of various types of so-called man-made or **artificially induced mutations** for studies on qualitative and quantitative genetics, gene expression, chromosome reorganization (chromosome engineering), evolutionary and taxonomic genetics, reproduction (or flowering) genetics, developmental genetics and physiological and biochemical genetics.

At another occasion, Nilan (1981a), in a treatise about mutation research in barley, mentioned four fields of research in which the use of mutations might be of interest:

1. to increase knowledge concerning the inheritance and genetics of a given trait;
2. to reveal the biochemical, physiological and anatomical processes that occur between the starting point of the gene and the ultimate phenotype of a number of traits and the genetic control of all these steps;
3. to analyze the nature of mutations;
4. to probe the genetic constitution of loci.

In a more general way one could also refer to the value of induced mutations when studying fields like plant evolution, plant physiology, symbiosis and pathogenesis and an increasing number of publications on these topics has appeared already. An early discussion on the value of biochemical mutants in higher plants was written by Rédei & Acedo (1976). Special conferences on the use of developmental mutants in higher plants were organized, for instance, in 1986 at the University of Nottingham, UK (Thomas & Grierson (eds.), 1987) and in Japan in 1991 within the frame work of the *Gamma Field Symposia* (Anon., 1992a). In addition, mutations, nowadays, are indispensable tools in several applications of 'biotechnology' or 'genetic engineering'. In Box 1.4 additional information about the use of mutations 'as tools' is given whereas in a later section (Section 1.3.10) and the accompanying box (Box 1.8) attention is paid to the relation between mutation breeding and modern biotechnology.

1.3.3. Spontaneous mutations

Part of the genetic variation of plants is derived from so-called **spontaneous mutations**: mutations which occur without intentional human intervention. In most cases the origin of spontaneous mutations is unknown. Recently, it has been suggested in several publications that spontaneous mutations often may result from the activity of so-called **transposons**: mobile genetic elements that can move within the genome (the complete set of chromosomes) from one place to another and affect the activity of the gene in which they are inserted. Evidence for this suggestion for instance has been derived from experimental work with *Drosophila*. In Chapters 2 and 3 more attention will be paid to transposon activity and a number of references will be presented.

Spontaneous mutations usually arise at low frequencies, e.g. during one (plant) generation not more than 10^{-5}–10^{-8} per gene or per locus. As a single plant cell may contain up to 100 000 genes (or billions of base pairs), it

can be envisaged that, even with such low frequencies of spontaneous mutations, and even for short-living species, each individual plant or even each plant cell may carry one or more spontaneous mutations which were collected during its lifetime. In vegetatively propagated crops such mutations may pile up throughout consecutive generations as a consequence of their mode of propagation. Like in cross-pollinated crops, recessive mutations may remain undetected for a long time when no easy distinction can be made between plants which are homozygous and heterozygous dominant for the specific trait. Mutations are often deleterious and it is estimated (Kondrashov, 1988) that at least one new deleterious mutation per individual per generation may occur. This may have important consequences for the **fitness** of the plant population in which such effects occur, in particular when obligately vegetatively propagated crops are involved. **Fitness** (or **Darwinian fitness** as it is sometimes called) is an expression to indicate the relative probability of survival and rate of reproduction of a phenotype or genotype (Suzuki *et al.*, 1989). It is a quantitative measure for the reproductive success of a given genotype. Some details will be presented in Chapter 7.

Mutant plants with an altered phenotype as the result of one or more spontaneously occurring mutations, most probably have been used by early men right from the moment that they started making use of plants for food, clothing, or housing. Our ancestors, undoubtedly, gradually learned to recognize the most favourable plants and to search for the most tasty fruits, the strongest fibres, etc., between and within – what we call now – a botanical species. They also have often learned to save the best seeds, etc., for future planting instead of eating them. Whether favoured plants have directly resulted from spontaneous mutations, or arose after natural crossing between genotypically different plants, is often difficult to retrieve. However, if at that time at a given location a fruit tree with spineless fruits would have shown up, surrounded by many trees of the same species with spined fruits, there is not much doubt that such spineless fruits were the result of an earlier spontaneous mutation for this trait. (Such details, of course, will have been of no concern to our ancestors.)

During the phase of domestication of plants different traits have changed by natural selection, or as a result of – intentional or non-intentional – selection by man. The basis for this selection, of course, was formed again by the occurrence of spontaneous mutations. Some examples of unintentional selection will be briefly mentioned here.

Preferential harvesting of cereal plants with heavy

Box 1.4 Mutations as tools in plant breeding research

Mutants are essential prerequisites for genetic studies, in particular for studying gene structure and functional relationships (Old & Primrose, 1989). Experiments in which mutations are used 'as a tool', often are performed with various prokaryotes (bacteria like *Escherichia coli*), lower eukaryotes (like the yeast *Saccharomyces cerevisiae*) or specific higher eukaryotes, like *Arabidopsis thaliana* (fam. *Cruciferae*).

Arabidopsis, also known as mouse ear cress, is a small, diploid, self-fertilizing plant with a short generation cycle of less than two months, tiny seeds and easy cultivation of up to 5×10^4 seedlings per m². It has 2×5 chromosomes in the somatic tissue and a small genome (haploid genome size: 100×10^6 bp or 100 Mbp), which enables efficient procedures, for instance when 'cloning' genes by molecular methods. An early review on the genetics and biology of *Arabidopsis* was presented by Rédei (1970). *Arabidopsis* is also an ideal model species for mutation studies, for instance because of its low sensitivity to X-rays and chemical mutagens, its small nuclear volume, the low number of so-called

genetically effective cells (GECN) after mutagenic treatment (e.g. 1.7 for X-rays and 1.5 for EMS) and the occurrence of stable autotetraploids. In recent years *Arabidopsis* has become a much used model plant for various genetic investigations. Until 1992 about 150 morphological/biochemical markers and some 300 RFLP markers have been identified. More references about *Arabidopsis* are presented elsewhere in this and other chapters. Another important species used in fundamental genetic studies is the fruitfly (*Drosophila melanogaster*). Pea (*Pisum sativum*), maize (*Zea mays*), barley (*Hordeum vulgare*), rice (*Oryza sativa*), tobacco (*Nicotiana tabaccum*) and tomato (*Lycopersicon esculentum*) are often used as 'model crops' for genetic studies, including studies of mutations.

Mutant collections have contributed significantly to the construction of genetic maps. Mutagenic effects can result in changes in the DNA which may affect the functioning of genes as well as the phenotype. Mutants of *Arabidopsis* are used to study processes related to flowering and plant architecture. In tomato, in 1986, some 1200

mutants of various nature were known (Rick, 1986). Starting with research on chlorophyll-deficient mutants many different traits have been studied in barley by research groups in Scandinavian and other countries for more than 50 years (N.B. barley mutants for various traits are described in Chapter 3). So-called viviparous mutants of maize, which fail to show an arrest of seed development, are an essential tool to identify the genes that are responsible for the process of maturation of maize seeds.

In addition to spontaneous mutants, use is made of artificially induced mutations for morphological, biochemical or other selectable traits to study gene expression. Such mutants are of particular use when, apart from the mutated gene or trait, no further disturbance of the 'genetic background' occurs. Some mutants are indispensable tools for specific techniques in the field of biotechnology, such as genetic transformation of plants or fusion of protoplasts from distinct species or genera.

grains automatically results in selection for late ripening. Also, by collecting only heavy cobs of maize, unconscious selection for low tillering capacity is performed. Ripe seeds of cereal plants with a stiff rachis, and mature non-dehiscent pods in legumes are more likely to be collected by the farmer and, in addition, loss during transportation to the storage place will be less. Seeds of those plants were probably used again next season for sowing and, in this way, unconscious selection for non-target traits may have taken place.

Some traits are valuable in nature to enable maximal dispersal of seeds, like the presence of small thorns, which would make those seeds stick to the fur of animals. Such traits lose their significance once plants are grown in fields, where farmers will prefer plants with thornless seeds (provided that 'thornless' is not linked to important negative traits). Other examples of early, conscious selection of spontaneous mutants are easy to

imagine: fruits with a better taste or a better keepability, non-bitter mutants of gourds from various species of the genera *Cucurbita* and *Lagenaria*, stronger fibres of cotton or bamboos, etc.

Records about the occurrence, frequency and use of spontaneous mutations at those early times, of course, are not available. The first extensive descriptions about economically useful mutants, to which publications attention will be paid in later chapters, were written by, for instance, Carrière (1865), Darwin (1868) and Cramer (1907). Although spontaneous mutants occur in low frequencies, they still may be of practical value these days and, occasionally, spontaneous mutants do result in cultivars of considerable economic importance. Recent examples include mutants of commercial interest in citrus, apple and different ornamental species. Several examples for vegetatively propagated crops will be given in Chapter 7.

Of particular economic significance in seed propagated crops has been the mutant dwarfing gene of the Chinese rice cultivar Dee-geo-woo-gen of rice (*Oryza sativa*), which, presently, is incorporated in many high-yielding, short-straw rice cultivars in tropical areas. Another spontaneous mutation in which breeders recently became interested, concerns the trait 'brown mid-rib', for instance in maize (*Zea mays*). Plants with this trait have a lower lignin content than normal genotypes, which positively affects the digestibility of maize leaves when used as cattle fodder. As is the case with many mutations, 'brown mid-rib' has negative characteristics as well, like delayed maturity, increased lodging, lower grain yield, etc. Negative effects may be overcome by further crossing and it is important to realize that even the most promising mutations may reveal their full value only after they have been submitted to further crossing programmes. In practice, still many (from an economic point of view) potentially useful mutations are discarded before they have been sufficiently investigated. This subject will be further discussed later in this chapter.

Nature's ability to increase genetic variation within a limited period of time may be demonstrated with many examples. Recently, Bozzini (1991) mentioned, for instance, that a single plant of the ornamental species *Kalanchoë blossfeldiana* (fam. *Crassulaceae*), collected in 1920 in Madagascar and afterwards introduced in the UK, has been the base of a great variety of cultivars which are grown now all over the world. Bozzini (loc.cit.) stated that all those cultivars had arisen from spontaneous mutations in the progeny of the aforementioned single plant. It is, however, rather doubtful whether Bozzini is fully correct in this respect. Selection work in the progeny of the collected plant has indeed resulted in new cultivars, e.g. with a compact plant type, and such compact types may have been the result of spontaneous mutant types, in particular when selection was performed in vegetative progenies (which has not been mentioned explicitly). In *Kalanchoe* vegetative propagation as well as propagation by seed is possible and intensive crossing work within and between different species has been performed already for many decades within Europe, for instance in Germany and the Netherlands (van Raalte, 1969; van Voorst & Arends, 1982). The last authors described the origin of *K. blossfeldiana* which, according to these authors, traces back to 1924 and no reference was made to the UK expedition in 1920, mentioned by Bozzini (loc.cit.). Whether the material from the UK was used also in the crossing work at the European continent has not been further checked. In later chapters earlier, and probably more reliable, examples will be given.

As most spontaneous mutations occur in low frequencies it is essential to develop efficient screening techniques. The German plant breeding scientist R. von Sengbusch found in populations of the common bitter and poisonous, alkaloid-containing, lupines (*Lupinus luteus, L. augustifolius* and *L. albus*, fam. *Leguminosae*) so-called sweet (alkaloid-free) plants. This discovery opened the possibility to use lupines as cattle fodder crop. The mutations occurred in low frequencies of 1×10^{-4} – 1×10^{-5}. The mass-screening techniques developed by von Sengbusch made it possible for one person to handle thousands of seeds in one day. Between 1927 and 1939 von Sengbusch and co-workers altogether screened 6–7 million seeds (von Sengbusch, 1942).

Parlevliet & Zadoks (1977) calculated for the pathogen leaf rust (*Puccinia recondita* f.sp. *tritici*) in wheat (*Triticum aestivum*), that if only 1% of the leaf area would be covered by sporulating uredosori, producing 300 spores per mm^2 per day, each day about 10^{11} spores would be produced. If the spontaneous mutation rate would be only 10^{-8} per locus, a production of 1000 mutants per locus per day (!) per ha would be the outcome. This astonishing result, according to the authors, is in agreement with the outcome of calculations for other plant/ pathogen systems.

For a final example reference is made to Redéi (1982a), who reported for maize (*Zea mays*) that a single maize-ear may contain 300–400 kernels. Dominant mutations – which are very rare – can be observed by surface examination of the endosperm cells per kernel, of which about 1400 are situated at the surface of the kernel. Since the endosperm is triploid, a single maize ear may reveal mutations in about $350 \times 1400 \times 3$ genomes = 1 470 000 genomes! This shows that situations exist where high numbers of genes or genomes can be studied relatively easily and where even mutations which occur in very low frequencies can be detected.

At present much attention is given to variation which is observed when plant parts are grown *in vitro*. This variation is classified nowadays as **somaclonal variation**. In many publications somaclonal variation is referred to as a new and – from a breeding point of view – very promising source of spontaneous genetic variation. Whether somaclonal variation indeed is a 'new' source of variation and whether it really is as 'promising' as some authors would like us to believe, will be discussed later in this book.

It may be useful to recall here that not all phenotypically observed variation refers to genetic changes. At the same time one should be aware of the fact that not all changes within the DNA, ultimately, do result in permanent changes of the DNA. Moreover, even if such changes in the DNA would be permanent, i.e. refer to 'real' mutations, they may not always result in visible – or in another way detectable – effects.

1.3.4. Induced mutations

Although selection for economically useful spontaneous mutants still takes place these days – and often with remarkable success – the purposeful induction of a specifically desired mutation at a specific time and place, and in a selected genotype, would be a much more attractive option. Another reason to consider using induced mutations is that breeding programmes could eventually be speeded up considerably. This, for instance, may be of particular interest in economically important cut flowers like chrysanthemum in order to reduce the costs of breeding (see also Chapter 7). As was mentioned in Section 1.1 of this chapter, Hugo de Vries, at the start of the twentieth century, was the first to suggest the artificial induction of mutations. But it took till 1927 before Muller, working with the small fruitfly *Drosophila*, proved for the first time that mutations could be artificially induced.

Various agents may artificially induce mutations; the most important methods at this time are the use of various types of **physical agents** (in particular different types of radiation) and **chemical mutagens** (for details see Chapter 4). Throughout the years many other methods have been tried as well, like the use of heat or temperature shocks, centrifugation of seeds, ageing of seeds and pollen grains and specific culturing practices in plants; but all those methods have proved to be by far less efficient than the use of radiation or chemical mutagens. In later chapters advantages and disadvantages of various methods will be discussed.

It is often found that induced mutations may occur in frequencies 10^3 higher than for spontaneous mutations, but considerable variation in frequencies can be observed between different traits. A clear distinction between mutations which occur as a result of exogenous agents or are caused by endogenous effects in the plant(cell) is difficult to make. Not all exogenous agents to which plants are submitted (like radiation from minerals present in the soil or from air pollution), are purposefully administered. On the other hand, the action of endogenous agents (often oxidants) may not be recognized as such. Moreover, mutations which have been present for a long time in a plant, but remained unobserved for some reason, at a certain moment may become visible, e.g. by a morphological change in a plant part, and then create the impression of a recent mutagenic event. And, finally, much damage to the DNA may be effectively – be it not always perfectly

– repaired, for which various systems have been detected.

Whether – apart from a higher frequency for induced mutations – real differences exist between spontaneous and induced mutations remains a subject of much discussion. Nilan (1981b) stated that 'there are actually no major differences between induced and spontaneous gene and chromosome mutations', but he added that older spontaneous mutations have been 'moulded' by recombination and natural selective forces into useful co-adaptive complexes. Konzak, Kleinhofs & Ullrich (1984), however, remarked that definite proof for the opinion that induced mutations and spontaneous mutations refer to the same kind of changes, has never been given.

Several mechanisms may be responsible for the occurrence of mutations after DNA damage. One well-known mechanism refers to an alteration in the specificity of so-called basepairing (e.g. de-amination of cytosine to uracil). The resulting mispairing does not lead to problems with replication and a mutation may be induced opposite to such lesions. In other cases the basepairing potential can be lost, which also may lead to mutations. Many such conclusions are the result of observations on lower organisms like the bacterium *Escherichia coli*. For higher plants still many details are not yet fully understood. In Chapters 3 and 4 more details will be presented about this and related topics.

1.3.5. Mutation frequency and mutation rate

Various methods are used to record how many mutations have been obtained, for instance in experiments in which the effects of different mutagenic treatments are compared. Common expressions like **mutation frequency** and **mutation rate**, often are used in a rather loose way, without making clear what exactly is meant by the figures concerned. In general, the mutation frequency is easier to determine than the mutation rate.

The notion **mutation frequency**, in its most simple form, could be defined as the observed (or estimated) number of mutations, divided by the population size. However, a statement like: 'The mutation frequency was 4.5%', is not very useful without informing the reader about which trait (with one or more genes involved), gene or category of mutations was studied. Moreover, it would be useful to know whether the mutation frequency determined was per unit of dose, per cell, per organism, per gamete, per plant(part) or per petri-dish and in which generation(s). It may be clear that a correct interpretation of mutation figures is impossible without having such details at hand.

Ake Gustafsson, an outstanding Swedish mutation

scientist, introduced for barley (*Hordeum vulgare*) a practical and unambiguous method to record mutation frequencies, indicated as the '**spike progeny**' method. According to this method, the number of spike progenies showing segregation for a given trait, is divided by the total number of spike progenies studied (Gustafsson, 1940). As an alternative, the mutation frequency could also be calculated from the number of mutated plants in the second mutated generation (indicated as M_2), e.g., per 100 or 1000 M_2-plants that have been analyzed. This method has been worked out by Gaul (1960), who worked on barley mutations in Germany. According to Blixt, Ehrenberg & Gelin (1963), this method has several advantages because variations in progeny size and the size of the plant area that is mutated do not affect the final result as would be the case when the spike progeny method would be followed.

In their textbook on genetic analysis Suzuki *et al.* (1989) defined **mutation frequency** as: 'the frequency at which a specific kind of mutation (or mutant) is found in a population of cells or individuals.' It should be noted that this definition only gives the proportion of mutants in a population and that the number of mutations that gave rise to the original ones is disregarded (Rédei, 1982a). Within one cell more than one independent mutational event can take place, but, because of the estimated number of genes per cell (e.g., between 10^4 and 10^5) and the common spontaneous mutation frequencies (e.g., 1×10^{-5} – 1×10^{-8} per gene), this is most unlikely in nature. Even in case of mutation frequencies which are a hundred times higher as a result of mutagenic treatment, the occurrence of two or more, independently induced mutations, statistically speaking, seems to be an exception. This may sound somewhat remarkable as, in case of 10^5 genes per cell and a mutation rate of 10^{-4} per locus, about ten nuclear mutations should be induced in each cell. But, on the other hand, not all genes are expressed in each cell at each moment and, moreover, many mutations, after having been induced may be repaired by various mechanisms to which more attention will be paid in Chapter 4. (N.B. When a specific trait is polygenically determined, this, of course, does increase the chance of mutations for this trait but not per locus.)

The **mutation rate** can be defined as: 'a number that represents an attempt to measure the probability of a specific kind of mutation event occurring over a specific unit of time' (Suzuki *et al.*, loc.cit.). In this context 'time' may refer to minutes or days, as well as to the length of the period of one cell cycle or a plant generation. Rédei (1982a), in a more simple way, refers to the mutation rate as: 'the overall frequency of detected *mutational* events per genome and per generation in large-scale

observations'. Different methods could be used to determine mutation rates, the most common being the so-called fluctuation analysis, developed by Luria & Delbrück (1943) for large populations of bacteria, grown from a single cell. Various methods to estimate mutation rates in experiments with bacteria were described extensively by Lea & Coulson (1949). Reference is also made in this context to a recent contribution by Stewart (1994) who, based on simulation studies, commented that the original averaging method used by Luria & Delbrück is better than has often been suggested. Stewart adds that when the expected number of mutations is small, estimation by the number of cultures without mutants is almost as reliable as the maximum likelihood method, but low expected numbers should be avoided when possible. Accordingly, for small values of the expected number Lea & Coulson's method is seriously biased, but for larger values it is almost as accurate as the maximum likelihood method.

Whether, in a specific population, the observed mutation frequency or mutation rate differs – statistically – from a population treated in another way, or from the control, can be easily checked by making use of probability tables (see, for instance, Kastenbaum & Bowman, 1970, and Bogyo, 1991). Such tables may give a minimum value at which, at a given level of probability, a specific hypothesis (H_o) – in which, for instance, is stated that mutation rates as a result of two different types of irradiation treatments are the same – is either rejected or not.

Estimates of mutation rates in populations of cells may show considerable variation, either depending on real differences in mutability between different objects, or in differences caused by different statistical procedures. Li & Chu (1987) evaluated various methods for the estimation of mutation rates in cultured mammalian cells and concluded that, for large populations, the method of maximum likelihood was to be preferred. (N.B. For details about this often used, but laborious method, the reader is referred to textbooks on statistical theory.)

In the previous examples mutation frequencies and mutation rates were determined in bacteria or mammalian cells. For fundamental studies, for instance when the efficiency of different mutagenic treatments is compared, higher plant systems in most cases are more complicated to work with, as such systems consist of groups of cells and tissues which, in various ways, are very heterogeneous. Moreover, even in case within a plant a relatively homogeneous group of cells would be the target (e.g., shoot tip meristems), such cells still might differ very much as to their stage within the mitotic cycle, etc.

Box 1.5 The mutation detection system in stamen hairs of Tradescantia spp.

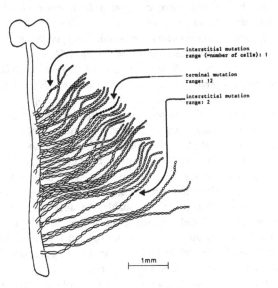

interstitial mutation
range (=number of cells): 1

terminal mutation
range: 12

interstitial mutation
range: 2

1 mm

Figure 1.3 Stamen of *Tradescantis* sp. with part of the hairs showing mutations.

In the *Tradescantia* mutation detection system mutations are studied which occur in cells of stamen hairs. One inflorescence of *Tradescantia* may contain 50 flowerbuds. Each flower contains six stamen, each with 70–100 hairs. Use is made of specific, well-studied plant material that can be vegetatively propagated. Somatic mutations in one or more cells of stamen hairs, which result in changes from blue (in certain clones purplish-blue) to red (or purplish-red), occur spontaneously and – with higher frequencies – after mutagenic treatments with physical or chemical agents. In diploid cells, mutations in general affect only one of the two corresponding alleles and, as the large majority of the mutations goes from dominant to recessive, the *Tradescantia* clones that are used

for mutation studies must be heterozygous (e.g., Aa) for cell colour. By analogy, triploid clones should be Aaa (not AAa!).

Each stamen hair, consisting of about 20 cells, predominantly originates by division of the terminal and subterminal cell. As a consequence, each hair could be compared with a cell culture of individual cells, whereas each individually mutated cell of a stamen hair can be considered as an individual mutagenic event. When within one hair two mutated cells are found which are separated by a non-mutated (blue) cell, this result is scored as two individual mutagenic events.

The simplicity of the *Tradescantia* system (although in some cases different interpretations of results may be given) enables easy screening of millions of hairs or cells for mutagenic events, even by inexperienced persons. The system is often applied to compare mutation frequencies resulting from different mutagenic treatments, but is, for instance, useful as well as a sensitive method to study mutagenic effects of air pollution.

For higher plants 'in vivo' model systems to study mutations and mutagenesis have been developed. The first one, set up by L. Stadler in the 1920s for quantitative studies of the chromosomal effects of X-ray treatments, was based on so-called mosaic endosperm: mutations for kernel colour in maize (*Zea mays*). The endosperm nucleus is triploid with two genomes from the female parent and one genome from the father. As a consequence, the loss of a dominant gene of maternal origin has no visible effect and studies of mutations should start from plants which are heterozygous for visible endosperm traits with the female parent recessive and the male parent dominant. Stadler (1930) mentioned that seven of the 10 maize chromosomes carried genes that could be identified in endosperm mosaics. Only one system carried more than one gene: the linked *C-Sh-Wx* complex (*C Sh Wx* grains are coloured, smooth

and starchy; *c sh wx* grains are colourless, shrunken and waxy). In Chapter 3 we will return to this system.

The best-known and probably most sensitive system in higher plants refers to (spontaneous or induced) somatic mutations from (mainly) dominant blue to recessive pink in the staminal hair cells of specific heterozygous clones of *Tradescantia* spp. In Box 1.5 a drawing of an anther of *Tradescantia*, showing some mutations in the stamen hairs is presented and some additional comments are given. The relatively high sensitivity of *Tradescantia* to radiation was established in the 1950s by the group of A.H. Sparrow at the Brookhaven National Laboratory, Upton, New York. Mericle & Mericle (1965; 1967; 1971), who also studied the genetics of the *Tradescantia* system, started with material obtained from Sparrow and Schairer and compared effects of 'background' irradiation and gamma rays. Underbrink *et al.*

(1970) studied the relative biological effectiveness (RBE) of X-rays and neutrons. Ichikawa (1972) studied the radiosensitivity of a triploid clone of *Tradescantia*. Underbrink, Schairer & Sparrow (1973) showed that the system is also applicable to chemical mutagens. Naumann, Underbrink & Sparrow (1975; with 126 references!) investigated the effect of dose rate on mutation frequencies (N.B. Mutation frequencies can be expressed in various ways, like the number of pink mutant events per 10^3 hairs, or per 10^4 divisions in hair-cells.) Sparrow & Sparrow (1976) compared differences in spontaneous mutation frequencies, flower colour and stamen hair cells of various clones of *Tradescantia* and hybrids, whereas Sparrow, Underbrink & Rossi (1972) analyzed dose–response curves after applying small doses of X-rays and neutrons. The *Tradescantia* system is still used on various occasions. Some recent publications from Japan, for instance, referred to gamma-ray induced mutation frequencies in stable and unstable clones (Ichikawa, Imai & Nakano, 1991; Ichikawa, 1994) and the effects of combined treatments of alkylating agents and X-rays (Ichikawa, Yamaguchi & Okumura, 1993).

In recent years the small weed *Arabidopsis thaliana* (mouse ear cress), fam. *Cruciferae*, has become the most important model species for advanced genetic studies on a wide range of traits in higher plants, Induced mutations are often used as a starting point for these studies. Some additional points are presented in Box 1.4 and elsewhere in this book.

In the last decades efficient *in vitro* systems have been developed for almost all kinds of organisms, in which homogeneous cell populations can be studied. Such systems have considerably facilitated fundamental mutation research work. However, for higher plants still many technical problems are encountered, in particular because the situation in a test tube or Petri dish seldom agrees with actual conditions in the field or in the greenhouse. The subject of *in vitro* mutation breeding of higher plants will be discussed at various places in later chapters, in particular in Chapter 5.

1.3.6. New plant cultivars from induced mutations

Early references

One of the first overviews about radiation-induced mutant cultivars was presented in 1962 by Singleton (1964) at the 3rd congress of EUCARPIA (which is the European organization for plant breeding). This list (slightly adapted) contained nine radiation-induced mutant cultivars for seed propagated crops [cv. Primex, white mustard (*Sinapis alba*), Sweden, 1950; cv. Chlorina, tobacco (*Nicotiana tabacum*), Indonesia, about 1950; cv.

Schaefer's Universal, bean (*Phaseolus vulgaris*), Germany about 1950; cv. Regina II, summer oil rape (*Brassica napus*), Sweden, 1953; cv. Weibull's StråI, fodder pea (*Pisum sativum*), Sweden, 1957; cv. Sanilac, navy bean (*Phaseolus vulgaris*), Michigan, USA, 1957; cv. Pallas, barley (*Hordeum vulgare*), Sweden, 1958; cv. N.C.4X, peanut (*Arachis hypogaea*), North Carolina, USA, 1959; cv. Florad, oats (*Avena sativa*), Florida, USA, 1960].

The list was not complete and the years mentioned, which (supposedly) refer to the year of release or registration of the mutant cultivar, are not always correct. The tobacco mutant cultivar Chlorina, for instance, was already released in 1934, and not 'about 1950', as Singleton erroneously mentioned. This mutant cultivar which, most probably, became the first radiation-induced mutant cultivar in the world, was grown already extensively in 1936 in Indonesia where it had been released (Tollenaar, 1934; 1938). Some details will be presented in Chapter 2.

Singleton (loc.cit.) also mentioned that also a number of successful mutant cultivars for vegetatively propagated crops, e.g. for fruit trees and ornamentals, had been obtained but – with one exception for cv. White Sim of carnation (*Dianthus caryophyllus*) – presented no cultivar names or other details. Singleton referred, however, to a publication on mutation breeding for vegetatively propagated crops by Nybom (1961), but this author did not mention names of mutant cultivars. For vegetatively propagated crops the mutant cultivar Faraday of tulip (*Tulipa* sp.), which was released in 1949 in the Netherlands by de Mol after X-ray treatment of tulip bulbs in 1936, apparently, is the first mutant cultivar in vegetatively propagated crops (Broertjes & van Harten, 1988). In Chapters 2 and 7 several (early) examples of induced mutant cultivars for this group of crops will be given.

Central registration of artificially induced mutant cultivars

In 1964 an important Technical Meeting on 'The Use of Induced Mutations in Plant Breeding' (Anon., 1965), jointly organized by the Food and Agriculture Organization of the United Nations (FAO), the International Atomic Energy Agency (IAEA) and EUCARPIA, was held in Rome. At that time about 30 (artificially induced) mutant cultivars were known. As a result of that meeting a small group of leading mutation scientists agreed to coordinate their activities in this field. Their efforts benefited from the establishment in that same year of the Joint FAO/IAEA Division of Nuclear Techniques in Food and Agriculture at IAEA headquarters in Vienna. Since then the FAO/IAEA Plant Breeding and Genetics Section in

Vienna and the Agricultural Section at the IAEA Seibersdorf Laboratory (situated near Vienna) have performed a wide range of activities to stimulate the development and application of mutation breeding. These activities among which include, starting from 1969, the registration of data concerning artificially induced mutant cultivars, have contributed significantly to our present knowledge of mutations and their application in plant breeding. Further details about the FAO/IAEA Plant Breeding and Genetics Section and the Seibersdorf group and their main tasks are presented in Box 1.6.

The year 1969 has been often considered the turning point from primarily fundamental research on mutations and studies on radiobiology to increased attention for applied mutation breeding for new cultivars (Micke et al., 1990). At a Symposium on The Nature, Induction and Utilization of Mutations in Plants, in Pullman, USA, Sigurbjörnsson & Micke (1969) presented a list of 77 mutant cultivars. Those mutants, predominantly, were still by-products of fundamental research, rather than results of purposeful mutation breeding programmes. When in 1990 another FAO/IAEA symposium was held in Vienna, mainly to assess achievements of mutation breeding work and to discuss persisting problems, Sigurbjörnsson (1991) could refer to some 1330 mutant cultivars that had been registered and released by the end of 1989. Most recently, during a FAO/IAEA meeting in Vienna on mutations in relation to molecular techniques, Maluszynski et al. (1995) reported that the FAO/IAEA Mutant Varieties Database at that stage included 1790 entries involving 154 plant species. The mutant cultivars had been released in more than 50 countries, of which the 'top six' countries were: China, India, the former USSR, the Netherlands, Japan and the USA. From the 1306 mutant cultivars for crop species, 1237 cultivars belonged to the seed propagated crops with in particular many cereals (322 for rice, 240 for barley and 140 for wheat).

Numbers of listed or released mutant cultivars for a given trait or a given country may differ considerably in different publications. In *Mutation Breeding Newsletter 38* (Anon., 1991a), for instance, a number of 82 mutant cultivars for seed propagated crops was given for the former USSR. Salnikova (1993a), however, mentioned 366 mutant cultivars from the former USSR; this last figure, most probably, refers to seed propagated as well as vegetatively propagated crops.

Nowadays, for a number of crops, one even has lost track of the newly induced mutant cultivars. Several reasons account for this. First, in particular for some important ornamental crops, mutant cultivars are not registered anymore. This, for instance, is the case for mutants in the cut flower chrysanthemum (*Dendrathema morifolium*) where it is certain that hundreds of not-registered, induced mutant cultivars for this ornamental have been commercialized. And this, probably, may be the case for some other ornamentals as well. Mutagenic treatment of cuttings of chrysanthemum, which is performed in order to stimulate and to speed up breeding for new flower colours and other traits, has become common practice in several countries, in particular in the Netherlands. Spontaneous mutations for such traits turn up as well and breeders often do not care to distinguish anymore between spontaneous and induced mutants. Part of the reason for this is that the public has become aware of – supposed or real – risks of what is commonly called 'biotechnology'. As a consequence, breeders may prefer not to mention anymore that their cultivars arose from mutagenic treatments, as this may negatively influence their customers.

A second reason why the listing of new mutants is not complete anymore, is that, nowadays, breeders pay much more attention to the use of mutants in cross-breeding whereas in the past, attention was mainly focused on so-called **direct mutants**. To obtain useful mutants directly from mutagenic treatment must be considered an exception rather than a rule and in most cases the induced mutations should be considered as 'raw material' or 'building blocks', of which the potential can be assessed only after further crossing.

A wide-spread lack of insight and knowledge among breeders and other scientists about the special features and requirements of mutation breeding has seriously hampered a realistic assessment of the potentials of mutation breeding for many years. To give just one example: it was often not realized that negative side-effects of valuable mutations in many cases can be considerably modified or even eliminated by additional crossing.

Despite the fact that, even these days, not all unfounded criticisms about the (potential) value of induced mutations have disappeared, the situation has changed considerably. Breeders with some basic understanding of mutation breeding have learned now to handle plant material that resulted from mutagenic treatments as a – potentially valuable – source of genetic variation that, in most cases, reveals its true potential only after additional crossing and selection work. An example, derived from mutation studies on barley (*Hordeum vulgare*), which were started in Sweden by Nilsson-Ehle and Gustafsson already before 1940, may further illustrate this point. Despite the fact that this work was primarily intended to study the process of mutagenesis for higher plants and not to produce new

At the International Atomic Energy Agency (IAEA) a Department of Research and Isotopes is operating, to which the Joint FAO/IAEA Division of Nuclear Techniques in Food and Agriculture and the group of 'Agency's Laboratories' belong. The IAEA itself was founded in 1957 to 'accelerate and enlarge the contribution of atomic energy to peace, health and prosperity'. The Joint FAO/IAEA Division was established in 1964. For mutation breeding in particular the Plant Breeding and Genetics Section and the Plant Breeding Unit in Seibersdorf are important. These two, relatively small, groups within the Department of Research and Isotopes are mainly involved in the following activities.

- Research programmes (which run for a number of years) are initiated for different topics, like for instance on the use of neutrons in seed irradiation (1966–9), on the use of induced mutations for rice improvement (1967–70), on nuclear techniques to improve crop protein (1969–79), etc. In 1989 the Section in Vienna fostered 70 research projects.

    Under the IAEA Technical Cooperative Programme assistance is given to plant breeders in developing countries in the form of equipment, supplies and expert advice in addition to training courses and fellowships (see later). In 1989 44 Technical Cooperation projects were under way in 35 different countries (Anon., 1990a).

- Publication of the *Manual on Mutation Breeding* (Anon., 1970 1st ed.; Anon., 1977a, 2nd ed.) as the outcome of another coordinated research programme. The Manual for years has proven to be a very useful source of information for different groups of users.

- Proceedings of many Symposia (often organized in combination with other organizations), Panels, Research Coordination Meetings, Seminars (often on a regional base), Consultants Meetings, etc., are published.

- In May 1972 a series, called *Mutation Breeding Newsletter*, was launched. The main aims of the Newsletter (which is issued free of charge) were to bring news of progress in the field, to announce coming events and activities, to mention useful publications, to publish research notes, to report on the activities of the Section and to produce lists of new cultivars obtained with the aid of mutations. Until 1994, 41 numbers were issued in this series.

- Registration of new mutant cultivars was set up right from the beginning. A preliminary list was presented in 1969 at a symposium on mutation breeding at Pullman, USA. The first complete list was published by Micke *et al.* (1985). A summary of the 'Mutant Varieties Database', referring to seed propagated plants, was last published by Maluszynski *et al.* (1991) in *Mutation Breeding Newsletter* 38. The part for vegetatively propagated crops, was last published in *Mutation Breeding Newsletter* 39 (Anon., 1992b).

- In 1982 a series of publications, called *Mutation Breeding Review*, was started. In each number one topic is high-lighted, like mutation breeding for a specific group of crops. In the numbers 3 and 7 of this series Micke *et al.* (1985) and Micke *et al.* (1990), of the FAO/IAEA Plant Breeding and Genetics Section in Vienna, presented useful reviews on the current state of mutation breeding, including quantitative information on numbers of mutant cultivars released, species and traits concerned, mutagenic treatments, examples, etc.

- In 1969 a First International Training Course on the Use of Radiation and other Mutagen Treatments for Crop Improvement was organized at the Casaccia Nuclear Research Centre near Rome, Italy. Consecutive courses were organized in different continents.

    The First International Training Course on Induction and Use of

Mutations in Plant Breeding at the IAEA Laboratories in Seibersdorf, Austria, took place in 1982. Since that time such training courses have been organized almost on an annual base. They are mainly intended to train junior scientists from developing countries. In addition, IAEA fellowships are made available for individual programmes.

- At the Seibersdorf Laboratory mutation breeding research, using nuclear and related biotechnological methods, is performed. Projects refer to priority areas defined by FAO and IAEA upon requests from member states. Much attention has been paid recently to *in vitro* mutation breeding technology for crops of importance to developing countries, like banana (*Musa* spp.), cassava (*Manihot esculenta*), yam (*Dioscorea* spp.) and cocoa (*Theobroma cacao*).

- At the Seibersdorf Laboratory technical advice on performing mutation research, in particular for developing countries, is given and radiation treatment services are provided for research by scientists working in member states.

Both '*in vivo*' and *in vitro* plant material can be exposed to ^{60}Co-gamma rays or to fast neutrons in the research reactor of the Austrian Research Centre at Seibersdorf. Between 1967 and 1990, 18 612 samples, concerning 217 plant species and 1134 cultivars were irradiated; in 13 265 cases by ^{60}Co-gamma rays, in 4892 cases by fast neutrons and in 455 cases by other treatments. Seed samples were treated in 16 984 cases, vegetatively propagated material 823 times, *in vitro* material 599 times and other material 206 times. Treatments were performed for 108 Member States. In recent years an increased interest in neutrons and in treatment of *in vitro* material could be observed (Brunner, 1992). (N.b. The mailing address of the IAEA and its sections is: P.O. Box 100, A-1400, Vienna, Austria.)

mutant cultivars, originally two well-known direct (or primary) mutant cultivars were obtained: cv. Pallas and cv. Mari, which were officially approved in 1958 and 1960 respectively. In addition, till 1982 eleven other (indirect) cultivars, derived from crosses of the aforementioned two mutants with another cultivar and from continued crossing, were obtained (Gustafsson, 1986).

A distinction between direct and indirect use of mutations is of particular relevance to seed propagated crops. Further improvement by crossing will not be possible in obligately vegetatively propagated crops, whereas crossing in other vegetatively propagated crops is often hampered. Specific problems of applying mutation breeding for vegetatively propagated crops will be discussed in Chapter 7.

Main sources of reference about induced mutants
For an inventory of mutant cultivars, obtained as a result of mutagenic treatments, different sources can be consulted. In particular for seed propagated crops often, but not exclusively, use is made of publications by co-workers of the IAEA. Much information can be found in Micke et al. (1987) and in *Mutation Breeding Review* 7 (Micke et al., 1990). The many tables are particularly valuable in providing a clear picture of the results obtained till then. Also very useful is a review article by Konzak et al. (1984). Data about more mutant cultivars can be found, for instance, in *Mutation Breeding Newsletter* 38 of December 1991 (Anon., 1991a) and *Mutation Breeding Newsletter* 39 of January 1992 (Anon., 1992b). These newsletters contain data about mutant cultivars recorded in the FAO/IAEA Database in Vienna (Maluszynski et al., 1991; 1992) and in the Symposium proceedings of the 1995 FAO/IAEA Symposium in Vienna (Anon., 1995), but these additional data do not essentially change the overall picture presented by Micke et al. (loc.cit.). For vegetatively propagated crops often use is made also of the publications by Broertjes & van Harten (1978; 1988) and van Harten & Broertjes (1989).

Table 1.1 shows the total number of commercial mutant cultivars released till 1989. The total number of mutant cultivars mentioned in this table should be considered an underestimation, in particular because many mutants in ornamental crops had not been listed yet in the IAEA data bank. For seed propagated crops some ear-

lier obtained mutant cultivars may have remained unnoticed as well at that time. Although, at that stage, still more 'direct' mutants were listed than 'indirect' ones (indicated in the table as: 'used in crosses'), a clear tendency could be observed already in favour of the indirect use of mutants. In the years to follow this impression indeed was confirmed.

Mutation breeding: not always cheap and easy!
The amount of work required to obtain a new mutant cultivar may differ considerably, depending amongst others on the crop and the trait involved. It is, therefore, necessary to take a stand against the idea that succesful mutation breeding implies that results can be obtained always in a fast and easy way. Mutation breeding, generally, in this and other respects does not differ from most other plant breeding activities and such fast and easy results are rather more exceptions than routine. And although knowledge of the genetics and breeding properties of the traits to be improved, in combination with some knowledge of mutation breeding, may increase the chance of success to a certain degree, the induction of specific desired mutations, basically, remains a random process. Optimal procedures to obtain the highest frequency of desired mutations have to be determined in an empirical way.

For a 'difficult' crop like groundnut (or peanut; *Arachis hypogaea*), an allotetraploid species, for instance, one million(!) so-called M_2-plants (i.e. plants of the second generation after mutagenic treatment of seeds) were grown by W.C. Gregory in 1959 in North Carolina, USA. This population, ultimately, resulted in only one commercial mutant cultivar, registered as N.C.4X (Singleton, 1964).

On the other hand, one example of fast and very successful mutation work should be briefly mentioned here. In chrysanthemum (*Dendrathema morifolium*) the period between irradiation of cuttings for mutation induction and commercialization of a new mutant cultivar may be limited to 1.5–2 years only and as for an adequate mutation experiment in this case not more than 200–500 cuttings are required, a relatively small investment in time and labour may do the job. A remarkable example has been the mutation breeding work in cv. Horim of chrysanthemum in the Netherlands. Within a few years after the irradiation project had started, a

Table 1.1. Number of mutant cultivars registered and released till 1990 (after Micke *et al.*, 1990)

Seed propagated crops			Vegetatively propagated crops			All crops
Direct mutants	Used in crosses	(Total)	Ornamentals	Fruits and others	(Total)	
567	285	(852)	409*	69	(478)	1330

* N.B. Estimation far too low!

whole range of flower colour mutants of cv. Horim were obtained, which quickly replaced other cultivars. At one stage this series of Horim mutants represented 35% of the (substantial) Dutch market (Broertjes, Koene & van Veen, 1980). More details on this example will be presented in Chapter 7.

The economic value of mutation breeding: some examples

Mutation breeding, in a number of cases, has proved to be very attractive from a commercial point of view, a fact that is not often mentioned in literature. In Italy, for instance, one mutant cultivar Creso of durum wheat (*Triticum turgidum* subsp. *durum*) was grown in 1990, about one-third of the total area for durum wheat. During a period of ten years an <u>extra</u> economic profit of 1800 million US dollars was obtained by growing this cultivar (Rossi, 1979; Scarascia-Mugnozza *et al.*, 1991; 1993).

Recently, Wang (1991) mentioned that in China over the period 1985–90 about 9 million hectares were planted with mutant cultivars of some 20 crop species, which is about 10% of the total area of cultivated land in China for these crops.

Kivi (1991) reported on three mutant cultivars of oats (*Avena sativa*). Seeds from the oat cultivar Sisu, which was released in 1948, were X-irradiated and from a cross between the induced mutant line and cv. Blixt the (indirect) mutant cultivar cv. Ryhti arose. This mutant cultivar was released in 1970 and occupied up to 41% of the total area for oats in Finland during the period 1970–80. Further crosses with cv. Ryhti resulted in two other mutant cultivars: cv. Puhti, released in 1970, occupying up to 30%, and cv. Nasha, released in 1979 with up to 15% of the total area during the period 1970–80. All three mutant cultivars shared the trait of stiff straw.

Mutant cultivars from various mutagenic treatments

Various mutagens can be applied to induce mutations. In 1972 the Joint FAO/IAEA Division undertook a preliminary survey to find out which mutagens were used by scientists working on induced mutations in crop plants. In total 217 answers were received from an estimated number of 300 scientists working in this field. Fifteen different mutagens were mentioned one or more times. The use of gamma rays and EMS (ethylmethane sulfonate) was reported more than 100 times, X-rays and neutrons about 100 and 80 times respectively. With respect to chemical mutagens four of them – ethylene-imine, diethyl-sulphate, colchicine and nitroso-methyl-urea – were mentioned 50 and 20 times. The use of UV-radiation, absorbed radioisotopes and five further chemical mutagens (isopropyl-methane-sulphonate, ethyl-ethane-sulphonate, methyl-methane-sulphonate, nitroso-ethyl urethane and nitroso-methyl-guanidine) was reported less than 20 times for each of the mutagens mentioned. More details on the results of this survey have been published in *Mutation Breeding Newsletter* 4 (Anon., 1974a).

Although recently new data have been presented by Maluszynski *et al.* (1995), the most detailed analysis of data concerning mutant cultivars is still based on the results published by Micke *et al.* (1990). As is shown in Table 1.2 (adapted from Micke *et al.*, loc.cit.), most mutant cultivars obtained so far, have resulted from radiation treatments, in particular from irradiation with gamma rays and X-rays which, gradually, have become the most important methods of treatment.

The overall number of the mutant cultivars (1028) to which is referred in Table 1.2 is lower than the overall number for all crops (1330) in Table 1.1. This discrepancy can be explained by the fact that exact mutagenic treatments could not be traced in all cases.

Micke *et al.* (loc.cit.), in their original table, for seed propagated crops, distinguished between cereals and other crops. Of the total number of 563 mutant cultivars for this category, 347 cultivars (or 60%) refer to cereals. This percentage is an average for all mutagenic treatments and does not exactly represent the data obtained for the various treatments. Total numbers are too low to justify any conclusion about the efficiency of specific treatments for different groups of crops. Nevertheless, it is interesting to observe that mutation breeding has resulted in high numbers of new cultivars in monocotyledonous crops like cereals, which often are found to be rather recalcitrant when some modern biotechnological techniques like genetic transformation are applied.

Mutants in different groups of crops

Plant breeding for vegetatively and seed propagated crops, in many ways, has developed along different lines. Authors of publications in which the results of different mutation breeding treatments or effects for different

Table 1.2. Number of mutant cultivars obtained from using different mutagens (after Micke *et al.*, 1990)

Mutagen	Seed propagated crops	Vegetatively propagated crops	Total
Gamma rays	366	204	570
X-rays	65	227	292
Neutrons	36	13	49
Other radiation	15	7	22
Chemical mutagens	81	14	95

traits or in different crops are evaluated, often have limited themselves to one of these groups only.

For vegetatively propagated crops Broertjes & van Harten (1988, Table 1, p. 4) referred to, in total, 311 registered and released mutant cultivars, of which 27 were fruit crops, 13 other crops (including root and tuber crops) and 271 ornamentals. The group of ornamentals contained 47 mutant cultivars for root and tuber crops, 90 for pot plants, 114 for cut flowers and 20 for other ornamental crops. The number of 114 for cut flowers – as has been explained before – must be considered a gross underestimation, as many mutants of cut flowers are not registered as such anymore.

Micke et al. (1990), who listed 465 mutant cultivars for vegetatively propagated crops, mentioned 407 induced mutants in ornamentals, 31 mutants in fruit crops and 27 in other crops. In both publications one is struck by the relatively high numbers of mutant cultivars that have been obtained in ornamentals. One reason for this is that directly visible mutations for traits like flower shape and flower colour in this respect are much easier to handle than, for instance, traits like protein composition in seeds or polygenically inherited resistances.

Many more examples for vegetatively propagated crops than can be presented in the present book have been documented by Broertjes & van Harten (1988) and in Mutation Breeding Newsletter 39 (Anon., 1992b). In this last publication a number of 523 mutant cultivars for vegetatively propagated crops, which were officially released (and about which mutants had been reported already in Mutation Breeding Newsletter, numbers 1–37), were mentioned. Most mutants, representing more than 60 different species (or genera), were produced in chrysanthemum (187, a significant underestimation!); Alstroemeria (35); Dahlia spp. (34); Streptocarpus (30) and Rosa spp. (27). Mutant cultivars from 19 countries were mentioned, with the Netherlands (173 mutant cultivars) and India (103 mutant cultivars) being the most important contributors.

Mutations for different traits

In Table 1.3 the different traits, improved by mutagenic treatments of seed propagated crops, have been categorized. Use is made again of the results collected and grouped by Micke et al. (1990). Cereals and 'other crops', this time, show some interesting differences and therefore both categories are mentioned separately. The distinction between direct and indirect mutants in the original table by Micke et al. has not been followed. In total 1019 mutant cultivars for in total 80 species of seed propagated crops were mentioned, from which 609 cul-

tivars originated in a direct way and 410 after crossing with a mutant.

For each trait mentioned in Table 1.3 several interesting examples could be given here, but we have to restrict ourselves at this place to some main points. Previous to Micke et al. (loc.cit.), numbers of mutants released and examples were published by, for instance, Singleton (1962; 1964), Sigurbjörnsson & Micke (1969; 1974), Micke & Donini (1982), Donini, Kawai & Micke (1984), Gottschalk & Wolff (1983), Konzak (1984), Konzak et al. (1984), Micke, Maluszynski & Donini (1985), in various issues of the Mutation Breeding Newsletter and Mutation Breeding Review. In Mutation Breeding Newsletter 38 (Anon., 1991a), data about officially released mutant-cultivars for seed propagated crops that were published in the previous issues (till January 1991), have been summarized.

Data about numbers of mutant cultivars do not allow conclusions to be drawn about the relative contribution of mutants to plant breeding activities in a specific country; nor do such data inform us about their economic value. In some cases almost all mutants, released in a specific country, may refer to one plant species only and sometimes a large majority of the induced mutants may have resulted from the work of one research institute or one person only. This last situation occurred for mutation breeding work on vegetatively propagated ornamentals in the Netherlands. Throughout the years private breeding companies have released and registered

Table 1.3. Improvements reported for mutant cultivars of seed propagated crop plants (after Micke et al., 1990)

Improvements	Cereals	Others	Total
Increased yield	248	131	379
Plant architecture:			
–reduced plant height	289	47	336
–other changes	94	84	178
Resistance:			
–pathogens	186	53	239
–pests	6	8	14
Earlier maturity	164	88	252
Seed traits:			
–morphology	40	29	69
–quality	115	61	176
Other improvements reported:			
–adaptability	82	39	121
–threshability	10	9	19
–easier harvesting	4	4	8
–cold tolerance and winterhardiness	25	6	31

N.B. It is not claimed that all these improvements resulted directly from induced (gene) mutations.

many mutant cultivars for this economically important category of crops and quite a few mutant induced cultivars have been – or still are – very successful from a commercial point of view. The basis for the success with vegetatively propagated ornamentals was laid in particular by the work of the late C. Broertjes, who, for many years, worked on mutation breeding for this group of crops at the former Research Institute ITAL in Wageningen and advised private breeders on their applied mutation projects. For seed propagated crops, where such a coordinator was lacking, almost no mutant cultivars have been produced in the Netherlands.

Some comments should be made concerning Table 1.3. First of all, mutations for different traits and occurring in different plant parts can not all be observed with the same ease. Screening for mutations for skin colour and skin structure in seeds, for chlorophyll deficiencies in young, small seedlings or for spines on stems and fruits, for instance, is much easier than for yield, protein content or composition in seeds, digestibility, alkaloid content in seeds, root tubers or leaves, etc. With respect to this second category of traits it is difficult to initiate adequate mutation breeding programmes unless efficient screening methods for those traits are available. Often such (large-scale) methods have to be developed before the breeding programme can be started. Another practical point to be considered (like in more 'conventional' crossing programmes) is the number of observations and the amount of time required when screening for desired mutations. In some instances it may take several years before it can be checked whether a specific improved trait (e.g. in fruit trees) is not accompanied by adverse side effects in the adult plant stage or when the plant is grown on a large scale. Moreover, it may take a number of years before the improved mutant plants are accepted by the growers and before, for instance, fruits with a different taste or colour, will be accepted by the consumer.

Mutations for increased yield. A general statement to be made with respect to the complex 'trait' yield is that increased yields often result as 'by products' of changes in plant architecture which, for instance, may have been induced by mutagenic treatments. The observation in Table 1.3 that relatively many examples of new cultivars with (genetically determined) higher yield have been obtained as a result of mutagenic treatment, probably much more reflects the permanent necessity to produce higher yields worldwide, than the degree to which mutation breeding programmes have been successful in this respect. It is, on the other hand, beyond doubt that many mutation programmes (and, maybe, other breeding programmes as well) for improved yield would have

been more successful if they had been performed in the proper way, i.e. if they had been executed on a scale large enough to detect positive mutations for yield characteristics and if selection had been carried out by plant breeding specialists who know how to handle polygenically inherited traits.

Nevertheless, several successful cases have been reported. The most spectacular example concerns the mutant cultivar Diamant from barley (*Hordeum vulgare*), which was released in Czechoslovakia in 1965. This new cultivar yielded 11% more than its – already high yielding – parent. From cv. Diamant the (indirect) mutant cultivar Trumpf arose in the former German Democratic Republic (GDR) into which, by further crossing, improved disease resistance had been introduced. The new cultivar Trumpf in 1973 produced 15% more than the best cultivars of that time and in 1975 it already occupied 70% of the total area for barley in the GDR. In later years cv. Trumpf became incorporated in many barley breeding programmes in a great number of countries (Bouma, 1967; 1976; Bouma & Ohnoutka, 1991; Konzak *et al.* 1984). More details are presented in Chapter 6.

In a previous section the successful mutation breeding programme for increased yield in durum wheat (*Triticum turgidum* subsp. *durum*) in Italy was mentioned (Rossi, 1979; Scarascia-Mugnozza *et al.*, 1991; 1993) and more examples will be discussed later in this book.

When interpreting figures in literature about realized increases in yield by mutation breeding or other breeding methods, it is useful to keep in mind that to add a few percents of yield to an already high-yielding cultivar is often more difficult than, for instance, doubling the yield of a primitive landrace in which, so far, not much breeding work was done. On the other hand, a simple calculation may show that an increase of yield by some percents in an already outstanding cultivar may add more additional kilogrammes than an – at first sight – spectacular increase in a landrace or primitive cultivar. The general advice to mutation breeders is always to start from the most modern and highest yielding cultivars that are available. Mutations for complex, quantitatively inherited traits, like increased yield, will be discussed again in Chapters 3 and 6.

Mutations for plant architecture. Mutants for changes in plant architecture are important from various points of view. Spectacular results have been obtained by induced and spontaneous mutagenesis with respect to lodging resistance in cereals, which trait is based on inducing short and stiff culm. Another advantage of short-straw mutants in cereals is that mutations for this trait may lead to a higher **harvest index**; this expression, for instance, signifies that the relative contribution of the ears – the most profitable and high-quality part of the plant – to the total biomass has increased. Short-straw mutants, in particular in barley and rice, will be further discussed in Chapters 3 and 6.

Konzak *et al.* (1984) described the (indirect) use of the radiation-induced trait for determinate bush habit in dry bean (*Phaseolus vulgaris*) in the USA, starting with the release of cv. Sanilac in 1957 (see also Singleton, 1962; 1964). Other examples refer to induction of a non-branching type of tomato (*Lycopersicon esculentum*) for outdoor growing which are better adapted to machine harvesting. Significant induced mutations for leaf and plant structure (so-called leafless and semi-leafless types) have also been obtained in pea (*Pisum sativum*).

Other interesting mutations in this category refer, for instance, to traits like increased branching (which may result in more seeds, pods or fruits), a more compact growth (for pot plants), different growth patterns (e.g. from indeterminate to determinate), shorter stolon length of potatoes, more concentrated flowering, etc. Mutation breeding for such traits in many cases has been successful and, if mutation programmes are properly designed and executed, many more results may be expected. More examples and additional details will be presented in Chapter 6.

Mutations for resistance. In most breeding programmes a major part of time and energy is devoted to breeding for resistance. Accordingly, mutation researchers in an early stage paid attention to this topic, but most workers were mainly interested in fundamental aspects, and whenever practical objectives (like inducing resistance against mildew in barley) were set, the approach in such programmes often was not well considered because of inadequate knowledge of the specific disease, of its interaction with the crop as well as of the possibilities of the mutation method. Freisleben & Lein (1942) were the first who published about mutation breeding for disease resistance. The Plant Breeding and Genetics Section at the IAEA headquarters in Vienna has set up many research projects on disease resistance, often in developing countries and organized several meetings on this topic. For proceedings of these meetings, reference can be made to, for instance, Anon. (1971; 1974b; 1976a; 1977b; 1983a and 1985). In 1977 an advisory group meeting was organized on the use of induced mutations for resistance to insect pests (Anon., 1978).

Results obtained so far have been rather disappointing, in particular when taking into account the large amounts of work that have been performed. Although mutations for improved resistance against various diseases have been reported for quite a few crop species,

only a few mutants have become of real practical value. Resistance to pathogens in plants can be complete or partial, durable or non-durable, etc., whereas its genetic control can be either dominant (the most common situation) or recessive, monogenic as well as polygenic. Testing for disease resistance, often, is not easy and in many cases the system of host–pathogen interaction is not well understood. Lack of knowledge of the epidemiology of the disease is another important limiting factor. When resistance should be monogenically controlled, new pathotypes often arise spontaneously within a very short time by mutation and selection among the millions of spores that are produced. Consecutively, those new pathotypes are able to break down the recently introduced defense mechanism. Protection by a monogenic system, therefore, often lasts for a short time only. There are, however, some very notable exceptions, which resistances have lasted already for many years (Micke, 1993a,b). The classical example is the *ml-o* resistance against all known pathotypes of powdery mildew (*Erisiphe graminis*) in barley (*Hordeum vulgare*), which monogenic resistance has lasted already for 50 years (a subject that will be discussed in Section 3.5 of Chapter 3). For the also long-lasting resistance of peppermint (*Mentha piperita*) against bacterial wilt (*Verticillium albo-atrum*) it has not been proven that this resistance is monogenic as well. This last example will be further treated in Chapter 7.

As breeding for resistance has already for many years been one of the main assignments for the breeder, the most common sources of resistance, based on simple monogenic or oligogenic systems of genetic control, often have been exhausted and therefore it may be difficult to obtain new resistance genes by mutation breeding. Induced mutations, predominantly, go from dominant to recessive, whereas most resistances are dominant. A complication of inducing mutations for resistance may be that negative side-effects (e.g. a linkage between increased resistance and a chlorophyll deficiency) occur which have to be eliminated again by further (back)crossing. Sometimes it is difficult or even impossible to break such linkages. Moreover, even if a new gene for resistance has been induced, it is most likely that this would be of temporary help only, as it may be anticipated that the new defence system will break down again within a limited period.

Although there are exceptions to this rule (e.g. the aforementioned *ml-o* resistance in barley), many breeders, at present, are not much interested in monogenic resistances anymore and, as will be explained later, in most cases mutation breeders can not offer other, more attractive options, like polygenically inherited resistances. In a few cases monogenic resistances based on dominant genes have been induced by mutagenic treatment and some examples will be discussed later.

In 1991, Micke (1991a), during a symposium in Japan, referred to three 'rather useful' examples of resistance, viz. *Ascochyta*-blight in chickpea (*Cicer arietinum*) in Pakistan; *Sclerospora graminicola* in pearl millet (*Pennisetum americanum*) in India, and the aforementioned *Verticillium*-wilt in peppermint (*Mentha piperita*) in the USA. In addition to the earlier mentioned reasons why mutation breeding for resistance so far has been of limited value, Micke (loc.cit.) states that many potentially useful mutations never will be detected, as traits like resistance or susceptibility will become expressed only in the presence of the matching pathotype. As a, preliminary, conclusion, it appears that mutation breeding for resistance should be performed only when other sources are not, or hardly, available or when a known resistance is very difficult to introduce into a specific crop or cultivar. It appears to be even more difficult to induce resistance against pests. The IAEA has also paid attention to this subject (Anon., 1978). In later chapters some other examples of the use of mutations for increased resistance will be discussed.

Mutations for earliness. Mutations for the next category of improved traits, early maturity, occur most frequently and can be found by very simple screening in the field or in greenhouses. As different systems of inheritance for this trait are known, further crossing work must reveal whether, in a specific case, monogenic (either dominant or recessive) or polygenic inheritance is involved. Many spontaneous mutations, as well as mutations induced by radiation or chemical mutagens are known for various crops. In 1960 the radiation-induced, early-maturing, semi-dwarf mutant cultivar Mari of barley (*Hordeum vulgare*) was induced in Sweden. This direct mutant has been often used in further crosses, from which several valuable indirect mutant cultivars were obtained (Gustafsson, 1986). Another example, reported by Hanna & Burton (1985) refers to induction of earliness in pearl millet (*Pennisetum* sp.) by treatment with 1% of the chemical mutagen ethylmethane sulphonate (EMS). Kivi (1981) selected from 350 000 M_2 plants of cv. Pokko, an early barley cultivar, 22 even earlier mutants. Two of these mutants – because of their overall performance – were worthwhile to be tested beyond the fifth mutation generation (M_5) for possible direct use. A new, early maturing mutant cultivar of banana (*Musa* sp.), cv. Novaria, was officially registered in 1993 in Malaysia. This triploid (of the so-called AAA group, derived from cv. Grande Naine (or Grand Nain) after mutagenic treatment *in vitro* with 60 Gy of gamma

rays at the FAO/IAEA Seibersdorf laboratories in Austria, in addition showed a stronger fruit stem (or peduncle). Some additional details are presented in Chapter 7, section 7.4.1.

More examples have been described extensively, for instance by Konzak *et al.* (1984).

Mutations for improved seed quality. In the previous paragraphs (and in particular in Box 1.1), some examples of mutations for improved seed quality were mentioned already. Breeders are very much interested in improvement of traits like content and composition of oils, proteins and storage carbohydrates, malting quality, digestibility, seed dormancy, etc., but these traits are by no means easy targets, neither for genetic improvement nor from the viewpoint of developing efficient screening methods. Most breeding programmes for quality traits suffer from these drawbacks and, because even for one specific trait within a crop more than just one genetic system may be involved, it is difficult to predict which traits can be improved in a relatively easy way. Particular interest has been paid to genetic changes in amino acids and increased protein content in barley and maize. The same applies to alterations in the fatty acid composition of crops such as sunflower (*Helianthus annuus*), soybean (*Glycine max*) and flax (*Linum usitatissimum*), in which considerable successes have been obtained.

Micke (1983a) and Konzak *et al.* (1984) have summarized results, Röbbelen (1990) has reviewed quality improvement by mutation breeding for oil seed crops and many additional results can be derived from reports in the proceedings of FAO/IAEA symposia and the like. In later chapters a number of experiments to induce mutations for such traits will be presented and results obtained will be discussed. One case will be briefly mentioned here. Rowland (1991) reported on a EMS-induced mutant in flax with seeds containing a stable low-linolenic acid content of 2% only, whereas in nature percentages may go up to 40%. This mutant was selected in the M_4, i.e. the fourth mutated generation after treatment. It was suggested that the low-linolenic trait is controlled by recessive alleles at two independent loci.

Mutations for seed morphology are of interest, e.g. in order to obtain seeds with a smooth skin instead of a rough skin; a soft skin instead of a tough skin; a better seed shape, which would make seeds more attractive for direct consumption (e.g. edible nuts) or a thin nutshell, like in oil palm (*Elaeis guinensis*); and many more examples could be thought of. A number of mutations have been induced for such traits and some examples will be discussed later.

Mutations for some other traits. This final category

in Table 1.3, which is in fact a 'rest group', includes various traits that are of interest to the plant breeder and for which mutations have been induced. Growers, continuously, would like to see certain genetic improvements in the crops they cultivate. Such wishes are not necessarily the same throughout the years, for instance because of changing systems of cultivation, increased interest in possibilities of mechanization and easy handling of crops, different storage systems, etc.

Many examples concerning different crops have been mentioned already by Micke & Donini (1982) and Konzak et al. (1984). Traits considered, for instance, are winter hardiness, heat and frost tolerance, adaptability to specific (acid, alkaline) soil conditions, increased suitability for mechanical harvesting, non-shattering fruits, threshability, tolerance to herbicides (also a 'topic' of modern biotechnology), milling or manufacturing quality and self-fertilizing or cross-fertilizing properties in plants (to be discussed specifically in Chapter 6). Recently, Forster et al. (1994) reported for cv. Golden Promise, a successful induced (direct) barley mutant from cv. Maythorpe, that this mutant has a reduced sodium content. The authors concluded that this favourable mutant trait, which is generally associated with salt tolerance, could be attributed to the mutagenic treatment.

An example of multiple changes in different traits is the aforementioned high-yielding barley mutant cv. Diamant. This mutant has shorter straw and more and larger spikes than the mother cultivar, cv. Valticky, a higher thousand-grain weight and an increased tiller number. But, according to Konzak et al. (loc.cit.) even more significant is the altered developmental pattern of this mutant cultivar for prolonged tillering, a longer juvenile phase, a better root system and a longer grain-filling period.

1.3.7. When to apply induced mutations in breeding of new cultivars?

In principle, two main situations can be distinguished where induced mutations are of value for the plant breeder. First, induced mutations may produce genetic variation to supplement the variation that was already available in the plant population studied. Second, plant breeders nowadays – in an increasing number of cases – apply advanced breeding techniques in the development of which induced mutations often have played a useful, if not indispensable, role. In this section the question of when to apply induced mutations as a source of genetic variation is briefly introduced. The use of induced mutations 'as a tool' will be discussed at a later stage.

When it is decided to start an applied mutation breeding programme, either to widen in general the spectrum of genetic variation, or to induce variation for a specific trait in a given crop, a greater or lesser part of the available genetic variation has been exploited already by the breeder. Various crops, in this respect, may have quite different breeding histories and the intensity of the previous breeding work determines how much of the available genetic potential variation has been exploited already. The need to make use of induced mutations, of course, depends on the level to which the breeder, in an effective way, has been successful in meeting his breeding goals.

Induced mutations, as said, may supplement the genetic variation that is already present within a plant species (or within a group of related species) in two ways. Induced mutations, in general, occur at much higher frequencies than spontaneous ones and for many traits (e.g. plant height, oil content, flower colour, leaf shape) new types may become observed. It is difficult – if not impossible – to determine whether such new forms, observed in the mutagenically treated population – on the long run and given the inspection of sufficiently large plant populations – would not have shown up as well as spontaneous mutations in the untreated plant material. Moreover, as processes of selection in nature and in man-made and man-controlled mutated plant populations differ considerably in most cases, spontaneous and induced mutation spectra, ultimately, may differ as well. This in particular may be the case when mutagenesis and the consecutive selection of mutants are applied in vitro. But, even if the final spectra of genetic variation would be identical, artificial induction of mutations may produce new alleles much faster than in nature and, thus, may speed up and economize breeding work considerably. The 'induced mutation approach' not only could be useful to keep the costs of a breeding programme low, but also – and this is of particular importance in ornamental crops – to produce within a very limited period of time, and ahead of other breeders, a range of novelties in a crop or group of cultivars, e.g. for colour of flower or leaf, flower shape, etc. Some very rewarding projects have been set up for this purpose and a few examples will be discussed in Chapter 7.

The aforementioned 'speeding up' of a breeding process by induced mutagenesis is not only of interest in crops in which much breeding work has been performed already, but also in relatively young crops that, still, are in an early stage of domestication. Now and then a (semi-)wild plant species is directly introduced as a new and promising crop, for instance because its seeds produce a valuable oil for industrial purposes or, in the case of grasses, because they make good lawn or are particularly useful as fodder. Two recently introduced oil crops are *Cuphea tolucana* (Campbell, 1987) and *Euphorbia lagascae* (Vogel & Röbbelen, 1989). Crops, when still in a semi-wild stage, often possess a number of traits, like shattering of seeds, undetermined growth patterns, sticky hairs or thorns, which are unfavourable to the grower, manufacturer or consumer and have to be eliminated in order to utilize these crops in an optimal way. Breeders, in such cases, may search for spontaneous mutants in which the unfavourable trait is absent. To try and induce artificial mutations is another possibility and, in fact, may be a quicker option. Cross breeding with genetically related plant material in which the undesired trait is absent would be another possibility. This, however, may be a risky affair, as by crossing not only the target trait (e.g. non-shattering seed) may be changed but many other traits as well, including the special feature trait (e.g. the specific oil composition) for which the new crop was selected.

This brings us to another, more common situation in which mutation breeding may offer better prospects than cross breeding. A breeding programme is often started from an already good and high yielding cultivar or breeding line in which, for instance, one or a few unfavourable traits, e.g. low lodging resistance, lateness or susceptibility to a specific disease, should be corrected, preferably without changing the genotype of the cultivar for other important traits. Crossing may offer good prospects only when the desired trait is present in closely related material, e.g. in near-isogenic-lines. However, when the required allele, for instance, should be present only in distantly related genotypes, in cultivars from abroad that are not adapted to growing conditions in a specific region, or in wild relatives, the result of making crossings will always be that a (large) number of unwanted alleles are introduced into the original cultivar. As a consequence, several generations of repeated backcrossing are needed before the required new trait and the good properties of the original cultivar have been combined again in a new, enriched cultivar of at least the same quality as the original one. Under such conditions the use of mutation breeding should be seriously considered as an alternative. This approach, for instance, has been successfully used for rice (*Oryza sativa*) by Rutger in the USA (Rutger, 1991; 1992). Details are presented in Chapter 6.

Finally, it is also conceivable that no crossable genotypes with the desired gene are available or that the desired allele simply is not known in nature. In that situation the use of mutation breeding even might be the only possibility that is still available. This, for instance, may be the case in obligate vegetatively propagated crops.

The most attractive option, of course, would be to apply real **site-directed mutagenesis**, aiming at changing or replacing only one particular gene. The common mutation breeding methods, as will be explained in Chapter 3, are not sufficiently precise to achieve this goal. Nowadays, it is possible, however, to apply sometimes one or a few of the new biotechnological techniques as a supplement, or sometimes even as an alternative to longer established breeding techniques (including mutation breeding) for this purpose. It should be realized, however, that most of these new methods – which will be briefly discussed and compared with mutation breeding methods later in this chapter – also do have their limitations and are not yet within easy reach for most higher plants.

A main point to keep in mind when comparing the advantages and disadvantages of various breeding methods, is to take into account the amount of work that is needed to screen for the desired trait in each specific approach. The breeder, further, has to check whether the mode of inheritance of the trait for which improvement is sought makes it accessible to various breeding methods. This subject will be further discussed at various occasions in later chapters.

1.3.8. Chimerism and competition between genetically different cells

Chimerism
Mutations, as a matter of principle, are single-cell events and – at least in case of gene mutations – concern only one of the homologous chromosomes and, as a consequence, also only one of the present two (or more) alleles. If a specific mutation occurs in one cell of an organism, it is – from a statistical point of view – most unlikely that an identical mutation in the same locus will arise within one of the neighbouring cells. Mutations with an increased ploidy level, in a way, are an exception to the rule of single-cell mutations, because this mutation may occur at the same time in many cells of a tissue. This topic will be briefly discussed in Chapter 3.

Plants, carrying genetically different cells or groups of cells within their somatic tissues, are called **chimeras**. As a consequence of the single-cell origin of mutations, chimera formation, or **chimerism**, is a very common phenomenon after a mutational event, irrespective of whether such mutations arose in nature or were artificially induced.

In seed propagated crops the genetic constitution of the next generation is determined only by the gametes and, as a consequence, mutations in the somatic tissue normally will be lost. However, in an increasing number of plant species it is possible to maintain this somatic tissue (including eventual mutations) and to induce its development into sporogenic tissue by various 'in vivo' or in vitro methods of vegetative propagation, as will be discussed in particular in Chapter 5. Such techniques are of growing importance, e.g. in relation to in vitro mutagenesis.

Male and female gametes ultimately trace back to a small number of so-called **initial cells**, situated in a strategic position within the meristem of shoot apices. The role of initial cells and related subjects will be described in particular in Chapter 7.

Let us take now a homozygous, self-fertilizing, diploid crop species – for instance pea (*Pisum sativum*) – as an example. It is assumed that this plant has a shoot apex in which two initial cells give rise to the germline and, thus, to the formation of the gametes. It is further assumed that for a specific, monogenically inherited trait: colour of the seed cotyledons (G for the dominant colour yellow; g for the recessive colour green), both initial cells originally have the genetic constitution GG (yellow). Seeds with the genotypes GG and Gg both show yellow cotyledons and can be easily distinguished from the gg genotype which has green cotyledons. Let us suppose now that one of the two initial cells produces (haploid) egg cells with the normal genetic constitution (G) and that a mutation has occurred towards recessive (g) in the other one of the two initial cells. As a result, 50% normal and 50% mutated egg cells are produced by the mutated initial cell. It is further assumed in this example that the corresponding locus in the (haploid) pollen remained unchanged (i.e. yellow; constitution G). As a consequence, after pollination, 50% of the seeds resulting from the mutated initial cell will be GG and 50% will be Gg. The non-mutated initial, of course, produces 100% GG seeds.

A special feature of cotyledon colour is that this trait, sometimes even without removal of the seed coat, can be observed already on the seeds produced on the original plant. Whereas the constitution of the cotyledon colour already refers to the next sexual generation, it must be noted that the seed coat itself consists of maternal tissue.

Plants that directly develop from a mutagenically treated seed – in accordance with the standard nomenclature for the consecutive generations in plant breeding programmes as F_1, F_2, etc. – are commonly called M_1 plants; plants of the next generation are called M_2 plants, etc. (for details see Box 1.7). In the example the M_1 plant, which is chimeric (partly GG, partly Gg) for cotyledon colour, is pollinated by non-mutated (G) pollen of the same plant and, as a consequence, the resulting seeds – which develop on the M_1 plant, but which (except for the seed coat) represent already the M_2 generation – will be either GG or Gg. The M_2 plants produce an M_3 generation which either will segregate for yellow and green cotyledon colour – in case the M_2 plant was been Gg – or remain all yellow – in case the M_2 plant was a GG plant.

The example illustrates that some knowledge of the basic phenomenon of chimerism is necessary to interpret the outcome of mutagenic treatments. Moreover, it is important to realize that not all plant species behave in a similar way. It should also be kept in mind that for most traits more genes are involved and that direct observation of an induced mutation in such situations may be much more difficult or even impossible.

During the sexual phase of seed propagated crops, mutations outside the gametes are eliminated, which makes it relatively easy to handle chimerism in this group of crops. The crucial point is to ensure that a mutation in an initial cell is transferred to the next generation and not lost halfway. In order to achieve this goal it is important to collect the necessary information about organogenesis and patterns of plant growth of each crop species. Monocotyledonous and dicotyledonous plants very much differ in this respect. Most information so far has been collected for (monocotyledonous) cereals, in particular on barley (*Hordeum vulgare*). For dicotyledonous crops some general recommendations were formulated during a consultants' meeting held in Vienna in 1981 (Anon., 1983b), but several aspects still need further clarification. Some additional information will be presented in Chapter 6.

For vegetatively propagated crops chimerism in many cases really does present major problems in different ways. However, the subject has been well studied and various methods to diminish the problem, or even to get completely rid of it, are available. The so-called adventitious bud technique, which originated from the research group of Sparrow in the USA and was taken up, perfected and successfully applied for a range of ornamental crops in particular by Broertjes in the Netherlands, deserves special attention. Theoretical and practical implications of chimerism for the group of vegetatively propagated plants, specific aspects for many different species as well as solutions and suggestions on how to deal with such chimerism have been described extensively by Broertjes & van Harten (1987; 1988). For further details see in particular Chapter 7.

Competition

Competition may occur between mutated and non-mutated plants at the population level, or within plants

Box 1.7 Terminology used to indicate different plant generations in mutation breeding programmes.

During a meeting organized by the Joint FAO/IAEA Division in Vienna in 1990, Sigurbjörnsson (1991) mentioned in his opening address that about 25 years earlier, a number of different notations were still used to designate different filial generations in mutation programmes, such as F_1, X_1, M_1, R_1, C_1, etc. Moreover, at that time a certain degree of inconsistency could be observed as to the question whether mutagenically treated seeds, plants, etc. should be considered as belonging to the first or second 'mutated' generation, etc. This caused unnecessary confusion, for instance when comparing methods and results of different authors with respect to treatments, optimal population size of different generations, preferred generation for selection for a given trait, etc.

It may be useful to summarize here the, in our opinion, most appropriate notation. It is advisable always to check carefully what was meant by different authors.

Plant breeders generally agree to indicate the generation following a cross between two parents (P_1 and P_2) as the F_1-generation, or, briefly, F_1. Accordingly the first mutated generation should be called the M_1. If seed is taken as the starting material for a mutagenic treatment, the original, untreated seed could be indicated as M_0 seed. After the treatment this seed should be called M_1 seed, whereas seedlings and adult plants, grown from M_1 seed are M_1 plants. Seeds which develop on M_1 plants are M_2 seeds. Each M_2 seed develops into an M_2 plant, which plants together represent the M_2 generation, etc. For vegetatively propagated crops a small addition to this notation is made. In accordance with M_1, M_2, etc. for seed propagated crops, here the notations vM_1, vM_2, etc. are used for the consecutive vegetative generations of a mutation programme.
(N.B.1. Some authors prefer V_1M_1, V_2M_1, etc., but there seems to be no special reason to do so.)
(N.B.2. In some earlier publications, and in particular in publications from the USA, the first mutated generation – M_1 in our notation – has been indicated sometimes as M_0. This notation does not make much sense and should not be used.)

It is further advised **not** to make use of special systems of nomenclature for mutagenic treatments with different types of radiation (like R_1, R_2, etc.), X-rays (X_1, etc.), neutrons (N_1) or with chemical mutagens. Other peculiar combinations, such as XM (plant material mutated by X-rays!) can be found in literature! This is highly confusing, in particular because each worker may like to develop his own system. Moreover, does the term R refer to radiation (in general) or to Röntgen- or X-rays, the term C to chemical or to colchicine, etc?

The kind of confusion, caused by such special notations, can be further illustrated by a system of notation that is used now and then to identify tissue culture regenerates and their progeny, in particular when studying 'somaclonal' variation. Orton (1983, 1984) described a system in which he indicated a regenerant, directly obtained from tissue culture as an R0 plant, the selfed progeny of an R0 plant as an R1 plant, the selfed progeny of an R1 plant as an R2 plant, etc. When 'somaclonal' variation – in fact, at least partly, a kind of spontaneous mutations – should have occurred as a result of tissue culturing, such mutations, according to this notation, were observed already in the R0 (or R_0). Another system, suggested to indicate the different generations after *in vitro* culture, makes use of the abbreviation SC_1, etc. (SC refers to 'somaclonal variation'; why not SV or SCV?), but whether SC_1 is identical to R_0 in the other system is not clear. In such cases it almost requires the special skills of a detective to draw justified conclusions when comparing mutation frequencies in a specific generation (*in vivo* or *in vitro*, spontaneously or after various mutagenic treatments).

Sometimes small differences in notation, like VM_1 (instead of vM_1) and M0 or M-0 (instead of M_0) are found in literature, but such differences are often the results of the limited skill of an author with word processing equipment, etc., and are not a cause of much confusion.

In conclusion it is advised to stick to a few clear rules concerning terminology and to avoid the use of non-standardized systems of nomenclature. The kind of mutagenic treatment that has been applied and the way consecutive generatons are studied in a particular case must be described adequately in the section 'materials and methods' of a publication.

at the level of mutated and non-mutated zones, tissues or cells. In this context some attention must be given to the concept of the **fitness** of a mutated cell, a mutated tissue, etc. The expression **fitness** can be defined as the selective value of an entity (a cell, a tissue, a genotype, a locus) in comparison with another entity. This selective value refers both to the survival value of, for instance, a cell, as well as to its reproductive capability. Fitness can also be expressed in terms of relative numbers of successful gametes contributing to the next generation. It is clear that the fitness of a mutated cell or – in later stages of development – of a mutated plant or plant part, will be decisive in determining whether the mutation will survive. Useful mutations, induced in cells which are drastically damaged by the mutagenic treatment, are not very attractive to the plant breeder, as the cells in which such mutations occur may have a very small chance of survival. Mutation breeders often try in an early stage to save, or put apart, plants which may contain useful mutations but, because of their lower fitness, probably would be lost when natural selection prevails. In recent times techniques to favour the further development of mutated cells, even in a case of a lower fitness, have been much improved, like the use of single-cell *in vitro* techniques. Selection of mutations under *in vitro* conditions will be discussed in particular in Chapter 5.

Sometimes favourable mutations for important traits are accompanied by negative side-effects: e.g. a desired compact plant type that resulted from a mutagenic treatment may be accompanied by a yield level that is lower than in the control. The expression of such negative side-effects may be considerably modified by growing plants under very different climatic conditions. This, for instance, was shown by Gottschalk (Gottschalk & Kaul, 1975; Gottschalk & Wolff, 1983) for mutated plant material of pea (*Pisum sativum*). When a specific pea mutant was cultivated in Bonn, Germany, an increased number of ovules per ovary was accompanied by a reduced number of pods per plant. In Kurukshetra, India, the mutant not only produced more ovules, but also more pods as a result of increased branching of the plant under those conditions.

At the population level an important question to consider is whether and how fast, in an – originally – homogeneous plant population, a mutated plant type, may replace the original plant genotype. According to Pickersgill & Heiser (1976) the general answer to this question is that the speed of such a shift largely depends on:

1. the prevailing system of reproduction (inbreeding, outbreeding, vegetative propagation),
2. the nature of inheritance of the mutation (dominant or recessive) and the degree of phenotypic expression,
3. the intensity of natural or artificial (human) selection for the new mutant.

With respect to this third point it should be kept in mind again that natural selection refers to fitness of the individuals concerned, in the sense that plants, tissues or cells with a low fitness have a lower competitive ability than the surrounding partners, which may result in a smaller offspring, etc.

In the early years of mutation breeding much attention was paid to the concept of competition between mutated and non-mutated cells in plant tissues. In fact the question is whether of two genetically different cells one cell has a selective advantage above the other. This competition may take place at the gamete level, leading to so-called **haplontic selection**, for instance between different pollen grains (certation) or in the female gametophyte. Competition between genetically different cells in somatic tissue during plant development, accordingly, could result in **diplontic selection**. The action of diplontic selection is best studied by making use of mutations which arise within shoot apices in the initial cells, which cells are the origin to which specific cell lineages can be traced back.

To give an example, let us consider the situation in the 'model crop' barley (*Hordeum vulgare*). In the embryo of a mature seed the main shoot and about three axillary shoots have been initiated already. Within a shoot apex a cell group – in fact a cell layer – is present from which the gametes arise and which consists of about four initial cells. These initial cells, present in each of the four shoots, give rise to the first four culms of a barley plant; later culms are the result of branching. Each culm bears an inflorescence consisting of a number of flowers and each flower produces a single seed. Let us assume now that mutagenic treatment of barley seed results in a mutation in one initial cell, giving rise to a visible mutation, for instance a chlorophyll-deficient stripe, that can be observed in plants of the M_1 generation. Starting from the apical area of a culm a long chlorophyll-deficient stripe (or sector) can be observed. The presence of long stripes demonstrates the stability of the shoot apex with its initial mutated cell (Dermen, 1947; Burk, Stewart & Dermen, 1964; Steffensen, 1968). Now and then stripes may taper off or even completely disappear somewhere along the plant. Such phenomena often are considered as indications for competition between the normal and the mutated tissue. In particular the German scientist Gaul strongly advocated the existence of such a selection mechanism and proposed the term **diplontic selection** that was mentioned before. Other, less commonly used names referring to the same process are **intrasomatic** or **intra-individual selection** (Kaplan, 1953). Gaul, on many occasions, has made a strong case of his views on the relevance of diplontic selection in relation to practical mutation breeding (see for instance Gaul, 1957; 1959; 1961a and 1964a,b) and quite a few leading scientists in the field of mutation breeding (e.g. D'Amato, 1965) have taken over and further disseminated Gaul's opinion. However, throughout the years, not everybody was convinced that diplontic selection indeed does play the important role that had been attributed to this phenomenon by Gaul. Conclusive proof as to whether, and to what extent, diplontic selection really does exist, is difficult to give, as most mutations are recessive and heterozygous in the M_1 generation and, therefore, phenotypically not expressed and not easily identified.

The disappearance of the aforementioned chlorophyll-deficient stripes was often used in the past as evidence for diplontic selection. However, as such stripes are the result of severe defects in the chlorophyll apparatus of a cell, they do not represent the common situation for mutational events at the gene level. Moreover, chlorophyll deficiencies are not caused only by changes in the nuclear DNA, which makes a correct interpretation of the observed phenomena much more complicated.

A category of mutations that can be used with relative ease to study the concept of diplontic selection are ploidy chimeras, in which for instance stripes or sectors, containing cells of various size as a result of different ploidy levels, can be observed within one plant or plant part. As was shown by Bain & Dermen (1944), Dermen (1945) and by many later authors, ploidy chimeras may be rather stable. Further details are outside the scope of this section.

Balkema (1971), who studied chimerism and diplontic selection in sunflower (*Helianthus annuus*) and *Arabidopsis thaliana* and, in addition, has provided the most complete review for different categories of mutations that have been put forward as evidence for the occurrence of diplontic selection, arrived at the conclusion that the behaviour of chromosomal aberrations in dividing cells of developing plants does not provide clear evidence in support of Gaul's concept of diplontic selection. She further concluded that loss of mutations or loss of chimerism often is not based on genotypic differences but can be explained on the basis of the normal development of a plant, for which phenomenon she suggested use of the expression **diplontic drift** (in analogy to the expression diplontic selection for loss of chimerism through genotypic differences). Some evidence for 'real' diplontic selection was found only when several potential meristems were present, of which only a few can develop, as may be the case for axillary or adventitious bud meristems.

Some other publications concerning diplontic selection should be briefly mentioned. Lindgren, Eriksson & Šulovská (1970) also concluded that any great effect of diplontic selection upon the sector size in their barley material looked highly improbable. Harle (1972) found for *Arabidopsis* no evidence for diplontic selection, and

Ukai & Yamashita (1974) concluded that chimerism does not cause particular problems for mutant selection.

With respect to the acting of selection, differences may exist between monocotyledous and dicotyledonous plants. In the report of a consultants' meeting on chimerism in irradiated dicotyledonous crops, organized in 1981 by the Joint FAO/IAEA Division in Vienna (Anon., 1983b), a very cautious standpoint is taken with respect to the possible action of cell competition within somatic tissues. There is no information as to the point whether identical competition patterns occur after treatments with (various) chemical mutagens and irradiation. Recently, it was reported (Cassells & Periappuram, 1993; Cassells, Walsh & Periappuram, 1993) that the use of the phenomena of diplontic selection and diplontic drift, in combination with *in vitro* mutagenesis in carnation (*Dianthus* sp.), by elimination of 'unfit' mutants, may considerably increase the frequency of flower colour mutants which are agronomically acceptable as well.

In conclusion, we believe that the opinion expressed by Balkema (1971) that genotypic differences between competing cells are not as important as was believed before by Gaul and many others, is still fully correct. Therefore, the often quoted statement by Gaul (1964a, p. 213) that 'the highest mutation rates are to be expected when there is no intercellular competition, or when it is still limited', needs reconsideration.

Based on the present views on chimerism and intrasomatic selection, Micke *et al.* (1990) have formulated some practical advice to plant breeders, which, briefly summarized, is as follows.

1. For seed propagated plants seeds should be harvested from the most chimeric part of the M_1 plants.
2. For dicotyledonous plants seeds should be harvested from different branches of M_1 plants because of the diplontic drift mentioned by Balkema (1971; 1972).
3. The ontogeny of the plant species should be taken into account when determining the most appropriate size of the M_1 population.
4. For vegetatively propagated crops chimerism either should be avoided or eliminated as much as possible. Nevertheless, so-called **periclinal chimeras** (see Chapter 7), because of their relative stability, often are useful, in particular in ornamentals.

These proposals will be discussed more into detail in later chapters.

1.3.9. Mutation breeding in practice: some main considerations

The expression **mutation breeding**, most probably, traces back to the 1940s, when Freisleben & Lein (1943a,b; 1944a,b) introduced the German equivalent ('Mutationszüchtung'). MacKey (1954a), for instance, already referred to 'mutation breeding'. Mutation breeding involves much more than simply searching for a useful spontaneous mutant or the artificial induction of a desirable mutation, followed by an immediate release of a new cultivar.

Mutation breeding programmes predominantly follow the consecutive steps of conventional schemes used in cross-breeding and for vegetatively propagated crops. Where needed, such schemes are adapted or improved for specific purposes, both in conventional (cross) breeding and in mutation breeding. In recent years, details of mutation breeding procedures have been worked out for a number of economically important crop species and for the most essential breeding objectives. Points deserving specific attention in mutation breeding are: choice of starting material, mutagenic treatments, screening and selection of promising mutants. The choice of the most suitable starting material for conventional breeding is not necessarily identical to that for mutation breeding, whereas mutagenic treatments, of course, are specifically developed for this breeding method. Mutagenic treatments will be discussed in particular in Chapter 4. Another important aspect to consider is how to increase the efficiency of selection of a desirable mutant, for instance by developing 'early' tests at seeds or young seedlings, by making use of specific tests for mutant traits (e.g. 'sweet' alkaloid-free genotypes in *Lupinus* sp.), by developing methods of indirect selection, by growing mutagenically treated cell populations on selective media *in vitro*, etc. The next steps, involving cultivar production, breeding for maintenance, etc., are identical to conventional breeding work.

For commercial plant breeders mutation breeding is just one of a number of methods with each method having its own advantages and disadvantages and requiring specific skills. Furthermore, for each breeding situation it should be carefully considered which approach may offer the best prospects in that specific case.

Mutation breeding in seed propagated crops and vegetatively propagated crops originally developed along different lines, with specific approaches and techniques for both groups of plants. With the increasing importance of a number of new breeding strategies, which in most cases are related to *in vitro* techniques, the differences between the two categories of crops are becoming less important. Most seed propagated crops now can be submitted to vegetative propagation and maintenance and for crops considered as strictly vegetatively propagating, in some cases new techniques have become available which allow one to overcome crossing barriers, like the use of protoplast fusion, irradiated mentor-pollen, etc.

In most present textbooks on plant breeding and applied plant genetics, mutation breeding is discussed in relatively few lines. Practical advice as to optimal procedures is often lacking, only in a few cases clear examples are presented, and a proper evaluation of the possibilities and limitations of this method is often lacking. On the other hand, in some other text books on plant breeding, examples of mutation work of relatively limited practical value have been worked out into detail because they were connected to the work of the authors. Useful reviews on mutation breeding in seed propagated crops have been written by Konzak (1984) and by Konzak *et al.* (1984). For vegetatively propagated crops methods and results obtained have been discussed extensively by Broertjes & van Harten (1978; 1988) and van Harten & Broertjes (1989).

Chimerism was mentioned already as one of the major technical obstacles that have to be dealt with in mutation breeding of vegetatively propagated crops. This problem can be adequately handled if appropriate methods for plant regeneration from single cells, are available. This, however, is not the case yet for all important crop species and, according to most authors, chimerism therefore remains the most important bottle-neck for quite a number of vegetatively propagated crops. On the other hand, specific, rather stable forms of chimerism do not cause any problem at all, or even may be of advantage to the breeder. Reference is made here to so-called **periclinal chimeras**, which are of particular interest in cut flowers like *Chrysanthemum*, some other ornamentals and some fruit trees. Examples will be discussed in Chapter 7. For seed propagated crops, the handling of chimerism causes considerably fewer problems, provided that the breeder bothers to get himself sufficiently informed about, in particular, initial cells, gamete formation, the population size required to detect induced mutations and the role of plant architecture with respect to chimerism. Such aspects are discussed in Chapter 6.

Other possible complications in mutation breeding programmes may be caused by space requirements (which is, of course, highly dependent on the crop: compare for instance *Arabidopsis* and big apple trees!) and the length of the period between induction of a mutation and its possible detection and selection. Another important aspect to consider is that after a desired mutation has been induced and detected, one must often still go a long way before the mutant is available as a cultivar. Such necessary further steps, probably after further crossing has been performed, include checking the mutant during a number of generations for stability of the mutated trait as well as for general agronomic value, propagation and multiplication, followed by registration and commercialization of the new mutant cultivar. As already stated: it is most advisable to leave those tasks to plant breeders.

When to apply mutation breeding?

During a symposium, organized by IAEA and FAO in Vienna in 1990, Knott (1991) discussed why certain mutation programmes were successful where others failed. Knott concluded that mutagens should be applied in specific situations, like the elimination of deleterious genes (provided of course that the induced deletion does not induce other deleterious effects). Mutation breeding, in his opinion, should not be applied when ample genetic variation for the desired traits is already present but, on the other hand, be a logical step if the desired genetic variation is not available in a crop after standard breeding work.

When considering in general whether to apply mutation breeding, several aspects should be taken into account. First, as mentioned already, an inventory should be made about the amount of variation that is available already for the relevant traits. Another important point of thought is whether the intended breeding goals are proper ones. It may be clear that a realistic answer to this second question can be given only if enough is known about the crop in general and the specific trait in particular, about possibilities and limitations of the intended breeding method and about possible alternative methods. As to this last point, not only the technical merits of a method, but – often neglected – also the costs of different methods in relation to the expected economic profit must be considered. Details can be found for instance in Brock (1971) and Brock & Micke (1979).

Minimum requirements for a commercially successful mutation breeding programme are: realistic goals, skilled personnel to do the job, an adequate stock – often of considerable size – of suitable plant material to start with (by preference including more than one cultivar), traits that can be easily recognized, or for which efficient screening procedures are available, adequate mutagenic treatment (which mutagenic agent, which dose or concentration?) and – last but not least – a certain amount of breeder's luck.

Limitations

Mutation breeding, of course, has its limitations. Basically, all genes may mutate, but not all genes or traits are equally mutable. As a consequence, mutations for some traits may be very rare or even may not be found at all. One could imagine in this respect that certain mutations might have such drastic physiological effects, that the cell in which the mutation occurs is unable to survive.

It is generally believed that mutations occur at random and cannot be 'directed' in the true sense of the word. Randomness of mutations is considered a fundamental tenet of evolutionary biology. This view, however, does not automatically imply that all parts of the plant genome show the same mutation frequency, nor does it mean that environmental factors should not affect mutation frequencies. For artificially induced mutations it has been suggested at various occasions that, in particular, radiation treatments should result in random 'hits' in the DNA or other cell contents of – randomly hit – cells, whereas it was initially thought that chemical mutagens might have a more specific action, for instance directed towards specific chemical bonds or specific molecules. In several publications (see for instance the discussion on mutations in barley in Chapter 3) it was also suggested that so-called 'hot-spots' for radiation-induced breaks occur as well, but very large numbers of mutants must be carefully screened before reliable conclusions can be drawn.

In recent years the prevailing idea of randomness of (spontaneous) mutations has been challenged on the base of results of practical work, for instance by Cairns, Overbaugh & Miller (1988), who worked with *Escherichia coli*. These authors suggested that bacteria, to a certain extent, could 'choose' which mutations – for instance mutations that are beneficial to the bacteria – they should produce. This opinion has raised much interest, but so far there seems to be no definite proof for this kind of 'directed' mutagenesis. Besides, the fact that such processes have been found for bacteria, does not necessarily imply that identical processes do occur in higher plants as well. Efforts to induce relatively high frequencies of specific, desired mutations in higher plants, for instance by choosing specific mutagenic treatments, do not necessarily represent a situation that is identical to the one that was described for spontaneous mutations in *E. coli*.

Time and exact place of occurrence of mutations (in which plant within a population and at which position within a plant?), to a large extent, are unpredictable. Sometimes, based on experience and knowledge, certain 'calculated guesses' can be made. Mutation frequencies for specific traits are also difficult to predict and, moreover, in particular spontaneous mutation frequencies often are too low to be easily determined, because of the number of plants or the size of the cell population that has to be studied in such a case.

Most mutations refer to a change in one allele only. If a recessive mutation occurs in one allele of a homozygous diploid plant, this results in a heterozygous plant (e.g. AA → Aa), which often can not be distinguished visually from the original AA plants and, hence, the mutant remains undetected in the generation in which it arose. Moreover, as has been explained before, mutations which arise outside the gametic tissue in most cases will not be detected at all, and even if they should be detected, such mutations will get lost unless adequate (in vitro) regeneration methods are available. Unfortunately, standard regeneration methods that can be used for different species often are not available and – moreover – genotypic differences within species also may strongly affect the efficiency of different methods.

The common situation is that most mutations go from dominant towards recessive, which implies that traits based on a dominant gene – for instance monogenic resistance against a specific pathogen – are much less attractive targets for mutation breeding than in the case of recessive inheritance.

Another complication is that most mutations are of no (direct) practical value from a breeding point of view. A score of reasons could be mentioned in this respect, such as the induction of a (potentially useful) mutation outside the initials from which the gametes arise in seed propagated crops or outside the (axillary) buds in vegetatively propagated crops, a lower survival rate or reproduction rate of the mutated cell in comparison with surrounding cells, negative pleiotropic effects, negative linkage effects, an unfavourable genetic background of the mutation, incomplete penetrance of the mutation (i.e. not all individual plants show the expected phenotype), etc.

A few examples of negative linkages are mentioned here. Efforts to produce mutant cultivars of barley (Hordeum vulgare) with a high protein content and an improved protein composition have been successful, but in practically all cases such improvements were accompanied by a low total grain yield (see for instance Doll, 1976; Doll & Køie, 1978; Munck, 1972; and Munck, Bang-Olsen & Stilling, 1986).

A spontaneous so-called brown mid-rib mutation – which segregated as a simple recessive – was found in maize (Zea mays) in 1924 at the University of Minnesota (see for instance Cherney et al., 1991). Brown mid-rib mutants, which (spontaneously arisen or induced) also

have been obtained in various such crops like sorghum (Sorghum bicolor) and pearl millet (Pennisetum glaucum), are characterized by a better digestibility, which is of considerable importance in forage crops. Unfortunately, this useful trait is negatively linked to yield, which results in reduced grain yield or lower yield of silage components. In such situations much further work is required to find out whether such negative linkages can be modified or broken. For sorghum reference can be made to a publication by Porter et al. (1978) who induced mutations for brown mid-rib in this crop by treatment of seeds with 0.1 and 0.2% of the chemical mutagen DES for three hours. As an additional reference to the use of brown mid-rib mutants in pearl millet, a publication by Degenhart, Werner, & Burton (1995) can be mentioned.

It is found sometimes that certain interesting mutations can be induced with relative ease in vitro, but another prerequisite for successful use of this mutation is that the specific cultivar in which the breeder would like to induce this trait, should be not recalcitrant with respect to regeneration, as otherwise the induced mutations still would get lost.

In a previous section of this chapter reference was made to experiments with pea (Pisum sativum) by Gottschalk & Kaul (1975). These workers studied the behaviour of an interesting mutant, which in Germany showed more ovules per ovary but a reduced number of pods when compared with the control, and also in India where, as a consequence of increased branching, the reduction in pod number did not occur. This work clearly demonstrates that in order to assess the potential value of mutants, it is very rewarding to test them under different environmental conditions. The same applies to changing the genetic background of a mutation. Transfer of a mutation to another cultivar of the same crop may result in a very different expression of that mutant, as can be demonstrated by the many recent examples of indirectly obtained mutant cultivars (see also in Section 1.3.6). Some examples will be presented in Chapter 6.

As early as 1954 Gustafsson mentioned that, even if large numbers of mutants are obtained, still only a very low percentage of all selected mutants are found to be of interest for further practical breeding work (Gustafsson, 1954). High frequencies of various kinds of mutations, in this respect, in most situations are not very useful. Plant breeders want useful mutations for agronomically important traits which mutations, preferably, should not disturb the further genetic set-up of the plant. In addition even in a self-fertilizing crop species – in which individual plants are predominantly genetically identical for important traits – one should attempt to obtain different plants in which a useful mutation for a specific trait is present, as even small differences in the genetic background may significantly affect the expression of the induced mutation.

When considering the wishes of the breeders and the limitations of mutation breeding, summed up in this section, it appears that the most essential point for the near future is to further increase the efficiency of methods for detection of favourable mutations and to develop efficient methods to propagate the mutant plants as fast as possible. The amount of work and time required – depending of course on the crop and on the proposed breeding goals – to a large extent determine whether mutation breeding is a realistic complement to other breeding methods or, may be, even an attractive alternative.

In addition to the aforementioned limitations of mutation breeding, one should always remain critical with respect to possible mistakes or unjustified claims about results of mutation breeding work. Remarkable or unexpected results may also be explained by outcrossing, contamination of the starting material or wrong interpretation of results. In some cases, even, the desire to please superiors (maybe beyond the knowledge of the 'boss'), excessive ambitions or a desperate need of successes in order to secure financial support for research may be the cause of exaggerated reports about the outcome of a mutation project. Mutation breeders, also in this respect, do not differ from other scientists.

1.3.10. Mutations and modern biotechnology

How do mutations and mutation breeding relate to what is, nowadays, commonly indicated as **plant biotechnology** or, somewhat more specifically, as **genetic engineering, genetic manipulation** or **genetic modification**? Two aspects should be considered in this respect: 1) in which way do mutations contribute to the field of biotechnology, and 2) how can biotechnological techniques contribute to plant breeding equally well or better than mutation breeding can do?

As to the first point, Old & Primrose in their book *Principles of Gene Manipulation* (Old & Primrose, 1989, pp. 87–9) state that mutations have been an essential prerequisite for any genetic study, in particular for those concerning gene structure and function relationships. Mutations have been successfully induced for this purpose with chemical or physical mutagenic agents. This approach, however, suffers from a number of disadvantages, such as a low probability of obtaining a specific desired mutant, the lack of proof that a certain phenotypic change indeed does refer to the gene in which one is interested, the difficulty to find out where in the gene

the mutation has occurred and, finally, whether for instance a single base change or an addition or deletion in the DNA has been the cause of the mutation. Old & Primrose refer in this respect to the possibilities offered by new molecular methods, which are indicated as *in vitro* **mutagenesis** or **site directed mutagenesis**. The use of the term '*in vitro* mutagenesis' is a rather unfortunate one, as this term is widely used already in a different way (see later), but apart from this, the new field of research indicated by the authors has become a highly valuable tool in genetic research.

We must give at this point a working definition of the term biotechnology. Since the early 1980s **biotechnology** in relation to plant breeding, most often, is used to indicate, in a rather general way, the application of a number of techniques and tools of cell biological and molecular biological origin in crop plants for economic purposes. Such techniques may help us to better understand genetic processes and, in this way, be of value for breeding research, as well as be directly utilized for the production of new, genetically improved cultivars. The terms **genetic engineering, genetic manipulation** or **genetic modification** are often used more specifically to indicate various ways of *in vitro* introduction of so-called **recombinant DNA** by asexual means (N.B. The expression 'recombinant DNA' will be explained later in this section).

Basically two major groups of biotechnological activities can be distinguished in plants: 1) all activities related to *in vitro* culture or *in vitro* propagation and, 2) the various 'recombinant DNA' methods. Techniques belonging to both categories are often combined and several ways of using recombinant DNA would not even be feasible without having suitable *in vitro* systems at one's disposal.

The history of applying *in vitro* techniques dates back to far before the advent of modern biotechnology. To mention a few important data: the first successful *in vitro* culture of root tips was reported in 1922; virus-free meristem cultures of *Dahlia variabilis* were obtained in 1952; the production of haploid plants from pollen grains of *Datura stramonium* took place in 1966 and the first protoplast fusions were reported in 1970 (Pierik, 1987).

In vitro culture, basically, refers to growing different plants, organs, explants, tissues, cell suspensions, single cells and protoplasts (cells from which the cell walls are removed) under controlled and aseptic conditions. The most important *in vitro* methods are as follows.

1. *In vitro* conservation of genetic variation (gene banks).

2. *In vitro* propagation and multiplication.
3. *In vitro* elimination of pathogens.
4. Production of haploid plants (from microspores, anthers or unfertilized egg cells).
5. Embryo rescue (e.g. after distant hybridization).
6. Regeneration of protoplasts or cell suspensions used for protoplast fusion or somatic hybridization.
7. *In vitro* generation of spontaneous (genetic) variation (or '**somaclonal variation**').
8. *In vitro* selection for mutations.
9. *In vitro* shortening of the breeding cycle.

Application of most of these methods may be relatively cheap as no expensive equipment is needed. On the other hand, considerable technical (often costly) skill is needed and much work may be involved. Although only some of the topics mentioned in the list are directly related to mutations and mutation breeding, application of *in vitro* methods has become common practice in many recent mutation breeding projects. We will return to this subject at later occasions.

The second group of biotechnological activities can be briefly indicated as '**recombinant DNA technology**'. Under this heading a wide range of techniques and tools are included, of which the most important ones are as follows.

1. Genetic transformation (directly or vector-aided).
2. Chromosome transplantation (N.b. This method is not much used anymore.)
3. Antisense RNA technique.
4. Transposable element techniques.
5. *In vitro* mutagenesis and 'somaclonal' variation.
6. Protoplast fusion. (N.B. Often included under pt 1.)

Particular reference must be made in this context to **genetic transformation**, which implies the transfer and incorporation of foreign DNA (e.g. a specific chromosome fragment or gene) into the DNA of a recipient plant cell, followed by recombination of that DNA into the cell's genome. For useful application of this technique in plant breeding it is essential that the newly integrated foreign DNA is expressed in a controlled way in the recipient cell and that transformed cells can be properly regenerated.

In Box 1.8 some additional facts about the aforementioned subjects are given, but for detailed information reference must be made to publications in various specialized journals on these topics.

The results obtained after applying some of these techniques may show a striking similarity with those produced by mutation breeding methods. This, for instance, is the case with the use of antisense RNA, a method by which the expression of genes is blocked or lost (see Box 1.8).

Once a plant has been transformed, its progeny in some cases may be directly useful as a new cultivar, but much more often it will be used as parent material for further cross-breeding programmes. An important characteristic of these modern techniques is that, at least theoretically, any identified gene – from microorganisms, other plant species or even from animals – can be transferred to plant cells. As a consequence, the access to genetic variation in nature has become much bigger, although the point that each desired combination of genes can be realized is still far beyond reach.

Another point should be mentioned here as well. Contrary to what is often believed and written, new methods of genetic manipulation are neither the first nor the only way to pass the 'species barrier'. In 1719, Fairchild had already made an interspecific cross between two ornamental crops *Dianthus barbatus* (vernacular name: Sweet William) and *Dianthus caryophyllus* (carnation), and more examples, mainly for ornamentals, are known. In some of these cases one still could doubt whether both parents are species <u>in the strict sense</u>, as crossing between them is possible. A better example is found in the successful crosses made late in the nineteenth century between wheat (*Triticum* sp.) and rye (*Secale cereale*), which resulted in the intergeneric hybrid *Triticale*. It may be expedient to keep in mind that it still took almost one more century before *Triticale* changed from a curiosity into a real new <u>crop</u>!

Before a start is made to compare the contributions made to plant breeding by modern biotechnological techniques and by 'conventional' mutation breeding methods, one general limitation of both methods should be mentioned here. Both categories, predominantly, contribute to the earliest stages of the breeding cycle, i.e. to the creation of genetic variation. Biotechnological methods may assist in, and sometimes speed up, the next step of the breeding programme: the selection phase. However, in the final, indispensable and time consuming steps, further perfectioning of the selected genotype into a cultivar, controlled multiplication and maintenance, official registration and commercialization, often not much room exists for applying modern techniques. *In vitro* multiplication may act as a time saving method but, on the other hand, the occurrence of uncontrolled mutations ('somaclonal variation') during the *in vitro* stage of a transformation programme may cause many problems and loss of time. As a consequence, the ultimate gain of time by applying new methods is often considerably less than was promised or anticipated.

Box 1.8 The most common methods of genetic manipulation.

Most of the methods described here refer to molecular genetic methods of DNA manipulation in plants or plant cells, often in combination with methods to transfer DNA in an asexual way and in combination with an *in vitro* regeneration method. A clear distinction between the earlier described methods of *in vitro* culture and **genetic manipulation** is not easy to make as various methods may serve more than one purpose at the same time. *In vitro* propagation may also lead to the occurrence of – wanted or unwanted – spontaneous mutations.

To avoid confusion, the expressions **genetic manipulation, genetic modification** and **genetic engineering** are considered here identical, although often 'manipulation' is considered to enclose more than 'engineering'. The word 'modification', nowadays, is often preferred above 'manipulation', which expression is believed to provoke negative reactions!

The most important methods are as follows.

1. *Genetic transformation.* This means the asexual transfer of defined fragments of foreign (donor) DNA into cells of the host recipient plant, followed by incorporation of (a part of) the donor DNA into the genome of the host. Introduction of DNA can take place in various ways: a) <u>by direct uptake</u>, b) <u>by making use of vectors</u>, c) <u>by protoplast fusion</u>.

ad a) <u>Direct uptake</u> of DNA in seeds, roots or inflorescenses, followed by expression in the resulting plants, appears to be an ideal method, but this is not feasible yet. Starting from protoplasts (naked cells), electric pulses may produce temporary pores in the plasma membrane, through which DNA fragments can enter. This process is called **electroporation**. Manual **micro injection** with a solution of DNA into the nucleus or organelles of intact plants/cells holds prospects.

By making use of the **particle gun bombardment** (or 'biolistic') **method**, small (i.e. 1–2 µm) tungsten or gold particles, coated with DNA can be shot into various types of cells.

So far, all methods of direct delivery of DNA are labour-intensive and not very efficient, but could be tried, for instance, when vector-aided transformation (see ad b) is not applicable. Another drawback is that often adequate regeneration systems for protoplasts are not available.

ad b) <u>Vector-aided transformation</u> or <u>indirect transformation</u>. The most widely used system is based on *Agrobacterium tumefaciens*, which is used as a vehicle to transfer DNA fragments (t-DNA) in which, for instance, a desired gene has been inserted, to the genome of plant cells, infected by the aforementioned bacterium. By using this method DNA fragments with a length between (theoretically) one and 20 000 base pairs (bp) can be transferred, even if such fragments are completely unrelated to the plant genome. The t-DNA may become stably established and expressed in the plant cells and, provided that suitable regeneration techniques are available, complete plants can be grown from those genetically transformed (or **transgenic**) cells. The first transgenic plants, obtained in this way, with resistance to the antibiotic kanamycine, were obtained in the early 1980s. The technique is not yet applicable in all important crop plants; in particular monocotyledonous crops (including all cereals) often are difficult to modify in this way. Alternative to the use of *A. tumefaciens* are other bacteria, like *A. rhizogenes,* and viruses, in particular the cauliflower mosaic virus (CaMV), but none of these systems has reached yet the state of perfection of the *A. tumefaciens* system.

ad c) <u>Protoplast fusion</u>. Various fusion techniques (in particular so-called **electrofusion**) can be followed to combine the genomes of different – more or less related – plant species. Often use is made of the chemical polyethylene glycol (PEG) that penetrates the membrane. The first successful fusion: the so-

called 'pomato', an intergeneric hybrid of potato (*Solanum tuberosum)* and tomato (*Lycopersicon esculentum),* has been described by Melchers, Sacristan & Holder (1978).

The fused cell, basically, contains a mixture of DNA from both 'parents' of nuclear (chromosomal) and extranuclear (plastid, mitochondrial) origin.

Selection of either **somatic** (or **nuclear) hybrids**, with the complete chromosome sets of both cells, or so-called **cybrids** (cytoplasmic hybrids), in which only one complete chromosome set is combined with the extranuclear DNA of both cells, is performed in many ways and with different degrees of success. Although, theoretically, the crossing barriers between species are eliminated by protoplast fusion, it appears that, still, distant parents often cannot be combined in this way and still many practical problems have to be solved.

Another goal of somatic cell fusion may be to combine the nucleus of one plant species with the cytoplasm of another plant species, for instance by killing the nuclei of cells of the latter species by irradiation.

Interest in protoplast fusions is declining now as methods of direct transfer of DNA are becoming more successful.

2. *Chromosome transplantation.* Starting from about 1983 methods became available for isolation and separation of individual chromosomes from complete genomes, e.g. by flow cytometric methods. After a successful transfer of, for instance, a single chromosome to a protoplast, however, the introduced chromosome is often rapidly eliminated during cell division and even when maintained in the recipient cell, stable expression of the desired genes does not always occur. Moreover, undesirable genes on the newly introduced chromosome will be expressed as well. It is tried sometimes to fragment chromosomes in small pieces by high dosages of, for instance, gamma-rays or X-rays, but this approach

also does not work properly for a number of reasons such as the random fragmentation effect of radiation and the fact that, by far, most fragments do not survive in recipient cells, unless they should be incorporated in other, complete chromosomes. An alternative way, studied by, for example, K.K. Pandey, is to irradiate pollen grains before pollination, in which case sometimes only a very few traits from the 'father' become expressed in the off-spring. Although in some cases the method may offer prospects, still many practical problems have to be solved. The 'Pandey' method in a way is comparable to direct gene transfer (see method 1 in this box). The method of chromosome transplantation is not much used anymore.

3. *Antisense RNA technique*. The antisense method – introduced for eukaryotes around 1988 – is based on the suppression of gene activity. The mechanism is known in nature where it is used, for example, in bacteria for gene regulation. For plants the main purpose of applying the method is to inactivate certain chromosomal regions or, in other words, to suppress the gene activity of specific, undesired genes. When the so-called target gene (in fact often only the coding part of this gene) can be isolated and re-inserted in the reverse sequence (as read from the starting point: the promoter) at the original place in the genome, a plant cell carries both the common (sense) and the reversed (antisense) gene in the homologous chromosome. The common gene will be the template for the (common) messenger (m-)RNA and the antisense gene for the reverse (antisense) m-RNA. The normal and the antisense m-RNA, which are complementary, associate and do result in double-stranded RNA. Normally the m-RNA will finally result in the production of a specific protein or enzyme, but because the RNA is double-stranded RNA the so-called translation process is blocked, in other words: the function of a

gene has been inactivated by the introduction of the antisense sequence of this gene.

The antisense method – provided that this method leads to complete and stable suppression – has some advantages, for instance when compared to the transformation method or 'classical' mutation breeding methods. The antisense method is site-specific for a specific gene and does not affect other genes. Another advantage is the 'dominant' expression of gene suppression, whereas most mutations are recessive. Moreover, the work can be targeted towards valuable traits. On the other hand, only inactivation occurs, but new genes do not arise.

Despite all these advantages several points still need further clarification before the antisense method will become an easily applicable and generally accepted breeding method for all crops and all traits. Moreover, it is known that application of the antisense method does not always lead to complete inactivation of the gene function. When that is the case the induction of a mutation that causes 'total loss of function' may be an attractive alternative again. A reliable system of regeneration after introduction of the antisense-gene by the *A. tumefaciens* method is (still) needed and whether this method, as is the case with several other methods of gene manipulation, will be readily accepted by society cannot be predicted.

4. *In vitro mutagenesis and 'somaclonal variation'*. In some publications treatments with mutagenic agents *in vitro*, as well as the occurrence of spontaneous (?) genetic variation under *in vitro* conditions, are also considered as 'biotechnological' or 'genetic manipulation' techniques, probably because they occurred as a side effect of genetic manipulation work. In our opinion it would be more logic to treat both subjects as belonging to 'classical' mutation breeding. Therefore both subjects will be discussed more in detail at other places in this book.

In 1994 the first genetically modified food crop, cv. Flavr Savr with an improved ripening pattern in tomato (*Lycopersicon esculentum*) was marketed. This topic is briefly described elsewhere in Chapter 1.

5. *Transposable elements*. Transposons are mobile genetic elements, discovered in the 1940s by Barbara McClintock, who became a Nobel prize laureate in 1983. Transposable elements can move within the genome from one place to another; they can insert themselves in or near genes and block or change the expression of that gene. This topic is often called **insertional mutagenesis** and will be discussed in more detail in Chapter 3 of this book. From a fundamental, as well as from a practical point of view, the use of transposons offers very interesting prospects.

Tobacco, in 1982, was the first crop to be transformed, followed in 1986 by tomato and sunflower (*Helianthus annuus*), and by rape seed (*Brassica napus*) and potato (*Solanum tuberosum*) in 1987. Monocotyledonous crops appeared to be less easy to transform; wheat (*Triticum aestivum*), for instance, could not be transformed until 1992. The fact that successful transformations are possible for a specific crop, does not imply that this approach has become a standard tool already, applicable on a routine base for all cultivars or starting material of that species.

The first approved field trials of crop plants in which recombinant techniques had been applied, started in 1986 in the USA and in Belgium. The most important crops investigated so far in this context are rape seed (*Brassica napus*), potato (*Solanum tuberosum*), tomato (*Lycopersicon esculentum*), tobacco (*Nicotiana tabacum*), maize (*Zea mays*) and flax (*Linum usitatissimum*). *Arabidopsis* is often used as a general 'model' crop. For experiments on flower colour expression, etc.; the ornamental garden petunia is often used.

A common (ultimate) goal of both mutation breeding and modern biotechnology is to perform **site-directed (or site-specific) mutagenesis**. The term **'site-directed mutagenesis'** refers to the induction of small changes within a specific gene, like addition, deletion or substitution of one or a few base pairs (often indicated as **point mutations**; for definition see Rieger *et al.*, 1991). Some modern biotechnological methods are much better suited for this purpose than mutation breeding ever will be. Biotechnological methods, in addition, enable the introduction of <u>additional</u> genes, which can not be achieved by mutation breeding methods. Until now it is impossible to predict the exact site at which mutations occur within a genome, whereas the exact site where transgenic DNA is incorporated into a chromosome is unpredictable as well. The same applies to the place within the chromosomes where transposons may become active and to the traits for which 'somaclonal variation' occurs.

Recombinant DNA methods, nevertheless, are already applied on a relatively large scale to introduce foreign genes or, in case of the antisense method, to block specific genes in cultivated plants. In 1982 tobacco (*Nicotiana tabacum*) was the first dicotyledonous crop in which genes from other species were incorporated by transgenic methods, whereas the first transgenic cereal, wheat (*Triticum aestivum*), was obtained only in 1992. In 1994 the first example of a genetically modified food crop: cv. Flavr Savr in tomato (*Lycopersicon esculentum*), with an aberrant ripening pattern and an improved firmness of the fruit skin, produced by the Calgene biotechnology company in Davis, California by means of the 'antisense' method, was marketed in the USA. At the same time more cultivars, genetically modified for various traits – in particular resistances to viruses and insects, herbicide tolerance and 'quality' traits – were about ready for official registration and marketing. With an estimated cost price of 25 million US dollars, cv. Flavr Savr probably has become the most expensive cultivar that has ever been developed, which makes its production an unattractive undertaking from a commercial point of view. Its release and marketing, however, should be considered, predominantly, as a test case to go through all procedures, required safety tests, etc.

Despite a number of apparent advantages of modern biotechnological methods (e.g. transfer of single genes and directional changes are possible; even genes from other organisms or completely synthetic genes can be used as well), the introduction and general acceptance of biotechnological techniques in applied plant breeding has not gone as fast as was anticipated first. One important reason is that, so far, very few genes for agronomically important traits have been recognized and cloned. This is not surprising as many of these traits are genetically complex and show polygenic inheritance. Next, after successful introduction, the newly introduced DNA is often not expressed at the proper place and time. Moreover, successfully transformed cells must be regenerated into normal plants without any major genetic change outside the target gene(s).

Until now new protocols had to be developed each time for the consecutive steps of introducing foreign DNA and regeneration of a specific plant species, with again further adaptations for each genotype. In order to reduce the costs of applying biotechnological methods substantially, it is necessary to develop standard protocols which can be followed irrespective of the genotype or crop species. Moreover, the fact that the gain of time in the breeding cycle as a result of biotechnologal techniques remained far beyond expectations, the high costs involved (salaries, expensive facilities, patent royalties), the (for non-plant breeders) unexpectedly long period between the moment of production of a new cultivar and its introduction to the market and, finally, specific problems with the introduction of genetically manipulated organisms, until now further limit the attractiveness of applying biotechnological methods for cultivar breeding. Since its start in the early 1980s, expectations of 'biotech companies' had to be adjusted downwards repeatedly and during the period of writing this book, as far as known, still none of agricultural biotechnology companies throughout the world had passed the stage from losing money annually and using up capital invested by mother companies or by share holders (including plant breeding companies who want to 'keep in touch'). Summarizing, it seems justified to conclude that much of the transgenic technology is still in its infancy and that we are only at the very beginning of the stage in which it may become possible for plant breeding companies to produce at a routine basis new, commercially attractive 'transgenic' cultivars. In the meanwhile, mutations play an important, and sometimes even indispensable, role in biotechnological research methods, e.g. for the localization of genes on chromosomes whereas, at the same time, mutation breeding – which is sometimes called 'poor man's biotechnology' – continues to play its role as useful additional technique in applied breeding, the significance of which, on the other hand, should not be overestimated as well.

A reason why mutation breeding still may remain attractive to plant breeders – probably even in the future again in an increasing number of cases – is that all biotechnological companies try to protect the use of methods developed or applied by them, as well as the results of their work, by patents or licenses, which may lead to considerably increased costs for the users. This in particular could be a serious threat to plant breeding in poor countries. The application of established mutation breeding techniques, on the other hand, is still free (unless, of course, it would be possible as well to include aspects of this approach in patents as well).

From the view point of mutation breeding, another recent, positive development is that, gradually, biotechnologists and plant breeders have come to understand that they may need each other. Apart from the application of biotechnological techniques for the production of genetically improved cultivars, the new methods are much applied to study structural and functional relationships in biological systems and to study the underlying mechanisms and processes (see for instance McPherson (ed.), 1991).

An interesting observation for the Netherlands has been that some breeding companies that, because of disappointing results, abandoned mutation breeding in the late 1970s and, consequently, switched to 'biotechnology', have returned now to some relatively simple and well established methods of mutation breeding. Examples are, for instance, mutation treatment (e.g. for flower colour) of seeds, cuttings, anthers, tissues or protoplasts with the chemical mutagens EMS or with X- or UV-rays, or selection for spontaneous mutations *in vitro* (which are often 'advertised' with the vogue term of **'somaclonal variation'**). The main reasons for this renewed interest in mutation breeding are that positive results may be obtained with relatively simple procedures and at low cost and, in addition, the interesting fact that the public acceptance of mutation breeding, suffers far less than 'genetic manipulation' from a negative image.

Commercial plant breeding remains a complicated activity, requiring both skill and an integrated approach. As a consequence, each new technique that may contribute to make the breeding process more efficient, should be warmly welcomed. This, indeed happened with mutation breeding in the 1960s and 'biotechnology' since the 1980s. Despite the fact that quite a few workers involved in mutation breeding had a background in (applied) plant genetics, plant breeding or related plant sciences, they still did not sufficiently realize that mutation breeding could be successful only when fully integrated with other breeding procedures. Moreover, the original euphoria which, apparently, goes with each new technique, made expectations rise too high, and problems often were very much underestimated. Despite some positive exceptions, which will be discussed later, the lesson had to be learned the hard way. And when it became clear that spectacular results were not obtained as easily as was anticipated, many scientists again moved away from mutations, in most cases without taking pains to analyze carefully why their projects had not been successful.

By comparison, biochemists and molecular scientists involved in plant biotechnology, almost without exception have considerably less affinity to crops and plant sciences than the aforementioned mutation breeders and, as a consequence, they may be even more inclined to overestimate the impact of their new tools on plant breeding.

In conclusion, it may take at least another ten years before a realistic assessment can be made of the contributions to plant breeding by biotechnology and to make a fair comparison of the relative role of mutation breeding and biotechnology in this respect.

2 History of mutation breeding

2.1. Introduction

Not much attention has been paid, so far, to the history of mutation breeding, and most information about earlier events and discoveries, achievements by various workers and the life of those workers themselves, is found scattered in literature. In more recent years some authors have contributed to this subject, in most cases on main lines only and often limited to some specific phenomena (e.g. mutations induced by chemical mutagens, mutations for a specific trait or gene) or to mutations for a specific crop (e.g. barley) or group of crops (e.g. vegetatively propagated crops). Some relatively recent studies may be mentioned in this context.

In 1976 Charlotte Auerbach, who became well-known for her research on chemical mutagenesis in *Drosophila* (a little fruitfly that about 1910 was found to be an excellent organism for studies on mutations and genetics), published a volume about mutation research, with special attention to fundamental aspects and a small final chapter on applied work, including mutation breeding (Auerbach, 1976). In the first chapter of her book Auerbach discusses the history of mutation research and we will return to Auerbach's views on this subject after briefly mentioning some contributions by other authors. Interesting information about earlier findings in mutation breeding has been presented on various occasions by the, also well-known, Swedish mutation breeding scientist Åke Gustafsson. In his last publication before his death he outlined in his lucid way the main achievements of mutation breeding until the 1970s and the contributions by a number of leading scientists in this respect (Gustafsson, 1986). Tilney-Bassett (1986) reported on historical data about chimeras. Broertjes & van Harten (1978; 1988) and van Harten & Broertjes (1989) briefly described the history of mutation breeding for vegetatively propagated crops.

Charlotte Auerbach (loc.cit.) distinguished five periods of mutation research; distinction was mainly based on the time of introduction of novel concepts and on new findings or fundamental break-throughs concerning mutation. Following this approach, in the first period (1900–27) basic concepts on mutation and (spontaneous) mutation rates were worked out. In the second period (1927–40) the mutagenic effects of X-rays were discovered and major attempts were made to throw light on the nature of mutation. Much impact was made in this respect by the so-called **target theory** (see later in this chapter). The third period which, according to Auerbach, started at about the beginning of World War II, was dominated by the studies on chemical mutagenesis and the introduction of micro-organisms in experimental analysis. The discovery of the double helix DNA model by Watson and Crick in 1953 inaugurated the beginning of the fourth period, during which period studies on DNA were the main topic. The fifth period started in 1965, during which period, again according to Auerbach, it gradually was realized that the phenomena of mutation could be understood only if mutation was considered as a biological process.

Although Auerbach's time divisions could be used as a starting point, these periods do not always coincide with the periods that can be distinguished for mutation breeding of plants. First of all, spontaneous mutations in plants have occurred as long as plants have been known and have been important for man for many centuries. The first European publication about a spontaneous mutant plant was written 400 years ago and reports about economically useful spontaneous mutants also were published already before 1900, for instance by Darwin (see later). Moreover, some findings and results from outside the field of biology signify important landmarks for the mutation breeder.

The discovery and the subsequent application of chemical mutagens from 1940 onwards, an important landmark in the opinion of Auerbach, on the other hand, did not lead to a major break-through in muta-

tion breeding until now, as it did not result in a mutation spectrum that significantly differs from that
induced by radiation. The same applies to the DNA
model of Watson and Crick from 1953 which, of course,
represents one of the most important scientific discoveries ever made, but did not significantly change methods in 'conventional' mutation breeding.

More important for mutation breeding were the
foundation of the IAEA in 1957 and the Joint FAO/IAEA
Division in Vienna, several conferences on the peaceful
application of atomic energy, including mutation breeding, the funds that were made available by the USA and
the IAEA for mutation research and mutation breeding
and for training in these fields (see also Chapter 1).

Based on a number of such – sometimes rather arbitrarily chosen – events, some main periods of mutation
breeding research will be discerned here. References in
the following sections concern in particular those subjects to which not much attention will be given anymore
in later chapters, like publications on mutagenic treatments during the first part of this century.

2.2. The first period: observation and documentation of early spontaneous mutants

300 BC – Early mutant crops in China
Based on data communicated by Professor Liu Zhongqi
of the Sichuan Academy of Agricultural Sciences to
Professor A. Micke (formerly Head of Plant Breeding
and Genetics Section, IAEA, Vienna), it appears that the
earliest description of the selection of spontaneous
mutants traces back to around 300 BC when selection of
mutants for, in particular, cereal crops, different for
maturity and other traits, were described in an ancient
Chinese book, 'Lulan'. Reference was made to a publication on plant breeding achievements in ancient
China by Huang & Liang (1980, in Chinese, original not
consulted).

1590 – The 'incisa' mutant of Chelidonium majus
It is generally accepted that the first reliable and
detailed description of a sudden genetic change in a
higher plant dates back to the end of the sixteenth century. In 1590, Sprenger, a pharmacist in Heidelberg,
Germany, discovered in his herb garden a strikingly
aberrant form of *Chelidonium majus* (greater celandine)
with deeply incised, laciniate leaves. According to
Stubbe (1963, p. 84), Sprenger named this plant
Chelidonia major foliis etfloribus incisis. This finding became

known to the botanist Gaspard Bauhin who, in 1596,
made a detailed description of this new type. Moreover,
Carolus Clusius in 1601 included the mutant plant in
his book *Rariorum plantarum Historia*, be it under a somewhat different name (*Chelidonium majus laciniato flore*). In
later years again other scientific names have been given
but there is no doubt about the origin of this new form
by a spontaneous mutational event.

Stubbe (1963; 1972) mentions that the new mutant –
which attracted wide attention all over Europe and
which was introduced in many botanical gardens – was
stable right from the beginning. This was confirmed by
all contemporary botanists who raised it from seed during a long range of years. In 1620 G. Bauhin reported
that in the Jardin Royal of Paris an even more extreme
form of the original mutant (indicated at that time as a
'sport of nature') had been generated, with an even
deeper incision of the leaves (and flower petals as well).

The name '**sport**' is a common way to indicate spontaneous mutants, in particular when occurring in ornamentals, fruit trees and (other) vegetatively propagated
crops. De Vries (1905, p. 310) described sports in the
following way:

. . . One of these doubtful terms is the word sport. *It often
means bud-variation, while in other cases it conveys the same
idea as the old botanical term of mutation. But then all sorts
of seemingly sudden variations are occasionally designated by
the same term by one writer or another. . . . If we compare
all these different conceptions we will find that their most
general feature is the suddenness and the rarity of the
phenomenon. They convey the idea of something unexpected,
something not always and not regularly occurring. But even
this demarcation is not universal and there are processes that
are regularly repeated and nevertheless are called sports.
These at least should be designated by another name. In
order to avoid confusion as far as possible with the least
change in existing terminology, I shall use the term*
eversporting varieties *(underlining van H.) for such forms as
are regularly propagated by seed, and of pure and not
hybrid origin, but which sport in nearly every generation.*

According to Tilney-Bassett (1986, p. 1) a **sport** can be
described as <u>a spontaneous change, or mutation, in a
part of the plant that creates a feature not previously
known within the species, their hybrids and descen-
dants</u>. In practice, only mutations that occur in the
somatic tissue of a plant are indicated as sports.
Such sports were considered very valuable sources of
new, genetic variation and were much appreciated by
breeders of ornamentals and fruit trees.

1667 – The first so-called graft-chimera

A plant consisting of two or more genetically different somatic tissues is called a **chimera**. The name **chimera** has been taken from the Greek mythological being with the head of a lion, the body of a goat and the tail of a dragon. Winkler (1907) proposed to apply the word 'chimera' in plants for a specific phenomenon, and according to his original definition (translated from German), chimeras are organisms which are composed of cells from different species ('Arten') and develop into a stable unity ('Individuum') <u>without</u> the occurrence of fusion between the cells of the different species. To this category belong the so-called **graft-hybrids**, which should be well distinguished from mutated plants that originate from single cell mutagenic events.

Baur (1909) extended Winkler's definition of a 'chimera' to plants consisting of somatic cells that are genetically dissimilar but in which the tissues all trace back to the same species. Over twenty years later Winkler (1935) defines 'chimeras' (again translated from German) as organisms ('Organismen') that develop from genotypically different cells without cell fusion into a whole, stable plant ('zu einem einheitlichen Individuum').

The first known description of a **graft-chimera** is the <u>Bizarria</u>-orange, known from Florence, Italy, since 1644 as the result of a scion of sour orange (*Citrus aurantium*) grafted on a seedling stock of citron (*Citrus medica*). The resulting plant showed a strange, mixed pattern with variable contributions of both 'parents' to foliage, flowers and fruits. The Bizarria-orange is a so-called **periclinal chimera** in which the tissue of one of both 'parents' is completely surrounded by that of the other 'parent' (see also Chapter 7). This case was published in 1667 in England in Volume II of the *Transactions* of the Royal Society. Details have been given, for example, by Cramer (1907), Neilson-Jones (1934) and Tilney-Bassett (1986). Recently, Spena & Salamini (1995) referred to even earlier publications concerning graft-hybrids in citrus. Some details are given in Box 7.1 in Chapter 7. In later years other interesting cases of graft-chimeras have been found – the most well-known being + *Laburnocytisus adamii* (originated in 1825 in Paris) and + *Crataegomespilus* – which will be briefly discussed in Chapter 7.

It should be emphasized here that graft chimeras – apart from the word 'chimera' – have no relation to chimeric structures which result from 'real' mutations. Mutations originate as single cell events and as the mutated cell and, eventually, a group of cells or a tissue, are surrounded by non-mutated cells, the plant (part) in which the mutation occurs, automatically, has a chimeric structure. The analysis of graft-hybrids, on the other hand, has helped us to better understand chimeric structures.

1672 – Variability in plants

One of the oldest and very detailed accounts on the variability of numerous groups of cultivars of trees, shrubs and herbaceous plants was published in 1672 by A. Munting under the title *Waare Oeffeninge der Planten* ('reliable study of plants'; original not consulted) (see Stubbe, 1963; 1972). Although other reasons – such as interspecific crosses – may account as well for part of the observed variation, it is most probable that many of the examples studied by Munting did refer to spontaneous mutations.

Seventeenth century – 'Imperial Rice' in China: a spontaneous mutant?

According to a citation that, so far, could not be traced down again, an early maturing, high yielding, aromatic mutant of rice (*Oryza sativa*) was discovered in the garden of Emperor Khang-Hi (1662–1723) and later on grown all over China. Because of its early maturity it was reported that this rice cultivar was the only one that could be grown north of the Great Wall, whereas in S. China two harvests per year were possible. This rice was called 'Imperial Rice' and history says that only the Emperor and his court were allowed to eat this rice.

In 1996 Professor Liu Zhongqi of the Sichuan Academy of Agricultural Sciences informed Professor Micke, that Huang & Liang (1980), whose work was mentioned before, refer to 'Imperial Rice', selected from a spontaneous mutant. According to these authors the selection procedure was recorded in a Chinese book 'Jixia Gewu Pian' which was published around 1691. (N.B. Original publications are in Chinese and have not been consulted.)

Late seventeenth century – Spontaneous mutants for the ornamental 'morning glory' (Ipomoea nil) in Japan

An account of early spontaneous flower mutants of the ornamental species 'morning glory' (*Ipomoea nil*) in the so-called Genroku period (1688 – 1703) in the area of Edo (now Tokyo), Japan, was given by H. Yamaguchi (Micke, personal communication).

According to this account, inhabitants of Edo looked for spontaneous flower mutants of this ornamental and competed for the most beautiful colours and flowering types. A white flowering type and dwarf mutants were described already in 1664 and 1709, respectively. Yamaguchi further referred to a book '*Honzokomokukeimo*' in which many early spontaneous mutants for flower type and leaf shape are described and adds that research in the 1990s has shown that a flower colour mutant with flecked petal colour resulted from transposon-mutagenesis (see for instance Chapter 1).

From the information provided so far by Yamaguchi it is difficult to judge whether the deviating flower types mentioned here all resulted from spontaneous mutations or may have been caused as well by normal segregation of genes.

1741 and following years – Description of various mutants by Linnaeus

The Swedish botanist Linnaeus (Carl von Linné, 1707–78) also described various examples of heritable variations in both wild and cultivated plants. It has been reported by Shamel & Pomeroy (1936) that already in 1741 Linnaeus was acquainted with the existence of a russet-type apple produced on a green-fruited tree. Smith (1958), in this respect, refers to a publication by P. Collinson in 1821 (original not consulted). According to Shamel & Pomeroy (1934), in 1765 the attention of Linnaeus was drawn to a tree on which a peach and a nectarine were growing which together arose from a single twig. Stubbe (1972, p. 210) mentions that a so-called peloric form of *Linaria vulgaris* (toad flax) was found in 1742 near Uppsala, Sweden. This mutant was described in 1744 by Linnaeus in his *Peloria* as well as by Rudberg in Linnaeus's *Amoenitates academicae*. The expression 'peloria' refers to the appearance of an actinomorphic flower in a plant that carries zygomorphic flowers.

1865 – Carrière publishes his book 'Production et fixation des variétés dans les végétaux'.

Starting from 1859 Carrière published on spontaneous mutations in various groups of crop plants, with many examples concerning conifers, but also for fruit trees and potato (*Solanum tuberosum*). Most examples have been published in Carrière (1865).

[Aside: *1865 – Gregor Mendel publishes his studies on inheritance, based on his results with pea (Pisum sativum).*]

1868 – Charles Darwin

Charles Darwin is in particular well-known for his book *The Origin of Species*, which was first published in 1859. In Chapter XI of another book: *The Variation of Animals and Plants under Domestication*, Darwin (1868) defines **bud variations** (or **bud sports**) as 'all changes in structure or appearance which occasionally occur in full-grown plants in their flower-buds or leaf-buds'. In this publication a number of cases of **bud variation** (or **bud sports**) have been described extensively for many crop plants as

well as for other plant species. Darwin attributed the observed changes in many cases to 'spontaneous variability', but expressed his ignorance of the laws of variation and did not speculate about the cause of this variability.

1901–5 – Hugo de Vries

The Dutch scientist Hugo de Vries, who lived from 1848 to 1935 and studied botany in Leiden, the Netherlands, and in Würzburg and Halle, Germany, became well-known for his work on plant physiology already before the turn of the century. Moreover, together with C. Correns and E. von Tschermak-Seysenegg, he was in 1900 one of the 'rediscoverers' of Mendel's laws on inheritance.

In an extensive publication *Species and Varieties: Their Origin by Mutation*, de Vries (1905) discussed various categories of abnormalities in plants that could be attributed to spontaneous (genetic) changes, not resulting from normal segregations or mutations. Some years before (in 1901) de Vries, whose work was mentioned already in Chapter 1 and to whom will be referred again in relation to 'the second period of mutation breeding', had coined the term '**mutation**' for sudden, shock-like changes of existing traits (de Vries, 1901; 1903).

1907 – P.J.S. Cramer on bud variations

In 1907 P.J.S. Cramer, who lived from 1879 to 1953 and who was a student of Hugo de Vries at the University of Amsterdam, published a large bibliographical monograph on bud variations. This monograph resulted from the prize winning answer to a question put forward by the Holland Society of Sciences at Haarlem (which is not identical to the Netherlands Scientific Society). The highly interesting book contains examples of 56 crop plants and hundreds of spontaneous mutants, which may differ from the original plant for a substantial number of traits, like (absence of) hairs or thorns, colour of vegetative plant parts, flowers or fruits, variegated leaf types, *incisa*-leaves, dwarfs, giants, 'weeping' varieties, broom-like types, monstrosities, etc.

It is not always easy – for the examples mentioned in this monograph, as well as in other (early) publications – to prove beyond doubt that all observed variants or aberrant types are caused by 'true' mutations and were not the result of epigenetic changes (N.B. The term 'epigenetic' refers to any change of activity of a gene during development of the organism from egg to adult; see Rieger et al., 1991). Evidence for the occurrence of a true 'genetic' mutation is in particular difficult to produce when plants can be propagated only vegetatively and not by seed.

In the aforementioned monograph Cramer (1907, p. 430) referred to several cases in which bud variations had led to cultivars of economic value. An old example concerning potato (*Solanum tuberosum*) was derived from the *Gardeners Chronicle* (1857, p. 613 and p. 629; original not seen) and refers to cv. White Fortyfold with white tubers, obtained from cv. Purple Fortyfold. As in this example only one eye from a purple tuber of cv. Purple Fortyfold had become white, contamination, often a source of wrong conclusions about the occurrence of bud sports, can be completely excluded. An interesting point is that cv. White Fortyfold was reported to be similar to the then known cv. Regent.

2.3. The second period: induced mutations; the mutation breeding concept

1895–1900 – The discovery of various kinds of radiation

In 1895 W.K. von Röntgen read a paper to the Scientific Society of Würzburg, Germany, in which he reported that he had discovered a new, penetrating kind of radiation that would darken photographic plates. In January 1896 a translation of this paper was published in *Nature*, where the term 'X-rays' was used for sake of brevity. Later, the expression 'Röntgen-rays' was used, after its inventor who, in 1901, became the first Nobel prize laureate for physics.

X-ray machines for medical purposes have been manufactured since 1896. The discovery of the ionizing properties of X-rays by J.J. Thompson and E. Rutherford traces back to 1896. More details and references about the discovery of X-rays are given by Maddox (1995).

Based on Röntgen's findings, H. Becquerel in 1896 became the discoverer of so-called natural radioactivity (emission of radiation) from uranium. Pierre and Marie Curie together discovered the radioactive element radium in 1897. In 1899 E. Rutherford and F. Soddy found that the so-called Becquerel radiation consisted of two types of radiation, which were indicated as α- and β-radiation (alpha particles (helium nuclei) consist of two protons and two neutrons, whereas a beta particle is an electron).

Important for mutation breeding was the discovery of a third type of radiation emitted by radioactive elements, and indicated as γ-(gamma-)rays in 1900 by P. Villard.

1901 – Hugo de Vries: 'Die Mutationstheorie'

In 1901 and 1903 de Vries presented (in two volumes) *Die Mutationstheorie* in which an integrated concept for the occurrence of mutations was outlined. This work has become a milestone in the study of 'mutations', evolution and genetics in general. In retrospect, *Die Mutationstheorie* refers more to evolution of species than to genetic variation useful to the plant breeder. Some details about the concept of de Vries and other aspects of his work on mutations have already been presented in Chapter 1, Section 1.1, of the present book.

The word '**mutation**', was coined by de Vries for sudden genetic changes, 'leaps', 'shocks' or 'saltations'. It may be interesting to mention that the term 'mutation' had been used before already in different ways, for instance in the seventeenth century to indicate changes in the life cycle of insects and in the nineteenth century by palaeontologists for clear abnormalities in fossils (Mayr, 1963).

1901 – S. Korschinsky

In an extensive contribution about sudden (genetic) changes in plants, Korschinsky (1901), who was director of the botanical gardens in St Petersburg, Russia, presented his Theory of Heterogenesis and gave many examples of 'mutation-like' changes in different plant species and crop plants. Korschinsky used the term heterogenesis, which had been coined in 1864 by A. Kölliker to indicate in a general way the occurrence of such sudden changes. In his book, Korschinsky described that in the progeny of completely normal plants, suddenly, some plants arise which differ more or less drastically for one or more traits from the parents as well as the other plants within the progeny. He also observed that such changes remained remarkably stable in later generations. Studying the catalogues of cultivars offered in the early 1800s by breeders of ornamentals, fruit trees, etc., he concluded that the changes occurred in all directions. Korschinsky further concluded that their frequency was much higher than was generally believed and that internal processes in the egg cells were responsible for the observed changes, but did not make any suggestion as to the possible cause of such processes.

As had been the case with Charles Darwin for plants and animals and with William Bateson for animals, an enormous wealth of information on 'sudden changes' in plants had been collected and studied by Korschinsky. Nevertheless, Korschinsky's work practically passed into oblivion because of the enormous impact of the – more complete and better founded – Mutation Theory of Hugo de Vries.

1904 – Hugo de Vries suggests artificial induction of mutations by radiation

In the introduction to the first volume of *Die Mutationstheorie* de Vries (1901) already predicted that it probably might become possible in future to <u>artificially induce mutations in (cultivated) plants and animals</u> .

In 1904, during a lecture presented in the USA, de Vries suggested producing artificially induced mutations by means of the recently discovered sources of radiation. More details about this lecture by de Vries can be found in Blakeslee (1935; 1936).

1897–1908 – Early work on irradiation of plants

Soon after the discovery of X-rays and various other types of radiation, emitted by radium, plant scientists started to irradiate plant material in order to investigate the biological effects of ionizing radiation. Most investigations were performed with X-rays (or Röntgen-rays) and most studies referred initially to physiological effects like germination of seed, retardation of plant growth, recovery after treatment, effects on protoplasma and the like.

Early reports about X-irradiation of (higher) plants, cited by Koernicke (1904a,b), trace back to Lopriore (1897; original not consulted), who studied effects on pollen germination, and to Atkinson (1897), who X-irradiated leaves of *Caladium*, flowers of *Begonia* and different seedlings and did observe break down of chlorophyll activity and growth retardation after extended treatments. According to Koernicke (1904b) probably the first studies on the effects of irradiation on germination of seeds were performed, soon after the discovery of radiation, in the laboratory of Henri Becquerel by L. Matout (Becquerel, 1901; original not consulted), but details are not known. Other early reports come from Nathanson and Perthes who studied cell division in eggs of the *Ascaris* worm as well as growth retardation of seedlings of *Vicia faba* (Perthes, 1904). Koernicke (1904a,b) irradiated seeds of faba been (*Vicia faba*), winter rape (*Brassica napus*) and *Vicia sativa* and compared effects of growth retardation after different treatments. Koernicke (1905) and Guilleminot (1908) studied effects of radiation on germination of several horticultural crops like radish (*Raphanus sativus*), turnip (*Brassica campestris*) and beans (*Phaseolus* sp).

Starting from about 1907 attention gradually became more focussed on radiation induced changes in cell nuclei, nuclear divisions and on growth aberrations in plants. It soon was proved that **physical agents** (radiation, heat, electromagnetic fields) and various **chemical agents** (salts, sugars, acids, inorganic bases, chemical stains, etc.) could induce (substantial) damage to cell nuclei and cell division processes. Often irradiation experiments were performed with other material than plants, like for instance frog eggs, testicles of rabbits and, in an increasing number of cases, with the fruit fly *Drosophila melanogaster*. For an extensive summary of this work see Hertwig (1927). Hertwig mentioned that authors like Koernicke were already of the opinion that all kinds of radiation that had been discovered so far, did affect the chromatine in the cell nucleus, in particular during mitosis. In addition, many more historical data and references about early efforts to induce mutations, either by radiation or by chemical treatment, can be found in an outstanding and very complete monograph on mutations by Stubbe (1938), published as a volume of the series *Handbuch der Vererbungswissenschaft* (Handbook of Genetics), Vol.IIF), which was edited by E. Baur and M. Hartmann.

In 1912 the Italian scientist Alberto Pirovano started experiments with electromagnetic treatments of plant material, in comparison with X-rays and UV-irradiation. His first results were published in 1922 (see Stubbe, 1938) under the title (roughly translated): '*Electric Mutation Induction in Plant Species*'. Pirovano treated germinated seeds of many crop plants with various doses in electromagnetic fields and looked for aberrations in mitosis and meiosis, for morphological changes in adult plants, sterility, etc. Particular attention was paid to treatment of pollen of F_1-hybrids in fruit trees. In the next generation several aberrations as well as useful new plant types were obtained, some of which later became new cultivars. It is in this case impossible to distinguish between mutations and the results of normal Mendelian segregation. Pirovano, who summarized his own work in 1957 (Pirovano, 1957), also believed that by using electromagnetic treatments cross-incompatibility could be overcome, e.g. in crosses between *Cucurbita pepo* x *C. maxima*. Finally, despite the fact that Pirovano referred to the occurrence of natural as well as induced 'mutations', it appears that most of the changes obtained in his experiments could be propagated vegetatively only.

Very few scientists have continued on the line of using electromagnetism to induce mutations. A curious exception was found in *Mutation Breeding Newsletter* 36 from July 1990 where S. Starzycki from the Plant Breeding and Acclimatisation Institute in Radzików, Poland, briefly describes a 'new' method of mutation induction, which method implies exposing germinated seeds to the action of an electromagnetic field. Starzycki (1990), who does not make any reference to the early literature on the use of electromagnetism for mutation induction, calls his method 'ionophoren'. Some details will be presented in Chapter 4, Section 4.2.5.

Much-cited is a major work by Gager (1908) in which he, in a meticulous way, describes his own research concerning various effects of radium rays on plants. This work is preceded by an extensive study about sources of radiation and the effects of radiation on plants. Gager's own experiments were started in 1904 and, as usual in those days, concerned in particular the effect of radiation (from radium treatments) on germination of seeds, plant growth, photosynthesis, plant respiration and other physiological effects. The final chapters of his book are devoted to histological effects of radium radiation, to changes observed in cell nuclei and patterns of nuclear division, and to the effects of exposing germcells to radiation. In the last chapter Gager describes many aberrations in plants following irradiation and, in this context, refers to the Mutation Theory of de Vries. Gager assumes that most of the variations obtained were not 'true' mutations and that further evidence (e.g. from progeny tests) is needed before it can be concluded that radium rays may induce mutations.

In the aforementioned experiments seeds, seedlings, leaves as well as pollen were irradiated. Unfortunately, information about plant material and treatments is very incomplete in most cases. Although in some cases reference was made to changes within nuclei, such effects were not clearly described and no convincing evidence for the induction of mutations was presented.

Stubbe (1938, p. 259) points out that already in many of the early experiments – most probably – induced mutations were obtained, but that in practically all cases the way of experimenting did not allow definite conclusions to be drawn. Most often use was made of genetically ill-defined (plant) material and of inadequate methods to compare the frequency of mutations induced by a certain treatment and the spontaneous mutation frequency. Moreover, most experiments were performed on far too small a scale to obtain (statistically) significant results.

1909 and 1911 – First proof of mutations induced by chemicals

From the beginning of the twentieth century, many workers tried to obtain chemically induced mutations in higher plants. Stubbe (1938), for instance, refers to the pioneering work by MacDougal, Humbert and Dewitz during the period 1909–13.

Experiments with chemical treatments remained without success until Wolff (1909) and Schiemann (1912), who were both students of Erwin Baur, showed that a range of chemicals could induce mutations in bacteria (in particular in *Bacillus prodigiosus*) and in fungi (*Aspergillus niger*). Although in both cases no evidence for

the occurrence of 'real' gene mutations could be derived, there is no doubt about the hereditary character of the observed changes as, in particular, the variations found in the material of Schiemann could be vegetatively propagated for more than 25 years without any change (Gustafsson, 1986). In conclusion, it is generally accepted now that the work of Wolff and Schiemann proved for the first time that mutations can be artificially induced.

Experiments to induce mutations in *Antirrhinum majus* (snapdragon) with various chemicals, started by Baur in 1916 were not successful. Earlier, in 1908, Baur had also been unsuccessful in inducing mutations by means of high temperature and of blue and yellow–red light (see Baur, 1932).

1910 – T.H. Morgan: first mutation experiments with Drosophila melanogaster

Morgan – who became well-known for his experiments with *Drosophila melanogaster* in which he proved that genes are located at fixed positions on chromosomes – started mutation experiments with *Drosophila* in 1910. He applied high temperatures and various chemicals. Irradiation experiments with radium, performed in 1911, yielded some mutations but these results were not obtained when the experiments were performed ('repeated') with X-rays. The negative outcome of numerous experiments that had been performed with *Drosophila* made Morgan and his colleagues Bridges and Sturtevant in 1925 still strongly to doubt whether it would be possible to artificially induce ('by altering the environment') mutations (see Stubbe, 1938, p. 262).

The early 1920s – From qualitative to quantitative radiation biology

Starting from the 1920s the use of mathematical and statistical methods to interpret results of radiation treatments came into fashion. The action of radiation was studied, in a simplified way, as a function of the amount of absorbed radiation energy, which was difficult to measure. Based on studies of dose-response curves it was tried to come to an understanding of underlying mechanisms. Early attempts to develop a mathematical analysis of such dose–effect relationships were performed by Blau & Altenburger (1922), Dessauer (1922) and Crowther (1924). These early studies, which – together with more recent ones – will be briefly discussed at a later occasion, eventually lead to the development of the so-called 'Hit Theory', which gradually was extended to the 'Target Theory'. Both 'theories' in fact are hypotheses, as was pointed out by Lawrence (1971). Good, general accounts and many references on this subject can be found in the aforementioned publication (Lawrence, loc.cit.) and in Dertinger & Jung (1970) or Lawrence (1991).

1920 – Vavilov's 'law of homologous series of variation'

N.I. Vavilov (1922) predicted that the appearance of a given type of mutation in a plant species may be expected when such a mutation has been found already in another species phylogenetically related to the first species.

2.4. The third period: proof of induced mutations; theoretical background; the first commercial mutant

1927 – Gager & Blakeslee report on induction of mutations in Datura stramonium

Gager & Blakeslee (1927) exposed flower buds of plants of *Datura stramonium* (thorn-apple, Jimson weed), which had been selfed for nine generations, to radium rays. The radium preparations used consisted of small sealed glass tubes of capillary bore containing radium emanation. According to the authors the α- and β-rays are largely or wholly absorbed by the glass walls of the tubes, whereas the 'physiological effects' observed in their experiments 'are to be attributed to the action of γ-rays'. Three different types of 'mutations' (non-disjunction of chromosomes, aneuploidy and two real 'gene mutants') were reported. One 'gene mutation', after selfing, segregated in 162 normals and 34 albinos. The authors mention that, although this trait is not convenient to work with as it is, when in homozygous condition, not viable till beyond the seedling stage, it should be classed as a simple gene mutant. The other 'gene mutation' referred to so-called 'Ilex individuals' in which a trait 'Swollen' was observed. The *Ilex* parent was heterozygous for the gene 'Swollen'. After selfing the character did segregate in a somewhat complicated way, which was explained by the authors by the fact that the 'Swollen' trait was located in the '*Ilex* chromosome'. As the authors, originally, were looking only for chromosome mutations and not for gene mutations, they did not save seeds of the normal offspring from the treated capsule to check if they were heterozygous for other 'new genes' (some additional data can be found in the aforementioned publication). According to Stubbe (1937a) this work on plants was performed independently from Muller's work on *Drosophila*.

Radium treatment of snapdragon (*Antirrhinum majus*),

performed by Stein, who was a student of E. Baur, produced heritable malformations in plant tissues. Stein, however, did not indicate these effects as mutations, but as 'radiomorphoses' (Stein, 1922; 1926; 1929; 1930). Treatments were performed by exposing seeds for nine or more hours to about 15 mg of radium. The radium was kept in a glass-platinum tube which absorbed all α-rays and 99% of the β-rays.

For more details and additional references see Baur (1930; 1932) and Stubbe (1937a; 1938).

1927 – Definite proof of mutation induction by X-rays by H.J. Muller. The ClB-Method

In 1927 H.J.Muller presented definite proof that X-rays could induce mutations within genes (sometimes indicated as **point mutations**; see also Chapter 3). Use was made of ingenious experiments to study mutation rates in the X-chromosome of *Drosophila melanogaster*. The so-called *ClB*-Method, which is based on the induction of lethal mutations on the male X-chromosome (i.e. gene mutations which are inherited in a Mendelian way), became a model method for quantitative studies on sex-linked mutations in populations of *Drosophila*. The *ClB*-method is explained in Box 2.1. It was found that X-rays could increase the mutation frequency with a factor of 100 or more.

In his classic paper Muller (1927), who was not a plant breeder, referred to the possibility of obtaining genetically superior plants, animals and man by applying X-radiation. Soon after the 6th International Congress of Genetics in Berlin, where Muller (loc.cit.) presented his results, mutation work was taken up at various places. Some additional details about Muller, who became a Nobel laureate in 1946, and his research are presented in Box 2.1.

Muller was not the first who succesfully applied X-rays for the induction of mutations in *Drosophila*. During the early 1920s James W. Mavor presented a clear case of a genetic effect, i.e. of a mutation, induced by X-rays (Mavor, 1925). The effect obtained by Mavor did not refer to a gene mutation but to a disturbance of a sex-linked character due to non-disjunction, which leads at meiosis to two chromosomes going to one cell and none to the other. As Singleton (1955; 1962, p. 303) points out, Mavor intended to determine if X-rays could disturb the inheritance of a sex-linked character and there is no doubt that a hereditary change was obtained. Moreover, many more effects which – correctly – have been registered as 'mutations', do not refer to gene mutations but to other hereditary effects (see also Chapter 3).

Box 2.1 H.J. Muller (1890–1967)

Hermann Joseph Muller (1890–1967) studied at Columbia State University, under the renowned geneticist T.M. Morgan. In 1920 he joined the Department of Zoology at the University of Texas. In particular his research on *Drosophila melanogaster* became famous. Important topics of his studies were the phenomena of 'crossing over', inheritance by means of multiple factors and, from 1918, mutations. In this last context he defined **'mutation rate'**. Initial experiments to induce mutations by treatments with various chemicals and high air pressure remained without success. Some modest results were obtained in experiments with high temperature treatments which he performed together with E. Altenburg.

In 1927 Muller surprised the audience at the Fifth International Congress of Genetics in Berlin with the definite proof that X-rays could induce mutations (Muller, 1927). It had taken Muller and co-workers a period of seven years to build up the genetic stock of *Drosophila* for the range of ingenious experiments in which it was shown definitely that irradiation with X-rays does lead to mutation rates that statistically differ from spontaneous mutation rates. In another publication Muller (1928, original not consulted) stated that he had not been able to observe qualitative differences between spontaneous and induced mutations.

The best-known method applied by Muller to detect all kinds of mutations on the X-chromosome of *Drosophila* (irrespective whether they arose spontaneously or

artificially, after chemical or physical treatment) is the so-called *ClB*-Method. A good description of this method (see also Fig. 2.1) was given by Singleton (1962). In the expression *ClB* the *C* stands for a long inversion which prevents crossing-over within the inverted part of the chromosome concerned. Gametes with cross-overs only very seldom survive. The *l* refers to a recessive lethality gene on the X-chromosome. Homozygous *l/l* females and (hemizygous) males with the constitution *l* are non-viable. The *B* indicates the so-called Bar eye, which trait is used as a marker trait.

In the *ClB*-Method a heterozygous *ClB* female is mated to an irradiated male fly with the intention to detect all lethal mutations that have occurred in the X-chromosome. Fig. 2.1 shows that in the so-called R_1 generation, (which, in fact, should be called M_2!) theoretically, four different types of flies could occur. So-called 'normal' females are discarded; normal males are saved for further testing; *ClB* males do not survive and *ClB* females (showing the Bar eye) are used for testing. The expression 'normal females' might be somewhat misleading, as such females do carry the X-chromosome from the irradiated male! The cross between normal males and the heterozygous *ClB* females again results in four possible categories of progeny: *ClB* females; 'normal' females, *ClB* males that will die because of the *l*-factor, and other males, carrying the *L*-factor on the X-chromosome

of the original, irradiated male. If a mutation for lethal zygote from *L* to *l* has been induced in the X-chromosome of the original (irradiated) male, no males at all will be present anymore in the generation studied now (called R_2 by Singleton in Fig. 2.1).

Singleton stresses the point that only one female fly must be tested in each M_2 (R_1 according to Singleton) culture. In this way it is possible to trace back the complete segregating generation (R_2 in the picture) to a single treated X-chromosome. If some of the 'R_2' culture bottles only contain females, evidence is provided that a mutation from *L* to *l* had been induced in the original treated chromosome. Normal females tolerate such induced lethal mutations in heterozygous condition (black bar in Fig. 2.1) and can be used for further studies. In addition, any visible mutation that has occurred on the X-chromosome can be found among the surviving males in the generation indicated by Singleton as R_2.

From 1933 Muller worked for three years at the Institute of Genetics in Moscow and thereafter until World War II at the University of Edinburgh, where he met with Charlotte Auerbach. From 1945 he was Professor of Zoology at Indiana University. In 1946 he received the Nobel Prize in physiology and medicine.

1928 – Report on the successful induction of mutations after irradiation of barley and maize by L.J. Stadler

In 1928 Stadler, a plant geneticist, published his first results on the effects of irradiating seeds of cultivated plants. In this and following years many reports on irradiation of barley (*Hordeum vulgare*), oats (*Avena sativa*), wheat (*Triticum aestivum*) and maize (*Zea mays*) with X-rays

and radium rays, were published (Stadler, 1928a,b; 1929; 1930 and later years). Stadler showed that dry seed can withstand 15–20 times higher doses of radiation than germinating seed and that irradiation of germinating seed gave eight times higher mutation frequencies than dry seed. He also found that the mutation rate is proportional to the doses with which the plant material was treated and that differences with respect to mutation

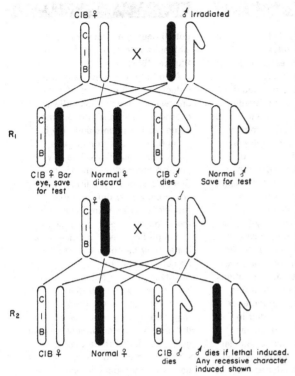

R₁

CIB ♀ Bar eye, save for test

Normal ♀ discard

CIB ♂ dies

Normal ♂ Save for test

R₂

CIB ♀

Normal ♀

CIB ♂ dies

♂ dies if lethal induced. Any recessive character induced shown

Figure 2.1 Muller's *CIB*-Method for detecting mutant genes in the X-chromosome of *Drosophila*. (From Singleton, 1962).

opinion of many plant breeders about the feasibility of mutation breeding. At a later stage Stadler also came to believe that X-rays cannot produce true mutations whereas Muller, on the other hand, claimed the opposite. Auerbach (1978) mentioned that, many years later, Muller's claim was proved to be correct by the discovery of molecular geneticists who showed that X-rays can induce base changes in the DNA.

At that stage already, but also in more recent years, strong objections arose against Stadler's view, in particular in Sweden by Nilsson-Ehle and his student Gustafsson, as well as in Germany by Baur and, later, by Stubbe. In particular Gustafsson, on many occasions, has expressed as his opinion that mutation <u>in the hands of skilful breeders</u> can be an important tool in plant breeding (see for instance Gustafsson, 1986), and results obtained in later years have clearly shown that he has been right in this respect. The mutagenic effect of UV radiation was convincingly demonstrated in *Drosophila* by Altenburg (1934). Some years later, studies on UV induced mutations in maize pollen were taken up by Stadler.

More details about Stadler's work and his opinions on mutations are summarized in Box 2.2.

1927 – Efficient mass screening techniques for spontaneous mutants in lupine developed by von Sengbusch

Based upon a suggestion by E. Baur, the German scientist R. von Sengbusch started in 1927 to develop efficient mass screening methods to detect alkaloid-free plants for the common bitter lupines (*Lupinus luteus, L. augustifolius* and *L. albus*) which could be used as cattle fodder. Spontaneous mutations for the trait 'alkaloid-free' occur at very low frequencies (see also Section 1.3.3 from Chapter 1) and, as a consequence, very large populations, consisting of hundreds of thousands of plants, have to be investigated in order to detect such rare plants. This work, summarized by von Sengbusch (1942), has been an important example to demonstrate the value of efficient selection methods, in particular for traits which occur at low frequencies.

1928 – Early practical projects on mutation induction in crop plants

Soon after the results obtained by Muller and Stadler became known, mutation breeding was taken up at several places. In 1929 the German (plant) geneticist H. Stubbe (1929) made an appeal to apply mutation research in plant breeding. Stubbe and co-workers performed mutation research on *Antirrhinum majus* (snapdragon), tomato (*Lycopersicon esculentum*), soybean (*Glycine*

rate exist <u>between</u> but also <u>within</u> plant species. The ploidy level of a crop was found to play a role as well. Stadler, whose experiments started simultaneously with those of Muller, but whose results could not be published at the same time due to the longer generation cycle of plants, indeed was the first to prove the feasibility of inducing mutations in higher plants and from this point of view his contributions to mutation breeding have been very significant. Stadler's methods have been followed for many years by other scientists.

Stadler, however, came to believe that induced mutations would benefit practical plant breeding only in very few cases. The main reasons for this opinion were, first, that he assumed that enough valuable genes (alleles or spontaneous mutations), well adapted by many generations of natural selection, would be present for further improvement of most cultivated plants and, second, that most newly induced mutations would refer to changes into a negative direction or would be strongly linked with other deleterious mutations. These assumptions made Stadler conclude that plant breeding would be more beneficial by making use of the variation that is already existing in the plant population than breeding by means of induced mutations. This pessimistic view by a leading scientist, has for many years affected the

Box 2.2 L.J. Stadler (1896-1954)

The first proof of considerably increased mutation frequencies in crop plants after mutagenic treatment came from Lewis John Stadler (1928a,b; 1929; 1930; 1932) who obtained his Ph.D. in 1922 at the University of Missouri and remained employed at this University throughout his academic career. In order to study whether spontaneous and induced mutations are identical, Stadler also determined in a very accurate way spontaneous mutation rates for various traits and in different crops, in later years in particular maize (Stadler, 1942).

Stadler's experiments, however, mainly referred to the self-fertilizing crop barley (*Hordeum vulgare*). After mutagenic treatment of resting or germinating seeds, planting took place at such distances from each other that sufficient tillers were formed to detect chimeras. The occurrence of mutations was detected by growing the self-fertilized progeny of each individual spike (or head) of a plant. A control, consisting for instance of 1500 spike-progenies, was included in the experiments in order to determine the spontaneous mutation frequency. Mutations in seedlings, mostly

referring to chlorophyll aberrations, were used as an index for the mutation frequencies obtained after different treatments. About 90% of the occurring mutations could be observed already in the seedling stage. Other mutations referred to seed traits or to developmental characteristics of the adult plant. The genetics of the obtained mutations was investigated by raising a selfed next (M_3) generation, or by back-crossing with untreated control plants.

All mutations tested by Stadler were found to be recessive and, as a consequence, Stadler believed that recessive mutations are the result of chromosome aberrations, or indicate the destruction of a gene. Comparable experiments were performed by Stadler in maize (*Zea mays*) and similar mutations were reported. Stadler was slightly more optimistic about the quality of UV-induced mutants which did not show the associated effects that were observed after X-ray treatment (Stadler, 1939; Stadler & Roman, 1943). Many of the collected UV-data have never been published. Many more details of Stadler's work can be found in his own publications (which have

not all been mentioned here), as well as in reference articles (e.g., Stubbe, 1938).

According to Stadler radiation-induced mutations in most cases are related to deleterious effects, caused by deletions and chromosome rearrangments and only in exceptional cases to real gene (intragenic) mutations. The few positive induced mutations, in Stadler's opinion, are likely to be accompanied by unfavourable mutations which were induced by the same mutagenic treatment. Although he agreed that desirable genes may be obtained in this way, he believed that promising genotypes can be obtained in an easier way by hybridization.

Stadler (1930) concluded that it might be preferable to make use of the great wealth of already existing genetic variation in nature, from which more valuable traits, without negative accompanying effects, may be derived than after irradiation. He argues that mutations observed in nature have survived selection pressure already for a very long time and, therefore, can be directly applied in breeding work. This opinion appears correct, but also emphasizes the point that

max) and several other crop species and demonstrated the role of mutations in domestication of crop plants. In Box 2.3 more attention is given to Stubbe and his important contributions to our present knowledge of mutations.

Two other pioneers in the field of mutation breeding, who must be mentioned here, are H. Nilsson-Ehle and Å. Gustafsson in Sweden, who started with mutagenic treatment of cereals (in particular barley) in 1928. This work will be briefly discussed in a later section of this chapter, and in more detail in Chapter 3. In 1954 Singleton (1955) presented an invitational paper before the general meeting of the Crops Science Divisions of the American Society of Agronomy. In this paper Singleton did credit to Nilsson-Ehle and Gustafsson who, in 1928, had already embarked on studies of agricultural plants by means of mutations induced by X-rays and UV-radiation. Singleton pointed out that these authors

were the first who accomplished in barley the production of genetically superior cultivars by the use of radiation.

In 1928/29 W.E. de Mol, a Dutch scientist, started X-irradiation of bulbs of hyacinths (*Hyacinthus* sp.) and tulips (*Tulipa* sp.). De Mol (1931; 1933 and at later occasions) reports that genetic changes in flower colour were obtained after X-irradiation of incised bulbs of hyacinths. Part of the mutation work of de Mol, who published hundreds of papers on this topic, will be discussed in Chapter 7. For references see for instance Broertjes & van Harten (1988).

1926/28 – Vavilov's theory on gene centres or 'centres of origin'

Vavilov developed the view that concentrations of genetic variation for cultivated plants occur in several regions of the world. According to Vavilov a centre of induced mutations in most cases should not be considered as 'final products', but as **raw material for further plant breeding work** (as was pointed out already in Chapter 1).

Stadler further mentions that induced mutations of course would be much more promising for practical breeding if mutations could be induced selectively. However, he adds that there are some cases in which induced mutations might offer better possibilities and in this respect, he refers to corn (or maize) breeding with continuous inbreeding, followed by crossing for first-generation hybrids. Induction of positive, non-deleterious mutations in those almost completely homozygous lines (in which no further progress can be obtained by selection) might lead to some further improvement.

The second promising application, mentioned by Stadler, concerns the use of induced mutations in breeding of (vegetatively propagated) fruit trees. Such crops predominantly started as heterozygous seedlings of unknown parentage, which were selected for their attractive characteristics. As propagation by seed immediately would lead to loss of the favourable combination of traits, they are vegetatively propagated.

Stadler was aware of the fact that some important cultivars have arisen from spontaneous mutations (**somatic mutations** or **bud sports**) and added, for this group of crops, that it does not seem unlikely that induced mutations may produce valuable new genetic variation.

Despite these positive suggestions, the predominantly negative opinion of Stadler about the practical value of induced mutations for plant breeding for many years greatly discouraged the initiation of further mutation breeding activities, in particular in the Western world. It appears, however, that not many people have endeavoured to carefully study Stadler's results and opinions, in order to try and find practical answers to some of Stadler's critical remarks about the limitations of the mutation method. It may be emphasized in this respect that the experiments performed by Stadler were excellent and that Stadler was completely right in pointing out that the majority of induced mutations consist of 'junk'. Positive results can only be obtained after this junk is eliminated and the rare, positive cases are picked out by effective methods of mass-screening.

In particular Nilsson-Ehle in Sweden in the 1930s, followed by Gustafsson, the German scientists Baur and Stubbe and the Russian wheat researchers Delaunay (1930; 1931) and Sapehin (1930) already expressed a much more optimistic opinion than Stadler about the potentialities of induced mutations for plant breeding (see for instance Gustafsson, 1947; 1986). Gustafson himself proved that favourable mutations can be obtained and are not necessarily accompanied by negative effects.

Stadler did not live long enough to see for himself that his pessimistic views on mutation breeding were unjustified as can been shown by the many mutant cultivars, some of which are of considerable economic value, which have been produced by now. Obituaries on Lewis John Stadler were published by Rhoades (1956) and by Rédei (1971).

origin is characterized by dominant alleles. Towards the periphery the frequency of recessive alleles increases and diversity decreases. The occurrence of many recessive alleles, ultimately, is the result of mutation, whereas an important further role is played by inbreeding, geographical isolation and genetic drift. Gene centres play an important role in our present thinking about the occurrence of spontaneous mutations and of so-called parallel (or homologous) mutations in various crops. More details can be found for instance in Zeven & de Wet (1982).

1928–34 – Continued studies on mutation theory and practical applicability

Following Muller's and Stadler's experiments, the genetic effects of irradiation, in particular with X-rays, were investigated in many different plant species. Different doses were applied under various environmental conditions and various plant parts and organs were studied in different stages of development. With respect to the genetic nature of the plant material, an increased interest arose for studies on the effects of irradiation on polyploids and aneuploids. A useful reference in this respect concerns the work of Ichijima (1934) on artificially induced mutations, in particular on aberrant chromosome numbers in rice (*Oryza sativa*).

Not much attention was paid yet to the fact that breeding work in self-fertilizing crops, cross-fertilizing crops and vegetatively propagated crops requires different approaches and that the same should apply with respect to mutation breeding. Effects of treatments were expressed in terms of the number of gene mutations obtained or of the relative frequencies of gene mutations and chromosome mutations.

Box 2.3 H. Stubbe (1902–89)

Hans Stubbe, one of the real pioneers of mutation breeding in plants, was born in Berlin, Germany, where he studied agricultural sciences at the Humboldt University. He started his scientific career as an assistant of Erwin Baur in 1927, first at the University Institute of Genetics in Berlin-Dahlem and later (till 1936) at the Kaiser-Wilhelm-Institute for Breeding Research in Müncheberg, with experiments on induced mutations in *Antirrhinum majus* (snapdragon). In the early 1930s, soon after Stadler's publication on induced mutations in barley, Stubbe published extensively about his efforts to induce mutations by X-rays, UV-light, temperature shocks, seed centrifugation and chemical treatments (Stubbe 1930a,b; 1932). His work yielded many valuable results. He showed that mutagenic treatments resulted in a significant increase of gene mutations in

Antirrhinum, and that a clear correlation existed between the dose applied and the mutation frequency, but he did not find that the height of the applied dose affected the relative frequency of different categories of mutations.

Stubbe observed the phenomenon of **reverse** (or **back**) **mutations** and was, with Baur, the first who paid attention to so-called '**Klein-mutationen**', or **micromutations** (which will be discussed in Chapter 3). He further described in a very detailed way many different types of mutation and changes for shape, colour and other traits in snapdragon.

In 1937 and 1938 Stubbe published two monographs on mutations, which are among the best publications that were ever written in this field (Stubbe, 1937a; 1938). Another very valuable publication concerned the history of genetics (Stubbe, 1963, in German), which covered the

period till the rediscovery of Mendel's laws in 1900. In 1972 an (extended) English translation (Stubbe, 1972) was issued.

In addition to his work on snapdragon, Stubbe also studied mutations in tomato, barley, soybean and faba bean. He contributed significantly to fundamental studies on the mutation process and on the role of mutations in evolution.

Apart from his great merits for the science of mutation, the activities of Stubbe as director of the famous Institute for Research on Cultivated Plants and director of the first genebank in Europe (outside the USSR) should be mentioned here as well, whereas many other scientific contributions must remain unrecorded here.

Valuable research work on mutations in higher plants during this early period, in addition to that performed by Stadler (and – indirectly – by Muller), was done for instance by Gager and Blakeslee (*Datura*), Goodspeed (tobacco, *Nicotiana tabacum*), Tollenaar (also tobacco), Nilsson-Ehle and Gustafsson (barley, *Hordeum vulgare*), Stubbe (*Antirrhinum*) and de Mol (hyacinths and tulips). We will refer to their work at several later occasions. In addition, Delaunay (1930; 1931) and Sapehin (1930) published about induced mutations in wheat (*Tritium aestivum*); Lindström (1933) and MacArthur (1934) on tomato (*Lycopersicon esculentum*) and Asseyeva & Blagovidova (1935) on potato (*Solanum tuberosum*).

More examples can be found in the monograph by Stubbe (1938) who, in one of the final chapters of that book expresses the opinion that at that time enough theoretical knowledge had already been collected on the induction of mutations to apply this technique in applied mutation breeding. Stubbe further advised practical plant breeders, when considering the use of mutations in their breeding work, to pay much attention to carefully determining the breeding goals in each specific case and to work on large plant populations in order to

be able to select for useful mutations. In addition, adequate screening methods must be available to select properly in such large populations.

1934/1938 – The first commercial mutant cultivar 'chlorina', obtained in tobacco after X-irradiation

It had been demonstrated in the late 1920s by Goodspeed and co-workers in the USA (Goodspeed, 1929), that mutations could be induced in tobacco (*Nicotiana tabacum*). Starting from 1929 and following Goodspeed's work, D. Tollenaar, a Dutch scientist, irradiated inflorescences of tobacco with X-rays with the purpose of obtaining genetically improved cultivars of tobacco (1934). One of the mutants was a chlorophyll-defective plant, characterized by a light-green leaf colour. This so-called 'chlorina-type', a trisomic **aneuploid** (with an extra chromosome) which was described in detail (Tollenaar, 1938), in 1936 already occupied about 10% of the total area grown with tobacco in the 'Vorstenlanden' area in the Dutch East Indies (now Indonesia). More details are given in Box 2.4.

1934 and following years – A physical mutation theory by Timoféeff-Ressovsky and co-workers

In 1925 the young Russian researcher N.W. Timoféeff-

Box 2.4 The first commercial induced mutant 'chlorina' in tobacco

D. Tollenaar, who worked at the Tobacco Research Station in Klaten, Java, in the former Dutch East Indies (now Indonesia), starting in 1929, irradiated inflorescences of a local, fully homozygous line of tobacco (*Nicotiana tabacum* subsp. *havanensis*) cv. Kanari, of which the origin is not exactly known, with X-rays (5 and 10 min. at 50 kV and approx. 3 mA, distance of object to source: 35 cm). According to most sources *N. tabacum* is an amphidiploid species with $2n = 4x = 48$ chromosomes.

In his first radiation experiments Tollenaar (1934) already obtained many mutants which, predominantly, were **factor** (**gene, locus**) mutations. Other mutations found were associated with chromosome rearrangements. Tollenaar (loc.cit.) reported that, after eight generations already

more than 100 different mutants had been obtained, which he studied carefully and described in a very precise and detailed way.

In a following publication Tollenaar (1938) describes different types of mutants including a mutant indicated as 'chlorina'. This 'chlorina type' is a **chromosome mutant** – in fact an **aneuploid** with the chromosome constitution $2n + 1$ – which also carries a gene mutation for a yellow-green leaf colour. Tollenaar reports that the chlorina mutant is grown as F_1 hybrid cross with the original ('typica') type of 'Kanari' – i.e. **as an indirect mutant** – in 'de Vorstenlanden' of Central Java, an area famous for tobacco growing. The attractive light leaf colour of the 'chlorina type' and resulting hybrids is of particular importance when the leaf is used as a 'wrapper' in cigars.

The economic significance of this indirect mutant can be derived from the fact that, according to Tollenaar, in 1936 already 10% of the total area grown with tobacco was covered with hybrid cultivars in which the 'chlorina' trait had been incorporated. In the discussion of a contribution about the later Pallas barley mutant, Borg *et al.* (1958) reported on some personal information obtained from Tollenaar concerning the chlorina mutant. Tobacco planting in 'de Vorstenlanden' area was terminated in 1939/40, since the market for this kind of tobacco lay exclusively in Europe, and collapsed because of the start of World War II. After this war, tobacco growing was not taken up again in this specific area of Indonesia.

Further details about the way of inheritance of this mutant, unfortunately have not been published.

Ressovsky was invited to move to Berlin to organize a new Department of Experimental Genetics at the Kaiser-Wilhelm-Institute. This move took place in 1926 and within a few years he became an important mutation geneticist. Timoféeff-Ressovsky's life and major achievements have been described by Paul & Krimbas (1992).

Together with co-workers, of whom in particular M. Delbrück, G. Zimmer and his wife (Helena Aleksandrovna) should be mentioned, Timoféeff successfully tried to provide a general, physical, explanation for the occurrence of both spontaneous and induced mutations, based on experiments with *Drosophila*; see, for example, Timoféeff-Ressovsky, Zimmer & Delbrück (1935) and Timoféeff-Ressovsky (1937). This work of the Timoféeff group resulted in the **target-** or **hit and target theory**, in which the genes were considered the **targets** for the **hits** which were produced by the radiation. Summarizing after Auerbach (1976; 1978), the target theory implies that mutations and chromosome breaks arise as a result of single hits. The hits are produced by one ion or a few ions which arise when cells are irradiated. The frequency of gene mutations (and small deletions) increases linearly with dose and independently of dose rate or dose frac-

tionation. If a treatment results in an exponential killing curve and a linear mutation curve, it is assumed that single 'hits' have occurred.

Radiation may also induce chromosome breaks which, in most cases, do not rejoin in the old way and, as a consequence, (larger) rearrangements like inversions or translocations may occur. It was proven that chromosome breakage always precedes rearrangement. The frequency of rearrangements and large deletions increases roughly as the square of the dose. The curve is bending down again at higher doses. Lowering the dose rate leads to a lower frequency of X-ray induced rearrangements. Cytological evidence that chromosome breaks and rearrangements occur in irradiated plant cells was provided by Sax (1938). Most details are outside the scope of this book, but some general principles of the interaction between living material and radiation are discussed in Chapter 4.

Some other researchers whose names are directly related to the target theory, are (i.a.) Lea and Catcheside (Lea & Catcheside, 1942; Catcheside & Lea, 1943; Lea, 1946).

In addition to the work on physical aspects of mutation, Auerbach (1976) mentions a number of other important landmarks with respect to fundamental mutation research obtained in this period, such as the difference between intergenic and intragenic changes, the relation between crossing-over and chromosomal rearrangements, the successful method of analyzing dose-effect curves, studies about the occurrence of spontaneous mutations and, finally, the specific mutagenic effect of ultraviolet light, as was demonstrated in particular for maize (*Zea mays*) by Stadler and collaborators in the 1930s.

1937 – The chromosome doubling effect of colchicine on plant chromosomes

In 1937 two important publications appeared in which the chromosome doubling effect of colchicine, an alkaloid derived from *Colchicum autumnale* (Autumn crocus), was described for plants by Blakeslee & Avery (1937a, b). The suggestion to Blakeslee to investigate the anti-mitotic activity of colchicine on plants came from O.J. Eigsti, a co-worker in Blakeslee's laboratory who, in 1936 had attended a lecture by E. Allen, G.M. Smith and W.U. Gardner in the USA on the various effects of colchicine. These speakers, for their part, had taken notice of earlier investigations in Belgium by A.P. Dustin, L. Havas and F. Lits on various effects of colchicine on animal tissues, which had been reported in 1934. In 1937 Dustin, Havas & Lits published on various effects of colchicine treatments on seedlings of wheat (*Triticum aestivum*) and on roots of *Allium* and *Tulipa*, including the anti-mitotic activity of colchicine. This original work, undoubtedly because of the nature of the journal in which it was published, remained unnoticed among botanists.

In addition to Eigsti's suggestion to Blakeslee in 1936, a similar suggestion was made in that same year by D.F. Jones to B.R. Nebel and M.L. Ruttle who, independently of Blakeslee, also started investigations on the effects of colchicine on plants. The first contribution on this topic was published by Blakeslee (1937) in a French journal, but the papers of Blakeslee & Avery (1937a, b) were those which, in particular, drew general attention. Eigsti's suggestions were acknowledged in a small footnote only. Publications by Nebel and Ruttle on the same subject (originals not consulted) followed only a few months later. More details about the discovery of colchicine and other interesting data about the early years of colchicine research, accompanied by an impressive list of references, including all early references to the work of the aforementioned scientists, were published by Krythe & Wellensiek (1942). Rieger (1963), in addition, mentioned that the effect of colchicine on cell division was discovered also in 1937 by Gavaudan & Pompriaskinsky-Kabozieff (for references see Rieger, loc.cit.).

Blakeslee and Avery tested the anti-mitotic effect of colchicine for a wide range of plants. It was found that increased ploidy levels could be obtained in a relatively easy way and may lead to an overall enlargement of plant organs (cells, stomata, leaves, flowers, fruits, seeds, etc.) but also to slower growth and reduced fertility.

Polyploidization of crops soon became an important research topic and expectations from the viewpoint of practical breeding rose high. It appeared later that results often were disappointing, in particular in seed propagated crops. Nevertheless, doubling of chromosome numbers, nowadays, is an often applied method.

More details about ploidy mutants, and how to obtain them, will be presented in Chapter 3.

The 1930s – Start of the Swedish mutation research programme with studies on chlorophyll mutations by Gustafsson

Throughout the years continued attention has been attracted by extensive projects on induced mutations which started in the early 1930s in Sweden. The main initiators of this work were the young Åke Gustafsson and his supervisor, H. Nilsson-Ehle, who studied, predominantly, mutations in barley (*Hordeum vulgare*). Because of the interesting and promising results that were obtained, Gustafsson was able to secure sufficient financial support to continue this work and after World War II he gathered around him a group of specialists from different disciplines. The field of studies was broadened and improvement of the methods to apply induced mutations in plant breeding also became a subject of research.

Although attention in particular went to fundamental aspects, the project also convincingly demonstrated the value of induced mutations for practical plant breeding and Gustafsson himself became the most influential advocate of mutation breeding. The work of the Swedish group and other scientists working in close relation with them, will be discussed in particular in Chapter 3. Åke Gustafsson's significant contributions to mutation breeding and some additional information about his person are presented in Box 2.5.

1941 – Chemical mutagenesis: Auerbach, Rapoport, Oehlkers and others

Efforts to obtain artificially induced mutations by treating *Drosophila* and seeds of various plant species with different chemicals started at the very beginning of this century. Stubbe (1937b; 1938) mentions several early reports on treatments of higher plants, for instance by MacDougal in 1909, by Humbert in 1910/11 and by Dewitz in 1913, but in none of these cases were positive

Box 2.5 Åke Gustafsson (1908–88)

Åke Gustafsson, who probably might be called 'father of mutation breeding', was a Swedish botanist and geneticist who, after his doctor's degree in 1935, became a university docent in Lund. He had a special interest in problems of plant taxonomy, apomixis in higher plants and cytogenetics. In 1947 he just missed the prestigious full professorship in botany in Lund and became professor of forest genetics in Stockholm. His work there was very successful and Gustafsson stayed in Stockholm for 21 years, where he continued to work on a broad range of research subjects within the fields of botany, plant taxonomy, genetics and forestry. In 1968, when he had reached the age of 60 years, he returned as a professor to Lund.

Already in the early 1930s, and like several others, Gustafsson, still a student under H. Nilsson-Ehle, became interested in the question whether artificially induced mutations were identical to spontaneous mutations and whether they could be of practical use in plant breeding in the same way as spontaneous mutations. This subject had been touched already by Hugo de Vries around 1900, but became topical again around 1930 when H.J. Muller expressed his rather optimistic expectations about the practical value of induced mutations and when L.J. Stadler, who was an eminent plant breeder, advanced the pessimistic opinion that induced mutations in most cases would be useless for plant breeding. Gustafsson was not convinced by Stadler's views and suggested his supervisor Nilsson-Ehle to start experiments on induced mutations in barley. After X-irradiating seeds of the cv. Gull, primarily chlorophyll mutations were induced and studied in detail

(Gustafsson, 1938; 1940).

Of more practical value were the so-called 'erectoides-types' with compact spikes and, often, stiff and short straw, which are the most frequently induced morphological changes in barley and had been obtained in the 1930s by Nilsson-Ehle (Gustafsson, 1947). Other interesting groups of mutant types followed, such as short awn types ('breviaristatum'), waxless mutants ('eceriferum'), six-row types ('hexastichon'), powdery mildew resistant mutants, anthocyanin-free types ('exrubrum'), etc. More references to Gustafsson's work can be found in particular in Chapter 3.

Throughout the years a unique collection of genetically and agronomically well-documented mutant genes in barley has been built up in Sweden and based on this collection exemplary studies on fundamental and practical aspects of mutations were performed. These studies made barley one of the few 'model' crops for mutation research in higher plants. In 1994 the Nordic Gene Bank contained about 10 000 barley mutants (Udda Lundqvist, 1994; personal communication).

Gustafsson, as was said before, at an early stage of his mutation research had already contested Stadler's pessimistic opinions and took it as his job to prove the practical value of induced mutations for plant breeding. Final proof came from the 'Gustafsson group' when, beginning in the early 1960s, a number of very good mutant cultivars of barley were obtained as a result of purposeful mutation breeding work. Several of these X-ray induced mutants, of which in particular the stiff straw mutant cv. Pallas became famous, have been very successful from an economic point of view in Scandinavia as well as in other countries, either as

'direct' cultivars (cv. Pallas and cv. Mari) or after further crossing work as 'indirect' mutant cultivars.

Some 240 scientific papers were authored and co-authored by Gustafsson. His last contribution about mutations was presented in 1986 on the occasion of the 100th anniversary of the Svalöf breeding company (Gustafsson, 1986).

Almost until his death in 1988 he continued to advocate the value of mutation breeding and, being a gifted speaker and debater, he probably became the most formidable adversary of those geneticists and plant breeders who, often without adequate knowledge of the facts, expressed their doubts or disbelief about the feasibility of mutation breeding. Gustafsson also openly attacked the pernicious doctrines of Lyssenkoism and stood up for the victims of the 'witch-hunting' caused by the Lyssenko adherents. Another, less known feature of Gustafsson has been that he was a writer of poetry and essays which, unfortunately, never have been translated from Swedish.

Åke Gustafsson has contributed greatly to a better understanding of mutations and the possibilities of applying them in plant breeding. His work on induced mutations in crop plants has proven beyond any doubt their value for plant breeding, provided that the mutation method is used in the appropriate way, i.e. by skilled plant breeders who know their material as well as the possibilities and limitations of different breeding methods. Gustafsson also showed what can be achieved in research when knowledge, skill, perserverance, leadership and a spirit of cooperation are brought together.

results obtained and the same applied to experiments with *Drosophila*. According to Ehrenberg (1960), the oxidant iodine was the first chemical demonstrated to induce mutations in *Drosophila*. This chemical was tested in 1932 by the USSR scientist W.W. Ssacharow (for reference see Ehrenberg, loc.cit.).

Gustafsson (1986) mentions that for higher plants 'the first more generally accepted paper on **chemical mutagenesis**' was published in 1943 by Friedrich Oehlkers in Germany. The work of this researcher concentrated on studying chemically induced chromosome translocations, originally in *Oenothera* (Oehlkers, 1943), but later also in *Vicia faba* (faba bean) and *Antirrhinum* (snapdragon). Oehlkers treated meiotic cells in developing buds of cut inflorescences of *Oenothera* with different chemicals – of which in particular urethane (+ KCl) and ethyl urethane became well known – and obtained various types of chromosome aberrations in frequencies which differed significantly from the spontaneous rate in the control material. Results, published in a range of articles (see also Oehlkers, 1946; 1949; 1956) were interpreted in terms of the so-called **breakage and reunion mechanism** in which two duplex chromosomes are broken at corresponding points and, consecutively, cross-wise rejoined.

In 1940 C. Auerbach started research in cooperation with J.M. Robson on chemical mutagenesis in *Drosophila*, in particular with 'mustards' of the sulphur and nitrogen type. It was known already that much similarity existed between the effects of mustard gas and X-rays and in 1942 an internal report was submitted to the British Ministry of Supply in which for mustard gas a mutagenic effect similar to that of X-rays was reported. Simultaneously, in the *Drosophila Information Service* the induction of mutations by treatment with chemicals was reported in a rather sketchy way (Auerbach, 1943). Details of this work could be published only after World War II, starting with a short note in *Nature* in 1946 (Auerbach, 1943; 1951; Auerbach & Robson, 1944; 1946; 1947). Soon afterwards Gustafsson & MacKey (1948) proved that mustard gas was mutagenic as well in barley (*Hordeum vulgare*).

At about the same time the USSR scientist Rapoport, who earlier had discovered already the mutagenic activity of formaldehyde mixed with the food of *Drosophila* larvae, published on the mutagenic effects of a variety of other chemicals (Rapoport, 1946; 1948). These chemicals, according to Auerbach (1960), subsequently, were detected independently also by 'Western' geneticists.

Although Oehlkers's contribution to chemical mutagenesis for higher plants may have remained somewhat underestimated – probably, as Auerbach (1976) did put

it, because urethane 'is a more "spotty" mutagen than mustard gas, acting only under certain conditions and in certain organisms' – it is indisputable that the impact of the work by Auerbach and Rapoport to the further exploration of this field has been much greater. More details about chemical mutagenesis will be presented in Chapter 4. The merits of both Auerbach and Rapoport are briefly discussed in Box 2.6 and Box 2.7 respectively.

1942 – First report on X-ray induced resistance in barley

Resistance to powdery mildew (*Erysiphe graminis* f.sp. *hordei*), obtained in cv. Haisa II of barley (*Hordeum vulgare*), after treatment with X-rays, is the first example of the induction of mutations for disease resistance. The German scientists Freisleben & Lein (1942) screened some 12 000 plant progenies (or 240 000 seedlings) from which 19 highly resistant plants were obtained. One mutant, indicated as 'M 66', possessed resistance to all different races of powdery mildew in Germany and it was further found that the observed resistance was based on a single recessive gene.

This mutant – later named *ml-o 1* (Jørgensen, 1976) – was the most spectacular result of a large scale mutation breeding programme initiated cooperatively at institutes in Halle and Müncheberg, Germany (Freisleben & Lein, 1943a,b; 1944a,b). More details about this – as it turned out – highly effective and durable mutant and similar mutants that were induced later as well as found in nature, will be discussed in Chapter 3.

1953 – The Watson–Crick model of the gene

In 1953 Watson and Crick presented their double helix model to explain the structure of DNA. At that stage the basic molecules or building blocks – four nucleotides, each with the same phosphate and sugar and with one organic base (adenine, guanine, cytosine and thymine) – had been known already for many years. In the early 1960s the work of various research groups made clear that a group of three nucleotide pairs (a triplet) codes for a single, specific amino acid. These amino acids are the building blocks of the proteins which, in fact, are the 'translation' of the genetic information in the DNA.

According to Auerbach (1976, p. 10) the acceptance of the Watson–Crick model of the gene implied that 'the central problem of mutation research was solved once and for all' and the new knowledge also provided explanations for many problems that could not be solved in previous years. Auerbach mentioned in this respect that, for instance, now final proof could be given for the occurrence of so-called intragenic mutations; the action

Box 2.6 Charlotte Auerbach (1899-1994)

Charlotte Auerbach was born in 1899 in Krefeld, Germany. She studied at various Universities and graduated in biology in 1924. She was a teacher in Berlin till 1933 when new laws prohibited Jewish citizens being employed in state schools and forced her to abandon this profession. The increasing uncertainty for Jews in Germany made her decide to leave the country and with the help of a friend, Professor H. Freundlich in London, she managed to come to Britain in 1933. After having been introduced to F.A.E. Crew, head of the Institute of Animal Genetics in Edinburgh, she was offered a modest position there to work on *Drosophila*. She obtained her Ph.D degree at the University of Edinburgh in 1935.

In Edinburgh Auerbach found much incentive in the presence of a strong international group of research workers. Some years before World War II, H.J. Muller, the later Nobel prize winner, also arrived in Edinburgh and he and Auerbach discussed at length problems in mutagenesis and related fields. Muller stimulated Auerbach to try and obtain induced mutations in *Drosophila* by making use of chemicals which were known to be carcinogenic. Until 1940, however, experiments remained unsuccessful. In that year Auerbach, in collaboration with

J.M. Robson from the Pharmacology Department in Edinburgh, started experimenting with mustard gas [$(CH_2ClCH_2)_2S$]. Treatments were performed (in a most primitive and dangerous way!) by Robson – who had some experience with mustard gas as a mitotic inhibitor in rats – whereas Auerbach performed the genetic analysis of the treated material by means of Muller's *ClB* method (see Box 2.1) by which X-linked visible and lethal mutations can be detected. By letter she informed Muller that after treatment with mustard gas 93 lethals were obtained from 1213 individuals tested, against only three lethals from 1216 individuals in the control.

Apart from publishing some results in internal reports to the Ministry of Supplies of the British Government; a short general note in *Drosophila Information Service* and a letter hinting at the mutagenic effect of mustard gas in *Nature* (Auerbach & Robson, 1944), publication of the results had to wait till after the war.

In 1946 Auerbach and Robson sent another letter to *Nature*, in which it was stated that chemical substances could be as effective as X-rays in inducing mutations and chromosome rearrangments. In the next years more details followed

(Auerbach, 1947; Auerbach & Robson, 1946; 1947; Auerbach, Robson & Carr, 1947).

Auerbach remained active in this field of research for many years and contributed may publications in congress proceedings and reviews. She was also author of several books (e.g. Auerbach, 1961; Auerbach & Kilbey, 1971; Auerbach, 1976; 1978). In the later stages of her career she was sometimes called the 'grand old lady' of chemical mutagenesis. However, as was stated before, her original work was performed together with Robson, whereas opinions differ as to the importance of the relative contributions by each of them. Beale (1993) who wrote a very interesting account on the work and persons of Auerbach and Robson, in this context mentioned that, according to Auerbach, the original idea to use mustard gas as a mutagenic agent was forwarded by the pharmacologist J.D. Clark in 1940. When referring to early major contributions to chemical mutagenesis, in any case the work of Auerbach and Rapoport and, to a smaller extent, Oehlkers, has to be bracketed together.

Charlotte Auerbach died in 1994. An obituary was written by Kilbey (1995).

– and sometimes the specificity – of certain mutagens could be explained now by their reactions with DNA and that the double-stranded nature of DNA accounted for the occurrence of so-called mosaic mutations.

It is noteworthy that only two years later, Auerbach (1978) presented a much more balanced view when she remarked that the Watson–Crick model did make it clear why mutated genes replicate as such, what kinds of changes can produce mutations and, for some chemicals, by what kinds of change they do produce mutations but, on the other hand, that the model did not provide answers to all scientific questions that could be raised with respect to mutation research in general.

Moreover, it has to be added that the new knowledge about DNA certainly did not solve any practical problem of the mutation breeder. Indeed, induction of mutations is only a small part of mutation breeding and the selection of valuable mutations in large populations of treated plant material is mostly a much bigger and more difficult part of the job. It would, indeed, have been a major break-through if our understanding of the DNA structure had made it possible to induce in a more specific way mutations for valuable traits but, so far, this goal was not achieved and induced mutations, until now, continue to arise at random.

Box 2.7 I.O. Rapoport (1912-1990)

Iosif Abramovich Rapoport was born in Chernigov, Ukraine. His work as a young scientist was already highly praised by the famous N.I. Vavilov. During World War II, while recovering from war injuries, Rapoport defended his Doctor of Science thesis in Moscow, after which he returned to the front. After the war he came back to Moscow to work at the Institute of Cytology, Histology and Embryology, where he discovered, in experiments with *Drosophila*, a number of interesting and powerful chemical mutagens, such as dimethyl sulphate, ethyleneimine, diazomethane and various epoxides. The first scientific paper concerning these discoveries, entitled 'Carbonyl compounds and chemical mechanisms of mutations' appeared in 1946. Many other publications followed.

In 1948 Rapoport was no longer allowed anymore to continue his work on genetics as a consequence of the absolute support enforced by the Soviet Government to the Lyssenko doctrine, according to which all scientific work which was not strictly in line with the 'only right theory' (i.e. the opinions of Lyssenko) had to be abandoned. It was not until 1957 that Rapoport could return to his research work, this time at the Institute of Chemical Physics of the USSR Academy of Sciences in Moscow, where he continued his research on chemical mutagenesis on

Drosophila till his death. A high mutagenic efficiency was found for certain chemicals, some of which are still often used as mutagens nowadays, such as several so-called **nitroso**-compounds (to be discussed in Chapter 4). Some chemicals belonging to this group were found to be two or three times more effective than for instance X-rays and other effective mutagens such as **ethyleneimine**. Rapoport called such mutagens 'supermutagens'.

(N.B. Zoz (1967) and Makarova (1967), both co-workers of Rapoport, reported on this topic during the Mendel Memorial Symposium in Czechoslovakia in 1965. According to Zoz, of all mutagens studied so far 'most effective was **N-nitrosoethylurea**, inducing up to 15% of dominant (!) viable mutations in wheat, 50–80% of dominant (!) mutations in pea and up to 100% of recessive mutations in wheat and pea'. The presented figures, however, look rather unlikely. A possible explanation may be (Micke, personal communication) that 'high doses' of such mutagens were used which caused 90% killing or more. Thus up to 100% of the surviving plants may segregate in M_2 or even show dominant changes in M_1 which, for wheat, may mainly refer to chromosomal aberrations and aneuploidy.)

In 1965 Rapoport set up the Center of Chemical Mutagenesis (in

some publications also called the Department of Chemical Genetics) at the aforementioned institute, at which place fundamental genetic research, methods to efficiently apply chemical mutagens to different plant parts and practical plant breeding were combined. This mutation breeding work was exclusively based on the application of chemical mutagens. Joint mutation breeding programmes were started with about 150 research stations in the USSR and other Eastern Block countries and at least 3000 cultivars of different crops were mutagenically treated.

Salnikova (1993a) reported that until 1993 these coordinated efforts had resulted in 366 mutant cultivars or cultivar lines, mainly for agricultural crops and a few for horticultural crops, of which already more than 134 had been 'introduced'. Salnikova further reported that in the year 1992 mutant cultivars of winter wheat (*Triticum aestivum*), spring and winter barley (*Hordeum vulgare*), white lupine (*Lupinus albus*) and maize (*Zea mays*) represented 25% of the total number of cultivars for these crops in the USSR.

An obituary on Rapoport by Salnikova was published in *Mutation Breeding Newsletter* 40 (Salnikova, 1993b).

The early 1950s – Barbara McClintock; mutations induced by gamma rays

In 1951 Barbara McClintock reported in a conference on quantitative biology at Cold Spring Harbor, USA, on the existence of unstable genes, a phenomenon caused by the movement of so-called **controlling elements** in maize (*Zea mays*). This discovery, based on findings which trace back to experiments started in 1944, for many years got by almost unobserved, but McClintock, convinced of the scientific importance of the subject and despite the lack of interest, continued her work with an almost unbelievable determination.

In later years, after similar phenomena had been found in micro-organisms and the scientific community started to better understand what McClintock's discoveries implied, her work made a considerable impact on genetic research as well as on related fields like evolution studies and mutation breeding research. The earlier mentioned **controlling elements** became recognized as **transposable genetic elements** (or **transposons**). In Box 2.8 some additional facts about McClintock's life and work are given.

We have been unable to trace in the literature when

Box 2.8 Barbara McClintock (1902-1992)

After having studied genetics and having obtained her Ph.D. degree in 1927 at Cornell University, USA, and after a number of short employments, McClintock in 1936 got a temporary faculty position at the University of Missouri, USA, thanks to L.J. Stadler. Although her exceptional talents were widely praised it was not easy for her as a woman and because of her 'difficult' character to secure a permanent position. In 1942 she moved to Cold Spring Harbor N.Y. to the Department of Genetics of the Carnegie Institution of Washington, where her main scientific work was done and where she was going to stay until her death.

In a small experimental garden McClintock grew her own genetic stocks of maize (*Zea mays*) and, for many years, performed in a meticulous way detailed cytogenetic work with a simple light microscope. In her studies on the behaviour of broken chromosomes in maize in the fall and winter of 1944 and 1945, she discovered 'totally unexpected segregants exhibiting bizarre phenotypes'; these segregants were 'variegated for type and degree of expression for a gene'. Part of the observed maize kernels showed unexpected changes in pigmentation. Soon it became clear to her that these results had to be explained by the action of a kind of **controlling elements** or **transposable (genetic) elements** (nowadays: **transposons** or, in a more popular way, **jumping genes**) which, simply said, are pieces of DNA that can change

their position in the chromosomes and can be inserted at a new location in the same or another chromosome. At this new location they may cause effects that can be observed in segregants of self-pollinated progenies (McClintock, 1984).

Starting from 1948 McClintock published her findings on mutable loci in maize extensively in various issues of the *Carnegie Institution Year Book* (McClintock, 1948; 1949; 1950a) and in *Proceedings of the National Academy of Sciences* (McClintock, 1950b) but did not get much reaction from the scientific community. As people did not grasp the significance of her findings, McClintock's work met the same lack of interest when she presented her results at the Cold Spring Harbor Symposium on Quantitative Biology in 1951. In her lecture she referred to 'controlling elements', controlling gene action as well as causing mutations by transposition in the maize genome. There was no discussion or arguing after her presentation (McClintock, 1951) but despite these negative experiences McClintock was convinced of the importance of her findings and, imperturbable, continued her work on maize, showing that transposable elements indeed are mobile and can move from one place to another within a genome and that they can produce a gene mutation caused by an insertion (McClintock, 1954; 1956).

The fast development of bacterial genetics in those years enabled

much faster proof for the stability or instability of a gene than before. By the mid-1970s it was proved that certain mutations in bacteria resulted as well from the insertion of a piece of DNA in a gene or, in other words, from the effect of transposable elements. A review by Fincham & Sastry (1974) in a leading serial, *Annual Review of Genetics*, finally, made McClintock's work accessible to a much wider scientific audience.

For her extraordinary contributions to modern genetics, made almost 40 years earlier, Barbara McClintock was awarded the Nobel prize in 1983. The lecture she delivered on that occasion was published in *Science* (McClintock, 1984). She remained scientifically active till the very last and she died at the age of 90 in 1992. An obituary was written by Fedoroff (1994).

More details about transposable elements and their importance in relation to mutation breeding, are presented in Chapter 3. Some easily accessible publications that have contributed significantly to a better understanding of the concept of transposons, showing that such elements in bacteria and maize are very much alike, were published by Cohen & Shapiro (1980), by Fedoroff (1984; 1989) and by Campbell (1993). These publications also give us a better insight into the person and the work of McClintock.

exactly the first evidence was presented that γ-rays do induce mutations as well. This evidence for plants, apparently, has been obtained only some decades later, namely after mutations were obtained as a result of **chronic irradiation**, produced by a ^{60}Co source in a so-called **gamma field**. Sparrow (1954), for instance, reported on visible colour changes in leaves and petals in a number of different plant genera after chronic radiation of dormant

seeds, bulbs, rhizomes and growing plants in a gamma field at Brookhaven National Laboratory, USA. This field, the first in the world, had been installed in 1949. Evidence for mutations in crop plants induced by **acute gamma irradiation** may be even of more recent date as this method, most probably, became fashionable later than chronic irradiation with γ-rays.

Starting from the early 1950s, experiments on

chronic irradiation of maize (*Zea mays*) by means of a ^{60}Co gamma field source were performed by Singleton and colleagues at the Brookhaven National Laboratory. In these studies mainly endosperm traits were studied because it was found easy to obtain large seed populations and because mutations for such traits could be easily detected with a high degree of certainty. From these and other studies, Singleton (1955) concluded that mutations can be induced most effectively by subjecting plant material to a relatively high dose for a short period, i.e. by acute treatments, during the 'sensitive' period of the plant. The author added that this 'sensitive period' for maize was about a week before pollen shedding, but 'definitely after meiosis, which is the pollen sensitive period'.

1956 – Sears transfers resistance from Aegilops to wheat (Triticum aestivum) by a radiation-induced translocation.

A very elegant way to obtain resistance to leaf rust (*Puccinia triticina*) in wheat (*Triticum aestivum* subsp. *vulgare*) was demonstrated in 1956 by E.Sears. Sears made use of a radiation-induced translocation following hybridization to transfer a resistance gene from a species of a related genus: *Aegilops umbellulata*. This subject will be further discussed in Chapter 3.

1958 and following years – Application of chemical mutagens on higher plants; ethyl methanesulphonate (EMS)

The pioneering work by Auerbach, Rapoport and Oehlkers (see earlier sections) was soon followed by many other workers. Systematic research on chemical mutagenesis in higher plants started in particular with work by the Swedish group (Å. Gustafsson, J. MacKey, L. Ehrenberg, A. Hagberg and U. Lundqvist and many others). Gustafsson & MacKey (1948) had already applied mustard gas as a mutagen in barley (*Hordeum vulgare*). References can be found in publications by Westergaard (1957), Röbbelen (1959) and Gustafsson (1960). In the hope of finding mutagenic specificity, hundreds of chemicals were tested for this purpose. Main attention went to the occurrence of a high mutation inducing **effectiveness** (i.e., a high comparative frequency of a given mutation at a given dose of treatment), a high **efficiency** (i.e. a relative high rate of desirable changes – like gene mutations – in comparison to the frequency of undesirable effects like gross chromosomal damage, sterility or lethality) and, in particular, the aforementioned differences in mutation spectrum, etc. Results, initially, were not very encouraging and mutation rates obtained remained low, often because the high toxicity of the mutagen allowed treatment with low concentrations only.

During a conference on chemical mutagenesis in 1959 in Gatersleben, Auerbach (1960) referred to the early discovery of ethylene imine (EI) a substance that produced high mutation rates in plants, by Rapoport. Auerbach also mentioned at that occasion the mutagenic action of the important group of alkylating agents to which also ethylene imine belongs. Many alkylating agents had been tested for their mutagenic activity at that time by Fahmy & Fahmy at the Chester Beatty Institute in London (see for instance Fahmy & Fahmy, 1956a, b). Tests for mutagenicity often were performed because certain substances had shown already that they were carcinogenic which, in case of alkylating agents, is mainly due to chromosome breakage. In Chapter 4 more attention will be given to the different groups of alkylating agents.

On the same occasion Westergaard (1960) reported on early studies concerning the mutagenic effects, induced by alkylating agents, on **reverse** (or **back)mutation** for two specific alleles (adenineless and inositolless) in *Neurospora crassa*. Westergaard, referring also to earlier results of Kølmark and some other workers, in an earlier publication (Westergaard, 1957) already included the compound ethyl methane sulphonate (EMS), which belongs to the group of alkylating agents and has become the most frequently and universally used chemical mutagen in plant breeding. (N.B. The full name for EMS is written in various ways.) In the experiments reported in 1957, however, EMS only showed an average mutagenic effect of 17 reverse mutations per 10^6 conidia for an adenineless strain of *Neurospora crassa*, whereas **diepoxybutane** gave the best results with 85 reverse mutations per 10^6 mutations.

Heslot and colleagues in France, who most probably were the first to test the mutagenic effect of EMS and other alkylating agents in a crop plant (barley, *Hordeum vulgare*), found that EMS was ten times as mutagenic as the most effective dose of gamma-rays in that crop. Details can be found in Heslot (1960), Heslot & Ferrary (1958) and Heslot et al. (1959; 1961). In addition, Swaminathan, Chopra & Bhaskaran (1962) reported at an early stage about the use of EMS in barley and wheat (*Triticum* spp.) and, accordingly, Neuffer & Ficsor (1963) for maize (*Zea mays*) and McKelvie (1963) for *Arabidopsis*.

2.5. **The fourth period: large-scale application of mutation breeding for many crop species; assessment of possibilities and limitations; international cooperative activities developed by the Joint FAO/IAEA Division in Vienna**

The first practical efforts to make use of induced mutations for plant breeding purposes started soon after the definite proof for the mutagenic effect of X-rays had been given. Nevertheless, not many applied projects on mutation breeding were started, most probably as a result of the pessimistic views expressed by Stadler in this respect, and a large majority of the work on mutations in plants in the following 25–30 years remained directed towards fundamental problems. Some exceptions to this rule have been mentioned by Gaul (1963), who referred to applied mutation work by Delaunay (1931) and Sapehin (1936) in the USSR, Freisleben & Lein (1944a,b) in Germany and Gustafsson (1947) in Sweden. Some other references to early practically oriented work of these and other investigators have been mentioned in this chapter.

Stimulated also by US efforts to compensate for the disastrous impact of the atomic bombs dropped on Japan by promoting peaceful applications, interest in the application of mutagenic techniques to improve genetic traits in crop plants quickly increased in the early 1960s. However, before it came to **mutation breeding** in the proper sense, still many practical questions had to be solved – such as the choice of the most appropriate starting material, the quantity of seeds or plants to be treated, the optimal dose for various mutagens, differences between crops and cultivars, optimal treatment conditions, the influence of modifying effects, etc. Many meetings, conferences, symposia and the like on these topics became organized by various organizations. A few early ones will be mentioned here: The Brookhaven Symposium on Biology No 8 in the USA (Anon., 1955), two International Conferences on the Peaceful Uses of Atomic Energy, organized by the United Nations in Geneva in 1955 (Anon., 1956a) and 1958 (Anon., 1958), the Symposium on Mutation and Plant Breeding in Ithaca, USA, organized by the National Academy of Sciences and the National Research Council (Anon., 1961a), the Symposium on 'Effects of Ionizing Radiations on Seeds', organized by IAEA in Karlsruhe, Germany in 1960 (Anon., 1961b), three sessions of the 'Erwin-Baur-Gedächtnisvorlesungen' in Gatersleben, Germany, respectively on chemical mutagenesis in 1959 (Stubbe, 1960), on radiation induced mutagenesis in 1961 (Stubbe, 1962) and on the utilization of induced muta-

tions in 1966 (Stubbe, 1967). In Japan the first *Gamma Field Symposium* was organized in 1962. A symposium on the induction of mutations and the mutational process was organized in 1963 in Prague, former Czechoslovakia, by Véleminský & Gichner (1965). In 1964 a meeting on the use of induced mutations in plant breeding, jointly organized by FAO, IAEA and EUCARPIA and very important for the future of mutation breeding, was held in Rome (Anon., 1965). This meeting was followed in 1969 by the equally important FAO/IAEA symposium 'Induced Mutations in Plants' in Pullman, USA (Anon., 1969).

Partly as a result of the views and results presented during these meetings, interest in mutations in plants and in their application for practical purposes further increased. Some major lectures on mutation breeding were presented in Köln, Germany, in 1959 during the general meeting of the second EUCARPIA congress (Anon., 1959) as well as in later years during several meetings of the Section Mutations and Polyploidy of EUCARPIA, that was founded about one year earlier in Svalöf, Sweden. Journals on mutations were launched, for example, *Radiation Botany* in 1961 and *Mutation Research* in 1964. Gradually, also the number of publications on the practical use of mutations in plant breeding in these and in general journals on genetics and plant breeding increased. Many publications were related to projects or other activities for which FAO and IAEA, through their Joint FAO/IAEA Division in Vienna, provided financial and logistic support. These activities (the first and second edition of the *Manual on Mutation Breeding*, the *Mutation Breeding Newsletter*, the *Mutation Breeding Review*, symposia, training courses, facilities for mutagenic treatments etc.) were mentioned already in Chapter 1.

Mention should be made also of the significant role played in those early years by several research laboratories in the USA, Sweden and Japan and various other countries but, with an exception for the Brookhaven National Laboratory, Long Island, USA, where – soon after World War II – many (cooperative) mutation projects started and many scientists got their first training in this field, no names of specific laboratories will be given here.

1964 – Establishment of Internationally Coordinated Mutation Breeding Research Programmes by the Joint FAO/IAEA Division of Nuclear Techniques in Food and Agriculture

Following the aforementioned FAO/IAEA/EUCARPIA symposium on induced mutations in plant breeding in Rome in 1964, it was decided by a group of leading scientists to coordinate their activities in this field. These efforts first resulted in the establishment of an Advisory Group on Mutation Breeding for the Joint FAO/IAEA Division, which was founded on 1 October 1964 in Vienna at the IAEA Headquarters by bringing together two units, working on the use of atomic energy in agriculture at FAO and IAEA respectively (see *Mutation Breeding Newsletter* 34, 1989). As a second step, with financial resources from both UN organizations, research projects all over the world could be initiated, supported and supervised, meetings could be organized, training could be given, administrative support could be provided, etc. This subject was discussed already in Box 1.6 of Chapter 1. In retrospect, the impetus of these activities to stimulate mutation breeding work and to evaluate the merits of the method, have been very significant.

1969 – The Pullman Symposium on Induced Mutations in Plant Breeding; first classified list of mutant cultivars

During the 1960s not only did the interest in mutation breeding – supported also by the easy availability of funds – increase, but so did the criticism about the use of public funds for research that had never shown its usefulness by developing improved cultivars of economic value.

The FAO/IAEA Symposium on Induced Mutations in Plant Breeding, organized in 1969 in Pullman, USA, from the view-point of plant breeding, marked the turning point from – predominantly – fundamental research on the phenomenon of mutation to mutation breeding of new cultivars. At this symposium a first classified list of mutant cultivars was presented by Sigurbjörnsson & Micke (1969). This list, containing 77 mutant cultivars obtained by mutagenic treatments, was the first of a series, produced later on a continuous basis by the FAO/IAEA staff in Vienna.

It should be realized that the majority of the mutant cultivars on this first list still were by-products from fundamental mutation research in higher plants. But in the years to follow, gradually, mutation work that was directly initiated for the production of new mutant cultivars, was taken up in an increasing number of countries.

1981 – First major symposium on the use of induced mutations as a tool in plant breeding

In the 1970s it could be proven beyond doubt that mutation breeding was able to contribute significantly to the production of new, successful cultivars, and now it was time to remind research workers, the public and donors, that induced mutations had more to offer than a quickly increasing number of mutant cultivars in agricultural and horticultural crops.

Right from the beginning of plant research, spontaneous and induced mutations have been used as tools for various research topics and their contributions have been acknowledged in many articles. Some main applications have been mentioned in Box 1.4 of Chapter 1. Attention had been paid already to this subject for instance during the annual *Gamma Field Symposia* in Japan, but it was not until 1981 that the IAEA and the FAO jointly organized a symposium called 'Induced Mutations – A Tool in Plant Research' to review this topic. It was mentioned in the foreword of the proceedings of this symposium (Anon., 1981a) that, apparently, this assembly was the first to concentrate fully on to the value of induced mutations as a research tool. Topics discussed during this meeting were the role of mutations in relation to studies of genetics, plant evolution, plant physiology, plant parasites, symbiotic relationships, *in vitro* culture, gene ecology and, of course, to their direct use for plant breeding research.

The timing of this meeting more or less coincided with the advent of the new research fields of applied molecular genetics in gene manipulation and biotechnology. In the years that followed, public attention (and funds) shifted more and more towards these new and challenging activities. It was suggested at various occasions that within a decade biotechnology would completely revolutionize plant breeding, and that before the turn of the century many (if not most) new plant cultivars would be obtained in this way, instead of by the current or 'conventional' breeding methods known (including mutation breeding). However, progress in these fields, as had to be admitted even by its most ardent proponents, worked out to be much slower than was predicted by them and until 1995 only one cultivar (cv. Flavr Savr in tomato; see Chapter 1), obtained by biotechnological methods had been officially registered and released. Moreover, it appears that Flavr Savr is the most expensive cultivar ever produced and it must be concluded that this cultivar was released not because of its expected economic profit, but to test the feasibility of the method, the official registration procedures and the acceptance of 'genetically engineered' products by the public.

During the period 1985–95 the great attraction of biotechnological approaches started to overshadow the value of the mutation work as a method of proven value for breeding new plant cultivars and as a distinct area of plant research. In some cases, long established mutational techniques, nowadays, are even believed to have arisen as a result of biotechnological work. This, for instance, has been the case with the 'rediscovery' of the potential of spontaneous mutations as a source of useful genetic variation by biotechnologists. This phenomenon, known already for many centuries, was studied now in combination with an *in vitro* phase of the plant material under the new name '**somaclonal variation**'.

The conclusion, at about the end of our 'Fourth Period of Mutation Breeding' is that, despite a certain risk that in future mutation breeding may not always be recognized anymore as a distinct discipline, the application of mutation techniques in (crop) plants will remain an attractive field of studies with proven merits, both for applied plant breeding and for fundamental plant science. Micke & Donini (1993) added that the potential of modern mutation technologies refers to both seed and vegetatively propagated crops and that such technologies should be combined with refined selection methods applicable to large populations. More in general, mutation breeding, preferably, should be performed by experienced plant breeders. Finally, a relatively new and attractive field of interest is the role of spontaneous and induced mutations as an important and sometimes even indispensable tool in various molecular and biotechnological approaches. Some examples were already preseted in Chapter 1.

1990 – Joint FAO/IAEA symposium in Vienna to assess results of 25 years of applied mutation breeding

In 1990 a symposium was held in Vienna to evaluate the practical results of the use of mutations for plant breeding purposes in the period following the symposium in Rome in 1964 that was mentioned in a previous section. The symposium in 1990, in our opinion, marks the end of the fourth period in which mutation breeding has clearly proved its worth, but also reached its peak as a subject of research.

It was mentioned before that, in 1962, Singleton (1962) could only refer to nine induced mutant cultivars of arable crops, whereas in 1969 the total number, including mutant cultivars for horticultural crops, amounted to 77 (Sigurbjörnsson & Micke, 1969). In 1990 the total number of mutant cultivars had increased to more than 1300, of which 850 represented seed propagated, mainly arable crops. The number of 480 that was mentioned for vegetatively propagated, mainly ornamental crops, in fact was a large underestimation (Anon., 1991b; for more details see also Chapter 1). Since

that time the number of induced mutant cultivars has continued to increase but, as was mentioned before, interest in the technology and in research for mutation breeding started to dwindle.

The decreased interest for mutation breeding, undoubtedly, was caused by the fact that many research institutes and commercial companies – with high expectations about quick and easy results from the new biotechnological techniques – started concentrating their main attention and research funds on this challenging, but expensive, new field. In addition, the turn towards biotechnology in plant breeding research may also have been caused by the gradually gained insight into limitations of mutation. For instance, by means of ionizing radiation it is not primarily the basic units of heredity, i.e. the genes, are altered, but chromosome segments. Indeed, non-ionizing radiation (UV) produces relatively more 'point mutations', but its application in higher plants is often hampered by low penetration. Some chemical mutagens, such as sodium azide (see later), have proved to be more effective for induction of 'point mutations' than radiation, but they are not effective in all crop species.

These (real or assumed) limitations of mutation methods applied so far, together with the expectation that by biotechnological methods in the near future any desired single gene can be replaced or changed in any desired direction, sufficiently explain why, in particular, many research workers on plant breeding subjects in recent years switched their main interest towards biotechnology. In practice, however, most plant breeders continue to produce new cultivars, predominantly, by conventional methods, including the use of mutation methods but, at the same time, keep an eye open for the opportunities offered by new, often molecular, methods which may contribute to the introduction of new useful genes for important traits, and which may facilitate breeding work and speed up the breeding cycle.

This chapter on the history of mutation breeding mainly looks backwards and is not the most appropriate place to speculate about the future. However, we may express here our conviction that mutations and mutation breeding, though in most cases in combination with other (current or future) methods, will always continue to play their role in plant breeding, as tools as well as in a direct way.

3   Nature and types of mutations

3.1.   Introduction

Although the mechanisms of spontaneous and induced mutagenesis, particularly in higher plants, are still not fully understood, it has become clear by now that errors in replication, recombination and repair of damage in DNA are involved. Other phenomena like the occurrence of so-called **transposition effects** caused by the transfer of a DNA-segment to another position, may result in modification or inactivation of gene action and, in this way, result as well in heritable changes (see for instance Peterson, 1993).

The four endogenous chemical processes which, most likely, occur in case of DNA damage are: oxidation, methylation, deamination and depurination, of which oxidation may be the most important one. A considerable part of the DNA damage may be counteracted by repair processes with the result that only part of the DNA damage leads to mutations.

As, for instance, was pointed out earlier by Drake (1969), mutation frequencies depend on 1) the amount of primary damage, 2) the degree to which the damage is repaired, and 3) the probability that the mutation produces a recognizable altered phenotype. Repair of DNA takes place along several biochemical pathways, which processes – to various degrees – are subject to errors ('error prone') and, consequently, also may lead to **mis-repair** of the DNA damage. Recently, interest in studying such DNA repair mechanisms has increased considerably (Marx, 1994). This subject will be also briefly discussed in Chapter 4.

Nowadays a considerable body of information on the nature of mutations – often based on studies of physical and chemical damage to DNA in lower organisms – is available. Many details, for instance, can be found in Friedberg (1985).

Plant breeders are particularly interested in mutations which concern single traits only. In practice this is often 'translated' as interest in mutations within single genes in which one allelic form is mutated into another. For such mutations in literature different words are used, like **gene mutations**, **intragenic mutations** or **point mutations**.

In addition to mutations within single genes, different kinds of other, heritable changes in the chromosomal DNA, like larger deletions, insertions, duplications, substitutions and translocations, may result as well in alterations of the genotype. The aforementioned **gene mutations** and the alterations of this second category, which are often called **chromosome mutations**, according to various authors, are mostly considered the two basic levels (or classes) of mutation (Rieger *et al.*, 1991; Russell, 1992; Suzuki *et al.*, 1989). Chromosome mutations (in the sense as mentioned before), together with changes in chromosome numbers, often are called **mutations in the broader sense**, whereas molecular changes within the gene may be called **mutations in the narrow sense** (Rédei, 1982a).

Classification systems

The aforementioned categories (gene mutations, mutations at the chromosome level and at the genome level) still do not cover all possible situations, as mutations also may have an **extranuclear** (or **extrachromosomal**) origin. It, therefore, would be very useful if an unambiguous and complete system to classify different categories of mutations were available but in practice several systems are used and it is not always made sufficiently clear which types of mutations are included in a specific category.

Stubbe (1938) proposed the following system.

A.   Mutations within the genome (= nuclear mutations)
   1. **gene mutations**
   2. **chromosome mutations** (fragmentations, deficiencies, deletions, inversions, duplications, single and reciprocal translocations, fusion of chromosomes)

3. **genome mutations** (changes in chromosome or genome number)
B. Mutations within the plastidom (= the total of the plastids of a cell)
  1. **plastid mutations**
C. Mutations within the plasmon (= the total cytoplasmic genetic system of eukaryotes)

Stubbe (loc.cit.) added that, hitherto, mutations within the **plasmon** were unknown. According to our present knowledge chloroplasts and mitochondria are the major constituents of the plasmon. (N.B. Definitions of 'plastidom' and 'plasmon' are after Rieger *et al.*, 1991.)

In the FAO/IAEA *Manual on Mutation Breeding* (Anon., 1977a, p. 107), Gustafsson & Ekberg distinguished the following main categories of mutations:

1. **genome mutations** (including the occurrence of differences in numbers of complete chromosome sets: **polyploidy** and **haploidy**, as well as differences in numbers of individual chromosomes: **aneuploidy**)
2. **chromosomal mutations** including
   - structural rearrangements
   - gene mutations
3. **extranuclear mutations.**

Although this grouping has the advantage that it includes – on main points – all the changes that are induced by common mutagens in plants, it suffers from some shortcomings. The expression 'genome mutation' is somewhat ambiguous, as it refers to changes in the number of genomes, as well as to changes in the chromosome number, i.e. within a genome. The expression 'chromosomal mutations' may be misinterpreted as well, unless it is made clear that all mutations within chromosomes, including mutations within the gene, are included and that changes in chromosome number are excluded. An advantage of putting structural rearrangements and gene mutations together under 'chromosomal mutations' is that the distinction between 'gene mutations' and other mutations within the chromosome is fluent. However, because of the importance of gene mutations from a plant breeder's point of view, it would appear to be useful to discern gene mutations as a special category. In addition, one could also argue that aneuploids should be added to the category of 'chromosome mutations' as well.

A possible alternative could be the system of classification used by Auerbach (1976) in which the following categories are distinguished:

1. **intra-genic mutations** (referring to all changes within individual genes)

2. **inter-genic** (or structural) **mutations** (within chromosomes)
3. **changes in chromosome number.**

The expressions used by Auerbach – which had already been applied much earlier, for instance by Muller in 1941 – are less ambiguous than those used in the FAO/IAEA system, which, from this point of view, would make the use of Auerbach's system recommendable. In Auerbach's system, on the other hand, the category of **extranuclear** mutations – of considerable interest to plant breeders – is missing.

In conclusion, we have a certain preference for a system that is, on main points, in line with the grouping proposed by Stubbe (1938) and that, first, makes a distinction between **nuclear** and **extranuclear** mutations and, secondly, distinguishes for the category of nuclear mutations between a) mutations at the level of (or within) **genes**, b) different types of mutations referring to individual **chromosomes** and c) changes in the number of **chromosomes** and genomes. This last group could, if so desired, be subdivided in **aneuploids**, **haploids** and **polyploids**. However, there is, in our opinion, no need to introduce yet another system of grouping the different categories of mutations as long as it is made sufficiently clear in each specific case to which type of mutation(s) is referred.

Another attractive way to elucidate different features of mutations is to make – in various ways – comparisons between two categories of mutations. In this respect one could refer for instance to the way of origination of mutations (**mutations of spontaneous origin** versus **induced mutations**), the important difference for plant breeders between **macromutations** and **micromutations**, **dominant mutations** versus **recessive mutations**, **positive** (or **progressive**) **mutations** versus **negative** (**destructive**) **mutations**, mutations *'in vivo'* versus **mutations** *in vitro* and, finally, groups of mutations for specific traits (for instance chlorophyll mutations, mutations for short straw, mutations leading to resistance to a specific disease, high-protein content mutants, etc.).

In the following sections of this chapter we will make use of several of these categories to further discuss the spectrum of mutational events and to illustrate their relative importance for plant breeding. The occurrence of various kinds of mutations in relation to the application of *in vitro* techniques, because of the particular interest nowadays in this subject, will be discussed separately in Chapter 5.

3.2. **Nuclear and extranuclear mutations**

Nuclear mutations may include genome mutations (polyploidy, haploidy); changes in the number of individual chromosomes (aneuploidy); structural changes within chromosomes; as well as mutations in the narrow sense at the level of (or within) genes. As mutation breeding is focussed in particular on mutations at the level of genes and their alleles and on smaller changes within chromosomes, emphasis will be placed on these categories and on their contrast with extranuclear mutations. We will start at the 'lowest' level with gene mutations.

3.2.1.   Genes and gene mutations

Introductory remarks

Before starting to discuss gene mutations, some words must be said about chromosomes, genes and DNA. In principle each eukaryotic chromosome is made up of a linear arrangement of genes and other DNA. Genes, the basic functional units of life, which also carry information from one generation to a next, in fact, are segments of DNA. A distinction can be made between **structural genes**, which encode the amino acid sequence of proteins 'required by the cell for enzymatic or structural functions', and **regulatory genes**, which 'are responsible for controlling the expression of the structural gene clusters' (Lewin, 1987, pp. 220 and 221). Regulatory genes usually code for a protein that controls the **transcription** of the structural genes. (N.B. Genetic **transcription** is the first step in the process of gene expression in which RNA is synthesized from a DNA template.) It should be further remembered that regulatory sequences also occur as part of each gene.

DNA (deoxyribonucleic acid) consists of two long, chain-like molecules, called **polynucleotides**, which are interconnected and wrapped around one another as a spiral ladder-shaped double helix. The building blocks of these chains or strands are the **nucleotides**. Four different nucleotides, each with a different nitrogenous base, but with an identical (deoxyribose) sugar and a phosphate group, are discerned. These four nitrogenous bases are **adenine, guanine, cytosine** and **thymine**, abbreviated as to A, G, C and T respectively. The bases A and G have a similar structure and belong to the group of the **purines**; the other two bases C and T, which are also similar, are **pyrimidines**. The individual strands of the aforementioned double helix structure are held together by hydrogen bonds, formed by pairing of two of the above mentioned bases. Each so-called **base pair** (abbreviated: bp) consists of one purine base and one pyrimidine base with G being paired to C and A being paired to T.

The sequence of the nucleotides in the DNA determines the genetic properties of any cell and organism. A combination of three nucleotides – a **triplet** or a **codon** – provides the code for the assemblance of a particular amino acid (e.g., AAG codes for the amino acid lysine), specific codons start or terminate the genetic message readable on the DNA molecule. In total 64 combinations of three nucleotides are possible but this number is far in excess of the 20 needed to code for the 20 known amino acids and for so-called **start/stop codons**. It was found that in a number of cases more than one codon can specify for the same amino acid.

The genetic set-up of prokaryotes and eukaryotes

Much fundamental information regarding chromosomes, DNA and mutational events has been derived from research on lower organisms (prokaryotes), in particular the bacteria *Escherichia coli* and *Streptomyces*, the alga *Chlamydomonas*, viruses and bacteriophages. The best studied (lower) eukaryotes at the DNA level are the (baker's) yeast *Saccharomyces cerevisiae* and the fungus *Neurospora crassa*. In the DNA of *S. cerevisiae* 12×10^6 bp or 12 mega base pairs (abbreviated: Mbp or Mb) are present, divided over 16 chromosomes. In 1989 a project got underway to unravel the molecular structure of the entire yeast genome. It was found that yeast genome encoded some 6000 proteins and in 1996 this job was completed.

The haploid nucleus of the fruitfly *Drosophila melanogaster* contains 165 Mbp. For the plant *Arabidopsis thaliana*, with five chromosomes in a single genome ($x = 5$), a genome size of 70–100 Mbp was reported; the corresponding figure for rice (*Oryza sativa*; $x = 12$) is about 450 Mbp, but for barley (*Hordeum vulgare*; $x = 7$) a much higher figure of 5000 Mbp is given. In diploid plant species like the three mentioned here a double set of one genome is present in the somatic tissue (abbreviated: $2n = 2x$).

A gene may be made up of several hundreds up to several thousands of base pairs. The number of (nuclear) genes for higher plants may be estimated at about 25 000, but figures in literature vary between 10 000 and 100 000. Williams (1995) reported for *Saccharomyces cerevisiae* that, on average, each stretch of 2000 bp contains a gene that encodes a protein made up of more than 100 amino acids. More complicated proteins may even consist of thousands of amino acids.

The *S. cerevisiae* mapping programme was the first completed project in which the entire genome of an eukaryote has been 'sequenced'. Other mapping projects for major crop plants, as well as for the human genome (3300 Mbp) are under way, but it is believed that in many other eukaryotic species relatively much DNA is present that does not code for proteins and, as a consequence, shows a much lower gene density than in *S. cerevisiae*.

Mutations

Theoretically, all changes which occur in the DNA sequence may result in changes in the genetic code i.e. **mutations**. Whereas a single plant gene may contain several thousand base pairs or, in other words, a thousand or more triplets or codons, a gene coding for a mutant trait may differ from the original gene ('wild type') for instance in one codon only, where such a change even could refer to the alteration of (only one atom of) a single base within one of the nucleotides.

A **gene mutation**, which term is used here to indicate all heritable changes within the limits of a gene, may be the result of alterations of the DNA sequence of the gene, like gain or loss (**addition** or **deletion**) of one or more base pairs, or **substitution** of one base pair by another.

It may be interesting to mention here that before 1985 it was already possible to screen for deletions, insertions and rearrangements in genes by methods like Southern blotting, but scanning for smaller changes, including single base changes was still beyond practical reach at that time. At present, the use of sequencing methods allows the detection of even single base substitutions in the coding sequences. This is of particular interest for human genetics because of the connection between mutations of certain genes and particular diseases. For details see for instance Prosser (1993).

A deletion or an addition (insertion) for any number of nucleotides other than three, leads to a **frameshift mutation** (or reading frame mutation). Such mutations may result in another starting point for transcription of the genetic code or in a reading shift within the code, both of which lead to 'misreading' of the genetic code. To illustrate these and other mutations, some examples are presented in Box 3.1.

After a frameshift mutation three different situations may occur: a) so-called **missense** codons arise which code for another amino acid, b) the original codon is changed into a **stop codon** – also called **nonsense** codon – which does not code for an amino acid but acts as a 'stop' or comma in the genetic message, and c) despite a change of one of the three nucleotides of the codon – often the last one – the same translation product (specific amino acid sequences or proteins) is obtained as before the frameshift mutation. This is called a **sense** codon. The expression **silent mutation**, **samesense mutation** or **sense mutation** refers to a base pair substitution that results in the production of the same amino acid as before, e.g., the triplets AAA and AAG both code for the

production of lysine. The expression **neutral mutation** is sometimes applied when there occurs a change of one amino acid to another which, however, does not lead to changes in the function of the polypeptide or protein concerned. However, **neutral mutations** are also referred to as changes that are not phenotypically expressed and therefore (supposedly) not subject to immediate environmental selection.

Suzuki *et al.* (1989, p. 340) make the comment that the expression **nonsense codons** has been a misnomer because of their important function as **stop codons**.

Substitutions of a single base pair can be either **transitions** or **transversions**. Transitions involve changes from one purine/pyrimidine base pair to the other purine/pyrimidine combination. This results in four combinations: AT → GC, GC → AT, TA → CG and GC → TA. Transversions involve the change from a purine/pyrimidine base pair to a pyrimidine/purine base pair at a specific site, like for instance the change from AT → TA, AT → CG, GC → CG and GC → TA. Base pair substitution also can result in the above mentioned **missense** and **stop codons**.

In many publications and textbooks the terms **gene mutation** and **point mutation** are considered as synonyms, like for instance in Suzuki *et al.* (1989), where in the glossary (p. 727) a **gene mutation** is defined as 'a point mutation that results from changes within the structure of a gene'. Lewin (1987), in his highly valued textbook *Genes* (p. 34), gives the following definition: 'A **point mutation** is a change affecting a single base pair in a gene (such as a substitution of one base pair for another)'. The expression 'point mutation', obviously, refers to the limited size of a mutation whereas the expression 'gene mutation' has to do with function.

A mutation for a single base pair may result in a substantial effect such as loss of function of the affected gene. Van der Leij *et al.* (1991) proved in a monohaploid clone of potato (*Solanum tuberosum*), that loss of a single base pair in the so-called GBSS gene was responsible for the occurrence of a mutant clone with amylose-free (*amf*) tubers.

Another example in this respect refers to loss of resistance. This was demonstrated for the pathogen *Cladosporium fulvum* in tomato (*Lycopersicon esculentum*) by a research group at the Department of Phytopathology at the Wageningen Agricultural University. A single base pair in the *Cladosporium* was changed in such a way that the avirulence factor could not be recognized anymore by the host plant, and as a consequence the host plant's resistance capacity became ineffective (Joosten, Cozijnsen & De Wit, 1994).

Box 3.1 Examples of some types of mutations

Frameshift mutations arise from insertion or deletion of one or more nucleotides – in numbers other than three (or multiples of three). Because the starting point of transcription of the genetic code is 'shifted' or displaced, the codons read differently and the genetic message of the mRNA will be different from the point of the frameshift mutation onwards and probably makes no sense.

When one or more groups of three nucleotides are inserted or deleted, no frameshift mutations occur and each individual 'word' of three nucleotides will continue to give a genetic message (in most cases: to code for a specific amino acid), but the contents of the old and the new message or 'sentence' (which probably consists of some hundreds of 'triplets') may differ to a smaller or larger extent, leading to a difference in the gene product (usually a protein).

It is not difficult to conceive what could be the result of different types of changes if a gene is compared with a sentence in a book or an article, consisting exclusively of words of three characters. As an example we take the sentence: 'DID THE OLD MAN SIP AND CRY' (and we forget about the questionmark). If, for instance, from two consecutive words of this sentence the final, and the first character, respectively, would be deleted, under the strict condition that after deletion the characters can only occur in groups of three, then the original sentence could be changed in different ways. If each deleted character is indicated by a down arrow (↓), two deletions at the positions indicated by such arrows, like for instance: 'DID TH↓ ↓LD MAN SIP AND CRY', would be read as: 'DID THL DMA NSI PAN DCR Y'. It is clear that this sentence has become incomprehensible.

If a deletion should have occurred at the position of the last character of the sentence only, the first part of the question still would make sense: 'DID THE OLD MAN SIP AND CR↓', would be read as 'DID THE OLD MAN SIP AND'.

If a deletion for instance would refer to two characters in one word and to one character in the other word, the message, although incomplete, still could be meaningful: 'DID THE OLD MAN S↓↓ ↓ND CRY' would be read as 'DID THE OLD MAN (SND) CRY'. Whether this 'SND' would still be a meaningful 'word' would depend on different aspects, like the language used by the 'reader' (RNA, mitochondria).

If one complete 'word' of three characters would be deleted or added, the message would be incomplete and therefore either could become incomprehensible or be changed considerably. This may be illustrated in the following examples.

1. (deletion of one word): 'DID THE ↓↓↓ MAN SIP AND CRY', reads as 'DID THE MAN SIP AND CRY', looks like a meaningful question, but a perhaps important part of the information of the original sentence (OLD) is missing.
2. (insertion or addition of one word): '↓↓↓ DID THE OLD MAN SIP AND CRY' could become: 'WHY DID THE OLD MAN SIP AND CRY', which is a different question.

Again a new situation would arise if the character P in 'SIP' would change ('mutate') to the character T. The whole sentence still would make sense ('DID THE OLD MAN SIT AND CRY') but the meaning of the question would partly differ from the original one.

Finally, it should be clear that not only the size of the deletion but also its position is of importance and that, probably, in a sentence certain parts or words could be omitted without losing the vital meaning of that sentence. The size and the position of a deletion or insertion may also play a role in determining whether the change will be – or can be – repaired or not.

In another experiment Hsu *et al.* (1993) showed that streptomycine resistance in *Nicotiana plumbaginifolia* might be related to an A to G transition (a point mutation which results in substitution of arginine for lysine) at a specific position of the chloroplast *rps12* gene.

The position of point mutations within the genetic material seems to be non-random as was shown already in 1961 by S. Benzer (for reference see Rieger *et al.*, 1991), for the phage T4 on which 2400 mutations were mapped on 288 different sites. Whereas most sites showed one or a few mutations only, two sites had mutated 312 and 615 times respectively. There may be a certain 'preference' for specific sites or regions in the genetic mater-

ial. The mutation type involved and the choice of specific mutagenic agents may play a role as well.

For higher plants Wessler & Varagona (1985) studied 22 mutations at the *Wx* locus of maize (*Zea mays*). They found that seven mutations were due to 'large' insertions, six mutations were caused by deletions larger than 50 bp and the other nine represented base pair substitutions and small deletions and insertions of less than 30–50 bp. Stadler, MacLeod & Dillon. (1991) investigated 135 strains of the fungus *Neurospora crassa* for independently arisen spontaneous mutations at the *mtr* locus. (The *mtr* gene accommodates the structural gene for amino acid permease.) It was found that 54% of the mutants were the result of single base pair substitutions and 34% were showing deletions, including either one or more genes. Most of the remaining 12% mutants were caused by insertions; several of them referred to so-called tandem duplications of 400–1000 bp.

Previously, in 1965, Freese & Yoshida (see Rieger, Michaelis & Green, 1976, p. 546) reported that different mutagens, or specific conditions, may favour certain transitional mutations. Alkylating agents, for instance, by preference, lead to changes from $G \rightarrow A$ or from $T \rightarrow$ C. These are also the most frequent directions of change for spontaneous mutations. Kohalmi & Kunz (1988) induced 318 forward mutations in the SUP-4 gene in Saccharomyces cerevisiae by treatment with the alkylating mutagens EMS (ethyl methanesulphonate) and MNNG (methyl nitrosoguanidine). Only base pair substitutions were found, of which more than 96% consisted of the transitions mentioned also by Freese & Yoshida (loc.cit.). A considerable similarity was found on the sites of the SUP-4 gene that were mutated by EMS and MNNG.

Old & Primrose (1989, p. 98) mention that in Escherichia coli point mutations can be obtained at specifically defined sites by making use of advanced methods of molecular biology, like enzymatic misincorporation of non-complementary nucleotides. For higher plants the combination of real site-directed mutagenesis for (or within) specific genes that code for important plant traits with, in addition, the induction of changes towards the desired direction (higher resistance, higher yield, different nutritional composition), still seems to be far beyond reach, at least as long as – for each specific situation – complete knowledge of the gene function, the role of all codons within the gene concerned and the interaction with other genes is lacking. In a later section of this chapter the Swedish mutation programme on barley will be discussed and at that point we will return to the subject of 'specificity' of different mutagenic treatments in relation to the relative frequencies of induced chromosomal aberrations versus gene mutations. To manipulate proportions of different mutagenic events, at this time, seems to be the best result that can be achieved with 'conventional' mutation breeding for higher plants.

Not all 'gene mutations' are really gene mutations
Gustafsson and Ekberg in the FAO/IAEA *Manual on Mutation Breeding* (Anon., 1977a, p. 113) already discussed three criteria which are used generally to determine whether a mutation is a 'real' gene mutation. These criteria are:

1. cytological irregularities do not occur,

2. heterozygotes show regular segregation patterns and lethal phenomena of extreme kinds do not occur,

3. the induced mutation can revert to the original ('wild') type.

In analogy with the comments made in the aforementioned *Manual*, some remarks are made here concerning these criteria. First of all, relatively small deficiencies and duplications which, nevertheless, do exceed the physical limits of a single gene (and therefore – according to our definition – should not be indicated as 'real' gene mutations), may not show any sign of cytologic irregularities. As a consequence, they can not be distinguished from 'real' gene mutations by cytogenetic methods.

A next point, mentioned as well in the *Manual*, refers to the fact that 'real' gene mutations – as is also the case with larger deletions and the like – often are associated with the occurrence of sterility and lethality in the mutated plant population. In particular meiotic processes like the formation of bivalents may be seriously affected. Homozygous mutants may show a lower competitive ability as well as a lower seed production. Nevertheless, segregation for monogenically inherited mutant traits does not differ significantly from the expected 3 : 1 ratio.

Another subject of discussion is that induced mutations sometimes seem to revert back to the original type, as, apparently, could be inferred from the re-appearance of the original phenotype for that specific trait. However, whether this indeed is the result of a true back-mutation (or reversion) at the mutated locus, or caused by a so-called suppressor mutation (sometimes also called second-site mutation), which suppresses the action of an earlier mutation without changing the genetic code of the original mutation itself, is often not easy to determine. The occurrence of suppressor mutations indeed could be easily mistaken for proof of the presence of a real gene mutation. Contrary to what is often thought, reversion by suppressor mutations may occur either outside (extragenic) or within the locus of the first mutation (intragenic). This makes the use of the other expression 'second-site mutation', instead of suppressor mutation, less desirable.

Lewin (1987, p. 35) points out that mutations for a single base pair (point mutations) can either revert by restoring the original sequence or by the action of a compensating mutation elsewhere in the DNA. Insertions can revert as well, either by deletion of the original insertion or by the action of a suppressor gene. A deletion, according to the same author, cannot revert and in this way both point mutations and insertions could be distinguished from deletions. In conclusion it may be said that it is extremely difficult to give final proof that, after reversion of the phenotype, the action and position of the new allele are completely identical to those of the originally non-mutated allele. Studies on the action of suppressor genes have become very popular in recent years, in particular for prokaryotes and lower eukaryotes, and many more details, which are outside the scope of this book, can be found in modern textbooks on molecular genetics.

A characteristic feature of point mutations and small deletions, which may be obtained after treatment with X- or gamma rays, is that their frequency – within a certain range – increases in a linear way with an increase in the radiation dose. This observation suggests that this mutation type results from a single event (a 'hit') and that the number of such 'hits' increases linearly with increasing dosages. In this way the 'real' gene mutations differ from chromosome mutations, like larger deletions, requiring more than one event so that the dose curves show an exponential pattern. We will return to this subject in Chapter 4.

It appears that the large majority of the induced 'real' gene mutations, in fact, do not result from point mutations but from two (or even three) hits very close to each other and which cause small deletions, i.e. the loss of, for instance, only 10 or 20 base pairs, after which the broken ends on either side of the DNA molecule are joined together by specific processes.

Recessive versus dominant gene mutations
A large majority of 95% or more of the group commonly indicated as 'gene' mutations shows recessive inheritance, which usually refers to non-functioning of the gene or absence of the gene product of the mutant allele (and which often is the proof of a deletion rather than a mutant allele with a new specificity). It has even been doubted for many years and on various occasions whether mutations towards dominant ($a \rightarrow A$) could be induced at all. Stadler (1944), for instance, reported that in X-irradiated populations of maize (*Zea mays*) which were large enough to yield 900 000 losses of the dominant allele A by deficiency or mutation towards a colourless allele, not a single mutation from $a \rightarrow A$ had been induced. Stadler referred in the same article to similar results with X-ray progenies of barley (*Hordeum vulgare*).

More recently, and in strong contrast to this result, induction of mutations with a dominant expression has been mentioned in several cases and at frequencies which may be in the range of one dominant for 100 or 200 recessives. Before discussing some examples, how-

ever, it must be stated that, as in the case when trying to prove that a gene mutation is a 'real' one, it is not always easy to discern – for instance for higher plants – between the occurrence of 'real' dominant mutations and other possible explanations (like uncontrolled outcrossing, impurity of the seed material or deletion of an inhibitor), for the dominant expression of a gene. From the point of view of plant breeding the relatively low percentage of mutations with a dominant expression, logically, makes the application of mutational techniques less attractive in case the desired trait requires the dominant allele of a gene.

Brock & Micke (1979) refer to Gaul (personal communication) who, based on 119 mutants reported by different sources, estimated that the proportion of dominant viable mutations in barley was about 6%. In the same publication it was calculated that the percentage of dominant visible mutations in mice was about 1 to 2%. Some references to dominant mutations in crop plants have also been presented by Konzak *et al.* (1984).

Neuffer (personal communication) of the University of Missouri who, starting in 1980, treated pollen of mays (*Zea mays*) with chemical mutagens (more details will be presented in Chapters 4 and 6), obtained until 1991 more than 750 different mutants of which 54 were reported to be dominant. Neuffer & Chang (1989) mentioned that dominant mutations are 200 times less frequent than recessive ones in maize. Coe (1993), in a table of defined and designated gene loci of maize, reported about twelve dominant mutations which had arisen either spontaneously, via transposons or after treatments with the chemical mutagen EMS.

Lundqvist & Lundqvist (1988a), who performed extensive mutation research on barley (*Hordeum vulgare*), reported on mutations for *eceriferum* (waxiness of the leaf surface), which, most probably, is the best investigated trait in cultivated plants. The 1580 mutants mentioned in that publication, induced by seven different kinds of mutagenic treatment, were localized to 79 so-called *cer* loci, of which for only one locus (*Cer-yy*) dominance was found. More details about *cer*-mutations will be discussed later in this chapter.

Lundqvist, Meyer & Lundqvist (1991) and Lundqvist & Lundqvist (1991) reported that after mutagenic treatment with various ionizing radiations and chemicals, 71 lines of barley had been obtained with resistance for powdery mildew (*Erysiphe graminis* f.sp. *hordei*). In 28 lines the resistance inherited in a recessive way and referred to the *Ml-o/ml-o* locus; the other 43 lines showed dominant inheritance of resistance. Ionizing radiation was reported to be significantly more effective for producing dominant inherited resistance to mildew than other mutagenic treatments. For the dominant mutants it could be proven that the vast majority represented genuine dominant mutations.

An interesting case of dominant mutations refers to the trait *Knotted* in maize, which leads to considerable changes in leaf morphology. The dominant gene action, in general, results in the formation of protrusions (knots) along the lateral veins and the veins themselves become more prominent because the trait is asscociated with non-green cells. The ligule, which is situated at the fringe of leaf blade and leaf sheath, is often displaced and some other phenomena can be observed as well. The observed symptoms may occur in different degrees of severity, depending on the genetic stock. The original *Kn1-o* mutation was discovered as a spontaneous mutant in 1941 and other knotted maize mutants were found afterwards.

Hake, Vollbrecht & Freeling (1989) were able to isolate the *Kn1* gene by making use of the technique of **transposon tagging**. Transposons will be discussed later in this section. The authors showed that for the occurrence of this specific knotted mutant type, called *Kn1-2F11*, the presence of an unlinked **transposable element** (see later), known as *Ac*, is required. According to Hake *et al.* their results suggest that this knotted mutant *Kn1-2F11* arose from the insertion of such a transposable element. However, they add that such insertions normally do not cause dominant mutations. Additional molecular data concerning this interesting 'gain-of-function' mutant type suggests that a tandem duplication of genomic DNA rather than an insertion may be responsible for the dominant mutations at the *Knotted* locus (Vollbrecht *et al.*, 1991; Hake, 1992).

In addition to the example of the *Knotted* mutants in maize, which suggests a relation between dominant mutations and transposable elements, other factors may be associated with or be mistaken for the occurrence of dominant mutations. Mention should be made in this respect of **antisense genes**, whose action may result in non-functioning of genes or blocking of gene expression (see also Chapter 1 for a brief description). The application of the antisense technique genes shows a certain similarity to site-directed mutagenesis and without going into detail it can be mentioned here that antisense genes act as **dominant suppressor genes** and, thus, could become expressed without further breeding work in diploid or even polyploid cultivars.

'Mutation-like' effects

Gene mutations were the main topic of the previous section and, in relation to this, the occurrence of so-called

transposable elements was briefly mentioned. More generally it can be said that – next to the afore discussed gene mutations and to structural mutations (see Section 3.2.2) – several other types of 'mutation-like' genetic changes are known.

As a first category so-called **unstable genotypes** or, better, **genotypes with one or more unstable alleles**, are known in cultivated plants. They refer to loci which produce mutations with a high frequency of reversion and which are often called 'unstable' or 'highly mutable'. Unstable mutations were studied in the first decades of the twentieth century, for instance in maize (*Zea mays*) by Emerson (see Emerson, 1914; original not consulted) and Emerson (1917; 1929)).

In early literature it was often not exactly known, or not specified, whether the unstable character of a specific allele was caused by the structure of this allele itself or by factors outside the allele. Demerec, in this context, proposed in 1937 the expression **mutator gene** to indicate 'any mutant gene that increases the spontaneous mutation frequencies of other genes' (Rieger *et al.*, 1991, p. 343). Robertson (1978), for instance, describes for maize a 'mutator system' that induced mutation rates of 40 times the spontaneous mutation rate. In the glossary by Rieger *et al.* (loc.cit.) it is further mentioned that some of the proteins specified by mutator genes may be concerned with DNA repair, DNA replication or precursor synthesis. Suzuki *et al.* (1989) use the expression **mutator**: a mutant 'strain' that has an increased spontaneous mutation rate, and add that this phenomenon, in many cases, is caused by a defective repair system. More recently, Prina (1992) reported on a mutator nuclear gene inducing a wide spectrum of cytoplasmically inherited chlorophyll deficiencies in barley.

Although plants carrying unstable alleles have often been described as 'mutants', such genetic changes, according to present opinions, most often, are not caused by common mutations but by other effects like changes in gene regulation.

Until a few years ago plant breeders did not pay much attention to phenomena like the ones mentioned here, probably because the ways they operated were not understood. As a consequence, it seemed doubtful whether such systems could ever be manipulated to the benefit of practical plant breeding. During recent decades, however, much more has become known about the action of, what are commonly called now, **transposable genetic elements** or **transposons**. Transposable elements, in popular publications sometimes and somewhat misleadingly called 'jumping genes', are DNA sequences that can move within or between chromosomes from one place to another. The merit for understanding the action of transposons, to a great extent, goes to Barbara McClintock, who, from the early 1940s, worked in this field with maize as a major crop. At the time that this phenomenon was discovered and explained by McClintock, the concept of mobile DNA pieces was considered highly improbable and was met with disinterest and even disbelief by almost the entire scientific community. McClintock, however, established the existence of 'mobile genetic elements' and, for the first time, clearly demonstrated the effect of **transposition** (i.e the process of transfer of a DNA sequence to another position at a chromosome), based on a system indicated as the 'Ac-Ds (*Activator-Dissociation*) element family' (McClintock, 1948).

What actually happens in the case of transposition may be illustrated with the following example, derived from Lönneborg & Jansson (1993). If a transposon should become inserted in a gene that is involved in anthocyanin pigmentation of the aleurone layer – the outer layer of the maize endosperm – this layer no longer expresses its normal (dominant, for instance purple) colour but becomes colourless. As transposon-induced changes are more or less unstable, the inserted element can be excised again from the gene and, as a result, the original purple colour can be restored. The occurrence of variegated seeds in this case demonstrates the action of transposons.

Only many years after their discovery have transposons become characterized in detail, both at the genetic and molecular level. Reviews on this subject, for instance, have been written by McClintock (1951), Fedoroff (1983; 1984; 1989) and Döring & Starlinger (1986). A special volume, written on the occasion of the ninetieth birthday of McClintock (Fedoroff & Bottstein, 1992) contains contributions by colleagues, several of her own papers as well as a bibliography of McClintock's work. For some additional details about McClintock and her work see Chapter 2, in particular Box 2.8, and Box 3.2 in this chapter.

It took decades before it was recognized that mobile DNA elements are ubiquitous in all kinds of living organisms and before the significance of the work of McClintock, from a fundamental point of view as well as with an outlook at its possible use for genetics and plant breeding purposes, became appreciated.

Before McClintock, M.M. Rhoades (1936; 1938), when analyzing ears of Mexican black maize (*Zea mays*), had already found a clear example of an unstable mutant allele (*a1*) at the *A* locus that reverted back at high rates, but only in the presence of another, unlinked, dominant gene, indicated as *Dt* (dotted). This case represents one of the first known examples of an unstable mutant allele, characterized by a very high rate of reversion.

McClintock first used the expression **controlling elements** in order to indicate that such elements have the capacity to control gene expression at various loci and produce unstable target genes (McClintock, 1956). In modern text books and glossaries on genetics there is still no unanimity with respect to definitions used for the terms **controlling elements, transposing elements** and **transposons**. If we follow Suzuki *et al.* (1989) in this respect, 'transposable genetic element' is 'the general term for any genetic unit that can insert into a chromosome, exit and relocate and according to these authors the term includes: 'insertion sequences, transposons, some phages, and controlling elements'. A **transposon**, according to the same authors, is 'a mobile piece of DNA that is flanked by terminal repeat sequences and typically bears genes coding for transposition functions', whereas a **controlling element** is 'a mobile genetic element capable of producing an unstable mutant target gene', of which 'two types exist, the regulator and the receptor elements'. According to these descriptions, transposons and controlling elements are specific groups of transposable elements. Russell (1992) also distinguishes between transposable elements and transposons, but in most other publications the term 'controlling element' is used in a rather general way, whereas often no distinction is made between the expressions transposable element and transposon. We will also use these words in a not too strict sense.

Nevers, Shepherd & Saedler (1986) mention four genetic criteria to identify **insertional mutagenesis**:

1. the occurrence of 'wild' (dominantly inherited) spots on a recessive background,
2. segregation according to Mendelian laws,
3. new alleles arise with heritably altered phenotypes,
4. sometimes a second factor is required for the observed instability.

Transposable elements are interesting in their own right, but also of importance as tools for studying the expression and functioning of genes. They are replicated, excised from a position and carry codons for enzymes by which they are causing chromosomal breaks and insert – in principle randomly – at a new location in the genome. The result of this action may be that the transposable element changes the expression and/or structure of a gene in or near which it was inserted. This explains why this topic is also called **insertional mutagenesis**. It was mentioned before that when the transposon is excised again from the 'mutant locus', the 'insertion mutation' may disappear and the phenotypic expression of the original gene (the 'wild type') may be revealed again. Normally, some sequences remain at the site of the former insertion which may further affect the expression of the gene concerned. The process of insertion and reversion (excision) may occur in frequencies of about $10^{-5} - 10^{-6}$, which is much higher than for random spontaneous mutations and also explains the term **mutator genes** which is still used sometimes.

Two types of transposable elements are normally distinguished: **autonomous elements**, which can move independently in the genome, and **non-autonomous elements** which can only transpose or induce chromosomal breaks in the presence of the autonomous element.

Because of their ability to disrupt gene action by their insertion and their ability to generate deletions, rearrangements and point mutations, transposable elements, in many publications, are considered as mutagens. A possible explanation for the occurrence of transposon induced mutations could be the presence of inverted sequences that stay behind when a transposon 'moves out' from a specific locus.

When a transposon with a known DNA sequence is inserted into a functional gene, in which the newly introduced DNA prevents or changes the gene expression, it can be used as a **molecular marker** or **tag** to isolate and clone that gene. This field is called **gene tagging** on **transposon tagging**. The method has been in particular successful when transposons endogenous to the species are used and when genes with an easily screenable mutant phenotype are isolated. The reader may be reminded that mutations provide another method of identifying genes through 'lack of function', which is usually an irreversible phenomenon.

Transposable elements have become important topics of research and are extensively discussed in practically all modern textbooks on genetics. In addition to the publications mentioned earlier in this section, reference is made, for instance, to contributions by Peterson, (1987, with 276 refs!), Nelson (1988), Burr & Burr (1989), Gierl, Saedler & Peterson (1989) and Gierl & Saedler (1992).

Finally, it is not clear yet whether transposons possibly could become a source of stable and (potentially) useful mutations for the production of genetically improved cultivars.

So-called **epigenetic effects** represent another category of 'mutation-like' effects, which are not always mentioned in textbooks on genetics and plant breeding. The expression 'epigenetic effect' was proposed by C.H. Waddington in the early 1950s to indicate 'any change of gene activity during development of the organism from the fertilized egg to the adult' (Rieger et al., 1991).

Epigenetic effects, according to this definition, are related to the regulation of the expression and interaction of the genetic material during development rather than to changes in the DNA itself. The term traces back to the German scientist K.F. Wolff who, in the mid-eighteenth century, suggested that organisms develop by a process that starts with an undifferentiated fertilized egg, followed by the formation and addition of new parts that were not present before. This process was indicated as 'epigenesis'. Rieger et al. (loc.cit.) add that 'epigenetic switches' turn particular genes on and off during development, producing either transient or permanent changes in gene activity. Molecular biologists and other scientists, nowadays, frequently use the expression 'epigenetic effects' to indicate, often in a rather loose way (visible and – predominantly – non-heritable) changes that may be observed for instance in transgenic plants and for which no clear explanation is available yet. Whereas some authors simply state that epigenetic variation is similar to non-genetic variation (see for instance de Klerk, 1990), it is often not made clear whether such effects are transient, permanent during life time, transferable to the next generation or really heritable.

Nowadays, the occurrence of epigenetic effects is frequently related to the mechanism of methylation, which is a chemical modification of the DNA (Holliday, 1987; Matzke, Matzke & Mittelstenscheid, 1994). An example is the introduction of a foreign gene for flower colour in *Petunia* by genetic engineering, resulting in the suppression of the action of both the newly introduced and the related endogenous gene or genes (van der Krol et al., 1990). The observed suppression of the endogenous gene appeared not to be simply the result of transgene expression and the authors suggested, among other things, the occurrence of a double epistatic effect, occurring as a result of methylation.

Holliday (1987) further reported that epigenetic effects may be heritable in somatic cell lineages and may result in a phenotype that is formally identical to that caused by a 'classical' mutation. Epigenetic changes, however, may occur at much higher frequencies than mutations. The occurrence of a considerable amount of instability or 'epigenetic' variation is, a rather unexpected, side-effect of modern biotechnology and *in vitro* work. With respect to this last point, it is often part of the so-called 'somaclonal variation' that will be discussed in Chapter 5. Breeding of new genotypes of lettuce (*Lactuca sativa*) in the 1990s in the Netherlands by applying modern methods of biotechnology, for instance, was seriously hampered by the observed (epigenetic?) instability of the new genotypes that had been produced in this way. Another well-documented

Box 3.2 Transposable elements and plant breeding

Transposable genetic elements (transposons), unstable mutations, insertion mutations, mobile controlling elements (as McClintock initially called them) or 'jumping genes' refer to mobile DNA fragments (which may vary in size from approx. 1000 to 17 000 or even more base pairs) that can move within the (plant) genome and alter or block the expression of a gene in which, or in the vicinity of which, they are inserted. The action of transposons, in a way, may be considered as a mutagenic action. When a mutation that resulted from an insertion reverts again in some cells of a tissue but not in others, genetically dissimilar tissues may arise and when such genetic differences are visible, so-called variegated tissues may be shown.

Variegation patterns in maize, caused by unstable mutations, were already described in 1914 by Emerson (1914; 1917; 1929). The pattern of visible somatic variegation depends on the size and frequency of the areas (sectors) in which the changed gene expression becomes visible. This again is determined by the moment during development of the plant or plant part involved and the rate of 'reversion' by excision of the transposon. Traits affected by transposons in almost all cases inherit in a Mendelian way. In diploid plants a transposon is situated on only one of the two corresponding genes and, hence, selfing in such a case will result in a 1 : 2 : 1 segregation in the absence of dominance. When variegation is the result of other processes it may not be inherited at all, or the mode of inheritance often is non-Mendelian.

Transposable elements are known now to occur in at least 75 species of higher plants, including both monocotyledonous and dicotyledonous species. Most examples concerning the action of transposons refer to maize (*Zea mays*), one of the best investigated crops in this respect. Other plant species that have been studied relatively well in this respect are *Antirrhinum majus* (snapdragon), the garden plant petunia (*Petunia hybrida*) and *Arabidopsis thaliana*.

Many excellent and very detailed articles about transposon studies in the aforementioned crops are available. For maize for instance Fedoroff (1989) reviewed the most important aspects in a lucid way. In maize two different types of transposons can be distinguished: the **autonomous elements** and the **non-autonomous elements**. Autonomous elements can move independently in the genome, whereas the structurally heterogenous group of non-autonomous elements can transpose only in the presence of the autonomous element.

Four so-called 'maize element families' have been studied rather in great detail on their genetic interaction. These groups are called *Ac/Ds* (*Activator/Dissociation (Modulator)*, *Spm* (*Suppressor/mutator or Enhancer/Inhibitor*), *Mu* (*Mutator*) and *Dt* (*Dotted*).

Colouring of maize kernels is caused by the dominant alleles in the *C* locus that codes for a purple pigment (see for instance Fedoroff, 1984). Transposable elements are often, but not exclusively, the cause of changes in kernel colour. Insertion of the *Ds* transposable element in this *C* locus inactivates the gene, which results in colour-less kernels or kernel stripes. When the *Ds* element in some kernel cells is transposed away from this locus (which occurs only when *Ac* is present as well), the dominant gene expresses again and sectors of pigmented cells become visible, which give the maize kernel a spotted appearance.

From 1942 onwards Barbara McClintock published frequently on her research on 'mobile controlling elements', in particular in the Year Books of the Carnegie Institution of Washington, but in many other scientific journals and in proceedings of symposia and congresses as well. Fedoroff (1989), for instance, mentions more than 30 publications by McClintock. McClintock reported that unstable mutations were the cause of changes from pigmented to unpigmented and *vice versa*. The aforementioned *Ac/Ds* system was the first one on which McClintock (1948) demonstrated transposition. Gene suppression caused by elements like *Ds* in the absence of *Ac* looks like a stable mutation and in such a case shows Mendelian inheritance. Another system of unstable mutations identified and described by McClintock was the *Spm* (Suppressor/mutator) group. In maize, nine so-called 'two-element systems' were already established by 1987.

Another example of transposition effects refers to the *waxy* locus in maize. The dominant allele in this locus encodes for an enzyme needed for the synthesis of amylose starch (granule-bound starch synthase), the presence of which gives the endosperm a translucent appearance. When the transposable element *Ac* is inserted into the *waxy* locus, gene function is suppressed and the endosperm will become opaque. Subsequent transposition of the *Ac* element out of the locus results in variegation for translucent and opaque sectors in the kernel. An interesting question is whether the locus is again completely identical to the original one.

The ornamentals *Antirrhinum majus* (snapdragon) and garden petunia (*Petunia hybrida*) are now also often used for transposon research. For snapdragon it was found that transposon activity, in addition to the induction of a high rate of genetic instability caused by gene suppression, may result in the generation of many new alleles which, for instance, give the *Nivea* and *Pallida* flower colour.

Starting from a white-flowering petunia, a salmon-red petunia was constructed by transforming the white plant with plasmid DNA, containing the *A1* gene of maize and a promoter, which together produce pigments unknown so far in petunia, like the aforementioned salmon-red colour. It was found in this petunia research programme that the frequency with which transposon-induced alleles can be recovered from genetic crosses may go up to 10^{-3} or 10^{-4} if the target gene is genetically linked to a transposon showing a high level of excision (Döring, 1989).

Transposon activity leads also to structural instability and by this may induce various types of aberrations, like duplications, deletions and inversions in both the host DNA and in the transposable elements themselves. In an experiment involving 30 000 transgenic, heterozygous petunia plants, carrying the *A1* maize gene,

four white-flowering plants were shown to be mutants in which part of the *A1* gene had been deleted. Other plants with aberrant flower colours, however, were caused by so-called hypermethylation of the promoter and therefore not stable (Meyer *et al.*, 1987; 1992).

Transposons are also of practical value for **tagging** of genes. In plant species like maize, snapdragon and *Arabidopsis*, a number of transposons have been molecularly cloned.

Such cloned transposons – after having been inserted into a specific gene indicated by absence of gene expression – can be used for isolation of that plant gene.

The transposon, of which the DNA sequence is known, is used as a **probe** or a **tag** to recognize the gene in which it has been inserted. This method is generally known as the **transposon tagging-method**. (N.B. For a useful review on this subject see, for instance, Walbot, 1992.)

It is possible to transfer a transposable element from a plant species in which this element is endogenous, to other species in which such transposons were not known before, e.g. from maize to *Arabidopsis*. For this transfer use is made of the *Agrobacterium* system. In 1993 it was shown by research workers from the CPRO-DLO (a government plant breeding research institute in Wageningen, the Netherlands), that a maize transposon can be used for gene tagging in *Arabidopsis*. Most details about molecular and other aspects of the transposon work are outside the scope of this book.

example concerns the occurrence in the 1980s of unexpected and undesired variation in oil palm (*Elaeis guineensis*) in Malaysia after a cycle of *in vitro* propagation from root explants. More details are given in Chapter 5.

Apart from the evidence for a role of DNA methylation in the occurrence of epigenetic modifications, the process is not well understood yet. In 1993 R. Jorgensen reviewed 'genetic variants' that arise 'by an imposition of epigenetic (developmental) information on the germ line'. He refers in this context to the term **'epimutation'** (Holliday, 1987) to indicate genetic variants due to DNA modifications. Jorgensen adds that the absence of suitable explanations for the occurrence of 'the parallel phenomena of directed epimutation and selection-induced mutation', demonstrates our present lack of knowledge about the functioning and organization of the nucleus.

In conclusion it still seems rather improbable that epigenetic effects will ever become of (much) direct value for applied plant breeding.

Paramutations represent another category of such mutation-like effects to which most (but not all) textbooks on genetics and related topics do refer. Combination of two specific alleles in a heterozygote plant may lead to peculiar, heritable changes in which the genetic constitution of one allele affects the expression of the other allele. Rédei (1982a) briefly describes **paramutation** as a non-Mendelian alteration, enhanced by the heterozygosity of certain alleles and which produces a characteristic, unidirectional change in one of the alleles.

Wisman (1993) and Wisman, Ramanna & Koornneef (1993) mention that paramutation was first noticed by Bateson & Pellow (1915, original not consulted) for the 'rabbit ear rogue mutation' of pea (*Pisum sativum*). Forty years later Brink (1956) found some unexpected results when studying progenies of maize (*Zea mays*) from crosses in which alleles of the R locus – which locus affects the seed colour pattern – were involved. The following analysis is derived from Suzuki *et al.* (1989, p. 385). Effects of the three different alleles: r^r (colourless), R^r (dark purple) and R^{st} (stippled with irregular spots on a light background) were studied. The cross $R^r r^r \times r^r r^r$ resulted in 50% dark seeds and 50% colourless seeds. In the cross $R^{st} r^r \times r^r r^r$ 50% stippled and 50% colourless seeds were obtained. Finally, in the cross $R^r R^{st} \times r^r r^r$ Brink got 50% stippled seeds and 50% seeds that showed a weak colouring. It was concluded that in the heterozygote $R^r R^{st}$ combination the R^r allele changes towards another, new and stable allelic situation, indicated as R'. In other words, apparently the activity of the **paramutable allele** (R^r) depends on a **paramutagenic allele** (R^{st}) that was present on the homologous chromosome in the previous generation.

Another maize locus for which paramutation has been studied is the locus *B*. A third, well-known example of paramutation, the *sulfurea* (*sulf*) 'mutant' in tomato (*Lycopersicon esculentum*), recognized by an aberrant leaf colour and obtained after X-irradiation of the tomato cultivar Lukullus, has been described by Hagemann (1958; 1969) and in Hagemann & Snoad (1971). Wisman *et al.* (1993; see also Wisman & Ramanna, 1994) reported on a new paramutagenic allele for *sulfurea* following *in vitro* culture.

Hagemann (1958) uses the expressions **'somatic conversion'** and **'paramutation'** as synonyms. Mention should be made in this respect also of the earlier expression **'gene conversion'**, which was introduced by Winkler (1930) and should be considered as a synonym as well. Because of the date of publication it seems to be justified to give priority to the expression 'gene conversion' as the only correct one for 'paramutagenic-like' phenomena. However, as 'paramutation' is generally accepted by now, it may be difficult if not impossible to gain general acceptance for this change.

Rieger *et al.* (1991), referring also to Brink (1958), mention that paramutation 'represents a kind of gene instability due to heterochromatization at or near a locus at which metastable repression of gene activity occurs'.

More details about paramutation can be found in, for instance, Coe (1966), Brink (1960; 1964; 1973), Brink, Styles & Axtell (1968) and Hagemann & Berg (1977). The molecular base of paramutation is subject to much speculation (see Dooner & Robbins, 1991, and Krebbers *et al.*, 1987). R. Jorgensen (1993) related the occurrence of paramutations to the process of methylation. Recently, the organization of paramutagenicity was discussed by Kermicle, Eggleston & Alleman (1995).

No reports are known which describe efforts to make use of paramutations for practical plant breeding purposes.

Another intriguing 'mutation-like' phenomenon which is often not mentioned in text books on genetics, is the occurrence of so-called **genotrophs**. The expression 'genotroph' traces back to Durrant (1962) who has produced a range of publications on this subject (1958; 1959; 1962; 1971; 1981). Seeds from selfed lines of flax (*Linum usitatissimum*), in particular of cv. Stormont Cirrus, were sown by Durrant on two soil types which strongly differed for the content of minerals like K and P. Plants resulting from the 'rich' and 'poor' plots were very distinct with respect to plant morphology and state of development and, for instance, showed differences in plant weight, plant height and degree of branching. If seeds were collected from both types of plants and grown separately, these differences were maintained for many generations of selfing. A period of five weeks was found to suffice for the occurrence of this phenomenon. It appears that the plants have been 'transformed' into new <u>heritable</u> types as a result of environmentally induced changes or, in other words, by 'nutritional conditioning' of the plant populations. A possible explanation has been sought in the remarkable difference in nuclear DNA content between both types, the large type, which originated at the rich soil, showing 16% more DNA per cell nucleus than the small type. The observed increase in nuclear DNA under more favourable conditions is assumed to come from <u>amplification of specific DNA sequences</u>. A series of publications on the molecular aspects of genotrophs in flax has been published by Cullis and colleagues, including a report on the observed differences in DNA amount between different lines (Cullis, 1973), some reviews on the subject (Cullis, 1977; 1986) and a contribution about the inheritance of variation in the proportion of the DNA, coding for ribosomal RNA (Cullis, 1979).

A more or less similar case was described for tobacco (*Nicotiana rustica*) by Hill (1967), Hill & Perkins (1969) and Perkins, Eglington & Jinks (1971). Here the expression **transmutation** was used to indicate '<u>environmentally induced heritable changes</u>'. (N.B. The word 'transmutation' does not occur in Rieger *et al.*, 1991.) Another example, described earlier by Highkin (1958) in pea (*Pisum sativum*), refers to temperature-induced variation for various traits, but the observed changes were not as stable as those observed at the same time in flax.

Cullis (1986) has suggested that the observed phenotypic changes are caused by alterations in gene control as a result of changes in the chromosomal structure in which heterochromatin might play a role. According to Fieldes (1994) the phenomenon of (environmentally induced) genotrophs may be related to the changes that have been observed as a result of **'somaclonal variation'** after the passage of plant cells through a cycle of tissue culture and also may be caused by **epigenetic effects** which could be the result of the aforementioned methylation of specific genes. It has not been determined yet whether or not the aforementioned phenomena are caused by similar mechanisms.

Genotrophs are not very common and until now no reports are known of purposeful use of such mechanisms in the Western world. May be part of the findings in the former USSR concerning 'environmental conditioning of crops' and some old literature in Germany (the so-called 'Herkunfts-Wirkung' of seeds of one cultivar propagated under (very) different conditions), the Netherlands and some other countries as well, could be explained by genotroph-like effects.

The last system that produces 'mutation like' effects and that is only briefly mentioned here is the **antisense RNA technique** which system was mentioned already earlier in this chapter and briefly discussed in Chapter 1. By making use of this system – in fact a gene regulation method – it is possible to suppress the gene activity of specific, for instance unwanted, genes. The antisense method, in a way, could also be considered a mutational method.

3.2.2. Structural mutations

Structural mutations arise as a result of a chromosomal break: a discontinuity within a chromosome which may either occur spontaneously or be artificially induced. Sometimes the expression <u>clastogenic effects</u> is used to indicate the effect of any agent that may induce structural changes in chromosomes (Rieger *et al.*, 1991). Four types of structural mutations within chromosomes are commonly distinguished: **deletions** (or **deficiencies**), **duplications, inversions and translocations**. Mutations of this group, together, are also called **chromosome rearrangements** or, sometimes, **chromosome aberrations**, although this last expression often also involves changes in chromosome number.

The large majority of structural mutations either result in loss of smaller or bigger chromosome segments (and, hence, the loss of genetic information on those chromosomal sections) or in rearrangements of genes on the chromosomes. Mutation of genes may occur as well, but this is not necessarily the case. As opposed to gene mutations, most structural mutations can be observed with a microscope. Detailed information on structural mutations can be found in most textbooks on genetics but, as they are not as valuable for plant breeding as 'real' gene mutations, structural mutations will not be discussed here as extensively as gene mutations.

Deletions (Deficiencies)

When a chromosome segment and the genetic information within this segment are lost, this is commonly indicated as a **deletion** or a **deficiency**. It is estimated that at least 90% of the radiation-induced mutations refers to (gross) deletions. The effect of a deletion of a complete gene is illustrated by the following example. Yatou & Amano (1991) reported for the *waxy* locus in rice (*Oryza sativa*) that entire deletion of the gene resulted in the complete absence of the (*waxy*) enzyme that codes for amylose synthesis in the endosperm. This effect could be shown by making use of the so-called Southern blot analysis method.

The word deletion, in practice, is commonly used for the loss of both terminal and internal (intercalary, interstitial) chromosome segments, but Rieger *et al.* (1991) suggest that for eukaryotes the word deletion could be applied to the former and deficiency to the latter group. We will use both words as synonyms. The difference between terminal and internal breaks is that terminal deletions may result from just one chromosomal break and that for internal deletions two breaks are required.

When deletions (or inversions) occur for any other number of nucleotides than three (or multiples of three), commonly, **frameshift mutations** occur; this category of mutations was discussed in Section 3.2.1.

Broken segments either may be lost (deletion, deficiency) or become attached anywhere else on a chromosome (translocation). The broken ends on both sides of the deletion normally are reunited again and in that case in meiotic pairing the homologous chromosome may show a so-called loop corresponding to the deleted part. A deletion may be associated with a duplication if breaks occur at the same time at the same position in two homologues. One of both homologues will then be supplemented with the chromosome segment deleted from the other homologue.

Deletions may cause the loss of a few genes only but also the disappearance of a complete chromosome arm carrying many genes. Whereas 'real' gene mutations can revert to the 'wild type', a characteristic feature of deletions is that the larger the lost DNA segment, the more unlikely it is that they ever revert. Ionizing radiation, like X-rays, neutrons and gamma rays, is an effective tool to induce deletions.

Radiation-induced deletions, basically, occur at random positions in any chromosomal region. It was mentioned in some early reports (see for instance Khush & Rick, 1968) that neutrons produce more breaks in euchromatin than X-rays, but reliable information about the relative frequencies of induced deletions in euchromatin, heterochromatin and the centromeric region is not available. It is, on the other hand, known that different loci may show differences in the number of induced mutations, as well as in the potential number of mutant alleles after different mutagenic treatments. This subject will be discussed later in this chapter in the section about the Swedish mutation research in barley.

The consequences of the occurrence of a deletion for a plant cell are predominantly determined by the role of the genes concerned. In case of heterozygosity the remnant dominant or recessive gene on the homologue will be expressed and there may or may not be a change in the phenotype of the cell. If the deletion involves the loss of the centromere, the whole chromosome may be lost. Deletions often are lethal for the plant, in particular when in the next generation both homologous chromosomes carry the same deletion.

Duplications

Duplications are very important from an evolutionary standpoint, because additional genetic material becomes available in this way which, potentially, could assume new gene functions (Suzuki *et al.*, 1989). Mutations in one of both duplicated regions (even deletions) would not lead to the loss of function of the gene(s) concerned, as their original function is secured by the other copy.

The position of duplications is random within a chromosome and the size of duplicated segments may differ considerably. Duplications mostly occur in homologous chromosomes but can also be transferred to non-homologous ones. In the latter case the duplication is not linked and therefore will segregate. At homologous chromosomes the duplication may be situated in the chromosome distantly, close together or even adjacent to each other. Duplicated segments can be inserted either in tandem-sequence (tandem repeats) with both segments showing the same sequence, or in reverse order (inverse repeats). In the *Manual on Mutation Breeding* (Anon., 1977a, p. 111), Gustafsson & Ekberg mention that duplications, and possibly tandem repeats in particular, are common in (crop) plants, for instance in maize (*Zea mays*), barley (*Hordeum vulgare*), pea (*Pisum vulgare*) and cotton (*Gossypium* sp.).

Effects of duplications are largely locus-specific. If a duplication involved one complete gene (or more genes) this may result in a double dose of the gene product(s) and therefore in a different phenotype. The behaviour of a duplication may be similar to a gene mutation. The classical example of the dominant *Bar*-eye mutation in *Drosophila* (a slit-like eye, differing significantly from the normal, oval eye) is caused by a tandem duplication that probably resulted from an unequal crossover (Suzuki *et al.*, 1989, p.179).

Inversions

An **inversion** is a change in the linear sequence of one or more genes in a chromosome segment, following two simultaneous breaks and caused by a 180° rotation of the segment concerned and a subsequent reunion of the inverted segment at the breakpoints. Inversions occur spontaneously or can be artificially induced. They are called **paracentric** when the centromere is not included and **pericentric** when the centromere lies within the segment. Crossing over within the inverted segment may result in the production of duplicated and deficient chromosomal strands, depending on whether paracentric or pericentric inversions are involved and whether single or double crossing over occurs (Rédei, 1982a).

Homozygous inversions in most cases do not much affect the phenotypic appearance of the plant. Significant cytogenetic effects may occur in heterozygotes with one normal chromosome and another with the inversion. Because there is no loss or gain of genetic material, plants which are heterozygous for inversion may be normally viable (Suzuki et al., 1989), but problems arise in meiosis and sporogenesis. Plants heterozygous for a pericentric inversion show both pollen and embryo sac abortion; paracentric heterozygotes show pollen sterility but little or no female sterility. Seed sterility in plants which are inversion heterozygotes is generally less than in heterozygotes which are the result of a translocation. It may be difficult to detect small inversions because meiosis and the consecutive gamete formation is (almost) normal.

Translocations

Translocations, which – both in positive and in negative ways – may be the most important of chromosome aberrations, are characterized by a change of position of chromosome segments within the genome without loss of genetic material. In a translocation, most often, part of one chromosome has broken off and becomes attached to another chromosome. Translocation chromosomes may be recognized cytogenetically or in suitable crosses. Translocations should not be confused with the result of (meiotic or mitotic) **crossing-over**, which takes place in the four-strand stage of chromosomes and refers to the exchange of corresponding chromosome parts between homologues by breakage and reunion (Suzuki et al., 1989; Rieger et al., 1991).

In nature translocations occur in low frequencies only, but they can be easily induced by ionizing radiation and chemical mutagens. According to the *Manual on Mutation Breeding* (Anon., 1977a, p. 109) different types of breaks and rearrangements in chromosomes may occur as a result of radiation treatment or by the application of chemical mutagens. It is possible to break or change in this way specific (for instance undesired) gene linkages. This subject was discussed in particular for barley by Hagberg (1986).

Following Russell (1992) a distinction is made between **intrachromosomal** (within a chromosome) and **interchromosomal** (between chromosome) translocations. Intrachromosomal translocations, which are also indicated as transpositions by Russell, are non-reciprocal and may involve a change of a chromosome segment within the same chromosome arm or from one arm to the other. **Reciprocal exchanges** (sometimes also called interchanges) are the most common type of translocation. This situation will occur when chromosome segments from two non-homologous chromosomes are exchanged, resulting in two different translocation chromosomes.

Gustafsson, Lundqvist & Ekberg (1966) report that in cases of heterozygous translocations (in M_1), with both translocated and normal chromosomes present, mention is often made of abortion of gametes and semi-sterility. This may become visible by the occurrence of a mixture of shrivelled, abnormal pollen and normal, viable pollen. Homozygous translocations, which occur in M_2, M_3, etc., mostly do not result in major changes in morphology or viability in diploid crops like barley.

The first and best known example of a successful radiation-induced translocation of agronomic importance, refers to the transfer of resistance to brown rust (in the USA: leaf rust) (*Puccinia triticina*) from a chromosome of a grass-like species called *Aegilops umbellulata* (syn. *Triticum umbellulatum*) to a chromosome of bread wheat (*Triticum aestivum* subsp. *vulgare*) by Sears (1956). Some details are presented in Box 3.3.

A sometimes heard, justified, criticism on this work is that Sears, who was a cytogeneticist and not a plant breeder by profession, performed this highly interesting work with cv. Chinese Spring, a cultivar with many agronomic shortcomings and, at that time, already unattractive from an economic point of view. As a result, the newly introduced resistance did not have much practical impact, probably opposite to what would have been the case if a modern cultivar had been used as a crossing partner. In order to use this valuable resistance on a broad scale in wheat, it would have been necessary to transfer this resistance by means of multiple crossing (in which the translocated gene may be lost again as well) to better, modern cultivars at the cost of much work during several years.

The 'translocation method', over the years, has been considered a most interesting and elegant demonstration of what can be achieved by using mutation methods in combination with other plant breeding techniques. The costs and the time involved in these activities and the fact that the translocation method is already laborious by itself and requires careful cytological investigation, made the use of induced translocations to be considered by most plant breeders not to be a very suitable method for practical genetic improvement of crop species. In addition, it must be mentioned

---

Box 3.3 The introduction of a gene for brown rust from *Aegilops umbellulata* (syn. *Triticum umbellulatum*) to bread wheat (*Triticum aestivum* subsp. *vulgare*) by means of an induced translocation[*]

The cytogeneticist E. Sears in 1956 provided a beautiful example of applying an X-ray-induced translocation for the genetic improvement of a crop plant. The wild species *Aegilops umbellulata* carries a resistance gene for brown rust (or leaf rust: *Puccinia triticina* = *P. recondita*), the presence of which gene would be highly desirable in bread wheat (*Triticum aestivum* subsp. *vulgare*). *Aegilops umbellulata*, unfortunately, is cross-incompatible to the hexaploid bread wheat. Sears, therefore, made a so-called bridge-cross between *Ae. umbellulata* and *Triticum dicoccoides* (wild emmer), and the resulting **amphidiploid** (see Chapter

1) could be crossed with cv. Chinese Spring of bread wheat. One plant was obtained that had 21 pairs of chromosomes plus a so-called **isochromosome** (i.e. with genetically identical arms) of *Ae. umbellulata*. Plants of this type, in fact **addition lines**, were X-irradiated before meiosis in order to translocate the area containing the desired gene of the alien chromosome to another chromosome, originally belonging to the wheat genome.

From 6091 plants tested, one single plant proved to be highly resistant to *Puccinia triticina*, and, moreover, showed no cytologically detectable anomalities and had a normal fertility. Rédei (1982a) added

that this specific plant carried a very small translocation containing the resistance gene and, in addition, some factors that slightly affected the flowering time. The plant, in nearly all other traits, was described as a 'perfect, resistant wheat plant without the undesirable properties of the wild plant'.

The induction of resistance for *Puccinia triticina* in wheat has been the first case of the transfer of a radiation-induced chromosome segment with a desired resistance gene.

[*]N.B. Different views on the rather confusing taxonomy of cultivated wheats and related wild species are not discussed here.

---

that the method is applicable only for simple (monogenically) inherited traits and that one cannot predict on which chromosome, and where on the chromosome, a translocation will occur. Moreover, meiotic irregularities and transfer of undesired genes, situated on the translocated part of the alien chromosome, may further complicate this approach.

In spite of these draw-backs several other examples of successful translocations have been reported, mostly in wheat with different sources of resistance from, for instance, rye (*Secale cereale*) (Driscoll & Jensen, 1963) and *Agropyron elongatum* (Smith et al., 1968). Other examples refer to oats (*Avena sativa*) with a translocation from *Avena barbata* (Aung & Thomas, 1976); sugar beet (*Beta vulgaris*) with resistance for nematodes from *B. patellaris* and *B. procumbens* (Savitzky, 1975, 1978) and tobacco (*Nicotiana tabacum*) with resistance for tobacco mosaic virus from *N. glutinosa* (Chaplin & Mann, 1978). The publication by Smith et al. (1968), for instance, referred to the official registration of cv. Agent as a new wheat cultivar, carrying a resistance gene to brown and black rust from *Agropyron elongatum*, whereas Chaplin & Mann (1978) reported for tobacco on the new cv. Samsun with resistance to tobacco mosaic virus.

3.2.3. Changes in chromosome numbers
The subject of variation in chromosome numbers and ploidy levels is of considerable importance for various fields of plant breeding, and plant breeders spend much

time in understanding these phenomena and making optimal use of them in their breeding programmes. Some of the most common terms are briefly introduced here.

The term **ploidy** is related to the number of sets of chromosomes or **genomes** in a cell, a tissue or a plant. The term **euploid** is used to indicate a plant (cell, tissue) with one complete genome (resulting in a **monoploid** plant) or with whole multiples (**diploid, triploid**, etc. or, in general, **polyploid** plants) of the complete genome of the species concerned. The basic chromosome number of a genome is indicated by the symbol x. For barley (*Hordeum vulgare*), for instance, $x = 7$; for snapdragon (*Antirrhinum majus*) $x = 8$ and for potato (*Solanum tuberosum*) $x = 12$. The chromosome number in the gametes is indicated as n; the number of chromosomes in the somatic tissues as $2n$. Consequently, the correct chromosome formula for the genetic set up of (normal) diploid barley is $2n = 2x = 14$.

Doubling of the two complete genomes of a diploid plant results in a **tetraploid** (or, more in general, in a **autopolyploid**). Halving of the diploid plant leads to a plant with a single genome or a **haploid**. Potato is an **autotetraploid** with the chromosome formula $2n = 4x = 48$. A haploid derived from this tetraploid potato, in fact a so-called **dihaploid**, is characterized by $2n = 2x = 24$. A haploid of a diploid, e.g. cv. Kleiner Liebling of *Pelargonium* has the formula $2n = x = 9$. Haploid plants derived from diploids are rather uncommon.

Allopolyploidy is the result of combination of the genomes of different (diploid) species, genera or families. An **allopolyploid** from two parents is also called a **amphidiploid**.

Aneuploids are plants which differ from a basic *euploid* plant because they have one (or a few) of the chromosomes of a complete set in excess or deficient.

In the following sections some additional information about the most common categories of plants with aberrant chromosome numbers is given.

Aneuploids

Aneuploids, in which one or several complete chromosomes are added or eliminated (indicated in diploid organisms for instance as $2n - 1$ or $2n + 2$, etc.), can be obtained in different ways. They either occur in nature or can be produced by man and may be detected phenotypically, since they often show a lower growth rate than normal (Avery, Satina & Rietsema, 1959).

Crosses between species which are not closely related (wide crosses), even with different ploidy levels, may account for aneuploids; the use of *in vitro* culture also appears to result in rather high frequencies of aneuploids (included in so-called 'somaclonal variation'). Application of radiation or chemical mutagens to induce aneuploids in general seems not very promising, although highly polyploid crops like for instance sugarcane (*Saccharum officinarum*) – where the induction of aneuploids by radiation may be rather likely – may easily tolerate losses of one or some chromosomes, which in some cases even could be beneficial.

A closer study of aneuploids shows that they will be mostly **monosomics** (with for diploid species one chromosome of a homologous pair missing) or **trisomics** (with one extra chromosome). In the 1930s E.R. Sears discovered monosomics (with $2n-1$ chromosomes) in wheat (*Triticum aestivum* subsp. *vulgare*) and started isolating all 21 possible monosomics and trisomics in cv. Chinese Spring. This material became very useful to assign genes to specific chromosomes and for other genetic studies. (See for instance Sears, 1944; 1954.)

Ahmad & Khawaja (1992) reported that irradiation of seeds of barley (*Hordeum vulgare*) even resulted in a **nullisomic** plant (with both homologous chromosomes of a pair missing), which is quite uncommon for a diploid species. The authors further mentioned that the obtained nullisomic plant was completely sterile.

Aneuploids are of value for cytogenetic studies and for plant breeding research but, so far, have been of rather limited importance for applied mutation breeding. A positive exception is the cv. Madame du Bary (or Madame du Barry, according to Darlington, 1973, p. 191) of the

ornamental bulb crop hyacinth (*Hyacinthus orientalis*). This cultivar has a red flower colour and is a (spontaneous) aneuploid mutant ($2n - 1$) from the light pink, diploid ($2n = 2x = 16$) cv. Distinction. Hyacinths appear to be an exception to the rule that aneuploids that are derived from diploids always show detrimental effects, as more cases of successful aneuploid cultivars of hyacinth are known. A possible explanation may be that hyacinths in fact are not 'real' diploids or triploids, but tetraploids or hexaploids (J.P. van Eyk, personal communication).

The other example refers to the aneuploid ($2n + 1$) mutant cultivar 'chlorina' of tobacco (*Nicotiana tabacum*) which, most probably is the first induced mutant cultivar ever obtained (Tollenaar, 1938). It remains somewhat doubtful whether the desired 'chlorina' trait (see also Chapter 2, in particular Box 2.4) is related to the presence of an extra chromosome.

Haploidy

Manipulation of the **ploidy level** of crop plants gives the plant breeder an important tool to increase or decrease genetic variation. The expression **ploidy**, as was mentioned before, refers to the number of **genomes**, for instance per cell. To indicate the basic number of genomes present in a cell, the notation *x* is used. The number of chromosomes in the gametes is the **haploid** chromosome number (notation *n*). Somatic cells which arise after gamete fusion contain $2n$ chromosomes.

Normally, the haploid phase is restricted to the gametophyte, but in exceptional cases and by *in vitro* culture, haploid sporophytes may be obtained which are viable but, of course, sterile. **Haploid plants** or **haploids** are sporophytes resulting from female or male gametes and, hence, with the gametic chromosome number. They may arise from diploid plants (so-called monohaploids or monoploids), as well as from, for instance, tetraploids (dihaploids).

In order to illustrate the potential value of haploids it may be useful to mention that, when a breeder wants to obtain homozygous lines, the use of so-called doubled haploids produces instantly the same effect as at least six or seven generations of inbreeding. A disadvantage of the method is that not all potential recombinants will be realized. Haploids produced by means other than by the meiotic process can be called **genome mutations**. Their occurrence has been known since the 1920s. Detailed investigations on their nature and origin, however, started after World War II. Haploids have been found in the meantime in about 250 species at least. Depending on the plant species and their mode of origin, they occur at different but low frequencies (e.g. 2×10^{-4} in tomato).

Haploid plants may occur 'in vivo' as well as in vitro. Different modes of origin in nature ('in vivo') are known. They may occur where microspores (male nuclei), embedded by anther tissue, develop into haploid plants (**pollen haploids**) by a process called **androgenesis** or, less frequently, where the haploid cell of the embryo sac becomes a full plant (**gynogenesis**). Other ways of origin are by semigamy (with both haploid egg cell and haploid male nucleus developing independently into haploid embryos), the occurrence of twin embryos and by natural genome elimination (which processes are not further explained here).

Use of irradiated pollen for pollination purposes may result in the production of higher frequencies of haploids. It is assumed that in that case the second mitotic division in the pollen is disturbed and, consecutively, that the division of the generative nucleus into two sperm cells is prevented. As a result the egg cell will not be fertilized but be stimulated into embryonic development. One single restitution nucleus with 2n chromosomes arises in the pollen tube, which may fuse with the secondary nucleus of the embryo sac and give rise to tetraploid instead of triploid endosperm development.

Lacadena (1974) reviewed literature concerning pollen irradiation and other mutagenic treatments, but did not arrive at clear conclusions as to the efficiency of such methods for the production of haploid plants. Most examples refer to pollinating normal plants with irradiated pollen. Logically, the opposite – irradiation of the embryosac, followed by pollination with normal pollen – is not often tried. Raquin (1985; 1986) reported that pollen irradiation with higher dosages (> 600 Gy) of gamma rays, in combination with in vitro ovary culture, should be considered a potentially efficient method for the production of haploids of Petunia. It should be noted in this respect that the doses of radiation that are applied in such situations are much higher than for mutation breeding.

Other less applied methods to increase the frequency of haploids are treating pollen with colchicine or laughing gas (nitrous oxide, N_2O) and some other chemicals, late pollination, the use of aged pollen and the use of so-called wide crosses (crosses between distant plant species). As to the use of chemical mutagenic agents, results in relation to haploid induction are very meagre. It is believed by several authors that specific chemicals can induce **parthenogenesis** (diploid embryo formation without participation of male gametes), but not haploidy.

In some early publications it was mentioned that colchicine treatments may result in haploid plants; e.g. Levan (1945) reported on a haploid plant of sugar beet (Beta vulgaris) obtained in this way. Suitable chemicals should be able to inactivate the pollen nucleus without disrupting the ability of the pollen tube to penetrate the egg cell wall, which appears to be a necessity to stimulate the egg cell to divide.

A succesful start of applying in vitro methods to produce haploids was made in 1964 when Guha & Maheshwari (1964) obtained haploid plants of Datura inoxia by in vitro anther culture. More details can be found for instance in Bajaj (1990a). In later years the use of anther culture has proved to be an efficient method for the production of haploids for many other plant species. But haploid production by anther culture is not always successful in practice as in some plant species it may be very difficult to produce some haploids in this way or because too low numbers of haploids are obtained, leading to the elimination of numerous, potentially valuable segregants. In such situations culturing of unfertilized ovules and ovaries offers an alternative.

Finally, in later chapters the value of using haploids in mutation breeding will be discussed as well.

Polyploidy

Polyploids are cells, tissues or plants with an increased genome (chromosome set) number. They often, but not exclusively, arise by doubling of the genome when doubling is not followed by cell division. The process of doubling can occur either spontaneously or, more efficiently, by special techniques to induce doubling. Polyploidy is a rather common phenomenon and plays an important role in the evolution of species, a subject that is not further discussed here. It is estimated that 50–70% of the flower crops are polyploids. The rate of polyploids in higher plants is correlated with the geographical latitude and altitude of an area. For instance, the percentage of polyploids in Sicily (geographical latitude: 36–38° N) is 37%; in England (50–61° N) about 52%, and at Peary Island (82–84° N) even 86%. The proportion of polyploids at a given geographical latitude also increases from lower to higher altitude.

Polyploids containing only homologous chromosome sets are called **autopolyploids** or **autoploids**, whereas polyploids originating from zygotes with different chromosome sets – which is most often the result of interspecific hybridization – are called **allopolyploids** (or **alloploids**).

Increased ploidy levels generally lead to an overall enlargement of cells, stomata, flowers, fruit, seed and plant size as well as to broader and thicker leaves but, as drawbacks, also to slower growth and reduced fertility. Seed propagated crops suffer more from these drawbacks than vegetatively propagated crops. Cross-pollinated crops which are grown for their vegetative parts and which, originally, are low in chromosome number, can better withstand increased ploidy levels. It is often mentioned that $3x$ is the optimal ploidy level and in vegetatively propagated plants such plants are favoured because of their exceptional vigour, whereas seedlessness (e.g. in water melon) as a result of the common sterility of triploids, is another attractive point. Triploids are not propagated through seeds, except when they are apomictic.

Since the role of haploids in plant breeding has become much more significant in recent decades, chromosome doubling of haploids in order to produce true (diploid) homozygotes, has become a subject of considerable attention. For details on this specific topic reference can be made to Jensen (1974). Spontaneous or induced doubling of chromosomes – which occurs frequently in differentiated tissues – can occur either somatically or meiotically. In the case of somatic doubling endoreduplication (in the interphase stage of the cell cycle), endomitosis, colchicine-induced (C-)mitosis or fusion may be the causes. Meiotic doubling is due to the formation of so-called unreduced gametes or $2n$-gametes by either incomplete first meiotic division ('first division restitution') or incomplete second meiotic division ('second division restitution').

Doubling of chromosomes induced by man can be obtained, for instance, by wounding of plants, followed by callus formation and regeneration of the tissues concerned. In recent years, the use of explant culture *in vitro* for (spontaneous?) doubling of the chromosome number has been advocated by different workers. The mechanism of this type of doubling is not known. Spontaneous doubling, frequently, occurs in explant cultures (tissues, androgenic cells). The disadvantage of such methods, however, may be that during cell culture the effect of doubling is accompanied by 'somaclonal (or gametoclonal) variation' for various traits. Protoplast fusion may lead to identical effects of doubling, but this method in particular offers prospects for allopolyploidization. Positive results have been mentioned with protoplast fusion of *Nicotiana tabacum* (Gleba, Butenko & Sytnik, 1975) and *Datura innoxia* (Schieder, 1976; 1977).

Treatment of various plant parts with colchicine, an alkaloid derived from *Colchicum autumnale* (Autumn crocus), has become by far the most applied technique. It was mentioned already in Chapter 2 that A.P. Dustin and some colleagues first reported on this method in 1936 (see Dustin et al., 1937), but publications by the well-known scientists Blakeslee & Avery (1937a,b) became the start of a real colchicine craze. Within a few years the chromosome doubling technique had been applied to at least fifty plant species and an impressive number of publications on various ways of applying this drug and on the effects of colchicine in plants arose. The contributions of Blakeslee and Avery in fact were based on suggestions and initial observations by Blakeslee's assistant O.J. Eigsti who, earlier, had become in touch with Dustin's work (see Eigsti & Dustin Jr., 1955). Further details about the early years of colchicine treatments can be found in Krythe & Wellensiek (1942).

Meristems of seeds, seedlings, plants or vegetative plant parts, pollen grains, zygotes, young embryos' cell culture, etc. can be treated with colchicine in concentrations between 0.005% and 1.5%, administered during 1–24 h or even up to 10 days. Treatments can be performed in many ways, e.g., via aqueous solutions, in alcohol, in combination with lanolin paste and agar, by direct soaking of the material in a solution, by brushing the solution over shoot tips, by smearing colchicine (in lanolin) on growing plant parts, via a cotton pad placed in a leaf axil, by injection, etc.

The effect of colchicine, of which the chemical formula is presented in Fig. 3.1a, is based on partial or complete inactivation of the spindle mechanism in the metaphase and prevention of the anaphase. The chromosomes stay at the equatorial plane but split

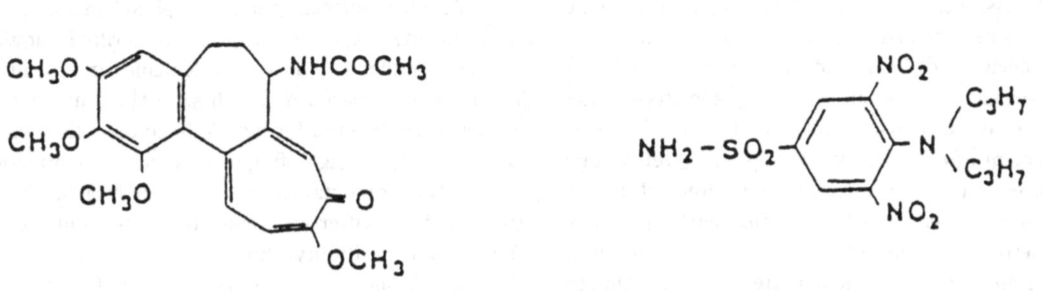

(a)

(b)

Figure 3.1a. The chemical structures of colchicine (a) and oryzaline (b).

longitudinally and when inactivation is complete, the divided chromosomes remain in a single restitution nucleus. As a consequence nuclei with a doubled genome number are formed. This effect of colchicine occurs in dividing cells only. The process goes on as long as the colchicine is present and therefore, after treatment, colchicine should be washed out again thoroughly. An optimal concentration of colchicine inhibits spindle formation but should not affect normal cell functions, but colchicine treatments produce negative side effects, such as irregularities in mitotic division, growth retardation and chromosal deficiencies. Frequencies of such events, however, are rather low. Application of dimethyl-sulphoxide (DMSO) at a concentration of, for instance, 2%, facilitates transport of colchicine through the cell membranes and, hence, allows the use of lower concentrations.

Colchicine also induces mutations other than doubling of chromosomes. In an extensive review on chromosome doubling techniques Jensen (1974) mentioned that in addition to the aforementioned negative side effects, like mitotic irregularities, etc., other mutagenic effects including even 'quantitative changes' have been reported for various crops in literature. Much information has been collected from experiments with sorghum (Sorghum bicolor) and several examples are mentioned (see for instance Franzke & Ross, 1952, and Barabas, 1962). However, Jensen (loc.cit.), who also gives many references, adds that the interpretations of some of these results are refuted by other research workers.

In 1994 the IAEA data bank on induced mutant cultivars contained, amongst a total of 141 records for chemically induced mutant cultivars, six colchicine mutants, induced during the period 1961–85 (Maluszynski, personal communication). Field crops (red clover, ryegrass and turnip) as well as ornamentals (Streptocarpus, Chrysanthemum) were involved and the mutated traits, amongst others, concerned yield, lodging resistance, flower colour and flower morphology. Nevertheless, according to Jensen (1974) colchicine should not be regarded as a 'true chemical mutagen'.

A problem of colchicine-induced increased ploidy levels, in particular in vegetatively propagated crops, is the occurrence of **chimerism**, caused by the simultaneous development of tissues of different ploidy levels in one plant or plant part. Many details about **ploidy chimeras** have been discussed for instance by Dermen (1940; 1960). As chimerism is of considerable importance in mutation breeding and deserves particular attention in relation to mutation breeding of vegetatively propagated crops, this subject will be treated extensively in Chapter 7.

In addition to colchicine a number of other chemi-cals with chromosome doubling effects (spindle poisons) are known (see for instance Jensen, 1974). Best known is nitrous oxide or laughing gas (N_2O), a relatively harmless gas with easy and quick penetration into cells when under high pressure. Compared to colchicine, the nitrous oxide produces higher frequencies of doubled plants and less lethality. Additional application of DMSO also in this case leads to an increased efficiency of the doubling treatment (Subrahmanyam & Kasha, 1975).

Other chemicals mentioned in relation to chromosome doubling are the herbicides amiprophos-methyl (APM) and pronamide (when applied at low levels in vitro; see for instance Stadler, Phillips & Leonard, 1989), trifluoralin [2,6-dinitro-N,N-dipropyl-4-(trifluormethyl) benzeamine] and acenaphtene. Other methods like the use of centrifugation, temperature shocks and hormones may also result in doubling of chromosomes, but all these methods are only seldom applied.

More may be expected from the use of oryzalin [4-(dipropylamino)-3,5-dinitrobenzenesulphonamide; see also Fig. 3.1b], because this chemical is considerably less toxic than colchicine and, hence, may cause no growth retardations and mutations. The use of oryzalin, a herbicide, as a polyploidizing agent in higher plants was proposed by Morejohn et al. (1987). Details about treatments with this rather new chemical can be found in Verhoeven, Sree Ramulu & Dijkhuis (1990) and van Tuyl, Meijer & van Diën (1992). Van Tuyl et al. described the use of oryzalin for two flower crops: Lilium and Nerine. Oryzalin was used as a stock solution of 20 mg ml^{-1} in water-free DMSO. Treatments were carried out in aqueous solutions at 20 °C for 4 h on lily scales and Nerine explants in vitro. In a comparative experiment with colchicine it was concluded that oryzalin much more effectively inhibits spindle formation than colchicine, and at lower concentrations (0.01–0.001% for oryzalin; 0.1% for colchicine). As opposed to colchicine, in the treatments with oryzalin no negative side effects were observed. Recently, van Duren et al. (1996) reported on the induction of autotetraploids in diploid banana (Musa acuminata) after treatment of shoot tips in vitro in liquid medium, supplemented with colchicine or oryzalin. Treatments were performed with 5.0 mM colchicine for 48 h or 30 µM oryzalin for 7 days, in combination with 2% (v/v) DMSO. High frequencies of non-chimeric tetraploids were obtained in the fourth generation (vM$_4$): viz. 23.1% after colchicine treatment and 29.1% after treatment with oryzalin.

Earlier, Awoleye et al. (1994) reported for cassava (Manihot esculenta) that oryzalin was less effective than colchicine. Some Dutch breeders of cabbage crops (Brassica sp.) also communicated that oryzalin did not give better results than colchicine whereas the frequency of plants showing spontaneous chromosome doubling was high enough to refrain from treatments with colchicine, oryzalin or other chemical substances.

Finally, Stadler et al. (1989) correctly pointed out that different plant species may require very different effective concentrations of colchicine, oryzalin, etc., to obtain maximum mitotic arrest.

3.2.4. Extrachromosomal mutations

After the rediscovery of Mendel's principles in 1900, it became generally accepted that genes are located on chromosomes which are situated in the cell nucleus. Most classical geneticists strictly adhered to this concept and did not believe that other regions of the cell might be concerned as well with heredity. For instance, Beale & Knowles (1978) mention that, despite the fact that it was known already in 1909 that in higher plants certain traits did not segregate according to Mendel's laws, T.H. Morgan (1926) still wrote that 'The cytoplasm may be ignored genetically'. Gradually, however, the number of examples grew in which reciprocal crosses between different plant species did not produce identical results and the impression arose that some traits were transmitted more through female than male gametes. This soon resulted in the notion of **extranuclear inheritance** and when it was discovered that chloroplasts and mitochondria do contain DNA as well, the belief in extranuclear inheritance became firmly established.

Extrachromosomal or **extranuclear mutations** have increasingly attracted attention in recent years. The expression 'extrachromosomal mutation' indicates that genes do exist also outside the chromosomes, and these genes can be mutated as well, whereas the expression 'extranuclear mutation' more explicitly refers to genes which are situated outside the nucleus, that is in the cytoplasm. For plants both expressions, most often, are considered as synonyms. In plant cells two extranuclear genetic systems exist: the **chloroplast** and the **mitochondrion**. Chloroplasts and mitochondria are mainly responsible for photosynthesis and ATP synthesis, respectively.

Although it gradually became clear that the organization of extranuclear genes and the mode of inheritance of traits determined by such genes differed considerably from nuclear genes, not much attention was paid to this subject for many years. Starting from the 1960s interest in cytoplasmic inheritance increased, as is shown by a number of useful reviews, for instance by Hagemann (1964), Walles (1971), Sager (1972), Grun (1976), Beale & Knowles (1978) and Hagemann (1982). The earliest reports about extranuclear inheritance trace back to 1909 and refer to studies on patterns of green–white leaf variegation observed by Baur (1909) on *Pelargonium* and by Correns (1909) on *Mirabilis jalapa* (four-o'clock plant).

Leaf variegation in ornamentals is a trait of considerable economic importance and, therefore, of interest to plant breeders. The leaves of *Mirabilis* are variegated with white tissue and green tissue, but branches with only white or green leaves are found as well. Aberrant leaf colours of this type are commonly explained by the occurrence of defective mutant genes in the chloroplast DNA. In variegated leaves both white and green (defective and non-defective) chloroplasts are present, whereas in whole green or white leaves only one of each type occurs. Crossings between flowers of all three different types of branches show two remarkable features. First, when reciprocal crosses are made, followed by selfing or backcrossing, the resulting progeny does not show the normal Mendelian segregation ratios that would have been obtained when nuclear genes had been responsible. A second interesting point is that the phenotype of the progeny is determined exclusively by the phenotype of the mother. This phenomenon is called **maternal inheritance**, a common, but not exclusive phenomenon in case of organelle inheritance.

According to Tilney-Bassett (1991), an authority in this field, the inheritance of mutant plastids that result from a non-nuclear mutation in the plastid DNA is always maternal in the broad sense of meaning. The author adds that in a more narrow meaning of the word 'maternal', however, plastids are inherited exclusively from the maternal parent, as distinct from **paternal** inheritance (plastids solely from the father) or **biparental** inheritance (plastids from both parents). Smith (1989), in a review about biparental inheritance of organelles, estimates that about one third of the higher plants at least occasionally show biparental inheritance for plastids, whereas for mitochondria only one example of biparental inheritance is known. Tilney-Bassett (1991) further mentions that about four-fifths of flowering plants are maternal in this respect, one-fifth biparental and none totally paternal. A few maternal species are occasionally biparental. Many examples for both maternal and biparental inheritance in higher plants and many other details about leaf variegation, chimerism and related topics, can be found in Kirk & Tilney-Bassett (1967; 1978) and Tilney-Bassett (1970; 1975; 1986; 1987; 1991).

Most studies about extranuclear inheritance, so far, have been performed with chloroplasts, and – because of complications when working with higher plants – in particular with chloroplasts in lower organisms like the single cell alga *Chlamydomonas* (Beale & Knowles, 1978). The inheritance of mitochondria until recently was an almost untouched topic (see for instance Reboud & Zeyl, 1994).

For studies of organelle inheritance use can be made of phenotypic markers, molecular techniques and cytological methods. Extranuclear mutants which arise in cell cultures (*in vitro*) can also be used in such organelle studies (see for instance Maliga, 1984). Methods like electrophoresis may also reveal differences between mutated and non-mutated extranuclear DNA. Chlorophyll aberrations, resulting from **extranuclear** deficiencies – sofar the most important source of information – are mostly associated with chloroplast changes, although yellow or light green striping in leaves may be caused as well by alterations in the mitochondrial DNA, as was reported for instance by Newton & Coe Jr (1986) for maize.

In addition to leaf variegation, extranuclear genes have been reported to cause dwarf growth and to affect flower morphology and induce other morphological aberrations, tolerance to herbicides (sometimes encoded by mitochondrial genes) and the well-known race-specific susceptibility to the toxin of Southern corn leaf blight (*Drechslera maydis*; earlier *Helminthosporium maydis*) in plasmatypes of maize (*Zea mays*) containing the so-called Texas cytoplasm causing male sterility. Probably some resistances to other fungal diseases are also determined by extranuclear genes, as has been suggested for instance for *Puccinia striiformis* in wheat (*Triticum aestivum*) and *Alternaria* in tomato (*Lycopersicon esculentum*). According to, for instance, Russell (1978) and various other authors, the importance of the cytoplasm for the expression of disease resistance has been seriously underestimated. Reliable conclusions about the occurrence of extranuclear influence on resistances, of course, can only be drawn after reciprocal crosses between resistant and susceptible lines have been made and carefully analyzed, which often has not been done.

Much attention is paid also to cytoplasmic male sterility (CMS), for instance in maize, because of its importance to the plant breeder in relation to the production of hybrid seed. Practically all authors agree that CMS is a trait encoded by the mitochondrial genome (Lonsdale, 1987). Mutation and recombination in the mitochondrial (mt) DNA may be the basis of most spontaneously occurring CMS.

Most knowledge about **extranuclear mutations** has been collected for the plastids (the plastome) through chlorophyll mutations. They may be useful for fundamental studies as well as for practical applications, like induction of aberrant leaf variegation patterns in orna-

mentals and their application as markers in somatic hybridization. Fluhr *et al.* (1985) mention that spontaneous plastome-dependent variegations in plants 'occur at a frequency of about 5×10^{-4}' (per cell per generation?) and that 'the spontaneous mutation rate for specific traits is estimated to be less than 10^{-9}' (per ?). The authors refer in this respect to Epp (1973), in which publication, however, these figures do not occur. Epp, for his part, refers to Michaelis (1969) who worked with *Epilobium hirsutum* and observed spontaneous plastid mutations 'in 0.01–1.3% of the plants within control populations.' Spontaneous mutations for plastids, as far as is known, are always related to deficiencies and non-functioning of genes.

Extranuclear mutations – as opposed to nuclear mutations – principally can be observed already in the M_1 generation and, hence, the presence of mutations like leaf variegation in M_1 may be an indication for the occurrence of extranuclear mutagenic events. (N.B. Severe chromosomal aberrations may induce similar effects!) Since a cell with a plastid mutation still carries mostly non-mutated plastid copies, cells with a mixed plastid population (mutated and non-mutated) must be sorted out by mitoses at the chloroplast level and, subsequently, when a still meristematic cell with only mutated plastids exists, many more cell divisions are necessary to develop a 'mutated sector'. But, nevertheless, young seedlings, after mutagenic treatment of seeds, may show already visible (chlorophyll-deficient) areas or sectors. In addition to mutations in the chloroplast DNA, chromosome mutations, dominant nuclear gene mutations, somatic crossing-over and transposable elements also may cause variegation in M_1 (see for instance Blixt, 1972, for *Pisum*), but most nuclearly encoded chlorophyll mutations, because of recessivity, are not observed before M_2. Examples of induced mutations for plastids as well as for mitochondria, referring to different species can be found for instance in Hagemann (1964), Kirk & Tilney-Bassett (1967), Epp (1973), Gleba *et al.* (1975), Hagemann (1979), Schieder (1977), Kirk & Tilney-Bassett (1978), Hagemann (1982), Hosticka & Hanson (1984), Fluhr *et al.* (1985), Börner & Sears (1986) and Tilney-Bassett (1991). The work on chlorophyll mutations by Scandinavian mutation researchers like Gustafsson, von Wettstein, Lundqvist and several others will be discussed in Section 3.5 of this chapter.

Epp (loc.cit.) mentions that nuclear gene-induced plastome mutations are characterized by three features: 1) a nuclear gene which increases the variegation frequency is involved, 2) the induced variegation is persistent through the sexual cycle as a cytoplasmically inherited phenomenon, and 3) continued expression of the induced variegation is independent of the nuclear gene's action.

It must be added that still many questions about plastids, for instance concerning their number in irradiated 'initial cells' as compared with differentiated cells, the division rate of mutated and non-mutated plastids, the phenotypic expression of a plastid when one of its genes mutates, etc., are not sufficiently answered yet.

Based on the fact that cytoplasmic male sterility, apparently, occurs in nature, it has been assumed occasionally that application of mutation breeding techniques for this purpose may be feasible as well. Kaul (1988) refers to nine different crop species in which CMS has been obtained by applying various mutagenic agents, such as X-rays, gamma rays, EMS, ethidium bromide (EB), streptomycine and acridine. Sager (1972) already reported that mtDNAs are in particular susceptible to acridine dyes under conditions that are reportedly not-mutagenic for nuclear genes and an identical effect was reported specifically for the effect of streptomycine in algae. There is sufficient evidence that in nature mitochondrial DNA is involved in CMS and, as a consequence, the availability of mutagenic agents that would induce specifically high mutation frequencies in mitochondrial genes would appear attractive.

Although it has been reported that some mutagens are in particular effective in inducing high frequencies of extranuclear mutations (see, in addition to Sager (loc.cit.), for instance Hagemann, 1964; 1976; 1979; 1982; Hosticka & Hanson, 1984, and Fluhr et al., 1985), several of such reports have to be considered with some doubt. A positive exception, probably, may be made for chemical mutagens belonging to the group of so-called nitroso-compounds which, however, produce high frequencies of nuclear mutations as well. Pohlheim (1974; 1981) and Pohlheim & Beger (1974) were successful with N-nitroso-N-methyl urea (NMU) in Saintpaulia ionantha (African violet). (N.B. In literature, for this and other chemical compounds various names and abbreviations may be used. For instance, NMU is also indicated as NMH.) This subject is further discussed in Chapter 4, Section 4.3.1. Hagemann (1976, 1982) mentioned positive results in Pelargonium zonale and Antirrhinum majus. Hosticka & Hanson (1984) referred to the induction of very high rates of variegation due to plastid mutations in tomatoes (Lycopersicon esculentum) by the use of nitrosomethyl urea. Most mutations were reported to represent cytoplasmically inherited plastid mutations, as they showed the typical sorting-out pattern of variegation (including mixed types) and appear visibly in the M_1 generation.

Sears & Sokalski (1991) studied the effect of UV irradiation and treatment with nitrosomethyl urea on plant material of Oenothera carrying the homozygous recessive plastome mutator (pm) allele; this allele itself produces spontaneous mutation frequencies which are 200-fold higher than (wild-type) spontaneous levels. The authors reported that the plastome mutator activity is UV-independent, but that chlorotic areas (or 'sectors') scored during the period of three to seven weeks after seed irradiation with UV, were earlier noticeable than in the 'control'. On the other hand, a 'dramatic increase' in the frequency of pigment-deficient areas was found after treatments with up to 10 mM of NMU during 30 or 60 min. in both 'wild-type' material and in material carrying the plastome mutator alleles in the nucleus in homozygous recessive condition. According to the authors the combination of the plastome mutator and the NMU treatment resulted in a synergistic effect, suggesting that the plastome mutator may involve a cpDNA repair pathway.

Fluhr et al. (1985) observed high frequencies of extranuclear, chloroplast-encoded mutations for resistance to antibiotics in Nicotiana after treating seeds with 5mM NMU for 2 h. Timmons & Dix (1991), however, who also studied mutations for resistance to antibiotics in different species of Nicotiana, which either occurred spontaneously or had arisen after treatment with 5 mM NMU, reported a much higher yield of resistant shoots from haploids than from diploid plant material. This result, probably, could indicate that not cytoplasmic but nuclear mutations have been involved! Successful plastome mutagenesis for streptomycin-resistance with another nitroso compound: nitroso guanidine, in protoplast cultures of Nicotiana plumbaginifolia, was reported by To, Chen & Lai (1989).

Although it has been established that the nuclear genome does have a regulatory function on extranuclear genes, not much is known about an interaction between nuclear genes and the extranuclear DNA. Nuclear genes may affect the frequency and expression of plastome mutations (Kirk & Tilney-Bassett, 1967; 1978; Rédei, 1973; 1982a). The action of the aforementioned so-called mutator genes in the nucleus may greatly enhance the mutability of the plastome and produce high rates of plastid mutations, as has been shown for instance in Petunia hybrida (Potrykus, 1970), Oenothera (Epp, 1973) and maize (Zea mays) (Thompson, Walbot & Coe Jr, 1983).

Probably the best-known example in this respect (see for instance Rédei, 1982a, and Coe Jr, Thompson & Walbot, 1988), concerns the (nuclear) iojap (ij) gene in maize that, when in homozygous recessive condition, causes the induction of (maternally inherited) modifications in the plastid DNA resulting in a specific type of plastid defect which occurs in high frequency. Genotypes of maize of which the *iojap*-gene is acting, are recognized by variable chlorophyll deficiencies and albino striping. Once the mutation has taken place, the mutant plastids are not dependent anymore on (multiply independently of) the action of the *iojap* gene. The *iojap* gene is known since 1924 (Jenkins, 1924) and was studied starting from in the 1940s by M. Rhoades (see for instance Rhoades, 1943). Coe Jr. *et al.* (1988) conclude that the action of the *iojap*-gene is position-dependent (i.e. the location of a cell within the plant meristem or plant organ) as well as differentiation-dependent (i.e. dependent on the attainment of a certain state of differentiation that is specific to the genetic background).

Rédei (1982a) refers to the effects of a mutator gene (*striata*) which appears to control the mutability of both plastids and mitochondria in barley (*Hordeum vulgare*) and to the *chm* locus that produces mutations on many sites within the plastids of *Arabidopsis*. Prina (1992) also published on cytoplasmically inherited chlorophyll deficiencies in barley as a result of the action of a (nuclear) mutator gene. Several other scientists have studied chlorophyll mutations, e.g. Röbbelen in Germany, who worked on *Arabidopsis* (see for instance Röbbelen, 1966; 1972).

It would be tempting to further speculate on the feasibility of a practical breeding programme, starting with a search for mutator genes which induce a high frequency of changes in plastids and mitochondria in some species, and followed by the transfer of such mutator genes – for instance by transformation methods – to important crops in which these genes have not been observed yet.

Finally, it may have become clear that still much is unknown about extranuclear inheritance and that (mutation) breeders, working with traits which show this type of inheritance, should expect to encounter unexpected or unexplainable results. In addition, experiments to determine to which degree a specific trait is under extranuclear control, are very difficult to perform. For instance, even cell fusion experiments, in which substitution of the nucleus does not result in changes in the expression of certain traits, would not provide conclusive evidence that these traits would be completely under extranuclear DNA control.

3.3. **Spontaneous and artificially induced mutations**

3.3.1. **Spontaneous mutations**

Possible causes of spontaneous mutations

Spontaneous mutations arise without human intervention whereas **artificially induced** (briefly: **induced**) **mutations**, on the other hand, are the result of the intentional application of a mutagenic agent to a cell, a tissue or an organism. Spontaneous mutations may occur without the ('*in vivo*' or *in vitro*) action of so-called 'natural' mutagens. However, there could be some disagreement as to the point whether mutations caused by, for instance, cosmic rays, or the emission of radiation from disintegrating elements like uranium, thorium, etc., should be called 'spontaneous' or not. For us the essential point of difference between spontaneous and induced mutations is the deliberate action undertaken by man.

For plant breeders, when comparing spontaneous and induced mutations, two points are essential: 1) does mutagen application lead to a considerable and predictable increase in **mutation frequency** as compared to the spontaneous mutation frequency, and 2) are the **mutation spectra** of both groups of mutations (phenotypically and molecularly) identical? Before discussing those points, some general information about the most typical features of both groups will be given.

Spontaneous mutations may be the result of events within the cell (endogenous or intracellular factors) or be caused by external (exogenic) factors. The most common causes of spontaneous mutagenesis are: errors in DNA replication, recombination and repair; spontaneous deletions; the action of transposons and/or mutator genes; damage to DNA by cosmic particles or UV radiation; temperature shocks; geomagnetism; radioactivity, e.g. the presence of radioactive isotopes in the soil; and chemical changes in plants and plant parts (seeds, pollen) caused by ageing or other types of stress. A general review of the various kinds of effects caused by 'environmental stresses' on plants was published by Levitt (1972).

As far as mutagenesis through stress is concerned, the following may be relevant. When plants are grown under stress conditions, for instance after heavy frost or after severe damage caused by insects, it has been found that in particular in the outer layers of several plant species investigated, like onion (*Allium cepa*), spinach (*Spinacia oleracea*), *Crocus* and different species within the family of the *Cruciferae*, chemical compounds are produced like isigrine, which is a precursor of mustard oil and related to mustard gas which was the first chemical mutagen, discovered in the 1940s by Auerbach and Robson. Such compounds do occur in healthy plants as well, but in much lower concentrations.

Spontaneous mutations may also be obtained when plant parts are grown *in vitro*. One could argue, however, that the *in vitro* conditions (often with growth hormones added) are deliberately created and not 'natural' in the true sense and that, as a consequence, such mutations, at present commonly called 'somaclonal variation', should be classified as 'induced' mutations. In addition, as will be discussed in Chapter 5, only a small part of the phenotypic differences observed after an *in vitro* phase, refers to <u>genetic</u> changes whereas, in addition, part of these 'real' genetic effects might refer to variation pre-existing in the cultivated explants and not be a 'de novo' effect of *in vitro* culturing conditions.

Errors in DNA replication, which are probably the most important source of mutations, include transitions, transversions, frameshift mutations and deletions (Suzuki *et al.*, 1989, p. 478–82). Drake (1991) described a number of recent advances in trying to understand the nature of spontaneous mutation at a more molecular level. Micke (1991a), referring to Haynes & Kunz (1988), mentions that errors in DNA replication may have a probability of 1 per 10^{-2} replicated genes, but that various repair systems, including post-replication repair, may reduce this potential mutation rate to the commonly known rate of spontaneous mutations which is between 1 per 10^{-8} to 10^{-9} loci. Haynes & Kunz (loc.cit.) further comment that 'if all sources of error, damage and decay had free rein, the mutation rate would be so great that the genetic integrity of cells could not be maintained.'.

The existence of DNA repair was discovered only in 1964 and, according to Micke (1991b), 'is still being viewed sceptically for plant material treated with mutagens other than UV'. Micke further pointed out that the question of whether all genes do mutate is more of theoretical nature and is not identical with the question of why particular mutations have not been obtained by plant breeders.

Whether all genes do mutate and whether in higher plants such repair processes occur completely at random, is not known. Many aspects of processes involved in DNA damage and repair, also referring to higher plants, have been discussed in a special issue of *Mutation Research* (Véleminský & Gichner, 1987).

Ames (1989), in a contribution about the relation between endogenous DNA damage and cancer and ageing, mentioned that <u>oxidation</u>, <u>methylation</u>, <u>deamination</u> and <u>depurination</u> are important (endogenous) processes that – most likely – lead to significant DNA damage (provided of course that such effects are neither repaired nor incorrectly repaired). Ames, who for instance investigated the level of sponatenous DNA damage caused by oxidation in normal rat liver, found that 1 per 130 000 bases in nuclear DNA and 1 per 8000 bases in mitochondrial DNA was damaged. The insertion of transposable elements resulting in spontaneous mutations has been discussed earlier in this chapter. The effects of 'spontaneous' radiation will be treated together with the intentional use of different types of radiation in Chapter 4.

Seed ageing, in the early years of mutation research, was considered a useful method to augment the spontaneous mutation frequency. Ageing, in this respect, in particular refers to physiological ageing, which may be caused for instance by bad storage conditions. It may be useful to add here that spontaneous mutations occur as well when seeds are stored for a long time under optimal conditions, like is the case in gene banks. Starting with de Vries (1901) for *Oenothera*, many authors referred to such phenomena, like Nilsson (1931) for *Oenothera*; Navashin (1933a,b) for *Crepis* sp.; Peto (1933) for maize and barley; Cartledge & Blakeslee (1934) for *Datura*; and Stubbe (1935; 1936) for *Antirrhinum*. Kostoff (1935a), in a short note, points out that the 'mutations' to which de Vries, Navashin and various other authors refer, obviously are due to chromosomal alterations, like segmental interchanges between non-homologous parts of the chromosomes, and not to gene mutations.

Ageing of pollen may also result in higher mutation frequencies. Stubbe (1936) stored pollen of *Antirrhinum* for 11 weeks and obtained 10 times as many mutations as compared to the number of mutations using fresh pollen.

Other exogenous factors like temperature shocks – in particular when high temperatures are involved (see for instance Navashin & Shkvarnikov, 1933) – and the absence of certain minerals (e.g. Mn) in the soil in which plants have been planted, also may result in increased mutation frequencies. For most exogenous factors – maybe except for cosmic radiation – it is almost impossible to determine how much of the observed, increased mutation frequency indeed has been caused by a specific factor.

It may be interesting to refer briefly in this context also to so-called **selection-induced mutations**, which were discussed, for instance, by Cairns, *et al.* (1988), Drake (1991) and Hall (1993, abstract only). Cairns *et al.* state that their 'belief in the spontaneity (or randomness) of most mutations is very insecure as several experiments with bacteria do show. Evidence suggests that bacteria can choose which mutations they should produce'. Drake adds that (different types of) adaptive bacterial mutations can occur at sharply increased rates, provided that growth is not limited by other constraints. Hall, at the 17th International Congress of Genetics in Birmingham, UK, in 1993, reports that such mutations have been found in different genes of bacteria and yeast and may involve base substitution, frameshift mutation and excision of mobile elements. They occur primarily in non-dividing cells and more often when the resulting mutant genes have a selective advantage than when they are neutral. The most interesting feature of selection-induced mutations may be that they occur specifically in genes responsible for traits that are under selective pressure, i.e. in those genes where the occurrence of mutations does help to solve the problem caused by the specific type of stress to which the cell population has been submitted. Whether the phenomena described here occur as well in cell populations of higher (diploid or polyploid) plants is not known yet, but there is no specific reason to assume that this would not be the case. It would also be interesting to know whether such phenomena occur only in suspensions or also in specialized tissue.

Recently, Rasmusson & Phillips (1997) suggested that newly generated and existing genetic diversity may be much more important than was previously believed, even in plants generated from narrow gene pools.

Spontaneous mutation frequencies

The number of spontaneously occurring mutations – mostly called the **mutation frequency**: the number of mutations at a given checkpoint, or the **mutation rate**; the number of mutations per time period; see also Section 1.3.5 in Chapter 1 – depends, among other factors, on the size and the DNA sequences of the gene involved and on the mutational mechanisms which cause the mutations. The specific spontaneous mutation frequency, that is the frequency at which a specific kind of mutation occurs in a particular locus (or, even more precisely: from one specific allele *A-1* to another specific allele *A-2*, or from *A* to *a*), for instance per 10^6 cells or gametes or per 10^6 plants studied, is rather low under 'in vivo' conditions, e.g. about 1×10^{-6} to 1×10^{-8} for a given locus per haploid genome per plant generation. On the other hand, assuming the presence of 100 000 loci in the nuclear genome of a plant and a mutation rate of 1×10^{-6}, a mutation would still occur in one among every ten cells. When unstable genes are involved a mutation frequency of up to 1×10^{-3} may be found.

Data about spontaneous mutation frequencies have been collected for different plant species, including crop species, by a number of authors. As the number of genes coding for a given trait is unknown in most cases, it should be realized that such data mostly refer to the spontaneous mutation frequency for a given trait multiplied by the number of controlling genes. In addition, it is not clearly indicated in all examples whether the collected data refer to <u>any mutation</u> at a given locus or to a <u>specific mutation</u> at that locus.

Interesting data concerning mutation frequencies for various traits in barley (*Hordeum vulgare*) will be presented in a special section (Section 3.5) on mutations in this 'model crop'. Many data have also been collected for *Antirrhinum* (snapdragon) by Baur, Stubbe and Döring (see Stubbe, 1966). Baur, in 1924, calculated on the basis of segregating F_1 plants, an overall and accumulated spontaneous frequency for gene mutations of at least 5%. Stubbe found, on the basis of haploid nuclei, on the average, about 0.6% spontaneous gene mutations per generation.

Mutation frequencies in the 'model' *Tradescantia* stamen hair system (see also Chapter 1, Box 1.5) have been extensively studied by Sparrow's group at The Brookhaven National Laboratory USA (see for instance Sparrow & Sparrow, 1976) and in later years also by several Japanese workers (see for instance Ichikawa et al., 1996). Spontaneous mutation frequencies, depending also on the clone used, for instance may vary from about 3.5 to 23 spontaneous mutants per 10^5 stamen hair cells. Calculations can also be based on the number of mutagenic events per given number of hair-cell divisions (Ichikawa, 1992). Use is often made of special, preferably diploid ($2n = 12$), hybrid clones that are heterozygous for flower colour (i.e petal and stamen hair colour). Pink mutant petal spots, which occur spontaneously at a frequency of 0.5 per petal, can also be scored for estimating mutation frequencies. Mutagenic treatments may lead to mutation frequencies up to about 100 times higher. For further reading references can be also found in Chapter 1, Section 1.3.5.

In Table 3.1 some data are presented for mutation frequencies at different loci in maize (*Zea mays*); these were originally collected by Stadler (1942) and have been discussed at many occasions (see for instance Suzuki et al., 1989, p. 156). The table shows that the average number of mutations, calculated in this example per million gametes for a specific locus, may be 500 times higher than for another locus. As is the case for most other examples, Stadler looked for mutations to a specific allele and did not check the overall number of mutations per locus. For his studies Stadler (loc.cit.) used a genetic stock, dominant for a number of genes, as female parent with all tassels removed well before any pollen-shedding could occur. Pollination was performed

Table 3.1. Spontaneous mutation frequencies at some specific loci of the genome of maize (*Zea mays*). Data collected by L.J. Stadler (1942)

Gene	Number of gametes tested	Number of mutations	Average number of mutations per 10^6 gametes
R	554 786	273	492
I	265 391	28	106
Pr	647 102	7	11
Su	1 678 736	4	2
C	426 923	1	2
Y	1 745 280	4	2
Sh	2 469 285	3	1
Wx	1 503 744	0	0

with pollen from a multiple recessive stock. In this way Stadler studied mutations in female gametes and was able to study large populations, consisting of millions of plants. He regarded all mutants showing no sterility (which is often considered characteristic of losses of a DNA fragment) as gene mutations.

In another study, Stadler (1951) reported on the behaviour and frequencies of allelic series of different gene loci, like the *R*-locus and the *A*-locus in maize (*Zea mays*).

In Chapter 1 it was mentioned that for some specific loci it has been investigated in detail which types of molecular mutagenic events do occur and in which relative frequencies. Wessler & Varagona (1985) identified among a total of 22 independently arisen, spontaneous mutations at the *Wx* locus in maize, seven mutations of the insertion type and six partial deletions, whereas the other nine were either base substitutions or small deletions and insertions. Stadler *et al.* (1991), who studied spontaneous mutations for the *mtr* locus in the fungus *Neurospora crassa*, found 54% single base-pair substitutions, 34% deletions and among the remaining 12% mainly insertions.

More recently, Dillon & Stadler (1994) described sequence analyses for 34 spontaneous *mtr* mutations which confirmed that the majority of those mutations were base substitutions (48%) or deletions (35%). Most of the observed sequence differences were found in the DNA (1205 base pairs) that is flanking the so-called 'open reading frame (ORF)' of 1472 base pairs. (N.B. According to Suzuki *et al.* (1989), open reading frames are long sequences that begin with a start codon but are uninterrupted by stop codons except at the 'termini'.) In a specific 'mutator strain (*mut-1*)' a much higher frequency of small frameshift mutations, involving loss of only one or two base pairs, was observed. As an explanation for this, the occurrence of small 'mistakes' during DNA replication was mentioned. Drake (1991) mentioned that for a range of lower organisms (phages, bacteria, yeast, fungi) the mutation rates expressed per genome per DNA replication are very similar (about 0.003). Whether this holds as well for higher plants is not known.

Significance of spontaneous mutations

Throughout the years spontaneous mutations have been a subject of many detailed studies in order to understand their nature. Spontaneous mutations – despite the fact that their frequency in general remains low – are, in the long run, an important source of genetic variation in plant populations as they occur in the many cells of all plant parts and may affect each trait. It is often stated that spontaneous mutations occur more often in dividing cells, including gametes or germline cells, than in somatic cells. It is, however, difficult to substantiate this opinion. It may be interesting to refer in this respect to a review on spontaneous mutations in plants by D'Amato & Hoffmann-Ostenhof (1956). These authors in particular pay attention to endogenous effects within the plants which may be responsible for the occurrence of spontaneous mutations. In the discussion, these authors mention that 'No doubt, different types of tissues and cells show a different degree of sensitivity to an identical mutagenic action'. (N.B. What is meant is endogenous mutagenic action.) According to them the 'much higher frequency of spontaneous chromosome aberrations in the root tips as compared to the shoot tips of *Crepis* and some cereal species' can be explained in this way and 'finds its close parallel in the similar behavior of the same organs in X-ray treated fresh dormant seeds'. This, of course, is not necessarily the whole story concerning possible differences in the frequency of spontaneous mutations in various tissues on different stages of activity of a cell.

Spontaneous mutations are of particular value as a source of genetic variation in vegetatively propagated crops where they are less submitted to selection than when they have to pass meiosis and a haploid phase. In obligately vegetatively propagated crops they even are the only source of variation. Many spontaneous mutants (sports) of direct practical value have been obtained in this group of crops, in particular for ornamentals and fruit crops.

An important advantage of spontaneous mutations in plant breeding could be that at the moment of their discovery they often have already gone through a phase of adaptation in which natural selection and eventually even recombination play an important role. As a consequence such mutations, if they refer to a useful trait, can be of direct practical value in plant breeding, whereas newly induced mutations did not have the chance yet to become adapted.

The most striking disadvantage of spontaneous mutations – from the viewpoint of plant breeding – is their low frequency of occurrence. In addition, they occur unpredictably at any time and any place within one cultivar or another, as well as in each cell, irrespective of its position. Initially the plant will be heterozygous for the mutation. Thus it will not be very easy to select them in a nursery. They also occur in cells in which the mutated trait cannot be expressed unless special measures are taken; they may be found in outdated cultivars without much commercial value or in plants that can only be crossed with other plants with much difficulty or maybe not at all. Therefore, many spontaneous mutations could be of considerable economic value, but are of no use to the plant breeder, either because they could not be detected or because of the problems to transfer useful mutant genes to cultivars of economic value. Certainly, much of this applies as well to induced mutations.

3.3.2.   Induced mutations

Methods to induce mutations

As has been mentioned before, for the (artificial) induction of mutations several methods can be followed. **Physical** and **chemical mutagens**, the only two groups of mutagens that are presently used worldwide, will be discussed in more detail in Chapter 4. Before efficient procedures with those groups of mutagens had been developed, several other methods were tried and applied, most of which, nowadays, are of value only from a historic point of view. One could mention in this respect the application of temperature shocks, centrifugation of seeds and the use of electromagnetic fields. Temperatures which differ considerably from – what are considered – the optimal ones may result in mutations. Earlier, reference was made to a publication by Navashin & Shkvarnikov (1933) on this subject. Another report refers to mutations induced in seeds which were submitted to hot water treatments in order to get rid of virus infection or some specific fungi like *Ustilago nuda* (loose smut) in barley (*Hordeum vulgare*). This subject was reviewed by Lindgren (1972). Unexpected and, in most of such cases, unwanted mutations may also occur as a result of the chemical disinfection of seeds.

Based on earlier experiences with centrifugation of eggs of silkworms (*Bombyx mori*) in Japan, it was found that centrifugation of seeds also led to the induction of mutations. Kostoff (1935b; 1938) in this way obtained for various plant species (*Vicia faba, Nicotiana langsdorfii*, F$_1$ hybrids of *N. rustica* x *N.tabacum, Crepis capillaris* and *Pisum sativum*) plants with chromosomal damage, aneuploids, polyploids, plants with morphological aberrations and plants which were variegated for plastids, probably as a result of sorting out of chloroplasts. Kostoff (1938) presented a detailed description of the observed events and a list of references on this subject.

The work by Yu & Dodson (1961) with cv. Montcalm and cv. Hannchen in barley became well known. During centrifugation (at 25 000 g) of seeds, which had been pregerminated for 24 h at 10 °C, germination stops, nuclei start to show 'plasma tails', the number of dead cells increases and, afterwards, many structural chromosome mutations are observed. Centrifugation at the aforementioned conditions for 20 h results in 94% dead seed.

The centrifugation method has no practical value, but in the past breeders of, for instance, tomato and cucumber were warned that mutations might occur when centrifugation was used to extract seeds from the fruits.

Another method for mutation induction that is not applied at a large scale, refers to the application of electromagnetic fields. This subject has been briefly mentioned already in Chapter 2.

Do induced and spontaneous mutations differ?

It has been suggested on various occasions (see for instance Sankaranarayanan, 1993) that spontaneous and induced mutations do have certain mechanisms in common. For instance, certain similarities are observed between endogenous mutagens produced by normal aerobic metabolism and the mutagenic effects of hydroxyl radicals (see also Chapter 4) that are produced by radiation. In both situations specific enzymes may repair the damaged organic bases. According to Sankaranarayanan radiation-induced point mutations may show a greater similarity to spontaneous mutations than deletions, because spontaneous deletions may occur at more specific positions and induced deletions will be evenly distributed all over the chromosomes because of the random distribution of the radiation energy in space. However, so-called hot spots for radiation-induced mutations may occur as well.

Stadler, one of the great mutation geneticists (see also Chapter 2), concluded in the 1930s that X-rays did not produce the kind of mutations that arise spontaneously. Some of the mutations induced by UV radiation, on the other hand, were indistinguishable from those arising spontaneously.

Many authors, nowadays, express the opinion that there is little reason to assume that the mutation spectra of spontaneous and induced mutations should be fundamentally different (see for instance Brock, 1980). However, mostly phenotypic comparisons were made and, for instance, Konzak et al. (1984) mention that definite proof for identical kinds of changes after spontaneous and induced mutagenic events has never been given. In this respect reference could be made to Ichikawa et al. (1991), who worked with heterozygous stamen hairs of *Tradescantia* (see also Box 1.5 in Chapter 1), and conclude from their results that mechanisms of initiation and repair for radiation-induced and spontaneous mutations – at least partly – are different. The majority of gamma ray-induced mutations are caused by chromosomal breaks which often lead to deletions, whereas spontaneous mutations are predominantly of the base-change type and result from errors, like somatic recombinations, during DNA replication. Another interesting observation is that in the Swedish barley mutation research programme (see later this chapter) more than 1300 induced mutants were obtained by various mutagenic treatments for the 'trait' *eceriferum* (*cer-*) – in fact a trait complex, related to the presence and structure of wax in various plant parts – whereas screening large untreated populations resulted in six spontaneous mutants for this trait only. A further comparison of the spectrum of mutations for *eceriferum* after various mutagenic treatments reveals even more interesting details, but this subject will be discussed in Section 3.5.

The aforementioned facts indicate that some doubts can be raised when it is too easily assumed that spontaneous and induced mutations are identical also with respect to their mode of origination and the action of repair mechanisms and not only the phenotypic mutation spectra.

Are (induced) mutations 'type' and 'site' specific?
Different mutagens may have a certain preference or **'mutational specificity'**, either for specific types of mutations or for specific mutational sites (hot spots), as was first noted by Benzer for the so-called *rII* region of the phage *T4* genetic map (see for instance Benzer, 1961). Suzuki *et al.* (1989, p. 482 onwards) summarize results from work by various authors with *Escherichia coli*, which show for the gene *lacI* that different mutagens may favour different types of substitution. The chemical mutagen EMS, in this respect, favours GC→AT transitions, whereas the chemical mutagens aflatoxin B$_1$ favours GC→TA transversions. This is explained by referring to different mechanisms of mutagenesis which may be involved. Moreover, even within the same category of effects sites with very high mutagen-induced mutation frequencies may be observed; these hot spots sometimes may be related to the presence of 5-methylcytosine residues, but in most cases the reason is unknown.

For a plant breeder it would be highly interesting if he could identify specific regions on the chromosomes where a certain mutagen could be applied with a high efficiency. This would be in particular the case when genes for important traits would be situated in such areas. To illustrate the present level of 'specificity' of different mutagens for a certain category of mutations in higher plants, reference is made to Lundqvist *et al.* (1991). These authors, taking the frequency of recessive mutations in locus *Ml-o/ml-o* of barley as a standard, reported that ionizing radiations produced significantly more <u>dominant</u> resistant genes to powdery mildew (*Erysiphe graminis* f.sp. *hordei*) than various chemical mutagens like EMS, NMU and sodium azide (NaN$_3$) that are among the most efficient mutagens known. Such results, however, have to be considered with caution, as irradiation also may result in a higher level of semi-sterility in M$_1$ and, thus, may favour outcrossing; this mechanism could be responsible as well for the results obtained. Recessive *ml-o* mutations may be mainly the result of deletions.

Experiments along these lines are also performed with *Tradescantia*, *Arabidopsis*, *Drosophila* and mice. To mention an additional example, Pastink *et al.* (1991) found for a particular gene (the vermilion gene for eye colour) in two different strains of *Drosophila* after treatment with EMS (5–20 mM for 24 h) overall mutation frequencies of 2.7 and 3.4 × 10^{-4}, which in the majority referred to GC→AT transitions. In an earlier experiment in which Pastink *et al.* (1989) applied a 1 mM solution of nitroso ethylurea (NEU) for 24 h, a rather similar spectrum was found but with a larger share of AT → GC transitions and, in addition, a more significant proportion of transversions. The overall mutation frequency for the vermilion gene in this case was 1.4 × 10^{-3}, which was 14 times above the background frequency of 1 × 10^{-4}.

In conclusion: although for higher plants considerably less is known about such 'specificities', there is enough evidence – as will be illustrated again when discussing the Scandinavian barley mutation research programme, later in this chapter – that a certain level of 'mutagen specificity' and 'locus specificity' can exist as well for higher plants. It must be admitted, however, that not all claims made in the past about 'specificity' of mutagenic treatments to higher plants are justified. Conclusions which are not based on careful analysis of results obtained from large populations, mutagenically treated and grown under strictly controlled conditions, with all (?) possible sources of error excluded, and with large control populations, should be always considered with suspicion.

3.3.3. Concluding remarks on spontaneous and induced mutations

When comparing spontaneous and induced mutations, two differences have been mentioned already. First, there was the observation that natural variants in higher plants – starting from a single-cell mutational event –

before their discovery, already may have passed many 'selectional sieves' and, therefore, often, can be of direct practical use in plant breeding. Many mutant plants which arose from mutagenic treatments, on the other hand, still are just <u>raw material</u> or <u>building blocks</u> in which – even when considerable genetic improvements for important traits are involved – much additional breeding work is required before their real value can be properly assessed. In addition, too many of such 'raw' mutants did not make it in spite of considerable breeding efforts, as has been the case with the aforementioned radiation-induced *ml-o* resistance in barley. However, many positive exceptions to this rule, leading to **direct mutant cultivars** without much additional breeding work, have been found as well and examples of such cases are presented at several places in this book.

A second point of difference is that, as a rule, induced mutations occur at considerably higher frequencies, for instance 10^2 or 10^3 times higher than spontaneous ones, but for induced as well as spontaneous mutations considerable differences for various loci may be observed. The estimate of spontaneous mutations, however, is biased by the fact that their moment of origination is unknown. Their situation with respect to chimerism and heterozygosity differs considerably from induced mutations.

Ichikawa *et al.* (1991) found after treatments of heterozygous, phenotypically blue, stamen hairs of different clones of *Tradescantia* with low doses of gamma rays, an increase of the number of mutagenic events towards pink cells of 10–20 events per 10^4 hair cell divisions per unit of dose (in earlier publications: rad; nowadays Gy or gray), but the rate of increase was also related to the spontaneous mutation frequencies which differed for the various clones as well.

Mutation frequencies for chemical mutagens given in literature are often higher (e.g. 5–10 times) than those induced by radiation. In general, however, too high mutation frequencies are not desirable as the occurrence of more than one or just a few induced mutations per cell may make it impossible to utilize a favourable mutation that has been induced in that cell. Irrespective of the mutagen used, a mutation frequency of about 1×10^{-4} per locus seems to be about optimal for plant breeding purposes.

One point that has to be briefly mentioned here is that induced mutation frequencies for diploid and tetraploid forms of one species may differ considerably. For instance, Müntzing (1942) reported on the frequency of chlorophyll mutations in diploid and tetraploid barley. Results showed that in the M_2 75 chlorophyll mutations occurred for about 5500 diploid plants and none

for tetraploids. Polyploids, on the other hand, also may have some advantages for mutation breeding, such as a higher mutation frequency per cell and a higher tolerance for drastic mutations as was discussed, for instance, by Hagberg & Åkerberg (1962). We will return to this subject in later chapters.

Are there other points of difference between spontaneous and induced mutations? A practical advantage of the artificial induction of mutations is that the breeder can choose <u>which</u> genotype(s) or which cultivar he wants to submit to a mutagenic treatment and <u>when</u> he wants to do this. Based on some knowledge about the mode of reproduction of the crop, the way of inheritance and the number of loci involved for the trait under study, the mutation frequency of those loci and the amount of plant material used in the experiment, and taking into account the selection procedures that will be applied (see for instance Chapters 6 and 7), an estimation can be made of the number of induced mutations that may be expected. This estimation, however, remains rather poor as our knowledge of various aspects is still very inadequate.

In case of spontaneous mutations the breeder can only search everywhere for a specific mutation to appear, but without knowing where and when this will happen. A very desirable mutation, for instance for a rare resistance, may be found only in very low frequencies and only in an outdated cultivar or in wild plant material that can not be easily crossed with the modern cultivar or genotype in which the breeder would like that specific mutation to be introduced. Moreover, a mutation for resistance may occur in the absence of the pathogen and therefore never be selected. Nevertheless, the rareness of such a favourable event and the aforementioned fact that a spontaneous mutant trait, at the moment of its discovery, often has already passed many selection barriers, may make extensive efforts to make those spontaneous mutant traits available for further breeding work very worthwhile.

It may be important to add at this place that selection of spontaneous mutant traits under conditions which are not representative for those under which a crop is normally grown may often be without value for a modern crop cultivar!

In conclusion, the opinion that spontaneous and induced mutations do not fundamentally differ, may not be accepted as a general working hypothesis. Not all different phenotypes of mutations can be simply packed together and, as a consequence, it may be anticipated that differences may exist between both categories. The considerations mentioned above always have to be taken into account.

3.4. **Macromutations and micromutations**

3.4.1. Introductory remarks

In genetics and plant breeding a distinction is made between discontinuous traits which display a few distinct phenotypes only and traits which show continuous variation and are characterized by the presence of a whole spectrum of phenotypes. Mather (1941) introduced the term **oligogenes** – nowadays more often indicated as **major genes** – for genes that determine the phenotype for discontinuous traits and **polygenes** – or **minor genes** – for traits with continuous variation, and this has been the starting point of much confusion about possible major and minor effects of genes (see also for instance Scossiroli in the *FAO/IAEA Manual on Mutation Breeding*, Anon., 1977a, p. 118). Mather (loc.cit.), who uses the terms 'polygenic' and 'quantitative (or metric)' variation as synonyms, refers to the existence of 'balanced polygenic combinations' among the genes, which effect 'the best compromise between the advantageous qualities of immediate stability and the ultimate variability necessary for a change in an altering environment'. It must be commented at this point that it would be preferable to avoid the use of misleading words such as monogenes, oligogenes and polygenes and, instead, refer to monogenic, oligogenic and polygenic variation.

Mutations, in principle, may refer to all traits and in practice often a distinction is made between mutations producing a striking change in a trait and mutations which, because of the continuous character of the trait involved, can be detected by statistical methods only. More or less in line with the distinction between major and minor genes, the terms **macromutation** and **micromutation** have been introduced in the past. In 1940 Goldschmidt defined a **macromutation** as 'any genetic change leading to a striking change of the phenotype' (see Rieger *et al.*, 1991). Or, more in general: **macromutations** could be characterized as distinct, qualitatively inherited genetic changes which can be observed at individual plants. Remarkably, Rieger *et al.* (loc.cit.) in their *Glossary of Genetics* do not give a definition for 'micromutation', whereas definitions for the terms macromutation as well as micromutation are absent in leading textbooks on genetics like those by Rédei (1982a), Lewin (1994), Suzuki *et al.* (1989) and Russell (1992). One could, at this stage, think of a **micromutation** as a genetic change resulting in a small effect that, in general, can be detected only by help of statistical methods. In many publications about mutation breeding reference is still made to the concept of macromutations and micromutations, although these terms sometimes are interpreted in different ways.

Before further elaborating on this rather controversial and confusing topic, some general remarks must be made about population genetics and quantitative genetics, most of which have been derived from Falconer (2nd edn, 1981) and Snedecor & Cochran (6th edn, 1976).

A continuous trait is encoded by many independent genes or small chromosomal changes. Most genes have small effects on the phenotype (Brock, 1967; 1970). Together, they produce many genotypes. In addition, environmental factors cause a range of phenotypes to be produced for each genotype. Continuous traits must be described in quantitative terms and, hence, are called quantitatively inherited traits. Situations in which both qualitatively and quantitatively inherited traits are involved, of course, do exist as well. When, for instance, a number of genes together determine the level of resistance in a specific case, each individual gene may contribute in a major or a minor way in this respect.

As quantitatively inherited differences usually depend on genes with relatively small effects and usually, but not always, are governed by involvement of many loci, Mendelian analysis, as a consequence, is inapplicable and normal segregation ratios are not found. Nevertheless, the fundamentals of Mendelism remain valid also as a base of quantitative inheritance.

When micromutations are involved, populations must be studied instead of individuals. Changes in the genetic composition of populations may be the result of too small population size, differences between individuals in viability and fertility, the mating system, selection pressure and by mutations. Mutations may occur in nature in a population at a frequency of 10^{-5} or 10^{-6} per locus per generation per haploid genome.

Quantitative traits in most cases show continuous variation and always a scattering around a phenotypic mean. Their variations are 'continuous' for two reasons, 1) because multiple genes are affecting the trait, 2) because of the superimposition of multiple variation arising from non-genetic causes. Most quantitative traits (sometimes also called 'metric traits') normally show a standard frequency distribution.

Most readers may be sufficiently acquainted with the common expressions used in quantitative genetics and terms such as the mean, the variance and the standard deviation. The (arithmetic) mean for a given trait equals the sum of all measurements, divided by the number of measurements. Variance quantifies the distribution of individual measurements around the mean and is calculated as the mean of squared values of the deviations from the population mean. The expression standard deviation (which is the square root of the variance) provides a measure for the variability and is a property of the trait involved and of the population under study. In practice a population (e.g. the collected values for plant length of an M_2 plant population) is characterized by the mean and the standard deviation. Various tests, described in many textbooks about statistical methods, can be applied to investigate whether the difference found between measurements for two populations (e.g. a mutagenically treated population and a control) is statistically significant.

3.4.2. Early reports

The aforementioned differences between traits for which segregation ratios strongly suggest qualitative inheritance, and traits for which quantitative variation is observed, have led to the concept of **macromutations** and **micromutations**.

According to Gaul (1965a) it was Johannsen (1903) who for the first time really proved the existence of spontaneous small mutations for traits with continuous variation (e.g. the width/length ratio for seeds) in bean (*Phaseolus vulgaris*). Baur (1924), who described small mutations in snapdragon (*Antirrhinum majus*), was the first to recognize the importance of small mutations in evolution. Stubbe and colleagues continued the work of Baur and his associates on this subject and described many micromutations ('Kleinmutationen') in detail (Stubbe, 1934; 1966). They further applied the term 'Grossmutationen' for major mutations or macromutations (Stubbe & von Wettstein, 1941). Scossiroli (1965), however, claims that the work by Buzzati-Traverso and by himself in the early 1950s on *Drosophila* and by Gregory on groundnut, or peanut (*Arachis hypogaea*), represented the real start of studies on (radiation-)induced mutagenesis for quantitative traits. Gregory, whose work will be briefly discussed later in this section, at an early stage had already strongly emphasized the (potential) practical value of 'small' mutations in plant breeding (Gregory, 1955; 1956; 1961; 1965; 1966). He confirmed that the term 'micromutations' is synonymous to the 'Kleinmutationen' of Baur (Gregory, 1965). Another early report on 'minor quantitative changes' comes from Nybom (1954), who described for barley (*Hordeum vulgare*) several types of induced mutations which only slightly deviate from the mother line for one or several quantitative characters. Traits mentioned by Nybom in this respect, for instance, were kernel size, yield, straw stiffness and straw length, but no exact figures were presented.

Starting from the mid-1950s, in particular H. Gaul in Germany has stressed, on many occasions and with great enthusiasm, the significance of micromutations – as opposed to macromutations – for plant breeding. Gaul

(1965a) underlined repeatedly that macromutations and micromutations are not distinct classes and that a distinction between both categories is relative and arbitrary. Moreover, this distinction does not have a molecular basis, as genes for quantitative traits do not differ in principle from common genes in oligogenically inherited traits. The significant point of making a distinction between both categories, according to Gaul (1961b; 1963), is that different methods are required to detect mutants of both groups and to utilize them in the most efficient way for plant breeding purposes.

Nevertheless, discrepancies can still often be observed between the interpretations by various authors of the aforementioned expressions and the various authors do not always make sufficiently clear what they mean in a specific case, for instance when it is stated (see for instance Stebbins, 1950) that 'micromutations' are corresponding to the 'small variations' mentioned by Darwin as the building blocks of evolution. To give another example, Gustafsson (1986) comments that major mutations 'in the sense of Baur and Stubbe', would be of interest for plant breeding, whereas minor mutations 'would then act as a sort of modifying genes'. It is not fully clear what Gustafsson exactly was aiming at.

On the other hand, one can only fully agree with another remark by Gustafsson in the same publication: 'There is in my opinion no doubt that both major and minor mutations are important in plant breeding and in the process of evolution as well'. It is remarkable, however, that Gustafsson in this context did not refer to Gaul's many contributions to this subject.

3.4.3. Some examples

As, so far, mainly examples for mutations which show discontinuous variation have been given, we may at this point pay in particular attention to some general characteristics of traits with continuous variation caused by 'micromutations' (in the sense as meant by Gaul). Most studies on quantitative inheritance concentrate on yield, plant height, 'heading date' and various resistances against pathogens in different crops, but continuous variation may be observed as well in many other situations.

Gaul (1961b) expresses the opinion that a mutation in one or a few genes may lead to, for instance, a 5% increase in yield in cereals, with only small or even without significant morphological or other (undesired) changes in the plant. An increase in yield of 5%, provided that this could be achieved without too much work, indeed would be very attractive to a breeder, but the correctness of Gaul's assumption has never been proved. It is beyond doubt, however, that mutagenic treatments do lead to an increase in genetic variance, for yield as well as for all other quantitatively inherited traits. Results of detailed studies on selection for such traits in barley (*Hordeum vulgare*) were published by Gaul & Mittelstenscheid (1960; 1961). Gaul (1961b), in addition to the work of Gregory on groundnut that will be discussed soon, refers to some early studies on induced mutations causing small changes, or quantitative effects. Oka, Hayashi & Shiojiri (1958), in this respect, mention that mutations in 'polygenes' for plant height and heading date in rice (*Oryza sativa*) result in about a symmetrical increase of variability in both 'high' and 'low' direction. Moës (1958; 1959) describes induced quantitative variability for different traits like number of tillers, seeds per spike, kernel size, lodging resistance, leaf size and straw height in barley. Cooper & Gregory (1960) discuss the quantitative expression of resistance to the two most common leaf spot diseases (*Cercospora arachidicola* and *C. personata*) in groundnut (*Arachis hypogaea*). In these experiments, selections were carried out until M_9 (or X_9).

Gregory provided for groundnut, a self-fertilizing allotetraploid (amphidiploid) crop, the first definite proof that numerous mutations for continuously varying traits in crop plants could be induced by mutagenic treatments. Preliminary experiments with X-rays started in 1949, followed by X-irradiation of 75 000 seeds with dosages of 10–18.5 krad (100–185 Gy). Results showed that the genetic variance for single plant yield, measured in normal looking, randomly taken M_2- and M_3-plants (X_2- and X_3-plants in Gregory's notation), equalled four times that of the control (Gregory, 1955). Singleton (1962) described that Gregory's experiments from an M_2 population of more than one million plants yielded a few beneficial mutants, from which only one commercial mutant cultivar resulted. The mutant cultivar NC4X (or N.C.4X), derived from cv. N.C.4 after X-ray treatment of dry seeds in 1949, was released in 1959 in North Carolina, USA. It has been entered in the FAO/IAEA lists of induced mutants (Sigurbjörnsson & Micke, 1974; Micke *et al.*, 1985) as cv. N.C.4-X, and is characterized by a tougher hull (which resists damage during harvest and transport), high yield and good quality. Details about the commercial significance of this mutant cultivar are not known.

During a FAO/IAEA Technical Meeting called 'The Use of Induced Mutations in Plant Breeding' in 1964 in Rome, several presentations concentrated on induced mutations for quantitative traits. In addition to the aforementioned lecture by Gaul (1965a), reference can be made for instance to contributions by Brock (1965), Ehrenberg *et al.*, (1965), Frey (1965), Gregory (1965),

Lawrence (1965) and Scossiroli (1965). Some other publications about this subject appeared at about the same time. Pate & Duncan (1963), for instance, irradiated pollen of cotton (*Gossypium* sp.) with gamma rays (400–3200 rad) and studied their material until M_4. These authors mention that 'changes in quantitative characters of economic importance may have been induced', but further results on this subject could not be traced in literature. Joshi & Frey (1969) reported for oats (*Avena sativa*) that alternating mutagenic treatments and selection for several cycles had proved to be an effective method for obtaining genetic gain for 100-seed weight. Gardner (1969; see also Gardner, 1961), on the other hand, after 10 cycles of mass selection in a population of maize that had been irradiated repeatedly, was unable to select higher yielding genotypes from the irradiated population than from the control populations.

In conclusion, and despite several statements in the aforementioned and other publications about 'promising results' or, more specifically 'promising mutant lines', no mutant cultivars for quantitative traits have resulted from this work. The only exception known in this respect is the aforementioned mutant cultivar N.C.4-X in groundnut (*Arachis hypogaea*).

3.4.4. Characteristics of induced variation for quantitative traits

It is commonly observed that the estimates of phenotypic variation for continuous traits are larger in irradiated populations than in the control populations. In other words, it may be anticipated that, after mutagenic treatment, selection for both 'plus' and 'minus' mutants outside the limits of the original frequency distribution of the untreated population should be feasible. Most studies also show that the mean values for quantitative traits in a (mutagenically treated) M_2 population shift away from the direction of previous selection in the control populations. If yield is taken as an example, this would imply that the mean value for yield, at least in the M_2, will be lower than in the untreated control. The increased variance and the shift of the mean are illustrated in Fig. 3.2. The observed differences between treated and untreated populations diminish in subsequent generations. The common explanation for this observation is that the majority of the induced mutations in the treated populations will be detrimental, but that in following generations selection takes place against such undesired mutations.

Brock (1967; 1971; 1976) confirmed the aforementioned findings for flowering time and seed weight in comparative studies in which seeds of *Arabidopsis thaliana* were treated with chemical mutagens and ionizing

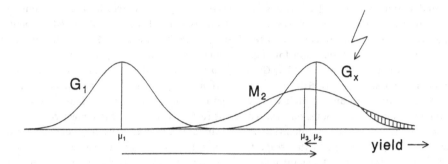

Figure 3.2. Mutagenic treatments lead to increased variation for quantitatively inherited traits and to a shift of the mean (μ) towards the direction of previous selection G_1 = yield (e.g. grain yield) for an unselected starting population; G_x = yield for an improved population (e.g. a cultivar) derived from G_1 (after x cycles of selection) and submitted to mutagenic treatment (indicated with ⚡); M2 = second mutated generation; $μ_1$, $μ_2$, $μ_3$, = mean of G_1, G_2, M_2, respectively.

radiation. The response for flowering time was skewed to lateness. Treatments with EMS and gamma rays showed comparable amounts of genetic variation but EMS treatments showed larger responses to selection and realized heritabilities than treatments with gamma rays. Brock (1970) also described effects of thermal neutrons for the inductions of mutations for quantitative traits, but did not find concluding evidence at that time about which kind of treatment should be preferred for this category of traits.

Brock (loc.cit.) states that the value of induced mutations for quantitative traits 'is largely determined by the importance of linked groups of genes and the degree to which natural selection has built-up linked gene complexes of adaptive significance in naturally occuring genotypes'. Based mainly on Brock's experiments, Brock & Micke (1979) further point out that the response of quantitative traits in self-fertilizing species to random mutation depend on 1) the number of genes involved, 2) the relative proportion of genes with positive (plus) or negative (minus) effects, 3) the extent to which genes of the parental genome operate as a balanced set, and 4) correlations as a result of linkage or pleiotropic effects. The effects of an altered genetic balance and of genetic correlations may disturb the expected pattern of an equally increased variation for all quantitatively inherited traits in the M_2 generation and lead to the aforementioned shift in mean away from the direction of previous selection, which normally implies a change towards a less desirable (in case of yield; lower) mean.

Earlier, Ehrenberg et al. (1965) reported for barley (Hordeum vulgare) that, following various mutagenic treatments, the mean values for tillering, grain weight (per row or per plant) and surviving plants in the progenies (up to M_4) decreased by up to 1–2% after treatments with X-rays and neutrons, and 5–10% after various chem-

ical treatments. No increased variance was observed for tillering per row. Increased variability and declined mean values in M_2 and M_3 were observed as well for the traits plant height and number of kernels in wheat (Triticum aestivum ssp. vulgare) by Borojevic (1969). In later generations, the mean value increased again and variability decreased. In M_5 mean values in treated and untreated populations had reached the same level again, but the treated polulation still showed a higher genetic variability, both in the 'plus' and 'minus' direction.

It may be concluded that for quantitative traits in general effective selection responses can be obtained in either direction (e.g. plant material with – genetically – higher and lower yield) because of the increased genetic variance. However, selection must be performed also for all other economically important traits, as they may carry genetic changes as well which, in majority, will be deleterious. If no selection is applied after mutagenic treatment, the average result will be a regression of the mean and a reduction in fitness for all adaptive and previously selected traits. As a consequence of the continued, purposeful selection against such detrimental effects (e.g., by discarding all aberrant-looking plants in subsequent generations), the difference between the means of treated and untreated populations, as was mentioned before, in later generations should become smaller and may even completely disappear. This, indeed was observed for instance by Gardner (1969) for maize (Zea mays), an open-pollinated crop in which, after ten cycles of mass selection for high grain yield, the treated population and the control showed equal population means for grain yield. However, according to Gardner, it appeared that the additive genetic variance for grain yield had decreased in the control series and was substantially higher in the mutant series. Based on these experiments with recurrent treatments with thermal neutrons, which started already in 1955 (see Gardner, 1961), Gardner predicted considerable future genetic improvement and higher selection limits for the mutant population, but according to a later, difficult traceable reference (Gardner, 1972), this goal was not achieved. Later reports about these experiments by Gardner are not known to us.

The observed lower means for quantitative traits (like yield per row or per plot) as a result of the mutagenic treatment and the lack of success in selection for such traits, often discouraged breeders, and in particular those who had no previous experience in mutation breeding, to continue their mutation projects. However, the increased genetic variance should enable further selection for outstanding (high yielding) plants, provided that selection is based on individual plant progenies.

3.4.5. Frequencies of mutations for quantitative traits

It may be expected a priori, that the probability of obtaining a mutation-induced change in a trait will be higher when more genes are involved (Gaul, 1961b). Evidence for the occurrence of this phenomenon was obtained at an early stage from experiments with Drosophila and from the aforementioned mutation experiments of Gregory. Mutations ('micromutations') for yield, according to Gaul (1963), occur at least 50 times as often as macromutations, but the number of genes involved in the 'trait yield' is not known, but certainly high.

In addition it is sometimes assumed – but never confirmed beyond doubt – that 'small mutational steps' do occur more often than large steps (which may be only the consequence from the above statement about the number of genes).

Most mutants show a reduced vitality, probably often as a result of the pleiotropic effects of mutated genes or because of additional mutations in the genetic background. It should be kept in mind however that, as was mentioned before, 'normal' mutations also may produce a shift of the mean for a specific trait towards a lower (undesired or negative) value. It has been speculated (see for instance Gaul, 1958; 1965a; 1967, and Gregory, 1965; 1966) that in case of drastic phenotypic changes (a 'large mutation step') caused for a given trait, a higher level of injury and a lower vitality is found than would be the case when smaller mutational events are involved. It should be commented, however, that a 'micromutation' is not identical to a smaller mutagenic event.

Gregory (1965; 1966) confirmed that the number of mutant plants decreases with increasing magnitude of the change and put forward the hypothesis, that the probability of obtaining improvement increases with decreasing magnitude of change to the limit of P = 0.5. Gaul (1967) also speculated that the frequency of mutants with a (slightly) improved vitality will be higher in case of smaller mutation steps. In other words: according to this view the magnitude of change is said to be negatively correlated with the probability for fitness. Gaul (loc.cit.) mentions that he started his experiments in 1956 with this concept and adds that Gregory followed and further extended this hypothesis. This extension refers to the concept of Gregory (1965; 1966) that the probability of genetic improvement of a certain effect by mutation should decline about exponentially with the magnitude of the effect. This resulted in the opinion that micromutations should be more important in plant breeding than macromutations. This view, so far, has not been confirmed by practical results.

3.4.6. Selection methods for quantitative traits

Selection procedures for mutations in qualitative and quantitative traits of course differ. Whereas selection for recessive, monogenically controlled, qualitative traits may start in M_2, selection for mutants in quantitative traits must be postponed at least till M_3 (examination of progenies of randomly harvested M_2 plants) and continued in subsequent generations. Because of the observation that mutations usually are associated with altered genetic backgrounds (i.e., with the occurrence of mutations in non-target genes) the newly obtained mutants either should be 'backcrossed' to the control, be submitted to further hybridization programmes, or the experiment should be large enough to induce so many mutants for that specific trait, that further selection for all important traits can be carried out (Brock, 1965; Brock & Micke, 1979).

Micke (personal communication) has expressed the suspicion that much of the so-called 'quantitative variation' may have resulted from outcrossing with partially sterile M_1 plants, leading to a heterozygous M_2 and segregation in M_3, M_4, etc. This point, in any case, underlines the importance of high quality experimental work and the need to draw conclusions with utmost care.

Selection procedures will be also discussed in Chapter 6. For additional reading about the theoretical background and experimental work on quantitative mutations, and in addition to already mentioned publications and their lists of references, see for instance Brock (1976), Dellaert (1979; 1982), Gaul (1964a, b) and Yonezawa (1975)

3.4.7. Concluding remarks

In retrospect, because of the importance of various quantitative traits in crop plants, the relatively high frequency of occurrence of mutations for quantitatively inherited traits and the assumed higher degree of fitness, it appears that mutation breeding should be succesful also in the improvement of such traits (Frey, 1965). However, until now most induced mutant cultivars represent changes for qualitatively inherited traits like flower colour, plant morphology, monogenically determined resistances, etc. (see for instance Konzak et al., 1984). This, partly, may reflect the much higher costs of selecting among large populations in $M_3 - M_5$ than in M_2 and M_3 for qualitative traits, but also the low efficiency and other problems encountered when selecting for quantitative traits. The low heritability of quantitative traits and a possible lack of genetic variation for polygenic traits, may further explain why in most cases no significant genetic improvement could be observed.

Accordingly, MacKey (1984), during a FAO/IAEA consultants meeting in 1982 in Vienna, pointed out that mutations for quantitative traits are more difficult to detect, evaluate and handle than those for qualitative traits. Because of their small phenotypic expression 'some kind of accumulation and rearrangement' is required in order to make optimal use of the relatively large number of 'micromutations'. MacKey (loc.cit.) adds that this is a relatively easy affair in cross-breeding species but, on the opposite, to perform an adequate programme for selection and accumulation of 'micromutations' in vegetatively propagated crops and strictly self-fertilizing crops, is very difficult. The crucial question, according to MacKey, is whether the percentage of deleterious changes by 'micromutations' is just as high as for 'macromutations'. As, in his opinion, there seems to be no reason why these percentages should differ (although the phenotypic expression for 'micromutations' will be smaller and less obvious), the induction of 'micromutations' should not only be useless but even more harmful as it will be much more difficult to get rid of the many negative ones, even by extensive cross breeding and recurrent selection. These considerations made MacKey strongly question whether induced 'micromutations' indeed may add significant useful variation to the existing natural variation.

The critical comments by MacKey certainly have to be taken very seriously indeed. It may be useful to recall that it is the 'genetic background' which largely determines whether a favourable (induced) 'major' mutation will become valuable from an agronomic point of view. It must be considered in this context that alterations in the genetic background, predominantly, are a kind of 'micromutations' and that many plant breeders have negative experiences with 'altered genetic backgrounds'.

In our opinion it is most doubtful whether, in the years to come, once more, extensive, time- and labour-consuming experiments, similar to those set up in the past by Gaul, Gregory and some others, will be initiated again. This, however, seems necessary to find additional information for a number of points that still are not fully clear. These answers which, probably, also may be partly obtained from simulation studies and from applying new molecular techniques, are needed to find definite proof whether 'micromutations' are as important for practical plant breeding as is still believed sometimes.

3.5. 'Phenotypic classification' of mutations

3.5.1. Introductory remarks

Detailed studies on mutations as observed by their phenotypic expression, so far, have produced a wealth of information useful for different fields of plant science and genetics, whereas mutants of practical importance are obtained as well in this way. By concentrating on a few plant species and by tuning the efforts of different scientific disciplines in well-organized cooperative projects, quick progress can be made, both with respect to understanding the fundamentals of mutation breeding as well as from a practical point of view, for instance when comparing effects of different mutagenic treatments for specific plant species or particular traits, etc.

The small weed Arabidopsis thaliana, for a score of reasons (see for instance the early review of the genetics and biology of this species by Rédei, 1970, and further Chapter 1, Box 1.4), is often used as an excellent model plant for many purposes, and many extensive and very detailed investigations have been performed already or are still under way. Genetic studies on chlorophyll mutations in Arabidopsis, for instance, were performed by Röbbelen in Germany since 1956 (see for instance Röbbelen, 1972). In addition to Arabidopsis thaliana, several crop plants are used as well as 'model plants' for mutation studies, like pea (Pisum sativum), tomato (Lycopersicon esculentum), snapdragon (Antirrhinum majus), maize (Zea mays), barley (Hordeum vulgare) and, nowadays, also rice (Oryza sativa) in Asia.

In 1928 the Swedish barley mutation research programme was initiated by H. Nilsson-Ehle and Å. Gustafsson. This programme will be discussed in detail in Section 3.5.2. Starting from 1940, gradually, the range of crop plants extended, and at various Swedish institutes and private breeding companies mutation studies were performed on, for instance, wheat (Triticum aestivum), oats (Avena sativa), lupines (Lupinus sp.), soybean

(*Glycine max*), flax (*Linum usitatissimum*), white mustard (*Sinapis alba*), pea (*Pisum sativum*) and fruit trees. The Wallenberg Foundation included mutation breeding in its research-supporting programme in 1948. For a review of the Swedish mutation breeding programme on the aforementioned crops, some early contributions about fundamental aspects of mutations and mutation breeding, and on the history of Swedish mutation breeding until 1954, reference is made to a series of articles, dedicated to H. Nilsson-Ehle (who died in 1949), published in 1954 in *Acta Agriculturae Scandinavica*, vol.4.

Gradually, mutation breeding programmes on various crop plants were started in many other countries as well, for instance on barley in Germany (Freisleben and Lein, Gaul, Hoffmann, Scholz), in the USA (Nilan) and in the former Czechoslovakia (Bouma); on pea in Sweden (Blixt and Lamprecht), Germany (Gottschalk) and Poland (Jaranowski); on wheat in the USA (Konzak), the former USSR (Skarnikov), Argentina (Favret) and on durum wheat in Italy (Scarascia Mugnozza and associates); on rice by different groups in China and Japan, in France, in California, USA (Rutger); on soybean in the former GDR, Poland, China and Japan; on pea, wheat, sunflower, cotton and various other crops in the former USSR and on (vegetatively propagated) ornamentals in the Netherlands (Broertjes). This list, of course, is far from complete and valuable contributions of many other research workers in various countries are mentioned throughout this book.

In 1974, over 2000 mutations for pea were already known (Blixt, 1975). About 400 mutations, for instance, refer to developmental and morphological traits, and were recently described by Murfet & Reid (1993). Studies of the many spontaneous and induced mutants in pea have resulted in the identification and symbolization of more than 500 'classical' genes. From the mutants obtained only about 30% had a sufficiently high fertility to be directly propagated by seed. According to Gottschalk (1987) only about 1% of all mutants selected are found to be of practical value for further breeding work. Similar information is also available for tomato and maize and many more interesting facts derived from mutation studies could be given for these and other 'model species'.

In the next section we will focus on mutation research in barley. Despite a number of evident disadvantages from an experimental point of view when compared with *Arabidopsis*, the diploid and self-fertilizing crop barley (*Hordeum vulgare*) with its few but large chromosomes ($2n = 2x = 14$), has proved to be a good model plant for mutation experiments. Barley is easy to handle in field and greenhouse, and produces sufficient seeds

from single plants. More or less concerted efforts of barley mutation research workers all over the world have provided us with a wealth of, often generally applicable, information about many aspects of mutations and mutation breeding. As the extensive Swedish barley mutation work (in which researchers from some institutes in other Scandinavian countries, like Denmark, participated as well) in fact has been one of the earliest and particularly well-organized and excellently documented mutation programmes, and has continued to be the most important focus point of barley mutation research for almost 60 years, we will concentrate here on this programme. This limitation, however, by no means implies that results of barley mutation work in other (aforementioned) countries like Germany, the former Czechoslovakia and the USA should not be worthwhile of being mentioned. It is also needless to say that valuable examples to illustrate phenotypic classifications could have been derived from other programmes as well as from other crops. More 'case studies' will be presented in particular in Chapters 6 and 7.

3.5.2. The Swedish barley mutation research programme

As was mentioned before, the Swedish (or Scandinavian) barley mutation research was initiated in 1928 by H. Nilsson-Ehle and Å. Gustafsson. Initial studies, in which X-rays and UV were applied, mainly dealt with fundamental aspects like types of chromosome aberrations, effects upon different cell stages, effects of treatment conditions like water content in seeds, radiation induced sterility in M_1 and lethality. An early account of the general biological aspects of this mutation work was published by Gustafsson (1941).

Experiments with X-irradiation of cv. Gull produced different kinds of chlorophyll mutations as the most common category of easily observable genotypical changes in seedlings. Attention to this category of mutations, which in general were sublethal, was soon followed by interest in other groups of more vital mutations like, starting in the mid-1930s, the so-called erectoides types. This mutant group, on which much research has been performed and which is also of much interest for practical plant breeding, is characterized by lodging resistance and compact spikes. Mutation frequencies and mutation spectra for various other groups of traits were studied and compared. Other methods of treatment, the use of different radiation sources and, later, of chemical mutagens, were gradually included in the research programme and mutagenic effects caused by the various treatments, as well as the behaviour of different categories of mutations, were carefully analyzed.

It was found for instance (MacKey, 1951; 1956; Lundqvist, 1992) that barley seeds were 20–30 times more sensitive to neutron irradiation than to X-rays when equal doses are given and that germinating seeds were only 2–3 times more sensitive to neutrons than dormant seeds. Neutrons are 10 times as effective as X-rays (of equivalent energy dissipation), in producing chromosome disturbances and about 50–100 times as effective with respect to inducing M_1 sterility and increasing mutation frequency in M_2.

Gustafsson's group was the first to approach the induction of mutations by chemicals in a systematic way, starting with mustard gas (see for instance Gustafsson & MacKey, 1948, and Gustafsson, 1986). Lundqvist (1992) mentioned that, after three decades of research, some chemical mutagens were found to produce up to 80% mutated plants for a given trait, which is 20 times higher than can be observed after most treatments with ionizing radiation. Some chemicals, like sodium azide, produced primarily gene mutations, whereas neutrons caused many more aberrations at the chromosomal level. To give one example from the USA about the specific action of sodium azide, reference can be made to Kleinhofs et al. (1978). These authors reported that treatment of germinated seeds of barley (Hordeum vulgare) and pea (Pisum sativum) for two hours with 10^{-3} NaN_3 (0.1 M phosphate buffer; pH 3) resulted in about 13 nitrate reductase-deficient mutants per 10 000 seedlings in barley and 3.5 per 10 000 seedlings in pea. Chlorate-tolerant mutants occurred at a frequency of 6 per 10 000 seedlings in barley. Waxy endosperm mutants (presumed single-locus mutants) occurred at a frequency of 2.7 per 10 000 seeds in barley.

Other groups of mutant types studied in Scandinavia, for instance, include changes with respect to spikes and spikelets, culm length and culm type, leaf blades, awns, seed size and seed shape, growth type, anthocyanin and resistance to barley powdery mildew. Lundqvist (1992) points out that in the 1930s a group of 'morphological mutations', as a rule clearly showing qualitative inheritance, had already been distinguished from the group of 'physiological mutations' in which, for a number of traits like culm-length, earliness, tillering capacity, seed size, etc., quantitative variation was observed. Lundqvist adds that no sharp distinction could be made between both groups 'as all mutations observed were pleiotropic'.

Studies on mutation frequencies were performed according to the so-called **spike progeny method**. This method was introduced by Gustafsson (1940) and became a standard method for measuring mutagenic effects. According to this method the number of segregating spike progenies is divided by the total number of spike progenies, which results in a figure that can be used to easily compare the effects of different mutagenic treatments, etc.

In 1953 Gustafsson, who had already brought together a number of colleagues in order to better coordinate the scattered efforts on mutation breeding, became the leader of the so-called 'Group for Theoretical and Applied Mutation Research' established by initiative of the Swedish Government. This did lead to a considerable extension of the (non-commercial) research activities, the peak of which was reached in the 1950s and 1960s (Lundqvist, 1986). Throughout the years about 9000–10 000 different barley mutants have been studied from a genetic and agronomic point of view. These mutants, which belong to some 10 main categories with 95 different types of mutants, were in particular studied and maintained at the experimental fields of the Swedish Seed Association in Svalöv. Lundqvist (loc.cit.), who worked at the Svalöf Plant Breeding Company (since 1993 Svalöf Weibull AB), housed in Svalöv, has presented a table containing the number of mutants and different loci for 11 groups of genetically investigated barley mutants (Table 3.2). Reference is made in this respect also to a publication by Gustafsson et al. (1969), in which a system of symbols for barley mutants is proposed. Most mutants are now stored in the Nordic Gene Bank, Sweden.

In a number of small sections we will discuss some major results of the Scandinavian barley mutation work. Not all mutant groups studied by the Scandinavian group will be discussed here. Most information traces back to reports of U. Lundqvist, in particular to her doctor's thesis (Lundqvist, 1992) and to personal communication. In addition, reference could be made to the many contributions on these subjects by the Scandinavian group in proceedings of many international conferences, in the earlier issues of the journal Hereditas and in the Barley Genetics Newsletter. Two topics from the Scandinavian barley mutation programme – the research on protein improvement by mutational methods, and a short description of the barley mutant cultivars obtained from the programme – will be discussed in Chapter 6.

Chlorophyll mutations

Chlorophyll formation in plants is the ultimate result of a long chain of steps in a biochemical pathway in which many loci are involved. Blakeslee (1936) already reported that for the normal development of chlorophyll in maize (Zea mays) the interaction of at least 65 'factors' is necessary.

Von Wettstein et al. (1971) reported that for barley at

Table 3.2. Survey of 11 genetically investigated mutant groups (according to Lundqvist, 1992, p. 38)

Mutant group	Number of mutants	Number of loci
Praematurum (early heading)	172	9
Erectoides	205	26
Breviaristatum (short awns)	140	17
Eceriferum (waxless)	1 580	79
Intermedium	103	11
Hexastichon (six-row)	41	1
Macrolepis (lemmalike glumes)	40	1
Bracteatum (third outer glume)	28	4
Calcaroides	18	4
Mildew resistance	77	several
Exrubrum (anthocyanin-less)	approx.500	approx.27

that time 198 recessive lethal mutants for chlorophyll had been assigned to 86 different loci by diallelic crosses and concluded that the nuclear genes hold a very tight control over plastid biogenesis. They mentioned for the same crop that for the category of albina mutants several hundred genes could inhibit chlorophyll formation in chloroplasts. Mutations in any of these loci may cause a deficiency in pigment and result in albina plants among which phenotypically different types (e.g. for chlorophyll distribution in different plant parts) can be distinguished. Von Wettstein *et al.*(1974) presented evidence for the control of chlorophyll(ide) synthesis in higher plants through the products of regulatory genes in the nucleus.

Chlorophyll mutations are useful markers in genetic studies and in physiological and biochemical research. Since the pioneering work of Stadler (1928a,b; 1930) chlorophyll 'aberrations' have been used in many cases to estimate spontaneous and artificially induced mutation frequencies in both germinal and somatic tissues. The spontaneous mutation rate is very low, e.g. 0.2–0.7 mutations towards recessive per 1000 M_1 spike progenies, as compared to 150 chlorophyll mutations after administering a modest dose of X-rays. Figures presented in literature for chlorophyll-deficient lethals range from 3.1×10^{-4} to 1.4×10^{-3} per diploid genome per generation in several annual crops. More details about mutation rates of deleterious alleles and further references can be found in Johnston & Schoen (1995).

Chlorophyll is nuclearly as well as cytoplasmically inherited. Many of the nuclear mutants show monogenic inheritance and were mutated towards recessive. The special features of extranuclear inheritance have been discussed earlier in Section 3.2.4. Mutations in nuclear genes may alter the genetic stability of extranuclear DNA in chloroplasts and, probably, also in mitochondria. Tilney-Bassett (1975; see also Kirk & Tilney-Bassett, 1978, p.387), basically, distinguished two

types of nuclear genes inducing plastid mutations which either produce a wide variety of mutant phenotypes or a single, specific mutant phenotype. Prina (1992), in a contribution about the effect of mutator genes (see earlier in this chapter), reported that mutations referring to the 'wide-spectrum group' occur only exceptionally in barley, as is the case for most other monocotyledons.

Gustafsson (1940) distinguished for barley nine different classes of chlorophyll mutations with a number of subdivisions.

1. Albina (no carotinoids or chlorophyll formed).
2. Xantha ('carotinoids prevail or chlorophylls are not even produced').
3. Alboviridis (with different colours at the leaf base and leaf tip), subdivided according to colour in alboxantha, xanthalba, viridoalbina and alboviridis (sensu stricto).
4. Viridis (a heterogeneous group, characterized by a uniform yellowish-green or light-green colour already at the seedling stage), with a subdivision in virescens (light-green gradually changing to dark green and often viable), chlorina (yellowish green; often viable), lutescens (leaves wither and turn yellowish; lethal) and albescens (like lutescens but more extreme and lethal).
5. Tigrina (transverse destruction of pigments; transverse stripes brown or yellow).
6. Striata (longitudinal white or yellow stripes).
7. Maculata (showing spots of chlorophyll and/or carotin destruction distributed over the leaf).
8. 'Undefined mutations'.
9. 'Plasm mutations'.

This classification system has proved to be also useful to describe chlorophyll mutations in other cereals.

The most important results obtained by Gustafsson after irradiation of cv.Gull with 50 or 100 Gy (5000 or 10 000 rad) of X-rays showed that albina, xantha, alboviridis (belonging to group 4), viridis and tigrina occur fairly commonly. The albina type arises in the M_2 generation independently of the degree of sterility of M_1 plants, whereas xantha, alboviridis and viridis are not randomly distributed over the different sterility classes. Xantha accumulates in the offspring of M_1 plants with a slightly reduced (70–90%) fertility and not above 90% fertility; viridis occurs in particular in highly sterile M_1 plants (0–30 and 30–70% fertility). Gustafsson concluded that viridis and xantha, apparently, are conditioned by chromosomal aberrations. In such cases increased dosages of irradiation result in proportionally increased mutation rates, correlated to a highly increased reduction in the fertility of M_1 plants. Mutations of the <u>albina</u> type behave like intragenic changes.

Gustafsson also found that, for instance, the <u>albina</u> type occurred in particular after treatments with relatively low dosages of irradiation and in water soaked seeds, whereas <u>viridis</u> and <u>xantha</u> occur mostly at high dosages, <u>tigrina</u> only after irradiation of dry seeds and in low frequencies only. Based on these results, Gustafsson (1940; 1941; 1942) assumed that the mutational process could be more or less 'directed'. As with the advent of modern molecular techniques the term 'directional mutagenesis' has obtained a rather different content, it would be better to interpret Gustafsson's results as indications for a certain degree of 'non random' or 'differential' mutagenesis.

Von Wettstein, Gustafsson & Ehrenberg (1959) showed that remarkable differences were observed when the relative frequencies of spontaneous and induced <u>albina</u>, <u>viridis</u> and other types of chlorophyll mutations were compared. These results, to which also was referred at the 1969 FAO/IAEA Symposium in Pullman, USA by Gustafsson (1969a) are shown in Fig. 3.3. A most interesting conclusion of this early work, in which for all groups together more than 2500 independent mutations were involved, is that – opposite to the prevailing opinion – it appears possible, at least to a certain extent, to influence the relative frequency of different mutation types by selecting a specific mutagen. However, as hundreds of loci are involved in chlorophyll formation (von Wettstein et al., 1971; 1974), it may be clear that such observations do not refer to real 'locus-specific' mutations.

In the aforementioned publications detailed information about the number of loci for the different groups of chlorophyll mutations and other details of chlorophyll synthesis was presented as well. The monogenic recessive inheritance of most chlorophyll mutations can be confirmed by selfing M_2 heterozygotes and by determining the frequency of homozygous recessives in the M_3. A proper-sized experiment, for instance, must at least involve 1000 M_1 spikes with 20–30 M_2 seeds collected per M_1 spike.

The prevailing lethal effect of chlorophyll mutations, as well as the fact that plastids are extranuclear organelles, were complicating factors with respect to fundamental research on mutations and, as a consequence, at a number of places the attention gradually

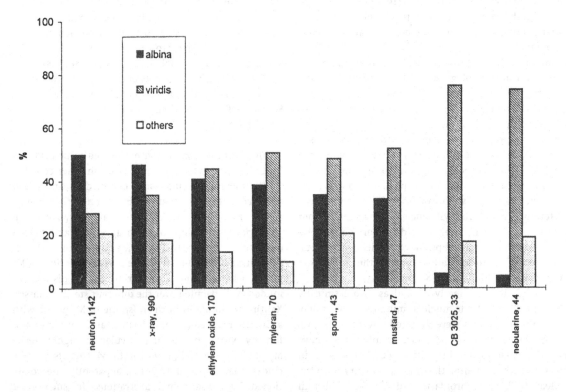

Figure 3.3. Distribution of three categories of chlorophyll mutations (albina, viridis, others) in barley, either obtained spontaneously (spont.) or after various mutagenic treatments (after von Wettstein et al., 1959, and Gustafsson, 1969a). (N.B. The figures following names of treatments refer to numbers of mutants studied.)

shifted to other groups of mutations. For an early review see Gustafsson (1941). Nevertheless, studies concerning chlorophyll mutants and nuclear genes affecting chloroplasts continue to be published. In *Barley Genetics Newsletter*, for instance, Simpson & von Wettstein (1991), report that among 84 non-lethal mutants of the chlorina type, 21 were localized by diallelic crosses to five loci; among 122 lethal mutants for albina, 37 were found to be located at 29 loci; eleven nuclear striata mutants at nine loci; 76 lethal xantha mutants at 20 loci; 48 lethal mutants of viridis at 36 loci and 27 lethal mutants of tigrina to 15 loci.

Erectoides (dense-spike) mutations and laxatum (lax-spike) mutations

Mutants of the erectoides (*ert*-)type are characterized by their spikes which are, in comparison to the control, dense (or compact) and, therefore, erect. In addition, erectoides plants often, but not necessarily, have increased tillering, shorter and stiffer culms (resulting in lodging resistance) and, sometimes, an increased wax coating (Lundqvist, 1992). With respect to culm length of the mutants it was found, that the most upper internode is generally longer and the basal internodes shorter than in the mother cultivar.

Erectoides mutants in barley were discovered as a very common group of viable morphological mutations at about the same time as the chlorophyll group. Their frequency of occurrence, after irradiation with X-rays, is about one erectoides mutation per 25 chlorophyll mutations. Detailed information on morphological and cytogenetical aspects has been presented in an extensive publication by Persson & Hagberg (1969). Of about 700 *ert*-mutants that had been collected at that time, it was determined for 182 mutants that they could be allotted to 26 different loci. Non-allelic mutants are indicated as *ert-a*, *ert-b*, etc.; allelic mutants of the *ert-a* group as *ert-a¹*, etc. In 1994, according to Lundqvist (personal communication) in total 1264 mutants were known of which at that time 205 mutants had been submitted to tests for allelism. As after the intensive studies by Persson and Hagberg not many additional tests for allelism have been performed. These 205 mutants, which were located on the 26 loci to which was referred before, mostly occurred among the aforementioned 700 *ert*-mutants.

A large majority of artificially induced *ert*-mutations, like most mutations, are recessive. In some cases simple phenotypic inspection elucidates to which locus-group certain *ert*-mutants belong, because of the specific phenotypic expression of mutations at certain loci. The heritability of spike internode length in nature appears to be very high which, according to Persson & Hagberg

(loc.cit., p. 134), implies that it is possible to handle many of the *ert*-alleles as 'qualitative' genes. Earlier in the publication (p. 116), however, these authors stated that the genetics of variation in this trait 'is, of course, quantitative in nature'. It is probably meant that mutations in (major?) genes for the erectoides trait – which itself is a quantitative trait – can be detected by qualitative methods.

It was known already from the 1920s that the internode length of the spike is determined by multiple factors, but the number of genes involved was not known at that time. Localization studies for the different *ert*-loci showed that these loci were distributed all over various chromosomes. It was concluded that most *ert*-mutants, once they have been carefully localized on the chromosomes, will be useful as genetic markers because they can be easily recognized also in segregating populations.

Mutations in three different loci, known as *ert-a*, *ert-c* and *ert-d*, have been indicated as 'mutagen specific', as more than 80%, 50% and 70% of the mutated alleles had been induced by treatments with X-rays. For the other 23 *ert*-loci in general only few mutations are known and (probably for that reason?), no effects of specificity for certain mutagenic agents have been found. On main lines no differences in spectra between spontaneous and induced *ert*-mutations have been found.

Many erectoides mutants are of direct interest for practical plant breeding purposes. One of the best-known mutants, with a mutant gene called *ert-k³²*, has been released as cv. Pallas and has become an outstanding barley cultivar. More details are given in Chapter 6.

The opposite of dense-spike mutants are so-called lax-spike mutants, called laxatum (or *lax-*). Persson & Hagberg (1969), in a special paragraph on this topic, reported that 201 *lax*-mutants had been collected, but that they had not been studied in detail yet. This group of mutants was studied very intensively in the 1980s by Larsson (1985a,b,c). Lundqvist (personal communication) mentioned that one specific set of mutant alleles, belonging to a single locus (*lax-a*), produces five stamens instead of three.

Short-awn (breviaristatum) mutants

The final grain size of barley is considerably influenced by the awns in the spike. Natural variation for the short-awned type appears to be rather limited and induced mutants in this respect, with often a good viability and fertility, can be obtained relatively easily. The induced mutants cover the whole range from very short to medium awn length relative to the normal length of the parent types. From the 470 different mutants obtained until 1992 (Lundqvist, personal communication), 140 have been localized to 17 different loci. Most of these loci can be easily distinguished by their specific phenotypic expression. Kucera, Lundqvist & Gustafsson (1975) have published extensively on breviaristatum mutants, artificially induced by several types of ionizing radiation and chemical mutagens in various cultivars of barley. The frequency of mutants for chlorophyll was about 25 times higher than breviaristatum; thus breviaristatum mutants occur about as often as erectoides and ecer-iferum mutants (this last category will be discussed in a later section). It was reported that chemical mutagens (in particular ethylene imine and sulphonates) were in general two to three times more efficient than radiation in the induction of breviaristatum mutants. The number of mutants that until 1992 was localized to different loci, may have been too small to conclude whether chemically induced mutants and radiation induced mutants occurred in ratios comparable to those for erectoides-, eceriferum- or other groups of mutants.

Intermedium and six-row (hexastichon) mutants
These mutants refer to changes with respect to the development of lateral spikelets in the spikes and a single mutational step may change two-row barley into six-row barley or into different mutant types which are inter-mediate (int-) between both extremes.

All known 41 cases of single-step mutations towards six-row barley (hexastichon or hex-) affected one single locus (hex-v) on chromosome 2 (Gustafsson, 1979; Gustafsson & Lundqvist, 1980). Lundqvist (1992) mentions that all the truly wild forms of the genus *Hordeum* are two-rowed, with only the central spikelet of each node producing a full-grown grain, the two lateral spikelets being sterile. Under domestication barley types have arisen with so-called six-rowed spikes, that is with all three spikelets at a single node producing grains. Studies on mutations for this trait are of particular interest from the point of view of plant evolution. Homozygous int-mutations show lateral spikelets with an increased size and fertility. Reference can be made to a whole series of publications on this subject, for example, Gustafsson & Lundqvist (1980), Lundqvist (1986), Lundqvist & Lundqvist (1987; 1988b,c; 1989), Lundqvist, Akebe & Lundqvist (1989) and Lundqvist & Lundqvist (1990). This work has been summarized in Lundqvist (1992).

A total of 126 int-mutants has been obtained, of which 103 were localized to eleven int-loci, whereas 69 int-mutants were studied more in detail in crossing experiments. Three loci were represented by more than

thirteen mutants and only one mutant could be assigned to each of five loci (Lundqvist & Lundqvist, 1988b). For eight loci all int-types that have been investigated were recessive to the two-row type. Thirteen mutants, which all referred to locus Int-d, showed a more or less clear tendency to be dominant to the two-row normal state.

Gustafsson & Lundqvist (1980) already concluded that hexastichon types either may arise by direct mutation in the two-row locus (V→v) of so-called distichon-types (*H.vulgare* convar. *distichon*), or as double recessives in the second generation of hybrid combinations of special int-types. Lundqvist (1992) adds that in one specific case, involving the six-row locus hex-v, semidominant inheritance was observed.

Further research has shown that crossing of so-called double int-mutants (int/int) with the hex-gene, produced very interesting six-row types with big spikes and thick culms. In addition, of 88 mutations induced for this trait in three different well-known two-row barley cultivars (cvs Bonus, Foma and Kristina), 76% had arisen after treatments with chemical mutagens, 12.5% after using X-rays and 9% after using densely ionizing (neutron) radiation. So far, no definite preference of specific loci to certain types of mutagen treatment has been observed (Lundqvist & Lundqvist, 1988b).

Wax (eceriferum) mutants
The outer epidermal (or cuticular) surfaces of most plants are covered by epicuticular waxes. In barley the synthesis and excretion of such waxes is controlled by eceriferum (cer-)genes for which many hundreds of induced mutants have been obtained, against a very few (so far only six) spontaneous ones. This mutant type, which has been studied for many years in large research programmes, most probably is the best investigated trait complex in higher plants. The cer-mutants may refer to wax composition and structure, and to the absence or reduction of the wax coating on three different plant organs: spike, leaf sheath and leaf blade. Mutations may affect a single organ or different combinations of organs at the same time.

Based on such distinctions five phenotypic classes are discerned. In 1988 in total of 1580 cer-mutants were localized by diallel crossings to 79 different loci, of which 78 showed recessive mutant alleles whereas dominance was observed for only one locus (Cer-yy) (Lundqvist & Lundqvist, 1988a). Often the expression 'complementation group' is used to indicate a series of mutations that, when tested pair-wise, are unable to complement and, hence, refer to the alleles at the same locus. In 1994 the number of complementation groups for cer-mutants had already increased to at least 85 and at regular intervals more loci and further details are reported, for instance in issues of the *Barley Genetics Newsletter*. The high number of mutants obtained and loci involved, make a detailed analysis of the results very attractive. Because of the many loci, gene localization of a newly found mutant is performed by making use of a simplified procedure, in which only all possible test-crosses within the phenotypic class to which the mutant belongs are made (Lundqvist & von Wettstein, 1962).

Based on a total of 1580 mutants, the largest phenotypic class, 'spike and leaf sheath mutants', is represented by 533 mutants, localized on (or in) 8 loci only, whereas the second largest class, 'leaf blade mutants' contains 390 mutants, localized on 25 loci, etc. (for details see Lundqvist & Lundqvist, 1988a). The mutation frequency for the different *cer*-loci shows considerable differences. Whereas for each of 44 different loci only between one and five mutants have been obtained, sixteen other loci together are responsible for 70% of all *cer*-mutants. Two mutant types of the class 'spike and leaf sheath' (*cer-c* and *cer-q*) together are represented by no less than 382 mutations. For the aforementioned locus *Cer-yy*, a so-called 'spike mutant', with the wax coating absent on the spikes, 18 mutant alleles were identified which all proved to be dominant and had been induced by mutagenic treatments with various types of ionizing radiation as well as with different chemicals (Lundqvist, von Wettstein-Knowles & von Wettstein, 1968; Søgaard & von Wettstein-Knowles, 1987).

According to Lundqvist (1992) mapping of the different *cer*-loci is under way and, so far, linkage studies by the Carlsberg research group in Denmark have revealed that, with the exception for chromosome 6 where only one *cer*-gene was localized, some 25 different loci are well distributed over the other six chromosomes of barley.

Another subject, studied in a detailed way for *cer*-mutants, has been the effect of different mutagenic agents on the distribution of mutations for the different phenotype classes and for the individual loci. It was calculated that X-ray treatments resulted in about one *cer*-mutant against three *ert*-(erectoides) mutants. As for *cer*-mutants 79 loci have been discovered against only 26 loci for *ert*-mutants, this result may be somewhat unexpected. Taken into account that both mutant types can be observed relatively easily and that both groups of mutants have been studied on a very large scale, and provided that such results cannot be explained by statistical errors, one can only speculate about a possible different 'sensitivity' of *cer*- and *ert*-loci to X-rays.

Distribution of the 1580 *cer*-mutants per mutagen over the five phenotypic classes, shows that very few mutations (altogether 24 only) in very few loci result in wax changes at the same time in spike, leaf sheath and leaf blade together. Other comparisons are less easy to make, but the number of mutants per phenotypic class differs considerably between the mutagen applications.

Results for the individual loci are presented in Fig. 3.4. Because of the aforementioned differences in numbers of mutations from different mutagenic treatments, the number of mutations per locus in Fig. 3.4 are presented as a percentage of the total number of mutants for that specific mutagen. Many details can be found in Lundqvist & Lundqvist (1988c) and Lundqvist (1992), but the most important results are briefly summarized here.

- Mutagenic chemicals and ionizing radiation show different effects with respect to *cer*-mutations. In particular sodium azide (NaN₃) differs from the treatments with ionizing radiation in this respect.
- Among the mutagenic chemicals, sodium azide differs considerably from the organic chemicals which, according to the effects on *cer*-loci, are relatively homogenous. Lundqvist (1992, p. 51) refers to Nilan (1981a, b), who has pointed out that the ratio of chromosomal aberrations to gene mutations should be about 1 : 1 or somewhat lower for organic chemicals such as ethylene imine and ethyl methanesulphonate (EMS), and 0 : 1 for sodium azide which, essentially, does not induce chromosomal aberrations.
- For physical mutagens relatively high frequencies of chromosome aberrations are found; more for densely ionizing radiation treatments than for sparsely ionizing radiation treatments.

With respect to effects on different loci the following main points can be noticed.

- For the two loci *cer-c* and *cer-q* (both of the class 'spike and leaf sheath mutants') the highest numbers of mutations (together 364) were obtained, especially by chemical mutagens. Two other loci, *cer-a* with 62 alleles and *cer-x* with 33 alleles, have also more mutations induced by chemicals.
- For locus *cer-b*, with 37 alleles mutated, on the other hand, a higher sensitivity for radiation is observed.
- Radiation is also more effective on the loci in which only low numbers of mutations are induced.
- Finally, the normalized results of Fig. 3.3 reveal

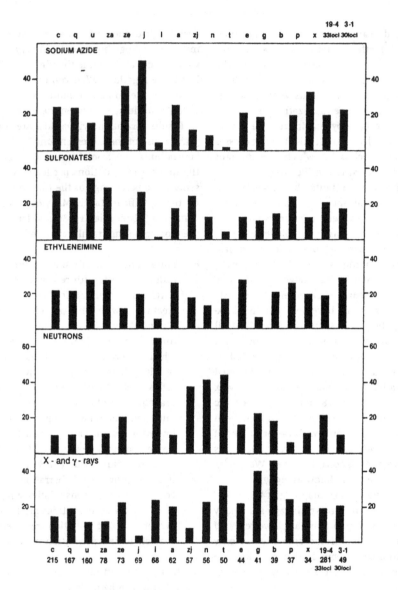

Figure 3.4. Percentages of induced wax (eceriferum; *cer-*) mutants at different loci on barley chromosomes, normalized for different mutagens (after Lundqvist, 1992). (Characters represent different loci; figures along the X-axis represent numbers of mutations per locus; figures along the Y-axis represent percentages.)

some strikingly different mutagenic effects for a few particular loci, e.g. the effect of neutron radiation on locus *cer-i* and the effect of chemicals, in particular sodium azide, on locus *cer-j*.

Early mutants (praematurum)

There has always been great interest, in particular in Scandinavia, in breeding for earliness in cereals and in the 1940s it was proven that X-irradiation could produce mutations affecting maturity time both towards later and earlier, although mutations towards later occur more often. Up to 1957, 20 early mutants, that could be

assigned to three or four loci, had been induced and selected in cv. Bonus (Gustafsson, Hagberg & Lundqvist, 1960). Screening for earliness is easy in cereals when instead of time of ripening the time of 'heading' is taken as the criterion. Earliness in most cases inherits in a recessive way, but could be dominant in other cases.

According to Lundqvist (1991), in the Swedish barley mutation project about 1100 different early mutants have been isolated, of which for only 172 mutants with easily recognizable improved earliness (i.e. more than a week earlier) gene localization studies have been performed. Those so-called *mat*-mutants are assigned to nine different loci. The phenotypically most prominent locus appears to be the *mat-a* locus, which controls the photoperiodic reaction. Mutation from the dominant allele to recessive alleles had the result that the mutants are heading at eight hours daylength and produce viable seed as well.

A mutant with the allele *mat-a^8*, derived from cv. Bonus, had become directly registered in 1960 as the succesful Swedish mutant cultivar Mari, from which in 1970, by crossing with the primitive cultivar Monte Cristo, another succesful (indirect) mutant cultivar, cv. Mona, arose. Cv. Mari, which was also characterized by its short and stiff straw, performed in particular well in Middle and North Sweden (where earliness is of special importance), but had 5–10% lower grain yield in South Sweden (Gustafsson et al., 1960).

Lundqvist (1991) reported that early maturity mutants may be produced relatively often after treatment with chemical mutagens of the sulphonate group, whereas sodium azide seems less effective in this respect.

Mutations for powdery mildew

The majority of the reported cases of induced mutations for disease resistance, as was pointed out by Jørgensen (1991), in fact may have been the result from outcrossing and other sources of contamination. This also applies to resistance for powdery mildew (*Erysiphe graminis* f.sp. *hordei*) in barley (*Hordeum vulgare*). A special type of resistance, the so-called *Ml-o* (or *Mlo*) resistance, may be the only significant exception to this rule. Resistance of this type has been studied at various places in Scandinavia and in particular at the Risø Research Laboratory in Denmark by J.H. Jørgensen.

Powdery mildew in barley has been the first disease against which X-ray induced resistance was obtained, namely already by Freisleben & Lein (1942) in cv. Heine's Haisa in Germany. The mutated locus was indicated as *Ml-o*. This mutant line was not included in breeding programmes because, in the absence of the pathogen, the grain yield showed a reduction of 5–10%.

In later years many more powdery mildew resistant mutants have been described. In 1992 already more than 150 *ml-o* gene mutations were known, mainly in two-rowed spring barley (Jørgensen, 1992). Jørgensen (1976) reported on studies with ten of these mutants which had been independently induced by various mutagenic treatments on the *Ml-o*-locus. Ten different alleles, which were all recessive, were assigned *ml-o 1* to *ml-o 10*. After the first discovery of artificially induced mutants in the early 1940s, and subsequently in numerous other muta-

tion experiments, *ml-o* resistance was detected in 1971 also in a German collection of Ethiopian landvarieties of barley dating back to 1937 (Jørgensen, 1976; Micke, 1992).

The *ml-o* type of resistance does not conform to the gene-for-gene system and appears to be monogenic and always recessive, non-race specific and very durable (Jørgensen, 1991; 1992; 1993). The *Ml-o* locus is located on chromosome 4. Resistance of the *ml-o* type has very attractive features for further plant breeding work but, until 1986 its further exploitation was severely limited by the occurrence of necrotic leaf-spots as a result of pleiotropy which, in the absence of the pathogen, cause a significant reduction in grain yield. In later years it has been found that this negative pleiotropic effect can be substantially reduced by altering the genetic background in which the *ml-o* alleles operate.

Since the negative side-effect of necrotic leaf spotting has been almost eliminated, the spontaneously arisen Ethiopian *ml-o* resistance, indicated as *ml-o 11*, has become much used in European barley breeding and starting with cv. Atem, released in the Netherlands in 1979, several mutant cultivars with this *ml-o 11* resistance have been produced. Micke (1992) reported that by 1991 at least sixteen cultivars which were released in the UK, the Netherlands and Germany, carried the *ml-o 11* resistance gene. More recent (scattered) information shows that in 1995 *ml-o 11* was present in (almost) every advanced European cultivar.

In principle, the artificially induced *ml-o* resistance should be at least equally useful or in fact even better, as in this way resistance of the *ml-o* type can be directly induced in each modern barley cultivar without having to go through the time and labour consuming system of making crosses, followed by repeated backcrosses, with the primitive Ethiopian cultivar. Crosses, on the other hand, have the advantage that undesired pleiotropic effects of a mutant allele can be removed.

As far as is known, so far only one cultivar of agronomic importance (high yield) with an induced source of *ml-o* resistance has been commercialized. This mutant cultivar, cv. Alexis, released in 1986 as a malting barley in Germany (former FRG), is based on an EMS-induced mildew resistant mutant (allele *ml-o 9*) from mutant cv. Diamant that was further crossed with several other cultivars and breeding lines. In 1990 cv. Alexis represented 35% of all spring barley grown in Germany. The corresponding figure for the same year for Denmark as well as Italy was 20%. More details can be found in Jørgensen (1992).

Lundqvist (1986) reported that in Sweden artificially induced mutants for *ml-o* resistance have been obtained since 1969 in different cultivars. In tests against susceptible material, 28 out of 77 mutants were found to be recessive and 49 dominant. All recessive alleles referred to the *ml-o* locus. In addition, Lundqvist (personal communication) reports that, so far, sodium azide has been the most efficient mutagen with respect to inducing resistance against powdery mildew.

Mutations for anthocyanin content (exrubrum)
Chemical reactions between proanthocyanidins (which are colourless compounds present in the testa layers of grains of the cereals barley and sorghum only) and specific proteins, which are present in malt and beer when produced from barley, cause the brewing problem of the formation of beer haze (or chill haze). Specific measures must be taken, e.g. chemicals added, to prevent this haze formation in beer. Brewers, therefore, are interested in barley cultivars in which the biosynthesis of so-called proanthocyanidins is genetically blocked (von Wettstein *et al.*, 1977; 1980; 1985). Until 1974 no proanthocyanin-free barley lines were known.

The research group of von Wettstein at the Carlsberg Research Laboratory in Copenhagen, Denmark, has studied this subject in a very detailed and rather original way, from a practical as well as from a more fundamental point of view. It should be mentioned in this respect that, in general, biochemically well-defined (induced) mutants are relatively rare in higher plants. As proanthocyanidins are colourless, they cannot be detected visually in the grains and, hence, a special chemical test (at present the so-called vanillin test) has to be applied.

Jende-Strid in 1974 found the first proanthocyanin-free (or *pac*-free) mutant (*ant 13-13*), an EMS mutant of barley cv. Foma, in a collection obtained from Svalöf, Sweden. Jende-Strid & Lundqvist (1978) reported on the presence of 42 anthocyanin-poor or anthocyanin-free mutants, either of spontaneous origin or artificially induced, present in the Svalöf mutant collection. On the basis of crossing experiments it was found that 31 of these mutants referred to 14 different loci. During scoring of mutant phenotypes the reduction or absence of anthocyanin was determined in stem, auricles, awns and lemma. Mutant plants were only scored as anthocyanin-free when anthocyanin was absent in all these four groups. Jende Strid & Lundqvist (loc.cit.) further proposed to change the originally used symbol *rub* for genes controlling anthocyanin pigmentation (Gustafsson *et al.*, 1969), to the symbol *ant* for all genes affecting the anthocyanin, anthocyanidin and proanthocyanidin. In later years the gene symbol *pac* was proposed for the proanthocyanidin group only.

Throughout the years a collection of more than 630 *pac*-free barley mutants has been built up at the Carlsberg Laboratory, for which purpose about 14 million (!) plants had been investigated. Of the obtained mutants 543 have been genetically defined by diallelic crosses and assigned to 27 different 'complementation groups' (Jende-Strid, 1988; 1990; 1991). Nine loci could be identified at that time and 475 *pac*-free mutants were localized to one of these nine loci. Mutations at these loci block the biosynthesis of proanthocyanidin in the seeds. For the three loci *ant 13*, *ant 17* and *ant 18*, a total of 139, 177 and 126 allelic mutants respectively were found of which, probably, some refer to the same mutational event. For other loci only few allelic mutants had been identified at that time.

More recently, Jende-Strid (1994) reported that in 1994 the Carlsberg collection of *pac*-free mutants contained 541 numbers, of which 516 had been assigned to eleven loci. Jende-Strid added that all *pac*-free mutants have been induced by treatments of barley seeds with sodium azide (NaN$_3$). Most mutants are the result of treatments with 1 mM NaN$_3$ in phosphate buffer; this dose also gives the highest chlorophyll mutation frequency on M$_1$ spike basis and M$_2$ seedling base without reducing the viability of the treated seeds. (N.B. More details about sodium azide will be presented in Chapter 4.) Mutants for the *pac*-trait are screened among M$_2$ plants by testing their M$_3$ seeds. In the early years anthocyanin-free mutants were selected in M$_2$ populations in frequencies of 3×10^{-3} (later 13×10^{-3}) on a M$_2$ plant basis.

The practical aim of the project is to breed barley cultivars which are high-yielding, *pac*-free and possess good malting properties. The first commercial result of the Carlsberg research programme, cv. Galant, has not been a great success, because of negative pleiotropic effects, due to deficiencies in the viability of the aleurone layer, resulting in a too low grain yield. During a FAO/IAEA Symposium on *'The Use of Induced Mutations and Molecular Techniques for Crop Improvement'*, held in 1995 in Vienna, von Wettstein (1995) reported on the release of two new commercial *pac*-free malting barley cultivars: cv. Caminant (*ant 28-48* (Grit) × Blenheim) and cv. NFC 8808 (*ant 27-488* (Zenit) × Sewa). Both cultivars at that time were on the European recommended list of barley. Cv. Caminant surpassed the Danish standard in the years 1991-4 by 4%, whereas a new introduction *ant 29-2110*, during the same period even outyielded the standard – including the excellent brewing cultivars Alexis and Canut – by 8%.

4 Induction of mutations

4.1. Introduction

For the purposeful induction of mutations generally two main categories of mutagenic agents are distinguished: radiation and chemicals. In addition, it has been suggested on some occasions, for instance by Bird & Neuffer (1987), that the intentional use of transposable elements (also called insertional mutagenesis) and callus cultures (because of the occurrence of so-called 'somaclonal variation') should be considered as well as suitable 'biological' techniques to induce mutations.

Transposable elements, according to some authors (see for instance Lewin, 1994), may be a major source of 'spontaneous' mutation in the genome. The application of transposon techniques, nowadays, is mostly considered part of the field of 'modern biotechnology'. Some basic information on this topic has been presented in Chapter 3. The subject of 'somaclonal variation', which refers to the phenotypic variation observed following *in vitro* culture and which is genetically determined only to a part, will be discussed in Chapter 5.

At this time it seems still somewhat premature to consider transposons and 'somaclonal variation' (in the latter case tissue culture is sometimes considered the mutagenic 'agent') as methods to induce mutations. It is not possible yet to manipulate both systems in such a way that, irrespective of the plant species, the genotype and the kind of (ex)plant material used, significant increases in overall mutation frequency can be obtained, let alone that this could be done in a predictable and controllable way for specific genes or traits. This, however, may become a reality in the future for transposable elements, in particular when the transfer of transposons at will to important crop species, in which they have not been identified so far, would become an accessible method. Despite some claims of successes with the use of 'somaclonal variation' for plant breeding purposes and for reasons which will be explained in Chapter 5, it is believed that the future role of 'somaclonal variation' will remain limited.

In the following sections we will concentrate on different types of radiation and on chemical mutagens. In some authoritative publications on mutation breeding, including the FAO/IAEA Manual (Anon., 1977a), many data about the properties of radiation facilities, various types of radiation, chemical mutagenic agents and their (known or supposed) interaction with plant genetic material are given. However, a considerable part of this information does not have much bearing on plant breeding. The practical breeder is mainly interested to know about the availability of certain radiation facilities or effective mutagenic agents, and in which way they should be handled in order to create optimal conditions for inducing the mutations he is looking for.

In the following sections a balance is sought between some useful or interesting background information and the essential basic information on mutagens and mutagenic treatments for breeders who may like to apply mutation breeding techniques. Many more technical details can be found in a score of textbooks and special publications, which will be referred to later.

4.2. The use of radiation for mutation induction

4.2.1. Kinds of radiation, type of action, factors influencing effects

Electromagnetic waves and corpuscular radiation
Various types of radiation, basically belonging to two different categories, are of interest for the induction of mutations: a) X-rays and gamma rays which normally are considered to travel in waves and belong to the **electromagnetic spectrum**, and b) radiation types consisting of moving particles like protons, neutrons and electrons; this category is also indicated as **corpuscular radiation**. Heat, which, like other radiation, is a release

of energy from an atom, is of no importance for the purposeful induction of mutations.

For mutation induction use is often made of high energy types of radiation, which produce discrete releases of energy, called ionizations, as they pass through matter. So-called 'ionizing radiation' may be radiation in the form of waves or particles. Charged particles (like electrons or protons) as well as uncharged particles (like photons and neutrons) can produce ionizations in matter. (N.B. The expression ionizing radiation is sometimes confused with radiation causing radioactivity. However, X- and gamma radiation, as well as beta radiation, do not leave any radioactivity in the irradiated (plant) material, whereas alpha and neutron radiation (which are not often used in mutation breeding) normally cause only a low level of temporary radioactivity.)

A wealth of literature on the effects of ionizing radiation on plants is available. Sparrow, Binnington & Pond (1958) produced a bibliography that covers the period from 1896 to 1955 and contains 2600 publications!

In practice, nowadays, most often use is made of sparsely ionizing radiation, to which category X-rays and gamma rays belong. Gamma rays and X-rays (or Röntgen-rays), as was said before, are electromagnetic radiations and have an energy level that is high enough to ionize atoms in molecules with which they interact. Electromagnetic waves belong to a continuous spectrum, including long wavelengths radiowaves, which are non-ionizing, just as is the case with infrared and visible light; ultraviolet, which is in the border area between non-ionizing and ionizing radiation; and finally X- and gamma rays, which have very short wavelengths and are of the ionizing type. Fig. 4.1 shows the spectrum of electromagnetic radiations. Gamma rays have shorter wavelengths, but otherwise are identical with X-rays with respect to their physical properties. By convention, although not completely correct, the expression X-rays is mostly related to photons from 'machines' and gamma rays to photons from radioactive sources (see later).

Ultraviolet light (or UV) is also a type of electromagnetic radiation used for mutation induction, but at the wavelengths that are employed, UV in most cases is non-ionizing. X-rays, gamma rays and UV will be discussed in Section 4.2.3.

In addition to the aforementioned types of radiation, use can also be made of (densely ionizing) alpha or beta particles and neutrons. All particle radiation can be identified as ionizing radiation. The aforementioned alpha and beta particles are emitted by radionuclides and may be artificially accelerated by machines.

During the process of irradiation, such densely ionizing, energetic particles (which usually carry an electrical charge) liberate electrons in collision with atoms of different molecules. When particles slow down, ionizations become less likely or even impossible. In biological systems electrons with an energy level higher than 10 electronvolts (eV) are considered to be ionizing. Below that level so-called excitations, transfer of electrons to higher energy levels, may occur.

Nuclides are atoms having a specified number of neutrons and protons in their nuclei. The expression iso-tope refers to atoms with the same chemical properties (= number of protons and electrons in the nucleus) but with different nuclear mass due to different numbers of neutrons, e.g. ^{37}Cl and ^{35}Cl. Radionuclides or radioisotopes are unstable and tend to become more stable by emission of alpha or beta particles or neutrons. The time required to become stable is defined by the 'half-life' of radioisotopes, which is the time for half of the atoms to be stable. Also gamma rays may be emitted without changes in the neutron/proton ratio. Nuclear radiation possesses high kinetic energy in the range of some thousands to several millions of electronvolts.

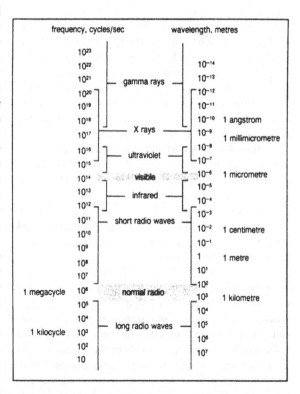

Figure 4.1. The electromagnetic spectrum. (From F. Bueche, *Principles of Physics* (1965); McGraw-Hill Book Company, publisher).

For ionizing radiation a further distinction can be made between directly ionizing particles, which are charged particles and require sufficient kinetic energy to produce ionizations by collision with negatively charged electrons or the positively charged atomic nucleus, and indirectly ionizing particles, which are uncharged neutrons and which can pass through matter even at slow speed and interact with the atomic nuclei producing unstable constitutions resulting in emission of high-energy protons, electrons, gamma rays and sometimes larger particles consisting of several nucleic elements (e.g. two protons and 2 neutrons = alpha particle).

Radiation in nature

Radiation is known as a natural phenomenon of geophysics. Man and all other living organisms are continually exposed to 'natural' radiation of various kinds. Some details are given in Box 4.1.

Dose, dose rate and some other important concepts and definitions

The number of mutations induced depends on the radiation type, the applied **dose** and the **dose rate**. The plant species, the ploidy level, differences in developmental stage, physiological condition, etc. may be the cause of differences in response to irradiation and, therefore, always have to be carefully recorded. Such differences are indicated with the term **radiosensitivity**, which is further explained in Box 4.2. Environmental factors affecting cell physiology like temperature, oxygen content, water content, etc., may have a modifying effect on the outcome of a radiation treatment. This underlines the importance of making detailed records when mutation experiments are performed. More particulars in this respect will be given at a later stage.

A distinction must be made between the term **dose**, which refers to the energy emitted by a source and the **absorbed dose** which is received by a plant, a tissue, etc. Only the latter is relevant. The **absorbed dose** is roughly defined here as the amount of energy that is transferred by the radiation treatment from the source to the irradiated object and that is absorbed by this object.

The **dose rate** according to the FAO/IAEA Manual (Anon., 1977a) is the rate at which a given dose is administered or, in other words, the quotient of the dose by the time (e.g. dose min⁻¹) and, when measured at the target level, indicates how much energy is absorbed by the irradiated material (e.g. seeds, cuttings, or *in vitro* cell suspensions) during a given unit of time.

The old, but still much used unit of exposure emitted by the source is **Roentgen (R)** or **Röntgen** (named after C.W. Röntgen, Germany), whereas the old unit for the emitted rate of exposure is **R.min⁻¹**. The old unit for the **absorbed dose** (per gram of tissue) is **rad**, or radiation absorbed dose (with 1 R = 0.877 rad in air). When X-rays and gamma rays are used, 1 R in plant tissue or human tissue corresponds to about 0.95 or 0.96 rad respectively, and in practice this conversion is often omitted.

The present unit of radiation, according to the internationally accepted SI system, is defined as **Gray** (named after L.H. Gray, a British radiobiologist). The corresponding symbol is **Gy**, with 1 Gy = 100 rad. These terms are also mentioned in Box 4.2.

The **dose rate** as well as the **dose** received (during a given period) by an object depends on the distance between object and source and the output from a physical radiation source shows a decrease in energy that has an inverse square relationship with distance from the source, or in an equation:

$$a_1/a_2 = (b_2/b_1)^2$$

when a = dose (rate) and b = distance.

To measure the (absorbed) radiation dose, various devices can be used. In this respect we can mention dosimeters connected to ionization chambers (in which the number of ionizations in air as a result of irradiation is determined), scintillation meters, thermoluminescense dosimeters, film dosimeters (via blackening of an exposed photographic film), so-called Fricke dosimeters (measuring the oxidation of Fe^{2+} to Fe^{3+} as a result of irradiation), etc. Technical details are not of particular interest to the plant breeder, but dosimetry is important for reproducibility of results and proper interpretation.

According to the FAO/IAEA Manual (loc.cit.) the frequency of gene mutations, small deletions and single-strand breaks – which are believed to have resulted from so-called 'single hits' – is directly (linearly) related to the dose and is not affected by dose rate. This may still be true for physiologically rather inactive tissue like dry seeds. There are, however, some recent reports on mutation experiments in lower organisms in which it was mentioned that also in case of 'true single hits' the dose rate did affect the mutation frequency for specific loci. As to a possible explanation for this unexpected effect (provided of course that results have been correctly interpreted) one could hypothesize that at high dose rates the system(s) that do repair radiation damage do not work anymore as effectively as at lower dose rates.

Chromosomal aberrations often require two or more breaks and their frequency, when caused by X- or gamma

Box 4.1 Some facts about 'background' irradiation to which man is exposed

All living organisms are exposed to radiation from the 'natural background' whereas most human beings may receive additional 'man-made' radiation as well, in particular from using ionizing radiation for medical purposes. The radiation dose to which an average person, for instance in the USA, is exposed from natural radiation sources, is estimated on an annual base to be 0.7 to 1.1 **millisievert (mSv)** (= 70 to 110 **millirem**) and, in addition, up to 1 mSv (= 100 millirem) mainly as a result of medical exposures. Thus it is estimated (see for instance Upton, 1982) that, in addition to the natural radiation, about the same radiation dose is accumulated by a person during their lifetime from man-made sources.

The unit of (absorbed) dose is **gray (Gy)** or in old units **rad** (1 rad = 0.01 Gy). The unit **sievert (Sv)** refers to the absorbed dose = amount of radiation absorbed per gram of human body tissue, multiplied by a weighing factor for the damage potential of a particular type of radiation (see also Box 4.2 for units).

The use of X-rays for diagnostic purposes is a common cause of irradiation for most people. Individuals, when treated for cancer may receive much higher doses; e.g. to control a cancer a total of 60 Gy may be required. Other sources like radioactive fallout, nuclear industry, consumer products and travel by air, on a population scale, contribute less than 0.01 mSv on a year's basis.

Other reports, also referring to the USA, mention that from a total radiation dose received by man, about 82% is derived from so-called 'natural sources', in particular gamma rays and heavy particles. Those sources are as follows.

1) Radon gas, a gas with a half-life of four days only, released in the earth as one of the radioactive products in the chain that starts with uranium-238 by decay of radium present in stones and bricks and infiltrating our houses). Radon is responsible for 55% of the total 'natural' radiation.
2) Radiation from inside human bodies, in particular from radioactive potassium in the muscles, with a contribution of 11%.
3) Cosmic and solar radiation (although the atmosphere protects us from most celestial radiation) which accounts for 8%.
4) Radiation from rocks, soils and groundwater (other than radon), which contributes also 8%.

The radioactive isotype ^{40}K represents only 0.001 18% of the natural potassium but, because of the ubiquity of this element, it contributes one-third of the external and internal dose from the 'natural radiation background'.

Man-made sources, which together account for about 18%, include

1) medical X-rays with 11%,
2) nuclear medicine (for cancer treatment) with 4%,
3) consumers' products, in articles as different as fertilizers, televisions, smoke detectors, false teeth and camping lights, with 3% and
4) some other sources, like fallout by leakage of nuclear reactors or from weapons, with less than 1%.

In tobacco and tobacco smoke ^{210}Pb and ^{210}Po occur. The amount of radiation for heavy smokers may be competitive with that of radon!

Papers, books, etc. may be (small and non-hazardous) sources of gamma radiation, produced by isotopes like ^{232}Th, ^{226}Ra and ^{40}K.

Radiosensitivity is expressed in **becquerel (Bq)** (desintegration.s^{-1}) or **curie (Ci)** = 3.7 x 10^{10}Bq. Small doses of radioactivity are expressed in **picocurie (pCi)** in which p = 10^{-12}, **nanocurie (nCi)** with n = 10^{-9}, or **microcurie (µCi)** in which µ = 10^{-6}. One kg of dry weight of paper, for instance, may produce 50–2000 pCi for each of the three isotopes mentioned.

The **dose** received by a person is expressed in **Gy. h^{-1}** or **rad. h^{-1}**. From a mutational point of view it has been calculated that about 0.2–0.5 **Sv** (20–50 rem) would be needed to double the (human) mutation rate per lifetime. Based on this calculation it is assumed that less than 0.2% of all new cases of genetically determined human diseases can be attributed to the current natural background radiation. For cancers, however, 1–3% can be attributed to natural background radiation. There is still some doubt as to the question whether low dose ionizing radiation is only harmful if it exceeds a certain threshold value or is cumulative in its effects.

A high dose of 10 **Sv** (10^3 **rem**) will lead to a quick death as a result of heavy damage to the central nervous system. A dose of 3 **Sv** (300 **rem**), when delivered to the whole human body, will be lethal in about 50%of the cases. This is recorded as: **LD 50** (LD = lethal dose) = 3 **Sv**.

Considerable differences may exist between different areas. Citizens of a town at sea level receive considerably less radiation than persons living in the mountains. People directly working with radiation, of course, run higher risks than an average citizen. This was in particular the case in the early 1900s when not so much was known yet of the effects of radiation. It is estimated that several hundreds of these early radiation workers died of diseases caused by X-rays. In the 1920s hundreds of girls were employed in the USA by companies to paint dials of watches with a mixture of radium and zinc sulphide. The girls moistened the hairs of the brushes by putting them in their mouths. By the end of the 1920s some of them had died from bone cancer.

rays, often shows a curvilinear expression and is proportional to about the square of the dose. If a time factor is involved, primarily physiologically active tissues in this situation are affected by the dose rate.

More recently, Gonzáles (1994a,b) outlined that it may be presumed that radiation acts through single track interactions which, in a homogeneous cell population, occur randomly according to a Poisson distribution. The theoretical dose–response relationship, i.e. the mathematical relation between the dose and the probability of expression of an attributable radiation effect, is mostly described by a linear-quadratic expression. When low radiation doses are administered it is unlikely that a single cell is traversed by more than one track of radiation and, as a consequence, the dose–response relationship most probably will be linear, without a threshold dose and independent of dose rate.

Different types of radiation may have a different mutagenic effectivity which is related to the so-called **relative biological effectiveness (RBE)**, as is further explained in Box 4.2.

The response to different dose rates is often used as a criterion to distinguish between events in which single hits or more hits are involved. Extensive research on dose rate effects has been performed by Nauman *et al.* (1975) on the basis of experiments with the stamen hair colour model system of a clone of *Tradescantia*, heterozygous for the trait 'colour of stamen hair cells', which system has been already briefly discussed in Box 1.5 in Chapter 1.

Some of the results, obtained by Nauman *et al.* (loc.cit.) in their Fig. 4, and which have been redrawn and converted into Gy. min^{-1} by Fendrik & Bors (1991), are presented here as Fig. 4.2. (N.B. Because the results obtained by Nauman *et al.* for dose rates of 0.05 rad. min^{-1} and 0.5 rad. min^{-1} appeared to be essentially identical, Fendrik & Bors have combined these results in one curve for 0.005 Gy. min^{-1} (= 0.5 rad. min^{-1}).)

For a given total dose within the range from 0.1 Gy to about 1.6 Gy, the highest dose rate (0.3 Gy. min^{-1}) also yields the highest mutation frequencies (measured as no. of pink mutant cells per stamen hair or, according to Fendrik & Bors in Fig. 4.2, per stamen (= Staubgefäß)). The lower dose rates (0.05 and 0.005 Gy. min^{-1}) invariably show lower mutation frequencies for each total dose and within the range studied all three dose rates show a linear increase in mutation frequency. For the dose rate of 0.3 Gy. min^{-1} a 'saturation effect' is observed above a total dose of approx. 1.6 Gy and a similar effect, probably, may occur for the lower dose rates, but these combinations of doses and dose rates were not included in the experiment.

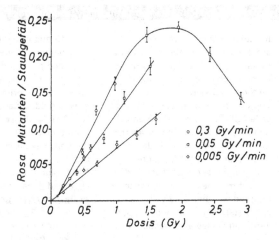

Figure 4.2. The production of mutations in relation to dose and dose rate in *Tradescantia* hair cells. (Rosa Mutanten = pink mutants; Staubgefäß = stamen). From: Fendrik & Bors (1991, p. 52). After Nauman, Underbrink & Sparrow (1975, adapted).

As is illustrated in Fig. 4.2, the effect of the dose rate on mutation frequency becomes more manifest as the dose is increased or, in other words, increasing the dose rate results in slopes which are progressively steeper. This observation, according to the authors, would be expected because the level of repair of sublethal damage increases during the longer periods that are required to deliver a given dose at lower dose rates.

According to Nauman *et al.* (1975) the majority of the radiation-induced mutations from blue to pink in the *Tradescantia* system may result from (small) deletions which, at higher doses and dose rates, presumably, result from so-called 'two-track processes'. The authors further state that the pink mutations produced in proportion to dose, probably, are a combination of 'true point mutations' (i.e., base changes in DNA which are not subject to dose-rate effects and small 'one-track' deletions, but they add that observation of a linear dose response by itself is not enough proof that the aberrations did not arise from 'two-track processes'.

Acute versus chronic irradiation

In irradiation programmes, commonly, a distinction is made between the – most often applied – **acute irradiation**, in which the total dose is administered at a high dose rate in a single exposure, administered in a number of minutes up to a few hours, and irradiation treatments which are administered over several weeks or even longer at a low dose rate (e.g. a total dose of 100 Gy at a dose rate of 1 Gy. day^{-1}), called **chronic irradiation**.

Until the 1950s mutagenic treatments, almost exclu-

Box 4.2 Common units and terminology, used in radiation biophysics and some related subjects

In radiobiology and related subjects, including the induction of mutations by radiation treatments, different units are used concurrently. It is generally advised to apply as much as possible the so-called SI (Système Internationale) units, but in quite a few cases the use of 'old' units or so-called 'special units' is still common practice. For some of these 'old' units it may be even very much doubted whether they will ever be completely replaced by corresponding SI units (see also Section 4.2.2).

Radiation units and definitions
* **joule** (symbol **J**), in fact the unit of energy or 'work', is also used as the standard unit for the radiation energy (which is present, emitted, transferred or received).
* **electronvolt** (symbol **eV**) is a unit for radiation energy and also for the penetration power of a specific type of radiation. According to different sources in literature, 10–30 eV may be required for one single ionization.
* **röntgen** or **roentgen** (symbol **R**) is the old special unit of exposure, or the quantity of ionizations in air. This unit is still used in reports on mutation breeding.
* **gray** (symbol **Gy**) is the SI unit

used to quantify the absorbed dose of radiation (1 Gy = 1 J. kg^{-1}). The absorbed dose is the energy transferred by radiation to the target and absorbed there. The unit **gray** should replace the old unit **rad** (1 Gy = 100 **rad**), but – apart from a factor 10^2 – no real difference exists between those units.
* **rad** (plural: **rad**) is the old unit of absorbed dose of radiation **(or radiation absorbed dose) with 1 rad = 10^{-2} Gy** (or 1 Gy = 100 rad). Often applied as well is the unit **krad** (rad × 10^3).
* **gray per second** (Gy. s^{-1}) is the SI unit for absorbed dose rate. The expression **Gy. min^{-1}** is used as well.
* **rad per second** (rad. s^{-1}) is an old unit for absorbed dose rate. Other units for time, like rad.min^{-1}, are used as well.
* **sievert** (symbol **Sv**) is the SI unit for comparing – for a given (absorbed) dose – differences in (biological) effects related to danger for man resulting from, for instance, neutrons against X- or gamma rays. The unit **sievert** refers to the (biological) effect caused to man by 1 **gray** and should replace the expression **rem** that is still used in most cases (see below).

* **rem** (röntgen equivalent man) is the 'old' unit to compare, in the same way as was described for the unit **sievert**, differences in danger for man caused by different types of irradiation. One unit **rem** – by agreement – refers to the effect on man resulting from one **rad** of gamma rays. The unit **rem** is related to **rad**. 1 rem =10^{-2} **Sv**.
* **becquerel** (symbol **Bq**) is the SI unit for the activity of a radionuclide (or radioactivity). One **Bq** equals one disintegration per second (dimension t^{-1}).
* **curie** (symbol **Ci**) is the old unit for radioactivity that should be replaced by the unit **Bq**. 1 Ci = 10^{10} **Bq. s^{-1}**.
(N.b. It appears to be rather impractical to define the strength of, for instance, a ^{60}Co source in **Bq**!)
* **LET** (linear energy transfer) is the energy per unit of track length and gives a measure for the density of energy dissipated by particles along a given track, for instance in plant tissue. **LET** has the unit J. m^{-1}. X-rays and gamma rays from ^{60}Co and ^{137}Cs are generally classified as low-LET radiations, which may induce relatively more gene mutations and deletions. Alpha rays and neutrons are high-

---

sively, referred to acute treatments. Some early reviews on acute treatments were written by Johnson (1936) and Catcheside (1948). Although Stadler (1930) used a 'transportable' X-ray machine for irradiating flower buds of orchards trees, exposure of growing plants for prolonged periods remained an exception until, starting from the late 1940s, so-called gamma fields became installed. Sparrow & Singleton (1953) mentioned that it would have been possible to perform prolonged treatments at an earlier date by using radium, a continuous gamma emitter, but that this was considered impractical because of the scarcity and high price of radium. Early treatments with radium of *Antirrhinum*, resulting mainly in growth disturbances, were performed for instance in Germany in the 1920s by E. Stein (for references see Baur, 1932).

The first reports about the installation of a ^{60}Co source

at Brookhaven National Laboratory and some preliminary results of chronic irradiation of different types of plant material from various plant species in gamma fields, came from Sparrow & Singleton (loc. cit.) and Sparrow (1954).

As is pointed out by Briggs & Konzak in the FAO/IAEA Manual (Anon., 1977a, p. 37), a comparison of acute and chronic irradiation usually implies a comparison of high versus low dose rates. A common observation is that low dose rates and extended periods of treatment in chronic irradiation may lead to the occurrence of so-called recovery phenomena. No clear borderline can be drawn between both types of irradiation and there is also no need to do so.

Extensive fundamental studies about differences between acute and chronic irradiation treatments administered to *Antirrhinum majus* (snapdragon) and

LET radiations and may cause relatively more chromosomal aberrations.

Half-life indicates the time needed to lose 50% of the radioactivity by radioactive disintegration (or decay). The half-life of ^{60}Co, for instance, is 5.3 years; for ^{137}Cs 30 years.

Some other useful units and definitions related to the induction of mutations

* **RBE** (relative biological effectiveness). This expression refers to the ratio for biological effects of the absorbed dose of a particular type of radiation (e.g. alpha rays) to a reference type (^{60}Co gamma rays) needed for the same effect. The **RBE** of neutrons is higher than is the case for gamma rays, which implies that in order to obtain the same mutational or biological effect, a lower dose of neutrons must be given. The **RBE** is mainly a function of the **LET** (see before).

* **interphase chromosome volume (ICV)** is an index correlated to somatic radiosensitivities of different plant species. The **ICV** is determined by dividing the nuclear size – or **nuclear volume (NV)** – by the somatic (2n) chromosome number of the species involved.

* **radiosensitivity** is a relative measure that gives an indication of the quantity of recognizable effects of radiation on the irradiated object (plants, plant parts, different developmental stages within mitosis or meiosis, young seedling stage versus adult plant, or specific molecular functions). Actively dividing cells, for instance, are more sensitive to radiation than non-dividing (resting) ones and therefore only tolerate about 10% of the radiation dose for non-dividing cells. Radiosensitivity can be measured for instance by determining a radiation dose giving a certain percentage of surviving cells. Other criteria for measuring radiosensitivity are: no. of cell divisions, mitotic delay, growth inhibition and pollen sterility. Often a relation is found between the size of the nucleus (or the **ICV** = interphase chromosome volume) and its radiosensitivity, but this does not hold for the majority of the polyploid species. Radiosensitivity is a property of the particular biological target, but is subject to several modifying factors.

* **LD 50** (LD = lethal dose) is the dose at which 50% of the irradiated objects (seedlings, plant parts or tissues, pollen grains, dividing cells *in vitro*) die as a result of the treatment. (N.B. There is no unanimity as to the time factor, i.e. the best moment of determining the LD 50.)

* **isotopes** are nuclides with the same atomic charge (= number of protons in the nucleus), but with different atomic mass (e.g. ^{37}Cl and ^{35}Cl).

* **wavelength**. Electromagnetic radiations are characterized by their wavelength whereas the biological effects of various types of radiation vary with their wavelengths. Visible light, for instance, has wavelengths between 4000 and 7000 Å; UV radiation between 2800 and 4000 Å. The unit of **wavelength** is nanometer (nm; n = 10^9) and, previously, Ångström (Å) with 1 nm = 10 Å.

* (N.B. Definitions for **effectiveness** and **efficiency** of mutagenic treatments are given in Section 4.2.1 of this chapter.)

---

other plant species, as well as about factors affecting the radiobiological response of plant species to both mentioned types of irradiation, were performed by Cuany, Sparrow & Pond (1958) and Sparrow *et al.* (1961). These authors studied various parameters for response to different radiation treatments, like chromosome breakage, growth inhibition, lethality and mutation frequency. With respect to this last category they calculated for instance mutation frequencies (e.g. per cell per rad) for different genes affecting flower colour in *Antirrhinum majus*. Details are outside the scope of this book.

In 1967 a conference on the use of chronic irradiation in mutation breeding was organized in the series *Gamma Field Symposia* at the Institute of Radiation Breeding in Ohmiya-machi, in Japan (Anon., 1967). Reference could also be made to several other reports on the effects of acute and chronic irradiation within the framework of the *Gamma Field Symposia* and voluminous other literature is available on the subject of comparing effects of such different treatments.

The present, prevailing opinion about differences between effects of acute and chronic irradiation, briefly summarized, is that such differences, from a practical point of view, appear to be less important for mutagenesis than was believed in the past. This is in particular the case when seeds are treated. However, differences may be more important for active cells, including *in vitro* cultures.

Dose fractionation

An alternative to chronic irradiation is to make use of so-called **dose fractionation** (or **split dose irradiation**) in which more than one irradiation is administered, interrupted by (one or more) time intervals, the length of which may be varied. In general it appears that the biological damage after fractionation of a total dose is less than when the total dose is applied at once or, in other words, that the recovery from radiation damage is higher than after application of an identical single acute dose. A small initial dose reduces the effect of a large later exposure to a level below that of the large exposure alone and fewer mutations are obtained in somatic tissue when an acute radiation dose is fractionated than when the dose is given in one, short treatment.

The subject of fractionated irradiation received considerable attention in the 1960s. Literature has been reviewed, for instance by Broertjes (1972a), who himself performed extensive experiments on dose fractionation in treatments of detached leaves of the vegetatively propagated ornamental *Saintpaulia ionantha* with X-rays or fast neutrons. From his experiments Broertjes concluded that in the case of dose fractionation a mechanism is induced by a low initial dose which protects the leaves against part of the effect of one or more subsequent radiation doses. It was possible to calculate the extent of the protection effect under various conditions, e.g. after an initial dose of 0.5 krad (5 Gy) of X-rays or fast neutrons the protection effect was equivalent to 3–4 krad X-rays or 1.5–2 krad fast neutrons. The magnitude of the protection effect is also demonstrated by the parameter of 'survival' of *Saintpaulia* leaves, which refers to the number of leaves (or the percentage of the control) which are able to produce adventitious plantlets. Broertjes (loc.cit.) found that normally a dose of 6.5 krad of X-rays was lethal, whereas after one initial dose of 0.5 krad 100% survival was obtained. An initial dose of 0.5 krad of X-rays, followed by a second dose of 10 krad, still showed 25% survival. Broertjes further found that the number of mutants per 100 irradiated leaves after certain fractionated irradiations (e.g., a pretreatment with 0.5 krad of X-rays, followed by a second dose of 6 krad after an interval of 8 h) can be higher than the corresponding number produced at the so-called 'optimum acute dose' for X-rays and fast neutrons (3 krad and 1.5 krad respectively).

Split doses of ^{60}Co gamma rays and fast neutrons were also given to protoplasts of *Nicotiana plumbaginifolia* by Devreux, Magnien & Dalschaert (1986). Details will be discussed in Chapter 5. In a report from Japan, Sakuramoto & Ichikawa (1996) describe dose fractionation experiments using the mutation detection system in stamen hairs of *Tradescantia*. Details of this system have been described in Chapter 1, Box 1.5. The authors concluded that results were in line with the most previous findings.

In conclusion, in the past much attention has been given in radiobiological research to a comparison of the effects of acute versus chronic irradiations, including also studies on the effects of fractionated radiation treatments. In general and mainly for practical reasons, it seems preferable to apply simple treatments with an acute dose in mutation breeding. The disadvantages of chronic irradiation including the expensive requirement of a suitable source (gamma-field or gamma-room) for extended treatments, the time (often many weeks or months) involved, the repeatability of the experiment, taking into account the physiological development of the plant material during a chronic treatment, are not repaid in most instances by a higher number of mutations or a more favourable mutation spectrum. Similar conclusions may be drawn with respect to the effects of dose fractionation. Studies about effects of chronic irradiation are perhaps of significance to investigate the effects of continuous irradiation by cosmic rays or to be prepared for catastrophes with nuclear energy, resulting from power plant accidents or nuclear war. Some additional details about acute and chronic irradiation will be given further on in this chapter when different types of radiation are discussed.

Effectiveness, efficiency and efficacy

In the past much attention has been paid to comparing the effects of different mutagenic agents, in particular at the time when it was still believed that particular chemicals might produce more specific effects on the genetic material (which, in a strict sense, could not be demonstrated until now for known mutagens).

Some important expressions that were already briefly mentioned before (e.g. in Chapter 2) are the **effectiveness**, the **efficiency** and the **efficacy** of different mutagens; these expressions may refer to radiation as well as to chemical mutagens. **Effectiveness** is defined as the number of mutations produced per unit of dose (e.g. dose rate in Gy or milligrammes per litre of chemical agent x duration of treatment). When two treatments with mutagenic agents induce the same frequency of (a given type of) mutations, these agents are said to show the same mutagenic effectiveness at a given dose.

Efficient mutagenesis implies the wish to produce desirable changes free from association with undesirable effects (Konzak et al., 1965). Hence, **efficiency** can be defined as the ratio of specific desirable mutagenic changes to undesired effects (like plant damage, sterility or lethality). Konzak et al. mention some parameters that can be used as 'reasonable quantitative measures' of the relative efficiency of different treatments, like the number of mutations (for which often the number of chlorophyll mutations is taken) per 100 M_1 spikes (or 100 M_2 seedlings), divided by the percentage of injury, lethality or sterility.

A third expression sometimes encountered in literature is **efficacy**. According to Gaul et al. (1972) – who refer in this context to Gregory (1961) – this term is less clearly defined and apparently refers to the power of a mutagen to produce useful mutations. Gaul et al. (loc.cit.) add that it is obviously more difficult and time-consuming to determine the efficacy and that any estimate of the efficacy is subject to the mutagenic treatment, as well as to the selection technique applied. The confusion about the exact interpretation of the term 'efficacy' may be further increased by the fact that in French literature sometimes the word 'efficacité' is used (see for instance Heslot et al., 1961) with about the same meaning as the word 'efficiency' in English.

In conclusion it can be said that, so far, the expression **efficacy** has not found general acceptance and that it would be advisable not to make use of this term.

Radiation sensitivity and modifying factors
The final outcome of a mutagenic treatment of plant tissue is determined by many factors of various nature, and the effectiveness and efficiency of the treatment are further influenced by so-called **modifying effects** caused by internal (biological) as well as external factors. The conditions under which the (plant) material is maintained before, during and after mutagenic treatment determine in which way irradiation induced effects are modified by the environment (Caldecott, 1956; 1958).

Radiation sensitivity or **radiosensitivity** – for a definition, see Box 4.2 – has been the subject of many detailed studies. During a conference on *Fundamental Aspects of Radiosensitivity*, organized at Brookhaven National Laboratory, New York, USA (Anon., 1961c), many aspects of this field of studies were discussed. Differences in radiosensitivity may occur as a result of the specific ontogenetic or physiological state of the irradiated plant material. Dividing cells, for instance, are more sensitive to irradiation than non-dividing cells. The effects caused by the aforementioned differences of the starting material indeed could be indicated as biological 'modifying factors'. More details on this subject will be given in later chapters.

In addition, differences in radiosensitivity between species and, to a lesser extent, between cultivars or breeding lines may exist. The occurrence of such differences, which are directly related to the genetic set-up of the starting material and, hence, are not 'modifying effects' in a strict sense, underlines the importance of performing preliminary experiments in order to deter-mine optimum mutagenic treatments before starting large scale mutation projects for practical purposes.

However, at this point we cannot elaborate on all aforementioned biological factors, but must limit ourselves to differences in radiosensitivity between species. Such differences often have been explained as resulting from differences in size of cell nucleus, or **nuclear volume (NV)**, and more precisely on the base of the so-called **interphase chromosome volume (ICV)** of different species. The ICV (see also Box 4.2) is the quotient of the average nuclear volume (determined by measuring the diameters of a number of nuclei of shoot apex cells during interphase under a microscope) and the somatic (2n) chromosome number for the species involved. Nuclear volume and ICV are genetically determined parameters for radiosensitivity of a plant species or a cultivar and it seems, therefore, also not correct to group these factors under the heading 'modifying factors' in a strict sense.

The ICV concept was developed and strongly advocated for chronic irradiation by Sparrow and co-workers in the early 1960s at the Brookhaven National Laboratory (for some early reports see Evans & Sparrow, 1961, and Sparrow & Evans, 1961) and is still used by many workers in mutation research as a suitable method to compare or predict radiosensitivities of different species. Sparrow, Price & Underbrink (1972) undertook a literature survey on DNA values for various taxonomic groups of organisms and estimated the nuclear volume size for almost 1000 organisms and species. They mentioned in this respect that nucleic acid contents varied from 1.3×10^3 nucleotides for the lowest viral RNA to 4.0×10^{11} nucleotides of DNA for the highest eukaryotic haploid nuclear volume. Average nucleic acid values per chromosome vary from 1.3×10^3 to 2.6×10^{10}.

A correlation between the ICV value and radiosensitivity was first reported for herbaceous plants in 1963 (Sparrow, Schairer & Sparrow, 1963) and later for woody species as well, with woody species being 2 to 2.5 times more sensitive than herbaceous species with comparable ICVs (Sparrow et al., 1965). It has been shown that in general (e.g. for a range of diploid species) a clear-cut negative correlation exists between radiosensitivity and the ICV value. To be precise: the lower the volume (for species with the same ploidy level) the higher the cellular radiosensitivity. Details and many references can be found in Underbrink, Sparrow & Pond (1968). The FAO/IAEA Manual (Anon., 1977a) mentions that, based on these relationships, it is fairly easy to predict for somatic tissue the 50% lethal dose (LD 50), which is an easy parameter to roughly determine the required radiation dose for plant species for which this has not been recorded yet.

Within one plant genus the nuclear volume increase is about proportional to the ploidy level. Conger et al., (1982) reported that in the range diploid (2x) to decaploid (10x) the radiosensitivity in terms of lethality is found to be about stable, as for such a range of species within one genus the sensitizing effect of increased target (= DNA) size is approximately compensated by the increase in genetic redundancy (more than two chromosome copies present in the nucleus). Still, it is often considered advantageous to irradiate, for instance, a tetraploid instead of a diploid plant. It is believed that the genetic redundancy in polyploids (to which group more than 35% of all angiosperms belong) allows a considerable rearrangment of the chromosome organization without loss of viability of the plant, as is often the case in diploids, where all chromosomal mutations are accompanied by reduced viability, a higher level of sterility or other adverse effects.

In addition to biological factors, a number of external (or environmental) modifying factors also influence the response of (plant) cells to a mutagenic treatment. The FAO/IAEA Manual (Anon., 1977a) mentions in this respect 1) oxygen, 2) water content, 3) conditions and duration of post-irradiation storage, and 4) the temperature.

For practical mutation breeders it suffices to know some basic principles only with respect to these factors, as for most types of treatments and crop species adequate procedures have been worked out and can be found in literature. These factors are important only when dormant seeds are irradiated. For irradiation of actively growing cells (or plant parts) or for in vitro treatments the situation is not comparable.

ad 1) Oxygen is considered the most important modifying factor, as the presence of oxygen could enhance biological radiation damage by a factor of up to 100. The subject of radiobiology will be discussed later in this chapter. The other aforementioned factors apparently all interact with the oxygen, and therefore play a more modest role. It appears that oxygen effects are much more important in sparsely ionizing radiation types (X- and gamma rays) than, for instance, when fast neutrons are applied. Antioxidants and anaerobic conditions may act protectively to limit radiation damage. So-called sulfhydryl or (SH-) compounds (which belong to the mercapto group), like 1,4-dithiotreitol (DTT), cysteine and cysteamine, are effective radioprotectors, as can be measured on the basis of seedling injury, M_1 sterility, embryonic lethality, ovule sterility, etc. For some details on DTT applications see van der Veen, van Brederode & Vis (1969) and Sree Ramulu & van der Veen (1973). DTT apparently does not improve the mutation spectrum in vegetatively propagated crops (Broertjes, 1976a).

ad 2) Moisture content or water content is also considered an important modifying factor – in particular in resting seeds – obviously in relation to respiration and gas transport. It has to be distinguished from the effect of soaking seeds in water, which initiates enzymatic processes prior to germination including pre-mitosis. However, it appears that moisture effects are not as clear-cut as has been thought in the past. In the FAO/IAEA Manual many details are given on effects caused by various water contents in barley seeds. In resting barley seeds with less than 14% water, the sensitivity to X- and gamma radiation increases as the water content decreases. One has to distinguish apparently the effect of higher water content as increasing respiration, and of very low water content as preserving radicals till the time of germination.

For many plant species the water content of seeds when stored under normal laboratory conditions appears to be in the range of 10.0 and 11.5% and a difference of 0.2 or 0.3% may already greatly affect the radiosensitivity of some species. For seed treatments with chemical mutagens, presoaking is common practice and important in relation to the uptake of the mutagen. This subject will be discussed later on.

The most important point for the mutation breeder to keep in mind is that, in order to perform reproducible results, the water content of seeds should be determined, controlled (e.g. no more than 0.1% deviation) and noted down, in particular when X- and gamma rays are involved.

ad 3) The effect of temperature is minor in comparison to oxygen and water content, but in interaction with these factors and storage (ageing), temperatures may have a recognizable modifying effect as well during and after radiation treatments. For chemical mutagens the temperature during the reaction is extremely relevant as it strongly affects the reactivity of the process!

ad 4) Seed ageing, in the early days of mutation breeding (see also Chapter 3), was considered a method to increase the frequency of 'spontaneous' mutations. Hugo de Vries (1901) in this respect already described the effects when using five year old seed of Oenothera. Seed ageing – in the physiological meaning – in particular during the post-irradiation phase (e.g. storage conditions before sowing the irradiated seeds) also may drastically increase the effect of the mutagenic treatment. It is therefore generally advised to sow seeds as soon as possible after mutagenic treatment and if storage for extended periods should be unavoidable, this should be done dry, at low temperatures, in the dark and as much as possible in the absence of oxygen in small air-tight tinfoil bags or in sealed boxes.

Keeping records

Earlier in this section the importance of keeping records (in addition to all relevant information about the radiation source and treatment conditions) of all kinds of details about the plant material that is submitted to irradiation was stressed. Such details are required to determine, in particular, which dose gives optimal results, that is the highest mutation frequency – preferably of the desired mutation type – in combination with a high survival rate, no reduced seed vitality and fertility and no unwanted other accompanying (genetic) changes, etc.

When a successful experiment is to be repeated such records are indispensable. Records, for instance, have to include the species name (in Latin), the name and history of the specific cultivar or genotype involved, the ploidy level, the growing conditions of the parent material (where and under which conditions grown, seed storage,etc.), the kind of material treated (seeds, tubers, bulbs, cuttings, cell suspensions, etc.), under which physiological conditions, etc. In Chapters 6 and 7 more specific points will be mentioned in this respect.

Before further discussing other relevant aspects of various radiation types and radiation sources (or facilities), some general information, including a small number of often used expressions and definitions, has to be given.

4.2.2. Radiation quantities, units and definitions

Throughout the years a range of different quantities, units and definitions has been used to describe various phenomena in the field of radiobiology and related topics and, in addition, now and then new units have been introduced to replace previous ones. This, often, has made the literature not easily accessible to non-professionals. In the following sections we will use as much as possible the so-called SI (Système Internationale) units, which – officially – were introduced in 1974 (but in fact later), and the recommendations by the ICRU (International Commission on Radiation Units and Measurements), which are published for instance in *ICRU News*. An important report of the ICRU with definitions of physical quantities pertaining to ionizing radiations, is ICRU Report 33, *Radiation Quantities and Units* from 1980, that should be replaced in the near future by a new report.

Various descriptions that will be used in the following sections have been derived from the authoritative book on radiation biophysics by E.L. Alpen (1990). Use is also made of a number of earlier, general references, like Lea (1946), Bacq & Alexander (1961) Dertinger & Jung (1970), Lawrence (1971), the FAO/IAEA Manual (Anon., 1977a) and Chadwick & Leenhouts (1981).

In radiation science many 'special units' are known

and in 1975 the ICRU recommended the use of SI units instead. Basic SI units, for instance, are kilogram (kg) for mass, metre (m) for length, second (s) for time and ampère (A) for electric current. A derived unit is for instance metres per second (m. s⁻¹). Within the SI system joule (J) is the unit for energy or 'work'(= 1 kg. m². s⁻²). It is rather doubtful whether the old 'special unit' for energy, electronvolt (eV), which equals 1602×10^{-19} J, will be completely replaced by J. The ICRU further (probably also for political reasons) proposed two new radiation units: becquerel (Bq), named after a French physicist who discovered radioactivity, instead of the old unit curie (Ci); and gray (Gy) to replace the units rad and Röntgen.

In Box 4.2 the most important and most often used definitions and units to describe and quantify different aspects of radiation facilities and radiation are compiled.

When radiation sources are used to induce mutations in plant material, a number of particulars have to be noted down carefully with respect to those sources and the way they are handled in order to be able to repeat an experiment in a completely identical way. Apart from the technical details about the radiation source and the conditions under which it has been operating (including adjustments like the use of special filters), the radiation dose and the dose rate, which have been defined already, must be always recorded.

4.2.3. Different types of radiation and radiation sources

The present-day use of radiation is limited to a few types only, which will be described in the next sections. An anecdotal illustration of the somewhat peculiar ways in which some radiation treatments used to be performed in the past is given by the following story. J. James, after World War II, started a series of experiments to induce mutations in roses by irradiation. In his own words (James, 1961): 'At first I collected every luminous dial clock or watch possible and scraped off the luminous material. This gave a minute quantity of radium to which I exposed the terminal buds of several roses. My first success was an induced sport of "Soeur Thérèse"...'! James continued to work on mutations in roses and later succesfully (and in a more cautious way, we trust) applied ⁶⁰Co gamma rays. Further references are given in Broertjes & van Harten (1988, p. 203).

Ultraviolet radiation

When ultraviolet light (UV) is used to irradiate pollen, more than 60% of its energy is already absorbed by the thin pollen wall before reaching the nucleus and when other plant parts are involved more than 90% may be absorbed by compounds of the epidermal cell layer.

Despite this limited penetration, UV radiation does have some features which makes its use attractive, at least when small targets like pollen grains, cell suspensions and spores of fungi are treated.

Mutagenic treatment of pollen grains for breeding purposes is of particular interest, both from a fundamental and a practical view point, because of the haploid character of the pollen grains. UV has also been used to increase for instance the low spontaneous mutation frequency (one mutation per 10^7 cell divisions) in baker's yeast (*Saccharomyces cerevisiae*).

The recent interest in *in vitro* methods in plant breeding has led to a further revival of the interest in UV irradiation for mutation induction.

In the 1930s Stadler & Sprague (1936a,b; 1937) and Stadler (1939) reported that UV induces mutations that closely resemble the spontaneous ones. Attractive features, mentioned in particular for UV radiation, are that it has a certain degree of specificity, is easily available, convenient and relatively safe. Stadler & Sprague (1937) reported that even after very high (unspecified) doses of UV radiation had been applied, only one translocation was found after examination of 100 M_1 plants which had been taken at random. For a 'moderate' dose of X-rays (exact dose not given), 44 out of 100 M_1 plants showed at least one translocation. The corresponding percentages of pollen lethality after UV radiaton and treatment with X-rays were 14% and 77%, respectively. A large part of the UV data collected by Stadler has never been published. Some experiments went on till after Stadler's death (Rhoades, 1956; see also Chapter 2, Box 2.2).

Nuffer (1957) showed that mutations which only affected locus β (the central gene of the close linkage group, indicated as 'α β *Sh*' on chromosome 3 of maize, *Zea mays*) could be induced after UV irradiation of pollen grains but not after X-irradiation.

UV radiation mainly results in unusual or very specific bonds (e.g. pyrimidine dimers) between DNA bases, in breaks or in pairing between complementary bases. It is often reported that UV radiation produces relatively many discrete mutations, both of the frameshift type as well as base pair substitutions, which are in particular of the transition type.

In a later section of this chapter we will discuss in a more general way how repair of radiation damage occurs. In repair of UV damage in pollen, a topic of practical importance, several mechanisms may be involved, including so-called light repair (photo reactivation) in which repair enzymes are activated, dark repair (excision repair) and recombinational repair. Mutations may arise as a result of the fact that repair processes are error-prone (Jackson, 1987).

Generally, three classes of UV radiation are distinguished: UV-A, UV-B and UV-C, of which UV-A (with wavelengths 320–400 nm) and UV-B (280–320 nm) are present in the sunlight. (N.B. See Box 4.1 for units used, etc.) No significant UV radiation below 295 nm reaches the Earth's surface, but because of the depletion of the ozone layer surrounding the Earth, the UV-B level at the Earth will increase, which results in damage to plants and in changes in plant morphogenesis. The 'natural' UV-C region includes wavelengths below 280 nm, which are highly energetic and, as they are absorbed by the aforementioned ozone layer, are not present in the sunlight that reaches the Earth's surface.

Shortwave UV-C is used as a mutagenic agent. The UV radiation used for this purpose, which mainly causes a so-called excitation effect in which orbital electrons are raised to higher energy states, is generated by mercury or cadmium lamps (so-called 'black' lamps, germicidal lamps for sterilizing purposes and 'sun' lamps). The minimum radiation energy required for a single ionization is more than 10 eV, which is about the case for the shortwave UV-C. (N.B. Visible light ranges from 2 to 3 eV.) Wavelengths of 250–290 nm are biologically most active since at about 254 nm maximal absorption by nucleic acids (DNA) is observed.

In the FAO/IAEA Manual (Anon., 1977a, p. 12) Briggs & Constantin give several early references which provide details about equipment and procedures for UV treatments. (N.B. Different units have been used over the years and it has not been tried in all cases to standardize them.)

Emmerling (1955) irradiated pollen grains of maize (*Zea mays*), which were deposited in a single layer on a quartz slide, with a Westinghouse Sterilamp (24 W, without filter), which emitted about 80% of its energy near 253.7 nm (2537 Å). The irradiation was performed from above and below for 10, 20 and 30 s. The method to irradiate both from above and below is still sometimes advised, as in this way different positions of the nuclei in the pollen grains may be reached.

Nuffer (1957) shook pollen from freshly opened anthers of maize on a small glass slide which was placed for 30 s between two 'steri-lamps' (further details unknown) with the lamps at a distance of four inches (about 11 cm) from each other. Immediately afterwards the irradiated pollen was evenly scattered on the receptive stigmas of the female inflorescences (silks) of maize plants. If pollen from plants possessing the dominant genotype is irradiated and recessive females are used, the effects of the mutagenic treatment can be easily studied by investigating the progeny, in particular when easily visible traits are used.

Cline & Salisbury (1966) studied for 67 species of higher plants lethal effects of UV radiation at a wavelength of 253.7 nm (2537 Å), derived from a lamp which consisted of four 15 W low pressure mercury tubes. The total emitted radiation by this lamp was 10^6 erg. cm^{-2}. min^{-1} (0.076 cal. cm^{-2}. min^{-1}) at a distance of 20 cm. Exposures were varied by altering the distance between plants and source and by varying the time of exposure. An enormous variation in relative sensitivity was found. Leaves of *Coleus blumei* and *Pharbitis nil*, for instance, were killed at very low exposures (less than 2 cal. cm^{-2}), whereas some conifers like *Pinus ponderosa* and *Agave* sp. survived exposures which were 10^3 times higher.

Ikenaga & Mabuchi (1966) also UV irradiated pollen grains of a maize line with a dominant marker gene in a single layer in a Petri dish at 253.7 nm, emitted by a Toshiba germicidal lamp at a dose rate of 90 erg. mm^{-2}. s^{-1} for 10–90 s. (N.B. As UV does not penetrate glass or plastic, the cover of the Petri dish must be removed before irradiation!) In this experiment half of the irradiated material was submitted to a post-treatment with 30 min visible light from fluorescent lamps whereas the other half was kept in the dark, also for 30 min in order to study repair of UV-induced damage by photoreactivation. The authors indeed found considerably decreased mutation frequencies after the additional treatment with visible light (N.B. Further details are outside the scope of this section.)

A recent example of treatment conditions for microspores (pollen grains) of rapeseed (*Brassica napus*) is derived from MacDonald et al. (1991). Relatively large numbers of haploid embryoids (i.e. 4.5 embryoids per 1000 microspores plated) can be obtained on a routine basis and 90% of the embryoids can be developed into plants when proper techniques are applied. This material is considered ideal for comparative studies of *in vitro* mutagenesis. Isolated microspores in the late uninucleate or early binucleate stage were irradiated in Petri dishes. A Philips germicidal lamp, model 57415/40 IC, 15 W, with the major emission at the wavelength of 254 nm, was operated at a dose rate of 33 erg. mm^1. s^{-1}. The Petri dishes were put in a laminar flow cabinet at a distance of 21 cm from the source. The LD 50 (see Box 4.1) is often used to decide on the proper dose. This LD 50 may depend on various factors including the genotypic constitution and the kind of explant material. In the experiment described here it was found that at an exposure time of less than 20 s the number of embryoids produced was halved but, in contrast to treatments with X-rays and gamma rays, the potential for plant regeneration did not decrease significantly when treat-

ment time was increased to 40 or 60 s. The experiments by MacDonald et al. confirmed that UV mainly induced gene mutations and deletions.

X-rays

The history of the discovery of X-rays by Röntgen in 1895 has been briefly discussed in Chapter 2. We will refrain from presenting at this point particulars about the well-known role of X-rays in medical diagnoses, treatment of cancer, etc., and limit ourselves to providing some details which are related to the use of X-rays for the induction of mutations in plants. X-rays, at present, are the second most important source for the production of mutant cultivars. For historic reasons they are discussed here before gamma rays, which, nowadays, are most often applied.

X-radiation, an electromagnetic type of radiation with wavelengths of 10^{-3}–10 nm (compare UV lamps with wavelengths of about 254 nm), is a form of cosmic radiation, but can also be generated in high vacuum in the cathode tube of X-ray machines where fast electrons are electrically accelerated and then abruptly stopped by hitting a material with a high atomic number (a heavy metal), e.g. molybdenum or tungsten. When the electrons hit the metal, their energy is released as X-rays. The energy released after slowing down high-energetic electrons is known in literature as 'Bremsstrahlung'. X-rays and gamma rays differ from each other in the way that X-rays originate in the electron cloud which surrounds the atomic nucleus, whereas gamma rays result from the emission of energy by the atomic nucleus.

The radiation energy of X-rays may vary from 10 to 400 keV (k = 10^3). X-rays cause ionizations in the atoms of the (plant) tissue they meet. So-called hard X-rays, with the shortest wavelengths, show a penetration in tissue of many centimeters, whereas soft X-rays do not reach deeper than a few millimetres. Soft X-rays (with longer wavelengths), on the other hand, are more densely ionizing than hard X-rays. The higher the so-called peak operating voltage (kVp) of the tube, the shorter the wavelengths that can be produced. The usual X-ray machines have peak energies of 50 keV or more. Combination of a high peak voltage and specific filters (which mainly absorb soft X-rays) results in the production of hard X-rays, which react in a way which is about similar to gamma radiation. According to the FAO/IAEA Manual (Anon., 1977a), records of experiments involving X-ray irradiations should always contain information about the peak operating voltage (often 250 kVp) of the tube, the electric current used (e.g. 15 milliamperes or mA), the use or absence of (e.g. Al or Cu) filters (type; thickness) and the so-called half-value layer (HVL), which is a measure of the effective energy of the X-ray machine or of the quality of the beam and usually expressed in millimetres of Al or Cu.

The **radiation dose** (preferably expressed in **Gy**) and the **dose rate** (in **Gy. s^{-1}** or **Gy. min^{-1}**), of course, always have to be recorded as well. Common doses in case of X- (or gamma) irradiation may range from a few grays for vegetative plant parts, for instance about 4–5 Gy for whole bulbs of an ornamental like tulip (*Tulipa* sp.); 5–100 Gy when irradiating explants before starting an *in vitro* culture and up to 500 or even a few thousand Gy for seeds and specific *in vitro* material. Dose rates mostly range from 1 to 20 Gy. min^{-1}, but dose rates of 100 Gy. min^{-1} are mentioned as well in literature.

In Section 4.1 it was already pointed out how important it is to note down – in addition to all relevant information about the source, the treatment conditions, the dose and the dose rate – all kinds of particulars about the plant material to be treated. This information should include information about the way of propagation (cross-pollinator, self-pollinator, vegetatively propagated?), details known about the cultivar involved (e.g. its genetic parentage and constitution), the kind of plant material to be treated (seeds, cuttings, tubers, bulbs, plants, pollen, various types of *in vitro* material), the ploidy level, the physiological state of the plant material, etc. It is, in this respect, much better to discover afterwards that too many details were noted down than too few!

It was also mentioned before that the **dose rate** can be manipulated by changing the distance between the tube of the X-ray apparatus and the treated plant material. With common X-ray equipment this can be done by moving the tube upward and downward whereas the plant material has been placed in a fixed position on a table, preferably a rotating table on which the plants, cuttings, etc., are arranged in circles in order to administer exactly the same dose to each object.

A dosimeter (for instance an ionization chamber, in which the number of ionizations is counted as a relative measure for the dose received by the objects) is placed amidst the treated material in such a way that it has exactly the same distance to the X-ray source as the objects. This goal is relatively easy to achieve when irradiating seed or Petri dishes with small *in vitro* plant material. However, when for instance whole pot plants are irradiated, it should be realized that the reproductive parts of the flowers are the real targets whereas, in case of irradiating cuttings, this may be existing axillary buds or a meristematic area from which adventitious buds may develop. In addition, other areas which are genetically irrelevant but may be in particular sensitive to radiation and of importance for the future development of the plant, like the rooting area of cuttings, should be protected against radiation damage, for instance by a lead shield.

Gamma rays

Gamma rays, at present, are the most favoured mutagenic agent. Gamma rays were discovered in 1900 by P. Villard, after alpha and beta radiation had been discovered the year before by E. Rutherford and F. Soddy (see also Chapter 2).

Gamma rays originate in the atomic nucleus and can be emitted by unstable nuclei. They are electromagnetic radiation with the shortest wavelength (shorter than X-rays), i.e. they represent electromagnetic radiation with the highest energy level. Gamma rays often accompany the emission of alpha and beta particles. The level of radiation energy of gamma rays produced in nuclear (atomic) reactors may be up to 10 MeV (megaelectronvolt; M = 10^6). Gamma rays are not particles and have no electric charge. Their great penetrating power makes them dangerous as they can cause considerable damage when they pass through the tissue. Because of the hazards involved, gamma sources must be well shielded, e.g. by a 2–5 cm layer of lead or by a metre of concrete, according to the energy (which depends upon the isotope that is used).

Gamma radiation is usually obtained by desintegration of the radioisotopes ^{137}Cs or ^{60}C. Cobalt-60, which has two peaks in its spectrum of radiation energy at 1.33 and 1.17 MeV (specific activity 1–400 Ci. g^{-1}), has a half-life of 5.27 years. Cobalt decays to stable tin (stannium). Cesium-137 is mono-energetic with an energy peak at 0.66 MeV (specific activity 1–25 Ci. g^{-1}). For plant breeding purposes a ^{137}Cs source is more practical and economic as its half-life time is 33 years.

A gamma source can be used in the same way as an X-ray machine, but is often preferred for prolonged (chronic) treatments of several months or more, for instance in growth chambers, greenhouses or in gamma fields with a gamma source in the centre and growing plants in concentric circles around it. For acute irradiations rotating tables are used as well. At the time of exposure a ^{137}Cs or ^{60}C rod is brought up from the shielded area. The distance between the source and the irradiated plants determines the dose rate that is applied. For instance, a gamma-ray source, used for radiation therapy in a hospital, may produce 2 Gy. min^{-1} at 1 m distance over a field of about 40 cm^2. Most sources have an automatic time set that switches off when the total pre-set dose has been applied, after which the rod drops back automatically in the shielding. Chronic irradiations in general lead to a somewhat lower mutation frequency than when the same total dose is administered as an acute irradiation. Gamma sources can be used to irradiate a wide range of plant material, like seeds, whole plants, plant parts, freshly picked flowers on agar, anthers, pollen grains, single cell cultures or protoplasts. Examples of suitable doses and dose rates will be presented in later chapters.

Of about 20 gamma fields which, mainly in the 1950s and 1960s, were constructed all over the world, only a very few are still operating. The gamma field at Brookhaven National Laboratory, USA, was the first one to be established in 1949; the one in Bålsgard, Sweden, followed in 1951. Nybom *et al.* (1956) already described initial experiments concerning the effects of chronic irradiation of barley at the Bålsgard Fruit Breeding Station. The gamma field at the Institute of Radiation Breeding in Ohmiya, Japan, which came into operation in 1962, is shown in Fig. 4.3 (courtesy E. Amano). The picture, taken in 1985, shows the location which is surrounded by a 8 m high shielding dike, which limits the direct gamma beams to the field area. The field is situated about 1 km from the nearest residential area. The

^{60}Co source (2400 **Ci**/8.88 × 10^{13} **Bq**) is renewed every two years and will be operational at least until 2000. In the gamma field a daily dose (20 h each day) of 2–150 Gy may be administered (depending on the distance), whereas a treatment may be continued for 100 days or more. Further details can be found in issue 6 of *Gamma Field Symposia* (Anon., 1967).

Each gamma source has its own specifications. As an additional example, some particulars about another source are given here. The facilities of the Pilot Plant for Food Irradiation Wageningen, The Netherlands, which were owned by the RIKILT-DLO, a research institute of the Dutch Ministry of Agriculture, were often used for inducing mutations in plants (at relatively low costs). It was operational from 1979 till 1996 with a comparably small ^{60}Co gamma source. The advantage of this source was that low dosages could be administered, whereas much more material could be treated in the same time than would be the case with X-rays, where only a few plantlets can be put on a single, small rotating table, directly <u>under</u> the X-ray apparatus. With the gamma source use could also be made of (a number of) rotating tables which are situated adjacent to the gamma source

Figure 4.3. Gamma field of the Institute of Radiation Botany, Ohmiya, Japan (courtesy E. Amano).

with the plant material to be irradiated placed in trays on top of those tables. The aforementioned gamma source in 1994 produced about 50 Gy. h^{-1}. The closing down in 1996 of this government-owned facility (an economy measure) may seriously hamper the continuation of the successful mutation breeding work in the Netherlands.

Cuttings of ornamentals, for instance, may be irradiated with a dose of 10–30 Gy. When several doses are given, the easiest (and cheapest) way is to start with the lowest dose, switch off the source, remove those plants and proceed with the second lowest dose, etc. After finishing all treatments, the source is stored again in the protected area in the floor. The aforementioned Pilot Plant in Wageningen had a large gamma source as well, but this source produced far too high dose rates for mutation induction. In such a case the total dose required for mutation experiments, for instance, would be administered in less than a second, which would make it impossible to exactly control and repeat treatments.

As has been said before, short definitions of the most important units used in radiation physics mentioned in this chapter, as well as definitions of some other units that are of importance in relation to mutation breeding, have been put together in Box 4.2.

Alpha radiation

Alpha particles (helium nuclei) consist of two protons and two neutrons. Such clusters of protons and neutrons can be emitted from isotopes which, by the ejection, are transformed into other chemical elements. Alpha particles carry a positive charge and are relatively slow moving because of their mass and because they are hindered by the negatively charged electrons surrounding each atom. They can be accelerated and may have an energy level of some megaelectronvolts (MeV). They are potentially very dangerous, e.g. when the alpha-emitting substance (e.g. Radium or Plutonium) is ingested or inhaled, but as their penetration power is low; they cannot penetrate a plant epidermis, a human skin or even a sheet of paper. Therefore alpha radiation, despite the fact that it is a high-LET radiation type (see Box 4.2), is hardly used for mutation induction.

Beta radiation

A beta particle is an electron (a negatively charged particle) which can be emitted when a neutron turns into a proton. (N.B. When a proton turns into a neutron by emitting a positron, the combination of the positively charged positron with the negative electron produces two gamma rays!)

Beta particles are emitted by isotopes like ^{3}H, ^{32}P or ^{35}S and produce effects which are rather similar to those of X- or gamma rays. Their penetration power, however, is lower than is the case for X- and gamma rays (but deeper than alpha rays). They can penetrate a few millimetres in plant tissue, human skin or paper. As is the case with alpha particles, they are much more dangerous when the emission takes place inside living tissue.

Electrons are not often applied in mutation breeding, but when accelerated they can be used, for instance, for food irradiation or surface sterilization. It is mentioned in the FAO/IAEA Manual (Anon., 1977a) that beta particles can be brought for instance in plant tissue by making use of ^{32}P labelled solutions to which the plant material is submitted. This overcomes the low penetration power and, when incorporated directly into the cell nucleus, may give a somewhat greater localization of the action of the radiation.

Beta particles move at nearly the speed of light and the energy level of beta radiation may vary from 10 keV to 3 or more MeV. They can, like all particles, be accelerated. Use can be made of so-called (linear) electron accelerators. A 10 MeV accelerator is situated at the Risø National Laboratory, Roskilde, Denmark. So-called beta-trons, machines which are particularly developed for accelerating electrons, essentially are high-voltage X-ray machines (however, without emitting X-rays).

Neutrons

Neutrons, uncharged particles, can arise in different ways, for instance by ejection from the nucleus during a nuclear fission. During nuclear chain reactions ejected neutrons cause a cascade of nuclear fissions. Through this, great numbers of neutrons are produced. When colliding with other atoms, they may enter the nucleus due to the absence of any electric charge. Thus radioactivity can be induced as a result of making nuclei physically unstable.

Neutrons may carry different kinetic energies and therefore may be called either fast (N_f), or slow, or – if the energy is as low as that of gas molecules at room temperature – thermal (N_{th}), and as such show very different interactions with matter. The use of moderators, like carbon, paraffin or water, can lead to reducing energy and slowing down of fast neutrons, which results eventually in the production of thermal neutrons: (uncharged) particles with mass 1, travelling at a speed regulated by the environmental temperature.

Neutrons of mixed speed (fission neutrons) are produced in atomic reactors from ^{235}U fission. In the mid-1990s about 75% of all nuclear power plants in operation were of the so-called 'light water reactor type', which type, operating for more than 35 years, together with a second category of 'gas-cooled reactors', make use of thermal neutrons, i.e. with an energy level of less than 0.01 MeV. Fast neutrons with an energy level of 1 up to 20 MeV are used for sustaining the fission process in a third category of reactors, the so-called 'fast reactors' or 'fast breeders'.

Neutrons of defined speed can be produced by various types of accelerators and appropriate filters. They release energy in a different way than X-rays or gamma rays. Neutrons may have an ability of penetrating into tissue for many centimetres. Thick shielding by material of light atomic mass, e.g. by concrete or water, is needed.

Neutrons are considered to be very effective for mutation induction, but may also cause gross destruction of the treated material, like the induction of many chromosome breaks. Often the **relative biological effectiveness** (RBE ; for definition see Box 4.2) for neutrons is higher than for X-rays or gamma rays. Broertjes (1972a), who applied fast neutrons and X-rays to detached leaves of the ornamental crop African Violet (*Saintpaulia ionantha*) for studying the effects of various mutagenic treatments upon survival, production of adventitious plants and mutation frequency, found that, when compared with X-rays on a rad basis, the RBE of N_f was 2 to 3, depending also on the dose rate that had been applied.

Only a small proportion of the biological action of neutrons is caused by the two primary processes of interaction of neutrons with atomic nuclei: a) elastic collisions between neutrons of moderate energy and the nuclei ('scattering') and b) capture ('absorption') of neutrons by the nuclei. Most radiation damage is caused by the action of recoil nuclei (H,C,N,O) and by the secondary radiation produced in neutron capture processes. The radiation dose of fast neutrons is for 90% produced by recoil nuclei, whereas in the case of thermal neutrons alpha and beta radiation are responsible for producing the radiation dose. More details, which are of no direct importance to mutation breeding, for instance, can be found in Dertinger & Jung (1970).

Till about 1970 many reports were published about the application of fast and thermal neutrons in mutation breeding of seed propagated crops as well as vegetatively propagated crops. References for instance can be found in the FAO/IAEA Manual (Anon., 1977a; for seed propagated crops see pp. 120–1; for vegetatively propagated crops see pp. 162–3). Reports on treatments with neutrons at that time remarkably often referred to mutation programmes aiming at the improvement of polygenically inherited traits (e.g. see Brock, 1970 for the use of thermal neutrons in this respect). Doses of

thermal neutrons are often given as the number of neutrons per cm² (e.g. $7 \times 10^{12}\,N_{th}$. cm⁻²) whereas for fast neutrons the expressions rad or Gy for doses and rad. min⁻¹ and Gy. min⁻¹ for dose rates are used.

Indeed, when compared to the 1970s the use of neutrons for mutation induction has decreased considerably. Several reasons account for this. Nuclear reactors, in which neutrons are produced, often are not available and, even if they should be available, are not always easily accessible to plant breeders. Moreover, fast neutrons are always accompanied ('contaminated') by gamma rays, and thermal neutron capture by alpha, beta and gamma rays and the effects by the different types of radiation cannot be easily separated. Part of the mutagenic events that in the past have been attributed to neutrons, in fact may have been caused by the accompanying gamma rays, which in particular might have been the case for the more favourable gene mutations. This view is supported by the fact that secondary high-LET radiation, such as is produced by neutron capture in tissue (see Box 4.2), is known to produce relatively many gross chromosomal aberrations, whereas gamma radiation (low-LET) may result in a higher proportion of gene mutations and small deletions.

4.2.4. Radiobiology

General remarks

The interaction of radiation with matter has been a field of considerable research efforts for many years and still many aspects are insufficiently known. In the context of this book there is no need to point out more than the main lines only. For additional reading one could for instance refer to a number of specialized publications, like Lea (1946), Bacq & Alexander (1961), Drake (1969; 1970), Dertinger & Jung (1970), Lawrence (1971), Anon. (1977a; a contribution by Ahnström in the FAO/IAEA Manual), Véleminský & Gichner (1978; 1987), Chadwick & Leenhouts (1981), Friedberg (1985), Kimball, (1987), Ahnström (1989) and Alpen (1990). In particular Dertinger & Jung (1970) and the small booklet by Lawrence (1971) give 'relative outsiders' an easy introduction to this subject.

Ionizing radiation as well as UV radiation induce a broad spectrum of physical and chemical events in the DNA and RNA of prokaryotes and eukaryotes. The general opinion is expressed that when X- and gamma radiation are involved, at least initially, physical effects (followed by chemical reactions) play a major role, whereas effects of UV radiation, the study of which is often considered as a different field called photobiology, primarily are said to be of chemical nature. Plants, according to prevailing present opinions, are aqueous systems in which ionizing radiation results in effects of various kinds in all molecules of the irradiated material. The biological effect of ionizing radiation depends primarily on the amount of energy that will be absorbed by the biological system of which, of course, the chromosomes are the most important target. Based on statistics and due to the specific properties of plant material, radiation doses below 1 Gy mostly are considered not to cause observable effects of any significance in plants.

Early concepts

In the 1920s the so-called Hit Theory was developed which, using a physical–mathematical approach, offered an adequate base to interpret dose–response curves (e.g. for 'survival') after different mutagenic treatments. (For some early references, see Chapter 2, Section 2.3.) This Hit Theory, in main lines, refers to the observation that ionizing radiation transfers its energy in discrete packets and that interactions (or 'hits') with the biological material occur independently of each other. In addition, it was postulated that in order to obtain a specific effect or response within a specified target, a defined number of hits (e.g. per cm^3) was required. The Hit Theory gradually was expanded and developed into, what became known as the Target Theory, to which model the name of Lea (1946) is firmly attached.

The essential point of the Target Theory, according to Dertinger & Jung (1970, p. 219), is that the radiation sensitivity of a biological system is positively correlated with the size of its sensitive 'target'. Such 'targets', for instance, could be the total amount of DNA in viruses and bacteria or the cell nucleus – including the nuclear DNA (see the work of Sparrow and associates that was briefly discussed in Section 4.2.1) – of higher plants. The interpretation of dose–response (survival) curves – which are either exponential or sigmoid – was, for several years, largely based on Lea's model. According to Alpen (1990) the Target Theory of Lea, in which also the so-called Breakage-and-Reunion Theory of Sax (1940; 1941) had been incorporated, would suggest that aberrations that require one hit, will follow a dose–response relationship that is first order, whereas those that require two or more hits should show a relationship of second or third order.

According to several authors, however, the Target Theory, still did not sufficiently explain the collected data. Revell in 1955 proposed the Exchange Theory, in which the hypothesis was forwarded that all chromosome aberrations, including deletions, arise from an exchange process that looks identical to meiotic cross-

ing over (Revell, 1959). Comprehensive reviews of these early theories have been given by H.J. Evans (1962) and Revell (1974). Lawrence (1971, p. 28) mentioned that, although the hypotheses by Sax in 1940 and by Revell in 1955 had been posed as alternatives, both hypotheses might be correct, depending on specific conditions, etc.; but we will not further elaborate on this theme. Heddle, Whissell & Bodycote (1969) in this context commented that the occurrence of chromatid deletions results from a combination of unrejoined single breaks and incomplete exchanges between sister chromatids.

In more recent years some alternative theories to explain the molecular base of radiation biology have been put forward, starting with the model of Chadwick & Leenhouts (1981), which is either called the Molecular Model or the Linear-Quadratic Model. In this model, in which it is tried to bridge the gap between physical processes and the biological outcome, it is anticipated that the integrity of the DNA is important for the functions as well as the reproduction of the cell. Radiation may lead to the rupturing of molecular bonds, but repair is sometimes, and to a certain degree, possible. Alpen (1990) mentions that the model of Chadwick & Leenhouts is not adequate in all situations. Details are complicated and outside the scope of this book.

Radiation damage in cells and plants

Nowadays, it is beyond doubt that within a cell the DNA is by far the most important target for radiation. There is also general agreement on the basic concept of the Target Theory that the radiation sensitivity of a biological system increases with the size of its sensitive target (Dertinger & Jung, 1970, p. 219). The relative sensitivity of cells to radiation differs at various stages of their life cycle (specific stages of the mitotic cycle, meiosis). Resistance to radiation between different organisms may differ dramatically. Daly & Minton (1995), for instance, mentioned that a human, exposed to 5 Gy of ionizing radiation most probably would die, whereas the bacterium *Deinococcus radiodurans* (!) is capable of surviving $5–30 \times 10^3$ Gy, because of its highly effective mechanism to repair DNA damage in which hundreds of DNA fragments are re-assembled into intact chromosomes.

Radiation sensitivity in higher plants for the same number of nucleotides, is higher than in viruses with single-stranded or double-stranded RNA. Cell cultures of higher organisms, consisting of haploid cells, show a ten times higher radiation sensitivity than when diploid cells are involved and several other interesting phenomena can be observed with respect to radiation sensitivity of cells, tissues, specific molecular functions, developmental stages, etc.

Dertinger & Jung (loc.cit) suggest several possible explanations for such repeatedly observed results, such as the operating of better repair mechanisms in higher organisms; the possibility that different mutational events (e.g., point mutations versus chromosome breaks) are playing a predominant role at the haploid and diploid level, the presence of redundancy of genetic information in specific situations and, finally, the simple fact that, for instance, diploid cells contain two copies of each DNA molecule. In addition, for higher plants the situation in polyploid species may differ again from that in diploids and, in this context, the many contributions by Sparrow's group to relate radiation sensitivity to the interphase chromosome volume (or ICV) that have been discussed earlier in this chapter, may be mentioned here once again.

The different stages of development of radiation damage

For plant breeders a basic understanding of the various events that happen between the moment that energy enters the biological system (the plant) till the stage of the definite biological effect, including the induction of mutations, may suffice. Use is made here of a figure (Fig. 4.4) derived from Alpen (1990). Similar figures, showing more or less details can be found for instance in Dertinger & Jung (1970) or Lawrence (1971). The figure in particular refers to damage induced by electromagnetic, ionizing radiation.

When radiation passes through plant tissue (like any other matter), interaction with DNA molecules as well as all other molecules present in the system may occur. The three basic processes involved in the absorption of X- and gamma rays by interaction with atomic electrons: the photoelectric process, the Compton scattering process and the pair production process (i.e. the production of electron–positron pairs) are only briefly mentioned here. In the energy range from 50 keV to 20 MeV the Compton effect (in which only part of the energy is transferred to the atomic electrons) is the dominating process for substances of low atomic number, like plant tissues. Pair production occurs when photons enter matter with energy levels in excess of 1 MeV. The contribution of the photoelectric process, in particular in substances with low atomic numbers, becomes relatively small at higher energy levels. Details are outside the scope of this book.

Absorption of (part of) the energy by atomic electrons by the aforementioned processes may lead to ejection of electrons from these molecules and result in their ionization, as well as to a process of raising electrons to a higher energy state, called excitation. These ionizations and excitations mark the end of a first, very short phase the length of which is estimated at 10^{-16} to 10^{-12}s, called the physical stage, during which ionizing radiation leads to a range of non-directional effects in the DNA and in all other molecules of the irradiated system.

The second, chemical stage, which lasts for about 10^{-6} to 10^{-3} s, starts with the formation of activated molecules: so-called radicals (OH$^{\bullet}$ and H$^{\bullet}$), which arise from OH^{-} and H^{+}. This stage ends when the chemical stability is restored. A significant distinction is made between so-called **direct action** caused by macromolecular radicals (referring to the important biological molecules) and **indirect action** as a result of the activity of so-called free radicals, which operate in the aqueous system. Ahnström (1989) mentions that indirect effects, and in particular OH-radicals, account for about two-thirds of the total effects observed in aqueous systems. Alpen (1990) distinguishes five types of radiation-induced damage in DNA, including transitions and transversions of purines towards pyrimidines and *vice versa*; single strand breaks and double strand breaks resulting from damage to the deoxyribose–phosphate backbone at one, two or more locations, respectively etc. This subject has been discussed already in Chapter 3.

Repair of (single and double strand) DNA damage

The damage to the macromolecules (read: DNA), in part, can be repaired. The first reports about DNA repair refer to studies concerning repair in *E. coli*. Early reviews on this subject, for instance, were written by Witkin (1976) and Walker (1985).

Repair of radiation damage may be either error-free, or faulty (error-prone) repair takes place. In other cases the damage may be irrepairable, which may result in the loss of the damaged DNA parts. According to Alpen (1990), the first two mentioned (physical and chemical) stages are followed by a third one, called the biochemical stage, which lasts a few minutes and during which by various processes (enzymatic) repair of molecular damage takes place. In other publications this repair period is considered part of the second (chemical) phase.

It has been reported (see Micke, 1991a), that under normal conditions, i.e. without mutagenic treatments, errors in DNA replication may have a probability of about one per ten replicated genes, but by post-replication repair this is reduced to 1 per 10^8 replications. Single strand breaks in a DNA molecule (which, in particular, may be produced by UV and chemical

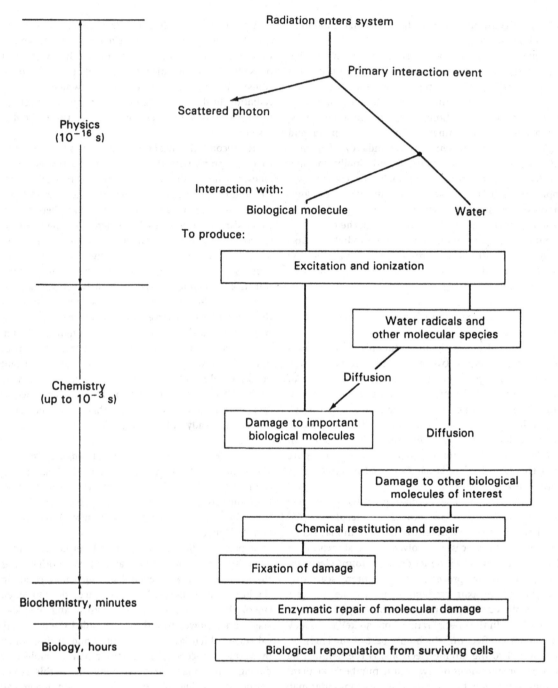

Figure 4.4. The processes from energy transfer to final biological damage. (From E.L. Alpen, 1990.)

mutagens), are believed to result in few cases only in loss of function of a cell. Usually, such breaks can be repaired fast and relatively error-free by enzymatic repair processes. The occurrence of double strand breaks has far more serious effects and repair is a much more complicated and error-prone process that often not only leads to mutations, but also to loss of reproductive capacity of the cell involved. Double strand breaks either may be the result of scission of both strands simultaneously or of two single strand breaks which occur almost simultaneously and not more than, say, ten base pairs apart.

The phase within the cell cycle in which the cells are at the moment of irradiation, determines which type of aberrations will occur. If double strand chromosome breakage occurs during interphase (before DNA replication) and if this breakage is not quickly repaired, the defect will either lead to non-replication or will also be replicated and be present in both daughter chromosomes in mitotic metaphase. If the break occurs after DNA replication, it exists in only one of the two chromatids and will usually result in a change in one of both daughter chromosomes only.

The final destiny of an induced aberration is difficult to predict. It is generally believed that ionizing radiation and some chemicals can induce aberrations by treatments in any phase of the cell cycle, with a different outcome. Other chemicals and UV irradiation are said to only induce aberrations in, or just prior to, the so-called S(= synthesis)-phase .

Modifying effects

Some additional remarks must be made about the period during which the damage to the DNA is administered. Alpen's figure, in this respect, does not show clearly that, during the phase in which direct and indirect action of the radicals lead to macromolecular changes (changes in the DNA), several protective and sensitizing systems and agents are active. They were already briefly mentioned in Section 4.2.1 and are called modifying effects. The action of such factors is not necessarily limited to this specific phase of the total process, but they could also act at other stages, i.e. in order to create radioprotective conditions before treatments, or during the prolonged final, biological or metabolic, stage.

Modifying factors either are of internal (mostly biological) nature, or refer to external influences. We have already mentioned in this respect the differences in radiosensitivity and mutability in relation to differences between species, genotypes, amount of DNA, interphase chromosome volume (ICV), ploidy level, etc. Therefore, at this moment we will limit ourselves to some comments concerning external modifying factors. The most important ones to be mentioned are water content, temperature, oxygen content, thiols and sulfhydryls, but some other factors play a (often modest) role as well. In general the presence of oxygen and water enhances biological damage by radiation. The effect of temperature remains relatively small. More details are presented in Box 4.3.

Contrary to many aspects of radiobiology which have been discussed before, this subject is, at least partly, under the influence of plant breeders and, therefore, deserves considerable attention. It is in each specific case worthwhile for the plant breeder to check literature (for instance by consulting *Plant Breeding Abstracts* and, over many years, in different issues of *Mutation Breeding Newsletter* or *Mutation Breeding Review*) whether something has been reported about the most advisable conditions of applying mutagenic treatments to a specific crop before starting his own experiments!

Again: repair mechanisms

Let us now briefly return to the action of repair mechanisms. Following Friedberg (1985) who, in nine chapters of his book *DNA Repair* has discussed the different types of DNA damage in a detailed way, repair – from a strictly biochemical point of view – is restoration of the original nucleotide sequence. An extremely efficient system of repair of DNA damage, reported by Daly & Minton (1995) was mentioned earlier in this chapter. This refers to the bacterium *Deinococcus radiodurans* in which the repair system is able to assemble hundreds of DNA fragments, which remained after treatment with 10×10^3 Gy of ionizing radiation, into intact chromosomes within a period of 12 to 24 h. Kimball (1987) who has given an interesting historical overview about evidence of repair of DNA damage in relation to the induction of mutations by radiation and chemicals, points out that cells have developed very complex mechanisms to protect themselves against damage to their genome, caused by internal and external factors. Those systems, however, are not perfect and those imperfections may lead to mutations and chromosomal changes. The fact that the length of the repair phase is of importance as well, according to Kimball, further supports the generally-accepted assumption that slowly dividing cells should be more resistant to permanent DNA damage than rapidly dividing ones.

The best known processes of repair of DNA damage, according to Friedberg (loc.cit.) are those processes in which the damage is repaired by specific repair enzymes or by a specific biochemical pathway. Friedberg distinguishes two fundamentally different systems of repair: 1) various types of direct reversal of the damage, and 2) excision of damaged areas.

A first example of direct reversal, given by Friedberg, refers to the enzymatic photoreactivation (or photorepair) of pyrimidine dimers, that are believed to constitute a major source of genetic damage after UV exposure at specific wavelengths. A second example concerns the repair of O^6 guanine alkylation occurring after treatment of cells with the monofunctional chemical mutagen MNNG (nitroso-guanidine). Other examples refer to the repair of sites where loss of one of the nitrogen bases in the DNA has occurred, by direct insertion of purines

Box 4.3 External factors modifying radiation treatments

Water content. The most important variable to modify the radiation response is the moisture of cells (tissues). Water plays also an important role as a medium for the production of so-called radicals, of which OH radicals are the most harmful ones. It can be concluded that the presence of water upto a certain level increases the effectiveness of ionizing radiation. In practice the water content of seeds should be calibrated by storage at a particular humidity before mutagenic treatments. Around 10% water content seeds are most radioresistant. Sensitivity increases up to ca. 20% H_2O. When dormant seed is soaked, e.g. for two or three days at room temperature, the seed may start germinating and its radiosensitivity of the seed will drastically increase as a result of mitotic divisions.

Temperature. Within certain limits temperature affects the outcome of a radiation treatment, i.e. in combination with increasing water content and oxygen availability, as it influences gas exchange and respiration. Temperature effects in general seem to be small, although not completely negligible. Temperature is an important factor during post-irradiation storage of seeds, etc.

Oxygen. The presence or absence of oxygen strongly influences the effectiveness of X- and gamma radiation in producing damage. (N.B. X- and gamma rays are so-called low-LET radiation, i.e. with a low linear energy transfer. See also Box 4.2.) Oxygen interacts with all chemical compounds and reaction mechanisms present in the system, e.g. with the formation of radicals and the repair of damage. For the same radiation dose, irradiation in anoxic conditions (e.g. in nitrogen) will result in more surviving cells or plants than in oxic conditions (e.g. in air).

Thiols and sulfhydryls. Thiols, organic compounds which are present in large protein molecules, can be effective protectants against radiation damage to DNA and other important biomolecules. They create anoxic conditions and carry off radicals.

Thiols either occur endogenous in biological material or are added from external origin, but have to act within the target cells. It has been known since 1949 that exogenous thiols may offer significant radioprotection. However, thiols have found little practical application as radioprotectants, partly because they are toxic, but also because they must be administered before irradiation in order to be effective.

Sulfhydryls, which are non-proteins, are better suited as radioprotecting agents. Well-known in this respect are glutathione, cysteine and cysteamine. Russian researchers in the 1970s reported that a direct relation was found between the sulfhydryl content in the shoot meristems of pea and wheat and the radiation sensitivity of the genotypes.

For plant breeding purposes use can also be made of 1,4-dithiotre-itol (DTT or Cleland's reagent), for which compound is claimed that it protects against chromosome breakage (two-hit events), whereas the frequency of 'real' gene mutations should not be affected. First proof of the protection by DTT against radiation damage came from research with animal leucocytes and early investigations with plants were performed at Wageningen University with seeds from Arabidopsis (van der Veen et al. 1969; Sree Ramulu & van der Veen, 1973). Broertjes (1976a), however, who treated freshly detached leaves of the vegetatively propagated ornamental species Achimenes, did not obtain an improved mutation spectrum and suggested that DTT may have no effect in vegetatively propagated crops. Recent reports which refer to the use of DTT in mutation breeding work are virtually absent. These and other substances, which also act as radical scavengers by creating anoxic conditions, are studied also in relation to cancer research and retardation of symptoms of senescence.

Other external factors. Seed ageing was tried in the early days of mutation breeding as a method to increase mutation frequency (see also Chapter 2).

Although radiosensitivity and mutation frequency in physiologically older seed indeed are higher, probably as a result of a less effective 'repair' system, this approach is not used as a suitable method to increase mutation frequencies.

It is, on the other hand, useful to realize that the physiological age of the seed does affect the outcome of an irradiation treatment. In the same context storage following mutagenic treatments affects the outcome of a mutagenic treatment as well. Low temperatures (e.g. −20 to −80 °C) and in particular a high moisture content may have a protecting effect against storage damage, but contradictory results are reported. When long term storage without the occurrence of mutations is the goal (e.g. in gene banks), anoxic conditions must be created in addition to dry and cold.

and to the repair of single strand breaks in DNA by direct joining of the ends.

The other repair system mentioned by Friedberg: excision of damaged nucleotides (or areas of DNA), followed by resynthesis of the excised parts, is the most general system of DNA repair observed in nature. Most evidence on this very complex system, which is very often acting on damage caused by ionizing radiation and chemical mutagens, has been obtained from studying prokaryotic systems, like the bacteria E. coli and the phage T4.

Friedberg adds that, in addition to these two systems, some other systems are studied as well in both prokaryotes and (lower and higher) eukaryotes and refers in this respect to the action of various endonucleases: enzymes that specifically recognize sites of base loss in the DNA. For UV, for instance, in addition to photorepair, also a system of dark repair, based on enzymatic activity, is known.

Much more could be said concerning the above mentioned repair processes (see for instance the table on different repair pathways in Suzuki et al., 1989, p. 495), but it may suffice to mention here that nowadays many studies are performed on repair processes in which, for instance, repair mechanisms that occur after treatments with different mutagens, like ionizing radiation, non-ionizing radiation (UV) and chemical mutagens, are compared. As most studies in this field are not performed on higher plants, with, for instance, a recent publication by Britt (1996) as an exception, still many questions remain open. It would, for instance, be useful for the mutation breeder to know whether (and to what extent) repair occurs in dry seeds before soaking and onset of germination, or refers only to physiologically active tissues.

The final stages of the induction of biological damage

The final stage of the whole process of irradiating plant material is called the biological, biochemical or metabolic stage, during which period the effects of the irradiation treatment become manifest. This stage, following the period during which repair of DNA damage by (bio)chemical processes has been completed, may last from a few minutes to many hours, days or even months. The cells with undamaged or faultlessly repaired DNA will develop into cells which are fully normal, whereas cells in which no repair or faulty repair has occurred, either will die or do survive as cells which either carry gene mutations or different types of chromosomal damage. In other words, the effects of the irradiation treatment become manifest at the cell level.

As has been explained at various previous occasions, it may take a considerable time before the effects of the mutagenic treatment become (ontogenetically) manifest in the plants, resulting from seeds or other plant parts which have been mutagenically treated.

Again, plant breeders should keep two points clearly in mind. First, that the damaging effect of most mutagenic treatments, basically, is non-directional, i.e. that damage may occur all over the (DNA of the) treated material. Ionizing radiation, in this respect, in most cases induces deletions of various sizes. Generally, small deletions are of much more interest to the plant breeder than large ones.

The second point is that mutations are single cell events but, at the same time, that in higher organisms damaged cells are part of a multicellular organism and that a whole range of factors – which have been discussed already in Chapter 1 – ultimately determine whether an induced mutation will become expressed in a plant (either in that generation or later).

4.2.5. Other sources of irradiation

On some occasions the use of other sources of radiation is mentioned. Nowadays, for instance, the application of lasers (e.g. helium–neon, argon and Co_2 lasers and Nd glass lasers) attracts some attention, in particular in some countries in Eastern Europe and in China (Khotyljova, Khokhlova & Kholkhlov, 1988; Vasileva et al., 1991, original not consulted; Wang, 1991; Rybínsky, Patyna & Przewózny, 1993). Lasers (light amplification by stimulated emission of radiation) refer to an intense, narrow beam of monochromatic light by amplifying radiation within or near the frequency range of visible light. By focusing a laser beam with a lens, a very high amount of energy per unit area can be concentrated within a limited area; this energy can be used for different purposes, like heating, melting, vaporizing, piercing, medical surgery, etc. Other interesting features of lasers are that they can penetrate to the interior of a cell without damaging the exterior, that they can induce chemical reactions, break atomic bonds, etc.

Vasileva et al. (loc.cit.) used lasers with wavelengths between 458 and 633 nm to irradiate seeds of pea (*Pisum sativum*) and reported on the occurrence of a wide range of chromosomal aberrations and mutations, including 'useful' mutations. Wang (loc. cit.), who reviewed mutation breeding work in China, referred to thirteen cultivars for various crops obtained in this way which had been released to farmers. Rybínsky et al. treated seeds of barley (*Hordeum vulgare*) with a helium–neon laser at 1 mW cm^{-2} for 0.5, 1, 1.5 and 2 h at a wavelength of 6328 Å (=632.8 nm). According to the authors, low doses of laser light stimulated shoot and root growth (by 5.7%

and 2.8% respectively), whereas higher doses resulted in reduced growth (10.3% and 10.8% respectively). The frequency of chlorophyll mutation in what was indicated as M_1, did not exceed 1% even at the highest laser dose.

As far as is known there is still no convincing proof for the induction of mutations by the use of lasers, and until now the opinion most often heard is that lasers either are not mutagenic at all or only have a very low mutagenicity. On the other hand, given the enormous power of lasers, one could also ask oneself the question why lasers shouldn't be suitable tools for mutation induction. As often much heat is generated, one should in this respect think about the mechanism by which lasers could transmit sufficient targeted energy to induce mutations in chromosomes without killing the cells involved. The use of specialized lasers and very short (repeated) treatments, e.g. less than one-thousandth of a second, may significantly reduce the heat problem and lead to a number of non-thermal effects which could be of interest to the mutation breeder, such as breaking molecular bonds.

Other possible treatments involve the use of ultra-sound, magnetic fields, electromagnetic fields, etc. The merits of several of these methods have been investigated already right from the beginning of this century, however without much success.

Mutagenic effects of magnetic fields have been reviewed by Tegenkamp (1969). It is known that magnetic fields may produce biological effects of various kinds. The field of study is often called biomagnetism and the observed effects refer to growth responses rather than to genetic effects. On earth a natural magnetic field or a 'geomagnetic background' with a full strength of 0.5 oersted (Oe) exists, whereas organisms also can be submitted for periods of, for instance, several weeks or even months to artificial static magnetic fields of up to 150×10^3 oersted.

Magnetic fields, like electric fields, can induce chemical reactions, but the amount of energy imparted by magnetic fields to biological systems is relatively small. According to Tegenkamp (loc.cit.) the amount of energy is ineffective below a certain threshold, e.g. $100 \times$ the field of earth magnetism. This author reported for pupae of Drosophila melanogaster that, after having been submitted to magnetic fields, morphogenetic abnormalities occurred, which lasted for 30 generations or even more. Tegenkamp mentioned that for plants not much evidence of 'lasting genetic effects' (mutations) is known and he concluded that the use of magnetic fields to induce mutations in plants neither seemed very promising nor practical.

With respect to the use of electromagnetic radiation, mention was made in Chapter 2 of early efforts by Pirovano (for a review see Pirovano, 1957). More recently, Starzycki (1990), who does not refer to the work of Pirovano or other early references, briefly reported about 'ionophoren' as a method of mutation induction in plants. The approach implies the use of an electromagnetic field between the anode and cathode of direct current at a voltage of 12 or 24 V, with the cathode permanently immersed in distilled water and the anode situated between germinating (not-immersed) seeds. Seeds of different crops like maize, barley, wheat and soybean were exposed for 2–10 min. According to Starzycki, distinct mutants were detected in M_3. It was announced that studies were to be continued in M_4, but more details and additional reports could not be traced.

On some occasions it has also been suggested to make use of internally applied alpha and beta emitters, for instance to induce mutations in trees. This, however, does not seem recommendable, because of problems with the dosimetry and reproducibility of such treatments.

Finally, when the positive results became known from treatments with UV, X- and gamma rays, neutrons and different chemical mutagens, most other efforts were abandoned. Although it cannot be excluded that some of the aforementioned methods may possess a certain ability to induce mutations, it is most unlikely that they will contribute significantly in this respect in future, mainly because of the low mutation frequencies induced. One of the very few reasons one could think of to modify this view, might be if it were discovered that by applying such methods more specific and desired mutation types could be obtained.

4.2.6. Other breeding applications of radiation

In plant breeding and genetics living organisms can be submitted to radiation for other purposes than for mutation breeding in a strict sense. One example: the use of pollen which is damaged by irradiation, in order to induce parthenogenesis – the formation of an embryo without the direct participation of a male gamete – and to obtain or increase the frequency of haploid plants, still could be considered as related to mutation breeding. It is assumed that the second mitotic division in the pollen is prevented as a result of the radiation treatment. As a result one restitution nucleus with 2n chromosomes is produced instead of two haploid sperm cells, which 2n nucleus may fuse with the secondary nucleus of the embryosac and give rise to the production of tetraploid endosperm and a haploid embryo of female origin only. This effect may be obtained as well after treatment of pollen with colchicine or laughing gas. The subject has been discussed already in Chapter 3, Section 3.2.3, and many examples could be quoted (e.g. Lacadena, 1974; Raquin, 1985; 1986; Pandey, Przywara & Sanders, 1990; Zhang & Lespinasse, 1991, and many others).

Pollen irradiation is also applied for other purposes such as a) to try and overcome incompatibility barriers (Pandey, 1974); b) to achieve limited DNA transfer from the pollen parent in crosses (Pandey, 1975; 1978); and c) to provoke nucleus substitution (Raquin et al., 1989).

ad a) When egg cells are pollinated with a mixture of pollen, one type compatible to the pistillate parent, but 'killed' by a high dose of radiation, and the other type containing alive incompatible pollen, the latter – occasionally – may be able to successfully fertilize the normally incompatible partner. This approach is commonly called the mentor pollen method, a term that is derived from Michurin. The radiation dose must be such that the mentor pollen should be still able to germinate and to grow a pollen tube into the style of its compatible partner, but cannot effect fertilization of the egg of the compatible partner. The rate of success of experiments with mentor pollen is often very low, but this would not matter as long as a few hybrid seeds are obtained in this way. Not many reliable reports on such experiments are available. Some positive results were obtained by Stettler (1968) in interspecific crossings of *Populus*, but in other genera, including *Lycopersicon*, *Nicotiana* and *Solanum* results were poor or even absent.

ad b) Fragmentation of pollen DNA by high dose irradiation prior to pollination can be a means to transfer to the recipient plant not the whole male genome but only a limited amount of donor DNA. The transfer of a single, useful gene would be ideal, but in practice always several chromosome pieces carrying a number of genes will be transferred. In this way so-called asymmetric hybrids arise, i.e. hybrids in which both partners do not contribute the same amount of genomic DNA. The chromosomes may be fragmented by the radiation treatment and the DNA may be considerably damaged, but pollen tube growth and discharging of the pollen nuclei into the embryosac may occur in a more or less regular way. The number of viable seeds produced in this way may be very low and the resulting hybrid will predominantly show maternal traits with a variable, but always low, portion of traits from the male genome. This method was called egg transformation by Pandey (1975), who used pollen irradiation before making interspecific crosses in the genus *Nicotiana*. Pandey assumed that the resulting progeny is of parthenogenetic (maternal) origin with some additional traits supplemented from the

fragmented pollen DNA that was inserted during early zygote chromosome replication. Definite proof for Pandey's hypothesis, so far, has not been given; many questions have not been satisfactorily answered yet and there is still considerable disagreement about possible mechanisms (see for instance Sanford, Chyi & Reisch, 1984a,b).

Caligari, Ingram & Jinks (1981) and Jinks, Caligari & Ingram (1981), who irradiated pollen of *Nicotiana rustica* immediately before pollination, explained their results by proposing that, despite the damage to the pollen DNA as a result of the radiation treatment at a sublethal level, biparental zygotes are formed following 'differential gene transfer' from the pollen parent. Powell, Caligari & Hayter (1983) studied – what were indicated as – M_2 derived from crosses between two cultivars of barley (*Hordeum vulgare*) with pollen irradiated at 5–20 Gy. With increasing doses the M_2 resembled more the maternal parent but a few traits from the paternal parent were still present. Snape *et al.* (1983), in comparable experiments with wheat (*Triticum aestivum*), used pollen that had been irradiated with 20–50 Gy and also concluded from cytological work that irradiation damage of the paternal genome in M_1 plants results in different degrees of preferential transmission of alleles of maternal origin.

ad c) X- or gamma irradiation is also used in somatic cell fusion experiments, either to limit the contribution of one of both partners (asymmetric fusion) or to mark one of the fusion partners for the purpose of recognition. The method of irradiating the donor parent in fusion experiments prior to fusion traces back to the 1970s, when it was found that foreign DNA fragments could be introduced in mammalian cells, where they replicated and expressed themselves (Schwartz, Cook & Harris, 1971).

For plants the picture is somewhat complicated as DNA occurs in the nuclear chromosomes as well as in the cytoplasm. When two complete protoplasts, i.e. somatic cells with nuclei and cytoplasms, are fused, it is anticipated that a fusion product originates with both genomes present – either in one (fused) nucleus or in two separate nuclei – and with a mixed cytoplasm including chloroplasts and mitochondria from both partners. When additive combinations of the complete genomes of the fusion partners arise, these are called symmetric hybrids.

All kinds of variations to this basic concept can be thought of. Cybrids (or cytoplasmic hybrids) are the result of mixing the cytoplasms of both fusion partners without fusion of the nuclei. This situation is believed to arise sometimes when the nucleus of one of both partners is completely 'inactivated' by irradiation prior to fusion. In that case the resulting 'cybrid' will carry only the unirradiated nucleus but cytoplasmic DNA from both the irradiated donor and the recipient.

Already many (asymmetric) fusions have been tried between a (recipient) crop species and a related wild (donor) species, for instance in order to try to introduce resistance from the wild relative in this way. When a donor is irradiated prior to fusion, it is impossible to predict which chromosome fragments will become transferred, but by inoculation with the proper pathogen, the 'carriers' of a desired gene for resistance could be easily identified. If it would be feasible to introduce the desired resistance in this way, fewer back-crosses would be required to eliminate unwanted genes! Doses of 300–500 Gy of X- or gamma rays often are found to be high enough for 'pulverizing' the donor DNA and for preventing division and callus formation of donor protoplasts. For references to irradiation prior to fusion, see for instance Gleba & Shlumukov (1990).

Gupta, Schieder & Gupta (1984) proved that it is possible to achieve by means of asymmetric protoplast fusion a stable intergeneric transfer of nuclear genes from *Physalis minima* into the genome of *Datura innoxia*. It was reported that the incompatibility barriers could be overcome by fusing *Physalis* protoplasts that had been 'completely killed' by a high dose (150 Gy) of X-rays. Another example of the fragmentation procedure described here is the fusion of protoplasts of potato (*Solanum tuberosum*) with protoplasts of the wild donor species *S. brevidens* which had been irradiated with 300–500 Gy. (Xu *et al.*, 1993). At the dosages mentioned, division of *S. brevidens* protoplasts failed and callus formation did not occur, but a *brevidens*-specific probe proved the hybrid character of regenerated shoots from the fusion product.

Other, often studied combinations are oil-seed rape (*Brassica napus*) with black mustard (*Brassica nigra*) and tobacco (*Nicotiana tabacum*) with *N. plumbaginifolia*. Bates *et al.* (1987) found that after gamma irradiation with 50 Gy (dose rate 4 Gy. min^{-1}) the protoplasts of *N. plumbaginifolia* were unable to divide more than once. In another example it was found that at a dose of 300 Gy of gamma rays, cell division in the wild species *Lycopersicon peruvianum* – an interesting donor species for tomato (*Lycopersicon esculentum*) – is prevented, but the functioning of the (extranuclear) chloroplast DNA is not significantly disrupted (Derks, Hall & Colijn-Hooymans, 1992).

Despite much research on fusions a number of problems must still be solved before significant practical results will be obtained. For example, it is difficult to determine an optimum dose of radiation, as too high a dose will result in much lower yield of fusion products and with too low a dose too much transfer of DNA takes place. It has been discovered, for instance, that after irradiating protoplasts of potato (*Solanum tuberosum*) before fusing them with protoplasts of tomato (*Lycopersicon esculentum*), much more potato DNA was transferred than was hoped for and that before the fusion was realized, different DNA fragments of potato, resulting from the irradiation, often had joined again already (Schoenmakers, 1993).

Some other applications of radiation to study various processes in plants are only briefly mentioned here. Cells could be marked, for instance, with a mutation, often leaf colour (chlorophyll) defects, to study plant development. Itoh & Kondo (1992) reported in this respect for seedlings of soybean (*Glycine max*), heterozygous for the marker trait and irradiated with a dose of about 2 Gy of X-rays, that the development of leaf primordia for this crop could be successfully studied two or three weeks after treatment.

Radioactive-labelled chemicals, for instance, can be brought into a cell culture or a specific plant section in order to study physiological processes. After exposing the labelled tissues to a suitable photographic emulsion and the following processing, the radioactive substances can be localized by the silver grains produced by the emitted radiation. The radiosiotopes used normally are electron emitters. This subject is not further discussed here.

Sterilization by irradiation of soil for potplants or for greenhouses (as an alternative to the use of methylbromide, a highly toxic chemical), is also performed at a relatively large scale. Also in this case the reproduction of harmful bacteria or arthropoids is prevented or the pathogenic micro-organisms are completely killed.

So-called low-dose stimulation of plants, for a number of years, has been a subject of some interest, in particular in Eastern European countries. It was strongly believed at some places that irradiation at low doses (e.g. a few grays only), would stimulate the physiological activity of plants and in this way, for instance, lead to better germination and higher yield. Until the late 1980s a special journal, called *Stimulation Newsletter* was issued in Hannover, Germany, but real scientific proof for low-dose stimulation has never been given. Most accounts of 'positive' or 'promising' results continued to be considered with much doubt and the field has never become popular, in particular not among research workers in the Western world. References can be found in the aforementioned *Stimulation Newsletter*.

Just for the sake of completeness some other applications of ionizing radiation, like its use to prolong the preservation time ('half-life') of food or to prevent reproduction or to kill pathogenic micro-organisms like *Salmonella* or *Listeria*, are mentioned here. The effect is based upon the same principles as mutation induction, but doses are so high that reproduction of harmful bacteria is prevented.

In order to prolonge the 'shelf-life' physiological processes are slowed down. For food preservation at present very different kinds of electromagnetic waves are applied, such as UV, infrared, gamma rays from ^{60}Co, X-rays, as well as accelerated electrons. Gamma doses below a certain threshold value are considered fully safe to human health as no radioactivity remains in the irradiated material and, according to the protagonists, significant amounts of toxic substances are not produced. Opponents, however, state that the vitamin content may go down and that toxic substances are produced and that (small) changes in taste, constitution and appearance may occur. Food irradiation, although for a limited number of food substances and with various regulations to inform the customer, was permitted in 1992 in more than 30 countries.

4.3.   The use of chemical mutagens

4.3.1.   A historical review of the application of chemical mutagens

The early years
Soon after the discovery of 'mutations' early in the twentieth century and after it had been suggested that man might be able to induce such genetic effects (see Chapter 1), the capability to induce mutations was tested for a wide range of chemicals. Wolff (1909) and Schiemann (1912), who both were students of Erwin Baur, induced mutations in bacteria and fungi by chemicals (for details and other references see Chapter 2). Starting from 1916 Baur himself tested many chemicals on their mutagenic effects in higher plants (Stubbe, 1937b). The oxidant iodine, applied in 1932 by the Russian scientist Ssacharoff to *Drosophila*, was the first chemical for which could be demonstrated that it induced mutations (Ehrenberg, 1960; Gustafsson, 1960).

A publication by Blakeslee & Avery (1937a,b) on the action of colchicine, an alkaloid derived from *Colchicum autumnale* (Autumn crocus), resulting in the production of polyploids, received much attention and the doubling of chromosome numbers in higher plants by colchicine became a frequently applied technique. One could take the position that this compound has been the first generally used and effective chemical mutagen ever.

However, the fact that this chemical does not affect the constitution of the DNA but the amount of DNA (or the genome number) within a cell, has placed colchicine in a somewhat special position and has meant, that colchicine, in most instances, is not considered as a mutagen in the strict sense. Additional details about colchicine, its discovery, dosages, methods of treatment, alternative methods to induce polyploidy etc., have been discussed in Chapter 3 and will not be repeated here.

The first generally accepted paper on successful chemical mutagenesis in higher plants was published by Oehlkers in 1943 (see Oehlkers 1943; 1946; 1949; 1953; 1956). This author treated cut inflorescences of *Oenothera* with urethane (+KCl) and ethyl urethane. The mutagenic action of these chemicals was proved cytologically beyond doubt by the increased numbers of chromosome breakages and translocations. According to present standards, the mutagenic power of these chemicals must be considered low. References to this and other early work on chemical mutagenesis have been presented already in Chapter 2. In addition, many references concerning chemical mutagenesis in higher plants for the period 1940–60, for instance, can be found in Prakken (1959), Röbbelen (1959) and Gustafsson (1960). The action of chemical mutagens, in the past, was often called 'radiomimetic', i.e. that effects were believed to 'mimic' or to be rather similar to those caused by radiation treatments, but this expression is not used anymore.

Auerbach and Robson discover mutagenic effects of mustard gas in Drosophila
Very significant for the further development of chemical mutagenesis, has been the work of Auerbach in England, which started in 1940. In 1942 Auerbach, together with the pharmacologist Robson – in an 'internal' report to the British Government – stated that mustard gas (dichloro-diethyl sulphide), a chemical used in World War I as a toxic weapon, was an effective mutagen in *Drosophila*. Publication in scientific journals had to wait till after the end of World War II. Details about the work and person of Auerbach are given in Chapters 2 and 3.

Conclusions about the effectiveness (the relation between effect and dose) of mutagens, in those days, were based on frequencies of chromosomal breaks and the number of dominant lethals, and not yet on the frequencies of gene (or point) mutations.

Effects of mustard gas on higher plants
The first studies on the effects of mustard gas on higher plants were performed soon after World War II on barley

(*Hordeum vulgare*) by Gustafsson and MacKey (see for instance Gustafssson & MacKey, 1948; MacKey, 1954b, and Gustafsson, 1960). When – soon afterwards – chemicals with higher mutation frequencies and a much higher efficiency became available, treatments with mustard gas were abandoned.

Rapoport discovers many new chemical mutagens
Also in the 1940s the Russian scientist Rapoport, like Auerbach applying the *Drosophila ClB*-technique to detect mutations, started to investigate mutagenic properties of a range of chemical compounds. Except for the mustards, it has been Rapoport who discovered the mutagenic effects of practically all other mutagens that became known at that time. Some of the first chemicals in this respect were formaldehyde (Rapoport, 1946; original not consulted); ethylene oxide (EO), diethyl sulphate (DES or dES); dimethylsulphate (DMS); ethylene imine (EI); diazomethane (DEB) and another epoxide: diepoxybutane (Rapoport, 1948; original not consulted). Several of these compounds were considered very powerful in comparison with mustards.

In 1948 Rapoport also described the alkylating effect of some chemical mutagens like ethylene imine on 'biologically important cell compounds' (which, in his opinion, were the proteins). It is generally agreed now that alkylating agents are the most important group of chemical mutagens.

After an unvoluntary break from 1948 till 1957, Rapoport could resume his research on chemical mutagens and, amongst others, showed that some chemicals, which belong to the group of the nitroso-compounds (a category of alkylating agents), were several times more effective in inducing mutations than both X-rays and the chemical mutagens known at that time. Nitroso-compounds, which will be discussed later, became mutagens of considerable practical importance and are still often applied, nowadays.

Studies on effects of chemical mutagens on higher plants by the 'Gustafsson Group' in Sweden
Gustafsson's group in Sweden was the first to systematically study the effects of many chemical mutagens in higher plants. Early results, for instance, have been published in Ehrenberg, Gustafsson & Lundqvist (1956), Ehrenberg & Gustafsson (1957), Ehrenberg, Lundqvist & Ström (1958), Ehrenberg, Gustafsson & Lundqvist (1959), Gustafsson & Ehrenberg (1959), Gustafsson (1960) and Ehrenberg, Gustafsson & Lundqvist (1961).

Ehrenberg *et al.* (1956), in a preliminary report on the effects of 'radiomimetic' chemical agents on barley (*Hordeum vulgare*), expressed the hope that with chemicals it would be possible to better direct the mutation process than is the case with ionizing radiations. In a series of publications relative percentages of sterility and the proportion of structural chromosomal mutations and chlorophyll mutations (which may be mainly the result of gene mutations), were determined for various chemicals and compared with spontaneous mutation frequencies and effects on chromosomes, induced by X-rays.

Results from earlier experiments with mustards (Gustafsson & MacKey, 1948; MacKey, 1954b) had shown already that, compared with the control, only a very small increase in the frequency of chlorophyll mutations was obtained, whereas the level of sterility increased to even 24%. For ethylene oxide (EO), chlorophyll mutations in those early years occurred in frequencies of 0.2%–1.8%, whereas the sterility levels were going up with increasing concentrations of the mutagen from 5.7% up to 22.1%. (N.B. In later years, as will be discussed further on, much higher percentages of chlorophyll mutations in M_2 were obtained after treatment of barley seeds with EO.) Formaldehyde, of which the mutagenic properties had been reported by Rapoport, and for which Favret in the 1950s claimed that it also induced mutations in barley, did not induce increased levels of mutations in the Swedish experiments. Swedish results further showed that chlorine and iodine (the mutagenic properties of which had been reported in 1932 by Ssacharow) produced only 0.3% and 0.6% chlorophyll mutations respectively, and 18% and 6% sterile plants, respectively.

A third group of chemicals studied belong to the purine derivatives, which specifically interfere with nucleic acid metabolism. Nebularine (= purine-9-D-riboside), a nucleoside not occurring in higher plants, produced only 0.8% chlorophyll mutations but, remarkably, showed no significant increase in M_1 sterility or structural mutations as compared to the control. Moreover, treatments with nebularine resulted in relative frequencies of different chlorophyll types, deviating from those observed for radiation and other chemicals. This observation led Ehrenberg *et al.* (1956) to conclude that some chemicals may induce more specific effects on the DNA than ionizing radiations. Alkylating and oxidizing agents, in this respect, are considered to act rather unspecific.

Another chemical, 8-ethoxycaffeine, which is known to break chromosomes and in particular to affect the heterochromatin, induced 0.1% chlorophyll mutations only, but a level of sterility up to 57%. Investigations by Ehrenberg *et al.* (1956) with acridines (acridine orange and acriflavine), the application of which compounds according to D'Amato (1950; 1951; 1952) induced increased levels of chlorophyll mutations, did not confirm the results of D'Amato.

Early results of comparing mutagenic properties of different chemicals showed that, so far, mutation frequencies induced by chemicals, in general remain lower than those obtained from X-ray treatments at comparable levels of sterility. Ehrenberg et al., (1956) pointed out that at that time for most chemicals optimal treatment conditions had not been worked out yet and, hence, that in a number of cases an increase in mutation frequencies may be expected in future. The authors further expressed their opinion that sterility (which they interpreted as 'dependent on chromosomal rearrangements') and 'mutation' were caused by different mechanisms.

New grounds for optimism about chemical mutagens

The overall not very optimistic conclusion about the mutagenic potential of chemical mutagens, which had resulted from the initial Swedish work, however, had to be adjusted after two other alkylating agents: ethylene oxide (EO; chemical formula CH_2OCH_2) and, in particular ethylene imine (EI; chemical formula CH_2NHCH_2) had been tested. It was shown (Ehrenberg et al., 1958; 1959; Gustafsson & Ehrenberg, 1959) that EI (of which the mutagenic properties had been established in 1948 by Rapoport) was 3–5 times more efficient as a mutagenic agent than X-rays. In treatments of dry seeds of barley with 0.04–0.07% EI, up to 30% of the M_1 spike progenies segregated for chlorophyll mutations, compared with 5–10% after treatments with X-rays. (N.B. Presoaked seeds are less sensitive!) In addition, whereas for most mutagenic treatments the relative contribution of structural mutations was found to be higher than of gene mutations (e.g. up to 65% or more), the percentage of structural mutations for EO and EI reached 45% only.

It is interesting to refer in this respect to Gaul (1958) who, at that time, expressed the opinion that the generally observed high toxicity of chemical mutagens prevented administering higher concentrations of chemicals or increasing the length of treatments beyond certain thresholds. This should explain why mutation frequencies obtained with chemicals remained far below those obtained from radiation treatments. Moreover, in case higher mutation frequencies had been reported, those, almost without exception, referred to (semi-)lethal chlorophyll mutations (e.g. in barley), which may not be gene mutations at all! The carcinogenic properties of many chemicals mutagens, which make treatments dangerous, were also mentioned as a serious drawback.

In the same publication, however, Gaul referred to two recent publications in which, as he called it, 'noteworthy progress' and 'exciting indications' of high mutation frequencies, comparable to those obtained by high X-ray dosages, had been reported, namely from treatments with ethylene oxide (Ehrenberg & Gustafssson, 1957) and with another new chemical, mentioned by Heslot & Ferrary, 1958) before, indicated as 'ethyl sulphate' by Gaul (but by which most probably EMS was meant). Apparently, Gaul was not aware yet of the Swedish work on barley with ethylene imine (EI), which had shown much better results than with ethylene oxide (EO).

Ethyl methanesulphonate

The late 1950s can be considered as the period of a growing awareness that chemicals, and in particular those belonging to the group of alkylating agents, might have great potential as efficient mutagens in higher plants, as had been shown already by Ehrenberg et al. (1959). This opinion soon was found to be not only justified for (semi-)lethal chlorophyll mutations, but also for viable mutations of the erectoides-type, for which category X-rays and neutrons, on the average, were reported to produce 4–5% mutations on M_1 spike-progeny basis, against about 9% for EO and even 20% for EI. The authors concluded that with such percentages a plant breeder could expect to find a viable mutant of potential interest to the breeder in at least every fifth or tenth spike progeny.

An important finding was that alkylating compounds could induce mutation spectra which differed from those obtained with ionizing radiation. This observation was based on the different ratios between so-called translocation sterility mutations, structural mutations and to differences within different categories of (both semilethal and viable) gene mutations.

Ehrenberg et al. (1959), in an addendum, mentioned that they recently had become aware of results from Heslot et al. (1959) with the alkylating agent ethyl methanesulphonate (EMS). Ehrenberg et al. added, that a comparison between myleran, a bifunctional alkylating agent (dimethanesulphonyloxybutane), and EMS, a monofunctional agent of which a molecule in fact constitutes half a molecule of myleran, shows a much higher mutagenic efficiency for EMS with 50% mutations against myleran with 2–6%.

Opinions differ as to who first demonstrated the special merits of EMS in inducing mutants. Shama & Sears (1964) mention that the first report, based on experiments with Neurospora, came from Kølmark (1953). (N.B. The Neurospora back-mutation test, which had been worked out in 1947, mainly by M. Westergaard, had been used already to test at an earlier stage many other chemical mutagens, like EO, EI, diepoxybutane, diazomethane, peroxides, dimethyl sulphate (DMS), DES, formaldehyde, phenols, penicillin, etc. See for instance Jensen et al., 1951; Kølmark & Westergaard, 1953; and Kølmark, 1953; 1956). According to other sources O.G. Fahmy & M.J. Fahmy at the Chester Beatty Institute in London were the first researchers to demonstrate around 1956 the mutagenicity of EMS on the basis of visible aberrations and lethals in the *Drosophila* ClB system (see for instance Fahmy & Fahmy, 1956a,b; 1957). Rapoport and associates also at an early stage studied the effects of EMS in *Drosophila*. In addition, Loveless (1958), for instance, reported on the mutagenic activity of EMS 'in vivo' and *in vitro* in bacteriophages. Gustafsson (1960, p. 25) mentioned that the proposal to apply EMS in higher plants had been made already in 1956/57 by Ehrenberg.

Anyhow, it appears that Heslot and colleagues in France, for the first time, successfully applied EMS to higher plants (see for instance Heslot & Ferrary, 1958; Heslot et al., 1959; Heslot, 1960; and Heslot et al., 1961). Their results showed that with EMS in barley chlorophyll mutation rates of 50% or more per M_1 spike-progeny could be induced, which made EMS much more effective than EI which, so far, had been considered the most effective mutagen.

Throughout the years EMS has been shown to be a very effective and efficient mutagen and has become the most frequently used chemical mutagen, whereas the application of most other available chemical mutagens drastically decreased. Gaul et al. (1972), for instance, showed for barley that EMS not only induced high mutation rates, but was also four to five times more efficient than X-rays, when assessing the frequency of chlorophyll mutations relative to the M_1 sterility and M_1 lethality, respectively.

The first meeting on chemical mutagenesis in 1959 in Gatersleben

In 1959 the German Academy of Sciences at Berlin organized in Gatersleben (former GDR) the first of a series of conferences to honour the German geneticist Erwin Baur. The topic of this first meeting with invited participants was chemical mutagenesis. Several leading European scientists in this field were gathered and the proccedings of this meeting provide a good view on the situation concerning the use of chemical mutagens at that time (Stubbe, 1960).

The conference itself may be considered the starting point of a quickly increasing interest in the subject of chemical mutagenesis and in the following years many

contributions at other conferences on genetics, plant breeding, etc. and in scientific journals appeared. There was a general feeling at that time that chemical mutagens might exert more specific effects on the DNA than X-rays, gamma rays and neutrons and, consequently, that it would soon become possible to 'direct' in this way the mutation process and to induce more specific categories of mutations at will. It should be added here that most details about the action of chemical mutagens were obtained from fundamental research with microorganisms like *E. coli*.

In addition to the remarks made already in this section, some main points, expressed by leading scientists like Auerbach, Gustafsson, Westergaard and Ehrenberg at the aforementioned conference are mentioned here. In her paper Auerbach (1960; see also Auerbach, 1961) pointed out that no specific mutagens for gene (or point) mutations were available yet and also expressed the view that the development of mutagenic purines and pyrimidines might be useful. Auerbach referred to Fahmy & Fahmy (see before) who, in addition to studying the effects of EMS, also had investigated the mutagenic properties of other alkylating agents, like methylmethane sulphonate (MMS) and triethylene melamine (TEM), which is the polyfunctional counterpart of EI. It was shown later that the mutation frequency induced by TEM remained far below those of EI or EMS. Auerbach also mentioned another potential chemical mutagen, called heliotrin, an alkaloid developed by Clark in Australia, but in later years no futher references to this chemical could be traced.

At the same conference, Gustafsson (1960) discussed the large scale comparative tests of different chemical mutagens and, in addition, observed that several mutagens might show a certain degree of specificity, for instance with respect to frequencies of different categories of chlorophyll mutations, *erectoides*-types and mildew resistance in barley. He further mentioned that frequencies of gene mutations and chromosomal breaks were not necessarily correlated and that, so far, practically no information had been collected on chemically induced extranuclear (plasmon) mutations.

During the conference in Gatersleben several authors elaborated on the, still largely unknown, mode of action of chemical mutagens. Reference was made to the different processes involved in case of the incorporation of, for instance, the base analogue bromouracil, ethylation (or alkylation), deamination by nitrous acid (HNO_2), etc. Much attention was also paid to the concept of 'specificity' of effects induced by different (groups of) chemical mutagens. In a contribution by Auerbach & Westergaard (1960), 'specificity', was considered both at a 'geographical level', that is at specific positions within chromosomes, as well as at a 'functional level', which designation refers to consecutive steps of the mutational process. However, the ways in which chemical mutagens interacted with the DNA, at that stage, were still completely obscure. In this respect mention should be made here also of another, interesting early article by Westergaard (1957) about chemical mutagenesis in relation to the concept of a gene as a unit of function, recombination and mutation, which publication was based on studies concerning chromosome breaks in plants and back-mutations in *Neurospora*.

Ehrenberg (1960), in his summarizing lecture, also underlined the lack in understanding at that stage of the processes involved in chemical mutagenesis. This concerned for instance the possibility of repair of damage, the question whether the induced damage in the DNA could be copied, the relation between chromosomal breaks and loss of reproductive ability, the stage at which specific processes occur, etc.

Some other early chemical mutagens

At the time of this first major conference on chemical mutagens, in addition to the group of alkylating agents, some other groups of chemicals with mutagenic potential were known. The observed mutagenic effects in most cases concerned chromosome breakage. Mentioned in this respect could be: antibiotics, like streptomycin and mitomycin; several so-called base analogues and related chemical compounds; hydroxylamine (NH$_2$OH) and acridine dyes, like acridine orange and ethidium bromide (EB). Some references concerning these compounds will be given at a later stage. Cytological studies revealed that ethylene diamine tetra-acetic acid (better known as EDTA), a so-called chelating agent, alters the activity of enzymes which are involved in oxidative or phosphorylative metabolism and increases the yield of radiation-induced chromosomal damage – assessed as reduced fertility and the occurrence of dominant lethals – in plant cells and *Habrobracon*, a small parasite wasp. Such observations, of course, do not necessarily imply that EDTA is mutagenic by itself. More details and some early references are given by Wolff & Luippold (1956) and LaChance (1958). Most of the chemicals mentioned in the previous sentences, have not contributed much to practical mutation breeding in higher plants and, therefore, their properties will not be discussed in detail, here or at a later stage.

Chemical mutagenesis in the 1960s; the FAO/IAEA Meeting in Rome in 1964

At the important FAO/IAEA Meeting on Induced Mutations in Rome in 1964 (Anon., 1965) several lectures on chemical mutagenesis were presented. Heslot (1965), who mainly referred to studies with lower organisms like bacteriophages, discussed the nature of mutations, induced by various groups of chemical and physical mutagenic agents, and pointed out that the way in which chemical mutagens act is largely determined by their physical properties.

Konzak et al. (1965), in a lecture about efficient chemical mutagenesis, stated that chemical mutagens may be attractive because of their potential for inducing interesting mutations, the relative ease of application and the low costs involved. They further mentioned the mutagenic effectiveness of EMS, which is much higher than that of any other mutagen known to have been applied to seeds. It should be remembered, however, that a high effectiveness does not necessarily imply a high mutagenic efficiency (i.e. the occurrence of a high ratio of 'desirable' mutations) as well, which is an essential requirement for mutation breeding.

Konzak et al. further referred to the fact that application of chemical mutagens is limited in particular by their solubility, toxicity (i.e. damaging effects on plants) and reactivity. The solubility of a chemical restricts the range of concentrations that can be applied. When a chemical exhibits a high level of toxicity, only low dosages can be administered, which may considerably limit the potential yield of mutations. A good system of post-washing may be another way to limit toxic effects. In addition, mutagens may quickly lose their mutagenic properties, for instance as a result of hydrolyzation in aqueous solutions. Concentration, temperature, pH, etc., may considerably affect the mutation frequency of a specific treatment. An increase of the concentrations of chemical mutagens normally results also in more desirable mutations but, in general, causes a proportionally even greater amount of (seedling) damage and lethality. Applying too high concentrations of EMS, for instance, results in a quick increase in the level of sterility. As a consequence, in order to obtain a high mutagenic efficiency in mutation breeding programmes, it is absolutely worthwhile to carefully work out optimal concentrations of chemical mutagens. Temperature may affect both the rate of diffusion of the mutagen in the plant tissue and the rate of hydrolysis. The pH may be an important factor, depending on the chemical. The rate of hydrolysis of EMS seems not to be much affected by a low pH, but the biological system itself may be rather sensitive to a low pH. Effects of hydrolysis can be prevented by the use of buffers which, for a number of chemical mutagens, has become a standard method.

Konzak et al. also confirmed a rule established by

Rapoport in 1948, that in a homologous series of chemicals (e.g. CH_3 – C_2H_5 – C_3H_7, etc.) the lowest and most reactive member (i.e. with an ethyl group) is the most mutagenic. In other words: monofunctional alkylating agents are the most efficient ones. In addition, it was found that, within the group of polyfunctional agents, the butyls and propyls are more efficient than the methylating agent.

It should be commented at this place that the observed higher efficiency of monofunctional alkylating agents does not imply that di- or polyfunctional agents should be less active at lower concentrations: the opposite is true. It depends, however, on the parameters studied, as lower concentrations of polyfunctional agents result in particular in a strong inhibition of growth and, therefore, should be called highly effective in this respect. Monofunctional agents, however, are active at much higher concentrations before cell death occurs; they produce higher mutation rates and, thus, are more efficient.

Further remarks by Konzak et al. (loc.cit.) about relevant specific properties of mutagenically treated biological systems will not be discussed at this place, except for some short remarks about the so-called pre-treatments and post-treatments. Pre-soaking of seeds in (distilled) water makes cell membranes more permeable to the mutagen and results in more uniform results of experiments. Post-washing in running water is often performed to remove remaining mutagen and its products (e.g. after hydrolyzation) from the treated material. Post-drying may facilitate further handling of the treated seeds.

The action of some other powerful alkylating agents, generally called nitrosamides or nitroso-compounds was reported in some other publications in the the 1960s, e.g. by Heslot, Ferrary & Tempé (1966) and by Rapoport et al. (1966). In particular nitrosoethyl urea and nitrosomethyl urea are still often used nowadays.

As chemical formulas, names and abbreviations for these and other mutagens of the nitroso-group (as well as for other chemical compounds) are written in various ways in different publications and in different languages, it should always be carefully checked which chemical compound was actually applied. For instance, in the FAO/IAEA Manual (Anon., 1977a) and other publications by the IAEA, arbitrarily, a distinction is made between compounds which contain urea (abbreviated as H, derived from the German word 'Harnstoff', i.e., literally, 'urine substance') and compounds with urethane (abbreviated as U, from urethane). According to this system, the compound nitrosoethyl urea (also written as nitrosoethylurea or nitroso ethyl urea) should be abbreviated as NEH, and nitrosomethyl urea (nitro-

somethylurea, nitroso methyl urea) as NMH. However, nitrosoethyl urea and nitrosomethyl urea are also often written as ethylnitroso urea and methylnitroso urea respectively, and the abbreviations, accordingly, would than be ENH and MNH.

Chemical mutagens containing nitroso urea are used much more often than mutagens which contain nitroso urethane and, consequently, several research workers (including the present author) – for the sake of convenience or out of laziness – apply in their publications the abbreviations NEU and NMU as (not unlogic) short for nitrosoethyl urea and nitrosomethyl urea respectively. As the official chemical formulas of these (and other) compounds often are more complex than has been mentioned here (e.g., N_1N-ethylnitroso urea or N-nitroso-N-ethyl urea instead of ethylnitroso urea) and, moreover, may undergo some changes – for instance in spelling – throughout the years, still other formulas and abbreviations can be found in literature for one specific compound. In situations where it really matters one should either consult the original publications, stick to one system – for instance the system used in the FAO/IAEA Manual – or try to avoid abbreviations as much as possible.

Keeping in mind that the aforementioned inconsistencies may also occur now and then in this book, in particular when the use of specific nitroso-compounds is discussed, we return now to some early applications. After it had been shown by Grant and Heslot (for references see Heslot et al., 1966) that nitroso-compounds induced chromosomal aberrations in root tips of field bean (Vicia faba), convincing evidence for their mutagenic potential was provided for barley (Hordeum vulgare) by comparing the mutagenic effects of these chemicals with those induced by gamma rays and EMS (Heslot et al, loc.cit.). Dry barley seeds were treated with different concentrations of NEU and NMU for 8 h at 25 °C. For NEU, depending on the dose applied, up to 45% chlorophyll mutations – calculated on spike progeny basis – were obtained, in comparison with about 20% for EMS. When comparing the effects of gamma rays 200 Gy (20 krad), EMS 0.4% and NEU 0.01%, which dosages refer to about equal levels of sterility, it was found that both EMS and NEU produced 4–8 times higher mutation frequencies than gamma rays. This led to the conclusion that NEU also appears to be a potent mutagen with a level of efficiency at least identical to that of EMS. In later years it has been claimed on some occasions that NEU and NMU may be of particular value for inducing mutations in the extranuclear DNA (Hagemann, 1976; 1982; Pohlheim, 1981). Some details will be discussed later in this chapter.

We may finish this section with a general remark. For the commonly applied chemical mutagens the efficiency of different mutagens at various concentrations and duration of treatment, as well as the effects of other factors such as pre-treatments, pH, temperature, post-washing, etc., have been determined in many cases and often can be found in literature, for instance in the many proceedings of meetings, etc. organized by the Joint FAO/IAEA Division in Vienna, in journals on plant breeding, in some specialized books and chapters on mutation breeding (several titles are mentioned for instance in Chapters 1, 6 and 7) or in the *Mutation Breeding Newsletter*.

Chemical mutagenesis in barley

Barley (*Hordeum vulgare*) has always been one of the most important 'model crops' for mutation breeding. Therefore, the proceedings of a series of conferences under the name 'International Barley Genetics Symposia', which are organized about every five years, offer an excellent standard to assess the relative importance of mutation breeding in relation to other breeding techniques, to consider the contribution of chemical mutagens in mutation breeding and to compare the contribution of various chemicals throughout the years.

It may be useful to add in this context that chemical mutagens have been subject to attention – and in a number of cases even <u>the</u> main subject – at several other meetings and in a number of reviews as well. Reference, for instance, could be made to a meeting called '*Induction of Mutations and the Mutation Process*' which was organized in 1963 in the (then) CSSR (Véleminský & Gichner, 1965). An early review on chemical mutagenesis, for instance, was written by Auerbach & Kilbey (1971). In the early 1980s, the IAEA organized two so-called Research Coordination Meetings on *in vitro* technology, during which the merits of chemical mutagens were also discussed (Anon., 1986a). In many other IAEA meetings effects of chemical mutagenesis were compared with those induced by radiation. In addition, Micke (personal communication) mentioned that several conferences on chemical mutagenesis were organized by the USSR Academy of Sciences in Moscow, e.g. in 1969 (report published in 1971), in 1981 and in 1986. The reports of these meetings, however, have not become widely known in the Western world.

At the First International Barley Genetics Symposium, organized in 1963 in Wageningen, the Netherlands, Nilan *et al.* (1964) concluded that for treatments with chemical mutagens optimum conditions still had to be established. This may also explain the considerable differences in mutation frequencies, observed for various chemicals in different experiments. The authors mentioned in this respect a number of modifying effects, that may considerably affect the outcome. The most important ones are:

- the reactivity of the agent, established by the hydrolysis rate,
- the solubility of the agent,
- the temperature and pH of the solution,
- the amount of (added) oxygen during the treatment,
- the size of the seed and (in case of barley) the presence of hulls around the seeds.

Another important subject, touched upon by Nilan *et al.* at that occasion, was that the degree of sterility in the M_1 is not automatically a reliable indication for the amount of chromosomal damage induced by the mutagenic treatment. The authors reported that EMS and DES induce relatively few chromosome aberrations, but a relatively high level of M_1 sterility.

The Second Barley Genetics Symposium, held in 1969 in conjunction with the FAO/IAEA Symposium in Pullman, Washington State (USA), did not yield significant new views or break-throughs with respect to the use of chemical mutagens. However, during the Third Symposium on Barley Genetics, in 1975 in Garching, Germany, several interesting papers were presented. On that occasion Sigurbjörnsson (1976) mentioned that, so far, only two direct mutant cultivars of barley had resulted from treatments with chemical mutagens: cv. Luther, obtained by R.A. Nilan, in the USA, after treating cv. Alpine with DES (conc. 3.8×10^{-4} M), which mutant was released in 1967, and cv. Betina, obtained after treatment of cv. Vada with EMS at the plant breeding company Ringot in France, released in 1970.

The action of a new inorganic chemical 'super mutagen' <u>sodium azide</u> (NaN_3), of which the remarkable mutagenic properties had been discovered in the 1970s, was reported by Sideris, Nilan & Konzak (1969) and Nilan *et al.* (1973). During the aforementioned symposium in Garching it was pointed out by Nilan, Kleinhofs & Sander (1976) that azide, a respiration inhibitor, had been studied since about 1964 in relation to its effects on radiation-induced chromosomal damage, whereas a weak mutagenic action of treating bacteria with azide had been observed already at an earlier stage. Many details about a wide range of effects in relation to azide mutagenesis also resulted from investigations with the bacteria *Salmonella typhimurium*. It was observed then, rather unexpectedly, that azide, in the presence of oxygen and at a low pH value (e.g. pH = 3), could be a potent muta-gen for higher plants as well. In barley, azide did induce chlorophyll mutations in frequencies that could go up to 46% on M_1 spike basis (which is about the same level as obtained after the most successful treatments with EMS or nitroso-compounds). A very spectacular addi-tional finding was that almost no chromosomal damage was found after treatments with sodium azide.

It was concluded that sodium azide, apparently, does produce point mutations only, which either are base substitutions or very small deletions within a locus. In other words: sodium azide should be considered a very valuable mutagen with not only a high effective-ness, but also with an exceptionally high efficiency. High mutation frequencies were obtained at relatively low dosages of, for instance, less than 1×10^{-3} M. Nilan et al. (1976) further pointed out that the relatively high degree of sterility, observed in M_1 barley plants after treatment with azide, at least in part, must be caused by (real) mutations and not by chromosome aberrations, because efficient excision-repair systems, essentially, did repair all azide-induced mutagenic chromosomal damage.

From 1973 onwards sodium azide was applied in many studies, both in order to produce new mutant cul-tivars, as well as to determine optimal treatment con-ditions (see for instance Nilan et al., 1973, and Sideris, Nilan & Bogyo, 1973). A good review of different aspects of azide application until 1978, with many references, was produced by Kleinhofs, Owais & Nilan (1978). The authors report that azide is not mutagenic in all plant species (or crop plants) studied and that even cultivars within one species may behave differently in this respect. This somewhat unexpected result may be partly explained by inappropriate treatments, in particular during the early years of application of azide to seeds, but the most important factor may be that azide itself is a so-called promutagen: a compound that is not muta-genic in itself but can be biologically transferred into a mutagen, so that an organic metabolite is the real muta-genic agent. (N.B. Another well-known promutagen is maleic hydrazide that is commonly used as a herbicide, fungicide, growth inhibitor and growth regulator. See for instance Véleminský & Gichner, 1988.)

Additional details about azide treatments, as well as about effects of promutagens will be presented later in this chapter.

Nilan (1981a), at the Fourth Barley Genetics Symposium, that was held in 1981 in Edinburgh, reported that, at that time, in addition to gamma- and X-rays, the chemicals EMS, DES, nitroso-compounds and sodium azide (of which the mutagenic action still was not well understood) remained the chief mutagens for barley. This observation is fully in line with the situa-tion for other crops as well. Nilan added, that by judi-cious selection of mutagens and manipulation of mutagenic treatment conditions, the breeder, to a cer-tain extent, is able to influence the kind of genetic changes he wishes to induce. However, this remark only refers to the possibility to influence more or less the rel-ative proportion of larger chromosomal aberrations against small changes. Myleran, for instance, induces high ratios of chromosome aberrations to mutations; EI induces about equal proportions of both; EMS, DES and some base-substituting nitroso compounds show rela-tively higher proportions of mutations, whereas sodium azide, apparently, almost exclusively produces gene (point) mutations.

Ukai & Yamashita (1981) reported at the same sym-posium that the ratio of early maturing mutants to chlorophyll mutations was lower for chemical mutagens than for gamma rays and, hence, they warned against too much optimism with respect to using chemical mutagens for practical purposes. Their results made them conclude that obtaining high frequencies of muta-tions for chlorophyll by applying chemical mutagens, does not necessarily imply the superiority of such treat-ments for other traits as well. The authors suggest that chemically induced early mutants may have an increased chance of being accompanied by undesirable mutations in comparison with early mutants obtained from treatments with gamma rays. Moreover, for some chemical mutagens, like EMS and certain nitroso-compounds, it was found that within the range of induced mutants for earliness, the relative number of very early mutants – which are the most favoured group – was lower. In addition Kivi (1981), who selected for ear-liness among 350 000 individual M_2 plants, after treat-ing six-row barley with sodium azide, also showed that high frequencies of chlorophyll mutations did not nec-essarily result in new cultivars in which the earliness of the best cultivar available is surpassed.

From the findings by Ukai & Yamashita and Kivi, it must be concluded that, when comparing efficiencies of different mutagenic treatments, an assesment of the rel-ative frequencies and spectra of chlorophyll mutations obtained after different treatments, in fact, does not provide sufficient information about the mutation frequency of other traits. In practice, however, most mutation breeders do limit their preliminary or pre-experiments to observations on mutation-induced chlorophyll aberrations and some other traits which can be observed relatively easily, because of the amount of work required when specific initial experiments should be performed for the particular trait in which the breeder is interested.

At the Fifth Meeting on Barley Genetics in 1986 in Okayama, Japan, it was again Nilan (1987) who reviewed the trends in barley mutagenesis during the past five years. Nilan concluded that no major developments with respect to further unravelling the mechanisms of chemical mutagens and ionizing radiation had been brought forward. He further observed that still a 'fairly steady flow' of mutation experiments, directed towards the development of new mutant cultivars, was set up, but that most publications did not come anymore from Western Europe, USA or Japan, but from countries like India, the (then) USSR and Eastern European countries. According to Nilan the period 1981–6 had shown an increased use of nitroso-compounds, but he stressed that, in his opinion, sodium azide should be in most cases the preferable mutagen, because of its high efficiency and effectiveness. This opinion, however, seems somewhat lonely, as for practical plant breeding purposes in most crop species – including barley – at that time (and still these days), the use of EMS in particular remained very popular.

Moreover, it had also become clear in the meanwhile that sodium azide could not be called a 'super mutagen' for all traits investigated and for all crops. This is further demonstrated by the fact that in 1994 – 20 years after the introduction of azide in mutation breeding – of the 141 direct, chemically induced and officially released mutant cultivars registered at the database of the IAEA in Vienna, only seven cases referred to azide-induced mutants (M. Maluszynski, personal communication, Oct. 1994). The best known azide-induced cultivar – at that time – was cv. Galant, a proanthocyanin-free mutant of barley, which was released in 1984 in Denmark (see also Chapter 3). In 1991 one azide-induced mutant cultivar of wheat (Triticum aestivum) with improved resistance for leaf rust was released in Pakistan, and between 1992 and 1994, five azide-induced registered mutant cultivars of sesame (Sesamum indicum) with various improved traits, were reported from Korea. As a comparison: the IAEA database reveals that, at that time, for instance six colchicine-induced mutant cultivar's were registered; five mutant cultivar's which originated from DES treatments; 11 from DMS; 18 from EI; about 50 from different nitroso-compounds (mentioned are ENH, MNH, NMH, NMU, NEU and NTMU) – which almost exclusively refer to mutants obtained in the former USSR (and North Vietnam) – and about 35 from EMS.

Finally, from studying the topics presented at the Sixth Barley Genetics Symposium, organized in 1991 in Helsingborg, Sweden, it must be concluded that at that time mutation breeding had lost its importance as a research subject of general interest and that, as could have already been anticipated in 1986, most attention had gone to subjects related to biotechnology and molecular genetics. Consequently, during the Helsingborg meeting no new approaches or major contributions concerning chemical mutagenesis were presented.

Chemical mutagenesis throughout the years: some conclusions

In summary it can be said that, after the definite proof in the 1940s that mutations could be induced by chemical mutagens, the discovery of EMS – which will be discussed in more detail later in this chapter – as a potent mutagen for higher plants around 1960 has been the next highlight in the history of chemical mutagenesis. EMS which, as a powder, can be easily catalogue-purchased, is cheap and does not leave toxic substances after hydrolyzation, and is still the most commonly applied and most successful chemical mutagen.

The discovery of EMS was followed by a period of many, and often very detailed, studies on the effects caused by different chemicals and on the most suitable treatment methods. Apart from some interesting results of fundamental significance, which often came from experiments with micro-organisms, not much progress was made with respect to the use of chemical mutagenesis for practical purposes during a period of one decade or longer.

Much interest and new hopes were raised again by the discovery of the strong mutagenic effect of sodium azide in the 1970s. Sodium azide – see also later this chapter – despite its proven unique ability to induce gene mutations almost without accompanying chromosome aberrations, and despite much research, in particular with barley, so far, has not fulfilled its expectations and did not become the supermutagen as was anticipated by its discoverers. In addition it should be mentioned that the use of sodium azide is also not without risks, for instance because it may become explosive under certain conditions!

In addition, some nitroso-compounds, introduced in the 1960s, are often applied as well, although they may be less easy to obtain in certain countries. It has been suggested at some occasions (see also Chapter 3), that nitroso-compounds – which will also be discussed later on in this chapter – may be useful for inducing extranuclear mutations.

Based on the present views on the possibilities and limitations of chemical mutagenesis in higher plants, the general advice to plant breeders of seed propagated crops at this stage should be that, in order to try and induce a high frequency and a broad spectrum of (useful) mutations, different chemicals should be tried side by side, e.g. EMS, sodium azide and nitroso-compounds. In addition, it may be advisable to try gamma rays and (in specific situations) UV as well.

4.3.2. Chemical mutagens: general characteristics and ways of application

General considerations

It may have become clear from the review in the previous section, that, despite the fact that throughout the years many chemicals with mutagenic properties have been discovered, relatively few of them, nowadays, are applied frequently for practical plant breeding purposes. Hence, in this section we will mainly pay attention to those chemical mutagens that are of some practical importance. Many details which do not seem to be of much use to plant breeders, will be just mentioned only or even may be completely omitted. For more details reference is made here in particular to the FAO/IAEA Manual on *Mutation Breeding* (Anon., 1977a), from which much of the information, discussed in this section, has been derived. Some additional comments about treatments and treatment modifications have been published, for instance, by Nilan (1987).

Most information about fundamental aspects concerning patterns of action of different groups of chemical mutagens, which almost exclusively has been collected in studies on bacterial and viral systems (see for instance Drake, 1969; 1970), will not be discussed here. Sometimes, mutagens of which the effectiveness has been proven for lower organism, are not as effective for higher organisms. Several mechanisms may account for this, like differences in repair mechanisms for various organisms. Finally, in this respect, for details on the chemical composition of the DNA and the subject of abnormal pairing behaviour as a consequence of the action of, in particular, chemical mutagens, reference can be made to most recent textbooks on genetics, like the ones by Suzuki *et al.* (1989), Russell (1992) and Lewin (1987; 1994), or more recent editions of these publications.

A summary of advantages and disadvantages of applying chemical mutagens

During two so-called Research Co-ordination Meetings on *in vitro* technology, organized by the IAEA in the 1980s, amongst others, advantages and disadvantages of chemical mutagens were discussed in this context (Anon., 1986a). As most points, mentioned in a list of pros and cons of chemical mutagenesis, do refer to *in vitro* as well as '*in vivo*' treatments, this list is also summarized here.

Advantages:
- mainly point mutations occur,
- less chromosomal damage is found than with physical mutagens,
- mutation spectrum probably differs from physical mutagens,
- high mutation frequencies may occur.

Disadvantages:
- penetration in multicellular plant tissues is often difficult,
- reproducibility of results is low, probably, because of lack of standardization of methods of treatment,
- treatments may be dangerous because of carcinogenic properties of chemical mutagens (biohazard boxes should be used).

Which parameters should be noted when treating with chemical mutagens?

With respect to standardization of experiments with chemical mutagens – a subject that deserves more attention than is commonly paid to it – the report of the aforementioned IAEA Research Co-ordination Meetings mentions that the following parameters should be taken into account:

- dose (concentration × duration of treatment) and amount of mutagen in relation to number and size of the treated objects (seeds, buds, etc.),
- pH,
- physical and chemical properties of the agent, e.g. the 'half-life' of the chemical,
- interaction with the culture medium (in case of *in vitro* treatments only),
- interaction with (in fact penetration in) specific plant tissues (seeds, seedlings, cuttings, buds, etc.),
- post-treatment conditions.

In addition to this list one could probably mention also, as a way to achieve standardization, the use of pre-treatment of the plant material. One could think, in this context, of standardization of pre-soaking methods of seeds, trying to solve problems with respect to semipermeability or impermeability of seeds coats (like may be the case in conifers and legumes) and of treatment of cells that have been synchronized at a specific stage within the mitotic cycle. Konzak & Narayanan mention in the FAO/IAEA Manual (Anon., 1977a, pp. 71–2) that the greatest efficiency (more mutations as compared to chromosomal aberrations) of treatments with chemical mutagens may be obtained in the mid to late G_1 stage of the mitotic cycle. For barley this may be achieved in practice by pre-soaking (immersing) seeds for 8–12 h at a low temperature in (de-ionized or distilled) running water.

The effective range of concentrations of a chemical mutagen is limited by its solubility and its toxicity. Despite the fact that methylating agents are more mutagenic than ethylating agents, the efficiency of a compound like methyl methanesulphonate (MMS) is lower than that of ethyl methanesulphonate (EMS) because of the higher toxicity of MMS. This higher toxicity results in a higher level of physiological damage to seeds and plants and a lower level of survival after the mutagenic treatment.

The concentration of the chemical mutagens and the duration of the treatment, in combination with the temperature during the treatment, are the most important factors which play a role when determining the dose. Prolonged treatments with high doses of mutagen result in relatively much physiological damage in relation to the mutation rate.

There is a tendency at present to treat with lower concentrations than was done in the past. It is assumed, nowadays, that about 20–30% growth reduction, which may correspond to a survival rate of 70–80%, may produce an optimal mutation yield, whereas in the past a lower survival rate of about 50% was thought to be an adequate dose for producing a maximum number of desirable mutations. Different chemical mutagens may require different doses in this respect. In addition as, according to the FAO/IAEA Manual, it has been found that the uptake saturation in dry seed embryos of rice (*Oryza sativa*) requires not more than 1 h, the use of shorter treatments is advocated. (N.B. For whole, hulled dry seeds of rice, however, about 5 h are required.)

Still following the FAO/IAEA Manual (Anon., 1977a, p. 68 and 69), pre-soaking, in this respect, may considerably shorten the treatment time as cell membranes are made ready permeable to the mutagen, and pre-soaking also results in reduced variability in the results of treatments.

When optimal treatment conditions for a specific crop or mutagen can not be found in literature, it is advised in general, to determine with a range of concentration x duration combinations in a seedling test at which treatment a growth reduction of about 20–30% is obtained. It has been found for barley, that at that level an optimal yield of mutations may be expected. Although this may be good advice in general, it would

be preferable to adjust these percentages by taking the nature of the mutagen into account as well, but for this purpose additional tests would be required. Furthermore, it should be remembered that differences between species, varietal differences, post-treatments and other factors may strongly affect the final outcome of a mutagenic treatment.

The half-life of the mutagen solution, that is the time during which half of the initial amount of the agent is hydrolyzed or degraded, also has to be considered in relation to a mutagenic treatment. In order to keep the concentration of the mutagen at a relatively constant level, mutagenic solutions have to be renewed with freshly prepared solution when the period of treatment is relatively long in relation to the half-life of the mutagen. It is advised in the FAO/IAEA Manual to renew the solution when about one-quarter of the solution has been hydrolyzed. Regular shaking or stirring of the solution is also advisable. Renewing the solution may have to be performed several times in the case of prolonged treatments. An alternative is to buffer the mutagenic solution. Both measures may not be necessary in short treatments with high concentrations, although for treatments with NaN_3 a pH 3 is required. When chemical mutagens with a short half-life are applied (e.g. DES, which has a half-life of 1 h only at 30 °C), it is advised to perform the treatment at low temperature.

Zimmer (1961) already showed that the dose–effect curves for 'various chemical reagents, e.g. poisons' may differ considerably from those for radiation treatments. This is shown in Fig. 4.5 (after Zimmer; derived from Dertinger & Jung, 1970). Chemical mutagens are characterized by the existence of a specific threshold-value, after which the survival rate may show a quick decline, whereas for radiation a more gradual decline in the number of surviving seedlings or plantlets can be observed. As a consequence, determining the optimal concentration for a chemical agent may be very critical from a practical point of view. Further treatment procedures, the choice of plant species and even cultivar, as well as growing conditions (field, greenhouse; '*in vivo*'/*in vitro*) may considerably affect the survival rate for a given dose.

Modifying factors for treatments with chemical mutagens

The FAO/IAEA Manual contains a list of factors that – before, during or after the treatment – may modify the outcome of a chemical mutagenic treatment. The most important factors are mentioned here:

1. pre-soaking (see before),
2. application of buffers (see before),

3. the use of de-ionized water in order to prevent undesired effects by metallic ions,

4. the use of so-called carrier-agents, like dimethyl sulphoxide (DMSO) is sometimes advocated, as it is believed that the uptake of chemical mutagens in multicellular systems may be increased by adding such carriers. DMSO, on the other hand may considerably decrease the mutation frequency, as for instance Ipser (1993) reported for treatments *in vitro* with DMSO calli of *Crepis capillaris* previous to application of MMS,

5. the occurrence of so-called after-effects, the level of which is significantly dependent upon the length of the period of storage and the temperature during storage.

The third point has not been mentioned before. It is reported in the FAO/IAEA Manual (p. 70) that certain metallic ions, like Zn^{2+} or C^{2+}, may affect mutation frequencies in various crops. Therefore, the general advice is given to prepare mutagenic solutions with de-ionized water.

In literature different opinions as to the significance of these modifying factors are expressed. Several of the

recommendations given here can also be found in an IAEA Training Manual on *in vitro* techniques, intended for participants of the training courses on mutation breeding at the FAO/IAEA Laboratories at Seibersdorf, Austria (Anon., 1990b).

Chemical mutagenesis of seeds, pollen, in vitro material, etc.

In most cases chemical mutagens are administered to seeds, although an increased interest in treating *in vitro* plant material can be noticed. Determining the most suitable dose and appropriate treatment conditions is not easy because of the many parameters involved. Fortunately, procedures have been worked out for most chemical mutagens by specialists and details about treatments often have been published in various publications by the IAEA or in specialized journals like *Mutation Research* or *Radiation Botany* (which is now called *Environmental and Experimental Botany*). A useful work of reference in this respect, in addition to the FAO/IAEA Manual, is a more recent publication, edited by Véleminský & Gichner (1987), in which the effects of several important factors have clearly been demonstrated.

Treatment of seeds can be done in different ways. Dry seeds can be exposed to different concentrations of a mutagen, seeds can be pre-soaked before being treated, etc. Satoh & Omura (1981) were much more successful in inducing endosperm mutations in rice (*Oryza sativa*) by treating fertilized egg cells with nitrosomethyl urea (1.0 or 1.5 mM for 3 h at 6 or 120 h after flowering) than by treatment of dry seeds.

Instead of seeds, other plant parts can be taken as well. So-called '*in vivo*' treatments can be performed by covering an inflorescence (or a bud) with a wad of cotton wool, soaked in the chemical. In the case of treating an inflorescence, a small polyethylene bag could be sealed around the inflorescence with tape and left for a certain number of hours, followed by some minutes of post-washing under running tapwater. Chemical mutagens could also be injected in several plant parts, for instance in nodes.

A very atttractive and effective option: treating pollen with chemical mutagens, has been described for instance by Neuffer & Chang (1989) for maize (*Zea mays*) (see also Chapter 6). An important advantage of treating gametes is that all recovered mutants have arisen independently of each other, so that M_1 plants are simple heterozygotes and are not chimeric. From treating maize pollen, 10 000 heterozygous M_1 seeds/plants can be easily produced. Dominant mutations are already detectable in individual mutant plants in the M_1 popu-

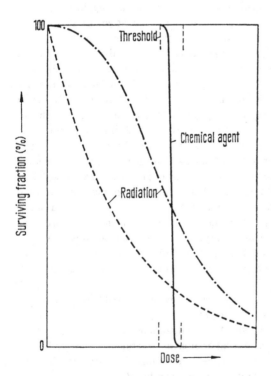

Figure 4.5. Dose–effect curves for the action of chemical mutagens and radiation (after Dertinger & Jung, 1970; original figure from K. Zimmer, 1961).

lation. An important point of attention should be to control pollen viability before and after treatment.

Results with 'in vivo' treatments of vegetatively propagated crops with chemical mutagens, so far, have not been very successful and it remains difficult to standardize treatment conditions. Positive results in vegetatively propagated carnation (*Dianthus caryophyllus*), for instance, were published by Hentrich & Glawe (1982). Some other examples and more details can be found, for instance, in Broertjes & van Harten (1988). The main points will be briefly mentioned as well in Chapter 7.

Application of chemical mutagens to various *in vitro* systems may offer several advantages, such as the possibility to apply mutagens under well-defined conditions. Indications about the physiological effects of chemical mutagens, for instance, may be studied on the base of survival of protoplasts, callus initiation, callus growth and survival, capacity of plantlets regeneration from regenerable callus, etc., but such results do not necessarily also provide information about an optimum yield of mutations *in vitro*. Treatments may be done before culture (whole plants, cuttings, bulbs, shoot tips, explants, single node cuttings) or during culture. It may take some time to determine optimal concentrations of the mutagen as well as optimal treatment conditions. The kind of tissue, type of mutagen, method of treatment and cultivar choice may cause differences in mutation response, but the role of the genotype in this respect seems to be relatively limited.

Omar, Novak & Brunner (1989), for instance, found for *in vitro* cultures of shoot tips excised from two clones of banana (*Musa acuminata*), that the best response to EMS treatment – measured by determining the number of new suckers – was achieved after incubation with about 25 mM (0.2%) EMS for 3h. It was established that the use of DMSO (4%) as a carrier agent also resulted in a much better uptake of the mutagen under *in vitro* conditions.

Masrizal, Simonson & Baenzinger (1991) concluded from EMS treatments (4 h at room temperature) of wheat (*Triticum aestivum*), that the doses required to achieve an LD 20 (about 0.35% EMS) or an LD 50 (about 0.80% EMS) when immature embryos and callus derived from immature embryos were treated, are practically identical to those required for seed mutagenesis (without pre-soaking). If this finding should have a general validity, this could save much time in designing *in vitro* mutation experiments.

Whether mutation frequencies observed from treatments of *in vitro* material indeed have been induced by the mutagenic treatment is not always clear, because sometimes genetic variation which was present already in the starting material but remained unobserved so far, may become revealed under *in vitro* conditions. In addition, the composition of the *in vitro* media may be responsible for the induction of mutations (see further Chapter 5).

The efficiency of the selection procedure is essential for mutation studies *in vitro*. Usually tolerances to certain metabolites and other compounds, supplied to the medium in toxic concentrations, are used for this purpose. In 1986 two special publications on *in vitro* mutagenesis were published by the IAEA in Vienna (Anon., 1986a,b). In Chapter 5 more details about chemical mutagenesis *in vitro* and about *in vitro* treatments will be presented.

4.3.3.   Classification of chemical mutagens

Introductory remarks

Depending on the authors – see for instance the IAEA Manual (Anon., 1970; 1977a); Suzuki *et al.* (1989) or Russell (1992) – chemical mutagens are classified in different groups. In a review of the main mutagenic compounds by H.Heslot in the first edition of the FAO/IAEA Manual (Anon., 1970), four different groups were distinguished: base analogues and related compounds, antibiotics, alkylating agents and miscellaneous. In the second edition of the Manual (Anon., 1977a) this division is maintained, whereas the number of compounds mentioned in the category 'miscellaneous' increased.

It has been mentioned before on several occasions that, for many years, by far the most important mutagens belong to the group of alkylating agents, with EMS and nitroso-compounds as the most commonly applied agents. Generally, other (new) mutagenic agents are of potential interest only, when they are able to induce a different mutation spectrum, e.g. by acting specifically on the extranuclear DNA. If this is the case for some chemicals described below, this will be specially mentioned.

The only other chemical mutagen that is often considered to be of about equal significance, because it is believed to produce primarily – or solely – point mutations by the induction of base substitutions, is azide. Colchicine, which was discussed already in Chapter 3 and which, apart from mitotic doubling of chromosome numbers, has produced a number of mutant cultivars as well, will not be treated again at this point.

Earlier in this chapter, mention was made of so-called promutagens. This term refers to chemicals that are not mutagenic by themselves, but can be biologically altered into mutagens by a process called 'plant activation' (Plewa & Wagner, 1993). The authors add that it is 'disconcerting that plants can activate promutagens and store the products in forms that may induce mutations in consuming organisms'. Probably the first report concerning promutagens was by Véleminský & Gichner (1968), who referred to the induction of mutations by several nitrosamine promutagens in *Arabidopsis thaliana*.

Sodium azide, which was briefly discussed before (see also later in this chapter) is also considered a promutagen, as is the case with, for instance, aflatoxins, furfurylamide, ethanol, aniline and a range of pesticides.

At this place it must be repeated that, for reasons which have been explained before, in the following sections several, sometimes unavoidable, inconsistenties may occur with respect to names, formulas and abbreviations used for various chemical mutagens.

Alkylating agents
Most often used and, by far, most successful from the point of view of producing new mutant cultivars, are a number of chemicals belonging to the group of the alkylating agents. The only compound outside this group that is rather frequently used nowadays is azide.

Alkylating agents are characterized by the presence of one or more alkyl groups, which react with DNA by alkylating the phosphate groups as well as the purine (A,G) and pyrimidine (C,T) bases of the DNA. When the organic bases are alkylated, the formation of 7-alkyl guanine is the most frequently occurring event. Fundamental aspects of alkylation mutagenesis, for instance, have been discussed by Lawley (1974).

Ehrenberg *et al.* (1956) had already pointed out that alkylating agents produce unspecific effects, i.e. that both small effects like point mutations and major chromosomal aberrations are induced. In this respect chemical mutagens like nebularine (which is not often used nowadays) and azide act in a more specific way.

Whereas Auerbach (1976) distinguished nine main classes of alkylating agents, in the FAO/IAEA Manual (Anon., 1977a) seven classes of alkylating agents are given. Each class is represented by a few up to many mutagens. The seven classes, mentioned in the Manual, are:

- sulphur mustards,
- nitrogen mustards,
- epoxides,
- ethyleneimines,
- sulphates and sulphonates,
- diazo compounds, and
- nitroso-compounds.

There is no need to mention here the names of all the (about 35) alkylating agents, given in the Manual as, nowadays, only a few of them are of practical importance. Those most important alkylating agents are: ethyl methanesulphonate (EMS), diethyl sulphate (DES), ethyleneimine (EI), nitroso ethyl urea, nitroso methyl urea and methyl nitro nitrosoguanidine (MNNG). These compounds, to each of which was referred to in Section 4.3.1, all are monofunctional compounds, i.e. they have only one functional alkyl group! The negative features commonly encountered when applying bi- or polyfunctional mutagenic compounds have been described before.

The relative effectiveness and efficiency of various mutagens has been compared in a number of studies. To give an example: Kaul & Bhan (1977) showed in rice (*Oryza sativa*), that for EMS both the effectiveness and the efficiency were superior to DES. The chemicals were for both aspects superior to gamma rays.

A general warning to be made about alkylating agents is that they are highly toxic and potential carcinogens and, hence, always should be handled with care, e.g. they should not be pipetted by mouth and use should be made of fume hoods. Nitroso-compounds are often considered to be particularly dangerous and carcinogenic but, when carefully handled, no greater danger applies to them over most other chemicals. Ethylene imine (EI), because of its volatility, is much more dangerous than EMS in this respect. Mutagenic chemicals always should be stored somewhere cool, preferably in special refrigerators.

Ethyl methane sulphonate (EMS)
As can be concluded from the information in the IAEA databank in Vienna (M. Maluszynski, personal communication), EMS is a very successful mutagen and most probably the chemical mutagen that is most often applied. This may be demonstrated by the fact that, by the end of 1994, 35 out of about 140 chemically induced, direct mutant cultivars were EMS-induced, and these mutants had been obtained in various countries, all over the world. Together with some nitroso-compounds, EMS also may be the most common mutagen for *in vitro* mutagenesis as can be concluded from an IAEA inventory in 1983, published in *Mutation Breeding Newsletter* 22 (Anon., 1983c).

Heslot *et al.* (1959) applied EMS for the first time in higher plants. On many later occasions it was reported that EMS not only produced higher mutation rates than irradiation treatments, but also was more efficient, i.e. produced more mutations relative to sterility. For results concerning treatments of barley (*Hordeum vulgare*), see for instance Gaul et al. (1972) and Constantin (1976). Zanone (1965), who compared the effects of EMS, EI and X-rays for frequency and spectrum of chlorophyll mutations in two types of vetch (*Vicia sativa*), concluded that EMS treatment resulted in more chlorophyll mutations than EI and X-rays. EMS treatments also produced a wider spectrum than the other treatments, but these differences were not statistically significant. Nilan (1981b) concluded that EMS induced higher proportions of mutations to chromosome aberrations. Kohalmi & Kunz (1988) mentioned that EMS induces high frequencies of base-pair substitutions, which were almost exclusively of the transition type at the G–C site, and little lethality. (N.B. This was also found for MNNG which mutagen was investigated in the same experiment.)

EMS is also a suitable mutagen for the induction of extranuclear mutations, as has been shown already by Favret & Ryan (1964) for barley and by Dulieu (1967) for tobacco (*Nicotiana tabacum*).

EMS is a colourless, liquid compound, with a molecular weight of 124, and 8% soluble in water. The half-life in water at pH 7 and 20 °C is 93 h. At 30 °C the half-life is 26 h. Information about sources of supply, health hazards, clean-up, disposal as well as references to methods of preparation can be found in the FAO/IAEA Manual (Anon., 1977a). Like all alkylating agents, EMS is very reactive and, as a consequence, solutions should be prepared just before use. For EMS the reaction with water is as follows:

$$CH_3SO_2OC_2H_5 + H_2O \rightarrow CH_3SO_2OH + C_2H_5OH$$

(EMS)   (methane   (ethyl alcohol)
                          sulphonic acid)

After hydrolysis the mutagenic activity of EMS and other alkylating agents has disappeared.

The FAO/IAEA Manual provides an example of applying EMS to barley seeds. If we omit at this place points like how to select the, genetically, most suitable starting material and how to determine the number of seeds to be treated (which subjects will be discussed in Chapter 6), the treatment procedure for this and other seed propagated crops could be as follows.

Store the chemical before use in airtight coloured bottles in the refrigerator. Put uniform seeds with a high level of germinability in quantities of no more than a few hundred seeds in small mesh bags from, for instance, plastic or nylon screening or – when treatments don't take too long – in small teabags. As uniformity and reproducibility of the treatments are important prerequisites, it is advised to keep the seeds prior to treatment in a

desiccator over a 60% glycerol solution in order to give the seeds a moisture of about 13%.

In order to realize optimal uptake of the mutagen in the seed embryo area, seeds should be pre-soaked by bringing the bags to beaker glasses with distilled or de-ionized water at room temperature for 16–20 h. (N.B. In other publications shorter periods of pre-soaking, e.g. 8–12 h are considered adequate.)

Temperature should be kept constant during treatment. Mutagenic treatments may be performed at room temperature or, by making use of a water bath, at a higher temperature of for instance 30 °C.

The actual treatment – when possible in a fume hood – with the mutagen may take from one half hour to a few hours, and for barley (*Hordeum vulgare*) and wheat (*Triticum aestivum*) the FAO/IAEA Manual recommends use of EMS solutions with concentrations from 0.05 to 0.1 M, whereas in other publications sometimes concentrations of up to 0.4 M are advised. (N.B. In many publications concentrations are still expressed as a percentage, say 0.2% or 0.2 g. per 100 ml. Given a molecular weight for EMS = 124, the figure of 0.2% EMS corresponds with 0.0248 M, because 0.2% = 0.248/124. Accordingly, mole concentrations of EMS can be converted into percentages by multiplying by 1.24.)

Following a 'standard recipe', used in the 1990s at several research institutes for treating pea (*Pisum sativum*), seeds which have been pre-soaked overnight, are treated with about 0.04–0.05 M EMS for 4 h, followed by 1 h post-washing.

According to another example, Grabau, Hanlon & Pesce (1995) successfully treated suspension cultures of soybean (*Glycine max*), two days after transfer, for 30 min with 1% EMS. As a next step, the suspension was washed twice with the medium (a supplemented Gamborg B5 medium), followed by resuspension in fresh medium and, after 24 h, addition of a selection medium which contained oligomycine (further details about selection procedures for EMS-induced oligomycine resistance are omitted here).

Depending on the crop and the specific cultivar selected, the optimal combination of treatment duration and concentration has to be found. Often the level of germination or the reduction of seedling length in M_1 is taken as a measure for the effectiveness of a mutagenic treatment, but determining the frequency of chlorophyll mutations in M_2 – despite an extra amount of work – has a considerably higher predicting value. For barley frequencies of more than 30% chlorophyll mutations have been observed in the M_2 but mutagenic treatments with high concentrations of chemicals are practically always accompanied by severe side effects, such as low survival rates and a high level of sterility in M_1 plants. As a consequence, most research workers prefer somewhat lower concentrations, i.e. they also accept that mutation frequencies are for instance 10–50% lower, in order to limit the occurrence of such undesired side effects and multiple mutations. Lower mutation frequencies may be compensated by a considerably higher number of seed bearing M_1 plants after mutagenic treatment.

In the FAO/IAEA Manual it is further pointed out that post-treatment drying is useful for convenient handling of the treated seeds and, further, that in case of treatments with alkylating agents like EMS, thorough post-washing is necessary in order to warrant maximal reproducibility and to limit the occurrence of undesired after-effects (which, apparently, can never be completely prevented). Post-washing for a period of four hours in running tapwater is often considered adequate. After treatment and post-washing have been performed, seeds should be surface dried for a short while, for instance at room temperature on blotting paper or – with the seeds still in the screen bags – with an electric fan or hair dryer (gently and in a fume hood). Immediate sowing is preferable but when this is impossible, the short drying period on blotting paper (or toilet paper) creates a situation in which the seeds sometimes may keep their germination power for several weeks, provided also that temperatures are not too high. For longer periods of storage low temperatures (e.g. 6–8 °C) are required.

After treatment the mutagen solution has to be disposed of, for which different methods are suggested in the Manual. A simple procedure is to mix the mutagen solution with excess sodium bicarbonate, for instance in a bucket, and to dump the mixture in a drum to which about 50 l of water is added afterwards. After two or three days the contents of the drum can be safely poured into a drain with excess water.

Other alkylating agents

Different alkylating agents have different physico-chemical properties, which may affect the methods of treatment, the safety measures that have to be taken, the way of storage, etc. DES, for instance, which was considered the most potent mutagen in the early 1960s, has a much shorter half-life than EMS, e.g. about 3.5 h at 20 °C and, as a consequence, solutions have to be often changed, e.g. every half hour or after 25% of the solution has been hydrolyzed. The concentrations at which DES is applied, according to Nilan et al. (1964), may be lower than for EMS, because DES reacts much faster. For DES treatment time is more critical than for EMS. A treatment with a low concentration of, for instance, 0.4 mM for 3.5 h, gives much higher mutation frequencies than higher concentrations for 1 h. Nowadays, EMS is preferred above DES, as EMS produces higher mutation frequencies and less physiological damage. In 1994 five cases of DES-induced mutant cultivars of different plant species were registered at the, earlier mentioned, FAO/IAEA database in Vienna (Maluszynski, personal communication). As has been mentioned before, the first chemically induced mutant cultivar (taken apart one or two earlier, colchicine-induced polyploid mutant cultivars), was cv. Luther in barley (Hordeum vulgare), which was released in 1967 and had resulted from treatment of barley seeds with DES by R.A. Nilan of Washington State University, USA, in 1960.

Another alkylating agent, ethylene imine (EI), which was considered as highly mutagenic in the early 1960s, also is not much used anymore. In 1994 the aforementioned FAO/IAEA data base contained 18 EI-induced mutant cultivars, which almost exclusively originated from the former USSR. Two other chemical mutagens which are not much used anymore are ethylene oxide (EO) and isopropyl methanesulphonate (iso-PMS). The database list of 1994 refers to two EO-induced cultivars, namely in barley and rice. Two mutant cultivars of rye (Secale cereale) with shorter culms were induced in 1981 in the former German Democratic Republic after treatment of seeds with iso-PMS (Mutation Breeding Newsletter 23; Anon., 1983d).

Nitroso-compounds are mainly solids, which are very soluble in organic solvents and in which the pH is important with respect to their reactivity. The three most important nitroso-compounds are: nitrosoethyl urea (NEH according to the FAO/IAEA Manual, but sometimes also abbreviated as NEU or, in German, ENH), nitrosomethyl urea (NMH, NMU or MNH) and MNNG (methyl-nitro-nitroso-guanidine).

Nitroso-compounds have to be stored in small quantities in a refrigerator. Temperatures above room temperature should be avoided by all means. These compounds, when not handled in a professional way, may cause considerable health hazards and cancer and, hence, treatments should be performed with very great care in a fume hood and care should be taken to prevent any waste. (N.B. The same precautions have to be taken when applying chemical mutagens of several other groups, like for instance imines and mustards, but compounds of those groups are not often used anymore.)

Müller & Gichner (1964) concluded that MNNG can

be classified as a mutagen with an extremely low relative toxicity – as is the case for EMS and some other nitroso-compounds – because the survival rate shows a decrease only at almost complete sterility.

Nitroso-compounds may be very useful mutagens for inducing nuclear as well as extranuclear mutations. Heslot et al. (1966) found after treatments of barley with nitrosoethyl urethane and nitrosomethyl urethane that the frequency of chlorophyll mutations – which may arise in different ways – could increase to 45% in the M_2 (on spike basis), which was about the same as for EMS and 4–8 times the frequency induced by gamma rays. These authors concluded that in particular nitrosoethyl urethane appears to be a very potent mutagen with a high efficiency. Zoz (1966, English summary only) reported that nitrosoethyl urea was the 'strongest' chemical mutagen, 'giving up to 80% recessive and 15% dominant mutations' in higher plants. Further details are unknown.

Nitroso-compounds have proved to be very effective mutagens for the induction of point mutations, as was published for instance for cereals in a range of articles by Malepszy, Eberhardt & Maluszynski (1973), Maluszynski, Eberhardt & Cudny (1974), Maluszynski & Maluszynska (1977), Maluszynski (1982), Maluszynska & Maluszynski (1983) and Maluszynski et al. (1987; 1989). Schy & Plewa (1985) reported that treatment of seeds with ethylnitroso urea was almost four times more effective than EMS in producing point mutations at the Yg2 (or Yg-2) locus of maize (Zea mays). A dose of 6.7×10^{-6} M of this compound administered for 8 h at 20 °C was found to induce the same mutation rate as one rad of gamma radiation (1 rad = 0.01 Gy).

Travis, Stewart & Wilson (1975) induced nuclear and cytoplasmic chloroplast mutations in Mimulus cardinalis by treating seeds with 0.01 mM MNNG for 30 min. McCabe, Timmons & Dix. (1989) described a simple and efficient system to induce extranuclear or plastome-related mutations (like white or variegated shoots or maternally inherited resistance to antibiotics) after treatment of several species of the Solanaceae family with methylnitroso urea. Seeds or shoot-cultures which had been derived from small leaf strips (0.2 × 1 or 1.5 cm), were incubated in liquid culture media containing 1 or 5 mM of this compound. The cultures were placed on a rotary shaker (120 rotations per min) for 90 or 120 min and washed three times with fresh medium, with or without the presence of streptomycine. Seeds were germinated and explants were placed on the surface of a culture medium. This procedure resulted in up to 3.5% of explants with albino or variegated adventitious shoots. At doses of 20 mM no more adventitious shoots were produced.

Hagemann (1976; 1982) successfully treated seeds of snapdragon (Antirrhinum majus) with concentrations of up to 17 mM for nitrosomethyl urea (mol. weight given as 103.8) and 31 mM for nitrosoethyl urea (mol. weight 117.1) for 2 or 3 h, with 6 h of pre-soaking, and obtained high frequencies of chlorophyll deficiencies, based on plastom mutations. In order to recognize extranuclear mutations Hagemann used three criteria: 1) the characteristic variegated pattern of the leaves, 2) cytological evidence of cells with different types of plastids, and 3) the occurrence of non-Mendelian inheritance for these traits. This last point provides the final proof, but this approach is not always possible, e.g. when the mutated sector does not give rise to flowers and seeds. The experiments by Hagemann (1982) – who uses the abbreviation NMU for nitrosomethyl urea, and NEU for nitrosoethyl urea – confirmed that nitrosomethyl urea showed a much higher mutagenic efficiency than nitrosoethyl urea, whereas the former compound was more toxic than nitrosoethyl urea. Hagemann further advised that because of an increasing level of sterility, one should not use too high concentrations of these compounds.

Extranuclear mutations may refer to a range of phenomena, including chlorophyll aberrations, cytoplasmic male sterility (CMS), resistance to antibiotics (e.g. to streptomycine), growth patterns, etc. (Razoriteleva, Beletskii & Zhdanov, 1970; Hagemann, 1976; 1979; 1982; Pohlheim, 1974; 1981; Pohlheim & Beger, 1974; Hosticka & Hanson, 1984; Fluhr et al., 1985; Davidson, Pertens & Armstrong, 1987; To, Chen & Lai, 1989; Walters et al., 1990; Sears & Sokalski, 1991; and Timmons & Dix, 1991). An extranuclear inherited trait of particular interest is cytoplasmic male sterility, which was claimed by Malinovskii, Zoz & Kitaev (1973, original not consulted) could be induced in sorghum (Sorghum vulgare) by nitroso compounds.

Nitroso-compounds (like other chemicals) are also used for treatment of vegetative plant parts, as was reported for instance by Pohlheim & Beger (1974) and Pohlheim (1981) with cuttings of African violet (Saintpaulia ionantha). (N.B. In the aforementioned publication Pohlheim abbreviates nitrosomethyl urea both as NMU (in the English summary) and as NMH (in the German text).

Nitroso-compounds are frequently used in in vitro mutagenesis. In addition to the aforementioned work by McCabe et al. (1989), reference can be made also to Fluhr et al. (1985), who observed a high rate of chloroplast-encoded mutations for resistance to antibiotics in Nicotiana spp., after treating seeds with 5 mM nitrosomethyl urea for 2 h. In addition, for instance, Swanson et al. (1989) treated microspores of Brassica napus with 20 µM nitrosoethyl urea.

For a number of alkylating agents the use of buffers and the choice of an appropriate pH is important. The production of acidic (and potentially toxic) compounds, which occurs when alkylating agents are brought in water, reduces the amount of mutagen available and, hence, the mutagenic effectiveness. Not all alkylating agents show similar reactions in this respect and the duration of the treatment may play a role as well. For EMS and DES buffering seems to be useful, although in practice this is often omitted. For nitroso-compounds a pH of 6 should be taken, as at that pH these compounds are most stable with a half-life of 24 h for nitrosomethyl urea and 31 h for nitrosoethyl urea at 20 °C (Hagemann, 1982). At pH 7 the corresponding values are in the range of about 1 h or less. This information is not found in the FAO/IAEA Manual (Anon., 1977a), where only mention is made (p. 57) of a half-life of 35 h for nitrosomethyl urea at 30 °C and pH 7 and, accordingly, 84 h for nitrosoethyl urethane.

As nitroso-compounds may break down in visible light, treatments and storage of the chemicals should take place under dark conditions. Dosages of up to 20 mM for 2–3 h yield best results. Details about preparing the solutions, buffers and treatment procedures – which do not differ significantly from those for EMS – have been published by Hagemann (1982).

The general remarks made about determining optimal concentrations and treatment periods for EMS, are valid as well for the other alkylating agents. The FAO/IAEA Manual, for instance, indicates ranges from 0.05 to 0.3 M for EMS, from 0.01 to 0.02 M for DES, from 0.85 to 9 mM for EI and from 1.2 to 14 mM for the nitroso-compounds. However, in the literature dosages well outside these ranges have been mentioned as well.

Methylating agents, as has been mentioned before, are generally considered more reactive on a per mole basis than ethylating agents, but more toxic as well. Plants, on the other hand, apparently can tolerate higher concentrations of ethylating agents, which makes them more efficient mutagens. According to the FAO/IAEA Manual this is true for sulphonates like EMS and MMS, but as far as nitroso-compounds are concerned, nitrosomethyl urea seems both more effective and efficient than nitrosoethyl urea.

Hagemann (1982), who provides many details about treatments with nitroso-compounds, cautions users of the risks when using these substances because of their carcinogenic and highly mutagenic character. Detoxification of all glass tools, etc. must be performed under an exhaust hood in a 20% sodium- or potassium-hydroxide solution. As an alternative treatment, glass tools can be immersed for 24 h in a diluted solution of sulphuric acid, again under an exhaust hood.

The figures on the number of direct, chemically induced mutant cultivars in the IAEA data base in Vienna show that, apart from the universally applied mutagen EMS, most of the chemically induced mutant cultivars are also derived from treatments with alkylating agents. Whereas EMS is used on a worldwide scale, the use of some other chemical mutagens like DMS, EI and nitroso-compounds has been mostly limited to workers on institutes in the former USSR and to students from other countries who studied there.

The information in the IAEA database further shows that by treatment with chemical mutagens like DMS, DES and EI, that are considered 'old-fashioned' nowadays, new mutant cultivars are still obtained occasionally.

Base analogues
Base analogues occasionally can be incorporated into the DNA because they are sufficiently similar to the DNA bases. They are able to replace the normal bases, without hindering replication, but pairing errors may occur. Base analogues induce transitions in both directions and, as a consequence, mutations induced by a base analogue can also be reverted by a base analogue. According to Drake (1970, p. 103), base analogues do not induce frameshift mutations and most probably also no transversions.

As a consequence of the aberrant pairing properties, mutations can be caused by the insertion of incorrect nucleotides opposite to the incorporated analogues. The base analogue 5-bromodeoxyuridine (an analogue of thymine) is effective only to replace the normal base when present during the S-phase of the mitotic cycle. Other well known base analogues are 5-bromouracil (also an analogue of thymine) and 2-aminopurine (an analogue of adenine). Suzuki et al. (1989), for instance, reported that 2-aminopurine – which very specifically induces mutations of the transition type (see also Chapter 3) – can pair with thymine but can also mispair with cytosine. As a consequence, incorporation of this analogue as adenine may generate AT → GC transitions by mispairing with cytosine during subsequent replications. When 2-aminopurine is incorporated by mispairing with cytosine, transitions of the GC → AT type will be the result. Transitions are induced as well by 5-bromouracil. It should be realized, however, that the 'specificity' of base analogues, to which is referred here, is not exactly the kind of specificity looked for by plant breeders who would like to change, for instance, one specific gene into a desired direction, leading to a higher plant length or better resistance, etc. Nevertheless, base analogues can be of much help to understand gene behaviour.

Base analogues, in general, produce a low mutation

frequency and the induction of commercial mutant cultivars in this way seems to be rather an exception. A combined treatment of 1 krad (lo Gy) of gamma rays and 1 mM of 5-bromodeoxyuridine (BUdR), applied for 1 h by Yamaguchi et al. (see Micke et al., 1985, in *Mutation Breeding Review* 3), in 1974 resulted in the release of the mutant cv. Fuji-2-jyo-II in barley (*Hordeum vulgare*) in Japan. Also in 1974 it was reported by Yamaguchi, Tano & Tatara (1974) that high frequencies of chlorophyll mutations and mutations for heading time and culm length had been induced in barley after treatment for 1 h with 1×10^{-3} M BUdR, preceded by 9 or 19 h presoaking. Earlier Zamecnik & Zamecnik (1967) had reported on defects in the chloroplast structure of *Ageratum corymbosum*, which were still present in the M_5 of this crop and had been induced by a seed treatment with 0.01 M of BUdR for 18 h. More details about the action of base analogues, and additional references, which mainly concern research with bacteriophages, can be found in Drake (1970).

According to the FAO/IAEA Manual, a number of compounds, related to the base analogues, but which are not incorporated into the DNA, show a completely different behaviour on chromosomes. The best known example in this respect is 8-ethoxycaffeine, that, like many N-methylated oxypurines, has strong chromosome breaking properties (Kihlman, 1950; Kihlmann & Levan, 1951; Ehrenberg et al., 1956). Kihlman & Levan (loc. cit.), after treating seeds of vetch (*Vicia sativa*) with 8-ethoxycaffeine, reported that chromosomal breaks in particular were induced in the heterochromatin. Maximum damage to the chromosomes in seedling roots of vetch was obtained in treatments with 7.5 mM for 6 h at 10 °C.

The only publications in which it has been mentioned that caffeine also may induce point mutations, as illustrated by the occurrence of chlorophyll aberrations, come from Vig (1973; 1975), who treated seeds of soybean (*Glycine max*) at concentrations of 0.05–1% for 4–24 h.

An extensive review on caffeine, its derivatives and their genetic effects, has been published by Kihlmann (1977).

Antibiotics

The first publications about the cytological and/or genetic effects of antibiotics on higher plants trace back to the early 1950s. Wilson (1950), who also referred to some earlier examples, reported on the effects of, amongst others, penicillin, streptomycin and neomycin in concentrations of 6 up to 800 p.p.m. on chromosomal aberrations in mitotic root cells of onion (*Allium cepa*). The FAO/IAEA Manual refers to the chromosome breaking properties of a number of antibiotics like azaserine, mitomycin C, streptomycin and actinomycin D and adds that their usefulness 'for practical purposes' is very limited.

Briggs (1974; 1975) described efforts to induce cytoplasmic male sterility in maize (*Zea mays*) by treatments with streptomycin 0.01% (and some other compounds as well). Although male sterile plants were obtained, it had to be concluded that this male sterility was not cytoplasmatically determined. More recently, Hu & Rutger (1991) reported on the successful induction of nuclear male sterility in M_1 in rice (*Oryza sativa*) after treatment with 800, 1600 or 2400 p.p.m. of streptomycin for 40 h at 15 °C. Out of in total 9000 M_1 plants, 24 plants with greatly reduced seed set were obtained. One plant from the 2400 p.p.m. series was completely male sterile. This sterility was controlled by a single recessive gene.

Sager (1960; 1972) had already described the specific effect of streptomycin in cytoplasmic genes in *Chlamydomonas*, whereas the occurrence of cytoplasmic mutants in higher plants has been described by Kinoshita & Takahashi (1969), Kinoshita, Takahashi & Mikami (1979; 1982) and Mikami, Kinoshita & Takahashi (1980) for sugar beet (*Beta vulgaris*); by Burton & Hanna (1982) for pearl millet (*Pennisetum americanum*) and by Jan & Rutger (1988) for sunflower (*Helianthus annuus*). Petrov, Fokina & Zhelagnova (1971) even tried to patent a method to induce CMS in maize (*Zea mays*) by applying streptomycin, but as far as is known this patent has never been granted or implemented.

In some publications, the induction of extranuclear mutations by other antibiotics like mitomycin-c has been reported as well (e.g. Jan & Rutger, 1988).

Acridines

Acridines, which could be classified as belonging to a group called 'miscellaneous chemical mutagens', according to the FAO/IAEA Manual, represent a group of heterocyclic dyes such as proflavine, acriflavine, acridine orange, ICR-170 and, the best known one, ethidium bromide (EB). Some acridines – in the presence of light – are mutagenic and may induce chromosome breaks. When acridines are applied to bacteriophages in the dark, they may induce deletions and additions in the DNA and, as a consequence, cause frameshift mutations (see for instance Lawrence, 1991). D'Amato (1950; 1951) claimed the mutagenic effect of three acridines (acriflavine, acridine orange and 9-aminoacridine) on the basis of chlorophyll mutations in the M_2 of cv. Aurora of barley (*Hordeum vulgare*), but this finding could not be confirmed by Ehrenberg et al. (1956) for the same crop.

Before the 1970s it had been discovered that such

acridines were highly effective in inducing plasmon mutations in, for instance, yeast and *Chlamydomonas*. According to Sager (1972), mitochondrial DNAs are in particular susceptible to acridine dyes under conditions that are reportedly non-mutagenic for nuclear DNA.

Mikami *et al.* (1980), in this respect, reported on the application of acridine dyes to induce cytoplasmic male sterility in sugar beet (*Beta vulgaris*). In a series of publications (Ashri, 1968; Levy & Ashri, 1973; 1975; 1978; Ashri & Levy, 1974; and Levy, Ashri & Rubin, 1979) procedures and effects of treatments of peanuts (*Arachis hypogaea*) with in particular ethidium bromide (which will be briefly discussed in the next section) have been described. Attention was focused on the induction of mutations for extranuclear traits.

Some other acridine dyes were tested as well. Ashri *et al.* (1977) induced trisomic mutants in peanuts affecting the branching pattern by treating developing embryos with acriflavine (1.15 mM = 300 mg. l⁻¹) for 72 h. Levy & Ashri (1978) treated seeds of peanuts with 0.11 and 0.38 mM acriflavine for 48 h and in this way obtained an extranuclear (plasmon) mutation affecting the growth habit of the plant. Mutant cultivars, obtained after treatment with acridines, so far, have not been registered.

Ethidium bromide (EB)

Ethidium bromide (EB, or in full: 2,7-diamino-10-ethyl-9-phenylphenantridum bromide), of which the mutagenic activity was mainly investigated in lower organisms and in cultures of animal cells, is of special interest to the plant breeder because it has been found to effectively induce mutations in mitochondria and plastids and to interact with respiratory enzymes. EB is a so-called intercalating agent which is able to slip between two adjacent base pairs along the double helix, thus causing errors at the next replication. As most often one or two base pairs are added and, less frequently, a base pair is lost, the result of an EB treatment normally will be a frameshift mutation (see Chapter 3). Intercalating agents, apparently, do not cause transitions or transversions.

The effect of EB on mitosis was shown already by Kihlman (1966) for vetch (*Vicia sativa*) and by Truhaut & Deysson (1972) for garlic (*Allium sativum*). In peanut (*Arachis hypogaea*) both nuclear and extranuclear mutations have been induced (for references see the previous section). EB-induced mutations for CMS have been reported for pearl millet (*Pennisetum americanum*) by Burton & Hanna (1976), for sugar beet (*Beta vulgaris*) by Kinoshita *et al.* (1982) and for maize (*Zea mays*) by Minocha & Gupta (1988).

In the research programme of Ashri, Levy and some other workers, which started in 1970 and in which particular attention was paid to the – genic-cytoplasmically controlled – 'bunch versus trailing growth habit' of peanut plants, seeds were treated, after 1 h of pre-soaking in de-ionized water, with a range of doses from 0.05 to 1.26 mM (or 500 mg. l⁻¹) in de-ionized water during 24, 48 or 72 h. It was found that the lower doses caused very little damage.

The compound EB is not degraded and, hence, does not have to be changed during treatment. Nevertheless, in some publications it is advised to change the solution when treatments take longer than 48 h. However, such prolonged treatments gave lower yields of mutations as well as a lower efficiency and effectiveness. On the other hand, treatments should not be too short either (Levy *et al.*, 1979). Levy & Ashri (1975) compared the mutagenic action of EB and EMS with respect to a number of traits, including chlorophyll, leaflets, plant size, growth habit and yield. It was found that EB had a much higher mutagenic efficiency than EMS and that both chemicals produced a different mutation spectrum. Taking into account also the good stability of EB, the authors concluded that EB has very good mutagenic properties.

Sodium azide

The only other chemical mutagen that, in addition to the alkylating agents, has been used on a relatively large scale, is sodium azide (NaN_3). This compound was employed originally as a respiration inhibitor to study effects of irradiation in barley (and also as a herbicide, nematocide, fungicide and preservative). Sodium azide was introduced as a mutagen in the early 1970s at Washington State University, Pullman, USA (Sideris, *et al.*, 1969; 1973; Nilan *et al.*, 1973; Kleinhofs *et al.*, 1974). Nilan (1987) claimed that, at that time, sodium azide should be the preferred mutagen, at least for barley. The successful work with sodium azide induce *pac*-free mutants in barley at the Carlsberg Research Laboratory in Denmark was described in Chapter 3.

The word azide refers to both inorganic an organic azides. For mutation breeding the simple inorganic metallic azides, in particular sodium azide and, to a lesser extent, potassium azide, are of interest. These azide salts are highly soluble in water and relatively stable. According to Kleinhofs *et al.* (1984) organic azides in fact may be better mutagens from a practical point of view as they are not volatile and, unlike the inorganic azides, do not require a low pH for effective uptake. Results of further studies about effects of organic azides in this respect are not known to us.

Azide often has been described as a 'super mutagen' in the sense that the ratio of mutations to gross chromosomal changes or large deletions is very high, that the toxicity of this compound is very low and, hence, that the mutagenic effectiveness is close to 100%.

Experiments based on the induction of clorophyll-deficient mutations in barley have proven that azide, under proper conditions (low pH, pre-soaking, room temperature, bubbling the solution with oxygen, etc.), may produce very high frequencies of such mutations (Nilan, 1981a). The effectiveness of azide in inducing mutations in, for instance barley, can be easily determined by recording the frequencies of chlorophyll-deficient 'striping' in M_2, and figures of 20–46% chlorophyll mutations have been reported in this respect. (N.B. Kleinhofs et al. (1984) mention that determining the seedling height in M_1 is not a reliable measure to check the effectiveness of azide.) In an earlier experiment Conger & Carabia (1977) compared the mutagenic effectiveness and efficiency of sodium azide and EMS on the basis of the induction of somatic mutations at the Yg-2 locus in maize (Zea mays) after treating dry or pre-soaked seeds. The dominant allele of the Yg-2 gene produces normal green leaf colour, while the recessive allele after mutagenic treatment of heterozygous stocks results in yellow-green stripes in the normal green leaf tissue. NaN_3 was found more effective than EMS (i.e. produced more mutations per unit of dose), but EMS was the more efficient mutagen (i.e.showed a higher yield of mutated sectors relative to other, undesirable effects). Both NaN_3 and EMS performed better in pre-soaked seeds than in dormant seeds, whereas in both cases the mutagenic effectiveness was not affected by concentration. The mutagenic efficiency in most cases tended to decrease at higher concentrations.

In another publication Kleinhofs et al. (1984) reported that for 'single locus traits' such as waxy and gigas in maize a mutation rate of roughly 1×10^{-5} per genome was obtained which, generally, is considered a very high rate.

Many aspects of azide mutagenesis and many references have been given by Kleinhofs et al. (1978). It was determined by the group of Nilan at Washington State University that it is not the azide that is the real mutagenic agent, but an organic metabolite that is synthesized in an azide-treated plant. This finding also explains why azide does not work as a mutagen in all plant species. The metabolite concerned was characterized in 1983 as β-azidoalanine.

In the FAO/IAEA Manual (Anon., 1977a) azides are described as mainly crystalline salts of which the alkali metal salts are relatively stable, but readily converted to HN_3, whereas the acid form is volatile and already boils at 36 °C. Conditions under which sodium azide is most effective – i.e., at a low pH and when solutions are bubbled with air or oxygen during treatments – strongly favour volatility. This may be dangerous for the laboratory personnel because at low pH the volatile hydrazoic acid is generated, in particular when the mutagenic solution is vigorously aerated. The exhaust air can be easily neutralized by passing it through a 0.1 N NaOH solution (Kleinhofs et al., 1984). When azide is spilled under acidic conditions, the area first should be ventilated and cleaned with water and soap or detergent. Disposal of azide can be done in several ways. For small quantities, mixing with excessive water and pouring down a drain, or absorption by paper towels or saw dust, which is burned afterwards, are acceptable solutions. When azide is disposed of in sinks with lead or copper pipes, these pipes must be decontaminated with strong NAOH as lead and copper azide residues may be explosive.

The FAO/IAEA Manual also presents the following treatment procedure for seeds of cereals with azide, developed by C.F. Konzak from Washington State University, USA.

No more than 250 seeds are put in small mesh bags . . . . These bags, consecutively, are presoaked in a container with flowing water at about 20 °C for 2 to 6 h, depending also on the temperature. The seed bags are taken out and the excess water is shaked off. A stock solution of, for instance, 1 M of sodium azide in water, should be prepared (Stock solutions can be stored almost indefinitely without losing their potency). The use of a buffer is essential and a phosphate buffer prepared to pH 3 is considered the best buffer available. In order to obtain 1 litre of approximately 0.1 M of a pH 3 phosphate buffer, 12 g KH_2PO_4 is used, to which H_3PO_4, as a powder or as concentrated acid, is added till the pH meter shows pH = 3. Glass or plastic containers are filled with a measured amount of buffer that is enough to cover all seeds (in general more than 1 ml. seed^{-1} will be needed) and bring this to a temperature of 20 °C. Consecutively, mutagen from the stock solution is added. By adding 1 ml of the 1 M stock solution, a 10^{-3} M solution is obtained.

The bags with the seeds are placed in the flasks with mutagen solution. Occasional shaking in order to keep the solution homogeneous is necessary and, as was said before, bubbling with air or oxygen may have a positive effect. An optimal combination of a high mutation frequency and a low level of induced sterility is reported to be obtained at NaN_3

treatments in the range from 0.001 M to about 0.003 M (Hodgdon, Nilan & Kleinhofs, 1979; Seetharami Reddi & Prabhakar, 1983). Standard treatment time at 20 °C is 2-4 h with higher temperatures allowing a shorter duration of the treatment. Much higher temperatures, e.g. when exceeding 30 °C, are not advisable and may be even dangerous. For low concentrations, prolonged treatment times are required.

After treatment the seed bags are removed and thoroughly rinsed off in tap-water. Some post-washing in running cold tap water for 30 min up to 2 h is advised, because without post-washing some azide may remain in the seeds and cause additional physiological damage. Afterwards, the seeds can be removed from the bags and dried on blotting paper. After about two days of drying the seeds can be stored in a refrigerator until sowing.

As has been said before, treatments with azide are not always successful. Kleinhofs et al. (1978) mentioned that azide, in a number of cases, was either not or only weakly mutagenic, which they believed to be due to inappropriate treatment conditions. Later, Kleinhofs et al. (1984) reported that azide is not effective in polyploid crops, which they explain by the fact that azide is not effective as a chromosome-breaking agent. Azide, moreover, appears not to be effective in particular plant species, for instance in Arabidopsis. The reasons for this are unknown and the aforementioned authors, at several occasions, have underlined the importance of performing initial experiments to check the effectiveness of azide before starting large-scale practical experiments. In addition the authors stress the importance of creating appropriate treatment conditions.

Crispi, Ullrich & Nilan (1987) found as a negative side effect of NaN_3 treatments in barley, that a relatively high degree of (partial) sterility was observed, which they suggested might have been caused by deleterious, recessive, minor gene mutations that were already present in heterozygous condition when the mutant lines were initially selected.

One final point to be briefly considered here is whether treatment with a so-called 'super mutagen' like azide is always desirable as, in addition to a high frequency of desired mutations, e.g. for short straw in cereals, many undesirable mutations for other traits will be induced as well in the same cell or tissue. Such undesirable mutations have to be got rid of, for instance by cross-breeding, which is costly and time consuming.

Hydroxylamine

Hydroxylamine, which is reported to primarily interact with cytosine and guanine, has been mentioned as a separate category in the FAO/IAEA Manual. Drake (1970) reports that hydroxylamine (NH_2OH) resulted from a search, initiated in the early 1960s by Freese, for powerful chemical mutagens which might act with a great specificity in non-replicating DNA, i.e. would mutate only a single base. Practically all research took place with lower organisms, such as bacteriophages. Effects of hydroxylamine on higher plants have been described in a few publications only, like by Prasad & Tripathi (1987) who applied this compound at 0.15 or 0.30% to barley (Hordeum vulgare). Mutation frequencies, e.g. on M_2 seedling base, remained lower than for DES and sodium azide, with which chemicals hydroxylamine was compared, but the level of seedling injury, chromosomal damage, etc. was less than for DES. There appears to be no specific reason why hydroxylamine should be applied in chemical mutation breeding.

4.3.4. Concluding remarks on chemical mutagens

The outcome of a treatment with a chemical mutagen, in rough terms, is determined by the properties of the mutagen, like its solubility, toxicity and reactivity; by the type and condition of the treated material before, during and after treatment, as well as by the treatment procedure.

One could wonder what gives chemicals their mutation inducing ability and why some chemicals do give rise to many more mutations than other chemicals? Unambiguous answers to these questions are still difficult to give. In the previous sections it has been pointed out that different groups of chemical mutagens can be distinguished which may affect the main targets for plant mutation breeding: the DNA in chromosomes, chloroplasts and mitochondria, in various ways, depending also on the way of treatment. By selecting a specific chemical mutagen and determining the optimal treatment method, in combination with the choice of the plant species, cultivar(s) and plant part (seeds, cuttings, 'in vivo' or in vitro) and selecting the method(s) which will be used to detect the desired mutant plants, the conditions of the experiment have been set.

Repair

As is the case with radiation, most, if not all, initial lesions in DNA caused by chemical mutagenic treatment are subject to repair. Repair processes are error prone and the degree of repair is known to differ with the kind of lesion, the plant species, cell type, rate of cell division, mitotic stage, etc. Based on results obtained from mutagenic treatments of slowly dividing cells – in this case the spermatogonial stem cells of the mouse – Kimball (1987) assumes that for some agents damage is already repaired before DNA replication occurs, whereas this may not be the case for other chemicals. In experiments with mice a mutagen like nitrosoethyl urea, for instance, acts very well in the G_1 stage, but this is not found for EMS. According to Kimball the reason for the different behaviour of both chemicals could be, that the damage induced by the nitroso-compound is not repaired in G_1 stage, whereas the EMS-induced damage may be repaired at that stage. Hence, it probably could be argued that it is not the mutation inducing ability of a (chemical) mutagen that determines the ultimate mutation frequency, but – at least to a considerable extent – the relation between induction and repair of mutations.

Specificity of chemical mutagens

Originally, it was believed – or at least hoped for – that chemicals would act more specifically than radiation, in the sense that either a higher proportion of 'real' gene mutations without much accompanying chromosomal damage could be obtained, or that high proportions of mutations for a specific trait or group of traits might be induced in this way. In later years the expression 'specific' was more considered in terms of a chemical mutagen affecting specific groups or sequences of the DNA or RNA. One could also distinguish between 'locus specificity' and 'site specificity' of certain mutagens.

The extensive Swedish work on specific groups of mutations in barley, that has been discussed in particular in Chapter 3, showed already that treatments with different chemicals, at least to a certain extent, may produce different mutation spectra, for instance for chlorophyll, wax, etc. In addition, Konzak et al. (1984), referring to previous research with micro-organisms, mentioned that EMS, DES and nitroso-compounds, for instance, produce relatively many mutations of the GC → AT transition type. To give another example of 'specificity', it is known that base analogues may act preferentially on specific sites of some nucleotides. But since the same nucleotides occur 100 or 1000 times in each locus, such 'specificity' (or 'differential behaviour') could never have any relation to the code of a gene. In addition, if ratio's of chromosome aberrations are compared to gene mutations, it has been proven that different chemicals may differ considerably in this respect.

Schubert & Rieger (1977) concluded from results of treating seeds of faba bean (Vicia faba) with different physical and chemical agents, that mutagens with so-called delayed effects, e.g. nitrosoethyl urea, mitomycin C,

tri ethylene melamine (TEM) and hydroxylamine (NH_2OH), showed many more 'hot spots' for chromatid aberrations than mutagens with non-delayed effects, like fast neutrons, X-rays and a chemical that has not been referred to before, called bleomycin. Non-delayed effects occur when the mutagenic agent produces aberrations also in cells that have completed the chromosomal DNA synthesis (or S-) phase, whereas agents with delayed effects are unable to do so.

However, from what has been discussed in the previous sections it must be concluded that most expectations as to the specificity of chemical mutagens have not been fulfilled. Auerbach (1976) already stated that chemical mutagens are unable to produce real 'directed' gene mutations for specific traits. She further suggested as the only realistic approach to obtain (a high proportion of) mutants of the desired type, to apply appropriate 'selection sieves' in later stages of a mutation induction programme. Rédei (1982a, p. 332) expressed the view that mutagenic agents do not have enough information in their molecular structure to recognize specifically one or only a few genes from the many thousand genes present in plants or other higher organisms.

It may be added here that in literature the expression 'directed mutation' is used in various ways. Mutations induced at specific positions at the DNA are sometimes called 'site-directed' (or 'site-specific'); some (chemical) mutagens may 'direct' the induction of relatively many mutations for a specific group of mutations for chlorophyll or wax as compared to frequencies of induced mutations for other groups; sodium azide 'directs' mutations towards a relatively high frequency of point mutations as compared to chromosome aberrations, etc. In addition, experiments with micro-organisms like E. coli, in the 1990s allow the interpretation that the use of selective media in vitro may also favour the induction (and not only the selection) of specific 'adaptive' or 'selective' mutations. This process, therefore, also could be called 'directed mutagenesis'. Why and how in such cases the induction of such specific mutations is favoured is not sufficiently understood yet.

Finally, real site-specific mutagenesis, i.e. alteration at a site within genes, nowadays, can be obtained in a number of cases by the use of recombinant DNA techniques. Genes can be cloned and relatively large amounts of DNA can be obtained in this way. Such DNA could be mutated in a test tube and re-inserted into (plant) cells by transformation or infection methods. The mutations produced in this way within a gene could be of various nature.

Chemical mutagens versus physical mutagens

In the past the question was often raised whether physical and chemical mutagens differ with respect to their mutation-inducing properties. According to Gaul in the FAO/IAEA Manual (Anon., 1977a, p. 95), chemical treatments in general gave more non-random distribution of chromosome breaks than X-rays, because of the preferential occurrence of breaks in specific chromosomes as well as in definite regions within chromosomes. For chemicals more breaks were reported in the heterochromatin than in the euchromatin. The breaks observed after X-radiation in general were at random, but relatively more chromosome recombinations and fewer fragments were reported than after chemical mutagenesis. Chemicals – still following Gaul – in general should produce more small structural changes, rearrangements and duplications. However, from what has been explained before, it seems most doubtful that the effects of chemical mutagens (and physical mutagens) can be generalized in this respect. Nowadays, these kinds of comparative studies are not often performed any more and, moreover, when comparisons are made, the interest of the investigators goes in particular to possible differences in the mutation spectra instead of to a detailed study of physical effects exerted by various mutagenic agents on the DNA. The various repair prosesses that may affect the final outcome of mutagenic treatments are also favoured objects of study.

4.4. Combined treatments of physical and chemical mutagenic agents

Soon after the discovery of the mutagenic effect of X-rays and chemical mutagens, plant material has been submitted to combined treatments, sometimes just to gratify the curiosity of the researcher, but in most cases with the purpose of increasing the mutation frequency as well as broadening the mutation spectrum. In practically all cases combinations of physical and chemical mutagens were applied (in this sequence). Examples of treatments with combinations of two different physical or two chemical mutagens are very rare. In practice, most often dry seeds are irradiated first, followed by a treatment with a mutagen in solution. Before discussing some results of experimental work, a few points must be further clarified.

Combined treatments, apparently, could be of value to the plant breeder, if such treatments would result in additional effects with respect to mutation frequency and mutation spectrum in combination with quicker positive results, less work and no decrease in rate of survival and fertility of the treated material in comparison with the sum of two separate treatments. Combined treatments would be even more attractive if synergistic effects would occur, either in a way that mutation frequencies should reach levels beyond the sum of both individual treatments, or if 'unique' mutations should arise in this way. In order to find out whether such goals could be achieved, it would be necessary to compare results of the combined treatment on a large scale with those from both separate treatments apart. In order to be convincing, results of such experiments should be compared at identical levels of survival and fertility, etc.

An ideal way to investigate the potentialities of combined treatments, for instance, would be to compare results of such treatments (and their reciprocals) with those for single treatments on the basis of traits which have been well-investigated, as is the case with the traits studied in the Scandinavian barley mutation programme (see also Chapter 3). Lundqvist (1992) in this respect referred to some results of combined mutagenic treatments for the *eceriferum* (or waxy trait), for which about 1600 mutants, localized to about 80 loci, are known. However, it should be commented that such combined treatments might be without any practical significance for traits mutating as frequently as the *eceriferum* character of leaves.

As a possible explication of increased mutation frequencies by combined treatments, one could think of less effective action of repair mechanisms when DNA, damaged by irradiation, is treated consecutively with a chemical mutagen, but no evidence is known for this assumption. However, small pre-treatments with radiation apparently do activate repair mechanisms.

Lundqvist (personal communication, 1994) mentioned that combined treatments of physical and chemical mutagenic agents had already been started in Sweden in the 1930s and 1940s. According to the field-books, these earliest treatments referred to combinations of X-rays and colchicine and some other chemicals, but further details are not known. In subsequent years the Swedish mutation breeding work on combined treatments involved studies on effects of X- or gamma rays combined with sulphonates and on combined treatments of neutrons and sulphonates (or reciprocal treatments). Most data collected on this work have never been worked out in detail but – still according to Lundqvist – Gustafsson, Ehrenberg and Lundqvist did not find significant differences when comparing mutation frequencies obtained from combined treatments with those from single treatments.

Before briefly discussing some recent reports concerning the 'standard type' of combined treatments (i.e. with an initial radiation treatment followed by application of a chemical mutagen), two other examples are mentioned here.

Favret (1964) reported for barley that a clear interaction was observed between treatments with EMS and X- or gamma rays with respect to efficiency and frequency of single gene mutations when EMS was used as a pretreatment. He concluded that application of the chemical gave protection against the effects of radiation in M_1, in particular when higher doses were used. In M_2 and M_3 pre-treatments with EMS diminished the frequency of chlorophyll mutations. This, of course, is not the kind of effect for which the plant breeder would like to perform combined treatments.

One of the few examples on combined treatments in which different chemical mutagens are used consecutively, was published recently by Konzak (1993). Seeds of oats (Avena sativa), which had been presoaked for 4–6 h, were treated with 0.1 and 0.2.% EMS for 2 h, followed by treatment with 1×10^{-3} sodium azide for 1 h. A large number of semidwarf, often dominant, mutants were obtained from M_2 and M_3 populations. Consecutively, it was possible to select mutants with an acceptable yield, in which only plant height had been reduced, whereas other traits remained unchanged. Konzak anticipated that mutant lines for commercial production would become available in this way.

Although, to the best of our knowledge, throughout the years, nobody has ever published convincing (and verifiable) data which clearly prove the superiority of effects of combined treatments above (the sum of) single treatments, still now and then publications appear in which it is mentioned that combining physical and chemical mutagens was an effective method of raising the mutation frequency and expanding the mutation spectrum. Wang (1991), in a review about mutation breeding in China, presented in 1990 during a symposium at IAEA headquarters in Vienna, reported that studies in wheat (Triticum aestivum) and rice (Oryza sativa) had shown that both additional and synergistic effects occurred, at least in M_1. Treatments were performed with gamma rays, in combination with EMS, DES or NaN_3. Biological injuries in M_1 increased with increasing doses of gamma rays and increasing concentrations of the chemical mutagen. Wang further mentioned that those studies had shown that seedling height, vitality index (?), root length and frequency of single micronuclei in root tips in M_1, could be used to predict the mutation frequency in M_2. For references to the original publications, see Wang (1991).

At the same symposium in Vienna, Mehandjiev (1991)

reported for soybean (Glycine max) about combined treatments with various physical and chemical mutagens. Three types of mutation effects were observed: synergistic (or 'superadditive'), additive and inhibiting. A synergistic effect was obtained after submitting seeds of the American cv. Beeson to a treatment with 100 Gy of gamma rays, followed immediately by a treatment with 0.2% of EMS for 4 h. This treatment also resulted in a greater frequency and a wider spectrum of induced mutations for morphological traits. The number of seeds treated in this experiment was not given. In an earlier contribution in the Mutation Breeding Newsletter 23, Gecheva (1983) reported on the release of the new soybean mutant cultivar Boriana, directly isolated in the M_2 after treatment of seeds from cv. Beeson with 50, 100 or 150 Gy of gamma rays (exact dose rate not mentioned), followed by 0.1% EMS. Early ripening and high protein content were mentioned as improved traits. Further details about the economic importance of this cultivar are not known. In addition, in Mutation Breeding Newsletter 28, Privalov (1986), working at Novosibirsk, former USSR, reported about combining gamma ray treatment (150 Gy) and nitrosomethyl urea (0.01%) for treatment of seeds of sea buckthorn (Hippophea rhamnoides). This work resulted in an improved type or 'cultivar' with a higher yield of the very nutritive fruits and an increased content of the valuable medicinal oil.

Reddy (1992), who combined gamma rays with EMS or NaN_3 treatments in barley (Hordeum vulgare), wheat (Triticum aestivum) and triticale, which is a crop derived from crosses between wheat and rye (Secale cereale), concluded on the basis of frequencies for chlorophyll mutations, that combined treatments were more effective (i.e. more mutations per unit of dose) than individual treatments. Chemical mutagens and combined treatments, on the other hand, did not differ significantly with respect to efficiency (i.e. frequency of chlorophyll mutations, e.g. in M_2, in relation to unfavourable effects such as seedling lethality in M_1). Many relevant questions, however, remain unanswered in the, somewhat scanty, discussion in Reddy's publication.

To mention one more recent example of studies on combined effects, Chauhan & Patra (1993) reported for poppy (Papaper somniferum) that combined treatments resulted in mutation frequencies superior to those of single treatments.

As there remains still considerable doubt among mutation breeders of, in particular, most western counries about claims concerning the 'added value' of combined treatments above single treatments, the method of combined treatments is not often followed for applied mutation breeding in those countries.

But, even if combined treatments should yield higher mutation frequencies, again, the question that was put forward previously when discussing the 'supermutagen' azide, presents itself, namely: do we really want higher mutation rates for applied mutation breeding? It should be realized that when mutation frequencies are increased, the frequency of undesired mutations increases as well and, may be, even more. In addition, undesired mutations and favourable mutations may be linked.

In conclusion, it appears that, also for combined treatments, the only point of real practical interest for the plant breeder is widening the mutation spectrum, and all efforts to do so should be taken seriously. Further steps of the mutation breeding programme, as has been mentioned before, are mainly a matter of selection.

5   *In vitro* techniques for mutation breeding

5.1.   Introduction

When, in the twentieth century, step by step, *in vitro* techniques came up, it was quickly realized that they could make very significant contributions to plant breeding, related directly to the genetic improvement of crops as well as to fundamental studies. In Chapter 1 (see in particular Section 1.3.10) the role of *in vitro* techniques as a supplement to the conventional way of vegetative propagation, and their more recent, indispensable contributions to genetic manipulation were outlined. Sybenga (1983), who reviewed the merits and limitations of different methods of genetic manipulation in plant breeding, made a distinction between 'molecular/*in vitro*/somatic' and 'plant level/generative' approaches. The author pointed out that techniques belonging to the first ('somatic') group were technically more demanding; that the 'generative approach' was often conceptually more difficult; and that techniques of both groups were laborious. Sybenga further concluded that many factors together determine which approach should be preferable and, further, made the important remark that traditional selection and testing procedures will remain indispensable in the final stages of any method of genetic manipulation intended for the production of new cultivars. In particular this last point has been often neglected by proponents of the application of 'new techniques' in plant breeding.

The advantages and limitations, or problems, of *in vitro* techniques for mutation induction and selection have been pointed out for instance by Constantin (1984). The most important <u>advantages</u> mentioned, phrased somewhat differently, were:

- uniform mutagen treatment and application of selective agents to homogenous populations are possible,
- in single cell systems the organized complexity of whole plants and seeds is circumvented,

- autotrophs can be easily detected on deficient media,
- anther cultures may speed up breeding considerably, and
- plant cells (unlike animal cells) may regenerate into whole organisms.

<u>Limitations and problems</u>, for instance, are:
- establishing cell cultures showing good regeneration may be difficult from a technical point of view,
- cultured cells and whole plants often express different sets of genes,
- selection at the cell level is not possible for many traits of agronomic importance,
- not all observed variation is genetic by nature, which complicates the selection process, and
- effective selection of desired mutants is often hampered by inadequate knowledge about biochemical pathways and developmental processes.

In the following sections a number of topics which are related to the use of *in vitro* methods for plant breeding purposes, and more specifically for mutation breeding, will be discussed in detail.

Throughout the years several conferences on *in vitro* mutagenesis and related subjects have been organized by the Joint FAO/IAEA Division in Vienna, but other organizations like EUCARPIA, the European Association for Research on Plant Breeding, have paid ample attention to such topics as well. Useful, comprehensive information on these subjects, for instance, can be found in a report (IAEA-TECDOC-392) of two IAEA research co-ordination meetings held in Vienna in 1983 and 1985 (Anon., 1986a), in the proceedings of a FAO/IAEA symposium on *in vitro* mutagenesis, organized in Vienna in 1985 (Anon., 1986b), and in the proceedings of a symposium entitled '*Genetic Manipulation in Plant Breeding*', convened in 1985 in Berlin by EUCARPIA, and edited by Horn *et al.* (1986).

In addition, mention can be made of a symposium 'Gene Manipulation for Plant Improvement in Developing Countries', organized in 1987 in Kuala Lumpur, Malaysia, by SABRAO, the Society for the Advancement of Breeding Researches in Asia and Oceania, edited by Zakri (1988). At this meeting Novak & Micke (1988) discussed mutations in relation to *in vitro* techniques. Much attention to the aforementioned subjects was also paid during an EUCARPIA meeting on 'The Methodology of Plant Genetic Manipulation', held in 1994 in Cork, Ireland (ed. Cassells & Jones, 1995), at which meeting Maluszynski, Ahloowalia & Sigurbjörnsson (1995) reviewed the state of applying '*in vivo*' and *in vitro* mutation techniques for crop improvement. Finally, *in vitro* mutation techniques were also discussed during the FAO/IAEA symposium, 'Induced Mutations and Molecular Techniques for Crop Improvement', organized in 1995 in Vienna (Anon., 1995).

Definitions of some important terms

The expression **mutagenesis** (see for instance Rieger *et al.*, 1991), refers to the whole mutational process and involves all mechanisms and different steps by which different types of mutations arise, either spontaneously or induced by mutagenic agents. Accordingly, **in vitro mutagenesis** may include all activities in which mutagenesis is combined with any *in vitro* method. (N.B. The – established – expression, '**in vivo mutagenesis**', in fact is a misnomer!) Following this line, the term **in vitro mutation breeding** may cover all activities in which mutational techniques are combined with *in vitro* methods for plant breeding purposes. However, the term **in vitro mutagenesis** is often applied in a much more narrow sense. Smith (1985), in a review on this subject (with about 300 references), has defined *in vitro* mutagenesis as a method by which specific DNA targets can be changed by chemical and/or enzymatic manipulation in a definable and often predetermined way. Despite the fact that in the aforementioned publication no attention is paid to agronomic traits and plant breeding, reading of this review is useful for all persons interested in fundamental aspects of *in vitro* mutagenesis.

Rieger *et al.* (1991) briefly define *in vitro* mutagenesis as 'mutagenesis of cloned genes'. They add that recombinant DNA technology is used and that three broad categories of *in vitro* mutagenesis are known: 1) methods that restructure segments of DNA, 2) 'localized random mutagenesis', and 3) 'oligonucleotide (site-)directed mutagenesis'.

Definitions for *in vitro* mutagenesis like those given by Smith and Rieger *et al.* are too narrow to be useful in relation to mutation breeding, because, for instance, mutations that either arise 'spontaneously' or after pur-

poseful (conventional) mutagenic treatments of *in vitro* plant material, as well as mutations that may arise as a result of the *in vitro* condition of the plant material (so-called 'somaclonal variation') would not be included. Therefore, in the following sections the term *in vitro* mutagenesis will be generally used in the broad sense as described earlier in this introduction.

5.2. From vegetative propagation to genetic improvement

5.2.1. Propagation and multiplication

An early and still very important application of *in vitro* techniques is the *in vitro* propagation and multiplication of selected plant material under controlled and disease-free conditions. This technique, in fact, is very similar to vegetative plant propagation, for which different techniques have been known for many centuries. Use can be made of natural organs like tubers, rhizomes, bulbs and bulbils, or techniques developed by man can be applied, like grafting of buds, rooting of cuttings, etc.

The essential point for propagation purposes is that the resulting plants should be 'true-to-type'. This implies all possible precautions to prevent the occurrence of genetic variation, as 'off-types' are not accepted by the customer, be it the potato farmer, plantation manager, florist or consumer. The availability of genetically uniform starting material in this respect is imperative. It is, however, a particular problem of *in vitro* conditions that they may affect the stability of the material and result in plants that are not 'true-to-type'.

For plant propagation *in vitro*, basically use can be made of various types of **explants**, i.e. any tissue, organ or plant part taken from an intact plant.

5.2.2. Other applications of *in vitro* methods

In addition to plant propagation, other *in vitro* approaches were soon found to be useful, for instance for eliminating diseases and for conservation and exchange of germplasm (see Chapter 1. Section 1.3.10). Gradually, more techniques were developed, such as production of haploids and 'cybrids' (a fusion product of an 'enucleated' cytoplast with an intact cell; see also Chapter 4, Section 4.2.6) or to rescue endangered embryos in immature seeds, that had resulted from difficult, e.g. interspecific, crosses and therefore often would not develop properly. This hybridization-cum-embryo rescue technique – despite the fact that often only a few embryos have to be rescued – is labour-intensive, and one wants to be assured that the *in vitro* phase does not affect the genotype of the hybrid.

Another *in vitro* application is the 'genetic transformation' of plants, e.g. by means of the *Agrobacterium tumefaciens* vector (see also Chapter 1, Box 1.8). For this, *in vitro* techniques are required, both to perform the gene transfer and, subsequently, to produce complete plants in which the newly introduced gene(s) are expressed and normally inherited. Also in this case it is essential that incorporation of the alien gene occurs without any genetic change in this or other genes of the transformed cell and the regenerated plants. The practical applicability of transformation work, however, is often seriously limited by instability of transformed cells.

Finally, *in vitro* techniques are useful as well in classical mutation breeding programmes, e.g. by vegetative propagation before or after a mutagenic treatment, by *in vitro* selection, or by clonal propagation of selected mutants.

5.2.3. Biotechnology for plant improvement

Under the title of 'biotechnology' (see also Chapter 1) nowadays most often attention is focussed only on the first step of a genetic engineering programme, i.e. on creating genetic variation, for instance by genetic transformation and protoplast fusion, or on simply searching for 'spontaneous' genetic variants which occasionally arise *in vitro* without intentional human action. The subsequent necessary phases of a programme aiming at genetic plant improvement, starting with the vegetative propagation of the transformed plant or the fusion product, unfortunately, often are considered as less important or scientifically less interesting. However, the choice of the starting material, the culture conditions *in vitro*, particularly the composition of the medium and other 'trivial' factors, to a considerable extent determine whether *in vitro* techniques are useful, and decide whether a specific *in vitro* method is suited for true-to-type propagation purposes or for provoking new genetic variation.

5.3. The starting material

Propagation by tissue culture, the simpliest 'biotechnology', involves growing different types of plant tissue on an artificial medium. This application of tissue culture emerged in the 1970s. Pierik (1987 and personal communication) estimated that in Western Europe in the late 1980s about 250 specialized laboratories together produced about 210 million *in vitro* plantlets. Of this number about 90 million referred to pot plants, 40 million to bulb and tuber crops and 13 million to orchids. In the Netherlands about 65 companies together produced 60 million *in vitro* plantlets.

In the 1990s it could be observed that much of the *in vitro* propagation work, in particular the mass-scale propagation, was transferred to countries with lower costs of labour. The remaining companies in, for instance, the Netherlands, concentrated on the propagation of 'more recalcitrant' crops, on small-scale propagation and on the application of more sophisticated tissue culture techniques for breeding purposes.

Different types of in vitro culture

In a comprehensive book on *in vitro* culture of higher plants Pierik (1987, p. 29) distinguishes the following types of *in vitro* culture:

- culture of intact plants (obtained from seeds),
- embryo culture (isolated embryo without seed coat),
- organ culture (meristems, shoot-tips, roots, anthers, etc.),
- callus culture (from plant material de-differentiated *in vitro*), and
- single cell culture and protoplast culture (from different sources).

Pierik, who refers in this respect to de Fossard (1976; 1977; original publications not consulted), further adds that for such cultures, three different degrees of tissue organization can be recognized.

1. 'Organized'. In this case the organizational structure of plants or of an individual plant organ can be maintained by the choice of the starting material: seeds, embryos or shoot-tips (in combination with a proper medium).
2. 'Non-organized'. Cells or tissues are taken from an organized plant part and become de-differentiated. Non-organized growth – especially callus growth – occurs, from which for instance suspension cultures of single cells, protoplasts or cell aggregates could be derived. Non-organized cultures generally show a low degree of genetic stability.
3. 'Intermediate'. (N.B. In fact more a consecutive, or neo-organized type.) Cells in an isolated plant organ or tissue first de-differentiate, followed by formation of a (callus-)tissue from which differentiated organs, like roots or shoots or even whole individuals (somatic embryos), arise. This development may either occur spontaneously or by providing special conditions.

For comparison: Novak (1991a) mentions that there are essentially two types of *in vitro* growth: 1) 'organized' and 2) 'unorganized'.

5.4. Kinds of variation *in vitro*

5.4.1. Introductory remarks

Several kinds of variation may be observed following cell and tissue cultures, but not all variation observed is genetic and only a part of the observed variation can be transmitted – either vegetatively or through a sexual phase – to later generations. Heritable changes may occur either spontaneously, i.e. without purposeful human action, or be induced by either mutagenic treatment or molecular genetic techniques. In order to determine with certainty that the observed variation – whether desired or not – is genetic in nature, a sexual phase is needed. However, when after several cycles of vegetative propagation a changed trait is still expressed, this, mostly, is also considered as proof of a genetic base for the observed change.

5.4.2. Epigenetic effects versus mutations

Much variation described is '**epigenetic**', which implies that the variation is not caused by a mutation but by a change in gene activity (Rieger *et al.*, 1991). Genes are continuously switched on and off, e.g. by environmental factors, in particular during the developmental process. Mostly this means temporary (transient) changes in gene activity, but often this also results in more permanent changes, e.g. in specializing tissues. When epigenetic changes are more or less 'stable', it is very difficult to distinguish them from real mutations as long as the plant material is vegetatively propagated (which may be also the case when plant material is cultured *in vitro*).

The fact that epigenetic changes, for instance DNA methylation, are reversible, according to Lörz & Brown (1986), allows an unequivocal distinction between such effects and mutations. However, in particular in many early publications the term 'epigenetic' to describe variation observed after an *in vitro* phase, has been applied rather liberally and the use of this adjective mostly implies that the nature of the (phenotypic) variation was still unknown.

The best criterion to distinguish between epigenetic effects and mutations is that the former cannot pass through the meiosis. If this test cannot be applied, some clues for differentiation from mutations can be derived from the much higher frequency of occurrence of epi-

genetic effects or the fact that they can be induced or even directed by specific selection pressure of the medium. However, there are always exceptions to 'rules' (see also the discussion in Chapter 4, Section 4.3.4., on '*Specificity of chemical mutagens*').

Epigenetic effects may be also mixed up with temporary physiological changes. Phillips, Plunkett & Kaeppler (1992), and various other authors, mention that epigenetic effects may be a factor underlying the induction of genetic variation *in vitro*. We will briefly return to this subject in Section 5.5.1 in which 'somaclonal variation' is discussed.

Whatever the cause of an observed change may be, when a change that was observed *in vitro* does not turn up again in later plant generations, the assumption that a genetic change has been obtained, has little support. However, it has to be remembered that 'true' mutations also may be lost during tissue or plant development as a result of intracellular competition or non-regeneration. As a consequence, one should be rather hesitant to classify potentially useful changes that have been observed *in vitro* as valuable – i.e. genetically determined and stable – for plant breeding purposes, before this has been checked carefully in a number of consecutive plant generations.

5.4.3. Genetic changes *in vitro*

Various explanations may account for (true) genetic variation that is found following *in vitro* culture. In general a distinction is made between so-called 'spontaneous' genetic variation and artificially induced genetic variation resulting from deliberate mutagenic treatments. For reasons that will be discussed later one might consider to distinguish a third category of *in vitro* variation, i.e. the genetic variation pre-existing in the explant material used to start the *in vitro* culture.

The word 'spontaneous' here refers to the absence of intentional human action, but this does not exclude that undeliberate human action has been the cause of this genetic variation. If *in vitro* cultures are started from explants derived from roots, leaves, stems, flower stalks, etc., it should not automatically be assumed that all cells of such different plant tissues are genetically identical.

This may be illustrated by one, often neglected, point. Root tips contain an area indicated as the 'quiescent centre', a reserve cell population which may contain (or is accompanied by) polyploid cells. If *in vitro* cultures are started from root tips, in which polyploid cells are present, it may be anticipated that such cells will be present as well in the resulting *in vitro* cultures, and thus explain why high frequencies of polyploid cells may be found.

How much genetic variation is pre-existing?

Thus, pre-existing genetic variation – that in most cases would have remained undetected in the mother material – may become revealed by the *in vitro* stage. In such cases the impression could arise that new variation has been induced by culturing the explants *in vitro*. This situation, most probably, occurs much more often than is anticipated, in particular in vegetatively propagated crops, where all kinds of mutations, including also aneuploids and polyploids, may be piling up during many generations and, because of the absence of a meiotic sieve, are not selected against. It may be justified to conclude that a considerable part of the total variation observed from *in vitro* cultures may be due to pre-existing variation in the starting material, but reliable data are lacking.

In order to avoid the starting material acting as an unintentional source of genetic variation, the utmost care must be taken when deciding which starting material should be chosen for *in vitro* cultures. Unfortunately, even nowadays, in many experiments not enough attention is paid to this point, which may result in premature or unjustified conclusions about the outcome of *in vitro* experiments.

The role of growth hormones

Further, in many *in vitro* cultures, growth hormones are purposely added to the medium in order to stimulate growth and differentiation of the explants in culture. These growth hormones (auxins/kinetins), in particular when added in high concentrations, also may cause genetic instability and it would not be correct to call the observed variation in that case 'spontaneous' in the strict sense.

Some tentative conclusions about genetic variation in vitro

It has been pointed out already that, if a change is observed *in vitro*, a sexual phase is required in order to give more definite proof for the genetic nature of the observed variation. In many cases, e.g. in case of *in vitro* propagation of (predominantly) vegetatively propagated crops like potato (*Solanum tuberosum*), cassava (*Manihot esculenta*), sugarcane (*Saccharum officinarum*) and many ornamentals, such proof has never been given.

All in all, it appears that, to a smaller or larger extent, part of what is called 'spontaneous *in vitro* variation' may not be 'spontaneous' and 'genetic' at all, and it can be said here already that many (if not most) claims made in this respect lack a solid base. This should be taken into account when considering the contributions of the so-called 'somaclonal variation' to plant breeding, which will be discussed in the next section.

At the same time, however, there can be no doubt that – similar to spontaneous mutagenesis in intact plants – genetic variation does arise under *in vitro* conditions and even may be considered a common feature of most tissue culture systems. Taking into account the numbers of individual cells, protoplasts, explants, anthers, etc., that are brought into culture, it would even be rather peculiar if not – frequently and for many traits – spontaneously arisen genetically determined variation of various nature would occur *in vitro*.

It is generally assumed that the *in vitro* process of spontaneous mutagenesis is enhanced by the absence of some controlling systems and regulating mechanisms that operate in complete plants. Phillips, Kaeppler & Olhoft (1994) in this respect refer to a 'breakdown of normal control' and report that for maize (*Zea mays*) single-gene mutations, chromosome aberrations, quantitative trait variation, transposable element activation and changes in DNA methylation have been documented. There is also strong evidence that the longer the time spent in a state of unorganized or disorganized growth, like in callus cultures, the more genetic instability is induced (Karp, 1995). Therefore, it can be anticipated that some of the genetic variation that arises *in vitro* is of potential value to the plant breeder.

Cell and tissue cultures, furthermore, sometimes offer opportunities by applying a specific agent in the medium to select on a large scale for cell lines in which a specific mutation occurred, for instance by growing cells of rice (*Oryza sativa*) *in vitro* at inhibitory levels of the amino acids lysine plus threonine to select for plants with altered seed proteins (see Schaeffer & Sharpe, 1990), or to perform selection for tolerance to salt or aluminium. Such cell lines, subsequently, may give rise to plants which possess the desired trait and may be useful as a 'parent' in further breeding programmes or sometimes be used directly as a new, improved cultivar (Maliga, 1984).

5.5. 'Somaclonal variation'

5.5.1. Historical background

The early years

It was discovered in the 1960s that when plant cells or tissues are brought into *in vitro* conditions, their genomes may show variation of different kinds for a wide range of traits. For the first report about this kind of genetic variability, Evans, Sharp & Medina-Filho (1984) refer to work by Murashige & Nakano (1967) with cultured cells of tobacco (*Nicotiana tabacum*). Evans *et al.*

(loc.cit.) add that Heinz & Mee (1969), who worked with sugarcane (*Saccharum* spp.) since 1962, were the first to report that the chromosome instability of cultured cells could be recovered in regenerated plants, and also the first who recognized the potential value of 'variant plants' regenerated from cell culture of sugarcane (Heinz & Mee, 1971). It should be added here that for instance Skirvin, McPheeters & Norton (1994) mention that, in the 1960s, hundreds of publications had already been written about somatic variation for animal cell cultures and refer in this respect to an entire book on this topic by Harris (1964).

The idea then took root here and there, that with the discovery of this *in vitro* variation for plants, a new and promising source of mutations for different traits of agronomic interest had been opened up that might be exploitable for plant breeding purposes (Morel, 1972; Nickell & Heinz, 1973; Heinz et al., 1977). At that stage particular reference was made to observations in some vegetatively propagated crops like sugarcane (*Saccharum* sp.) and potato (*Solanum tuberosum*). It was suggested that in this way useful genetic variation might become available, for instance for resistance to plant pathogens, for tolerance against stress by abiotic factors, for resistance against herbicides, for desired morphological traits, etc. and, moreover, that this variation perhaps might be different from the variation obtained by conventional breeding work. A number of early reports, for instance by Liu & Chen (1976), Skirvin & Janick (1976a,b), Behnke (1979; 1980a,b), Shepard, Bidney & Shahin (1980), Bidney & Shepard (1981), Larkin & Scowcroft (1981; 1983), Secor & Shepard (1981), Shepard (1981), Scowcroft & Larkin (1982) and Thomas et al. (1982), seemed to confirm that the goals mentioned were within easy reach. According to a proposal by Larkin & Scowcroft (1981), the aforementioned *in vitro* variation was called now **somaclonal variation**.

At that stage most practical plant breeders and growers did not pay much attention to this new subject, mainly because *in vitro* cultures were primarily used by them as a method for true-to-type vegetative propagation. As a consequence, the occurrence of unexpected variation – as is the case with 'rogues' or 'off-types' in multiplication fields of seed potatoes – was mainly considered a nuisance leading to financial losses. In addition, most breeders were already aware of the fact that much of the variation observed after an *in vitro* stage, was not transferable to later generations, in particular not by sexual propagation but often not by vegetative propagation either. Such untransferable variation, apparently, had no bearing on 'real' mutations, and mostly was labelled as 'epigenetic' (Meins, 1983). In addition, it was found that it could never be predicted for

which traits and how often unintentional *in vitro* variation would occur and in which direction such changes would go. Skirvin (1978) already reported that the amount of variation may differ among various clones and, in addition, may depend on the age of the culture, on the composition of the culture medium, on the selection pressure applied, etc. Another limitation to the use of promising variants from *in vitro* origin also soon became obvious: not all genotypes – even within the same species – can be regenerated easily. This implies that it may not be possible to regenerate genotypes in which valuable changes have been found. There is, however, no correlation between the occurrence of (useful) mutations and the regeneration ability of a mutated genotype. Until now no adequate solutions have been found for several of the problems mentioned here.

According to Maliga (1984) lines selected in tissue culture for a specific trait should be termed <u>variants</u> until the cause of the phenotypc change (mutation or epigenetic change) has been clarified. The term <u>mutant</u> should be used only when genetic or molecular evidence for mutation has been provided. Indeed, much confusion in literature would have been avoided if all authors on *in vitro* variation had followed this advice.

Undesired in vitro variation
One of the most striking examples in the 1980s of the problems that growers and breeders may meet as a result of unexpected and undesired variation after an *in vitro* cycle, refers to oil palm (*Elaeis guineensis*). Vegetative propagation of high-yielding genotypes of this cross-pollinated crop is very attractive from the view-point of starting genetically homogeneous plantations. Suitable *in vitro* methods in this respect offer several striking advantages, like the relatively small size of the propagules and the possibility to propagate free of disease. However, abnormal fruits and bunches were observed after *in vitro* propagation from root explants (Corley et al., 1986; Soh, 1987).

It was found that in the inflorescences of specific clones the sex of flowers changed, that fruit fertilization did not take place and bunch abortion occurred. These phenomena were observed in one clone that had been propagated by a tissue culture laboratory of a multinational company in the UK. They occurred after large-scale planting on oil palm plantations in Malaysia with 25% of all palms being affected during the second year and 90% during the third year after planting. Similar problems were met by another company when using leaf explants as starting material. Although exact figures have never been published, it is obvious that the financial losses caused by large-scale planting of such unsuitable clones must have been considerable, in particular because it takes a number of years in the field, depending also on methods of cultivation, before the oil palm plants have reached the stage at which their yield potential can be properly assessed.

Another example of undesirable variation, observed after *in vitro* propagation, concerns horticultural crops in the Netherlands. In total some 80 million plants (mainly ornamentals) were propagated *in vitro* in this country in about 1990, and at that time it was recorded that of a certain batch for a given crop sometimes up to 50% did not fully comply to the official cultivar description and, hence, had to be discarded. De Klerk (1990) in this context reported that the annual financial loss caused by off-types in ornamentals as a result of *in vitro* propagation, was equivalent to about one million US dollars for the Netherlands alone.

Another situation in which the occurrence of unwanted variation in the plant material should be prevented by all means, as was mentioned already earlier in this chapter, is when various *in vitro* culture methods are applied in relation to genetic engineering, as is the case when transforming plants with the *Agrobacterium* method. Karp (1991), in the introduction of a review on somaclonal variation, stated that the occurrence of somaclonal variation is one of the current problems facing the advance of genetic engineering in crop plants.

The 1980s: somaclonal variation 'rediscovered'
Despite the fact that the aforementioned risks and limitations of *in vitro* culture methods were already well-known, at least by the common growers and plant breeders, the 'advent of biotechnology' in the 1980s and the fact that large amounts of money from private industry as well as from governmental sources became available for this new field of research, resulted in a fast growing interest in *in vitro* culture as a source of genetic variation. At that time, many cell and molecular biologists, biochemists and other scientists entered the field of plant breeding and, despite the fact that most of them had no specific knowledge or previous training on this subject, their ideas and opinions received much (and often even disproportionate) attention. As a consequence, practical problems and limitations of various 'modern methods', in many cases already well-known to practical breeders, were neglected. The term 'somaclonal variation', following its introduction by Larkin & Scowcroft (1981), became generally accepted for *in vitro* generated variation and for a number of proponents of this method there seemed to be almost no limits to the virtues of this 'method' for plant breeding. Later in this chapter and in Box 5.1 further attention is

paid to the definition of 'somaclonal variation' and some other related terms and abbreviations.

Within a few years hundreds of publications appeared on this subject. Lal & Lal (1990) mention 353 publications! A cross-section of less than 10% of useful and interesting research contributions and reviews may include D'Amato (1975); Skirvin (1978); Shepard *et al.* (1980); Bidney & Shepard (1981); Larkin & Scowcroft (1981); Scowcroft & Larkin (1982); Evans & Sharp (1983); Meins (1983); Reisch (1983); Scowcroft, Larkin & Brettel (1983); Sree Ramulu, Dijkhuis & Roest (1983); Engler & Grogan (1984); Evans *et al.* (1984); Orton (1984); Maliga (1984); Sanford, Weeden & Chyi (1984); Creissen & Karp (1985); Karp & Bright (1985); Scowcroft (1985); Ahloowalia (1986); D'Amato (1986); Daub (1986); Evans (1986); Evans & Sharp (1986); Gill, Kam-Morgan & Shepard (1986); Lörz & Brown (1986); Sree Ramulu *et al.* (1986); Evans (1987); Gavazzi *et al.* (1987); Larkin (1987); Lee & Phillips (1987); Orton (1987); Scowcroft *et al.* (1987); Evans (1988); Evans & Sharp (1988); Lee & Phillips (1988); Scowcroft & Larkin (1988); Widholm (1988); Buiatti (1989); Evans (1989); Jackson & Dale (1989); Larkin *et al.* (1989); Phillips (1989); Bajaj (1990b); van den Bulk *et al.* (1990); de Klerk (1990); Wenzel & Foroughi-Wehr (1990); Phillips, Kaeppler & Peschke (1990); van den Bulk (1991); Karp (1991); Peschke & Phillips (1992); Buiatti & Gimelli (1993); van den Bulk & Dons (1993); Smith, Duncan & Bhaskaran (1993); Skirvin *et al.* (1994); Phillips *et al.* (1994); Ahloowalia (1995) and Karp (1995). Several of these publications will be discussed in the next sections.

When considering the enormous number of publications on 'somaclonal variation' that have been produced within a few years and assessing the funds going into this subject in relation to other plant breeding activities, a few question marks are justified.

5.5.2.   **Which variation is 'somaclonal variation'?**
In a recent publication Skirvin *et al.* (1994) state that somaclonal variation is 'variation among regenerated plants that occurs as a result of tissue culture of any type'. The authors – following Evans *et al.* (1984) in this respect – add that this variation may have been pre-existing or arose during the tissue culture phase, and may be either heritable, epigenetic or non-genetic. Earlier McPheeters & Skirvin (1983; 1989) had concluded that some of the variation observed in 300 '*ex vitro*' plants from shoot tip cultures of a periclinal chimera from blackberry (*Rubus laciniatus*) could be explained in terms of chimeral segregation (for pure thornless vs chimeral), but most other variation was assured to be of tissue culture induced ('somaclonal') origin.

It was mentioned before that the term 'somaclonal

Box 5.1 Somaclonal variation: some definitions

Larkin & Scowcroft (1981) proposed to use the expression '**somaclonal variation**' to indicate – in a rather general way – the phenotypic variation that is observed in plants that have been regenerated after an *in vitro* phase. The suggested term quickly became widely accepted. The authors, however, did not provide an unequivocal definition and soon different interpretations were given. It is therefore necessary to critically consider the definition and to investigate the real nature of the observed changes and their prospects for plant breeding in more detail.

The word '**somaclone**' refers to the old and commonly used expressions <u>soma</u>, the Greek word for body (tissue), and <u>clone</u> to indicate the vegetative progeny of a genotype. Related expressions are '**calliclone**' which, according to Skirvin & Janick (1976b) refers to clones originated from callus, and '**protoclone**' for clones derived from protoplasts (as defined by Shepard *et al.*, 1980). These last two expressions, however, did not find general acceptance. In addition, Evans, *et al.* (1984) proposed the word '**gametoclone**' – as a contrast to somaclone – for clones regenerated from gametes, i.e., anthers or ovaries but, again, this expression is not generally used. Another, related expression for clones developed after colchicine treatments of *in vitro* plant material, is '**colchiclone**', but this term also did not find acceptance.

Many authors, nowadays, use the expression 'somaclone' for all clones developed by an *in vitro* phase and 'somaclonal variation' for all phenotypic variation of those clones, often without having convinced themselves of the genetic nature of the observed variation.

In a number of publications the genetic variation that has been induced before or during the *in vitro* stage by intentional mutagenic treatments is included as well in the term. In addition, often no attention is paid to the presence of pre-existing genetic variation, i.e. the variation that was present already in the (cells of) starting material from which the *in vitro* propagation material has been derived. Although in many cases <u>chimerism</u> indicates newly arisen mutations, especially also in calli, it may be, in practice, often difficult to distinguish between pre-existing variation and variation that has newly arisen as a result of *in vitro* culture, since the pre-existing variation is only uncovered through *in vitro* cultures. In order to avoid confusion, authors using the term 'somaclonal variation' and other terms mentioned here, should always specify what exactly is meant.

In literature two different systems are used to name the various generations following the *in vitro* phase. Larkin & Scowcroft proposed to use the expression SC_1 for the first generation of plants following the *in vitro* phase. The next generation is named SC_2, etc. This system matches to a certain extent with the terms used for the different generations used in general plant breeding and mutation breeding where the suffix '1' indicates the first heterozygous generation.

According to another system, proposed by Chaleff (1981), primary regenerants are named as R_0, whereas R_1 represents the seed progeny of the primary regenerated plant. Chaleff's system, at least when applied in a strict way, does not fit in the notation that is generally used by plant breeders for the consecutive generations of a breeding programme (F_1 for the first (heterozygous) generation resulting from the crossing, etc.; M_0 and vM_0 for the starting material of a mutation progamme, M_1 or vM_1 for the first mutagen treated heterozygous generations, etc. The parents are normally called P_1 and P_2).

Hence, as plant breeders are the intended users of 'somaclonal variation', Chaleff's system should be rejected. (N.B. In Chaleff's system the first vegetatively propagated generation after the mutagenic treatment is indicated as R_0; the progeny of a (back)cross between a primary regenerant (R_1 according to Chaleff) and the (homozygous) parent cultivar (P) as F_1.)

Maliga (1984) pointed out that modifications to Chaleff's system have been suggested, for instance by Hibberd & Green (1982) and by Larkin *et al.* (1984), but he did not see the advantages of these modifications and, therefore, unfortunately, recommended to continue using the terminology of Chaleff on the basis of priority. Chaleff's system, indeed, is still often applied, e.g. by Evans (1986).

Finally, it has been proposed by the IAEA (see for instance Anon., 1986a) to use the symbol M_x for the consecutive generations produced after conventional mutagenic treatment as well as after *in vitro* mutagenesis. The x indicates the specific generation with x=1 denoting the treated (heterozygous) generation and x=2 the first segregating generation. Accordingly, it is advised to use V_x for vegetative 'generations', irrespective of whether the material is obtained by vegetative propagation or *in vitro*. It is added in the IAEA report that 'in special situations, SC for plantlet regeneration from somatic cells and GC for plantlet regeneration from gametophytic cells can be judiciously used'.

The conclusion is that the terminology concerning 'somaclonal variation' and the naming of different generations remains rather confusing, and authors should always indicate what exactly is meant by the symbols used by them.

variation' was coined by Larkin & Scowcroft (1981), but it is not fully clear whether, at least at that stage, the first authors of this term had in mind to imply all phenotypic diversity exhibited in plants regenerated from tissue cultures, as their early description of somaclonal variation is interpreted by for instance Creissen & Karp (1985) and Skirvin et al. (1994), or as Larkin et al. (1989) put it, to limit the use of this term to: '.. genetic, and therefore heritable, variation that arises following cell culture and plant regeneration'.

From various publications by Larkin & Scowcroft it can be concluded that these authors are of the opinion that pre-existing variation and variation generated in vitro cannot be separated and together account for the occurrence of 'somaclonal variation'. Consequently, according to this opinion the composition of the culture medium, including the action of growth hormones added to the medium, should also be accepted as contributing to 'somaclonal variation'. Of course, if one follows these lines one could even argue that if mutagenic agents are added to the medium on purpose (in addition to growth hormones which unintentionally may induce mutations) this, likewise, could be called a contribution to 'somaclonal variation'. For instance, Shen et al. (1995) reported on 'somaclonal mutants' in rice (Oryza sativa), obtained after combination of in vitro culture and gamma ray mutagenesis. Accordingly, in this way all phenotypic variation observed in regenerated plants and irrespective of its origin could be called 'somaclonal variation'.

In a more narrow and, from a practical (plant breeding) point of view, more useful definition of somaclonal variation, of course, the non-genetic component should be clearly excluded, because this variation is of no significance for plant breeding. For the time being, epigenetic variation should also be considered 'non-genetic', because it is still impossible to use this type of variation as a stable source of genetic variation.

In addition, in a narrow definition, it would be more logical to apply the term 'somaclonal variation' – or as a more restricting alternative, excluding the variation that is not genetically determined, 'somagenetic variation' – only for the genetic variation that has arisen during the in vitro stage and, thus, to exclude – if possible – pre-existing variation, as well the variation induced by mutagenic agents that have been added intentionally. (N.B. Genetic effects induced by hormones added to in vitro cultures, in fact, also should be excluded, but in that case no regeneration may take place after in vitro culture.) A serious practical objection against such a narrow definition would be the difficulty to agree which aspects should be included and, in addition, to prove in

quite a few cases that the observed variation does refer to 'real' somaclonal variation (or if our airy suggestion would be taken up: **somagenetic variation**).

The conclusion must be that, from a plant breeding point of view, the term 'somaclonal variation <u>in the wide sense</u>', is not very useful, but in practice botanists will continue to use this term for <u>all</u> variation observed in in vitro culture. Plant breeders should be well aware of the – predominantly – non-heritable character of most of this variation. The term 'mutation', of course, should be used only after the genetic nature of the observed variation has been established.

Finally, it is not difficult to agree with Wersuhn (1989), who stated that the choice of the word 'somaclone' (or 'somaclonal') has been a rather unfortunate or even bad one, and who further regretted that the whole terminology that has been developed for this field is very inappropriate. However, as we are discussing here the present role of – what is generally known now as – 'somaclonal variation', there is almost no escape from using this term, which will be applied in a rather loose way, unless authors have made very clear what this term implied in their work.

5.5.3. Some opinions about origin and possible causes of 'somaclonal variation'

In one of the earliest contributions about variation observed in cell suspensions and tissue cultures Nickell & Heinz (1973) working with sugarcane (Saccharum sp.) observed that this variation differed from clone to clone and speculated that it arises from 1) variation present in the original plant material, like aneuploidy, polyploidy and 'chromosomal mosaicism', 2) mutations caused by the medium (nutrients, auxins and other chemicals), and 3) the disintegration of intact tissues and organelles to form cell suspensions.

Novak (1991a; see also Karp & Bright, 1985) also distinguishes three different mechanisms that may explain the origin of 'somaclonal variation': 1) pre-existing genetic variation in the explant tissue, 2) genetic variation induced by mutagenic action of chemical compounds in the culture media, and 3) variation as a response of the plant genome to stress induced by the condition of the tissue culture. This distinction is generally accepted now. With respect to the last category, which may be the main source for the observed variation and is sometimes referred to as 'chaos in the test tube', the following may be added. It is believed that this 'mechanism' is related to or may be caused by epigenetic variation which, according to Novak (1991a) is characterized by non-permanent changes that sometimes may involve non-heritable traits that, nevertheless, may be transmittable through a limited number of sexual or vegetative generations. Novak adds that all responses to *in vitro* stress may involve modification of gene expression by DNA methylation, amplification of DNA sequences and transposition of mobile elements.

Skirvin *et al.* (1994), in a recent review, discussed a range of factors that may play a role in provoking variation during the tissue-culture phase. Mentioned amongst others are: the kind of explant, cultivar choice and cultivar age, ploidy level; method and specific conditions of culturing, including for instance the amount of growth regulators and whether specific selection pressure is applied; the length of the period *in vitro* and the proliferation rate. This summing-up more or less corresponds with four (groups of) factors, mentioned by Karp (1995), which influence whether or not variation is generated and how much. These four factors are: 1) the degree of departure from organized meristematic growth, 2) the genetic constitution of the starting material, 3) the growth regulators in the medium, and 4) the tissue source.

As has been shown on various occasions, for instance by Dolezel, Novak & Havel (1986) for garlic (*Allium sativum*), chances to generate 'somaclonal variation' generally become greater when the degree of departure from organized growth becomes greater and when the time spent in this disorganized state becomes longer. To explain such results two working hypotheses are generally proposed. The first implies that one or more 'control systems' which eliminate (the major part of) genetic changes in organized meristems (e.g., when shoot tips are cultured *in vitro*), break down in disorganized culture. The nature of this 'control' is still unknown. As an alternative, equally unknown mechanisms which provoke genetic instability could be released in such disorganized systems (Phillips *et al.*, 1994).

According to Karp (1995) an increasing amount of evidence becomes available for a certain genotype-dependency of 'somaclonal variation'. Almost no details are known yet, but the ploidy level certainly plays a role. However, even within one ploidy level, genotypes may show considerable differences as to the amount of 'somaclonal variation' observed. Karp speculates that differences between species may be partly related to the breeding system (outbreeding, inbreeding) or to what she calls 'differences between DNA sequences'.

Generally it has been found that the concentrations of growth regulators in the medium exert an effect on somaclonal variation. However, no alterations were found in the frequency of specific somatic gene mutations and small deletions in stamen hairs of *Tradescantia* hairs (an outstanding model system for determining

mutation frequencies that has been described in Chapter 1) as a result of growth hormones in the medium in which inflorescences of *Tradescantia* were cultured (Dolezel & Novak, 1984). Therefore, growth hormones, probably, may not be able to break chromosomes, but interfere with regulator genes. Species or genotypes also may show different sensitivities to growth regulators. Endogenous levels of growth hormones may also differ between cultured cells derived from different species and genotypes. This makes it difficult to establish general rules.

The use of various tissue sources may affect frequency and nature of 'somaclonal variation', and perhaps not only due to pre-existing genetic variation. In general, more variation is found when cultures are started from older or more differentiated sources. It is also known that certain types of mutations (chromosome losses, deletions, aneuploids), even if they should be detrimental to the plant, have a much better chance to be tolerated in polyploid crops.

Still too little is known of the cellular and molecular mechanisms behind 'somaclonal variation'. Many possibilities have been suggested in literature, amongst which are mitotic irregularities leading to chromosomal instability, the occurrence of gene amplification or deletion, as well as gene-inactivation or reactivation of 'silent' genes. In addition, transposition and somatic crossing over are mentioned as possibly significant factors. It has been further proposed that heterochromatine (chromosome segments that remain condensed during interphase) should be among the most important factors. DNA methylation, which is related to gene regulation and, hence, to the occurrence of epigenetic effects, has also received some attention as a possible explanation for 'somaclonal variation'. Phillips *et al.* (1992; 1994), referring also to earlier work, for instance stated that the mechanism responsible for the heritable variation observed among tissue culture regenerants is still largely unknown, but suggested that methylation may be an underlying factor in many of the tissue culture-induced phenomena (see also next section). The authors further reported that there is evidence for maize (*Zea mays*) that the tissue culture process induces changes in DNA methylation that can be detected several seed generations later, but added that mechanisms responsible for such changes, including the frequent occurrence of homozygous changes (in diploid tissue), are not understood. Moreover, they did not make clear how, for instance, effects of methylation can pass meiosis.

In the years following the aforementioned publication by Phillips *et al.* (1992) the relative role of the different factors mentioned before with respect to the generation of 'somaclonal variation', has not been further elucidated (see for instance Phillips *et al.*, 1994). As a consequence, it is still too early to reach definite conclusions as to the nature of the variation resulting from stress induced by the conditions of tissue culture.

5.5.4. Characterization of 'somaclonal variation'

In recent years gradually more insight has been gained into the various forms in which 'somaclonal variation' can be manifested. This is shown in the following overview which, roughly, follows the *FAO/IAEA Training Manual on Tissue Culture Techniques* (Anon., 1990b, p. 68).

- Monogenic and polygenic traits may be involved.
- Heterozygous (AA→Aa) and homozygous (AA→aa) changes occur.
- Mutations may show recessive and dominant effects. Sometimes remarkably high numbers of dominant mutations are observed.
- Changes occur in individuals and – simultaneously – in whole populations.
- Variation for morphological and physiological traits refers to changes in structural genes as well as in genes which regulate gene expression.
- Often aberrant chromosome numbers (aneuploids, polyploids) are found.
- Many structural chromosomal aberrations occur.
- Exchanges of sister chromatids are observed.
- Changes in DNA content of cells do occur.
- Changes within genes and in number of gene copies occur.
- Cytoplasmic mutations (in plastids and mitochondria) are observed.
- Increased transposon activity is found.

Some – from the view point of plant breeding important – additional comments should be made here. Most 'somaclonal variation', as is the case with (other) mutations, shows effects, opposed to previous selection; in other words: the large majority of the observed variation is useless. This is specifically the case for polygenically inherited traits (e.g. yield). Moreover, it seems that this variation cannot be directed and it is impossible to predict for which traits and how much variation will occur. Consequently, variation can not be limited to 'target genes' for which useful variation is desired. Like in conventional plant breeding, further selection work is required to get rid of undesired variation and to identify the (few) desired variants. Finally, the amount of 'somaclonal variation' may be highly variable, even

when starting from genetically homogenous plant material.

On the other hand, it is claimed in some publications that certain features should make 'somaclonal variation' particularly attractive for plant breeding. Of course, if 'somaclonal variation' should be able to produce changes which are not observed after application of other breeding methods (including 'classical' mutagenesis), this would be a valuable addition to existing plant breeding tools. The aforementioned list of characteristics of 'somaclonal variation' indeed does show several features which raise the particular interest of plant breeders, such as the occurrence of homozygous (AA→aa) changes and dominant mutations in sometimes high numbers; changes occurring simultaneously in whole populations; changes in chromosome numbers; changes in DNA content and number of gene copies; and an increased transposon activity. In the following sections we will further discuss several of the aforementioned points.

Evans & Sharp (1986) mention that 'somaclonal variation' shows a certain similarity to induced mutations, but the authors state that mutation frequencies *in vitro* are much higher. These authors further claimed that mutated plants derived from tissue cultures should be superior to plants resulting from mutagenic treatments. They reasoned this by assuming that the frequency of deleterious mutations should be less than after mutagenic treatment, because such *in vitro* mutations would be effectively selected against during plant regeneration. Whether this is really the case seems rather doubtful. When seeds, plants, pollen, etc., are mutagenically treated, each individual plant is (temporarily) affected by the stress caused by the mutagenic treatment. A cell of such plants carrying a desirable mutation also undergoes selection and has to compete with surrounding cells which may carry different mutations or no mutations at all. Moreover, when plants are propagated by seed, mutated cells also have to pass the meiotic sieve during which stage deletions, etc. are selected against. The only difference, when for instance suspension cultures are involved, is that during the initial stages stress conditions may differ considerably from those which occur when seeds, plants, etc., are submitted to mutagenic treatments. This view is supported by the observation that 'somaclonal variation' when starting from organized shoot apices, axillary buds, etc., is much lower than when 'unorganized' cultures are used.

The frequency of 'somaclonal variants'

Spontaneous 'sports' in plants commonly occur in low frequencies. One plant with aberrant (partly) flower colour, for instance, may be found among 10^4 to 10^5 non-mutated plants, depending on the genetic structure of the clone. Some plant traits show a higher rate of spontaneous phenotypic variation. Heiken (1960), for instance, has reported that the spontaneous frequency in potato (*Solanum tuberosum*) for so-called 'bolters', a physiological instability resulting in a strikingly aberrant plant growth – most probably an epigenetic effect and therefore not heritable – may go up to 1.5×10^{-3}. It is interesting to observe in this context that it has been just the high frequency of 'bolters' in potato that was mentioned in the early 1980s by Shepard – one of the pioneers on 'somaclonal variation' – as a 'proof' for the excellent prospects of this variation for plant breeding (Shepard *et al.*, 1980; Bidney & Shepard, 1981; Secor & Shepard, 1981; Gill *et al.*, 1986; Sanford *et al.*, 1984).

In addition to the 'high mutation frequencies *in vitro*' claimed by Evans & Sharp (1986), to which was referred in the previous section, Larkin *et al.* (1989) for instance reported that the proportion of <u>dominant</u> mutations *in vitro* could be ten times higher than among mutagen induced mutations, and that relatively many homologous mutations should occur in diploid tissue (AA→aa), leading to homozygous mutants. Explanations for these observations as well as quantitative evidence with adequate statistical proof, so far, have not been given. Provided that these observations are correct this result, perhaps, may be explained by changes in gene regulation (e.g., elimination of inhibitors) leading to the expression of pre-existing but silent dominant or homozygous recessive genes.

Indeed, very few results mentioned in literature concerning mutation frequencies as a result of 'somaclonal variation' are based on reliable experiments (with adequate numbers of plants) in which frequencies of proven mutations – either spontaneous or induced by radiation or chemicals – and proven 'somaclonal variation' have been compared. The shortcomings of the method that is usually followed to determine the extent of somaclonal variation, i.e. by determining the percentage of aberrant plants without a reliable point of reference, have been discussed in detail by de Klerk (1990). In this publication the use of another method – to determine the standard deviation of phenotypic variation of quantitative traits – has been advocated.

In this section we can only briefly mention some of the many examples that have been given in literature about frequencies observed for various traits of 'somaclonal variation'. This subject has been treated in several reviews, e.g. by Karp (1991) and by Skirvin *et al.* (1994). Skirvin *et al.* refer to publications in which, after biochemical and chromosomal analysis, the frequency

of changes was as a high as 100% (Orton, 1983) and to the 'more common' rates of 15–20%, reported for instance by Evans & Sharp (1983). For single-gene mutations the occurrence of one mutant in every 20–25 regenerated plants was reported by Evans (1988) and, according to Orton (1987), all lines of celery (*Apium graveolens*) that had been produced *in vitro*, varied from the chromosomal constitution of the parent. Oh *et al.* (1995) reported that 85% of the plants regenerated from protoplast cultures of diploid petunia (*Petunia hybrida*) were tetraploid, for which growth hormones in the culture medium were mainly held responsible. However, as Skirvin *et al.* (1994) rightly point out, many 'somaclones' may be identical and be derivatives of one and the same mutational event. This must, however, be distinguished from the claimed phenomenon, that whole populations change in the same way simultaneously, a subject matter unexplainable till this time.

Skirvin *et al.* state that, for instance, not 1–3% of the regenerants should be expected to mutate at a particular locus, but that it would be more realistic to expect that 1–3% of the regenerants will vary from their parent 'in some physical or biochemical manner'. This, of course, still would represent a considerable amount of variation. It goes beyond saying that such percentages do not reveal anything about the amount of variation in regenerated plants that could be useful for plant breeding purposes.

Gavazzi *et al.* (1987) reported for tomato (*Lycopersicon esculentum*) that by *in vitro* culture higher mutation frequencies and a different mutation spectrum had been produced than by pollen and seed treatment with EMS. They also mentioned the unexpected, frequent recovery of homozygous 'solid' mutants for leaf shape and plant vigor as a result of 'somaclonal variation' in the so-called R_2 (according to the system proposed by Chaleff, 1981; see further Box 5.1). We will return to this work later in this chapter when discussing chemical mutagenesis *in vitro*. The results by Gavazzi *et al.* were in line with the positive views expressed by Evans (1986; 1987; 1989). Maliga (1980; 1983), on the other hand, has pointed out that mutagenic treatment increases the frequency of 'variants' in the surviving cell population and is a key factor to recover rare phenotypes.

The spectrum and frequencies of putative mutations after regeneration of plants of tomato (cv. Moneymaker) from leaf, cotyledon and hypocotyl explants were compared by van den Bulk *et al.* (1990) with the results of EMS treatment of seeds. Phenotypic alterations which were observed in the first vegetative generation (called R_1) were not transmitted to the progenies, except when ploidy mutations were involved. Several monogenic mutations, allelic to known, recessive, single-gene mutations, were found in R_2. For some mutant types that could be scored unambiguously, results of treating tomato seeds with EMS (60 mM for 24 h) were compared with those obtained in the second vegetative generation of tissue culture (called R_2). The EMS treatments clearly produced higher mutation frequencies than were observed in tissue culture. No clear differences were observed between the mutant spectra (in M_2 and R_2) of both categories of plants. The only exception was the frequent occurrence of polyploid plants after regeneration through tissue culture. Such plants were not found after seed treatment with EMS. Contrary to Gavazzi et al. (loc.cit.), van den Bulk et al. did not find a different spectrum of mutations for EMS-induced mutants and mutants resulting from tissue culture. However, because of the use of different genotypes, different procedures, etc., a good comparison of the results obtained by van den Bulk et al. and by Gavazzi et al. (1987) is not possible.

Buiatti (1989) concluded that the frequency of mutagenic events in the case of 'somaclonal variation', when calculated on a cell base, did not differ in a positive way from the usual mutagenic experiments and further doubted whether the mutation spectrum of somaclonal variation indeed would include new mutations not observed in other kinds of experiments. However, it should be commented that certain mutagenic changes, for instance duplications, normally do not result in phenotypic changes. Buiatti further added that only in relatively few cases was the observed 'somaclonal variation' still present after one or two generations. However, Buiatti & Gimelli (1993) also reported on 'astonishingly high' frequencies of variants (e.g., up to 80% in rice, Orizya sativa) and a mutation spectrum that differed somewhat from recovered spontaneous mutations. These authors also gave a number of references in which mutation frequencies in regenerated plants of different species between 2% and 47% were reported. The relevance of such figures, however, remains low unless exact information is provided about the way these figures were obtained, the number of traits studied, the way of scoring, etc.

It is at this stage difficult to reach an unequivocal conclusion about the frequency of in vitro derived genetic variation. It is generally accepted that after an in vitro phase high levels of phenotypic variation may be observed and that this variation for a certain part refers to genetic changes induced during the in vitro phase. In addition, it is often believed that the frequency of such mutations should be high in comparison to the frequency of artificially induced mutations in plants.

Whether this is really true has not been proved yet. A further complication is that various methods can be used to determine mutation rates, for instance 'on a cell base' (for cultured cells or regenerated cells) or by determining the number of segregants after selfing regenerant plants; but often the description of the methods used is confusing or incomplete.

Sufficient and convincing statistical evidence for the claims of high frequencies of heritable variation from in vitro cultures has not been provided yet. A positive exception may be made for the occurrence of aberrations in chromosome number (aneuploids and polyploids), but even in such cases where screening is relatively easy, it can not be excluded that other factors than the in vitro phase itself – e.g. the occurrence of endomitosis, which is a fairly common phenomenon in differentiated and differentiating tissues – add significantly to the number of aberrant plants and thus confuse the picture.

In conclusion, much more statistically confirmed evidence will be required before claims about high frequencies of desired, heritable variation, obtained for a wide range of agronomically useful traits by bringing plant material under in vitro condition, are sufficiently substantiated.

'Somaclonal variation' for various traits

As has been mentioned before, 'somaclonal variation' has been reported for virtually all kinds of traits in many different crops. At the same time, detailed information about the nature of the observed variation and, in particular, convincing evidence about the stable genetic nature of this variation in regenerated plants during a number of (sexual) generations in most cases is lacking.

Maliga (1984) already correctly pointed out that the early positive results reported for potato and sugarcane referred to disease resistance in highly heterozygous, vegetatively propagated crops and that the factors conferring disease resistance had not been characterized genetically. He also mentioned that the spectrum of 'somaclonal' variation is limited, and that disease resistance is not the best trait to use for estimating genetic variability because the expression is dependent on the presence of the matching pathogen, on environmental factors and on the physiological stage of the plant. Maliga, nevertheless, expressed a qualified optimism about the potential of (the genetically determined part) of 'somaclonal variation', also for seed propagated species.

Evans (1986), during a FAO/IAEA Symposium in Vienna in 1985, mentioned that prior to work by Evans & Sharp (1983) on tomato (Lycopersicon esculentum) – in which the presence of single gene mutations in regenerated tomato plants was revealed – virtually nothing was known about the genetic base of somaclonal variation. Evans & Sharp (loc.cit.) reported for tomato (*Lycopersicon esculentum*) that among 230 regenerants from tissue culture 13 Mendelian (single gene) mutations were identified. These mutants carried recessive mutations for male sterility, jointless pedicel, tangerine fruit, virescent leaf, orange flower and fruit colour, chlorophyll deficiency and dominant (!) mutations for fruit ripening and growth habit.

In addition, D. Pratt (1983) found in *Nicotiana sylvestris* plants regenerated from protoplasts, that 9 out of 21 clones segregated for traits like leaf shape, pigmentation and flowering time. Somewhat later reference was also made to dominant resistance to the tomato disease *Fusarium oxysporum* (Evans *et al.*, 1984; Evans, 1986).

Such detailed genetic studies on agronomically important traits in relation to 'somaclonal variation', however, are an exception and it is in particular regrettable that details on the genetics of observed 'somaclonally induced resistances' are virtually lacking in most publications concerning this subject. (N.B. It may be added here that the optimistic view on the practical usefulness of somaclonal variation, expressed by Evans during the aforementioned FAO/IAEA symposium, was not shared by many participants, as can be concluded from a summary of the discussion on this topic; see Anon., 1986b, p. 367.) In addition, Vuylsteke, Swennen & De Langhe (1991; 1996) concluded from extensive experiments that so-called 'somaclonal variants' derived through micropropagation are of little value as a source of useful genetic variation for the plantain or cooking banana (*Musa* sp., group *AAB*).

In the first reports which mention positive results from 'somaclonal variation', reference is made in particular to increased levels of resistance to various diseases and to altered plant types in potato (*Solanum tuberosum*) and sugarcane (*Saccharum officinarum*). Reports for potato, for instance came from Gengenbach, Green & Donovan (1977), Behnke (1979; 1980a,b), Shepard *et al.* (1980) and Secor & Shepard (1981), whereas for sugarcane in particular the work of Heinz and colleagues (Heinz & Mee, 1969; Heinz, 1973; Heinz *et al.*, 1977) and of Jagathesan (1982) from the Sugarcane Breeding Institute Coimbatore, India must be mentioned.

It must be said here once again that potato and sugarcane, the two crops most often referred to, are vegetatively propagated, highly heterozygous and polyploid crops which can tolerate relatively easily the occurrence of aneuploidy or chromosomal deficiencies. As a result, such aberrations may have been piled up already and remained undetected during many vegetative genera-

tions. Therefore a considerable part of the variation observed in the regenerated plants in the aforementioned cases may have been pre-existing. Admittedly, it is somewhat difficult to give definite proof for the genetic nature (whether nuclear or cytoplasmic) of observed variation in vegetatively propagated crops. However, evidence for the cases mentioned here concerning potato and sugarcane, to the best of our knowledge, is limited to reports about stable <u>vegetative</u> propagation.

Increased levels of resistance after *in vitro* culture have been mentioned as well for several other crop plants, including much grown crops like maize (*Zea mays*) and tomato (*Lycopersicon esculentum*). For maize reference is often made to *in vitro* generated/selected resistance to the soil-borne disease *Helminthosporium maydis* (or *Drechslera maydis*). Details can be found for instance in contributions by Brettell & Ingram (1979), Brettell, Thomas & Ingram (1980) and Ingram & MacDonald (1986). An extensive review with many references concerning *in vitro* selection for resistance against various diseases in a range of crop species was produced by van den Bulk (1991).

5.5.5. Selection *in vitro*

At this place we will only briefly introduce the subject of selection *in vitro* and return to this subject in particular in Section 5.6.4. Selection *in vitro* for agronomically important traits would be very attractive, if it were possible to select on an individual cell base. However, this approach seems to be realistic only, for instance, in case of abiotic stresses, tolerance to specific chemical compounds and perhaps to certain pathogens. In the last case selection with phytotoxins and culture filtrates appears to be more effective than the use of the pathogen itself (van den Bulk, 1991). For practical purposes one should distinguish between selection based on whole plants *in vitro* and selection of cultured tissue, callus or single cells.

As an example, Blonstein, Stirnberg & King (1991) referred to selection for auxin-resistant mutants of *Nicotiana plumbaginifolia*. Auxin-resistant mutants (following mutagenic treatment of seeds with 5–50 mM EMS) were selected as M_2 seedlings *in vitro* in the presence of relatively high concentrations of growth hormones like IAA or NAA. By the over-dose of hormones growth of the normal (wild-type) seedlings was inhibited, but not of the mutants. Blonstein *et al.* selected nine independent (single locus, nuclear) mutants after screening 10 000 M_2 families (a family consisting of the plants grown from the seeds of one capsule per M_1 plant).

Duncan, Waskom & Nabors (1995) described *in vitro* screening and field evaluation for multiple soil-stress tolerance – involving salt, acid (high aluminium) and drought-stress tolerance – in sorghum (*Sorghum bicolor*) and concluded that *in vitro* selection was neither effective nor necessary to obtain improved stress-tolerant 'somaclones'. The authors advised exploiting 'somaclonal variation' through short-term culture (less than 12 months) with no *in vitro* selection.

Van den Bulk & Dons (1993), who studied the potential of 'somaclonal variation' as a source for resistance to bacterial canker (*Clavibacter michiganensis* subsp. *michiganensis*) in tomato, reported that the phytotoxic extracellular polysaccharides which are produced by the pathogen, were unsuitable for use as selective agent at the cellular level. Evaluation of 279 progenies at the plant level did show some variation for severity of wilting of the foliage but 'somaclones' with a major increase in resistance level were not found.

In addition to some publications mentioned already by van den Bulk (1991), some more recent contributions have appeared in which successful selection for disease resistance after an *in vitro* phase – without the application of mutagenic agents – has been reported. Song, Lim & Widholm (1994), for instance, described for soybean (*Glycine max*) that resistance to brown spot disease (*Septoria glycines*) was obtained. According to these authors this resistance had not been obtained before, that is since 1915 when the disease was discovered. In this experiment soybean plants, tolerant to a host-specific pathotoxic culture filtrate of *S. glycines*, were regenerated from *in vitro* selected immature embryos. Evaluation of the progeny in the field showed that until the R_6 (the sixth selfed generation of regenerated plants according to the system of Chaleff, 1981) the lines did not develop disease symptoms.

This result looks promising and it may be hoped that new, resistant soybean cultivars can be obtained in this way but, so far, no confirmation of the release of such resistant cultivars has been obtained. The same applies to the example that is most often quoted, i.e., the release of 'cultivar'(?) DNAP-17 of tomato (*Lycopersicon esculentum*) with *Fusarium* resistance (see for instance Evans, 1989). This last case and a few other examples will be briefly discussed in the next section.

5.5.6. Some results

General remarks

When evaluating the achievements of breeders using 'somaclonal variation', a realistic approach would be to make a comparison with the results of conventional breeding in reaching the same objectives. Breeding problems in principle can be tackled in different ways, which makes the choice of the most appropriate method important from an economic point of view. Therefore, 'somaclonal variation', in order to become a suitable alternative to other methods, should produce 'better' results, i.e., primarily not a higher mutation frequency but a different mutation spectrum. It would be useful as well if in this way results could be produced similar to those by other methods but in a cheaper, more simple or quicker way, at least for certain crop plants. In particular the claim that 'somaclonal variation' should produce quick results has often been made, e.g. by Evans *et al.* (1984), but still needs to be proven.

Although contributions may be made by 'somaclonal variation' to the generation of useful genetic variation, and in agronomically important traits, the ultimate proof for the usefulness of somaclonal variation for plant breeding, of course, is provided by the number of released or approved cultivars – which for sake of brevity may be called 'somaclonal cultivar' – and by the economic contribution of such cultivars as compared to other cultivars. Before discussing this subject, some other remarks should be made here.

Scowcroft, one of the leading forces in the field of 'somaclonal variation', advised biotechnologists in 1988 not to try anymore to look among 'somaclonal variation' for alleles that are known already, but to concentrate on other tasks. He referred in this respect to the 'introgression of alien genes' – for instance for disease resistance. In his opinion, this should be a relatively easy task with the aid of 'somaclonal variation', while – still according to Scowcroft – other methods often fail. Mention was made of efforts of his company (Biotechnica, Canada) on the introduction of (alien?) genes for resistance to *Phoma* into *Brassica* species, but results of practical significance were not presented.

At that time the first euphoria about the potential of 'somaclonal variation' had already largely disappeared (see for instance Wersuhn, 1989) and even the most ardent supporters had to admit that reliable *in vitro* selection for agronomically important traits often was not feasible. They also had to admit that, in most cases, by the use of 'somaclonal variation' the time required for breeding of new cultivars was not shortened, because first confirmation of expression and transmission is necessary, after which the usual several generations of testing of the improved plant material under 'field conditions' are needed before final proof of a useful genetic change in a still suitable genotype, from an agronomic point of view, can be given (Lee & Phillips, 1988). As with induced mutants, it may be expected that in many cases further cross breeding will be required.

One also gradually realized that the production of

'somaclonal variation' just as a source of non-specific genetic variation is not very attractive as this can be easily achieved as well by classical mutagenesis, and that *in vitro* approaches should be considered only if novel specific traits can be obtained. Moreover, it was also found that many 'somaclonal' changes for target genes were undesirable, while unwanted changes in non-target genes occurred as well and also undesired epigenetic effects (Phillips, 1989).

When scrutinizing the literature for reports about new cultivars which are claimed to have originated from 'somaclonal variation', one quickly learns that references to 'positive results' are often not correctly quoted. Sometimes, for instance, the original results had a bearing on preliminary small-scale and short-term experiments only and such results often just represent new 'selections' or lines with some potential instead of registered cultivars.

Another point to consider is that the use of a broad definition of 'somaclonal variation' does imply that more results will be indicated as 'somaclonal cultivars' than in the case of a narrow definition. For instance, a considerable part of all cultivars grown for vegetatively propagated crops are known to be periclinal chimeras, although the growers may be unaware of this. If such cultivars are used as starting material for an *in vitro* culture this may result in a process called 'uncovering' of periclinal chimeras and produce directly visible or cytogenetic changes. Under field conditions 'new' cultivars are produced in this way at regular intervals, in particular for ornamentals or when visible traits are involved. In the case of *in vitro* culture the frequency of 'uncovering' periclinal chimerism in the starting material will be even 100% if protoplasts or suspension cultures are started. The effect of 'uncovering' is often mistaken for a mutation and it can be shown from literature that this mistake is made also in case of 'somaclonal variation'. One example concerning blackberry (*Rubus laciniatus*), derived from McPheeters & Skirvin (1989), has been mentioned already in Section 5.5.2 and we will return to this subject later in this section. More details about periclinal chimerism, of which even plant breeders are often ignorant and which may remain unnoticed for many generations, are discussed in Chapter 7.

Reports about 'cultivars' attributed to 'somaclonal variation'

Buiatti (1989) mentioned that he was not aware at that time of any officially registered cultivar of 'somaclonal origin'. Vasil (1990), also stated that not a single example of any such new cultivar of some significance was known (to him) at that time.

Nevertheless, some named 'somaclonal cultivars' (which term, again, is used here for reasons of brevity only) had been reported before. Daub (1986), for instance, mentioned in this respect cv. Ono of sugarcane (*Saccharum* sp.), with resistance to *Sclerospora sacchari* (Fiji disease) and derived from cv. Pindar (see Krishnamurthi & Tlaskal, 1974), and cv. Scarlet of sweet potato (*Ipomoea batatas*) with a darker and more stable skin colour than the original cultivar (Moyer & Collins, 1983). Both examples refer to vegetatively propagated (polyploid) crops which, as has been said before, makes it difficult if not impossible to prove, at least in the narrow sense, the 'somaclonal origin' of the observed changes.

Another early and much quoted example of a 'somaclonal cultivar' refers to cv. Velvet Rose, a scented type of the ornamental crop *Pelargonium graveolens* (Skirvin & Janick, 1976a,b). The authors mention that the original cv. Rober's Lemon Rose, from which cv. Velvet Rose arose after an *in vitro* callus phase, was a periclinal chimera and they add that 9% of the 'calliclones' and leaf cuttings from the original cultivar spontaneously change to a polyploid form ($2n = 144$ instead of $2n = 72$). The fact that leaf cuttings taken from cv. Rober's Lemon Rose also show this high percentage of 'spontaneous' doubling of the chromosome number, makes it rather doubtful that the new 'somaclonal cultivar' indeed has arisen as a result of the *in vitro* phase per se. It is quite common in vegetatively propagated species that in particular L-3 tissue contains a certain percentage of polyploid cells (note that endopolyploidy is a fairly common feature!), and by making root or leaf cuttings it may be expected that some of these cuttings and *in vitro* propagated plantlets – partly or completely – trace back to such polyploid cells. It seems that this 'example' of successful variation 'of somaclonal origin' is much less convincing than is often believed and that this case may also refer to the dissociation of an already existing (periclinal) chimera.

Several references to 'somaclonal cultivars' have been mentioned in a number of review articles and some other publications (see for instance Karp, 1991; van den Bulk, 1991; 1995; Skirvin *et al.*, 1994). A list of 'somaclonal cultivars' for horticultural crops with the original literature references, produced by Skirvin *et al.* (1994), is presented here in a slightly different way, namely in chronological order. In addition, some other references and some comments will be given.

The list presented by Skirvin *et al.* includes the following.

* *Pelargonium* Velvet Rose (ornamental); source: Skirvin & Janick, 1976a,b.

* *Ipomoea batatas* Scarlet (root crop); source: Moyer & Collins, 1983.
* *Rubus* Lincoln Logan (fruit); source: Hall, Quazi & Skirvin, 1986; Hall, Skirvin & Braam, 1986.
* *Eustoma grandiflorum* (ornamental); source: Griesbach & Semeniuk, 1987; Griesbach, 1989.
* *Paulownia tomentosa* Somaclonal Snowstorm (ornamental); source: Marcotrigiano & Jagannathan, 1988.
* *Hemerocallis* Yellow Tinkerbell (ornamental); source: Griesbach, 1989.
* *Torenia* UConn White (ornamental); source: Brand & Bridgen, 1989.
* *Apium* UC-T3 Somaclone (vegetable); source: Heath-Pagliuso, Pullman & Rappaport, 1989.

Not all cases given here, again, represent convincing evidence for the occurrence of 'real somaclonal cultivars'. Cv. Lincoln Logan of *Rubus* undoubtedly refers to uncovering of a pre-existing periclinal chimeric cultivar. The example of *Apium graveolens* (celery) certainly does not represent a new cultivar, but only a 'selection' or a line, as was mentioned also in a review about 'somaclonal variation' for disease resistance by van den Bulk (1991). Van den Bulk also referred to a tomato variety of 'somaclonal origin', to which was referred in a previous section: cv. DNAP-17 with dominant monogenic *Fusarium* wilt race 2 resistance, originally reported by Evans (1989). This last author, employed by the biotech-company DNAP in Cinnaminson, New Jersey, USA, also reported another example: tomato cv. DNAP-9, with a high content of 'solids', and a cultivar of 'somaclonal origin' for bell pepper or paprika (*Capsicum annuum*) with very few seeds. It must be remarked, however, that these and some other positive results, claimed by Evans, did not show up again in later overviews on 'somaclonal cultivars'. Also Mutschler (1990), in a review on the use of biotechnology in tomato, did not refer to Evans's work. A possible explanation may be that the mentioned 'somaclonal cultivars' still were 'selections' or lines and had not yet reached the stage of an officially registered cultivar. Other explanations may be that the breeder of a new 'somaclonal cultivar' decides that this newly registered cultivar, apart from the improved trait, still carries too many negative traits to be able to compete adequately with other already existing cultivars. It is also possible that a good 'mutant line', or 'somaclonal line', was purchased by a commercial company which does not want to disclose the origin, fearing actions by radical environmentalists.

Karp (1991), in a review on 'somaclonal variation', also mentioned the example of *Paulownia* given by

Skirvin *et al.* (1994), and added that 'new crunchier varieties of celery and carrot have also been produced from tissue culture and are currently sold as 'Vegisnax'. Karp refers in this respect to a short report by Springen in '*Newsweek*' **109** of 26 January, 1987, p. 3, which is rather a commercial advertisement instead of a reliable source to substantiate scientific achievements. The example of celery probably refers to the same breeding line mentioned before. In a later review, Karp (1995) repeated the previous claims about *Paulownia tomentosa* and celery and added, without giving further details, tomato, sugarcane and *Sorghum* (with a reference to Duncan *et al.*, 1995).

A report about a 'somaclonal cultivar' not mentioned in the aforementioned reviews refers to cv. Andro of linseed or flax (*Linum usitatissimum*), officially registered in Canada in 1988 (Rowland, McHughen & Bhatty, 1989). This 'tissue-culture-derived cultivar' arose as a line from the only colony of callus cells derived from a single seed of cv. McGregor, which survived a highly saline culture medium. The cultivar was described as moderate yielding and early maturing, immune to all North American races of rust caused by *Melampsora lini* and moderately susceptible to wilt (*Fusarium oxysporum* f.sp. *lini*). The authors felt that cv. Andro would replace the rust susceptible cv. Noralta. Rowland *et al.* (1995) report that cv. Andro is best suited to the northern growing areas of Saskatchewan, but details about the acreage grown are not given. In the same publication additional work on 'somaclonal variation' in linseed is mentioned. A large range of variation was found in the 'somaclonal lines' of in particular cv. McGregor; this range was reported to be much larger than found in the breeder lines. Heritabilities for the 'somaclonal lines' varied from zero for yield to 0.37 for seed weight and 0.43 for oil content. A large search for novel fatty acid profiles was not successful. As far as is known, since 1988 no new 'somaclonal cultivars' for linseed have been reported.

Kleese (1993) reported on mutants for high-tryptophane content in seeds of maize (*Zea mays*), which had been patented as U.S. Patent 4 581 847. In order to detect such mutants, a tryptophane analogue (5-methyl tryptophane) was used as a selection agent in the growth medium.

At the 17th Symposium of EUCARPIA on creating genetic variation in ornamentals in San Remo, Italy, in 1993, Buiatti & Gimelli (1993) presented a table with a number of commercial cultivars which arose in twelve different species of ornamental crops after *in vitro* culture, with and without treatments with mutagenic agents. This list includes examples of new cultivars for seven species of ornamentals: chrysanthemum (1x), daylily (1x), *Ficus benjamina* (1x), marguerite (1x), orchids (1x), *Pelargonium* sp. (1x) and *Ranunculus* sp. (4x), which apparently arose without the purposeful use of mutagens. Although the original references are mentioned, it is sometimes difficult to check the exact background of the mentioned 'somaclonal mutants'.

Smith, Duncan & Bhaskaran (1993), presented a table concerning mutant cell lines of different origin resulting either from specific *in vitro* screening procedures or from 'somaclonal variation', possessing resistance or tolerance to various diseases and stresses, but admitted that most plant material generated in the aforementioned way was not utilized in crop improvement programmes. According to these authors the main reasons for this were, that the selected resistances were not stable or that other agronomic traits had been negatively changed.

Arihara *et al.* (1995) reported on the release of 'somaclonal variant White Baron' of potato (*Solanum tuberosum*) in Japan. This new cultivar (?) was obtained after protoplast propagation of cv. Danshakuimo, the predominant potato cultivar of that country. The Japanese name in fact refers to cv. Irish Cobbler, an old cultivar originating from the USA and introduced to Japan in 1907. Finally, Day (1993) stated that the work at the Plant Breeding Institute (PBI) in Cambridge, UK, on 'somaclonal variation' in potato (*Solanum tuberosum*) had been unfruitful as a source of stable and useful variants.

In most if not all of the aforementioned cases there is no definite proof that the obtained breeding products arose as a result of the *in vitro* phase and whether really commercialized cultivars were obtained. Because of lack of details it is also difficult to determine the exact story of a tomato cultivar of somaclonal origin, called 'VineSweet' which apparently was released in 1993 by the US biotech-company DNAP.

At the aforementioned symposium of EUCARPIA in San Remo, Buiatti & Gimelli (1993) evaluated the significance of 'somaclonal variation' for breeding of ornamentals and came to a somewhat more optimistic conclusion. They reported that, at that time, a number of private breeding companies already did introduce cultivars of 'somaclonal origin' on a routine base and do not consider the origin of such cultivars as a novelty anymore. If this statement should be correct, the situation for such 'somaclonally arisen mutants' would not differ from that for spontaneous mutant cultivars that frequently arise '*in vivo*', in particular in vegetatively propagated crops. Whether the number of mutant cultivars that arise from both systems differs, is difficult to check. When at this stage – about fifteen years after

'somaclonal variation' got the general attention of biotechnologists – the number and economic importance of 'somaclonal cultivars' is taken as a measure for the success of this approach, it must be concluded that results are very disappointing, at least for those who expected so much from this 'method'. A positive exception, for instance, may be cv. Andro, obtained in linseed (*Linum usitatissimum*) but, as was said before, economic data are not available (Rowland *et al.*, 1989; 1995). Even if some other successful examples should have escaped our attention, it appears that only very few cultivars of 'somaclonal origin' have been officially registered and released and, so far, none of these has become a cultivar of substantial economic importance. In addition, the prediction that by applying 'somaclonal variation' new cultivars can be produced much faster than by other breeding methods, has not been fulfilled.

On the other hand, given the significant role of *in vitro* techniques in plant breeding nowadays, new and sometimes useful genetic variation will continue to be produced as a result of the *in vitro* phase. Thus, there is also no doubt that in future new 'somaclonal cultivars' will be submitted at regular times for registration, in particular when ornamentals and some other vegetatively propagated crops are involved. As *in vitro* culture, predominantly, also implies vegetative propagation, this situation appears to be fully in line with the common occurrence of new spontaneously arisen 'sports' and the – nowadays common – induction of mutants in vegetatively propagated plants. Such sports and mutants have been successfully incorporated for many years in breeding programmes which, for this category of crops, sometimes may result in cultivars which can be released within two or three years of their discovery. We will further discuss this subject in Chapter 7. If the contribution of 'somaclonal variation' is also considered in this way, its role for breeding new cultivars has been brought back to a much more realistic perspective than before.

5.5.7. Concluding remarks on 'somaclonal variation'

It has been explained before that plant breeders should be interested at any time in ways to obtain new and better genes and gene combinations for important traits. 'Somaclonal variation' in this respect would be of considerable interest if better genes could be obtained easier or faster than by other methods or if desired genetic variation could be obtained for traits in which other breeding methods, so far, have not been successful.

In the previous sections it has been tried to characterize the phenomena of 'somaclonal variation', to indicate possibilities and limitations and to discuss results obtained so far. The most often claimed advantages of somaclonal variants are summarized here once again.

1. The high frequency with which such variants should occur.
2. The possibility to obtain variation for agronomically useful traits.
3. The expectation that new cultivars may be obtained faster than by other breeding methods.
4. The expectation that selection for useful traits is possible at the cell level.

It has been shown in the previous sections that in this way indeed variation for agronomically useful traits can be obtained, but that there is much doubt as to the other claims that have been made.

The possibility of using single cells as the unit for manipulation and very large numbers of single cells for selection, of course, has a lot of attraction but appears not to be feasible for all plant species and not for most traits of agronomic importance. A relative advantage of 'somaclonal variation' at least in comparison to the variation obtained by (other) methods of cell and molecular genetics, is that the general public aversion to most of such methods has not been extended towards cultivars originated from 'somaclonal variation'. As a consequence, such cultivars, as opposed to cultivars that have resulted from 'genetic manipulation', until now can be introduced without problems to the market.

Nowadays, a rather extensive list of potential complications and disadvantages of applying 'somaclonal variation' for plant breeding work can be compiled. The most important of these draw-backs – some in common with mutation breeding – are as follows.

1. The variation is completely unpredictable and uncontrollable.
2. The observed variation is predominantly negative since most of it goes opposite to the direction of previous selection.
3. Different species and cultivars do not show the same amount and type of variation.
4. Much variation observed is not genetically determined.
5. Only a limited part of the genetic variation is able to pass the meiotic sieve.
6. Variation does not occur for all traits.
7. Variation may involve more traits than one, and therefore can be used only via extensive cross-breeding, which may be very troublesome or even

not possible in most vegetatively propagated plants.
8. *In vitro* selection is not possible for most traits of agronomic importance.
9. Regeneration of genotypes possessing a desired genetic change may be difficult.
10. The resulting progeny may remain unstable when selfing or crossing takes place after the *in vitro* phase, may be as a result of increased transposon activity.
11. Somaclonal variation is a nuisance for gene banks and multiplication firms, but also limits the process of success for gene technology to develop good cultivars.

The first disadvantage mentioned, the unpredictable and uncontrollable character of the variation, is the most important disadvantage of 'somaclonal variation'. The same phenomena are often attributed to induced mutations, but the variation obtained there – after some experience with the plant material and treatment conditions – can be predicted at least to a certain extent, e.g. how many mutations may be expected for a given trait. Repeated experiments under similar conditions will produce comparable results, which is much less the case for 'somaclonal variation'. Disadvantage 2 also refers to mutation breeding, whereas the impression exists that, with respect to point 3, different cultivars show less variation in mutation breeding than in case of 'somaclonal variation'. This may be explained by the fact that the part of 'somaclonal variation' that is non-genetic is larger than is the case after mutation breeding (point 4). As to point 5 it may be observed that 'somaclonal variation' shows a certain similarity with mutation breeding for vegetatively propagated crops and less with seed propagated crops where all mutations must pass the meiotic sieve before they can even be observed. As a consequence, most large chromosomal deletions, etc., will not even reach the next generation. With respect to point 6, 'somaclonal variation' probably may offer somewhat better prospects than mutation breeding as changes towards dominant expression of genes seem to occur more often than after mutation breeding. Disadvantage 7 refers to the general finding that too much variation occurs under *in vitro* conditions, whereas breeders prefer to have only one genetic change for a given trait and leave the remaining non-target genes unaffected. When *in vitro cultures* are started from stable structures like shoot apices, the observed variation indeed appears to be much lower. Therefore, mutation breeding *in vitro*, starting from shoot apices, axillary buds, etc., may be more attractive than from protoplasts, etc. After the use of shoot apices and axillary buds, *in vitro* (pre-)selection may be possible for more traits than when protoplasts are involved (see point 8) and regeneration may be much easier as well (point 9). The use of only a few culture cycles may lead to less 'somaclonal variation', but it is still doubtful whether this is always the case. Singh (1993) observed that cells in ageing cell cultures may show high frequencies of aberrations, and gradually lose their capacity to regenerate plants. Transposon activation (point 10) does not play an important role in mutation breeding, at least not yet, and mutation frequencies in nature, in particular at low temperatures, are too low to be a nuisance in gene banks (point 11).

An important but often neglected, practical consideration is that after an *in vitro* phase in which 'somaclonal variation' has occurred, in most cases, all consecutive steps of a common breeding programme have to be taken. Selection for agronomically important traits often cannot be performed on single plants in the segregating generation (sometimes indicated as SC1), for instance when traits with a low heritability are involved. As a result, not much is left of the often predicted gain of time, when making use of the somaclonal variation method in comparison to conventional breeding work.

As a point of special importance it should be mentioned that use is made of *in vitro* methods also in practically all techniques of genetic engineering, like cell transformation and protoplast fusion. Undesired and unpredictable variation must be avoided in such experiments by all means. It seems that by making use of 'classical' mutation induction methods it would be easier to provoke genetic variation when wanted and to avoid or limit the occurrence of mutations when not desired.

When considering all aspects mentioned above, one wonders again why so much was expected from 'somaclonal variation', even by persons who should have known better. Skirvin *et al.* (1994), in this respect, concluded that 'somaclonal variation' will only become more useful for breeding if researchers learn to control and direct this phenomenon. There are, in our opinion, indeed good reasons to believe that 'somaclonal variation' only at rare occasions will be a practicable or better alternative to other sources of genetic variation, e.g. in plants with a narrow genetic base and limited possibilities to induce new variation by crossing (e.g. apomicts). In such rare cases where 'somaclonal variation' should be used it is absolutely imperative to work only with the most superior genotypes. Easy regeneration should be possible. A quick and reliable selection procedure for the desired trait should be available. Selfing and crossing should be possible to take care of changes in non-target genes.

5.6. Artificial induction of mutations and selection of mutants *in vitro*

5.6.1. General considerations

In situations where classical methods to obtain the desired genetic variation have not been successful or even are not available, one should consider whether induced mutations and/or methods within the realm of modern biotechnology could provide the desired genetic variation.

With the advent of *in vitro* techniques the interest in the use of *in vitro* plant material for mutation breeding started. Combination of mutagenic treatment and *in vitro* techniques is possible now for many seed propagated as well as vegetatively propagated crops. In particular for the latter group of crops the *in vitro* approach has certain advantages and may be an effective method – or, in the opinion of some – even the only effective method available (Novak, 1991a; Maluszynski *et al.*, 1995).

Sometimes the size of the common vegetative propagules (suckers, cuttings, rhizomes, etc.) makes it difficult if not impossible to handle the large numbers required in mutation programmes. Under such conditions the use of *in vitro* methods may be very attractive, as in this way often large numbers can be handled in a limited space within a short span of time. In addition, the use of *in vitro* methods facilitates keeping the plant material free of diseases and, also depending on the type of *in vitro* culture, may sometimes help in breaking up the chimerism that commonly occurs when multicellular organisms are subjected to mutagenic treatments. On the other hand, it cannot be said often enough that the *in vitro* phase of a breeding programme, without exception, must be followed by a number of decisive and often time-consuming steps 'in the field'.

The stages in a plant breeding programme at which *in vitro* mutagenesis (and 'somaclonal variation') and *in vitro* selection can be utilized, have been nicely outlined by Novak (1991a, p. 328) and are shown in Fig. 5.1.

Mutations *in vitro*, in fact, may occur under the following circumstances: 1) in organized structures (like shoot apex cultures) without mutagenic treatment (i.e., 'spontaneous'), 2) in organized structures, 'spontaneously' + mutagen induced, 3) 'somaclonal variation' resulting from disturbed cell physiology, and 4) combination of (3) 'somaclonal variation' + (2) mutagen induced.

After a reliable procedure for culturing and regeneration has been established – which subject will be discussed in a later section – according to Novak (1991a), the following three preparatory steps must be taken.

1. The establishment of a dose–response curve from measuring growth *in vitro*, using for instance dry weight, volume, cell number or plating efficiency. Such curves are established by dividing the average value for a given treatment (minus the initial value) by the average value for the control (minus the initial value) × 100. Similarly, dose–response curves for survival rate (or viability) and regeneration ability should be developed. This leads to a decision on the (supposed) optimal mutagenic treatment.
2. A suitable experimental design be set up for tissue culture by which chimerism can be reduced or even avoided.
3. *In vitro* screening for desired mutants be set up when possible. Otherwise one has to plan that numerous plants must be regenerated and then screened in the field or in the greenhouse. For this, the appropriate tests for screening and verification must be prepared too.

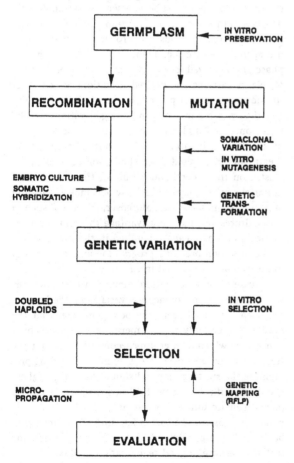

Figure 5.1. Application of biotechnology in the different stages of plant breeding (after Novak, 1991a).

This short description includes only the preparatory steps. The actual mutagenic treatment and subsequent handling of populations are not mentioned. These subjects will be discussed in Section 5.6.3.

If we return now to the steps proposed by Novak (1991a), step (1) has been briefly discussed already in general in Chapter 4 and some additional comments will be made at later occasions. With respect to the subject of step (2): why and how to limit or avoid chimerism, we may refer at this point to a detailed discussion of this subject in Chapter 7. An exception is made here for an interesting publication by Conger, Trigiano & Gray (1986), who developed a system of somatic embryogenesis for dealing with chimerism in *in vitro* mutagenesis of cereals and grasses, using cocksfoot (*Dactylis glomerata*) as a model.

Finally, a number of considerations in relation to the subject mentioned under step (3): *in vitro* screening and plant selection, will be discussed more in detail in Section 5.6.4.

How to assess results of mutagenic treatments in vitro

For an assessment of the results of mutagenic treatments *in vitro*, one should take for comparison exactly the same procedure excluding the mutagen, which will reveal spontaneous mutations as well as any 'somaclonal variation' that may occur. More difficult to compare are the results of a conventional mutation breeding programme in which for instance seeds or cuttings are treated. In any case it is useful to indicate what the expression '*in vitro* mutation breeding' implies in a specific situation. Strictly speaking, when the stage at which the mutagen is applied is considered the determining factor, only treatments <u>during</u> the *in vitro* phase, including treatment of explants at their initiation, should be coined as such. More often, however, any combination of a mutagenic treatment and a culturing phase is considered as such; for instance when only selection is performed *in vitro*, or even when *in vitro* work is limited to propagation of the mutagenically treated material after selection was performed already before the *in vitro* phase (see for instance Novak et al., 1986a).

Induced mutations versus 'somaclonal variation'

When deciding whether *in vitro* cultures should be submitted to mutagenic treatments, three different aspects must be considered: a) the mutation frequencies that are wanted, either in total or for specific traits, b) the spectrum of variation that can be obtained, and c) additional effects that may result from the mutagenic treatment (e.g. reduced regeneration ability).

Spontaneous mutation frequencies – e.g. between 10^{-6}–10^{-8} per generation on a cell basis – in combination with sometimes very large populations in which selection can be performed, seems to make it attractive to look for spontaneous mutants only. However, as has been pointed out before, much higher levels of 'spontaneous variation' are observed in particular in undifferentiated systems like callus or suspension cultures and, consequently, it may be anticipated that high frequencies of multiple mutations including unwanted changes for non-target traits arise. To get rid of this unwanted variation extensive recombination by crossing and selfing would be required, which would make this approach unsuitable for most vegetatively propagated crops.

There are, however, exceptions to the rule that undifferentiated systems are always unstable and produce much variation. Schieder (1976), for instance, reported on the irradiation of haploid protoplasts of *Datura innoxia* with doses from 2.5 to 15 Gy of X-rays (LD 50 = 10 Gy). From these protoplasts 2.5×10^4 calli were obtained, in which about 3×10^{-4} chlorophyll mutations were found, whereas according to this publication no chlorophyll mutations were found in a control population consisting of 105 calli from unirradiated haploids.

On the other hand, Crinò *et al.* (1990) reported for tomato – a self-pollinating crop – that the mutation spectrum derived from gamma irradiation of seeds, followed by *in vitro* culture of cotyledons did not differ significantly from the unirradiated cultured cotyledons, except that the plants derived from the irradiated material did show fewer chlorophyll mutations and less sterility. The authors suggested that such unwanted mutations are related to severe chromosome aberrations which, in case of gamma ray treatment, may be eliminated.

Other draw-backs of 'somaclonal variation', like the induction of relatively many non-genetic changes, epigenetic effects, unstable transposon-induced changes, etc., should be taken into account as well.

In only a few cases have reliable large-scale experiments been performed in which the effects of mutagenic treatments in combination with *in vitro* methods and 'somaclonal variation' have been compared. A positive exception in this respect is the work of Novak *et al.* (1986a,b; 1988) who compared effects of 'somaclonal variation' and variability induced by treatment of immature embryos of maize (*Zea mays*) with 5 or 10 Gy of gamma rays (dose rate 8.2 Gy. min^{-1}). Treated and untreated embryos were excised at the same time immediately after mutagenic treatment and cultured *in vitro*. The resulting plants are chimeric and, after selfing, the

next generation is made up of genetically different but heterozygous plants in which only dominant mutations can be recognized. After a second selfing the resulting plant material is segregating for dominant and recessive mutations.

When comparing these two groups of plants, a similar spectrum of chlorophyll and morphological deviants was observed for radiation induced and somaclonal variation. As these results refer to only one experiment with one (rather special) crop, limited numbers of plants and a limited number of traits studied, it is not justified to draw general conclusions about mutation spectrum and mutation frequencies obtained after mutagenic treatment in combination with *in vitro* culture and *in vitro* culture only.

5.6.2. Starting material and regeneration after the *in vitro* phase

In order to select the most suitable starting material for *in vitro* mutation breeding, the merits and disadvantages of different options must be assessed. Two examples will be given in this respect. If, for instance, small shoot cuttings with axillary buds are properly treated with a mutagenic agent, either before or during the *in vitro* phase, this may well result in the production of chimeric M_1 plants carrying the desired mutations which, after the normal selection procedures, most likely will give genetically stable mutants in later generations for the trait concerned and may show little or no variation in the first generation(s) of vegetative propagation. Treatments of haploid microspores (or haploid protoplasts), on the other hand, result in the induction of mutations at the haploid level, which makes selection of mutations much easier. Moreover, the use of such *in vitro* cultures also has the advantage that much higher numbers of 'units' can be handled in the initial stages of the mutation breeding programme and, thus, many mutants may be obtained in this way.

In case of mutagenic treatments *in vitro* (as well as when focussing on 'somaclonal variation'), the breeder should always start from the best genotypes that are available, preferably from superior and well-accepted cultivars in which further improvement for one or a few traits only would be desirable. Like in any breeding programme the starting material should meet the highest standards with respect to health, germination power, purity, genetic uniformity, etc. (see also Chapter 1). It makes no sense to use particular cultivars or lines, just because they are outstanding for *in vitro* culture, for instance because of good regeneration quality.

Another important, but often neglected, aspect concerns regeneration of plants <u>after</u> the *in vitro* phase.

When searching for 'somaclonal variation' often – and preferably – use is made of callus cultures, suspension cultures and the like. In practice, regeneration of vigorous and normal plants from such cultures, irrespective of the cultivar or genotype involved, is often much more difficult than when starting from relatively stable structures like (mutagenically treated) explants with existing apical or axillary buds. The latter, on the other hand, may have the drawback of showing significant chimerism. This subject will be discussed in detail in Chapter 7.

Haploids

A subject of interest in relation to *in vitro* culture and to mutagenesis, is the use of haploid plants. Various ways to obtain haploids and their value for breeding were briefly discussed in Chapter 3. For cereals, for instance, anther and microspore culture are the most applied methods. A disadvantage of anther culture is that both haploids and diploids may be produced, as plants may regenerate not only from microspores but also from the somatic cells of the anther tissues and, consequently, further screening for haploids is required. Pollination with irradiated pollen may also lead to haploids through development of unfertilized egg cells.

It may be remembered that especially selection of recessive mutations at the haploid level may have considerable advantages but that the final 'products', the crops that are grown in the field or the greenhouse, should contain again the normal somatic number of chromosomes, like diploid barley (*Hordeum vulgare*), with the chromosome formula $2n = 2x = 14$, or tetraploid potato (*Solanum tuberosum*), with $2n = 4x = 48$ chromosomes. In Chapter 7, for instance, an *in vitro* mutation breeding programme for potato, based on monohaploid ($2n = x = 12$) starting material, followed by the (time-consuming) return to the tetraploid level, for the production of so-called amylose-free potato cultivars will be discussed.

Another example refers to tobacco (*Nicotiana tabacum*), a functional diploid or allotetraploid crop [$2n = 2.(12 + 12) = 48$] which is highly susceptible to potato virus Y. In the past, irradiation treatment had resulted in a monogenically inherited, recessive mutant of cv. Virgin which was resistant to the highly necrotic strain NN of the virus. Witherspoon *et al.* (1991) inoculated 554 anther-derived haploids, obtained from cv. McNair 944 with the aforementioned strain NN to select for 'gametoclonal variants'. One haploid plant was found which showed a high level of resistance to strain NN and to three other strains of ten strains tested. Chromosome doubling and further crossing demonstrated the resis-

tance to be controlled by a single gene exhibiting incomplete dominance. The mutant is further used for breeding resistant cultivars.

References to a number of useful publications about the use of 'doubled haploids' in mutation breeding can be found, for instance, in Szarejko *et al.* (1991).

Culturing and regeneration

The prerequisite for *in vitro* mutagenesis programmes is to set up an effective, preferably cultivar-independent protocol for culturing and regeneration and the same, of course, applies to biotechnological techniques involving *in vitro* culture, such as transformation methods and the antisense method (see also Chapter 1).

It is imperative to control as much as possible all conditions that may influence the effects of mutagens on biological tissues. The most important factors are partly identical to those for common mutagenesis starting from seeds, seedlings, explants, etc.: ploidy level, phase within the DNA cycle, dose rate for radiations, and pH, concentration, temperature and duration of treatment for chemical mutagens. Others are specific for *in vitro* experiments: explant type, culture medium and physical culture conditions (including temperature, light and shaking).

The procedures that can be followed for *in vitro* mutagenesis with different types of starting material have been described in detail by Constantin (1984). Because the series 'IAEA-TECDOC' in which this contribution is published is not widely available, the general outlines, provided by Constantin, are summarized in Box 5.2. However, it must be realized that these protocols, in most cases, are incomplete and, therefore, many questions still remain unanswered.

A somewhat different categorization, in which five different systems of *in vitro* culture in combination with a mutagenic treatment are distinguished, is given in '*Plant Tissue Culture Techniques for Mutation Breeding: A Training Manual*', prepared by Novak and co-workers of the Plant Breeding Unit at the IAEA Laboratories in Seibersdorf, Austria (Anon., 1990b). The first group refers to the conventional mutagenic treatment of seeds. From the resulting M_1 plants generative explants for *in vitro* culture (e.g. anthers, pollen) are taken. This approach is applicable for plant species like barley, wheat, rice, various *Brassica* species and tobacco. Segregation of mutants occurs at the haploid level; regenerated gametes represent the M_2 generation. If microspores or anthers are mutagen treated instead of seeds, no chimerism occurs in haploid M_1-plants derived from anther culture as long as the mutagen treatment was finished before culture initiation.

Box 5.2 Protocols for in vitro mutagenesis starting from different types of material (after Constantin, 1984).

1. Adventitious buds from callus

a. Start from vigorously growing, healthy, disease-free plants.

b. Try explants from roots, stems, leaves, etc. or anything that may form callus.

c. Wash plants well with a mild detergent solution, rinse well, surface sterilize with hypochlorite (household bleach), alcohol, etc.

d. Rinse several times with sterile de-ionized water to remove all traces of the sterilizing agent.

e. Prepare explants aseptically and place in appropriate media which support callus induction (various options in literature).

f. Subculture frequently to establish a rapidly growing callus.

g. Apply mutagenic agents (after preliminary experimenting to establish dose–response curves) with various mutagens, concentrations, etc.

h. Subdivide the mutagenized callus (immediately after treatment or after a recovery period) on fresh medium. Apply *in vitro* selection pressure if a resistant auxotrophic or autotrophic mutant is sought. Otherwise, subculture the callus on a shoot-inducing medium.

i. Transfer shoots on rooting medium and complete plants to the greenhouse.

j. Test mutants for their performance for all relevant traits. Micropropagation may rapidly increase the number of genetically identical plants for testing.

2. Cell suspension culture

Follow steps 1a–1f to establish a rapidly growing, friable callus for starting a cell suspension culture.

a. Apply pre-selected mutagenic agent based on pre-established dose–response curves.

b. Following mutagenic treatment, cell cultures are washed and resuspended in fresh medium. When possible, apply selection pressure *in vitro* or wait till plants have been regenerated and can be observed in the greenhouse or field.

c. (*1*) Assuming selection for stress <u>tolerance</u> in cell suspensions, simply wait for some weeks and identify flasks with dividing cells. Those cells are plated on callus induction medium and transferred to shoot promoting medium; later to rooting medium.

(*2*) Assuming that selection pressure is applied at the time of colony formation, mutagenized cells are plated on a medium containing an inhibitory concentration of the selection agent. Tolerant cells will produce a colony from which eventually shoots can be derived.

(*3*) Another alternative of *in vitro* selection is to induce somatic embryogenesis in the cell suspension cultures with or without selection pressure.

d. Once mutant plants are discovered, they can be increased in number and tested for their performance.

3. Axillary bud proliferation

Shoot tip culture for rapid micropropagation via axillary bud proliferation is a standard *in vitro* method for many plant species. A multiplication factor of 10 or more can be realized at 4 to 8 weeks intervals.

a. Establish rapidly growing cultures and subject them to mutagenic treatments, based on dose–response curves. The survival after mutagenic treatment should be at least 50%.

b. Mutagenized shoot tip cultures can be used for several cycles for eliminating chimeras or repeated removal of axillary buds from the same M_1 shoot.

c. Axillary shoots are transferred to rooting medium and later to pots in the greenhouse.

d. Selection in most cases must take place on complete plants under field or greenhouse conditions. Since true axillary buds are of multicellular origin, much chimerism may occur.

4. Anther culture

Such cultures, predominantly, produce haploids.

a. Collect flower buds with anthers at the proper developmental stage, usually uninucleate pollen.

b. Pretreatments, like exposure to cold or centrifugation, may be applied.

c. Expose anthers to irradiation before the flower-bud is opened or to chemicals after anthers have been removed from the flower bud.

d. Plate on 'anther culture medium' until plantlet development occurs, either in the presence of selection pressure or with selection at a later stage.

e. Plants with desired mutant traits are 'doubled', either by treatment with colchicin or oryzaline, or by spontaneous duplication.

In the second category of *in vitro* mutation breeding systems which, according to the aforementioned publication (Anon., 1990b) is mainly applied for vegetatively propagated crops such as potato, cassava, yam, fruit trees and strawberry, use is made of node cuttings and shoot tips as explants. Mutagenic treatment is performed again before the *in vitro* phase, which aims at axillary or shoot tip meristem proliferation. Chimerism is to be expected.

In the third group, explants like petioles, leaf cuttings, shoot tips, bulb scales and segments of immature cotyledons of a wide range of crops, are subjected to mutagenic treatments before culturing. This treatment is followed ideally by direct organogenesis through adventitious shoot formation. In many cases chimerism may occur.

In category four, mutagenic treatment of explants like in group three is performed, followed by the initiation of callus, cell suspensions and protoplast cultures. Hypocotyls, pedicels, leaf segments, young zygotic embryos, nucellar tissue, etc., are used as starting material for such *in vitro* cultures. Subsequently, (indirect) organogenesis or somatic embryogenesis takes place. 'Somaclonal variation' is likely to occur in addition to mutagen induced variation. A positive feature may be here that *in vitro* selection is possible in some situations.

The fifth group mentioned in the IAEA 'Training Manual', refers to mutagenic treatments of already established *in vitro* cultures, followed by various ways of plant regeneration. Like in group four, it is, of course, crucial to have good *in vitro* regeneration. The treatment may technically be more cumbersome than treatment before culturing. The effect on the medium cannot be neglected. Therefore the medium has to be changed completely after the treatment. For this category of treatments *in vitro* selection may follow *in vitro* mutagenesis. If useful 'somaclonal variation' occurs, this may add to the mutagen-induced variation.

In all five categories mutagenic treatments may change the growth characteristics, cause delay of mitoses, reduce callus growth, as well as reduce the regeneration ability.

The doses, dose rates, concentrations of chemical mutagens, etc., indicated in the 'Training Manual' for the different systems described here, are mentioned in the next section.

5.6.3.   Mutagenic treatments

General remarks

Treatment of *in vitro* tissues with physical or chemical mutagens, according to a range of publications, may considerably increase the frequency of 'spontaneous' genetic variation which occurs *in vitro* and may also result in a different mutation spectrum (Novak *et al.*, 1986a,b; 1988; Cheng *et al.*, 1990). It appears that there is no general answer to the question of which kind of starting material and which method of *in vitro* mutagenesis should be preferred, as the species involved, the available methods of vegetative propagation, breeding objectives, crossing possibilities, selection methods, etc., play an important role in this.

Another aspect to be considered is the convenience of handling. Nickell & Heinz (1973), for instance, mentioned their preference for chemical mutagens because they considered chemicals much easier to handle in the liquid media in which the treated cells are grown and, in addition, such treatments do not require the special equipment needed for radiation treatments. But the preference of these authors is not shared by everybody.

A useful list with data concerning *in vitro* mutagenic treatments and numbers of frequencies of resulting mutants for a number of crop species has been compiled by Mathews & Bhatia (1983). The authors present data for mutagenic treatment of cells and protoplasts of tobacco (*Nicotiana tabacum*), potato (*Solanum tuberosum*), tomato (*Lycopersicon esculentum*), carrot (*Daucus carota*), soybean (*Glycine max*), rape (*Brassica napus*), sugarcane (*Saccharum officinarum*), maize (*Zea mays*) and rice (*Oryza sativa*). A list of 44 references is added. Another useful general contribution on *in vitro* mutagenesis has been made by Negrutiu (1990).

In practice it appears that the choice of mutagenic agents used for treatments *in vitro* does not differ much from that for the conventional treatments of seeds, seedlings, shoots with apical and axillary buds, etc. In most cases treatments are performed with X-rays or gamma rays, UV, EMS or nitroso-compounds. For a number of reasons irradiation treatments may be preferable, because application is fast and, in contrast to chemical mutagens, there is no risk that residues remain in the medium. In addition, most chemicals must be handled with much care as they may cause health hazards, their penetration in multicellular systems is more difficult and the reproducibility of the experiments may be less. When radiation types with low penetration power (in particular UV) are applied, open Petri dishes or vessels must be used! This, however, affects the sterile condition inside the vessel. Hard X-rays and highly energetic gamma rays can be applied on closed vessels. Further particulars of these agents can be found in Chapter 3. It should be further remembered that mutation induction, predominantly, is a random process and the choice of the mutagen, together with other factors, only to a very small extent may influence the spectrum of the obtainable variation. This subject has been discussed extensively in Chapter 4.

In addition to their genetic effect, different mutagens may exert various somatic effects on the *in vitro* culture, for instance influencing regeneration. Whereas higher doses of irradiation and chemical mutagen may have inhibiting effects, lower doses of irradiation may be stimulatory. Sangwan & Sangwan (1986), for instance, reported on the stimulatory effect of treatments with up to 10 Gy of gamma rays on somatic embryogenesis in carrot (*Daucus carota*) and on androgenesis in tobacco (*Nicotiana tabacum*) and *Datura*. Doses of 20 Gy and higher inhibited *in vitro* embryogenesis. One should therefore consider very carefully whether mutagenic agents should be applied and up to which dose or concentration.

Constantin (1984) has pointed out that it is safe to assume that increasing the dose of mutagen also leads to an increased amount of genetic damage, up to a point that it should be impossible to preclude the isolation of a desirable mutant trait without concomitant deleterious ones.

In Section 5.6.2 reference was made to various groups of *in vitro* plant systems in combination with mutagenic treatments before or during the *in vitro* phase. In the aforementioned IAEA 'Training Manual' (Anon., 1990b) for each system some information was given about doses and dose rates commonly applied in case of irradiation treatments, and about concentrations and treatment times for chemical mutagens.

The preferable dose, dose rate or concentration, treatment time, etc., of course, should be determined by preliminary tests which, as much as possible, should be performed with the same plant material and under conditions that resemble as much as possible the situation in the final experiment.

Two main groups of treatment procedures can be distinguished: treatment before *in vitro* treatment, and treatment when the *in vitro* culture has already been established. Some general information on doses and dose rates in case of irradiation, and on concentrations and treatment times when chemical mutagens are used, is presented in Box 5.3.

According to the IAEA report on '*In Vitro Technology for Mutation Breeding*', covering two consultants' meetings on *in vitro* technology for mutation breeding (Anon., 1986a), it would be preferable to perform mutagenic treatments before culturing, except in order to avoid chimerism to treat single cells during culturing, if such single-cell methods are available. However, as is shown by Deane, Fuller & Dix (1995; see also Box 5.5), treatment *in vitro* is also feasible for small multicellular explants.

The ways of handling different categories of *in vitro* starting material have been summarized in Box 5.2. In addition, it can be said that each category has its own merits. It is, for instance, the advantage of suspension cultures that such cultures enable starting from a homogenous group of cells, which even could be synchronized, whereas an advantage of obtaining mutations in haploid cells is that, after diploidization, homozygous mutants may be obtained.

Some parameters

When results of mutagenic treatments are presented and when the number or frequency of mutations in tissue cultures are determined, different methods are followed and, therefore, it is very difficult to compare results of experiments. Two expressions (see also Chapter 1, Section 1.3.5) are often used and, in many cases, confused: the **mutation rate** and the **mutation frequency**. The **mutation rate**, the frequency of (detected) mutational events per unit (genome) and per generation, from a practical point of view, is difficult to determine because the term 'generation' is dubious in this context. This subject, including the application of the so-called fluctuation test to calculate the mutation rate (Luria & Delbrück, 1943), was briefly introduced in Chapter 1. Details are outside the scope of this book but can be found in most textbooks on genetics, for instance in Suzuki *et al.* (1989).

Results of mutagenic treatments of seeds, whole plants, etc., are often presented as **mutation frequencies**: the frequency at which mutational events (all mutations or specific ones) are detected in a population of cells or individuals. Various systems are applied. For instance, the mutation frequency after mutagenic treatment of diploid seeds could be determined on the basis of the number of detectable dominant mutation events in the M_1 generation, either as a percentage of independently mutated cells or as a percentage of the number of plants studied. For diploids, recessive detectable mutations can be observed in M_2 either as a percentage of segregating/non-segregating plants or spike progenies, or as a percentage of mutated/non-mutated plants.

In the case of *in vitro* treatments, figures concerning mutation frequencies, for instance, may refer to the number of mutational events detectable in haploid cells in M_1, or detectable in regenerated plants. When regenerated plants are considered one could, for instance, determine the percentage of mutated/non-mutated regenerated plants, the percentage of calli producing mutated regenerants, or the number of mutated regenerants per (approx.) 1000 cultured cells. All these methods produce approximate results, which are quite

---

Box 5.3 Mutagenic treatments in combination with in vitro culture

1. In case of seed treatment <u>before</u> <u>in vitro</u> culture, the standard procedures for mutagenic treatments of seeds can be followed. Details can be found in Chapters 4 and 6. When explants are treated before the *in vitro* phase, irrespective of whether subsequent regeneration takes place via already existing shoot tips or axillary buds, by adventitious shoot formation, or via callus from cell suspensions or protoplast cultures, the same mutagenic treatments are advised. Irradiation can be performed within a wide range of doses and dose rates of X-rays or gamma irradiation, e.g. 10–100 Gy at dose rates of 1–20 Gy. min^{-1} depending upon the species. For treatments with chemical mutagens also a wide range of concentrations and different treatment durations are mentioned. For EMS, for instance, treatments with 0.1–1.0% for 2–6 h are suggested; for nitroso guanidine (NMMG or NG) 5–20 mg. l^{-1} for 0.5–4 h; for nitrosoethyl urea and nitrosomethyl urea 0.1–1 mM for 6–24 h.

2. When mutagenic treatments are given to already established *in vitro* cultures, somewhat different doses are mentioned in the IAEA '*Training Manual*' (Anon. 1990b). For treatments with X-rays and gamma rays doses may vary from 20–2000 (!) Gy, whereas dose rates range from 5–100 Gy. min^{-1}. UV treatments can be performed as well (for more details see also Chapter 4). For chemical treatments concentrations of 0.1–1.5% EMS applied for 2–8 h at pH 7 are mentioned. For NMMG treatments doses range from 5–30 mg. l^{-1} applied for 0.5–6 h at pH 6, and for nitrosoethyl urea and nitrosomethyl urea doses of 0.5–5 mM for 6–72 h at pH 6 are given. DMSO (2–4%) may be added as a carrier. Higher concentrations of chemicals and higher doses of radiation may negatively affect the regeneration ability. Low doses of radiation sometimes stimulate callus and cell proliferation.

---

acceptable if detailed searches for mutations are performed, but poor if only 'visible' or somehow obvious mutants are counted. Exact frequencies can <u>never</u> be determined if figures depend on human detection in phenotypes.

Different parameters can also be used to assess the **radiosensitivity**, e.g. of a specific kind of plant material (for details see Chapter 4, Box 4.2.). For practical purposes, an easy parameter that can be observed at an early stage is required. In the case of seed treatments, germination figures, growth retardation, inhibition of shoot growth, etc., are useful parameters. For *in vitro* mutagenesis, such parameters can be very different and are less easy to find.

Douglas (1986) exposed stem explants of poplar (*Populus* sp.) to gamma rays. Cuttings of about 6 mm in length, which had been derived from actively growing shoots, were irradiated with doses up to about 30 Gy within 24 h before being cultured on an MS medium. Formation of adventitious buds, common in poplar, decreased with increasing dose. At 30 Gy of gamma rays, which was the dose at which later on the highest figure of mutants per cultured explants were observed, bud formation showed a considerable inhibition (i.e. from 7.3 normal + mutant adventitious shoots per explant in the control series to 0.6 in the 30 Gy series). MacDonald *et al.* (1991; see also Box 5.4), who studied *in vitro* mutagenesis on (haploid) microspores of *Brassica napus*, found that the frequency of cell division gave a good and early picture of the number of embryoids produced later on from microspores. For a comparison of effects caused by different mutagenic treatment, e.g. with X-rays, gamma rays and UV light, the relative yield of embryoids, and the relative number of plant regeneration were used. Such data, of course, give only an indication of mutagen effectiveness and do not provide information about expected mutation frequencies, but become more useful if it is possible to determine, for instance, at which rate of survival and regeneration the optimal number of induced mutations can be expected.

In Chapter 4 the most important advantages and disadvantages of physical and chemical mutagens have been mentioned already and at this place only some special features, of relevance for *in vitro* mutagenesis, will be summarized.

Physical mutagens

Treatments with ionizing radiation and UV are relatively easy to perform and, in general, show a high degree of reproducibility. Most ionizing radiations easily penetrate multicellular tissues as well as tissue culture vessels.

Based on recommendations, formulated by the IAEA (Anon., 1986a) during two special meetings on *in vitro* mutagenesis, Novak (1991a) mentions the following general points to be kept in mind when ionizing radiation and UV are used.

Box 5.4 Examples of procedures with physical mutagens for various types of starting material in *in vitro* mutation breeding programmes

1. Roest & Bokelmann (1980), in a project aiming at the development of *in vitro* adventitious bud techniques for mutation breeding, developed a system to produce adventitious plantlets from explants of rachises, petioles and the upper pair of leaflets of cv. Désirée of potato (*Solanum tuberosum*). This *in vitro* experiment resulted in the production of 3167 plantlets originating from 457 explants with tall shoots ($\geq$ 1 cm). Explants had been irradiated with different doses of X-rays prior to placing them on a culture medium. In all cases callus was produced. The resulting plant material (with unirradiated controls) was further investigated by van Harten, Bouter & Broertjes (1981) for mutation frequency and chimerism in subterranean and aerial parts in three vegetative generations, of which the vM_1 and vM_2 were grown in the greenhouse and the vM_3, consisting of more than 13 000 plants, in the field.

The doses applied, determined after preliminary experiments, were 15–20 Gy of X-rays for explants of rachis and petiole and 22.5–27.5 Gy for leaflet-discs. Starting from 660 explants (including 40 controls), 534 explants produced adventitious

sprouts *in vitro*. These adventitious sprouts were subcultured for rooting. After rooting, the plantlets were transplanted to a soil mixture and transferred to a greenhouse. The time between excision of the explants and transfer to the greenhouse was five to eight months. On average 6.7 adventitious plantlets were obtained per irradiated explant and 10 adventitious plantlets per explant in the controls. The number of fullgrown vM_1 plants obtained from irradiated explants (vM_0) was about 2600; the corresponding number of mature 'groups of plants' in vM_3 was about 2400. Further details about mutation frequencies and rate of chimerism are discussed in Section 5.6.5. of this chapter.

2. Devreux *et al.* (1986) irradiated large populations of protoplasts of *Nicotiana plumbaginifolia* with ^{60}Co gamma rays (treatments in Petri dishes at 5–10 Gy. min^{-1}) and fast neutrons (from a Van de Graaf linear accelerator at 16 MeV and $\pm$ 4–6 x 10^{-2} Gy. min^{-1}). Dose–effect relationships and split-dose effects (e.g., for gamma rays 3 x 2 Gy with long intervals instead of 1 x 6 Gy) were investigated. Results

are discussed in Section 5.6.5. of this chapter.

3. MacDonald *et al.* (1991) developed microspore culture techniques for *Brassica napus* with the aim to produce large numbers of haploid embryoids. Per 1000 microspores plated, 4.5 embryoids were obtained in this way, 90% of which developed into plants. Incubated microspores in sealed Petri dishes were subjected to X-rays at a dose rate of 30.6 Gy. min^{-1} and to ^{60}Co gamma rays at 20.14 Gy. min^{-1}. UV light, with the major emission at 254 nm was administered at a dose rate of 33 erg. mm^{-1} s^{-1} to microspores in Petri dishes of which the lid was removed. UV treatment took place within a laminar flow cabinet with a distance of 21 cm between the lamp and the petridishes. Embryoid production and percentage of regeneration of the embryoids with increasing doses of X-rays and extended periods of UV treatment (e.g. 20–120 s) were studied.

---

1. The high penetration power required in multicellular systems (which makes gamma rays very suitable but UV unsuitable in most cases).

2. Experiments with radiation are highly reproducible because radiation dose and dose rate can be easily measured.

3. As radiation is suspected to produce chemical (toxic, morphogenetic) effects in the medium, transfer to fresh medium after irradiation is advised.

4. Fractionated radiation (see Chapter 4) may lead to better recovery from radiation damage.

5. Repair and recovery (see Chapter 4) may be affected by the physical parameters of the culture (temperature, oxygen, composition of the medium).

6. Ionizing and UV radiation may induce relatively many chromosome aberrations.

7. For UV treatments open Petri dishes should be used under aseptic conditions and cells, preferably, should be arranged as monolayers.

It must be noticed here that in various publications insufficient data are presented concerning the physical parameters of the culture (point 5), which should make it difficult to repeat experiments under identical conditions. Some examples concerning treatments of different kinds of starting material with physical mutagens are given in Box 5.4.

One additional aspect of mutagenic treatment of *in vitro* material with physical and chemical mutagenens must be mentioned here. It is essential to keep in mind that irradiation of culture media may have stimulatory or deleterious effects on cultured cells and tissues. Radiation, in particular when high doses are applied, may result in the production of radiolysis products (e.g. from the sucrose in the media) that may have indirect effects on cell growth and differentiation. Photo-oxidation, e.g. by UV light, also may result in toxic products. Howland & Hart (1977), for instance, already discussed starting material for *in vitro* work, treatments with ionizing radiation and UV, doses, dose rates, modifying factors, effects of radiation on cultured cells, repair mechanisms, etc.

Chemical mutagens

According to Novak (1991a), the most important points to consider when chemical mutagens are applied for *in vitro* mutagenesis are as follows.

1. The physical and chemical properties of the mutagenic agent in the culture medium, which depend on dose (concentration × time) and pH.
2. Mutagen solutions for *in vitro* treatments should be filter sterilized.
3. Pre- and post-treatment conditions (in particular pH, temperature and light) – which are highly variable when different experiments are compared – must be controlled (and noted down).
4. Uptake in multicellular systems is difficult. Carrier agents (e.g. DMSO, see Chapter 4) may facilitate penetration in such situations.
5. Application methods should be standardized in order to increase the reproducibility of the experiments.
6. Biohazard boxes or fume hoods should be used because of the carcinogenic properties of most chemical mutagens.

We will not discuss here again the specific merits of different categories of *in vitro* material as objects for treatments with chemical mutagens or in relation to selection of mutations.

Some examples of treatments with chemical mutagens are given in Box 5.5.

Which treatment?

It is not possible to provide general directives for the most suitable treatment. The crop, the cultivar, the ploidy level, the type of explant, the stage within the mitotic cycle and other factors together determine the appropriate mutagen and dose. A point of particular interest is, that enough cells or cultures survive after mutagenic treatment and that enough plantlets are regenerated. The regeneration capacity, therefore, should not be reduced below 60–70% of the control, although in many publications a more severe reduction by about 50% is considered acceptable. A higher regeneration capacity, of course, could be obtained by keeping (maximum) doses lower.

It is sometimes anticipated that radiation doses for *in vitro* treatments, as a rule of thumb, may be about 10% of the doses applied to dry seeds. Following this 'rule', treatments of vegetative material or *in vitro* material before culture should require about the same doses, whereas during culture slightly higher doses are used. To give an example: the optimum dose of X-irradiation of rooted cuttings of *Chrysanthemum morifolium* is about 15–20 Gy, whereas for irradiation of pedicel explants before culture a dose of 5–10 Gy is advised. In experiments related to a search for cold tolerance, cell suspensions of *Chrysanthemum* were irradiated by de Jong, Huitema & Preil (1991) with 15 and 20 Gy of gamma rays.

On the other hand, as was mentioned already in Chapter 4 (see Box 5.5. for treatment conditions), Masrizal et al. (1991) found for wheat (*Triticum aestivum*), that the EMS concentrations (applied for 4 h) at which the LD 20 and the LD 50 were reached by treatment of immature embryos <u>before</u> the *in vitro* phase, and after treatment of callus, derived from untreated immature embryos, <u>during</u> the *in vitro* period, were about identical to the ones when treating mature seeds. Also, no significant differences were observed between the two different cultivars (cv. Angus and cv. Pavon 76) which were used in this experiment and between the two different tissue sources. For instance, the predicted LD 20 was 0.35% ± 0.08% for the EMS treatment of immature embryos and 0.36% ± 0.10% for the callus treatment. When the mentioned similarity between results of EMS treatments for seed and *in vitro* material – calculated in this example on the basis of percentages of callus initiation and callus survival – could be confirmed for other crops, much experimental time could be saved as the appropriate concentrations for seed mutagenesis could be applied as well for *in vitro* mutagenesis. Of course one should keep in mind that in the case of chemical mutagens the doses applied are usually not identical with the dose that does reach the target cells in explants, calli or cell suspensions.

To be on the safe side when *in vitro* mutation induction programmes are planned, initial experiments to determine, for instance, the LD 20 or LD 50 for callus initiation, callus survival and regeneration of plantlets are recommended before large scale *in vitro* mutation breeding experiments may be initiated. In addition, it should be kept in mind as well that figures about

Box 5.5 Examples of procedures with chemical mutagens for various types of starting material in *in vitro* mutation breeding programmes

1. Bouharmont & Dabin (1986) reported on the treatment of explants and calli obtained from interspecific hybrid cultivars of the ornamental crop *Fuchsia* sp. Treatments of explants were performed with solutions of 80 mg. l^{-1} MNNG, 0.5% EMS, 0.1 M NaN$_3$ and 0.05 M N$_2$H$_4$, whereas for treatment of calli in the dark 80 mg. l^{-1} MNNG and 0.5 or 0.1% EMS was used. (N.B. This may be the only report on the application of sodium azide (NaN$_3$) *in vitro*.) Some callus subcultures were maintained for several months at 13 °C in the dark for the selection of cold-tolerant lines. A number of morphologically variant plants, including some polyploids, were observed, but no further reports about the progress of this work have been found.

2. Masrizal *et al.* (1991) collected immature seeds from spikes of two cultivars of wheat (*Triticum aestivum*) 14–16 days after anthesis. Per treatment about 40 immature seeds were placed in 40 ml EMS in a test tube at concentrations of 0.2, 0.4, 0.8, 0.12 and 1.6% v/v EMS, which figures correspond with 0.019, 0.038, 0.075, 0.113 and 0.151 M EMS. Treatments were performed for 4 h at room temperature, followed by rinsing the immature seeds five times with sterile distilled water and surface disinfection for 5 min in 0.52 sodium hypochlorite solution. After rinsing once again with sterile distilled water, the immature embryos (1–1.5 mm) were excised

and placed with the scutellar side exposed in Petri dishes containing a callus initiation medium. The objectives of the investigations were: to compare the ability of immature embryos to initiate calli and to study their survival and growth after treatment with various doses of EMS.

In addition, 28 day old calli which had initiated from immature embryos not treated with EMS, were treated with the same concentrations of EMS as mentioned before. About 20 calli per test tube were used. Treatments were performed with filter-sterilized EMS aseptically in a sterile hood.

After rinsing five times with sterile distilled water, the treated calli were placed in petridishes containing regeneration medium. Consecutively, the petridishes were sealed and incubated at 24 °C under a 12 h photoperiod for 30 days. Surviving calli, growing calli and/or regenerating calli were scored 28 days after EMS treatment. Some details about the results of this study are presented in Chapter 4 and in Section 5.6.3 of this chapter.

3. In mutation breeding experiments in which radiation as well as chemical mutagens were used, nucellar calli of different citrus species (e.g., *Citrus sinensis, C. reticulata, C. grandis, Fortunella hindsii* and *Poncirus trifoliata*) were treated *in vitro* with 0.3–0.5% EMS for 12–24 h (Wan *et al.*, 1991). Habituated embryogenic calli were selected among nucellar calli and

protoplasts were isolated from them, which after two to three months began to regenerate into plantlets. Some results and additional details are presented in Section 5.6.5 of this chapter.

4. Deane *et al.* (1995) described an interesting and very simple system of administering NEU (nitrosoethyl urea) to so-called 'curds' (small parts cut from the edible florets) of cauliflower (*Brassica oleracea* var. *botrytis*). Pieces of curd of about 3 mm in diameter were cut from freshly collected, surface-sterilized florets of 3-4-cm, and placed onto regeneration medium in Petri dishes. Petri dishes which each contained 5–20 curds were sealed and incubated. Curds became green after about one week and shoots were separated as soon as they appeared, to prevent crowding. After 4–5 weeks the shoots were transferred to another medium on which they could be maintained almost indefinitely. For rooting purposes shoots were cut and transferred to a rooting medium. Mutagenic treatment with the nitroso-compound was performed during the first stage on regeneration medium. One drop of a 0.3 mM solution was administered with a sterile pipette on each piece of curd, after which the plates were sealed and placed in the culture room. The NEU was not washed out. Results are briefly discussed in Chapter 5 (Section 5.6.5).

survival of callus or regeneration don't give very reliable information about mutation frequencies that may be expected!

5.6.4. Selection *in vitro* after mutagenic treatments

Earlier in this chapter it has been indicated that large scale selection *in vitro* for agronomically useful, genetically determined traits would be very attractive. For a number of traits this type of selection indeed seems to be feasible and some examples will be discussed in this section. No distinction will be made here between genetic changes which have been 'artificially induced' and those of 'spontaneous' origin.

Before discussing some selected examples, it should be pointed out that each system of culturing has its own potential and restrictions in relation to selection of mutants. For instance, the main advantage of the use of haploids in *in vitro* cultures is that, theoretically, all nuclear mutations, including recessive ones, could be recognized, but most agronomically useful traits are not expressed in single cells or callus and, in addition, reliable haploid systems have not been developed yet for all important crop species. Also one cannot be sure, that gene interactions are the same in haploid and diploid plants. Suspension cultures, although they mostly do not consist of single cells, may possess a certain degree of uniformity, and therefore promise a reasonable selection efficiency, but only for genes that are expressed in undifferentiated (single) cells. Mutagenesis in cultured protoplasts has more promise than suspension cultures as they certainly consist of single cells and therefore are physiologically and genetically more uniform, in particular when they are isolated from relatively uniform plant tissue. The small microcalli that derive from single protoplasts are relatively uniform, physiologically and in size. Finally, working with calli, embryos, explants, etc., requires more time and space than suspension cultures, and chimerism may cause major problems in the former. Advantages of the former also are the often higher probability of successful regeneration and the greater chance that some agronomically important traits may be expressed. In order to limit chimerism use could be made for instance of a single cell adventitious bud method *in vitro* (see for instance Broertjes, 1982, and Broertjes & van Harten, 1988). The merits of adventitious buds for mutation breeding *in vitro* as well as under field conditions are discussed more in detail in Chapter 7. In Section 5.6.1 reference was made already to the work of Conger *et al.* (1986), who studied *in vitro* mutagenesis in cocksfoot (*Dactylis glomerata*) and demonstrated a system to avoid chimerism by making use of somatic embryos which directly arise from single meso-phyll cells, but it seems that this system, so far, was not repeated in any other species.

For the selection of mutants under *in vitro* conditions, usually one creates a specific selection medium, for instance by adding a certain amount of herbicide, salt or aluminium or one exposes the cultures to physical stress such as heat or cold. Under such conditions all cells that do not possess resistance or tolerance to that specific stress factor will be killed. The surviving cells, subsequently, could be transferred to another medium for regeneration or, in the case of cultured tissue, be reproduced by adventitious shoot production. It is, of course, also possible to select for mutants at a later stage, for instance by administering a herbicide to seedlings from seeds of plants which had regenerated from *in vitro* plant material subjected to a mutagenic treatment.

Examples of in vitro selection

The first paper on *in vitro* selection of mutants was published by Melchers & Bergmann (1959) who reported on selection in a suspension culture of *Antirrhinum majus* for (spontaneous) mutants with tolerance to extreme temperatures. In this publication also some important attractive aspects of cell cultures for mutant selection were pointed out, such as the possibility to apply mutagens, the large number of cells involved, which allows real mass selection, the possibility to screen at the haploid level without bias by dominance or recessivity and – ideally – to regenerate only selected cells, showing for instance a specific tolerance. Ten years passed, before others continued along these lines (Binding, Binding & Straub, 1970; Carlson, 1970, 1973) and from then on many publications on this subject appeared.

Melchers (1974) discussed the potential of haploids in relation to *in vitro* mutagenesis. He pointed out for instance that the use of *in vitro* techniques makes it possible to plate about 10^5 cells of haploid plant material (e.g. protoplasts obtained from mesophyll cells of haploid plant material) on only 10 cm^3 medium on a small Petri dish. Melchers added that in this way it would be possible, after a mutagenic treatment, to select the 'few surviving cells and regenerate them to plants', followed by testing these mutants by conventional genetic methods. In addition, reference was made to the possibilities offered by haploids with respect to taking up foreign DNA by transformation and habituation, and to the possibility of obtaining fusion products from haploids.

In the following years extensive reviews about mutant selection in plant cell cultures were published for instance by Maliga (1980; 1983; 1984), Chaleff (1981), Bright *et al.* (1986), Duncan & Widholm (1986) and

Widholm (1988). Before discussing a number of examples, we will first mention here some main points treated in these reviews.

Chaleff (1983), who reviewed selection methods for agronomically interesting mutants by screening populations of regenerated plants and by direct selection *in vitro*, comments that not all traits expressed by the whole plant are expressed as well by the cultured cell (and *vice versa*) and adds that the *in vitro* selection technique is suitable only for alterations of basic cell functions. Chaleff indicated that perhaps some positive results could be obtained with *in vitro* selection for certain types of salt tolerance, resistance to toxin-producing pathogens, isolation of plant mutants altered in the control of amino acid biosynthesis and tolerance to herbicides. Positive achievements of *in vitro* selection for agronomically useful traits, however, could not be given at that stage.

Maliga (1984, with 140 references) pointed out that the early 1980s were marked by three main developments: 1) the introduction of haploid systems in *in vitro* research projects and the isolation of a variety of auxotrophic cell lines (note that auxotrophs refer to mutants that require a specific compound in the medium, which is made by a wild-type itself), 2) the increased attention to selection of mutants of (potential) agronomic value, like amino acid overproduction, herbicide resistance and salt tolerance, and 3) the fact that 'tissue-induced genetic changes' were considered now as an 'opportunity to obtain valuable genetic variations'. This last point has been discussed before. Concerning *in vitro* reactions to pathogens, it should be noted that such reactions are not typical for many diseases (see for instance Anon., 1985).

In an aforementioned report on '*In Vitro Technology for Mutation Breeding*' (Anon., 1986a) it was concluded that selection in *in vitro* systems had proved to be effective for a number of traits. Nevertheless, it was advocated to pay more attention to the development of more effective screening methods, especially for other traits of agronomic importance, including the use of markers, and to increase the search for correlations between *in vitro* recognizable traits and useful plant characters under field conditions.

In a review paper on *in vitro* selection, presented at a symposium in Vienna in 1985, Ingram & MacDonald (1986) indicated as important potential targets of *in vitro* selection, among others, efficiency of nutrient uptake and utilization, nutrient quality of the crop and tolerance to various stresses, like toxic metals and herbicides, and resistance to pathogens and pests. The imposition (or 'mimicking') of various types of stress and the recognition of biochemical markers were mentioned as major *in vitro* strategies. The suitability of different types of cultures for *in vitro* selection, in particular when selecting for resistance to pathogens, was discussed as well. Ingram & MacDonald further pointed out that for potato from 1 g of potato leaf 20×10^6 protoplasts can be obtained, which could be screened *in vitro* in one single operation. In comparison: the maximum number of plants that a potato breeder can handle in one year in the field is about 50 000 to 100 000 plants!

We will turn now towards a number of examples to illustrate in particular treatments and selection methods. Carlson (1973), who was the first to report on the application of selection techniques *in vitro* for the production of disease resistant plants, treated haploid plant cells and protoplasts of tobacco (*Nicotiana tabacum*) with 0.25% EMS for 1 h. Methionine sulphoximide (MSO), an analogue of methionine which itself is a structural analogue of the toxin produced by wildfire (*Pseudomonas syringae* pv. *tabaci*), was added to the medium. Surviving calli were plated on a medium without MSO where they remained for several months. Subsequently, the calli were divided and retested for resistance to MSO (10 mM). Calli which had retained the resistance were diploidized (by further culturing them as calli) and regenerated into whole plants. From in total 4.5×10^7 viable cells and protoplasts, 52 presumed mutant calli were obtained. From the only three calli which did not produce any tissue in which the MSO resistance had disappeared, plants were regenerated. Mutant plants that had been inoculated with *Pseudomonas* showed less or no chlorosis. These were crossed with non-mutant plants. Genetic analysis showed different patterns of inheritance of resistance for wildfire for the three mutants. No confirmation of these results in later years is known to us.

Colijn, Kool & Nijkamp (1979) reported for *Petunia hybrida* a method for positive selection of drug-tolerant mutants, which consisted of growing calli on media containing drugs at concentrations that do not allow growth of 'wild type' cells. From 29 drugs tested at various concentrations, ten compounds led to complete growth inhibition even at 'moderate' concentrations from 5 to 100 µg. ml^{-1}. Experiments were continued with two drugs (HgCl2 and 6FT) which were arbitrarily chosen from the aforementioned ten drugs. Cell suspension cultures of *P. hybrida* that had been subjected to treatments with 5–100 µg. ml^{-1} of the mutagen nitrosoguanidine (MNNG) were plated on drug-enriched media four days after the mutagenic treatment. Results showed that the MNNG treatment resulted in a strong increase in the number of drug-tolerant calli, but only at mutagen concentrations that do not cause a significant killing of the cell cultures of *Petunia*.

Sacristán (1982) performed *in vitro* selection for reduced susceptibility against the so-called black leg disease (*Phoma lingam*) in haploid plant material of spring rape (*Brassica napus*). Callus cell suspensions were produced from leaf and stem explants and in addition stem embryo cultures were used in the experiments. The cultures were subjected to mutagenic treatments, consisting of either 1–2% EMS or 20–100 µg. ml^{-1} nitrosoguanidine (MNNG) for 2.5–5 h. Treatment time and mutagen concentration were adjusted to reach at least a 40% 'level of inactivation' (which expression may refer to the percentage of non-dividing calli). Two *in vitro* selection systems were used: a) checking for absence of visible fungus growth on cultures inoculated with spores of the pathogen, and b) checking for tolerance to the toxic filtrate. Only the second system was found to be effective. In addition, some plants showing an increased resistance to *Phoma lingam*, were found in (untreated) controls. Preliminary studies were performed to elucidate the mode of inheritance of the observed resistance, but results from continued experiments are not known.

Selection for chloroplast-encoded (maternally inherited) antibioticum resistance was performed after NMU treatment of seeds of tobacco (*Nicotiana tabacum*) during a tissue culture phase by Fluhr *et al*. (1985). Seeds, soaked in 5 mM NMU for 2 h at room temperature, surface-sterilized with hypochlorite and rinsed with sterile water, were germinated and rooted in a Nitsch medium which contained an antibiotic (e.g. streptomycin 1 mg. ml^{-1}) for selection purposes. For regeneration, cotyledon sections with 'green islands of cells' were transferred to an MS medium with antibiotics. After shoot regeneration from explants, rooting was induced without the presence of antibiotics. The authors reported a high number of plastome-encoded mutations induced in this way.

Grunewaldt (1988) described a method to select for low temperature-tolerant genotypes in the ornamental African violet (*Saintpaulia ionantha*) after mutagenic treatment of small pieces of leaf discs with 15 Gy of gamma rays or 10 mM of the chemical mutagen NMU. Regeneration *in vitro* was performed at three different temperature regimes: a) 15 °C throughout the experiment, b) seven days at 26 °C followed by 15 °C, and c) 26 °C throughout the experiment. In Section 5.6.5 we will briefly discuss the outcome of this work.

Swanson *et al*. (1989) reported on the induction of mutants which showed 'field-tolerance' or 'field-resistance' to herbicides of the imidazoline group after treatment of microspores of *Brassica napus* with 20 µM of

nitrosoethyl urea. Mutagenically treated and untreated microspores were cultured in a microspore medium to which 40 µg. l^{-1} of the herbicide (tradename 'Pursuit') had been added. Small regenerated haploid plants were transferred to soil and after removal and rinsing with water, treated with 0.2% colchicine for 6 h in order to obtain doubled haploids. After repotting, selfing and backcrossing took place and plants grown from the obtained seeds were tested in the field for tolerance or resistance to the herbicide.

De Jong *et al*. (1991) reported on *in vitro* selection for cold tolerance in *Chrysanthemum morifolium*. After cell suspensions of cv. Parliament, developed according to a method described by Huitema *et al*., (1986; 1989), had been gamma irradiated with 15 and 20 Gy, the plated cell aggregates were kept for 90 days at 6 °C (control at 24 °C) in order to select for cold tolerance. Results will be also discussed later in this chapter.

To obtain aluminium tolerance, Matsumoto & Yamaguchi (1991) irradiated pieces of protocorms of a Cavendish type cultivar of banana (*Musa* sp.) with 20 Gy of ^{137}Cs gamma rays. At this dose protocorms normally maintain more than 80% of their growth. The treatment was performed after pieces taken from protocorms had been cultured in test tubes for one week at 28 °C. The irradiated protocorm pieces were left for one more week on the same medium and, subsequently, were transferred to a liquid selection medium with 10 mM aluminium chloride and some other components added. The growing pieces of protocorms underwent several successive phases of selection during three months by subculturing in the liquid selection. Subsequent testing for aluminium tolerance was performed in media containing different levels of aluminium chloride. The procedure resulted in a number of 'mutant' plants, regenerated from the irradiated material, which rooted more vigorously than the control plants in the rooting medium that was supplemented with high levels of aluminium chloride.

Syukur, Jacobs & Negrutiu (1991) irradiated protoplasts of *Nicotiana plumbaginifolia* with UV light at a dose rate of 22.5×10^{-6} J. mm^{-2}. s^{-1} for 20 s, one day after they had been isolated from haploid mother plants. Consecutively, irradiated and non-irradiated protoplasts were cultured in a K3M medium and, after some intermediate steps during which the selection agents NaCl, KCl, polyethyleneglycol and two proline analogues were added to the medium, selection for tolerance to salt, water ('osmotolerance') and proline analogue stress was performed. Results will be briefly discussed later in this chapter.

In order to select for salt tolerant mutants of sweet

orange (*Citrus sinensis*), Wan et al. (1991) exposed nucellar calli to 50–70 Gy of ^{60}Co gamma rays or 0.3–0.5% EMS and screened *in vitro* with 0.8% sodium chloride.

Cassells *et al.* (1993) and Cassells & Periappuram (1993) described for an interspecific hybrid of carnation (*Dianthus barbatus* × *D. caryophyllus*) a system of X-irradiation of nodes *in vitro*, followed by serial subculture of nodes of secondary shoots. This approach resulted in about 2% 'horticulturally acceptable' mutants, which figure – apparently – was considerably higher than after irradiation of cuttings. It is somewhat doubtful whether the explanation forwarded by the authors – *in vitro* elimination of unfit mutants through so-called **diplontic selection** – is correct, given the fact that this phenomenon operates as well, and maybe even to a greater extent, '*in vivo*'. Diplontic selection was discussed already in Chapter 1, Section 1.3.8 and earlier in this chapter.

Current opinions about in vitro selection

The traits so far under study for *in vitro* selection (see also Anon., 1986a) have been mainly related to biochemical characteristics, such as deficiencies for specific auxins, enzymes (e.g. nitrate reductase), tolerance to toxins like herbicides, antibiotics and pathotoxins, tolerance to various kinds of stress, like those caused by low temperature, frost, salt, heavy metals and drought.

Maliga (1980) already pointed out that a combination of tissue culture with continued genetic work at the plant level, despite the fact that this method is more time consuming, should be preferred above limiting the whole procedure – including selection – to the *in vitro* stage only. At the level of regenerated plants changes and mutations affecting growth and development, which almost certainly are not observed at the cellular level, can be easily detected. In addition, still according to Maliga, single new traits can be easily combined with others in crosses whereas maintenance of mutants by seed is easy. Plant breeders, of course, may comment on this opinion, for instance by pointing out the practical complications that may be met when making crosses, in particular when (often polyploid or aneuploid) vegetatively propagated crops are considered. But they will wholeheartedly support the view that *in vitro* work should be always followed by additional work at the level of adult plants and plant populations under 'natural' conditions which, in fact, will be decisive for the outcome of the *in vitro* work. This view, for instance, is expressed also by Ahloowalia (1995).

5.6.5.   **Results obtained by *in vitro* mutagenic treatments**

General remarks

In many recent publications about induced mutant cultivars no distinction has been made between such cultivars induced by treatment of seeds, plants, etc., or by treatment of *in vitro* material. There is no doubt, however, that only very few of the many '*in vitro* breeding projects' that have been performed until now, have resulted in new, officially approved and registered cultivars, let alone that mutant cultivars of considerable economic significance were released in this way.

Buiatti & Gimelli (1993) have presented a table containing a number of names of 'mutant cultivars' for ornamentals, obtained after an *in vitro* stage, in combination with or without the application of mutagenic treatments. This table, to which reference has already been made when discussing results of 'somaclonal variation', contained seven cases of cultivars which were attributed to genetic changes induced by mutagenic treatments. The species mentioned are: *Begonia masoniana*, *Begonia rex*, *Chrysanthemum* sp., carnation (*Dianthus* sp.), *Kohleria* sp. and *Weigela* sp. In all examples the plant material had been submitted to a treatment with gamma rays, with only one exception for the hybrid of *Kohleria*, where 500 mg. l^{-1} NMU was administered for 1 h at 20 °C to internode explants from *in vitro* grown shoots (Geier, 1983; 1989).

Unfortunately, several keywords and references in the table of Buiatti & Gimelli (1993) are not (fully) correct or are incomplete. In the original publication of Geier (1989), for instance, reference is made only to an interesting 'mutant clone', that still must be further evaluated before it can be released as a new cultivar. It is interesting to add here that the genus *Kohleria* belongs to the family of the *Gesneriaceae* and that, as with several other members of this family, use can be made of adventitious sprouts which often arise from single epidermal cells (Geier, 1983). In particular when the starting material has been vegetatively propagated for a number of generations, new cultivars may frequently arise as well as a result of dissociation of an already existing periclinal chimeric situation in the starting material. It cannot be excluded that the same applies to the mutant clone on which Geier reported. This situation does not differ from what is common experience when conventional methods of vegetative propagation, e.g. by making cuttings, are applied. Many details, in particular about the 'example crop' *Dendrathema* ('*Chrysanthemum*') are discussed in Chapter 7. For further reading, see Broertjes & van Harten (1988).

Another case to which is referred by Buiatti & Gimelli (1993) in the aforementioned table: the mutant cultivar (?) of *Begonia rex*, resulted from a treatment with 100 Gy of gamma rays (dose rate 2 Gy. min^{-1}) of adventitious buds which arose after one month from *in vitro* cultured leaf fragments (Shigematsu & Matsubara, 1972). Two out of 30 plantlets which developed from the irradiated adventitious buds showed chimeric leaves with sectors displaying a green leaf colour with silver white spots, whereas the original leaf colour was silver white. Small sections cut from the mutated leaf areas were cultured *in vitro* and from the adventitious buds plantlets were produced. Three plants were obtained in this way, all of which carried mutated leaves. The summary of the article by Shigematsu & Matsubara does not make clear, however, whether the mutant has been officially registered as a novelty and has been introduced on the market as a new cultivar. This, of course, also depends on the (commercial) attractiveness of the mutant. Again, in this case it cannot be excluded that the original 'mutant plants' with chimeric leaves were not caused by a newly induced mutation, but resulted from a histogenic effect of the radiation treatment, i.e. the dissociation of an already existing periclinal chimera.

Further examples given by Buiatti & Gimelli (loc.cit.) include one mutant variety induced by 10 Gy of gamma rays in carnation (*Dianthus* sp.) and two mutant varieties of *Chrysanthemum*, obtained also after treatment with 10 Gy of gamma rays. These examples have been derived from some short remarks found in a list of mutant cultivars in *Mutation Breeding Newsletter* 34 of 1989, where it was mentioned that the flower colour mutants were produced in the 1980s by workers of the Kasetsart University, Bangkok, Thailand. It was not mentioned, however, whether these mutants had been officially tested, approved and/or commercialized.

Duron & Decourtye (1986; 1990) confirm that several 'true new cultivars' of the ornamental shrub *Weigela* – dwarf types and plants with different leaf or flower colour – have been obtained after mutagenic treatment (e.g. with 40 Gy of gamma rays) of 3-year-old plants, followed by induction of bud neoformation *in vitro*. One of the obtained mutant cultivars, cv. Courtadur, with a very compact growth and a long flowering period, was mentioned by Buiatti & Gimelli (loc.cit.). Details concerning another example in the table, a dwarf type mutant of *Begonia masoniana* with aberrant leaf shape and leaf colour, obtained after a treatment with 100 Gy of gamma rays, could not be further verified, as this case refers to a Japanese patent (Suda, Matsubara & Kudo, 1982; original not consulted).

In conclusion, the lack of detailed information, the sometimes limited scale of experimenting and the fact that results of *in vitro* mutagenic treatments often have not been evaluated during some generations in the field, makes it very difficult to assess the real significance of various claims about the induction of higher mutation frequencies or different mutation spectra obtained *in vitro* after mutagenic treatments and the practical value of such observations for plant breeding.

Results from treatments with physical mutagens
A list of more than 25 examples of radiation treatments for *in vitro* plant material (cell suspensions, callus culture and protoplast cultures), including many references, was presented by Constantin (1984). Much work is performed with so-called 'model plant species', like *Haplopappus gracilis*, *Petunia hybrida*, *Nicotiana tabacum*, *N. sylvestris*, *N. plumbaginifolia* and *Datura innoxia*.

Tobacco (*Nicotiana tabacum*) has been one of the very first crops in which *in vitro* mutation research was performed on haploids. Nitsch, Nitsch & Péreau-Leroy (1969) irradiated haploid microspores or plantlets from cultured anthers of tobacco in test tubes with 15–30 Gy of gamma rays and obtained plants showing mutations for various leaf and flower traits.

In Box 5.4, the treatment procedures applied in an extensive *in vitro* mutation experiment on potato (*Solanum tuberosum*) by Roest & Bokelmann (1980) and van Harten *et al.* (1981) are briefly described. Plants resulting from adventitious sprouts which developed from X-irradiated explants of rachises, petioles and upper leaflets, were observed during three consecutive vegetative generations (called vM$_1$, vM$_2$ and vM$_3$). The first two generations were raised in the greenhouse and a vM$_3$ based on 13 380 tubers was grown in the field. Scoring of various (pre-established) groups of mutations was performed, following the procedures that had been used by the authors during more than ten years of field experiments after irradiation treatments of 'tuber eye pieces' of potato. Based on an 'overall mutation frequency' (i.e. the sum of mutations in all categories for which scoring was performed), up to 91% mutated clones were scored for the higher doses. When results for all groups of irradiated explants were taken together, an 'average mutation frequency' of almost 75% was obtained. In comparison, in a range of field experiments which had been performed in earlier years and in which mutation frequencies and percentages of chimerism had been determined on the basis of vegetative progenies from irradiated tubers and tuber eyes of different cultivars, a maximum mutation frequency of 38% mutated plants had been obtained. Without going too much into detail it must be mentioned here that the way of experimenting and the crop involved do not allow us to draw conclusions on the basis of 'independent mutational events'. In addition, and also despite the relatively large scale of the experiments and the care taken by very experienced field assistants when setting up field experiments and determining mutation frequencies (e.g. by excluding any 'mutation like' result which could have been caused by virus symptoms in the foliage), one could also, rightly, raise the question whether it is fully justified to compare data from 'explant'-experiments and from 'tuber'-experiments. It is, however, very doubtful whether in most comparable experiments by other workers more precautions to exclude unjustified conclusions were taken.

In the experiment by van Harten *et al.* (1981), the use of the adventitious bud technique (see also Broertjes & van Harten, 1988) resulted in a very low percentage of chimerism, namely 1.7% for the total experiment as well as for irradiated material only. A somewhat unexpected observation was the relatively high mutation frequency observed in the (small) unirradiated control groups, namely 12.3% for the group of leaflet-discs and even 50% when rachis and petiole explants were considered. The observed variation, nowadays, would be simply classified as 'somaclonal variation'. In this case, however, one should expect more variation among plants that trace back to leaflet-discs than to rachises and petioles, as for the last group adventitious sprouts arise almost directly, i.e. practically without a callus phase, whereas regeneration from leaflet-discs mostly requires the presence of a prolonged callus phase. Because in the experiment only a limited number of control plants had been included, a fair comparison between results from the much larger X-irradiated series and the controls and between the groups of controls is not possible. The experiment, anyhow, shows that *in vitro* mutagenesis of potato in combination with an adventitious bud method may result in a very high mutation frequency, a very broad mutation spectrum and a very low rate of chimerism. (N.B. The aforementioned experiment did not aim at the production of new mutant cultivars.)

Studies on *in vitro* mutagenesis of potato were also performed by Sonnino, Ancora & Locardi (1986). These authors cultured buds of cv. Desirée *in vitro* on a solid medium and irradiated the obtained plantlets with 30 Gy of gamma rays, at which dose about 40% of the micropropagated buds survived. Single-node pieces were cut off and transferred to a fresh medium; this process was repeated once. Afterwards the resulting plants (vM$_1$ up to vM$_4$) were potted and later transferred to the field. The authors commented that the method followed by them is simple and fast and that in this way after only

two cycles of *in vitro* propagation more than 75% of the detected mutants were 'uniform', i.e. non-chimeric. Among almost 1100 plants observed, 158 mutations were found for various morphological traits, including for instance changes in leaf size and shape, dwarf types, leaf colour, flower colour and shape, tuber skin colour and anthocyanin on stems. None of the 200 control plants showed any phenotypic variation when compared with the mother clone. It should be commented that these 158 mutation types, most probably, do not all refer to independent mutational events as several groups of mutations may (at least partly) refer to the same genetic event (e.g. pigmentation of flower, tuber skin and stem).

In Chapter 4, Section 4.2.1, brief mention was made of the work by Devreux *et al.* (1986) who irradiated protoplasts of the 'model species' *Nicotiana plumbaginifolia* with different doses and split doses of 60Co gamma rays and fast neutrons. Treatment procedures have been outlined in Box 5.4. When 6 Gy of gamma rays were split in three doses of 2 Gy, the survival rate strongly depended on the time intervals. When 'long' intervals (note that further details were not given) were taken between two consecutive split doses, a higher survival rate was observed than when a single full dose of 6 Gy was administered. In the case of intervals of just a few minutes between two split doses the survival rate was even lower than for a single full dose of 6 Gy. For treatments with a dose of 0.3 Gy of fast neutrons, administered in various ways: 2×0.15 Gy, 3×0.1 Gy and 6×0.05 Gy with intervals varying from 4 to 128 min, very surprising results were reported. Split doses with short intervals resulted in a radiosensitivity that was higher than when a single full dose was given, whereas for intervals of 32 and 128 min results were about identical to those for the single dose. Studies on dose–effect relationships were performed for haploid and diploid cells. Protoplasts derived from differentiated leaf cells showed a much higher radiosensitivity than protoplasts from very young calli.

Kleffel, Walther & Preil (1986), starting from embryogenic suspension cultures of poinsettia (*Euphorbia pulcherrima*), obtained a mutation frequency of 8.9% at a survival of 11% for the X-irradiated embryoids (against 100% survival in the control) at the highest dose applied, i.e., 60 Gy of X-rays. The highest dose corresponded with the lowest survival. A comparison of the 'relative effectivity' of different treatments showed that the optimal treatment was a (split) 2×20 Gy treatment with an interval of 1 h, administered to the cell suspensions.

Similar tests of radiosensitivity of *in vitro* plant material have been performed for other species. Protoplasts and embryogenic cell suspensions are often treated with doses in a range 10–60 Gy of X- or gamma rays at dose rates of 1–10 Gy. min⁻¹. Within one species sometimes remarkable differences in radiosensitivity between different cultivars may be observed. It is in most cases not easy to explain such differences. Walther & Sauer (1986a,b) irradiated small shoots, which had been taken as axillary shoots from *in vitro* grown cuttings of the ornamental crop *Gerbera jamesii*, with 10, 15, 20 and 25 Gy of X-rays respectively. The irradiated axillary shoots were X-irradiated and placed again on an MS medium. Axillary shoots which developed after the treatment were cut off four weeks after irradiation and three more times at intervals of four weeks each. All shoots were placed back on the MS medium for two weeks, rooted on a modified MS medium during three weeks, planted into a peat–soil mixture for 10 days, transferred to another peat–soil mixture and cultivated up to flowering. The resulting potted plants were checked for changed traits as compared with the control plants. In total 622 vM₁ plants were regenerated which traced back to the cuttings taken on four occasions (see before) from 30 explants per dose. Of the plants that were studied up to flowering, 90 plants (i.e. 14%) showed at least one 'mutative change' that remained stable throughout the period of observation. The highest number of changes was found at a dose of 20 Gy, at which dose about 38% of the plants showed at least one 'mutation'. The authors did not mention exactly whether the observed changes occurred throughout the whole plant or were found as chimerical structures in one or a few branches only. Moreover, it was not checked whether the observed changes (or mutations) manifested themselves as periclinal chimeras in later generations (see also the next example concerning the ornamental *Weigela*, described by Duron & Decourtye, 1986). It is, of course, somewhat premature to decide on the basis of observations during only one cycle of vegetative propagation whether such changes indeed represent 'real mutations'. This may refer in particular to plants with reduced stalk length, which account for 42% of all changes observed. This phenomenon, probably, may be explained as well by a transient growth retardation as a result of the mutagenic treatment. In addition to 42% changes for flower stalk length, 30% of the observed changes referred to changes for flower (petal) size and 19% to petal colour. In later studies Walther & Sauer (1991) obtained highest numbers of 'mutants' after fractionated X-ray treatments of *in vitro* derived microshoots of *Gerbera* in Petri dishes with doses up to 45 Gy (3×15 Gy with intervals of 4 h).

Duron & Decourtye (1986; 1990) studied the effects of *in vitro* irradiating small shoots with three internodes

of the arboreous shrub *Weigela* sp. on survival, root formation, growth of cuttings and the induction of mutations. Doses of 20, 30, 40, 50 and 60 Gy were given to 28 (2 × 14) shoots for each dose at a dose rate of 10 Gy. h⁻¹. From each series of 14 plants, theoretically, 14 apical shoots and 28 axillary shoots could be obtained. At 60 Gy almost no buds survived (LD 50 = 40–50 Gy). Reduction of rooting started when doses higher than 30 Gy were applied. After two or three years of testing in the field about 40% of the observed mutants proved to be periclinal chimeras.

An efficient system for mutagenesis using microspore cultures of rapeseed (*Brassica napus*) was developed by Beversdorf & Kott (1987). Uninucleate, potentially embryogenic microspores were irradiated. After four weeks of incubation the generated embryos were studied *in vitro* for chlorophyll mutations. Chlorophyll synthesis was taken as a visual selection criterium to identify viable embryos. About 8% of the microspores developed into embryos. The system described here is sometimes called 'isolated microspore culture (IMC)'. Polsoni, Kott & Beversdorf (1988) showed that in this way a couple of hundred thousand embryos can be produced within three weeks. In order to retain a sufficiently high number of embryos after irradiation, doses should not surpass 15 Gy. The observed relatively high sensitivity of haploid microspores to radiation was confirmed by the studies of MacDonald *et al.* (1991), which have been briefly discussed in Box 5.4.

Szarejko *et al.* (1991) also described the specific properties of mutagenized single, totipotent haploid cells (like isolated microspores) in combination with a system of rapid differentiation for obtaining chimera-free M₁ plantlets that can be screened for resistance or tolerance against particular selective factors. The authors reviewed the potential of the 'IMC-method' for a number of crops, like rice, wheat, barley, maize, rapeseed and potato. Results of several successful applications of this method for rapeseed were presented, including one case where even 50% of the isolated microspores had developed into embryos.

In an earlier section of this chapter in which *in vitro* selection was discussed, reference was made to work by Grunewaldt (1988) on *in vitro* selection for low temperature tolerant African violet (*Saintpaulia ionantha*). Grunewaldt reported that a good correlation was found between the low temperature tolerance during the *in vitro* stage and in regenerated plants. De Jong *et al.* (1991), who investigated whether the application of 90 days of cold treatment resulted in an increased number or different types of low temperature tolerant (LTT) mutants for *Chrysanthemum morifolium*, on the other hand, concluded that their method did not result in mutants which showed early flowering under a low temperature regime. The authors assume that the processes of rapid callus growth *in vitro* under cold conditions and early flowering induction under cold conditions '*in vivo*' do not have a common genetic base.

Novak et al. (1988) studied somatic embryogenesis in maize (*Zea mays*) and compared the genetic variability induced by gamma radiation and by tissue culture techniques. For this purpose, twelve-day old zygotic embryos, resulting from a selfed inbred line, were irradiated *in situ* with 5 and 10 Gy ⁶⁰Co gamma radiation. In another series, immature zygotic embryos were excised from caryopses and cultured *in vitro*. A third group consisted of immature zygotic embryos treated with 5 and 10 Gy ⁶⁰Co gamma irradiation and cultured *in vitro*. The authors concluded that, based on screening for chlorophyll and morphological offtypes, 'somaclonal variation', most likely, does not differ from the variation resulting from gamma irradiation. Combination of explant irradiation and *in vitro* regeneration was most effective for the manifestation of the aforementioned groups of mutations and, in addition, drastically increased the frequency of early flowering variants. However, as the authors correctly pointed out, mutagenic treatments may result in a decreased regeneration ability *in vitro*. The authors further mentioned that their results were in agreement with observations concerning 'somaclonal variation' in maize, made by Zehr et al. (1987).

Novak and associates (see for instance Novak et al., 1990; Novak, 1991b; and Novak et al., 1993) discussed gamma irradiation of *in vitro* cultured shoot tips of diploid, triploid and tetraploid cultivars with different genomes of banana and plantain (*Musa* sp.). Radiosensitivity was assessed by determining fresh weight of cultures and the degree of shoot differentiation. From seven clones tested a diploid clone (genomic constitution: *AA*) was most sensitive to radiation; a tetraploid clone (*AAAA*) showed the lowest level of radiation damage. These observations are according to expectations concerning the protective effect of higher ploidy levels in case of mutagenic treatments. For diploid clones doses of 20–25 Gy are advised, for triploid clones (*AAA, AAB, ABB*) doses of 35–40 Gy, and for tetraploid clones (*AAAA*) doses of 50 Gy. This work, among others, resulted after selection in vM$_4$ in a vigorous mutant that flowered after only nine months instead of the common 15 months under greenhouse conditions at the IAEA greenhouses in Seibersdorf, Austria.

In 1995 a mutant cultivar of banana, named 'Novaria', that resulted from the aforementioned work by the IAEA, was registered in Malaysia (Mak et al., 1996). This triploid (*AAA*) mutant, derived from cv. Grande Naine (to which was briefly referred already in Chapter 1) flowered in Malaysia on average 10 weeks earlier than its 'parent' and, in addition, showed some other favourable traits, such as a stronger fruit stalk. In 1990, 27 individual suckers, derived from the originally selected mutant plant, were introduced in Malaysia and, after micropropagation *in vitro*, planted at the United Plantations of Jenderata, state of Perak. In total 2000 plants were grown under commercial conditions, from which material, after continuous selection, in 1993 mutant cv. Novaria was selected. The new mutant cultivar appeared to be fully stable and non-chimeric. Introduction in Malaysia took place in cooperation with the Malaysian Institute for Nuclear Technology. Although no exact figures are known, the total number of hectares planted with the new mutant cultivar in 1997 in Malaysia (still?) seemed to be rather limited (van Zanten, IAEA, personal communication).

Gao, Cheng & Liang (1991) reported on irradiation of two-month old calli from wheat (*Triticum aestivum*) with 10 Gy of gamma rays which, after irradiation, were transferred immediately to fresh medium. The frequency of induced variants rose to 25%, which was about three times the frequency of spontaneous ('somaclonal') variation. Irradiation treatment also broadened the mutation spectrum. It was found, however, that a 10 Gy treatment causes much sterility and a poor regenerability, whereas it was concluded from another experiment that a much reduced level of sterility was found at a dose of 5 Gy.

Wan et al. (1991) discussed *in vitro* mutation experiments with calli of different citrus species in China (see also Box 5.5). Doses of 50–70 Gy gamma rays – which according to the authors approximately correspond with the LD 50 – produced about 6.5% callus with chromosome aberrations. The authors reported that the frequency of structural mutations in chromosomes after mutagenic treatments was 300 times higher than 'in nature' (this expression may be identical to 'somaclonal variation').

Syukur et al. (1991), who irradiated haploid protoplasts of *Nicotiana plumbaginifolia* with UV light (see earlier this chapter), selected for tolerance to salt or water stress and to proline analogues. Tolerant cells which are all characterized by overproduction of proline and which all carry the same single dominant gene, were obtained at a frequency of about one per 10⁻⁵ or 10⁻⁶ 'lines'. The tolerance is expressed more strongly in lines selected for salt and water stress than in those selected by proline analogues. (N.B. A mutation frequency of

1×10^{-5} for a mutation towards dominant is rather high, at least under 'in vivo' conditions.)

Finally, in Japan, in 1996 Nagatomi et al. (1996) briefly reported on three mutant cultivars of the ornamental *Eustoma grandiflorum* (lisianthus) which had been induced through *in vitro* culture of chronically irradiated floral parts and leaf blades. The mutant cultivars, indicated as 'spray types' with many small flowers and a favourable plant type, were entered for official registration in Japan.

Effects of treatments with chemical mutagens

Chemical mutagens have been frequently applied at an early stage to induce mutants in plant cell cultures (Chaleff & Carlson, 1974). Handro (1981), for instance, presented a list of treatments with chemical mutagens and of responses for various cultures involving different species. Treatments, as has been described earlier in this chapter, can be performed on seeds or other plant material before the culturing phase or mutagens can be added to the *in vitro* medium.

Another method is to deposit with a pipette a drop of the mutagen (in a predetermined concentration) on single explants in Petri dishes, as was done for instance by Deane et al. (1995) with small pieces of so-called 'curd' (explants of about 3 mm taken from detached surface-sterilized 3–4 cm florets) of cauliflower (*Brassica oleracea* var. *botrytis*) which had been transferred to Petri dishes before treatment. Some details about this easy but not very precise treatment procedure are given in Box 5.5. The most important results of this experiment are briefly given here. The low concentration of nitrosoethyl urea used (0.3 mM) meant that shoot regeneration from the mutagenized explants was not significantly affected. At the same time, from 50 leaves which were selected at random from shoots from each of ten batches of treated 'curd', more than 40% showed changes for chlorophyll content and distribution, and morphological changes were observed in more than 26% of the leaves studied. In comparison, less than 2% of the leaves collected from the (water-treated) control series showed abnormalities. From more than 6000 explants which had been mutagenized on a regeneration medium supplemented with 3 mM hydroxyproline, 31 hydroxyproline-tolerant shoots were obtained, while none was found in the control series. The authors reported that the hydroxyproline-tolerant shoots will be further evaluated for frost resistance.

The optimal concentrations depend on the kind of starting material and on the plant species. For treatments of explants and calli of various crops with EMS, often concentrations of about 0.5% are administered.

Nitroso-urea compounds, e.g. at concentrations of 0.1–0.5 mM, are often used for the induction of cytoplasmic mutations, as has been demonstrated in *Saintpaulia* sp., *Nicotiana plumbaginifolia* and other crop species.

Omar et al. (1989) dipped longitudinally dissected shoot tips of a diploid (*AA*) and a triploid (*AAA*) cultivar of banana (*Musa* sp.) in aqueous solutions with up to 0.8% (about 100 mM) EMS for 3 h at 28 °C. Dimethyl-sulphoxide (DMSO) 4% was added as a carrier agent. Fresh weight was determined at regular intervals and the number of initiated shoots was determined after 30 days. The number of adventitious buds decreased with increasing concentrations of EMS. The optimal dose of treatment with EMS for diploids and triploids was about 0.2% (about 25 mM) EMS during 3 h of incubation.

Earlier in this chapter when discussing 'somaclonal variation', reference was made to the work on tomato (*Lycopersicon esculentum*) by Gavazzi et al. (1987). These authors treated seeds and mature pollen of tomato with 0.1% or 0.2% EMS. Seeds were soaked for 18 h in an aqueous solution containing 0.1% of EMS. Pollen was suspended for 1 h 40 min in paraffin oil with 0.1 or 0.2% EMS and pollination was performed immediately afterwards by bringing the pollen suspension on the emasculated flowers with a fine brush. For more details on mutagenic treatment of pollen see Chapter 4. Double treatments were also performed, i.e. pollen was treated first and the resulting seeds (M_1) were treated again in the abovementioned way. Control seeds and seeds treated with EMS were brought on an MS medium and after 8–10 days of growth at 26 °C cotyledons were excised, sectioned with two diagonal cuts and plated on another MS medium (with IAA and zeatine). Calli were produced from the injured sites on the cotyledons, from which calli within 8–12 days shoots differentiated. Consecutively, these shoots after having reached a length of about 1 cm were cut off and transferred to a rooting medium. Controls were included as well. After 30 days on the rooting medium, the resulting plantlets were transferred to 'jiffy-pots'. From each cotyledonary section only one regenerated shoot was transferred to soil. The resulting M_1 (or R_1; see also Box 5.1 for nomenclature of the different generations) plants were selfed and the resulting M_2 (or R_2) plants were phenotypically examined for mutations. Scoring took place for a number of predetermined characteristics in seeds (seed colour, germination), seedlings (number of cotyledons, seedling size, lethals) and grown up plants (male sterility, changes for lateral shoot formation, height reduction, early ripening, leaf morphology, fungus resistance as judged by visual inspection). Chromosome aberrations were not determined. After comparing mutation frequencies and the mutation spectrum for the plants obtained from EMS-treated and untreated seed and pollen, Gavazzi *et al.* (1987) concluded that more mutations were obtained by 'somaclonal variation' in the control than by adding EMS to seeds, pollen or seeds and pollen. In addition, different spectra of mutations were observed for the control group and the EMS-treated group. For instance, one category of morphological mutations ('potato leaf') was observed only in the control series.

Van den Bulk *et al.* (1990), to whose work was also referred earlier in this chapter, also compared the frequency and spectrum of 'somaclonal variation' with the results obtained after treatment of tomato seeds with 60 mM EMS (in which the seeds were submerged during 24 h under dark conditions at 24 °C). These authors, who scored high numbers of M_1 (R_1) and M_2 (R_2) plants for changes concerning a number of predetermined traits, concluded that higher mutation frequencies were obtained after EMS treatments than only as a result of the *in vitro* phase. No clear differences were found between the mutation spectra of both groups, except for polyploids which did not occur after the EMS treatment. These results are not in agreement with the conclusions of Gavazzi *et al.* (1987) and earlier results reported by Evans & Sharp (1983). Van den Bulk *et al.* believed that their work and that by Gavazzi *et al.* are the only two reports aiming at a direct comparison of effects of 'somaclonal variation' and EMS-induced variation.

The different outcome of both studies, according to van den Bulk *et al.*, may be explained by the choice of different genotypes, different procedures followed during mutagenic treatments, different tissue culture procedures and the way of scoring based on different groups of traits. The most striking difference between both approaches seems to be that Gavazzi *et al.*, after having treated seeds and pollen with EMS, introduced an *in vitro* phase, whereas in the experiments by van den Bulk *et al.*, the seeds, after having been washed, were immediately sown, M_1 plants selfed and the M_2 evaluated in the greenhouse. A better comparison between both experiments would have been possible if van den Bulk *et al.* had included in their experiment a third group, consisting of EMS-treated seeds of which, subsequently, cotyledons had been transferred to an MS medium for callus development. Their main purpose, however, was to make a comparison between the mutagenic events produced in a 'conventional' mutation programme for tomato, starting from EMS-treated seeds, and 'somaclonal variation' resulting from *in vitro* plant regeneration from leaf, cotyledon and hypocotyl explants.

Another example of *in vitro* mutagenesis with EMS was reported by Wan *et al.* (1991). These authors, who also studied effects of gamma irradiation, applied EMS in concentrations of 0.3–0.5% for 12–24 h at 23 °C to calli from different species of citrus (*Citrus* sp.). (N.B. Further details about the way of treating calli with EMS were not presented.) The experiments, which started in 1979, had been set up to investigate methods of avoiding chimerism and diplontic selection and to obtain mutants for salt tolerance, low temperature tolerance and disease resistance. The LD 50 (which term in this case refers to the number of regenerating calli) for EMS was 0.3%, and the percentage of chromosome aberrations at that concentration in regenerated shoots resulting from the treated calli was about 6%. This last figure is slightly lower than the corresponding figure found for treatment with 70 Gy of ⁶⁰Co gamma rays. In this publication no results concerning induced stress tolerance or disease resistance were presented.

A comparison of the mutagenic effect of X-rays and the chemical mutagen nitrosoguanidine

Very few experiments have been performed in which effects of mutagenic treatments with physical and chemical mutagens have been compared on an adequate scale. An example in which some effects were compared which had been induced by X-rays and the chemical mutagen nitrosoguanidine (MNNG) on haploid and diploid protoplasts of *Datura innoxia* and *Petunia hybrida*, was described by Krumbiegel (1979). The author, who studied 'survival rates' (based on observations on division rates of protoplasts after seven days of culturing) and mutation frequencies for pigmentation patterns, mentioned that such comparative studies concerning survival of protoplasts of different ploidy levels after mutagenic treatment are rare. Protoplasts were obtained from (2n and n) shoots of *Datura*. The shoots were obtained from anther cultivation, with consecutive chromosome doubling of the haploid strain. For *Petunia*, shoots were used as well. For *Datura* doses of 15–30 Gy of X-rays were applied; for *Petunia* 2.5–30 Gy. The MNNG-treatment of protoplast suspensions was for both species performed with 5–50 μg. ml⁻¹ for 30 min in the case of haploid protoplasts, and with 5–120 μg. ml⁻¹ for diploid protoplasts. X-ray treatments as well as MNNG-treatments showed an exponentially decreased survival with increasing doses of mutagen for protoplasts of both species but the slope of the survival curves was steeper for haploids than for diploids. Based also on results from earlier experiments, the LD 50 for both haploids and diploids of *Datura* was determined at 7.5–10 Gy for the X-ray treatment and 10 μg. ml⁻¹ for the MNNG-treatment;

the corresponding values for *Petunia* were 5 Gy and 10 µg. ml⁻¹. Only at higher doses (e.g. 17.7 Gy or 30 µg. ml⁻¹ for *Datura* and 10 Gy or 20 µg. ml⁻¹ for *Petunia*) diploid protoplasts of both species are more tolerant to the mutagen than the haploid protoplasts. Finally, it was mentioned that treatment of about 105 haploid protoplasts of *Datura* with 10 µg. ml⁻¹ MNNG (about LD 50) resulted in the production of four chlorophyll mutants from 2.1×10^4 calli.

5.7. Some final remarks on 'somaclonal variation' and *in vitro* mutagenesis

It must be concluded here that 'somaclonal variation', *in vitro* mutagenesis and *in vitro* selection, so far, have not produced results which had a really significant impact on plant breeding. More specifically, as was mentioned already, *in vitro* mutagenesis in only few cases has resulted in the release of new mutant cultivars and none of these mutants, as far as is known, has had a really big impact from an economic point of view. Based on the number of newly released cultivars of 'somaclonal' origin, results are even more disappointing and in big contrast to the exaggerated expectations raised in the 1980s by its proponents. It is worthwhile to keep in mind that already for several years, and unnoticed by most biotechnologists, 'spontaneous' mutant cultivars (or 'sports') represent at least 2–5% of the acreage occupied by cultivars for (vegetatively propagated) root and tuber crops like potato, cassava, etc. For other vegetatively propagated crops, in particular ornamentals and some fruit crops, such percentages even may go up to 50% or higher and all the time new 'spontaneous' cultivars obtained in this way are added to the variety lists (more information on this subject will be presented in Chapter 7). Therefore, plant breeders will not be highly impressed when in the coming years, now and then, useful 'somaclonal cultivars' are introduced. Whether this is the result of spontaneous variation observed in plants or a result of an *in vitro* phase is not of major concern to them.

It has been pointed out before that also the high expectations with respect to selecting *in vitro* for disease resistance or stress tolerance and, thus, to the possibility of speeding up or facilitating the process of breeding new cultivars, so far, have not been fulfilled. Apart from

the impression that many putative mutants in fact may be epigenetic variants, and that more genes may be involved in stress tolerance than is generally assumed, there still appears to be a considerable discrepancy between many effects that are observed in *in vitro* experiments and those that occur at the level of intact plants. If reliable correlations could be established between certain *in vitro* expressed markers and desired traits in field plants, this indeed would greatly facilitate and speed up the use of *in vitro* techniques in plant breeding, but markers with such correlations are hardly available. The most probable reason for the lack of success with *in vitro* mutagenesis techniques may be the fact that selection for the great majority of agronomically useful traits can take place only at the level of adult plants under field or greenhouse conditions.

In conclusion: although it cannot be excluded that the use of selective media may create a different 'selection environment' and, therefore, could lead to mutations for specific traits which would not occur or not be observed under 'field conditions', there is, so far, no definite proof that *in vitro* mutagenesis does produce a mutation spectrum that differs from the spectrum resulting from 'conventional' mutation breeding. The same, in our opinion, applies to the use of 'somaclonal variation'. Another point, made before, is that the often claimed 'high frequencies' of mutation *in vitro* not necessarily are as beneficial to plant breeding as is suggested sometimes.

It seems highly appropriate to repeat also at this place a statement made already in Chapter 1. It appears that the number of professional, university-trained plant breeders with a liking for working in the field, quickly declines and that in particular large breeding companies – maybe temporarily – often choose to employ scientists with a strong background in molecular biology or biotechnology, who mostly prefer to stay far away from field work. Unfortunately, with the still increasing interest in molecular work, it is doubtful whether this situation will change considerably in the near future. Nevertheless, and despite the fact that contributions by molecular scientists by no means should be underrated, it may be predicted that the large majority of new, successful cultivars will continue to be produced by skilled plant breeders who know their crops and who are well acquainted with large-scale experimenting in the field and in greenhouses.

6 Mutation breeding in seed propagated crops

6.1. Introduction

General remarks

Mutation breeding programmes for seed propagated crops can be started by subjecting seeds, whole plants or gametes (pollen, egg cells) to mutagenic treatments. The aforementioned types of starting material, in principle, could be used also for mutation experiments using *in vitro* techniques. Vegetative propagation methods are available now for an increasing number of seed propagated crops, which offers additional possibilities, not only for breeding work, but also for the application of mutation methods, either by radiation treatments or by administering chemical mutagens. In this chapter we will concentrate on mutagenic treatments of seeds and gametes and, for this second type of starting material, in particular on treatment of pollen.

Another aim of this chapter is to discuss the consecutive steps of a practical mutation breeding programme for seed propagated species. As a starting point, a relatively simple situation is taken. First, it is assumed that the breeder is primarily interested in inducing mutations from dominant to recessive for a trait that is governed by one gene. In addition, a diploid, annual, self-fertilizing crop is considered where seeds or pollen are mutagenically treated with X- or gamma rays or EMS. Such a 'model crop' could be barley (*Hordeum vulgare*). For practical reasons, many fundamental studies on mutagenesis are performed with *Arabidopsis thaliana*, a small self-fertilizing, diploid weed that is often used for fundamental studies in genetics. Background information on *Arabidopsis* as a suitable 'tool' for this kind of investigation for instance, can be found in several research contributions and reviews by Rédei (1970; 1975; 1982b; see also Chapter 1, Box 1.4).

The application of mutation methods becomes more complex or less promising when cross-fertilizing crops are involved instead of self-fertilizing crops, when muta-

tions from recessive to dominant alleles would be desired, or when traits under polygenic control would be the target. Such situations will be discussed as well. Attention will be paid also to specific aspects of mutation breeding in polyploid crops.

Self-fertilizing and cross-fertilizing species

Two groups of seed propagated species are commonly distinguished: 1) **self-fertilizing** (often called **self-pollinating**) or **autogamous** species and 2) **cross-fertilizing (cross-pollinating)** or **allogamous** species. In plant breeding this distinction is generally very useful, although it should be realized that even in so-called 'strictly self-fertilizing crops' a small percentage of spontaneous cross-fertilization may occur. In barley (*Hordeum vulgare*), for instance, this percentage may be up to 2–5%, but in another self-fertilizing species, faba bean (*Vicia faba*), up to 20% outcrossing may be found. In text books on plant breeding, for species where hybrid seed production is practicable, hybrids are sometimes treated as a special, third group of seed crops. Hybrid seeds may be produced in self-fertilizing as well as in cross-fertilizing species. From the point of view of mutation breeding, there is no particular reason to treat such species as a separate group.

Self-fertilizing species. Cultivars of self-fertilizing species consist in principle of homozygous, identical genotypes among which, to a certain degree, aberrant genotypes may be present as a result of spontaneous mutation, by some pollinations in which foreign pollen is involved, or by accidental contamination with seeds of other cultivars or genotypes. In practice, plants in F_7 or F_8 are considered homozygous, despite the fact that 100% homozygosity can never be achieved. Progenies obtained after selfing – in a strict sense only after selfing of homozygous plants – are called <u>lines</u>. Actually, the word <u>line</u> is often used in a less restricted way and commercial cultivars of self-fertilizing crops are often composed of similar 'multilines' rather than represent strictly pure lines.

Various selection methods can be applied for self-fertilizing crops, e.g. mass selection, line or pedigree selection, back-cross selection, recurrent selection, etc. Often, various methods are combined or applied consecutively in different stages of a breeding programme. Selection can be performed relatively fast and easily when a trait can be assessed already in the vegetative stage. Stem length or fibre length (e.g. in flax, *Linum usitatissimum*) and stem colour, for instance, may be determined already in the vegetative stage whereas for seed yield or oil content of seeds one has to wait till seed set has been completed. As a consequence, plant breeders make a distinction between traits of which the phenotypic value can be determined before or after flowering and different selection schemes must be followed for both categories.

The selection programmes are often preceded by and interspersed with a crossing programme in order to create genetic variation by recombination of genes. Mutagenic treatments offer another possibility to create genetic variation. Plants with recessive mutations can be recognized relatively easily and early in self-fertilizers as segregation for recessive mutations in diploid plants normally can be observed already in M_2, provided the number of offspring is large enough.

Cross-fertilizing species. Populations of cross-fertilizing plants generally are very heterogeneous as well as heterozygous. For each locus for which two or more different alleles are present in the population, heterozygosity will occur in a certain percentage of the plants. In elder populations of cross-fertilizing species grown in isolation, each individual plant is a relative to all other individuals of the population by previous recombination, which may have taken place during many generations. The sum of genes within such a population is redistributed each generation over the individuals of that population. Genes that are situated close to each other on a chromosome may remain linked for a long time. Mutagenic treatments may break such linkages.

A range of mechanisms may promote cross-pollination and cross-fertilization, e.g., gametic incompatibility, heterostylism in flowers, protandry and protogyny, the occurrence of dioecious and diclinous plants, etc. Cross-fertilization is the normal way of propagation for a number of very important crop species, including for instance maize (*Zea mays*), rye (*Secale cereale*), oil palm (*Elaeis gilinensis*), cacao (*Theobroma cacao*), many species and interspecific hybrids within the genus *Brassica*, grasses, clovers (genus *Trifolium*), sugar beet (*Beta vul-*

garis), onion (*Allium cepa*), carrot (*Daucus carota*), and several species of citrus.

In cross-fertilizing crops variation not only exists between cultivars but also within cultivars. The progeny of an individual cross-fertilizing plant is called a family. Various selection methods, such as mass selection, family selection or progeny testing, recurrent selection and selection for hybrid vigour can be followed in cross-pollinated crops. Hybrid cultivars are common within this group. In cross-fertilizing species considerable genetic variation may be maintained for most traits. As a consequence, the main reason for applying mutation techniques is not to induce additional genetic variation in general but for specific traits.

Mutagenesis may be as useful in allogamous as in autogamous species, only for allogamous crops the application of mutation methods may become more complex, success is more difficult to verify and recessive alleles are considered less useful. Mutations are needed when the previous trend of selection has to be reversed. We will return to these subjects later.

Chimerism

Mutations are single cell events and, as a consequence, mutagenic treatment of seeds and other multicellular organs will result in plants which may carry one or several mutations, each one occupying a small part only. Such plants are called **chimeras**. Despite the fact that chimerism may be in particular troublesome for mutation breeding in vegetatively propagated crops (see also Chapter 7), breeders of seed propagated crops also have to be aware of the complications that are caused by chimerism. Therefore a special section of the present chapter (Section 6.2.8) is devoted to this subject.

For a majority of plant species it is not possible yet to regenerate plants *in vitro* from unicellular material like cell suspensions or protoplasts, in any case not on a routine basis and not without occurrence of 'somaclonal variation' (see Chapter 5). There are on the other hand, nowadays, quite a few seed propagated plants where various types of explants can be mutagenically treated. Use is made in such cases of immature embryos or mature ones from ripe (but not yet dry, and often presoaked) seeds, calli derived from embryos, young inflorescences or parts of them, etc., with subsequent *in vitro* culture and formation of adventitious shoots. Such modern methods could help to resolve chimeras or avoid them, but do not solve the problem of chimerism completely.

Only mutagenic treatment of unicellular plant material would – with a few exceptions – exclude the occurrence of chimeric plants. Gametes, especially pollen grains, are therefore of particular interest for mutation breeding of seed propagated crops. Their use and mutagenic treatment are treated extensively in this chapter.

Another, possibly alternative approach to avoid or limit chimerism is the use of adventitious shoots of single cell origin. As most knowledge concerning this subject has been collected for vegetatively propagated crops, this subject will be discussed in particular in Chapter 7.

6.2. Starting the programme: initial steps and theoretical considerations

6.2.1. Plant breeders and breeding goals in mutation breeding

The starting point of any breeding programme should be to formulate a (set of) well-defined breeding goal(s) and its basis should be the commitment of a well-trained plant breeder. This applies as well to mutation breeding. To start with the second condition, and as was said before: the production of new cultivars is a job for experienced plant breeders with a profound knowledge of their crops and who know the demands of 'the market'. Many mutation breeding programmes have failed because they were executed by the wrong persons. The plant breeder could be supported – but never replaced – by other specialists, for instance phytopathologists and molecular biologists.

Before starting any mutation breeding programme it should then be carefully considered whether this approach indeed may offer realistic prospects to achieve the pre-set breeding goals. As was mentioned before, in most cases a breeder has to choose among different sources of starting material and between various methods and his choice will be largely determined by economic considerations like: which approach is the quickest, the cheapest or the most easy or certain one. In a number of situations, when other methods that were tried so far have failed, only one suitable alternative may be available to reach the goal and this might be distant hybridization, genetic engineering or mutation breeding.

While considering the mutation breeding method in order to improve a specific trait (or set of traits), a number of questions should be raised. The answers will contribute essentially to making a realistic decision. Such questions are, for instance, the following.

– What is known with respect to kind and number of genes controlling the trait(s) in question? Is there anything known about linkages or pleiotropic effects?
– What is known of this crop or cultivar with respect to its system of generative propagation (self-fertilizer, cross-fertilizer), the ploidy level and kind (auto- or allo-), the degree of heterozygosity? Can it be vegetatively propagated 'in vivo' or in vitro?
– What would be, from a genetic point of view, the best plant material (or cultivar) to start with?
– Which material should be treated (seeds, gametes) and how much?
– Which method of treatment should be preferred and which doses or concentrations should be taken?
– How should the treated material be handled and how large should the selected population be?
– What is the most efficient procedure of selection for the target trait(s) of the mutation programme? Is the procedure already established? Could it be afforded?

This list, of course, is not complete and it may be clear as well that in many cases not all of these and other relevant questions can be answered. In several publications guidelines for the implementation of mutation programmes in seed propagated crops can be found. Very useful in this respect are the IAEA *Manual on Mutation Breeding* (Anon., 1977a) and a treatise by Konzak (1984). A similar list of questions for vegetatively propagated crops will be discussed in Chapter 7 (see also Broertjes & van Harten, 1988).

6.2.2. Preliminary experiments

Well-designed mutation breeding programmes should be preceded by preliminary experiments with various (but always the best available) cultivars of a crop species and the application of a small range of doses (because cultivars may differ with respect to their sensitivity to mutagens). For economic reasons such initial experiments in most cases will be limited to M_1 observations of not more than a few hundred objects per dose and cultivar.

When in literature treatment conditions have been given for previous successful experiments with similar targets, the best approach is to start – on a moderate scale – preliminary experiments with the most appropriate doses used in such experiments as a point of reference and to extend the programme with, for instance, treatments with doses 25% to 50% higher and lower. If figures for survival rate or growth reduction differ considerably from the results in previous experiments, it would be advisable to repeat the test series on the basis of the newly gained information, keeping all other treatment conditions as much as possible similar to those in the first test series. When in literature no previous figures can be found at all for the crop of interest or related crops, it is advisable to test a much wider range of doses and on a larger scale.

6.2.3.   The starting material

For mutation breeding always the best available and well-adapted material or cultivar(s) should be taken, i.e. the cultivars with the highest yield, the highest level of resistance against important diseases, etc. Moreover, efforts should be concentrated on one target only.

Negligence of the aforementioned points quite often has made that the practical value of an improved trait – even in the case of a spectacular increase in for instance level of resistance or increase protein content – remained very limited or that a lot of, in fact, unnecessary, additional work had to be done to round off the project with a satisfactory end result. A deterrent example in this respect was discussed already in Chapter 3, Section 3.2.2. If the famous work of E. Sears to transfer resistance for brown rust (*Puccinia triticina*) from *Aegilops umbellulata* to wheat (*Triticum aestivum*) by an induced translocation had been performed with another cultivar of wheat than the – at that time already outdated – cv. Chinese Spring, the practical impact of this work, undoubtedly, would have been much higher.

Another significant point to keep in mind is that use should always be made of seed that is of high quality and genetically as uniform as possible, e.g. from an advanced line or elite seed stock, grown in isolation. Seed impurity in the starting material, e.g. from uncontrolled outcrossing, which occurs also in predominantly self-fertilizing crops, may account for quite a few unjustified 'claims' of unusual, dominant genetic variation that was observed after mutagenic treatments. Well-designed, large-size control plots allow one to check whether the starting material was pure.

It is further strongly advocated that in the cultivars or genotypes used in a mutation programme, the improvement of only one or very few traits at a time is attempted. The question may arise whether it would be advisable to try to change for instance some yield components or to induce short-straw plants by mutagenic treatment in an already highly disease-resistant cultivar, or to start from a high yielding cultivar and to improve one of the other traits. The answer to this question depends in particular on the mutability of the various traits and – even more important – on the efficiency of screening. Throughout this book many contributions

concerning the efficiency of screening for mutants are discussed. In addition, mention is made at this place to a rarely cited publication on this topic, i.e. the proceedings of a special meeting on the efficiency of various breeding methods for screening of mutants, organized within the frame work of the *Gamma Field Symposia* in 1975 in Japan (Anon., 1975b).

It is sometimes observed that a specific genotype is more difficult to mutate for a given trait than another one or even that no mutations occur at all for the given trait. A possible explanation for this observation could be that the locus concerned in that specific genotype does not exist anymore, i.e. that a deletion is present at that position of the chromosome (Konzak, 1984). In such situations the possibility to obtain mutations for that trait would be higher if one would start with several genotypes.

6.2.4.   Which doses?

The ultimate aim of a mutagenic treatment is to induce mutations leading to genetic improvement of a specific trait. It has been mentioned before that selection of the desired traits normally starts not earlier than in the M_2 generation. Nevertheless, decisions about the 'most appropriate dose of mutagen' are often taken on the basis of observations on M_1 plants. The question whether it is justified to assume a correlation between such findings in M_1 and mutation frequencies in M_2 and later generations, will be discussed in Section 6.3.

In practice, for radiation treatments often a growth reduction for M_1 seedlings of 30–50% or a survival rate of 40–60% in comparison to the control plants is taken as a criterion for a promising treatment. Radiation doses commonly used in case of seed treatment, for instance, are 300–450 Gy of gamma rays or X-rays for barley (*Hordeum vulgare*), 450–600 Gy for tomato (*Lycopersicon esculentum*) and 100–350 Gy for pea (*Pisum sativum*). More details are presented in Section 6.3.3. Suitable radiation doses for treatment of pollen are much lower than the doses that are commonly applied to seed and may be in the range of a few grays only. Examples for pollen will be presented in Section 6.3.4.

On empirical grounds, for chemical mutagens in general much lower percentages of lethality, growth reduction or sterile M_1 plants are advised, but for various mutagenic chemicals such figures are not necessarily equally correlated to mutation frequencies and treatments with different chemicals may produce a very different outcome. Some practical examples have been discussed already in Chapter 4. For treatments with chemical mutagens it is always advised that if seeds are sown immediately after mutagenic treatment (followed in most cases by some hours of post-washing), gloves and protective clothing should be used to avoid all health hazards.

In recent years there has been an increasing awareness among plant breeders that their aim cannot be to induce high overall frequencies of mutations, but to obtain a reasonable number of useful mutations for a specific trait, without considerably changing the genetic background and without diminishing the fitness of plants in which useful mutations may have been induced. As a result, treatments are performed nowadays with lower doses than used to be recommended in the past.

Some authors have tried to define an 'optimum mutation rate'. Based on frequencies observed for chlorophyll mutations Yonezawa & Yamagata (1977) calculated that, in order to obtain directly useable mutants (i.e. without too many background mutations) in crops like barley and rice, mutation rates of not more than about 10^{-4}–10^{-5} per allele are to be preferred. Such rates, which are expected to coincide with about 10% chlorophyll mutants among the M_2 seedlings, may be obtained after seed treatment with doses below 50 Gy of X- or gamma rays (which is considerably less than the abovementioned doses of 100–600 Gy), or with concentrations below 2% during 1 h for EMS solutions. But, assuming that a plant genome comprises more than 50 000 genes and that (only) 200 loci are involved in chlorophyll biosynthesis, Micke (personal communication) points out that at the mutation rates suggested by Yonezawa & Yamagata, each cell would still contain up to ten different mutations (in the background). In cases of lower mutation frequencies, of course, more plants have to be treated and their offspring screened in the experiments. On the other hand, Yonezawa & Yamagata suggest that more powerful mutagenic treatments could be used for producing new, raw genetic variation for breeding stocks and hybridization. In that case a mutation rate up to 10^{-2} could be acceptable, which rate, according to these authors, is expected to be realized when about 65% of the M_2 seedlings emerge as chlorophyll mutants. Micke, in this context (personal communication), comments that at such high mutation rates each cell could contain up to 1000 mutations in the background which, of course, would be far too much, even for 'germ plasm'! More details will be presented later in this chapter.

6.2.5. First steps after mutagenic treatment

After mutagenic treatment, the resulting M_1 plants, as much as possible, must be protected from outcrossing, for instance by raising them on plots isolated by a wide border of control plants. Bagging of all flowers or inflorescences of M_1 plants, of course, would be the safest method, but this procedure often is far too laborious and costly given the required scale on which a properly designed mutation breeding programme should be performed. The use of recessive marker genes could help to detect and consecutively eliminate contamination by outcrossing.

In order to create reasonably uniform experimental conditions and to limit the effect of chimerism as much as possible, it is advised – at least for cereals – to sow at high seed density in order to obtain all of the progeny from spikes of the main culm and primary tillers.

Harvesting seeds from M_1 plants requires special attention. The appropriate way is determined by considering the specific properties of the plant species, the planned method of screening for desired traits and the generation in which screening will take place (see for instance Gaul, 1965b). For monocotyledons (cereals) a maximum level of genetic variation is obtained when seeds are harvested only from the main culm and primary tillers. For dicotyledons, seeds should be sampled from the main shoot and from each main (= early developed) branch. In each case the number of seeds harvested per tiller or per branch should be high enough to enable detection of the recessive genotypes with a high probability, i.e. 10–20 seeds.

6.2.6. Initial cells and the germ line

The role of **initial cells** must be briefly explained before it can be discussed how many plants have to be examined per M_2 family in order to detect segregants at a predetermined level of probability. Shoot development is usually directed by only a few meristem cells which are situated in the most distal part of the shoot apical area. Usually three cell layers (**histogenic layers**) are represented in the shoot apex and each layer contains one or a few of such meristematic cells, the so-called **initial cells**, situated in the most distal positions of these layers.

When plants are propagated by seed in which mutations are present – either resulting from a mutagenic treatment or of 'spontaneous' origin – those mutations are transmitted to the next generation only when they were induced in the initial cells of the sporogenic layer of the embryo. The somatic offspring of such cells are called **germ line cells** or **genetically effective cells** (GECs). Following Rieger et al. (1991), the term **germ line** had been coined in 1885 by A. Weismann to indicate a lineage of 'generative' cells that are ancestral to the gametes. These ancestral cells, together with the gametes are called **germ cells**, as opposed to somatic cells. (N.B. In dicotyledonous crops the germ line cells are part of the so-called L-2 tissue; in monocotyledons

L-2 and L-3 tissue may be involved. More details about histogenic layers and related topics will be discussed in Chapter 7.)

Depending on the seed, which is characteristic for a plant species, an average of 2–5 initial cells are estimated to be present for the germ line within the embryonic shoot apex. Single pollen grains and egg cells, basically, consist of one GEC. The **number** of **genetically effective cells (GECN)** can be determined by genetic methods. If among M_2 plants the segregation ratio is 3 : 1 (3 normal, 1 recessive mutant), the GECN in the treated material (M_1 seed) must have been one. In case of a 7 : 1 segregation two apical GECs must have been present in the embryo of which only one mutated. Kranz (1984) estimated for *Arabidopsis thaliana* that in the seed embryo of this small weed several thousand cells are present, of which less than 20 cells are meristematic shoot apical cells. During the first 36 h after imbibition, the germ line of the seed only consists of two initial cells, but within a week, according to Kranz, the GECN has grown to a dozen. This indicates that the GEC concept only works reliably in dry resting seeds but becomes uncertain after soaking the seeds, when the GECN may quickly increase by apical divisions. Presoaking of seeds before the mutagenic treatment could result in an increasing number of smaller mutated 'sectors'. On the other hand, high mutagenic doses may inhibit meristematic cells and could lead to a reduction of the GECN.

If a mutation occurs in an initial cell, this cell will produce a mutated cell lineage in the M_1 plant and if more than one initial cell would be present, the other initial cell(s) would not carry the same mutation, because mutations are single cell events. In other words, a mutated M_1 plant will be mutated for a part unless only one initial cell was present.

Li & Rédei (1969) have described what will be the fate of a mutation in a genetically effective egg cell of a self-fertilizing species (e.g. *Arabidopsis*). It is assumed that two haploid egg cells (note that two haploid egg cells equal two initial cells), both with genetic constitution A for a given trait, are present. If one egg cell mutates towards recessive ($A{\rightarrow}a$) and both cells are pollinated with non-mutated pollen (A), the resulting M_1 population will consist of 1 AA and 1 Aa plant, provided that no selection against the recessive allele has taken place. Even if the traits A and a can be discerned already on the somatic tissue (e.g. hairs on leaves present or absent), and if dominance is complete, both M_1 plants will look identical. Selfing of the M_1 plant with constitution Aa will result in segregation in M_2 in 1 AA + 2 Aa + 1 aa. Accordingly, selfing of the AA plant will result in AA plants, i.e., 4 AA. Consequently, the segregation of recessives in M_2 will be

one out of eight (12.5%). Continued strict selfing of each individual plant will show that, in addition to the recessive aa plants that had been detected already in M_2, more recessives will show up in the M_3 as a result of the segregation in the progeny of the two heterozygous (Aa) M_2 genotypes which, at that stage, could not be phenotypically distinguished from the AA plants. This is shown in Fig. 6.1, derived from Li & Rédei (1969).

Accordingly, when four initial cells are involved, the M_2 will show one recessive plant out of sixteen and, in addition, two out of fifteen normal looking M_2 phenotypes will segregate in M_3.

Yonezawa & Yamagata (1975a,b) already emphasized that delayed selection of mutants, e.g. not before M_3 to M_5, is a necessity (for statistical reasons) for quantitatively inherited traits, but also has practical advantages for single-locus mutants, when either the mutant character is not easily detectable or when the population size per generation must be limited to for instance less than 10 000 plants. But to benefit from a rather small increase in homozygous mutants in M_3, a population much larger than the M_2 has to be grown (See also Chapter 3 and Section 6.4.2 of the present chapter).

6.2.7. Generation size

How much material should be treated and of how many plants should a progeny (i.e. the M_2) of each mutagenically treated plant consist in order to detect segregants with a reasonable probability? This last figure is important also in order to determine the M_1 size. If a breeder, for instance, could handle up to 100 000 plants in the M_2 generation, but if from each individual M_1 plant 100 M_2 plants would be required, the maximum number of surviving and seed producing M_1 plants could be only about 1000, which would be far too low. Although the required number of M_2 individuals will be discussed later in this chapter, it can be said here already that this number is determined by the segregation ratio for a single recessive gene and the probability of occurrence of homozygous mutants. If pollen was treated, the expected mutant segregation ratio would be 3 : 1, but if seeds were treated often more than one cell of an embryo contributes to the next generation requiring a doubling of the plant progeny size.

A practical point to consider is that the number of seeds per plant or per fruit and the number of fruits per plant may be very different. To illustrate this point, two (rather extreme) examples are given. In peanut (*Arachis hypogaea*) 2–3 seeds may be present in up to 40 underground pods, which brings the total number of M_2 seeds per plant up to about 80. For jute (*Corchorus olitorius*) a single plant may carry 100 fruits with each fruit containing more than 150 seeds. In this last example it would be possible to produce an M_2 of 100 000 plants from about ten M_1 plants only, but this certainly would not be a suitable method in mutation breeding.

The size of the M_1 generation

Let us try to answer the question of what should be the proper M_1 size if one would like to obtain at least one mutagenic event after mutagenic treatment with probability P.

Several authors, for instance Rédei (1974a,b; 1982a), Brock (1977) and Brock & Micke (1979), have discussed this topic and related questions. Basically, they applied the following formula (in our notation).

$$n = \frac{\log(1-P)}{\log(1-f)} \qquad \text{Formula 6.1}$$

In this formula n is the number of treated 'units for mutagenic event' (or 'treated cells' according to Brock, 1977), and f is the theoretically expected fraction of mutations.

The number of mutations – events that happen with a low probability – in a population follows a (discontinuous) Poisson distribution. On the assumption of such a Poisson distribution for the number of mutations, probability tables are available from which it can be derived how many treated seeds or how many progenies (plants) are required to obtain at least one mutation or mutational event for a trait (or any other number of mutations) with a level of confidence of, for instance, 90%, 95% or 99%. Such tables can be found in many books on statistics like Snedecor & Cochran (1976) or in a contribution by Bogyo (1991) during a FAO/IAEA Symposium in Vienna in 1990.

If, following Rédei's example, a mutation frequency of 3×10^{-4} per 'unit' is assumed for a specific trait and if the P value would be set at 99%, this would result in

$$n = \frac{\log 1 - \log 100}{\log 9997 - \log 10\ 000} = 15\ 348 \qquad \text{Formula 6.2}$$

The resulting figure of 15 348, for several reasons, must be corrected in order to provide us with reliable

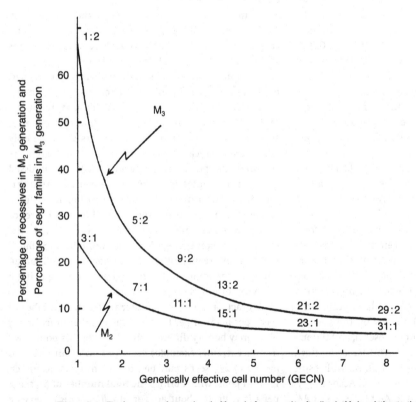

Figure 6.1. The frequency of recessives in M_2 and of segregating familes in M_3 in relation to the genetically effective cell number (after Li & Rédei, 1969).

information about the actual number of seeds required in M_1.

First of all, use is made of diploid seeds, in which two (allelic) genomes are present in each somatic cell. This implies that the number of seeds to be treated could be reduced to half, say to about 8000. Moreover, if more than one genetically effective cell (GEC) is present in the germ line and contributes to the gamete formation (and produces a sufficient progeny of M_2 seeds), this number of 8000 could be further reduced. For instance, if two (diploid) GECs would be present in the embryonic shoot apex in a seed, mutagenic treatment of in total about 4000 seeds would be adequate. However, when the number of GECs in the treated seeds goes up, for instance from one to two GECs, the progeny size in M_2 should be doubled to account for chimerism. More details will be presented further in this and later sections. In addition, as not all mutagenically treated material survives and produces seeds, a further correction has to be made. In order to obtain an M_1 consisting of 4000 surviving plants, and assuming a degree of sterility of 20%, the number of seeds to be treated should be increased to 5000 plants to compensate for sterile plants. These figures are in agreement with previous calculations by Rédei (1974a).

As an example, Kranz (1984) reported for an 'in vitro' experiment that, after mutagenic treatment of *Arabidopsis thaliana*, one independently segregating (i.e. single-locus) mutation for a specific nutrient-deficiency may be found in the progeny of 1000 surviving M_1 plants. Because of the presence of two genetically effective cells and two genomes in each cell, the mutation rate per genome for such traits would be 2.5×10^{-4}. This figure is well in agreement with the assumption made in the calculation of a mutation rate of 3×10^{-4}. For other mutation frequencies, of course, figures concerning the number of required plants would have to be corrected.

Most induced mutations are accompanied by other mutational events, either because more than one (independent) mutation has been induced in a germ cell or because of pleiotropic effects expressed by the mutated gene. Independent mutations will segregate in later generations, whereas real pleiotropic effects will last. Pleiotropic effects may be detrimental for the outcome of a mutation programme, for instance when a (desired) mutation for male sterility should be accompanied by growth retardation of the seedlings. Practice has shown that when, for instance, ten mutants for a desired trait are independently produced, these ten mutants may differ in their expression for this and other traits. This observation is very useful for practical mutation breed-

ing as among the ten mutants there is a chance for a mutant in which the negative effects caused by pleiotropy (or close linkage) are less pronounced. To use this option the M_1 population, in this example, must consist of ten times the size that was calculated before, or about 50 000 plants.

Despite the fact that this figure may put some breeders off, this required number is a realistic estimate. In practice, the M_1 generation usually is sown in bulk and, hence, does not require much space and effort until harvest and for safety reasons one should even consider, when possible, to grow more M_1 material than is strictly needed.

Some authors claim that under certain conditions the number of genomes that actually can be studied, may be much higher than is often anticipated. Rédei (1982a) exemplified that a medium sized cob (or ear) of maize (*Zea mays*) carries about 350 triploid kernels and that each kernel shows about 1400 cells at the surface in which dominant mutations for cell colour (which may have been induced after mutagenic treatment of pollen), should be readily revealed. Hence, a single maize ear may show mutations in $350 \times 1400 \times 3 = 1.5 \times 10^6$ gametes. Rédei further claims that, at least theoretically, it even would be possible to trace a rare dominant mutation for a visible trait, occurring with a frequency of only 1×10^{-6} within one single cob! This example, although interesting, may be somewhat out of touch with reality, as it refers only to a set of visible endosperm colour genes.

Finally, for additional details about handling the M_0 and M_1 population, reference can be made for instance to the IAEA *Manual on Mutation Breeding* (Anon., 1977a) and Dellaert (1982; 1983).

The number of plants to be examined per M_2 family

Before discussing this question it should be clear that the progeny of each (selfed) M_1 plant is called an M_2 family.

In order to calculate correctly frequencies for recessive mutations for diploids – again on a genome basis – Li & Rédei (1969) propose the following formula:

$$R = \frac{M}{S \times GECN \times 2 \times D} \qquad \text{Formula 6.3}$$

in which R = mutation frequency; M = the number of M_2 families segregating for recessive mutants; S = the number of surviving plants; GECN = the genetically effective cell number of the mutagen-treated tissues (germ line); 2 = the correction factor for diploidy. In addition, a factor D was added which apparently signifies a value for the dose of the mutagen treatment to which the plant material was subjected. The factor D is sometimes omitted, for instance by Kranz (1984) in his calculations for *Arabidopsis*, since the number of segregating M_2 families already is a measure of the dose of the mutagenic treatment. But the GECN may vary with the applied dose.

How many plants must be examined per M_2 family? This number (m) is, in our notation, determined by the desired probability (Q) of occurrence of or, better, detecting (= not let escape unnoticed) at least one induced homozygous recessive mutant (Q) and the mutated fraction (a) for a single recessive gene according to the formula:

$$m = \frac{\log (1 - Q)}{\log (1 - a)}$$
\hfill Formula 6.4

Earlier in this section reference was made to several research workers who have described approaches for calculating the optimal size of M_1 and M_2 populations and published additional details. Micke (personal communication) comments that one has to aim at the optimal combination of a high probability of not missing an induced mutant (by taking a large number of individuals per progeny), and the highest permissible number of progenies (up to a chance of getting 10 independent mutations in the same locus).

It was explained before that the assumed presence of two GECs in the treated embryo (in the absence of any discrimination) results in a segregation ratio of one to seven (or a = 0.125). Given this ratio the (theoretically) required number of offspring of each M_1 plant (= the size of the M_2 family) is 34.5 plants when Q = 0.99 (and 17.2 when Q = 0.90). In this case a population of, say 5000 fertile M_1 plants would result in 5000 progenies × 34.5 seeds = 172 500 plants required in M_2.

As an alternative one could plan on 10 000 fertile M_1 plants with 10 000 progenies in M_2 but only 17.2 plants per progeny. This would reduce the probability of detecting a recessive mutation in a particular locus to 0.90, but would increase the overall chance of obtaining induced mutations by doubling the number of 'M_1 cell progenies' examined.

It may be clear that it will often be difficult or even impossible to handle such large numbers of plants in the form of plant progenies, but these figures indeed indicate the optimal scale of performing mutation breeding programmes. The degree of difficulty depends upon the plant species and the objective of the experiment. For instance, in a cereal an M_2 size of 50 000 plants, according to several researchers, can be handled

well in practice – even in progenies – if appropriate machinery is at hand. Even in this situation the question remains, however, whether one should reduce the number of progenies or the number of plants per progeny. In general, it is considered better to maintain or even increase the number of progenies at the expense of the number of plants per progeny (Brock, 1979).

6.2.8. Chimerism in seed propagated crops

Chimerism – which phenomenon was already briefly introduced in Section 6.1 – most often, is considered a complicating factor in mutation breeding, and although chimerism is much easier to handle in seed propagated crops than in vegetatively propagated crops, this subject still deserves attention in this chapter. Chimerism is not always an obstacle. In fact, induced chimerism if properly handled can be even advantageous to the plant breeder as it allows the sampling of a larger cell population per mutagenically treated (M_1) plant (Constantin, 1983).

Mutations in initial cells

After mutagenic treatment of seeds a number of cells in the shoot apical area of embryos may be mutated, possibly including the few particular meristematic cells (= particular initial cells) that give rise to the pollen and egg cells of the M_1 plants. For strictly seed propagated crops only mutations in those cells (GECs), which normally belong to the sub-epidermal or sporogenic layer, can be of value. (N.B. One should keep in mind that sub-epidermal cells that are not 'real initial cells' may give rise to gametes in flowers that develop from axillary secondary meristems or from tertiary meristems on branches. For larger embryos, the most apical initial cells are only responsible for the flowers in the upper part of the plant; for small, undeveloped embryos such most apical cells account for all flowers.)

If a mutation is induced in one of these initial cells and if the mutated cell is able to divide and grow in a more or less regular way, this mutated cell will produce a mutated zone or 'sector' within the M_1 plant. A recessive mutation in one gene A (in diploid tissue from AA → Aa) normally will remain unobserved in the heterozygous M_1 plant but the mutation will segregate in the M_2. If we assume for instance the presence of four sub-epidermal initial cells (= 4 GECs) with the genotype AA, from which only one is mutated, the mutated heterozygous (Aa) area – given equal vitality, equal division rates, etc. – would represent only 25% of the total cross section of the sub-epidermal layer of the M_1 plant and homozygously mutated (aa) plants in the M_2 would segregate 1 : 3 only from 25% of the seeds (= 1: 15). In an

M_1 plant, however, several mutations may have been induced at the same time and, as a consequence, the M_1 plant can be multiple chimeric by several, genetically different, heterozygous sections.

Chimerism in relation to ontogenesis and
architecture of the plant: principles and examples

Mutagen induced chimerism in both monocotyledonous and dicotyledonous seed propagated plants is still not completely clarified. This statement was already made, for instance, by Balkema (1971) and in IAEA TECDOC 289 (Anon., 1983b) and not much has changed since. Much more than mutagen used, dose and duration of the treatment, it is the stage of development of the plant (seed embryo) during treatment and the typical ontogenesis and organogenesis of the species that play an important role. Studies, for instance at the IAEA Laboratories in Seibersdorf, Austria, on faba bean (*Vicia faba*), green pepper (*Capsicum annuum*) and flax (*Linum usitatissimum*), confirm the conclusion from many earlier investigations that mutated areas or 'sectors' in M_1 plants do not occur at random throughout the plant but follow a pattern typical for the species (Hermelin et al., 1983). This underlines the importance of applying appropriate harvesting methods for obtaining high frequencies of independent mutants in the most economic way. As an example, this subject was discussed for maize (*Zea mays*) for instance by Coe & Neuffer (1977) and Walbot (1983).

The mutated zone in an M_1 plant in fact is 'labelled' by the mutation, which allows a topographical analysis of the distribution of mutations in the M_1 plant. By harvesting the seeds of individual M_1 flowers separately and by sowing them according to their identified location within the plants, the chimerical structure of the M_1 plants can be reconstructed on the basis of evaluating the pattern of mutant gene segregation in M_2 and later generations. As the M_2 generation only partly reveals the induced mutations, depending on progeny size per fruit (e.g. pod), it is in most cases necessary to study later generations as well. For pea, for instance, 60% of the mutations were not detected before M_3 (Weiling & Gottschalk, 1961).

Nybom (1956) already reported that after X-irradiation of seeds of barley (*Hordeum vulgare*), a monocotyledonous crop, the mutated M_1 plant is chimeric for induced mutations and the heterozygous mutant tissue is mostly restricted to one spike or even less than a spike. For the same crop Frydenberg, Doll & Sandfaer (1964) determined frequencies of chlorophyll mutations in M_2 and found that the frequency of mutated X_1 (= M_1) spikes is higher with a factor 1.5–2 among spikes of which the initials were already present ('pre-formed') in the seed before mutagenic treatment (M_0) than in 'post-formed' spikes. The authors added that the frequency of mutant X_2 (M_2) plants in mutated X_1 spike progenies was lower by a factor of almost 2 in 'pre-formed' than in 'post-formed' spikes. Osone (1963) similarly found for rice (*Oryza sativa*) that the highest mutation frequencies (= number of different independent mutations) can be recovered in panicles from the main culm and earlier formed primary tillers.

Frequencies of chlorophyll-deficient seedlings in segregating (spike) progenies, for instance after radiation treatment of a cereal, often was less than the 3 : 1 ratio that was expected if a single mutated initial cell had been mutated. However, when species contain more inital cells in their apical meristems, treatments of seed with increasing doses of radiation was found to lead to more normal M_2 segregation ratios, i.e. to a relative increase of the percentage of recessive mutants, up to the regular 3 : 1 ratio. This is often explained by killing or inactivating some of the functional initial cells. Beard (1970), for instance reported for flax (*Linum usitatissimum*) that from a total of 65 M_1 branches that showed chlorophyll mutants in M_2 and M_3 after treating seeds with 300 Gy of X-rays, 69% were non-chimeric (i.e., were derived from a single initial cell), whereas 3% of the branches had a 'sector' size of less than 20%. This could be interpreted as flax normally having five inital cells in the sporogenic layer but that damage induced by such a high irradiation dose reduced in most apices the number of still functioning initial cells to one or zero.

According to Beard the method applied by him, to study chimeras by determining the limits of mutated 'sectors', appears to be more accurate than to use results from histological and morphological studies, which is the first and most common method. More details and further references can be found for instance in several publications by Gaul (1959; 1961b; 1963). A third approach: to study the behaviour of specially marked or mutilated meristematic cells of the apical zone (see for instance Ball, 1974), is not further discussed here. Some other early contributions about estimating the number of meristem initial cells, for instance, came from Weiling (1960) and Siddiq (1968).

After treatments with chemical mutagens, segregation ratios in general tend to show a larger deficit of recessive mutants than after radiation treatments (i.e., less than 3 : 1 segregation of recessives). This may be explained by a higher 'survival rate' of initial cells after treatments with chemical mutagens or by a higher 'starting number' of initial cells due to the presoaking that is commonly practiced for chemical mutagen treatment. For a good comparison with X- or gamma rays only presoaked seeds should be irradiated.

For a branched cultivar of pea (*Pisum sativum*), a dicotyledon, Blixt found that – as a rule – the mutated (heterozygous) area in M_1 plants was restricted to only one branch or part of a branch (Blixt, Ehrenberg & Gelin, 1958; Blixt, 1961). It was concluded that in order to study mutation frequencies in an accurate way, such frequencies should be related to the number of M_1 branch progenies and not plant progenies that have been investigated. In a later report Blixt (1972) analyzed for pea a chimeric M_1 plant, obtained after treatment with EMS, up to the M_5 generation. A picture of the aforementioned chimeric situation, derived from Blixt & Gottschalk (1975), is presented in Fig. 6.2.

In three separate branches (A,B,C) eight different mutant seedling traits were found. The number of mutant seeds for different seedling traits varied from six (for the trait 'narrow leaves') to one only (for the trait 'chlorotica') on a total of 83 seeds counted on the original M_1 plant. The mutations observed in the central (A-)branch and one side branch (C) did not overlap. According to Blixt these results indicate either that the branches A and C originate from separate initial cells each, or that the mutational events occurred later in ontogenesis, but the author added that his results were not conclusive on these points. From branch B a double-mutated ('narrow leaves' + 'chlorotica') plant was obtained in a later generation, indicating that either both mutations occurred in the same initial cell or (rather unlikely in late generations) that the flower in which the double-mutated seed had originated had been chimeric, i.e., developed from separate initial cells. In this experiment the position of mutant seeds within a pod was not exactly established. Blixt further concluded that the size of a mutated 'sector' obtained after EMS treatment was up to eight seeds (for the highest EMS concentration) whereas for gamma rays larger 'sectors' with ten or more seeds were found. Larger 'sectors' in this respect imply fewer active initial cells after the mutagenic treatment. An important practical conclusion from Blixt's work is that for investigations concerning the mutated 'sector' in *Pisum*, the best unit to handle are pods and not stem-branches including secondary branches as was suggested on earlier occasions. According to Blixt (1972) the position of each seed within a pod should be ascertained as well. However, before justified conclusions about mutated 'sectors' can be

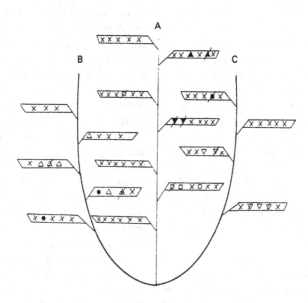

A, the main stem, B and C different stem branches. The symbols denote the following mutants: chlorotica I □; chlorotica II ○; chlorotica III ●; chlorotica IV ■; narrow leaves △; grey leaves ▲; chlorina ▽; chloroticamaculata ▼. The heterozygous seeds are denoted by a slash /.

Figure 6.2. Chimeric situations of an M_1 plant treated with EMS, as revealed by analyses from M_2 through M_5 (after Blixt & Gottschalk, 1975).

drawn on the basis of seed position within the pod, detailed information is required about the way in which for a specific plant species the ovules are initiated within the ovarium.

Hildering & Verkerk (1965) reported on the chimeric structure in plants of tomato (*Lycopersicon esculentum*) after treatment of seeds with EMS or X-rays. (N.B. It was not mentioned whether for seed irradiation use was made of dry or presoaked seeds which, as was mentioned earlier in this chapter, could be of interest with respect to the number of initials present at the time of treatment.) Seeds were harvested from fruits of the first five clusters along the main stem of M_1 plants and sown apart for each fruit. Visible mutations were scored on the base of M_2 progenies per fruit for, altogether, 200 000 M_2 seedlings, grown from in total 320 EMS- and X-ray-treated M1 plants. It was concluded that 1–3 cells do give rise to the sporogenic tissue of a tomato plant and that treatment with X-rays leads to a more severe elimination of initial cells than is the case for EMS. Balkema (1971) in this respect commented that ionizing radiation induces more chromosomal aberrations than EMS, which seems to result in more mitotic delay or elimination of cells and, hence, in fewer effective or a lower number of initial cells and an increased size of mutated sectors. Within the group of chemical mutagens, as has been discussed in Chapter 3, of course considerable differences may exist with respect to the amount of chromosomal aberrations. In the work by Hildering & Verkerk (1965) chimerism, predominantly, was found in the lower part of the tomato plant up to the second truss. The characteristic sympodial growth pattern of tomato causes that chimeric structures resolve between the first and the second truss. From a practical point of view, the most important, still preliminary conclusion of this work was that by sowing seeds of the first two trusses (clusters) of fruits, all mutations induced in the sporogenic tissue of the irradiated seeds are sampled. The position of the fruits within those trusses only rarely seemed to be relevant in this respect.

This work on the topography of induced mutations in tomato was continued and extended by Verkerk (1971) who, in a model study with detailed recording, treated seeds with 30 and 40 Gy of fast neutrons. In the experiment by Hildering & Verkerk (1965) harvesting of M_2 seeds was done only from fruits of two trusses of the main stem of all M_1 plants. In the later experiment many lateral shoots were produced by topping all M_1 plants above the second truss. These lateral shoots were cut off at a length of 15 cm, rooted and consecutively planted out in a greenhouse. Subsequently, in addition to the seeds from the first five fruits of the lowest two trusses, seeds of 3–5 fruits were harvested together from each of the first two trusses of these rooted lateral shoots. In planting the M_2, seed lots from different fruits were kept apart. The cotyledons and the next two or three leaves were scored after 10 and 20 days, respectively, for visible (morphological and colour) mutants. Progenies from 167 irradiated mother plants and eight control progenies were investigated. A distinction was made between 'singles', mutations found at a single sampling-point at an M_1 plant only, and 'sectors', mutations with a so-called vertical distribution, observed at various sampling points along the stem. Verkerk concluded that at the moment of seed irradiation the initial cells of the sub-epidermal layer of the first 5–7 lateral shoots are laid down already (in axillary meristems) in the embryo, which results in independent mutations ('singles') in only one side shoot. The initial cells of the terminal shoot apical area form all plant parts above about the sixth–ninth leaf onwards. The number of initial cells of the sub-epidermal layer of the terminal apex probably is limited to three cells only.

The results obtained confirmed what probably is true for most dicotyledonous species, namely that the highest number of independent mutations is recovered in seeds harvested from the lower part of the plant and that fewer but larger mutated sectors persist higher up in the plant, i.e. uniformity in that area increases in tomato, because of the aforementioned sympodial growth.

Earlier in this chapter brief reference was made to large-scale and very detailed experiments by Beard (1970) to estimate the number of initial cells and to study chimerism in seed flax (*Linum usitatissimum*), a self-fertilizing dicotyledonous crop. Seeds were irradiated with 300 Gy (30 kR) of X-rays and M_2 seeds from the five most basal branches were harvested separately for each M_1 plant without making a distinction between primary and secondary stems. Seed flax is recognized by short, coarse stems which are extensively branched. After the primary stem has elongated 10–15 cm, the first two (secondary) basal branches appear, one in the axil of each cotyledon. Later, other basal (secondary) branches originate, also around the cotyledonary node. In M_2 almost 50 000 'branch progenies' were studied for chlorophyll mutants.

Somewhat simplified, the screening procedure involved that if M_2 segregation ratios did not differ significantly from a 3 : 1 ratio (75% : 25%), the M_1 branch to which such results traced back, was considered to be not chimeric and derived from a single initial cell.

Recessive mutant frequencies in M_2, however, ranged from 4 to 30% (and in M_3 from 8 to 20%). It was found that in almost 70% of all cases studied only one initial cell had given rise to a particular branch of a flax plant; a two-cell and a three-cell origin was concluded in 12% and 11%, respectively, of the cases, and the presence of four or five initial cells was concluded only in 8% of all cases studied. Beard assumes that the primary stem may contain more initials than (the majority of) the basal secondary branches, but in the experiment no distinction was made between both groups. In addition, no distinction was made for the various positions of the secondary stems on the primary stem. In practice, this may be difficult, as all secondary stems – maybe except for the first two branches that originate at the cotyledonary node – originate very close to each other 'at any point around the cotyledonary node'.

Krausse (1982) studied for soybean (*Glycine max*, fam. *Leguminosae*) the induction of chlorophyll mutations by treatment with the chemical mutagen nitrosomethyl urethane (optimal treatment: 0.5–2 mM for 5 h). The author reported that M_2 plants referring to those regions of M_1 plants which were already determined by initial cells present in the dormant embryos, produced two or three times more independent chlorophyll mutations than M_2 seeds obtained from plant parts that were ontogenetically younger. The 'early determined regions' include the axillary meristems of the first (and probably the second) full leaf and the cotyledons, whereas higher situated leaves and their axillary branches originated later, i.e. after germination.

During the FAO/IAEA meeting on problems of chimerism in irradiated (mainly dicotyledonous) plants (Anon., 1983b), Kawai (1983) reviewed earlier literature on M_1 chimerism and advised for rice (*Oryza sativa*), which is a monocotyledonous crop, to harvest M_2 seeds from a small number of early developed panicles in order to obtain many mutants of independent genetic origin in a finite-sized M_2 population.

In the recommendations of this meeting it was concluded that, in particular with respect to the phenomena of chimerism in dicotyledonous M_1 plants, still much work would have to be done before adequate advice could be given to plant breeders. However, since that meeting not much additional information has been published. During the aforementioned meeting the architecture of shoot apices in dicotyledonous crops and their behaviour after irradiation was briefly reviewed by Davies (1983). This subject will be discussed more in detail in Chapter 7. However, some important aspects,

based on exemplary research on *Ligustrum* by Stewart & Dermen (1970a), will be briefly mentioned here already. Studies on (**mericlinal**) chimeric stripes in a **periclinal chimera** showed that one to three stable, but not permanent, initial cells were responsible for the growth of vegetative and generative shoots. This implies that no real permanent initial cells exist but that there are always some initial cells present and active at the most apical position of the shoot apex. Those initial cells divide at a very low rate: one division per twelve days being sufficient to maintain growth in *Ligustrum*. Following division, one of the 'first generation daughter cells' takes over the position and the function of the 'parent initial cell'. The authors further reported that one division of an initial cell, followed by further (and quicker) divisions of daughter cells, may be responsible for the production of cells that give rise to three to five nodes of the plant.

In dicotyledonous plants the organization of axillary buds plays an important role. Saccardo (1983a), during the aforementioned FAO/IAEA consultants' meeting in Vienna, reported on studies concerning induced chlorophyll and morphological mutations in order to elucidate chimera formation in M_1 plants and to study the ontogeny of sporogenic tissue in pea (*Pisum sativum*). Seeds were treated with either 0.2% DES for 15 h at 20 °C or 8 Gy of X-rays. In particular after DES treatment a positive correlation was found between M_1 injury (growth inhibition and leaf spots) and chlorophyll mutation frequency in M_2. For X-ray treatments this correlation was less clear but for both mutagens there was a clear correlation between M_1 damage and frequency of morphological mutations found in M_2. Results further showed that 50% of the mutated M_1 main stem but 77% of the branches are derived from only one initial cell, the 27% remainder from two or more (up to five) initial cells. As it was found that a pea seed embryo possesses already the main apex and several axillary apices, it is necessary to harvest the main stem and each branch separately. The results of Saccardo, on the whole, confirm earlier results obtained by Weiling & Gottschalk (1961) and Blixt (1972).

In order to determine the number of initial cells involved in the development of an M_1 plant Saccardo used two different methods a) the so-called 'topographical method' which was first described by D'Amato (1965) and b) a X^2 analysis for segregation of monogenic recessive chlorophyll mutations in M_2 with 1, 2 or 3 initial cells present in the respective M_1 main shoot or branch apex, described by Weiling & Gottschalk (1961).

Hermelin *et al.* (1983) found for faba bean (*Vicia faba*)

that most M_2 mutants were recovered when M_2 seeds had been collected from the second or third pod-bearing node at the M_1 plant but did not report on the number of independent mutations recovered from various nodes. For sweet pepper (*Capsicum annuum*) the highest level of segregation of mutants was observed in M_2 seeds from the fruits situated along the shoot up to the main bifurcation of the M_1 plants. For flax (*Linum usitatissimum*) the data confirmed that the number of sectors decreases, but that the sector size increases during ontogenetic development. Therefore, the easiest way to recover mutants may be to harvest M_2 seeds at the top position of the stem, as they will show more accurately a 3 : 1 segregation ratio (25% mutants). But in order to obtain as many different mutants as possible, the lower capsules should be preferred.

It may be useful to end this section with three general conclusions on (M_1) chimerism, drawn by D'Amato (1992) in a review on induced mutations for crop improvement of (mainly) seed propagated plants. These conclusions, which are of importance both from a theoretical and a practical point of view, are: 1) the size of a mutated area ('sector') in an M_1 plant increases with increasing dose of radiation and concentration of chemical mutagen; 2) because of the small size of the mutated 'sector' in many M_1 plants, only part of the actual number of mutations is recovered in the M_2 generation. The other mutations remain in heterozygous state and may not be observed before M_3. The proportion of mutations observed in M_3 may equal or even exceed the proportion recovered in M_2; and 3) the mutated 'sector' in M_1 after radiation treatment is in general larger than after treatment with chemical mutagens, probably because of the greater genetic damage (more chromosomal aberrations) caused by radiations.

Two comments should be made about these conclusions. With respect to conclusion 1) by D'Amato, one should distinguish between two different phenomena: a) the number of mutated cells and consequently the number of mutated 'sectors' will increase with increasing mutagen dose, and b) a further dose increase may cause an increase in 'sector' size at the expense of 'sector' number. In relation to conclusion 3) of D'Amato, Micke (personal communication) suggested as an additional explanation for the smaller 'sector' size in M_1 after treatment with chemical mutagens that such mutagens, as a result of 'delayed action' of chemical mutagens, still could be effective during or after first division or that subchromatid mutations could be induced (see also Blixt, 1972).

6.3. Mutagenic treatment

6.3.1. Introductory remarks

Different plant parts

For starting material, a breeder has several options. In general, mutagenic treatment of seeds and pollen is favoured. General aspects of seed treatment are discussed in Section 6.3.3 and, in addition, throughout this chapter most examples refer to mutation breeding programmes starting from seeds. Information on mutagenic treatment of pollen, egg cells, etc., is presented in Section 6.3.4 and as treatment of pollen, etc., is less common in practice, relevant data about achievements and other aspects of pollen mutagenesis are also brought together in this section.

Whole plants are relatively difficult to handle and require much space unless they are still in the seedling stage. As a consequence, treatment of whole plants in most cases is not considered attractive. It has been advocated on several occasions that treatment of whole plants with chronic irradiation (versus acute treatments) may have some attractive features. This subject was discussed in Chapter 4. The availability of a gamma source in the field or in a growth room facilitates treating whole plants but such facilities are expensive and, nowadays, not often available anymore. One principal problem of handling whole plants is that meristems in different stages of development are present, which makes the effect of mutagenic treatments less predictable and considerably increases the complexity of chimerism. Finally, it is often said that treatments of whole plants have been less successful than treatment of seeds and pollen, but it is doubtful whether enough experiments have been performed which allow for a valid comparison.

Treatment of leaves, explants, inflorescences, cuttings, etc., is not very popular for seed propagated plants; this is despite the fact that in vitro techniques have made this approach more attractive. This subject has been treated already in Chapter 5 and will not be discussed here again. Details about chimerism following treatment of whole plants, cuttings and other explants, will be discussed in relation to mutation breeding of vegetatively propagated crops in Chapter 7.

Diploids and haploids

Homologous genes are not mutated simultaneously and even if so, not in the same direction. Hence, in diploids a useful mutation towards the recessive allele still would be phenotypically hidden in a heterozygous plant. Heterozygous diploids might provide interesting starting material for mutation induction, but self-fertilizing diploids are homozygous for most of their genes. F_1 seeds may be easily produced, but their heterozygosity, of course, depends on the genes of the parents. Their advantage as starting material would be the exposure of two different genomes to the mutagenic treatment and not the immediate expression of a 'recessive mutation'. According to K. Borojevic (personal communication), irradiation of F_1 plants had been performed by Gregory in the 1950s (original publications not consulted) with peanut (*Arachis hypogaea*), a tetraploid self-fertilizing crop, and genetic variation was induced in addition to the genetic variation that resulted from hybridization. The main disadvantage of irradiating heterozygous diploids or tetraploids is the segregation in the F_2 (M_2) which does not allow us to distinguish clearly between recombinants and mutants.

The aforementioned problems do not occur when haploids are used. The two most important advantages of haploid mutagenesis are the early detection and easy isolation of recessive mutations and the possibility of obtaining homozygous diploids after doubling. Major disadvantages of using haploids are that many traits are not, or not optimally, expressed (especially grain characters) and that gene interactions and regulation of gene expression may be biased. A good introduction to the use of haploid cultures in relation to mutation breeding has been produced by Bajaj (1990a).

Saccardo (1983b), in a more general and somewhat different way, mentioned as most important advantages of mutagenic treatments of <u>gametes</u>, subsequently used in crossing with non-treated germ cells (see also Saccardo & Monti, 1984): 1) all induced 'visible' mutations that are not associated with haplophase lethality or embryonic lethality are detectable; 2) seeds for the M_2 can be harvested on any part of the M_1 plant (no chimerism) and a sufficiently large offspring can be obtained even in the case of severe sterility in the M_1 generation; and 3) the large numbers of M_2 seeds that can be obtained allows the phenotype to be observed in somewhat different genetic backgrounds by independent segregation of other mutations induced in the same cell.

Saccardo also mentioned a fourth advantage, implying that a relatively low number of – what were erroneously indicated as – M_1 plants should be required after mutagenic treatment of gametes. Saccardo in fact means M_0 plants on which M_1 seeds are produced by crossing between treated and non-treated gametes. Moreover, each M_1 seed produces one M_1 plant, representing only one GEC and, as a result, more M_1 plants would be needed than in the case of chimeric M_1 plants.

In a review about mutation breeding in sweet pep-

per (*Capsicum annuum*), Daskalov (1986, p.5) more precisely states the advantages mentioned by Saccardo (1983b) and also adds the following disadvantages of gamete mutagenesis: 1) one additional generation (= M_0) – albeit a small one – is required; and 2) the stage of development of the gametophytes and the treatment conditions are more difficult to control than with seeds.

When haploids are not available as starting material, mutated recessive genes first have to be made homozygous (by selfing or doubled haploid technique) to be detectable. The only exception is when rare mutations show dominant expression and when an efficient system of screening for such rare mutants (or mutant cells *in vitro*) is available.

Another point to be briefly mentioned in this section is general advice concerning mutagenic treatments. Brock & Micke (1979) suggest performing treatments with at least two different mutagens at doses which permit about 50% of the treated seeds to produce M_2 progeny. This may be a good starting point for a mutation breeding programme, in particular when experiments can not be based on results from previous experiments. It may be also useful to keep in mind a statement by the same authors, saying that the phase of inducing mutations is less critical than the selection phase that follows afterwards, provided of course that the mutagenic treatment has been effective at all.

Inducing mutations in polyploids
The subject of polyploidy was briefly discussed in Chapter 3, mainly in relation to techniques to double chromosome numbers. Polyploid plants have more than two genomes. They are called <u>autopolyploids</u> when they only contain homologous chromosome sets and <u>allopolyploids</u> when chromosome sets are involved which may be traced back to different parent species.

For theoretical and practical reasons, mutation breeding in autopolyploids has not received much attention, but much work was done on allopolyploids like wheat (*Triticum* sp.), oats (*Avena sativa*), cotton (*Gossypium hirsutum*, *G. barbadense*), peanut (*Arachis hypogaea*) and many ornamentals. Stadler (1929), in a publication on mutation rates in oats and wheat, remarked that only very few chlorophyll mutations could be observed in M_2. MacKey (1954a) reported on the basis of experiments with cereals that, after mutagenic treatment, allopolyploid species better tolerated chromosomal aberrations than diploids. Swaminathan (1965) who, for a number of crops, compared mutation induction in diploids and (auto- and allo)polyploids, concluded that polyploid species in most genera are more resistant to radiation than the corresponding diploids. Some short useful notes on this subject can be found also in the IAEA *Manual on Mutation Breeding* (Anon., 1977a). In 1972 a meeting on '*Polyploidy and Induced Mutations in Plant Breeding*' was jointly organized in Bari (Italy) by the FAO/IAEA Division in Vienna and EUCARPIA (Anon., 1974c), but in only a few contributions was attention paid to the specific problems of applying mutation breeding methods to seed propagated polyploids.

With respect to mutation breeding some features of polyploids deserve attention. When in a plant more genomes are present, the number of 'target-loci' per trait that are submitted to mutagenic treatments is accordingly higher. As a consequence, when mutation frequencies in diploids and polyploids are calculated on a 'per locus' basis, the frequency of mutations per locus should be the same, but mutations per trait should be higher by a factor 1.5, 2, 3 etc. This aspect was also introduced in some of the formulas that have been presented earlier in this chapter to calculate the number of seeds that should be treated mutagenically and to estimate mutation frequencies that may be expected in later generations. Unfortunately, many of the mutations induced in autopolyploids are not expressed as they more rarely become homozygous in all genomes. The important exception are allopolyploids and 'diploidized' genes in autotetraploids. For these, mutation rates induced by treatments with radiation or chemical mutagens may not differ from those in diploids, and the ratios with which induced mutants segregate as well as the generation in which the homozygous mutants are revealed may not differ between diploids, allopolyploids and 'diploidized' genes in autotetraploids. More recently it has been found, however, that in many allopolyploids sets of genes are homologous, i.e., behave like genes in autopolyploids.

The subject of radiosensitivity in various crops, including polyploids, has been studied for many years by the group of Sparrow at Brookhaven National Laboratory, USA, using chronic irradiation of plants. Sparrow and associates on many occasions expressed the opinion that the ICV, the average meristematic interphase chromosome volume (i.e. the volume of the interphase nucleus (NV) divided by its number of chromosomes), is an important determinant of radiosensitivity (e.g. as determined by plant lethality), in particular for diploid species. According to these authors the ICV yields a more consistent relation to radiosensitivity than the NV, at least at low ploidy levels.

In Chapter 4 reference was made already to a publication of Sparrow's group (Conger *et al.*, 1982) in which, for a number of polyploid series – consisting of different species within one genus or cultivars of different ploidy levels within one species – correlations between NV, ICV and radiosensitivity at different ploidy levels were studied. It was found that in general, within a genus, the NV increased proportionally to the ploidy level and with this the target size of radiosensitivity. It was further reported that for several ranges from diploid $(2n = 2x)$ to decaploid $(2n = 10x)$ plant species the lethality remained about the same, although the NV increases proportionally to the ploidy level. This was explained by assuming that the – expected – increased radiosensitivity as a result of the larger NV must have been about compensated by a higher 'genetic redundancy'; this term (see for instance Rieger *et al.*, 1991) refers to the presence of genes in multiple forms in the DNA.

Based on this result one could, as far as radiosensitivity is concerned, conclude that it would not make much difference whether mutation breeding should start from diploid or polyploid species, at least within the range from diploid to decaploid. Still higher ploidy levels show an aberrant behaviour. It is, in this context, sometimes advised to start – if possible – from tetraploids instead of diploids because of the higher buffering capacity of polyploids which allows the presence of a higher level of chromosomal aberrations in plant cells without affecting too much the viability of such cells and plants. According to this opinion, polyploids may carry more mutations which can be transferred to later (sexual) generations. This view, for several reasons, is questionable. First, buffering of chromosomal aberrations, although interesting in radiobiology, is of no value for plant breeders. In addition, selection of mutants is more difficult in polyploids as a) somatic mutations remain hidden, b) mutant segregation occurs in later generations, and c) relatively fewer mutations are obtained. It must also be remembered that the genetics of autopolyploids and allopolyploids – which latter may behave as functional diploids – differs considerably.

One other additional comment must be made here. In earlier publications by Sparrow's group the 'genetic redundancy' of autotetraploids was not taken into account. This point should be taken care of by dividing the NV by the number of chromosomes.

In journals that could be consulted by us practically nothing is published about effects of treatments with chemical mutagens on seed propagated polyploids. Swaminathan (1965) concluded, on the base of the few available data, that the frequency of mutational events, e.g. for a specific locus on chromosome 5A in different cultivars of bread wheat (*Triticum aestivum*), was lower after chemical treatments than when radiation treatments with similar survival rates were administered. He adds that this, probably, is a result of the low frequency of chromosome aberrations (in particular after EMS treatment) and of an assumed 'non-random localization' of chemically induced mutations.

Kleinhofs et al. (1984) reported that sodium azide is not an effective mutagen in polyploids, such as the hexaploid crops wheat and oats (Avena sativa). The reason given for this is that gene mutations and small deletions are not expressed in polyploids and that treatments with azide (practically) do not produce chromosome breaks (see also Sideris et al., 1973).

6.3.2. Predictions concerning the outcome of mutagenic treatments

Earlier experiments

As was mentioned already in Section 6.2.2, when determining the optimal scale of a mutation experiment and treatment (dose, concentration, dose rate, etc.) for a specific mutagenic agent, the best way, of course, is to assess results from previous experiments. Unfortunately, such information is not always available or treatment conditions are not described in detail.

Predictions on the base of observations in the first generations

When data of previous experiments are not available, at least M_1 and M_2 would be required before reliable mutation frequencies can be determined. For seed mutagenesis a general impression of the effectiveness of a mutagenic treatment (i.e. the number of mutations per unit of dose) can be obtained from studying the frequencies of easily detectable mutational events – like changes for chlorophyll or some morphological traits – in M_2 seedling populations. Based on such results the main experiment is started from exactly the same starting material and under identical conditions, and the best treatment method is chosen. In practice (see also Section 6.2.2) decisions about 'suitable' mutagenic treatments are often based on extrapolation from various effects of such treatments using parameters that can be observed already in M_1, like reduction of germination, seedling height, survival, seed set, degree of pollen sterility, or the occurrence of chlorophyll defects in M_1.

Mutagenic treatments may have a considerable detrimental impact on a wide range of characteristics of the treated (M_1) plant material. Such phenomena are often called 'physiological effects' or 'physiological damage', which imply that the observed phenomena cannot be related to any specific genetic aberration, although DNA damage and modification of gene expression are involved. Important for the breeder is that such phenomena will not be transferred to the next generation.

In particular the meiosis and the haplophase act as an effective 'sieve' in this respect. The 'reduced plant height' usually observed in M_1 after mutagenic treatments, could be used to illustrate this point. A progeny grown from seed, collected from small M_1 plants, probably segregates for various induced mutations, but selection of genetically shorter plants could be done with the same probability of success in the progenies of taller M_1 plants. Apparently, the M_1 plants with reduced plant length remained short because of inhibition of DNA synthesis or other 'physiological damage' as a direct effect of the mutagenic treatment. On the other hand, for instance, some positive correlation may be found between reduced plant height in M_1 and the frequency of chlorophyll mutations in M_2.

Dominant mutations could already be noticeable in the heterozygous chimeric M_1 and extranuclear mutations in plastids or mitochondria may become expressed already in M_1 if a sufficient number of mitotic divisions led to somatic segregation. If suitable methods to recognize such changes are available, scoring of somatic mutations in the treated generation is possible. However, the observed mutations might not be part of the germ line (= 'generative tissue') and therefore vegetative propagation would be required to obtain mutated offspring. Only in subsequent generations will it be possible to distinguish such desired mutations from temporary physiological effects. In addition, a useful dominant or cytoplasmic mutation, for instance for male sterility in leek (Allium ampeloprasum), may be present only in (part of) one floret (out of at least 200 florets that together constitute the flower head of leek) as a result of the occurrence of chimerism and, therefore, remain undetected.

As was said already, predictions on expected mutation frequencies in later generations, based on observations made in M_1, are often attempted. There is a correlation for chlorophyll mutations, but it has never been proven for other specific traits that a strong correlation exists between certain parameters that can be determined in M_1 and the frequency of mutations detectable in later generations. Therefore, all such predictions should be treated with caution. This may be illustrated by the following example.

Partial sterility is a common phenomenon after mutagenic treatments, which can be observed in M_1 as well as in later generations. It may be the result of quite different events such as point mutations, chromosome aberrations, cytoplasmic mutations or 'physiological damage'. Reference is made in this respect to a contribution of Gaul (1977). Problems with mitotic divisions in the treated material result in elimination of part of the causes of induced (male and female) sterility (diplontic selection; see Chapter 1, Section 1.3.8). Another part of mutagenic effects potentially leading to sterility, will be eliminated during gamete formation and as a result of pollen competition (haplontic selection). The causes of sterility that pass those sieves and enter M_2, may be related to observed M_1 semi-sterility, for instance reciprocal translocations or inversions without lowered fitness. On the other hand, the male sterility that segregates normal in M_2, M_3, etc., shows no relation to M_1 sterility at all.

The frequency with which male (and female) sterility occurs may be strongly dependent on the mutagen used. Ekberg (1969) distinguished five hereditary types of sterility in barley: 1) caused by translocations, 2) caused by inversions, 3) recessive lethal mutations causing abortion of 25% of the seed of heterozygous plants, 4) dominant female-lethal mutations causing abortion in 50% of the gametes or seeds in heterozygous plants, and 5) viable recessive mutations causing a more or less pronounced sterility in homozygous plants. Sterility types 1 to 4 are observed in the (heterozygous) M_1 generation (including M_2 seeds on M_1 plants), whereas type 5 is observed in homozygotes and does not appear until flowering in M_2. Radiation treatments in 82% of the cases induced sterility types 1 and 2 (note that types 1 and 2 both represent chromosome rearrangements). Alkylating agents mainly induced seed lethality (types 3 and 4) and only 27% of the lines after chemical treatments showed chromosome aberrations. Type 5 ('viable recessive mutations') was observed only after chemical treatments, but Ekberg commented that research had been concentrated on types 1 to 4, and that only few examples of type 5 had been selected for continued studies and no details were given. For further reading, reference is made to the original publication of Ekberg (1969).

Studies on sterility induced by X-rays and chemical mutagens in barley were performed as well by Prina, Hagberg & Favret (1986). These authors found that X-rays produce a low level of heritable sterility, this despite the many reciprocal translocations that still may be observed in M_2. Treatments with sodium azide, on the other hand, resulted in a much higher level of (heritable) sterility in M_2, but a very low level of chromosomal aberrations. Moreover, for various experiments with sodium azide applied under different conditions a very high correlation ($r > 0.95$) was found between the frequencies of sterility and the frequency of chlorophyll deficient mutations. In addition, studies based on observations on 68 000 M_2 seedlings showed that when different classes of sterility in M_1 spikes were distinguished

(< 25%; 25–49%; 50–75%; > 75%), a higher class number correlated well with an increase in chlorophyll mutants in M_2. These latter results, according to Prina et al., resembled those obtained by Künzel & Scholz (1971) for treatments of barley seeds with EMS and nitroso-compounds.

An additional comment to be made here is that due to the occurrence of semi-sterility in mutagenically treated populations there is an increased risk of cross-pollination, even in so-called obligate self-fertilizing crops. In order to exclude this possibility, all flowers or inflorescences should be bagged, but in practice this may be too big and too expensive a task in a large M_1 field and in that case other forms of isolation have to be used.

To present another example for correlation with M_1 damage: it was found for pea (*Pisum sativum*) that a positive correlation exists after seed treatment between the frequency of yellowish leaf spots on the first leaves of M_1 seedlings and the mutation frequency observable in M_2 (Blixt, Ehrenberg & Gelin, 1960; Blixt, 1972). The earliest leaves may show many small spots, whereas in later leaves fewer but larger spots appear, and no spots can be seen on leaves developing in later stages of ontogenesis. A few chlorophyll deficient spots which were part of a meristem may develop into chlorophyll deficient 'sectors'. There appears to be a striking correlation between such spots in M_1 and the frequency of maternally inherited chlorophyll mutations in M_2. In another publication in which the effects of EMS on leaf spotting and mutation frequencies were described, Blixt et al. (1964) concluded that in leguminous species when measuring the effects of mutagenic treatments, leaf spotting may be a much more reliable indicator of (or parameter for) effective mutagenic treatment than growth reduction.

According to Blixt (1972) the leaf spots in *Pisum* are somatic mutations (whether in the nucleus or the cytoplasm is not certain). The frequency of leaf spotting was found to increase with increasing doses of neutrons, gamma rays, X-rays and the chemical mutagen EI. The author mentioned that selecting seedlings with the highest degree of 'spottiness' and growing them to maturity offered the best prospects for the detection of high mutation rates.

In summary it must be concluded that some kind of correlation between observations in M_1 and the mutation frequency in M_2 may exist and therefore such M_1 observations can be used as a first indicator of an effective mutagen treatment. However, for different species and different mutagens one cannot take it for granted that such correlations do exist for a particular M_1 effect.

6.3.3. Mutagenic treatment of seeds

Mutagenic treatment of seeds – by irradiation or with chemical mutagens – indeed, is the easiest and therefore standard method in seed propagated crops. Although seed size may differ considerably, depending on the species, seeds in general do not require much space during storage, transportation and mutagenic treatments and, in dry condition, are easy to store during prolonged periods; they can be easily desiccated, soaked, heated, frozen, etc., and, as a consequence, mutagenic treatments and various pre- and post-treatments are easy to perform. An important advantage is also that it is relatively easy to reproduce treatment conditions and, hence, to obtain reproducible results.

The structure and state of development of the seed embryo, in particular of the main shoot apex and axillary buds present in the seeds, is very relevant for the outcome of a mutagenic treatment. In Section 6.2.6 of this chapter the terms initial cells – which in the case of the sporogenic layer are called genetically effective cells (GECs) – and germ line (formed by these GECs) have been explained. Independent initial cells are present in the meristematic areas of the future main shoot and of the branches (or tillers).

Each initial cell by its position is determined to perform a specific role in plant ontogenesis and organogenesis. As a consequence, a mutation induced in one initial cell destined to form a branch will result in a plant that carries the mutation only in that particular branch. Mutations induced in the initial cells of the outer germ layer, called L-1, may result in visible effects in M_1 plants like colourless areas or 'sectors' and, thus provide some information about the effectiveness of a mutagenic treatment for dominant mutations (recessive mutations are for self-fertilizing crops not visible in the M_1) and chromosomal aberrations. The only mutations, however, that are of real interest to breeders of seed propagated plants are those induced in the initial cells of the sub-epidermal layer (L-2), as they are the only ones that can normally be transferred by seed to the M_2.

The number of initial cells in L-2, present during and functional after the mutagenic treatment – predominantly – determines the pattern and degree of chimerism in the developing plant. Examples about numbers of initial cells in different plant species have been given by Jacobsen (1966). He, for instance, came to three to six functional initial cells in the L-2 of the main shoot apex and one or two for each of the – about – five axillary shoot apices in a barley embryo. The number of functional initial cells will be reduced by high doses of mutagen and probably increased by presoaking for chemical mutagen treatment (see Section 6.2.8).

In (dry) seeds all cells are – more or less – inactive and in this state they are less sensitive to radiation than physiologically active or rapidly dividing cells. In order to activate cells, seeds should be soaked at an appropriate temperature. Soaking before treatment with chemical mutagens may facilitate the uptake of the mutagen. The stage within the mitotic cycle (G_1, S, G_2,) also affects the sensitivity of cells to mutagenic treatments. In late S and in G_2 phase DNA replication has taken place, which implies that the 'target' for mutagenic treatments has doubled and that chromatid mutations may occur. As a result, only one of the two daughter cells will carry the mutation (chimerism!). Repair mechanisms may also act differently in different cell stages, but knowledge is insufficient to draw conclusions.

For the sake of reproducibility of mutagenic treatments, it is necessary to carefully record the condition of the seeds (age, moisture content, previous storage), as well as the treatment conditions, including all factors that may lead to modifying effects. The most important 'modifying factors' have been mentioned already in Chapter 4.

6.3.4. Mutagenic treatment of pollen

General aspects

Mutagenic treatment of pollen may serve several purposes. First of all there is the induction of mutations after which the mutagen treated pollen could be used to fertilize untreated egg cells and, in this way, to introduce mutated traits into the progeny. This approach accounts for the main part of this section.

Another option could be to try to grow plantlets directly from the mutated microspores or pollen, either as haploids or, after duplication of the genome, as diploids. Irradiation of pollen could also be performed for a number of other purposes, like the use of killed pollen as so-called mentor pollen for interspecific crosses. Such methods have been briefly described already in Chapter 4, Section 4.2.6, and will not be further discussed here.

In addition to the fact that large numbers of pollen grains can be easily treated with mutagen, the most attractive features of pollen for mutation breeding are the haploid nature and the single cell situation during mutagenic treatments. After pollination this may result in heterozygous diploid M_1 plants which are non-chimeric. When only pollen is mutagenically treated, the female partner remains free of somatic damage. As there is no chimerism, no diplontic selection occurs. Because of the absence of chimerism fewer M_2 seeds are needed per plant but, on the other hand, many more M_1 plants are required.

For some plant species, especially wind pollinators, pollen occurs in very high numbers. For maize (Zea mays), for instance, one single plant may produce about 18 million (!) pollen grains. However, most insect pollinated plant species are not as abundant in producing pollen as maize and, as a consequence, it is often difficult to collect sufficient anthers with pollen of good quality for large-scale mutation experiments for each cultivar and at the desired moment. Another problem of pollen treatment may be its short life time. After pollination pollen tube competition will occur which may eliminate a considerable part of the mutations ('haplontic selection').

Some of the advantages mentioned before with respect to pollen, in principle, can be attributed also to unfertilized egg cells but the number of female gametes is mostly much too low and it is much more difficult to determine the proper stage of mutagenic treatment for egg cells. As a consequence mutagenic treatment of egg cells is less practicable and, hence, not often performed in practice.

This section mainly concerns the mutagenic treatment of mature, dry pollen. In Chapter 5 attention has been paid already to the use of (haploid, uninucleate) microspores and of pollen grains (with one vegetative and one generative cell or nucleus) for in vitro work, for instance as starting material for the production of embryoids. This can be easily combined with mutagenic treatments. Reference is made in this respect to the experiments reported by MacDonald et al. (1991) with (late uninucleate or early binucleate) microspores of rape (Brassica napus). Some main points of the treatment procedures followed in this work have been summarized in the previous chapter, in particular in Box 5.4.

What happens during pollination?

When mature pollen is brought on a receptive pistil, pollen tubes start growing and penetrate the style. For most diploid plant species the generative pollen cell (and nucleus) divides in the pollen tube, after which the first generative nucleus (or, according to some interpretations, cell) fertilizes an egg cell to form the diploid zygote. The second generative nucleus fuses with the two polar cells (or secondary embryosac nucleus) to form the endosperm which is triploid. It should be realized that when mutations are observed in the kernels (in fact in the endosperm) of a crop like maize (Zea mays), such mutations will only be carried forward to the next generation when the same mutations have been present as well in the first generative pollen cell that contributes to the formation of the embryo. This will only be the case when such mutations were induced before division of the generative cell, which is not always the case. For maize, which is a somewhat exceptional case, Neuffer

(personal communication) explains that the second pollen division, in which two generative cells or nuclei are produced, occurs several days before shedding and, as a result, at the moment of mutagenic treatment of mature pollen two generative cells are already present. Consequently, an induced mutation can occur only in one or the other generative pollen cell, not in both. Thus, if a mutation towards <u>dominant</u> ($aa \rightarrow A$) should be induced, either a mutant (M_1) endosperm (Aaa) with a normal (M_1) embryo is produced, or the embryo is mutated (Aa) and associated with a normal (aaa) endosperm. (N.B An Aaa endosperm will not always express the 'dominant' mutation.) When the embryo should be mutated in the above described way, a normal-appearing kernel is planted but the resulting M_1 plant has the genetic constitution Aa and, after selfing, would produce M_2 seeds that would be either AAA, AAa, Aaa or aaa in their endosperms. This matter is particularly relevant if endosperm mutations are desired (e.g. with higher protein value) and M_2/M_3 seeds have to be chemically analyzed (see Micke, 1970).

Handling pollen in practice

A good way to separate pollen from anthers is by gently rubbing dry anthers with a small brush over a sieve. When possible, pollination should be performed with fresh pollen and take place as soon as possible after mutagenic treatment. Between collecting and mutagenic treatment or when pollination cannot be performed immediately after mutagenic treatment, pollen may have to be stored, but this period should be kept as short as possible, depending also on the species, unless special conditions are provided. A good way to store pollen, also for prolonged periods, would be in small glass vials over silica gel at a temperature of about 3–5 °C. Pollen can also be shipped (e.g. by letter in a sealed plastic cover with silica gel) with relative ease. As pollen from different species may react in very different ways to storage, etc., optimal conditions to maintain pollen vitality must be tried out in preliminary experiments. Another practical point that must be mentioned is that, often, pollen is shed already before flowers open.

Mutagenic treatments

In most cases dry pollen grains are irradiated or treated with a chemical mutagen as soon as possible after having been collected. An alternative approach is to irradiate inflorescences, flowers or tassels when they are still on the plant – for instance in a room with a gamma source – and collect pollen after this treatment, but in that case it is impossible to determine precisely the stage at which the mutagenic treatment is performed.

Gamma rays and X-rays are commonly used for pollen treatment. In addition, it has been reported occasionally that UV irradiation should be particular useful for pollen irradiation, but results of practical importance are not known. Examples of treatment procedures for pollen with UV have been presented already in Chapter 4 (Section 4.2.3).

Pollen can be irradiated, for instance, in small glass tubes by gamma rays. Depending on the species, doses may differ considerably and go up to 500 Gy or even higher, but usually pollen is more sensitive. Pollen grains may show very different sensitivities to irradiation (see for instance the review by Brewbaker & Emery, 1962) and as a consequence, initial experiments have to be performed in order to determine the optimal dose, i.e., the dose at which a maximum number of mutations is induced, whereas at the same time enough pollen has survived the mutagenic treatment to perform the required pollinations, and a sufficiently high pollen viability is maintained. One example is given here. Daskalov (1986) mentioned that for sweet pepper (*Capsicum annuum*) pollen, 'both gametes' or zygotes are irradiated with doses of X-rays or gamma rays within the range from 5–15 Gy at dose rates which vary from about 1.3 Gy. min^{-1} (in the case of gamma rays) to 1.8 Gy. min^{-1} (for X-rays). As a comparison, for irradiation of seed, doses from 60 Gy and higher are given. For 'both gametes' and zygotes the use of dose rates up to 10 Gy. min^{-1} was mentioned. In a comparison of the LD 50s, when different stages of cv. Albena were irradiated, Daskalov mentioned an LD 50 of 1.29 Gy for pollen, but this must be a printing error (probably 12.9 Gy is meant), because for 'both gametes' and zygotes an LD 50 of 8.5 and 9.0 Gy, respectively, was given, whereas the range of doses applied ranged from 5 to 15 Gy. (N.B. A publication by Auni, Daskalov & Filev from 1978 could not be consulted by us). For seeds the LD 50 may be in the range of 70–200 Gy. Finally, Daskalov emphasized that mutagenic treatment should be followed immediately by pollination on emasculated, non-irradiated plants of the same species.

Pollen viability can be measured for instance by scoring the percentage of germination (e.g. *in vitro* on an artificial medium), by determining pollen growth or pollen tube growth, by various staining methods (e.g. by carmine acetic acid, tetrazolium salts or lactophenolacid fuchsin), by specific enzyme tests (e.g. with fluoresceindiacetate = FDA), or by seed-set relative to non-irradiated pollen.

Neuffer & Coe (1978) and, in particular, Neuffer & Chang (1989) described an interesting method of treating maize pollen with the chemical mutagen EMS.

Tassels on superior inbred plants are bagged the day before pollen shedding. A stock solution of 1 ml EMS in 100 ml of paraffin oil is prepared (remember that several precautions have to be taken when working with dangerous chemicals like EMS; see Chapter 4). On the day of treatment – after repeated vigorous stirring – a treatment solution of 1 part stock solution and 15 parts paraffin oil is prepared and stirred again. Consecutively, good pollen free of anther rests is collected and less than one part pollen is mixed with more than 10 parts of the mixture in a small bottle (with cap) and again shaken with intervals for at least 45 min. The pollen in oil suspension is then spread on good female inflorescences (silks) of the mother plants (M_0) with a small brush. Stirring of the mixture must be continued between performing pollinations of different ears. Protective clothes (gloves, glasses) should be used and untrained persons should remain out of the field for 30 days, after which ears and kernels are safe for handling. Bird & Neuffer (1987) report that maize pollen may be kept in paraffin oil for well over 1 h and still give good seed set.

After 30 days in the field the ears and M_1 kernels will be safe for handling and harvesting is done by standard procedures. Dominant mutations will be visible already on the M_1 plants and their seeds and selfing of (heterozygous) M_1 plants will result in M_2 plants on which recessive mutations will become revealed.

Some practical examples

It has been pointed out before that the thread of this chapter is to follow in particular the effects of mutagenic treatments of seeds and the consecutive pattern of selecting favourable mutant plants. Moreover, most examples presented at the end of this chapter concern results obtained by mutagenic treatment of seeds. Therefore, we will discuss at this place a number of examples concerning mutagenic treatment of pollen.

Mutagenic treatment of pollen or egg cells or, more in general, gametophytes – a term coined already in 1851 by Hofmeister which, following the definition by Rieger *et al.* (1991), refers to the haploid generation which produces the gametes – so far, has not produced convincing results for plant breeding. Micke *et al.* (1990), in this respect, remarked that pollen irradiation has been used 'here and there', but that a controversy persists whether this method was superior or inferior from the view point of producing usable gene mutations. The review by the aforementioned authors does not give specific information about mutant cultivars that have resulted from treatment of gametophytes.

An early report on pollen treatment by X-rays and the chemical mutagen diepoxybutane for maize (*Zea mays*) was presented by Bianchi, Mariani & Uberti (1961). The X-ray dose used was 10–20 Gy and diepoxybutane was administered as a 0.2% solution for 8 h by a cotton wick which introduced the solution into the plant through a hole which was made in the maize stalk about 15 cm below the lowest tassel branch.

Contant et al. (1971) concluded from exemplary large-scale experiments with tomato (Lycopersicon esculentum) that irradiation of dry pollen or (premeiotic) pollen mother cells (present in flower buds of 3–4 mm) seems to hold little promise for mutation breeding purposes because of the 'extremely low' frequencies of recessive 'visible' mutations in the M_2 generation. Moreover, deleterious genetic effects caused by the irradiation treatment are not efficiently eliminated.

In these experiments irradiation of 'both gametes' (24 h after pollination of emasculated flowers and at least 20 h before gametic fusion) resulted in much higher mutation frequencies in tomato plants than irradiation of pollen mother cells and pollen, but the mutation frequency for the first group still remained considerably lower than after irradiation of dry seed. Irradiation of 'both gametes' stage resulted in a much higher load of sterility (indicating chromosomal aberrations) than treatment of pollen or pollen mother cells (PMCs).

Irradiation with ^{137}Cs gamma rays and fast neutrons was performed with a wide range of doses, depending on the kind of starting material, for instance for gamma rays with up to 4 Gy for PMCs, 50 Gy for dry pollen and 30 Gy for 'both gametes'. It was found for gamma rays that the maximum tolerable dose (= almost no seed set in M_0) in the case of mature pollen was higher than 100 Gy, whereas the comparable figures may be about 40 Gy for 'both gametes' and probably 10–20 Gy for the very sensitive PMCs. In comparison, dry seed of tomato may tolerate doses of gamma rays of more than 800 Gy. This, according to the authors, is also the main explanation for the very high mutation frequencies that can be observed after irradiation of dry seed and for the low mutation frequencies in pollen and PMCs, where severe dose restrictions are imposed by the low seed set in M_0.

Another result to be mentioned from the publication by Contant et al. (1971) concerns the important subject of the occurrence of chimerism after irradiation of prezygotic stages and zygotes. In general no chimerism is expected in those cases but the authors mention that some exceptions to this rule were found. Contant et al. reported that between 5 and 16% of the M_1 plants which were obtained after irradiation of prezygotic stages (PMCs, pollen, 'both gametes') were chimeric. According to the authors, the observed frequencies of chimerism are relatively low but nevertheless in contrast to some

results published for barley and durum wheat (to which publications the authors refer), 'where chimerism was completely avoided'. Among others, Devreux, Donini & Scarascia-Mugnozza (1968, abstract only) indeed reported for barley (Hordeum vulgare) and durum wheat (Triticum turgidum var. durum), two monocotyledonous crops, that chimerism could be avoided by irradiation of the 'gametophyte' stage. Lundqvist (1964), on the other hand, mentioned that chimerism occurred in the so-called F_1X_1 (= M_1) plants of barley after pollen irradiation. In this experiment pollen in the spikes of cv. Bonus was irradiated with 4–80 Gy of X-rays (optimal dose below 12 Gy). For UV radiation pollen from shedding anthers was treated in Petri dishes with 0.25–5 min of UV radiation (at 2537 Å). Pollination was performed on emasculated spikes. Chimerism was observed after both treatments. Lundqvist explains that one of the two haploid generative nuclei of a mature barley pollen grain fertilizes the egg cell and that, as a consequence, it would be expected that mutations in the generative nuclei would give rise to heterozygous non-chimeric 'F_1X_1 plants' and that these plants would segregate in the 'F_2X_2 generation'. Three possible explanations for the observed chimerism in the M_1 are suggested: 1) the mutations are induced in chromatids only instead of in whole chromosomes, 2) a so-called delayed mutagenic event occurs, i.e., mutations become established only after the first division of the fertilized egg cell, and 3) unusual processes of mitotic recombination or somatic segregation or repair may occur during embryo development and give rise to 'wild type tissue'. Lundqvist did not express preference for any of these possible explanations.

The size and structure of the mutated/non-mutated 'sector' could give some clue as to the time of the event during embryogenesis but this was neither studied by Lundqvist nor by Contant et al. A further discussion of this subject is outside the scope of our publication.

Contrary to the negative view expressed by Contant et al. (1971) on the usefulness of pollen and PMC-irradiation for mutation breeding in tomato, Saccardo (1983b), Saccardo & Monti (1984) and Saccardo et al. (1991), believe that pollen and gamete irradiations may be at least as efficient as or even better than seed treatments to obtain valuable mutations for plant breeding.

Saccardo & Monti (1984) also compared effects of irradiating gametes in different ontogenetic stages: pollen mother cells (PMCs), pollen and 'both gametes' (egg cell and growing pollen tube) for a few crops. For sweet pepper (Capsicum annuum) they found after irradiation of gametes with 7.5 Gy gamma rays, that the mutation frequency gradually increased from the pre-meiotic stage over the mononucleate to the binucleate pollen stage (including egg cell). The authors concluded that, from a practical point of view, mutagenic treatment of pollen or 'both gametes' offers the best opportunities for obtaining economically useful mutants for this species. The authors further reported that by X-irradiation of pollen of cv. Sprinter of canning pea (*Pisum sativum*) with 7.5 Gy of gamma rays, three new cultivars (cvs. Esedra, Navona and Trevi) and several other mutants with agronomically interesting features (e.g. for protein improvement) were obtained. The three cultivars – directly obtained or after further crossing work – showed improvement for various traits, like a longer cycle, a shorter plant height, more seeds per pod, smaller seed size, good properties for mechanical harvesting and earliness. To our knowledge, however, these pea mutant cultivars are the only examples of crop cultivars that resulted from mutagenic treatment of pollen. Saccardo *et al.* (1991), in addition, mention was made of some mutants with 'agronomically valuable characters' for tomato (*Lycopersicon esculentum*) and some 'advanced mutant lines' for sweet pepper.

In the same publication Saccardo *et al.* (1991), who referred in this respect to Contant *et al.* (1971), briefly mentioned that after gamete irradiation of tomato chimerism could not be completely avoided.

Earlier, reference was made already to mutagenic treatment of 'gametophytes' in a review on mutation breeding in sweet pepper (*Capsicum annuum*) by Daskalov (1986). The data mentioned in this review are mainly based on a publication by Auni, Daskalov & Filev (1978, original not consulted). From this work it was concluded that the highest frequency for chlorophyll and morphological mutations had been obtained after treatment of 'both gametes' or dry seeds. Daskalov's review also provides detailed information on how the mutagenically treated material should be further handled and presents data on agronomically useful traits obtained after mutagenic treatments. None of these cases, however, refers to traits induced by treatment of pollen, etc.

In the previous sections we have referred several times to mutagenic treatments of pollen of maize (*Zea mays*). Some major points will be summarized here and some additional points briefly touched upon.

Maize, a natural cross-pollinator, represents a very special case for mutation induction in seeds as this crop has separate male and female primordia in the kernel (see Neuffer & Chang, 1989, for some references about this subject). Much of the work on treatment of pollen, in particular with chemical mutagens, has been done in the USA. Many interesting details, which refer especially to endosperm traits, can be found in Neuffer & Coe (1978) and in an extensive review by Bird & Neuffer (1987).

The relative proportion of gene mutations to chromosome rearrangements after UV treatment (preferably from below as well as from above) of pollen grains of maize, as in various other crops, is much higher than when ionizing radiation is used. For various chemical mutagens it is also known that they have a lower proportion of chromosome rearrangements than ionizing radiation as well as a high rate of 'point mutations', but Bird & Neuffer report that several chemical mutagens have not been very successful in maize. Whether maize is compared in this respect only with prokaryotes, or also differs from other eukaryotes is not mentioned. Most effective compounds on maize pollen appear to be alkylating agents, in particular EMS, and nitroso-compounds like nitrosoguanidine (MNNG or NG) and, for maize to a lesser extent, nitrosomethyl urethane and nitrosoethyl urea.

After EMS treatment of maize pollen, chromosome aberrations and point mutations occurred in about equal numbers. An appropriate pollen treatment with chemical mutagens may involve mixing maize pollen for 45 min in a paraffin oil 0.06% EMS suspension or for 3 min in a 3.8% dilution of a saturated stock solution of MNNG. The use of higher concentrations in the latter case was reported to kill pollen.

In their review, Bird & Neuffer (1987) described types and frequencies of mutants induced in pollen by EMS, but they did not compare mutations induced by treatment of seed, pollen or otherwise. The induction of over 150 dominant mutants was reported, which was three times the number of dominant mutants known before applying chemical mutagens. The authors further mentioned that in particular EMS treatment of pollen resulted in a consistently high frequency of mutants. For instance, in an earlier experiment it was found that after treatment of pollen with EMS, at least three out of four pollen grains carried at least one mutation. In one case a single grain was reported to exhibit seven mutations! It was further mentioned that 535 loci had mutated and that the average mutation frequency per locus was about 1 per 1000. A later calculation, based on kernel mutants, for estimating mutation frequencies per locus showed that at least one mutation per locus occurred per 500 M_1 plants. The ratio of dominant to recessives varied for different phenotypes but on the average the rate was 1 to 200.

Mutagenic treatment of maize pollen, as far as is known, has not resulted in new mutant cultivars for this crop. Of course, if useful new traits should be induced in already adapted inbred lines that are further used in crosses for hybrid breeding, such results might remain unnoticed.

We end this section by briefly referring to a publication by Vizir *et al.* (1994), who irradiated pollen of *Arabidopsis* with gamma rays in order to induce deficiencies in the genome. For this purpose cut inflorescences of so-called 'wild-type plants' of *Arabidopsis* (in plastic tubes to avoid desiccation) were irradiated with up to 1500 Gy gamma rays from a ^{137}Cs source at a dose rate of almost 10 Gy. min^{-1}. Pollen from mature flowers was used immediately after irradiation to pollinate unirradiated mother lines ('multi-marker lines') that were homzygous recessive for several traits at a number of loci distributed over all five chromosomes. It was reported that pollen irradiated with doses between 300 and 600 Gy of gamma rays at 10 Gy. min^{-1} resulted in 50% aborted seeds in the siliques. All seeds aborted at 1000 Gy. At a dose of 600 Gy, 20% of the M$_1$ seed proved to be viable. It is remarkable in this respect that the authors do not consider the amount of pollen used per female as a factor that may significantly affect the percentage of abortion.

From the work by Vizir *et al.* it was concluded that pollen irradiation may be a good method to generate a large collection of different chromosome deficiencies which can be used for instance for fine-scale genetic mapping and various other advanced molecular-genetic approaches.

6.4. Selection methods

6.4.1. General approach

Following Dellaert (1980; 1983) the degree to which a mutation breeding programme will be successful depends to a great extent on a) the type of mutations desired, b) the ability of a specific treatment to induce a maximum number of independent mutations (indicated by Dellaert as the efficacy of the applied mutagenic treatment), and c) on the selection methods applied to detect desired mutants.

As has been mentioned in Chapter 4, Section 4.2.1, the ability to induce many useful mutations is sometimes called efficacy, but it was advocated at that place to avoid the use of this word. In addition, it should be remembered that induction of mutation is a random process and that it is the selection procedure that determines whether and how many desired mutants are selected, provided that such mutations have been induced by the treatment.

It was mentioned already in Chapter 1 that different traits have a different heritability, which implies a smaller or larger contribution of the environment to the observed variation. Selection for traits with a high heritability, of course, will be much more efficient (in a general sense of the word) than for low heritabilities. It is also easily understood that the efficiency of selection differs depending on the detectability of the trait concerned.

Selection for polygenically controlled characteristics like yielding ability, of course, is far more difficult than for flower colour or dwarf type and may not be started before M$_5$ when the plant population has stabilized. This subject will be discussed in particular in Section 6.4.2. Yield in various crops, moreover, may concern a wide array of different plant parts such as seeds, fruits, leaves, stems or tubers and may refer to such different products as starch, protein, sugar, fibres, latex or wood.

In order to determine – in a qualitative and quantitative way – the presence of desired products special techniques have to be applied. Desirable would be tests which, at an early stage, could provide reliable information about yield or level of resistance in field or greenhouse, but such tests often are not available. The limitations of *in vitro* techniques in this respect were mentioned already in Chapter 5. Useful would be in particular so-called non-destructive tests such as the 'half-cotyledon method' used for gas-chromatographic analysis of fatty acids in rape seed (*Brassica napus*).

When performing chemical analysis of seeds to screen for mutants with altered chemical composition, one should not forget the different genetic structure of seeds. Cereal seeds, for instance, are made up of a large triploid endosperm and a much smaller diploid embryo besides maternal tissue such as the seed coat. Seeds of leguminous species or Crucifers, by contrast, consist only of a diploid embryo within the maternal seed coat. A heterozygous cereal plant produces seeds with, genetically, four different endosperms and three different embryo genotypes. If whole seeds are analyzed, it will be rather difficult to clearly identify mutant seeds. Quantitative inheritance may be suspected, even if only one gene mutated. On the other hand, in seeds where no endosperm exists and the nutrients are stored in the cotyledons, only three types will occur (Micke, 1970). This may be one of the reasons why mutation breeding for seed quality was much more successful in oil seeds and legumes than in cereals (Micke, personal communication).

Another matter often forgotten in screening seeds for mutant phenotypes is that the developing seed is physiologically dependent on its mother plant and therefore the mature seed may contain compounds that are not in agreement with the embryo's genotype. As an example: it was not possible to screen seeds or cotyledons of sweet clover (*Melilotus* sp.) for low content of coumarin, because some quantity of this secondary metabolite in the mature seed was derived from the mother plant. A clear distinction of low-coumarin mutants was however obtained when – after germination – a piece of the primary leaf was subjected to the test. When (in the same plant species) mutant screening for permeable seed coat was undertaken, this could of course first be performed only on M_3 seeds, where the (maternal) seed coat belongs to the M_2 generation (Scheibe & Micke, 1967).

Sometimes, for complex traits with low heritabilities, so-called indirect selection is possible. An example concerns efforts to improve grain yield in oats (*Avena sativa*) by selection for harvest index, i.e. the grain/biomass ratio (Rosielle & Frey, 1975). Selection for increased harvest index could be applied already in early generations of mutation breeding programmes (M_2, M_3), but care should be taken not to end up in lower biomass rather than in higher grain yield.

We may proceed now to the third point and main question, i.e. which selection method offers best prospects to obtain high frequencies of desired mutants. Only in exceptional cases (e.g. after pollen irradiation) selection may start in the M_1. Following treatment of seeds, tubers, whole plants, etc., various non-inherited effects of the mutagenic treatments (mitotic inhibition, morphological deformations, reduced growth rates, non-flowering, sterility, etc.) are typical in the M_1 generation and mutations, with the exception of the very few dominant mutations, are not expressed yet. Therefore most breeders handle the M_1 generation as a bulk generation in which only few observations, e.g. on general health and growing conditions are made. An important point is that 'natural' selection among M_1 plants must be prevented. Anyhow, the size of M_1 in most cases is not really prohibitive, which means that normally sufficiently large M_1 populations can be grown without too much labour. Correlations which may exist between observations in M_1 and expected mutation rates in later generations were discussed in Section 6.3.2.

Sometimes selection can be practiced already on M_2 seedlings, e.g. for disease resistance, cold tolerance and N2-fixation. In addition, observations in the seedling stage of M_2 may help to determine which M_2 plants should be grown to maturity and, thus, which plants could be dismissed because it may be anticipated that they will contain no, very few or detrimental mutations. Blixt *et al.* (1963), in this context, concluded for pea (*Pisum sativum*) that scoring of mutations in the seedling stage of the M_2 generation was a good approximation for the chlorophyll mutation frequency and also provided a gross indication of the expected frequency of morphological and other mutations.

How to proceed from M_1 to M_2

The central question that must be answered, often will be how to proceed from the M_1 to M_2. If no limitations exist with respect to the number of plants that can be grown and screened in M_2, the breeder can follow the calculated plant numbers that were advised before, but this would result in enormous populations to be grown and studied. In practice, however, the size of an M_2 that can be properly handled is limited and will be fixed before. This M_2 size and the effort involved in screening it, indeed may be the true bottle-neck of a mutation breeding programme.

For self-fertilizing crop species several methods of harvesting M_1 and screening mutants in consecutive generations may be used. It has been explained before that in many cases not all seeds from each M_1 plant will be taken to start an M_2. Dellaert (1979) in this respect distinguished four alternative methods: a) to harvest one seed per M_1 plant, b) to harvest one reproductive organ (fruit, pod, capsule, spike, ear) per M_1 plant, c) to harvest several of such organs per M_1 plant and d) to harvest seeds from several areas (or 'sectors') of an M_1 plant and put them together. Dellaert also discussed the merits of these methods.

The purpose of all procedures, basically, is to establish for different situations an effective screening, based on probability theories, by which a desirable mutant in M_2 can be detected in a minimum number of M_2 or (M_1 + M_2) plants. In the so-called 'M_1-plant progeny method' which was described already by Stadler (1930) and Nybom (1954), either all seeds or up to a fixed number of seeds (e.g. 25) are taken from each surviving M_1 plant and sown in rows to produce the M_2 generation. Except for the fact that after mutagenesis culms of the same plant may be genetically different (chimeras!), this method, in fact, is similar to the classical 'plant-to-row' or 'ear-to-row' method.

The use in mutation breeding of an alternative method, the one-plant-one-grain method or single-seed descent (SSD-)method, was described among others by Yoshida (1962). Because of the importance of the SSD-method for mutation breeding, its main features have been put together in Box 6.1.

According to Yoshida the conventional plant progeny method using all seeds should not be used in mutation breeding. The SSD-method, in his opinion, is also inferior, except for situations when a mutant is easily distinguished in an M_2 field. When the discrimination of a mutant is difficult it is advocated that a so-called 'improved ear-to-row method' is used. It should be added here that these studies aim at detecting a particular desired mutation.

Box 6.1 The single seed descent (SSD-)method and some alternative options

Application of the SSD-method was proposed in 1939 by Goulden during the 7th International Genetics Congress in Edinburgh for wheat (*Triticum aestivum*) as an economic way to obtain homozygous inbred lines from as many segregating lines as possible (Goulden, 1939). For small-grain crops, where it is possible to grow two or even more generations within one year, the SSD-method combines the advantages of a quick generation advancement, almost no administration (no labels, no seedbags, no numbering needed!) and a high degree of accuracy in estimating the genetic progress.

The SSD-method is particularly useful for fast generation advance in a greenhouse. Since from each plant only one seed is harvested, optimal plant growth for producing a lot of grain is not required which makes it possible to grow plants in dense stands and to harvest at a young plant stage. Nevertheless, at least one spike should carry mature seeds at that stage. A first selection for yield may be performed in the F_6 (or M_6), although reasonable homozygosity is reached only in F_8–F_{10}. For this purpose individual spikes, ears, panicles, plants, etc., are then harvested and each spike (etc.) is sown in a separate row or

plot for detailed observation. Selected lines are harvested for further seed increase and testing.

Yonezawa & Yamagata (1975b) mention that, in addition to the above advantages, the SSD-method in the greenhouse is most useful in avoiding negative effects of natural selection and random genetic drift. One could argue, on the other hand, that the adaptive selection pressure to which segregating populations are exposed in the field is lacking in the greenhouse, but this disadvantage could be made up after homozygosity has been reached and large numbers of progenies are field tested.

One warning must be given. It is unavoidable that a SSD population shrinks by 5–10% each generation. This must be taken into account at the beginning. Instead of proceeding with only one seed per plant, often two or a few seeds are taken as bulk. This adjustment reduces the risk of genetic losses and does not fundamentally affect the principle of the single seed method as long as equal numbers of seed per plant are taken; but more space and labour are required.

Several alternative methods for fast generation advancement towards homozygosity exist, such as the random bulk (RB) method and the doubled haploid (DH-)

technique. A point of difference of the RB-method with the SSD-method is that in RB natural selection plays an important role.

With the DH-technique a fully homozygous situation can be achieved in only one step which, in theory, makes this a very efficient one to create inbred lines. Recombination, however, remains very limited because only a very small percentage of gametes regenerates into plants, leaving more than 90% of the potential recombinant variation unused. As a consequence, the genetic variance often is small as compared to other methods. Following mutagenesis, recombination plays a minor role and the DH-technique is more acceptable in that situation.

As Yoshida does not present a clear picture of his 'improved' method, we follow the explanation given by Gustafsson & Gadd (1966). According to these authors Yoshida's method implies that the probability of finding one M_2 line with at least one desirable mutant (P_m) and the probability of detecting at least one desirable plant among n plants in an M_2 line with one or more desirable mutants (P_n) are taken into account. Gustafsson & Gadd add that the product m × n (lines × number) is the total number of M_2 plants used for the detection of at least one desirable mutant and that the best result is obtained when P_m = P_n. The method of determining the values of m, n and m × n when P_m = P_n is called the 'improved ear-to-row method.'

According to Gustaffson & Gadd, in rice (*Oryza sativa*) the use of three seeds per ear progeny would be the cheapest and the use of 10–20 seeds the safest method in mutation breeding, considering the total costs for the M_1 and the M_2 generation (e.g. with in total 200 000 plants in both cases).

Rédei (1974a), who studied effects of various selection procedures in large populations of *Arabidopsis*, also mentioned that the costs of a mutation programme are determined by the combined costs of the M_1 and M_2 generations. These studies aim at a maximum of different mutations, and not necessarily desirable ones. Rédei found that, given a constant number of M_2 individuals, the effectiveness of recovery of different mutations may increase over sixfold if only one seed is taken from each M_1 plant. As has been explained as well in the IAEA *Manual* (Anon., 1977a), application of this SSD-approach will produce the maximal number of independent mutants in the M_2 population. According to Rédei (1974a,b) the mutation recovery is economically most effective when the maximal number of M_1 families are screened in minimal size M_2 (see also the table by Gustafsson & Gadd, 1966, p. 335). Rédei (1974b) opposes in this respect Harle (1972; see also reaction on Rédei's comment by Harle, 1974), who advocates to grow all seeds from the main inflorescences of M_1 plants.

Rédei (1974b) further states that the best way is to make an M_1 large enough to permit the use of a small M_2, but adds that when M_1 should be too large, the gain in number of different mutations recovered in M_2 would be nullified by the extra costs made to grow the M_1. This last statement, however, is open to question.

Other methods mentioned by Rédei to save money are – apart from the common practice of sowing the M_1 in bulk – growing M_1 plants at maximum density, delaying mutagenic treatments until the germline is multicellular (which results in more chimerism and more but smaller mutated 'sectors') and using pollen for induction of dominant mutations.

The following – theoretical and simplified – examples may further demonstrate that, given a fixed M_2 size, applying the SSD-method may result in a maximal number of <u>independent</u> mutants in the M_2 population. Assume that a breeder can handle 200 000 plants in M_2; further assume that in a segregating M_2 progeny one out of four plants is homozygous recessive for a target gene (e.g. $1AA + 2Aa + 1aa$) and that the mutation frequency for a specific, desired trait is 5×10^{-4} (i.e. that five out of 10 000 M_1 plants carry a mutation for the desired trait in heterozygous condition). It is further neglected that chimerism occurs, since in the SSD-system each M_2 seed refers to <u>one</u> GEC, no matter how many were present in the M_1.

EXAMPLE 1

Starting from, for instance, 20 000 seed-bearing M_1 plants, from each of which 10 seeds are taken, 200 000 M_2 plants arise which together contain 25 homozygous recessive mutant plants (= 200 000 $\times$ 0.0005 $\times$ 0.25). As the mutation frequency was 5×10^{-4}, no more than 10 M_1 parent plants are involved in the production of the 25 mutated plants. This means that only 10 independent mutations are detected among 220 000 ($M_1 + M_2$) plants grown.

EXAMPLE 2

Starting from 200 000 M_1 seed-bearing plants and taking one seed only from each M_1 plant, again 200 000 M_2 plants are produced and, given the same mutation frequency and segregation ratio, again 25 mutated plants would be present. These 25 mutants, however, must all have arisen from different parent plants, so they represent 25 independent mutations from 400 000 ($M_1 + M_2$) plants.

The biggest advantage of using the SSD-method is that it yields only independent mutations. Even in crops which are predominantly self-fertilizing, small variations in genetic background may exist, which may cause differences in the expression or penetrance of identical 'major genes'. On the other hand, every mutagen treatment causes background mutations, which segregate more or less independently. Therefore, from a practical point of view larger plant (spike) progenies could be more interesting than SSD, as in this way a wider choice of similar mutants may be found in some of which the expression of pleiotropic effects or close linkages with simultaneously induced, undesired mutations which accompany a promising mutation, may be less pronounced. This is the reason why plant breeders would prefer to have several parallel mutants for a desired trait at their disposal instead of one mutant only. However, one may argue, that the main bottle-neck is to find the particular rare desired mutant. Its pleiotropic problems might be solved afterwards by crossing, provided of course that crossing is possible.

In some specific situations, for instance when the technique of selecting in M_2 is particularly expensive (e.g. in case of amino acid determination), the application of more sophisticated selection systems could be of interest, for instance the use of the SSD-system in combination with a '<u>remnant (or spare) seed</u>' method (Dellaert, 1979; 1980; 1983). According to this method M1 plants that are heterozygous for the trait aimed at, are detected by sowing one of their seeds as M_2 and by analyzing its progenies (M_3 seeds or M_3 plants) for segregation. (N.B. It should be realized, however, that in 25% of the cases the heterozygous M_1 plant will <u>not</u> be detected.) From the original M_1 plants which were found to be heterozygous for a desired mutation, remnant seed (M_2) is then sown from which a larger number of M_3 seeds or M_3 plants can be tested again in a more significant manner. Following this approach desired mutations for delicate traits (e.g. protein quality) are selected in a more economic way than would be the case by immediate costly selection of all M_2 progenies or by analyzing a large number of M_1 progenies but without a chance for confirmation by making use of remnant seed. (N.B. Confirmation of the inheritance in M_3 is necessary anyhow!)

This subject was further discussed by Bhatia & Abraham (1983) who in principle supported the approach suggested by Dellaert, but commented that applying this amended SSD-method requires much extra work in harvesting, registration and planting (of the remnant seed) and this takes away most of the unique economic advantage of the SSD-method (no labels, no handling of seed samples). Moreover, they stated that it is not always known from which position at the M_1 plant the single seed should be picked in order to sample with the highest probability a heterozygous mutation. (N.B. From the previous discussion on mutagen induced chimerism it should be clear however that the most chimeric plant part is the best place in this respect.) For okra (*Abelmoschus esculentus*), a vegetable crop in the tropics and sub-tropics, Bhatia & Abraham advocate in this respect to harvest from each M_1 plant only the first fruit (containing about 50 seeds) on the main stem which is expected to have the highest probability to carry different mutations. Single fruit progenies or single seed bulks could be used to grow the M_2. In both cases record keeping would not be necessary, as identical mutants in the progeny of one fruit most probably trace back to the same mutational event, whereas in case of single seed bulks all mutants derive from independent mutational events.

6.4.2. Selection for quantitatively inherited traits

In Chapter 3 (Section 3.4) considerable attention has been paid to differences between mutations for so-called qualitatively inherited traits and quantitatively inherited traits. In the previous sections most attention was given to selection for qualitatively inherited traits. At this place we will limit ourselves to some additional comments concerning selection for traits of the other group.

Mutations for quantitatively inherited traits cause differences of degree (e.g. in yield or plant height) rather than of kind. Quantitatively inherited traits express themselves as a continuous spectrum and can be assessed only in populations instead of individual plants. Therefore improvements can be identified and confirmed only – with various degrees of difficulty depending on the trait – by statistical methods. Moreover, as the phenotypic value for quantitative traits is more variable by the interaction between genotype and environmental factors, the detection of useful mutations (with small effects per locus) is difficult.

Brock & Micke (1979) and Brock (1979; 1980) theoretize that the reaction of such quantitatively inherited traits to a mutagenic treatment depends on: a) the number of genes involved in the trait, b) the relative proportion of genes with positive and negative contributions to the trait, c) to what extent the involved genes in the parental genome act as a balanced set, and d) correlations due to linkage or pleiotropic effects.

Selection methods for quantitatively inherited traits of course differ from those for discontinuous (black/white variation or major gene) mutations. Mutagenic treatments were found to result in increased variation for such traits, both towards higher and lower values and at the same time, particularly in M_2 and M_3, in a shift of the mean, away from the direction of previous selection (see also Chapter 3, Section 3.4). In later generations this shift of the mean is compensated to some extent by selection pressure. Stabilization of variability and mean may not be reached before M_5 or M_6. Mutation breeding may have chances if the aforementioned shift of the mean goes in the direction of the breeding objective, as is the case when short straw mutants are favoured in an originally tall cultivar.

It has been emphasized earlier that mutation breeding programmes should always start from the best available (e.g. high-yielding) cultivar. In that case it can be assumed that selection for the most important traits has been performed already for generations and mutagenic treatment may have been chosen as a 'breeding method' because in practice a certain ceiling (e.g. for yield, quality or resistance) has been reached whereas further improvements are still desired. As a consequence of the aforementioned shift in the population mean after mutagenic treatment one would get for instance lower yield in most but not all of the mutants. This, however, is not always a disaster as new breeding goals may be set, including for traits that are genetically determined by multiple loci, which have not been subject to previous selection pressure. One could think in this respect for instance of resistance to a pathogen that was introduced only recently in a country or a new quality trait wanted by industries.

As population means and standard deviations must be studied for several generations in order to determine whether genetic variation for the desired traits has increased, larger numbers of plants are required in mutation breeding programmes for quantitatively inherited traits. It may take several generations before all the negative variants induced by the mutagenic treatment have segregated and have been eliminated and, consequently, before a correct impression can be obtained of positive genetic effects induced by the mutagenic treatment. Selection cannot start before M_3 under proper statistical design (with replications). Moreover, because of undesired pleiotropic effects or close linkages it is advisable to select in several independent populations displaying the desired genetic variation. Selection, for statistical reasons, should not start earlier than for instance in M_3 and should continue for several generations (see also Yonezawa & Yamagata, 1975a,b), but opinions differ as to the number of generations. Moreover, it is sometimes suggested to delay selection till as late as M_7. However, not enough evidence is available to convince the breeder in this respect.

Finally, the fact that after mutagenic treatment the population mean initially may show lower values, for instance for yield, should not make breeders believe that their efforts are in vain and that the experiments should be cut short prematurely. Plants with genetically improved genotypes may become revealed in later generations. Selection, ultimately, must lead to families that, during a range of years surpass the most outstanding families selected from the control population. Reference is made in this respect to a study started in the 1950s by Gardner (1961; 1969; 1972) who irradiated maize seeds with thermal neutrons and studied the effects of mass selection on grain yield. In the first of these publications Gardner advocated the use of mass selection based on grain yield of individual plants as a possible method of yield improvement. More details on the experiments of Gardner were discussed already in Chapter 3.

6.4.3. Selection in cross-fertilizing crop species

Cultivars in allogamous species (if they are not hybrids) are heterogenous populations in which all loci are to 50% heterozygous if open pollination is allowed and, therefore, there is not a single plant which will not be heterozygous for at least 50% of its loci.

Mutation breeding in cross-fertilizing, or allogamous, crop species, in general, offers fewer prospects than in self-fertilizing crops for the following reasons: 1) selection in heterogenous populations in most cases is more difficult, as it requires more time, labour and larger experiments, 2) recessive alleles play a minor role in allogamous plants but are the majority among induced mutations, 3) when self-pollination is performed in allogamous species to detect recessive mutations, one can hardly differentiate between induced mutation segregants and pre-existing recessive alleles, 4) selfing in such crops may be difficult or sometimes even impossible, which makes it not easy to get hold of induced recessive mutations in a homozygous state, and 5) use of recessive alleles by inbreeding is often hampered by

inbreeding depression. Otherwise it is possible to achieve homozygosity for a particular locus by eliminating the dominant allele from the whole population.

Based on his own, long-term mutation experiments with maize (*Zea mays*), involving up to 15 generations of selection, Gardner (1972) concluded that mutation breeding for increased grain yield in this crop, was not likely to be the most successful approach. On the other hand, an example where mutagenic treatment may be useful in this group of crops could be the induction of mutations in inbred lines which are built up for heterosis breeding. Other, potentially useful mutants in cross-fertilizers may refer to traits which – so far – are unknown in the crop species studied (like the non-shattering trait in new oil-crops) or may concern the flowering system, e.g. the incompatibility system or the induction of male sterility for the production of hybrid seeds. Some examples of successful mutation projects in allogamous crop species will be discussed in Section 6.6.

The selection methods most commonly applied are mass selection, family selection and selection for hybrid vigour, which all could be recurrent. The efficiency of various methods depends on factors, such as the (narrow sense) heritability of the trait involved, the frequency of dominance relationship among different alleles involved in a trait, whether the trait can be observed before or after pollination, etc. As mass selection does not provide reliable information about the genotypic value of the selected plant(s), often a system of family selection allowing progeny testing is practiced in cross-fertilizing crops. (N.B. A family, in this context, is the progeny of a single plant after open pollination.) For this purpose in practice often a number of (mother) plants are selected which are open-pollinated. The progenies from these plants are called half-sib families (a half-sib family has one parent in common).

A relatively efficient method for selection of mutants in cross-fertilizing species, called 'crossing within spike progeny method' (or CSP-method) was published in a short note by Ukai (1983) and in a more extensive publication (Ukai, 1990). Seeds are taken separately from each spike of a mutagenically treated population. These seeds are sown as single spike-progenies in separate 'hill plots'. Each hill is isolated from the other hills by bagging the plants of a hill together at flowering time to prevent outcrossing and to get intercrossing within the plants of each hill. Finally, seeds taken from each hill, so-called 'hill-progenies' are planted for selection of mutants. Ukai mentioned that this method – in fact an improved method of family selection with modest inbreeding to obtain mutants as recessive homozygotes – was successfully applied for rye (*Secale cereale*) and

Italian rye grass (*Lolium multiflorum*), but data were given only for Italian rye grass after treatments with gamma rays and ethylene imine (EI).

A comparison of the CSP-method and the open-pollination method showed that in M9, after six successive generations of recurrent irradiation with 300 Gy of ^{60}Co gamma rays, with the CSP-method up to 70% chlorophyll mutations per hill and 1.87% mutants on a plant base were obtained, as compared to 10% and 0.12% after open pollination. Further details about a comparison of the effects of gamma rays and ethylene imine are not given here. Ukai added that, unlike in self-fertilizers, seed sterility did not increase with recurrent irradiation and that the CSP-method can also be applied for other wind- or insect-pollinated crop plants.

In a short report of the Institute of Radiation Breeding in Ohmiya-machi, Ibaraki, Japan, Iida & Amano (1987) described how pollen irradiation (optimal dose 20–40 Gy of gamma rays) was used to induce mutations in cucumber (*Cucumis sativus*), which is an outbreeding species. The authors mentioned that the aforementioned CSP-method is difficult to apply in *Cucurbitaceae*, but that pollen irradiation may be a good alternative because male and female flowers are developing separately on the plants. Some mutant phenotypes, mainly concerning cotyledon colour and shape of leaves, were observed.

6.4.4. Selection in polyploids

Segregation ratios for polyploid and diploid genes differ and to detect a gene mutation in a polyploid plant will be more difficult than in a diploid plant. Let us first consider the induction of a recessive mutation in an autotetraploid plant (e.g. *AAAA* → *AAAa*). It is assumed that a plant with the genetic constitution *aaaa* can be distinguished from plants with the constitution *A*... and that all *A*... plants look alike. If only one 'initial cell' is supposed to have been present during mutagenesis and if random chromatid segregation is excluded (i.e. no double reduction: $\alpha = 0$), selfing of the M_1 plant (*AAAa*) will produce an M_2 which consists of one quarter with the quadruplex constitution (*AAAA*), half of the plants triplex (*AAAa*) and one quarter duplex (*AAaa*) plants. Hence, in the M_2 generation there will not be any plant with the *Aaaa* or *aaaa* genotypes. The recessive phenotype will appear not earlier than in the M_3. If one only considers the duplex (*AAaa*) plants present in the M_2, these plants will produce gametes with the ratio 1 *AA* : 4 *Aa* : 1 *aa* and selfing in this case will result in the production of only one visible *aaaa* mutant plant in an M_3 of in total 36 plants. Since *AAA*, *AAAa* and *AAaa* plants cannot be distinguished phenotypically, all M_2 plants would have to be self-pollinated. Therefore in M_3 only

one *aaaa* mutant can be expected among 144 M_3 plants. As a result, in comparison with mutation breeding of diploid plant species, additional costs are faced because the M_3 population must be 24 times larger than the M_2 generation. In M_4 and later generations, of course, additional mutant plants will occur, but it is clear from this example that homozygous recessive mutants in autopolyploids will not turn up before M_3 and that because of the low segregation ratio (one homozygous mutant in 144 plants for a duplex tetraploid in M_3 instead of 6 homozygous (*aa*) mutant plants against 10 *A* plants for a self-fertilizing diploid), much larger populations are required to recover these mutants in autopolyploids than in diploids. The conclusion must be that mutation breeding for autotetraploid crops is possible, but generally more generations and larger populations will be required than in the case of diploids.

However, for allopolyploids the situation differs much from that for autopolyploids. In 'ideal' allopolyploids the chromosomes of different genomes (e.g. for wheat: *AABBDD*) do not pair and such allopolyploids act like functional diploids. They are therefore amenable to successful mutation breeding just like diploids (see also the comments made earlier on this subject in Section 6.3.1).

6.5. Development of mutants into cultivars

The final steps of practical mutation breeding programmes are to produce improved cultivars, to commercialize them and to multiply and maintain these new cultivars. Most of these steps are rather similar to those for all new breeding products as a result of cross-breeding or otherwise. The importance of this stage and the amount of work required to perform these steps in an adequate way are most often underestimated, in particular by persons who are not breeders by profession.

It can not be sufficiently stressed that the production of useful genetic variation by mutations (or in any other way) does not automatically imply that new and better cultivars will be obtained. Although the degree of difficulty may vary considerably with the crop species, it is predominantly the breeder's skill that determines whether an induced useful mutant will become a valuable cultivar as well.

The mutation breeder, in most cases, aims at the improvement of one trait only. It, nevertheless, often will happen that, after a mutagenic treatment, out of the many thousand genes present in a plant cell several are changed. Some such changes may be noticeable, whereas other changes can not be easily observed but, nevertheless, may have a certain impact on the crop behaviour. In order to reach the desired situation, one improved genetic trait in an undisturbed genetic background, it is advised – when possible – to make backcrosses between the selected mutant plant(s) and the original cultivar and to re-select for the mutant trait in plants which are 'true-to-type' for all other essential traits.

Slight differences in genetic background may considerably affect the expression of the mutated gene, as well as of its accompanying pleiotropic effects. To illustrate this point, an example is given of a favourable mutation for resistance to mildew (*Erysiphe graminis*) in barley (*Hordeum vulgare*), a subject to which some attention was given already in Chapter 3, Section 3.5.2. In at least ten cases resistance of the so-called *ml-o* type has been obtained after mutagenic treatments in different cultivars of barley but until 1991 only one case was reported in which such an induced *ml-o* mutant had become a new cultivar. In the other cases the induced mutants showed too much yield reduction to be sufficiently attractive to the breeders. The only positive exception, mentioned by Micke (1992), refers to an EMS-induced mutant allele, called *ml-o 9*, obtained in cv. Diamant by Schwarzbach, which allele became incorporated in cv. Alexis.

It may be expected that continued studies, in due course, will result in more mutants in which this negative combination of *ml-o* resistance in barley and low yield is sufficiently modified or even completely absent. (N.B. Micke also reported that until 1991 at least 16 barley cultivars had been released in the Netherlands, the UK and Germany, in which the 'spontaneous' mutant allele *ml-o 11*, found in an Ethiopian barley collection, had been incorporated and where also intensive cross breeding had been necessary to remove undesired pleiotropic effects). Some additional details on *ml-o* resistance will be presented later in this chapter in a section concerning induced mutations for disease and insect pest resistance.

Promising mutants should be always tested under various environmental conditions. This may also show that some mutants which, at first sight, looked less suitable, could be of use in other climatic regions. This subject has been described, for instance, in detail by Gottschalk & Wolff (1983) for pea (*Pisum sativum*). Several examples for different pea mutants which were tested for yield potential in Germany and in different regions of India and Africa have been given. To mention just one example: pea mutant 68C was characterized by an increased number of ovules per ovary but a reduced number of pods per plant when cultivated in Middle Europe. When grown in North India the mutant maintained its increased number of ovules per ovary, but the reduction in pod number per plant, observed under German conditions, had disappeared as a result of increased branching of the pea plants under those climatic conditions.

6.6. Some achievements of mutation breeding in seed propagated crops

6.6.1. Introductory remarks

In Chapter 1 general data have already been presented to illustrate the practical significance of mutation breeding. The contribution of induced mutations to the improvement of various traits and the role of different mutagens in this respect have been briefly outlined; numbers of mutant cultivars obtained so far for various groups of crops were given and for some selected examples the economic importance of mutant cultivars has been indicated. In Chapter 3 different systems of classifying mutations have been discussed and as an illustration several mutation types in barley (*Hordeum vulgare*), studied in Sweden and Denmark, have been described. This Scandinavian work on barley in particular made this species the first and for many years the most important model crop for mutation breeding work.

Interesting examples of successful mutation breeding work in many other crops are available as well. Easily accessible information can be found in an array of publications by the IAEA in Vienna – in particular in the *Mutation Breeding Newsletter*, in the series of proceedings of symposia, etc., and in 'Technical Documents', most of which were mentioned already. The FAO/IAEA *Mutation Breeding Review*, either containing summary information on mutant cultivars – like for instance in Micke *et al.*, (1990) – or comprehensive reports about specific (groups of) crops, should be mentioned here as well. Valuable information concerning induced mutations for various important traits in seed propagated crops has been presented in a chapter on induced mutations by Konzak (1984) in a general book on crop breeding edited by Vose & Blixt (1984) and in a review by Konzak *et al.* (1984).

In 1990 a symposium was organized in Vienna jointly by the IAEA and the FAO to assess the contributions made by mutation breeding to crop improvement and the proceedings of this meeting (Anon., 1991b) also provide a wealth of information on practical achievements in many countries. More recently, two useful reviews on the use of induced mutations for crop improvement were published by D'Amato (1992) and Maluszynski et al. (1995).

In this chapter we have selected – from a much wider range of interesting topics – a limited number of case-studies of mutation breeding work for seed propagated crops to illustrate which kind of results can be and cannot be achieved.

It should be remembered as well that the indirect contributions by spontaneous and induced mutations as tools for the advancement of science, e.g. in genetics, evolution research, plant physiology, plant development, phytopathology, molecular genetics and genetic engineering, had much more impact than the direct contributions made by improved cultivars. Attention was paid to this subject, for instance in 1976 when, within the framework of the *Gamma Field Symposia*, a conference with the title '*Mutants in Physiological Research of Crop Plants*' was organized by the Institute of Radiation Breeding in Ohmiya-machi, Japan (Anon., 1976d). In 1981 indirect contributions by mutations were reviewed for the first time in a comprehensive way during a FAO/IAEA symposium in Vienna: '*Induced Mutations – A Tool in Plant Research*' (Anon., 1981a).

During a recent joint FAO/IAEA symposium in Vienna on '*Induced Mutations and Molecular Techniques for Crop Improvement*', von Wettstein (1995) discussed both fundamental and applied aspects of mutation induction and protein engineering for barley. As this contribution provides an excellent example of how fundamental research and applied mutation breeding can be combined, some comments are made here. Von Wettstein described among others the induction and detailed analysis of proanthocyanidin-free malt barley mutants which result from mutations occurring at different steps of the so-called flavonoid pathway. The role of proanthocyanidins for beer quality has been discussed already extensively in Section 3.5.2 of Chapter 3. Briefly, when proanthocyanidins are present in barley, so-called beer haze (or chill haze) occurs during the brewing process. To prevent this (unwanted) haze formation chemicals must be added. If a mutation blocks the aforementioned flavonoid pathway before it branches into a route producing anthocyanins and another route that synthesizes proanthocyanidins, the mutant plants are free of red anthocyanin pigmentation in various plant parts.

Von Wettstein mentioned that 18.5 million M_2 plants (!) have been screened for the absence of anthocyanin, from which 560 mutant plants were obtained which were further tested for the absence of proanthocyanidins by using the vanillin test and liquid chromatography. Recently, more efficient non-destructive tests for lack of proanthocyanidins were developed.

In this project use was made of treatments with NaN_3, which is described as the most efficient mutagen in barley; it induces very few, if any, chromosomal aberrations. Analysis of four independent NaN_3 (*ant-18*) mutants showed that three mutants referred to missense mutations in the respective genes and that one mutant had resulted from a GT to AC transition at the 5' splice site of intron 3 of the gene. More details about this research project have been given already in Chapter 3.

In Chapter 1 (see also Box 1.1) it was also pointed out that in plant breeding literature often some main categories of breeding goals are distinguished, related to yield, quality, time of maturity, resistances and reducing costs of cultivation. Mutations could be classified according to these categories of goals. Konzak et al. (1984), however, described mutants according to six different phenotypic groups for which rather frequently useful mutants had been obtained:

- mutants for growth habit, especially semi-dwarf stature mutants in cereals,
- mutants for disease and pest resistance,
- mutants affecting protein content, nutritional value or processing properties,
- mutants affecting cross- or self-fertility,
- mutants affecting other physiological traits.

In the following sections some, rather arbitrarily chosen, examples of (more-or-less) useful mutant genotypes or cultivars are discussed. These examples may represent achievements in a specific country only, or may refer in particular to the economic profit of mutation breeding. Despite the fact that much has already been said about mutations in barley, in particular in Chapter 3 (Section 3.5), mutation breeding research in this crop has been very extensive and a number of very interesting results have been obtained which deserve additional attention. We will return, for instance, to successful mutant cultivars of barley and to induced mutations for resistance to powdery mildew (*Erysiphe graminis* f.sp. *hordei*) in this crop, which is one of the best studied systems of resistance derived from induced mutations. Mutations for protein content and protein composition, also subjects to which considerable attention was paid for barley in Sweden and Denmark, will be discussed in this chapter as well, together with some particulars about mutations for protein in maize. Mutation studies concerning protein in other cereal crops, such as wheat, rice and sorghum do not differ fundamentally from those in barley. Several other aspects of mutation breeding in maize have been discussed already earlier in this chapter as well as in previous chapters.

Pea (like common bean, soybean, groundnut, lupin, mungbean, cowpea, faba bean, chickpea and pigeonpea) belongs to the group of grain legumes, in which many mutant cultivars have been released with many different improved traits. Several reviews on mutation breeding in grain legumes have been published for instance by Micke (1983a; 1984; 1988a,b; 1993b) and by Jaranowski & Micke (1985). Although mutation breeding in grain legumes will not be discussed in this chapter as a separate group, mention is briefly made at this place of mutants affecting legume plant–Rhizobium interaction. In a thesis of the University of Groningen, the Netherlands, Postma (1990) described studies on symbiotic nitrogen fixation for pea (Pisum sativum) and Rhizobium leguminosarum in which use was made of induced (monogenic, recessive) mutants obtained by Jacobsen (1984; see also Feenstra & Jacobsen, 1980, and Jacobsen & Feenstra, 1984) after treatment of seed presoaked for 8 h (seed coat perforated with sandpaper), during 4 h in a 800 ml solution containing 0.3% EMS. The treatment was performed in a glass beaker of 2 l under continuous aeration and occasional stirring in the presence of a 0.01 phosphate buffer (pH 5.7) at 20 °C. After EMS treatment, post-washing in running tapwater was performed for 1 h. In Postma's thesis many other references to earlier reports on this subject can be found.

Duc (1995), in the introduction of a report concerning mutation experiments on nodulation in faba bean (Vicia faba), briefly summarized results of previous experiments on this subject for pea (Pisum sativum), soybean (Glycine max), common bean (Phaseolus vulgaris) and chickpea (Cicer arietinum). Three main types of mutants affecting nodulation were distinguished: a) mutants showing no nodulation, b) mutants with ineffective nodulation, and c) mutants with increased nodulation. Mutant phenotypes of all three categories have been reported in literature for the four species mentioned, except for mutants of the 'ineffective nodulation type' in soybean and mutants for increased nodulation in chickpea. In his own experiment Duc (1995), after treatment of 1000 seeds of cv. Ascott of faba bean with EMS (0.1%, 6 h at 24 °C), also obtained the three aforementioned types of mutants. The 'supernodulating' mutant had 3 to 5 times more nodules than the control in a nitrogen-free flowing nutrient solution. Based on the analysis of 20 000 M_2 plants it was concluded that all mutant phenotypes were controlled by distinct single recessive genes. This work was also discussed earlier by Sagan et al. (1991) during

the FAO/IAEA symposium in Vienna in 1990 (Anon., 1991b).

A relatively new and attractive topic for breeders of, for instance, a crop like pea (Pisum sativum), is to increase the portion of certain starch components in the seed cotyledons from which biodegradable 'plastics' can be made. Mutagenic treatment, for instance with EMS, appears to be a relatively easy method to obtain the desired genetic improvement for such traits and the public acceptance of the 'mutation approach', nowadays, seems much better than the application of methods of genetic engineering (e.g. 'antisense' methods; see also Chapter 1) for similar purposes.

One example about which insufficient (easy accessible) information is present to discuss the subject more in detail, is the induction of a mutant cultivar in sea buckthorn (Hippophaea rhamnoides) by combined treatment with radiation and chemical mutagen. Privalov (1986) briefly reported about this small fruit tree that is grown in Siberia and valued for its highly nutritive fruits and valuable oil with medicinal properties. A new mutant cultivar was obtained after treatment of seeds with 150 Gy of gamma rays and treatment of the resulting M_1 seeds with 0.01% nitrosomethyl urea. The new cultivar had higher fruit yield and an increased oil content. More details, unfortunately, have been given only in publications in the Russian language.

As was also mentioned already in Chapter 5, a search in the literature for commercially successful cultivars from seed propagated crops that originated from in vitro mutagenic experiments did not reveal clear examples of any economically important cultivar. For some countries like China such positive results may have remained unnoticed by us, and in other cases insufficient details were available to be able to trace back the exact life history of such putative mutant cultivars of in vitro origin. One apparent positive exception for China mentioned by Wang (1991), is the successful rice mutant cultivar, R462, that resulted from irradiated anthers in vitro and has been released to farmers. Wang (1991) also reported that some other valuable mutants had been obtained from cultured tissues and cells, but further details were not given.

More recently, a summary from Wen & Qu (1996) about the achievements of mutation breeding in China came to our notice. In this report it was mentioned that up to 1993 more than 345 mutant cultivars from 31 plant species had been developed. Almost 80% of the mutant cultivars were directly obtained and about 20% resulted from further crossing work. In the publication no distinction was made between mutants produced via conventional ('in vivo') mutation breeding and in vitro mutation work. More than 75% of the mutant cultivars referred to food crops, 12.5% to oil crops, and the remainder to fibre crops, vegetables, food crops, forage crops and sugar crops. Most mutant cultivars concerned rice (*Oryza sativa*) with 122 mutant cultivars, and wheat (*Triticum aestivum*) with 91 mutant cultivars. Mutant cultivars for ornamentals were not reported. The authors further mentioned that over the period 1990–5, about 10^7 ha were covered by mutant cultivars, which area represents more than 10% of the total area cultivated with these crops in China.

6.6.2a.   Mutation breeding in rice

Rice, in particular Asian rice (*Oryza sativa*), together with wheat, is the most important staple food for man. World production in 1990 reached 520 Mt (Chang, 1995). Rice is mainly eaten in the humid tropics and subtropics. The most important rice producing countries are China and India.

Some early contributions on mutation breeding in rice were presented by Bekendam (1961), Li *et al.* (1961) and Matsuo & Onozawa (1961) during a symposium called '*The Effects of Ionizing Radiations on Seeds*' in 1960 in Karlsruhe, Germany, that was jointly sponsored by the IAEA and the FAO.

In a paper by Rutger on mutation breeding of rice in the USA (Rutger, 1991) and in a more extended review with worldwide coverage for this crop (Rutger, 1992), the merits of this approach and the results obtained have been clearly outlined. Rutger stresses that mutation breeding in rice, a self-fertilizing crop, has resulted in more cultivars than in any other crop. Within a period of about 30 years, in total 198 (directly obtained) mutant cultivars had been released worldwide. The use that was made of induced mutants in cross-breeding programmes during the period 1970–90 resulted in an additional 80 cultivars.

According to the FAO/IAEA database (see Maluszynski *et al.*, 1991) 114 mutant cultivars for rice had been released at that time in China, 31 in Japan, 24 in India and 17 in the USA. The most important improved characteristics are earliness (70 times), short straw (63 times), various traits concerning grains (27 times), resistances for various diseases, etc. (21 times) and yield (17 times). Most mutant cultivars were obtained by gamma irradiation of seeds with doses between 150 and 300 Gy. Population sizes used for M_1 and M_2 differed considerably between experiments, but Rutger himself was successful with irradiating about 5000 seeds at 250–300 Gy, from which 2000 panicles only were harvested to grow a 'panicle-to-row' M_2 generation.

Rutger (1992) reported that, from a total of 198

mutant rice cultivars, 159 cultivars arose from gamma ray treatments, 26 cultivars from other radiation sources and only 11 cultivars from chemical mutagenesis. Rutger pointed out that chemical mutagens produced higher mutation frequencies but strongly advocated the use of ionizing radiation because of practical problems (like soaking and redrying of seeds, safety of handling and disposal) encountered when chemical mutagens are applied. The same author mentioned that 'few if any' rice cultivars had arisen as a result of genetic variation from *in vitro* cultures.

The main reason why mutation breeding in rice has been very successful worldwide is probably the enormous attention of many researchers to this most important food crop of the world, its diploid way of inheritance and its self-fertilizing character. Rutger explains that positive results from conventional breeding work as well as from mutation breeding in particular have been realized – as expected – when only one or a few simply inherited traits were aimed at.

The first widely known rice mutant cultivar has been cv. Reimei, a short-straw (semi-dwarf) mutant from cv. Fujiminori which was released in 1966 by Futsuhara *et al.* in Japan (Futsuhara, Toriyama & Tsunoda, 1967; Futsuhara, 1968; Maluszynski, Micke & Donini, 1986). The mutant cultivar carries an allele that is allelic (or probably even identical; see for instance Kawai & Amano, 1991) to the well-known *sd1* (semi-dwarf) allele in cv. Dee-geo-woo-gen (DGWG), a spontaneous mutant discovered by Chinese scientists. This mutant DGWG possessed as its most important traits dwarfness, stiff-straw, fertilizer responsiveness, non-lodging, daylength-insensitivity and no seed dormancy, and has been the forerunner of the so-called 'Green Revolution'. It has given rise (after crossing) to the important cv. Taichung Native-1 and, after a cross with the tall cv. Peta from Indonesia, to the even more famous cv. IR 8 from the International Rice Research Institute in the Philippines. The aforementioned *sd*-allele(s) are frequently used at the IRRI and it appears that with the exception of cv. IR 5 (which is not a true semi-dwarf) virtually all other semi-dwarfs in major rice-growing areas carry the DGWG gene. So far, no other *sd*-alleles have been found as suitable from an agronomic point of view as *sd1*, and its frequent and worldwide use causes concern about genetic vulnerability.

If we briefly return now to the first artificially induced mutant cultivar for rice, cv. Reimei, it can be mentioned that in 1979 about 120 000 ha of this short-straw cultivar were grown in Japan, and that at least 33 other cultivars – including several important ones -have been developed from cross-breeding with this cultivar (Kawai & Amano, 1991).

Mutation breeding work on rice in the USA, mainly performed in California during the 1970s, has also been very successful, and Rutger (1991) calculated that this mutation work earned Californian rice growers an additional 20 million US dollars per year in the early 1980s. Three reasons were mentioned that have been mainly responsible for this success. First, genetic corrections for only one or two agronomic traits, e.g. for reduced plant height, were induced in high-yielding cultivars that were already widely grown and well-adapted to local conditions. Moreover, an alternative approach, crossing with semi-dwarf rice genotypes from Asia in order to introduce this trait to USA germ plasm, according to Rutger, appeared to be less promising because such genotypes were of the so-called *indica* type and did not have environmental adaptation required for California (i.e. a certain degree of cold tolerance) and a grain quality acceptable to US markets. In addition, back-crossing to US cultivars, which are mainly of the *japonica* or *javanica* type, is costly and time-consuming. Back-crossing programmes, nevertheless, have been successfully performed at research stations of the states Texas and Louisiana.

For California, mutation breeding was chosen as a quicker alternative. The most successful mutant cultivar, cv. Calrose 76, was selected in 1971 as a single short stature M_3 plant after treatment of seed of cv. Calrose with 250 Gy of gamma rays from a ^{60}Co gamma source. The new mutant cultivar was released in 1976 and registered in 1977 (Rutger, Peterson & Hu, 1977). The new cultivar is similar to its parent cultivar except that its straw is 25 cm shorter at maturity and the panicle has more awns. A single recessive gene has been found to be responsible for this short stature mutation.

Another significant factor for the positive outcome of this work that is worth mentioning, was that selection of promising mutants was done by rice experts and was strongly focussed on agronomically useful mutants. The third important point was that useful mutants were immediately integrated into conventional cross-breeding programmes.

In his reviews Rutger (1991; 1992) also paid considerable attention to so-called 'breeding tool mutants' in rice, like mutants for male sterility, mutants with aberrant hull colour (which may be of advantage as markers in crosses) and herbicide-tolerant mutants. Hu & Rutger (1991; 1992) analyzed 23 genic-male sterile (ms-)mutants of rice, ten of which had been induced by treatments with EMS, one by IE treatment and seven by gamma rays, whereas one was obtained after treatment with streptomycine (see also Hu & Rutger, 1991), one during tissue culture and three had arisen spontaneously. Four different pollen abortion types were found, but all mutants were monogenic recessive. Diallelic crosses showed that some mutants were allelic. Such ms-mutants are not of direct agronomic importance, but may become very useful in hybrid seed production.

In his review, Rutger (1992) also briefly discusses some other aspects of mutation breeding in rice. He refers in this respect to selection for early maturity, grain size (and kernel colour) and glutinous endosperm in Japan. For China (see also Wang, 1991) mention is made of the two most widely grown mutant rice cultivars in the world. Cv. Yuanfengzao, an early maturing mutant which was directly obtained from gamma irradiation of seeds from cv. IR 8, was grown in China in the early 1980s on over 1 million ha. Cv. Zhefu, also an early maturing directly obtained cultivar with in addition a broad disease resistance, a wide adaptibility and a high yield potential of almost 6400 kg. ha^{-1} on average, was grown in 1989 on more than 1.4 million ha. Wang (1991) further reported that seven other mutant cultivars of rice occupied areas exceeding 100 000 ha and that at that time from a total of over 110 rice mutant cultivars which had been released in China, 94 were direct mutants and 16 referred to crosses with mutants.

Rutger (1991) further mentions that 37 direct mutant cultivars and 18 mutant cultivars obtained after crossing, mainly with reduced plant height, had been produced for release in tropical countries in cooperation since 1961 between the agricultural research institutes IRAT in Montpellier, France, and national institutes in Cote d'Ivoire, Guyana, Senegal, Burkina Fasso, Cameroon and Brasil. Besides the famous semi-dwarf mutant IRAT 13, which was used in many crosses, several genetically semi-dwarf rice mutants are available from this programme (Jacquot, 1986; Clement & Poisson, 1988).

Rutger also discussed some examples for India and Pakistan. For Pakistan reference is made to agronomically promising mutants in the so-called Basmati rices: rice types that are highly favoured for their excellent aroma and specific cooking characteristics, but that suffer from some serious agronomic draw-backs such as a very tall stature, lateness and low yielding capacity.

6.6.2b. Mutant cultivars for durum wheat in Italy

During a symposium, jointly organized by the IAEA and the FAO in 1990 in Vienna in order to assess results of purposeful mutation breeding, Scarascia-Mugnozza *et al.* (1991) reported on the achievements obtained in this respect for durum wheat (*Triticum turgidum* subsp. *turgidum*) by a joint group of Italian researchers. This report was also published as FAO/IAEA *Mutation Breeding Review* 10 (Scarascia-Mugnozza et al., 1993).

A programme on experimental mutagenesis was started in 1956 in order to increase the yield of durum wheat, which had remained low compared to bread wheat (*Triticum aestivum*), in particular because of its lodging susceptibility due to straw weakness. The programme resulted in the production of some 1000 induced mutants for various traits. Eleven registered mutant cultivars were released up to 1989, five of which resulted from direct selection of mutants, whereas the other six came from continued cross-breeding work with these mutant cultivars or with other mutant lines derived from the programme. Between 1968 and 1971 four direct mutant cultivars – cv. Castelfusano and cv. Castelporziano (both derived from cv. Capelli), cv. Casteldelmonte (from cv. Grifoni) and cv. Castelnuovo (from cv. Garigliano) – were released. These mutant cultivars resulted from treatments with thermal neutrons (2x), fast neutrons and X-rays respectively. In 1988 another direct mutant cultivar, cv. Icaro, obtained after a fast neutron treatment of cv. Anhinga, was released. From the parent cultivars, according to Lupton (1992), cv. Capelli occupied 59% of the total area for durum wheat in Italy in 1947 and remained the most important cultivar till 1968. Cv. Capelli, a selection from a N. African population called 'Jan Retifah' (Lupton, loc.cit.), was still grown (3% of the total area) on a modest scale in 1986. The other parent cultivars mentioned before, cvs. Garigliano and Grifoni, during the period 1955–74 each occupied never more than 5–10% of the total area.

Data were presented on culm length, lodging resistance, heading date and yield for cv. Castelfusano, cv. Castelporziano, cv. Castelnuovo and cv. Casteldelmonte. The four mutant cultivars mentioned were characterized by short straw (between 73 and 92% of their mother cultivars), high lodging resistance, a slightly later heading date and a yield that was 7–18% higher than the yield of the mother cultivars. It was further reported that the fifth direct mutant cultivar, cv. Icaro, in particular had a good yielding ability, but as this mutant cultivar was released in 1988, yield figures for a range of years, apparently, were not available yet.

Cv. Castelporziano and mutant line Cp B144 proved to be particularly successful as parents in the production of indirect mutant cultivars because of a dominant mutation for short culm. The six indirectly obtained mutant cultivars (cvs. Creso, Mida, Tito, Augusto, Ulisse and Peleo), in addition to short straw and lodging resistance, all showed a good or – in the cultivars that were released most recently – even high yielding ability.

For one of the earlier indirect mutant cultivars, cv.

Creso, which was released in 1974 and can be characterized by a constant high yielding ability, a good adaptability and a good grain quality, detailed economic data have been collected for several years (see also Lupton, 1992). In 1990 cv. Creso had become the leading cultivar in Italy, representing 58% of the total amount of certified durum seed and occupying a growing area, rising during the 1980s to 460 000 ha, i.e. about one third of the total area under durum wheat in Italy. Its mean yield was estimated at 3.16 t. ha^{-1}, against 2.12 t. ha^{-1} and 1.75 t. ha^{-1}, respectively, for the two next most widely grown cultivars: cv. Patrizio and cv. Capeiti.

Scarascia-Mugnozza et al. (1991) calculated that an (even underestimated) benefit of 0.9 t. ha^{-1} over the other cultivars resulted in a total extra production for Italy of 450 000 t, which was equivalent to an additional economic benefit of about 180 million US $ at an annual base, or 1800 million US $ during one decade. It is interesting to add in this respect a remark by the authors, that the total costs of the mutation breeding programme on durum wheat, carried out at the Casaccia Research Centre near Rome during a period of 15 years, amounted to some 3.5 million US $.

6.6.2c. Mutation breeding in cotton

During the 1990 Symposium '*Plant Mutation Breeding for Crop Improvement*' in Vienna, jointly organized by the IAEA and the FAO, Iqbal et al. (1991) reviewed mutation breeding in cotton (*Gossypium* sp.) in Pakistan, a country with more than 70% of its population living as farmers, for which the cotton crop is a major source of income.

Because of the need for high yielding cultivars that can withstand large fluctuations in weather conditions, and the limited genetic variation in the available cotton germplasm, a mutation breeding programme was initiated by the Nuclear Institute for Agriculture and Biology in Faisalabad, Pakistan. In 1971 an M_1 population was grown from 1000 hybrid seeds of a specific cross between a local and an exotic cultivar (AC-134 × cv. Deltapine) that had been irradiated with 300 Gy (30 krad) of gamma rays. An M_2 population of about 11 000 plants was raised from five seeds per boll from each M_1 plant and selection for earliness, short stature, yield components and fibre properties resulted in 40 plants with promising properties. Consecutively, from the best family five plants were taken and selection was carried on to the M_4 generation.

In M_5 a selected line – later named NIAB-78 – bred true for earliness, short internodes and some other useful traits. Its hairiness limits insect pest infestation and because of its earliness the peak of bollworm attack is evaded. The new cultivar NIAB-78 can be recognized by its short stature, a high yield potential, the shorter period to mature (150–160 days instead of 190–210 days for other commercial cultivars) and a better tolerance to bacterial blight and salinity than other commercial cultivars.

In 1983 the new cultivar NIAB-78 was officially approved for commercial cultivation and became quickly accepted as the most productive cultivar in the country at that time. Seven years later it was still successful particularly in the Punjab area, with 65–70% of the total area for cotton being the predominant cultivar. The total production of cotton by then had been almost doubled to about 7.2 million bales in the Punjab and to about 8.5 million bales for the whole country. Depending on the location and the way of farming, yields between 2250 and 2600 kg. ha^{-1} were reported. The authors further mentioned that since the introduction of this cultivar farmers in the Punjab area had earned a total of Rs. 71 000 million, which is at least 15% higher than they would have been earning, if other recently released cultivars had been grown.

The release of a new mutant cultivar for cotton, NIAB-92, was described by Iqbal et al. (1994). Pre-soaked seeds were treated with 300 Gy of gamma rays (dose rate 400 Gy. h^{-1}) and sown in the field in 1984. In comparative field trials in 1990–1, NIAB-92 outyielded NIAB-78 by 15.3%; in farmers' fields the increase was only 8.8%. A number of other favourable traits, such as high resistance to leaf curl virus and several characters contributing to superior fibre quality, were mentioned as well.

Mention should be made also of a highly successful mutant cultivar of cotton, cv. Lumian No. 1, released in the Peoples Republic of China (see List of Mutant Varieties in the FAO/IAEA *Mutation Breeding Newsletter* 19; Anon., 1982b). Cv. Lumian No. 1 was released in 1976 by the Institute of Atomic Energy Application in Agriculture, Shantung Province, of the Academy of Agricultural Science, after treatment in 1971 of F$_9$ seeds from the cross Zhong No.2 × 1195 with 400–450 Gy of gamma rays. According to the aforementioned FAO/IAEA list, this mutant cultivar showed good vigour, an attractive plant architecture and a higher boll production and was grown in 1981 on 1.24 million ha in China. According to Wang (1991) later in the 1980s its peak was even 2 million ha.

6.6.2d. Mutations affecting protein content and composition, in particular in maize and barley

In maize (*Zea mais*) spontaneous mutations affecting the endosperm were detected already in 1912 in the USA.

The discovery in the early 1960s that in the spontaneous *opaque-2* mutant of maize the content of the amino acids lysine and tryptophan in the endosperm was increased (Mertz, Bates & Nelson, 1964), led to worldwide breeding efforts to convert normal maize to *opaque* maize and to induce more mutations for such traits. Lysine and tryptophan are so-called essential amino acids which, when present only in low quantities in the seed endosperm, limit the nutritional value of most cereals for monogastric animals and man. A second mutant gene affecting the amino acid pattern of maize endosperm proteins (with a lower level of the unfavourable zeins and a lysine concentration about equal to that in *opaque-2*), called *floury-2*, was reported in 1965 by Nelson, Mertz & Bates (1965).

Mutant genes like *opaque-2* and *floury-2* reduce the level of zeins which represent about 50% of the storage proteins in maize and are deficient in lysine and tryptophan. Unfortunately a number of negative phenotypic effects, such as low yield, reduced protein content and the presence of soft kernels, prevented *opaque-2*, *floury-2* and other mutations from becoming widely used. In order to overcome such problems, plant breeders utilized many and complex methods to modify this trait in order to develop *opaque-2* mutants with a hard, vitreous endosperm, a normal protein content and an enhanced percentage of lysine.

Micke (1983a), who reviewed the international research programmes that were started for the genetic improvement of grain proteins, mentioned some positive results of enriching maize with the *opaque-2* mutant. He reported that feeding programmes of pigs with *opaque-2* maize instead of normal maize grain did result in a gain of weight that was up to three and a half times faster than with the normal maize grain. Micke also mentioned that children in Colombia suffering from severe protein deficiency were brought back to normal health within 2–3 months on a diet of *opaque-2* maize.

Despite all efforts and the aforementioned encouraging results from the viewpoint of human and animal nutrition, the soft endosperm texture, the susceptibility of the seed to pathogens and mechanical damage and the lower seed yield, so far, have greatly limited the use of mutants with an improved protein composition and increased protein content (Nelson, 1981; Micke, 1983a; Mertz, 1992; Villegas, Vasal & Bjarnason, 1992).

The discovery of the aforementioned mutants in maize also led to a search for genotypes with an improved quality of the seed proteins in other cereals, including barley which, like most cereals, has a relatively poor protein quality because the major part of the seed protein consists of prolamines: proteins with a low lysine content. In 1968 a spontaneous high protein and high lysine barley line of Ethiopian origin, which became known as Hiproly, was discovered in barley collections in Sweden (Munck et al., 1970). The grain protein quality of this line was found to be based on a recessive gene, called lys-1. The finding of Hiproly was the cause of an extensive programme in Scandinavia as well as in other countries to produce artificially induced mutants in barley (and in other cereals as well) for protein content and protein composition. The Swedish/Danish barley work in this respect is well documented (see for instance Munck et al., 1970; Munck, Karlsson & Hagberg, 1971; Munck, 1972; Doll, Køie & Eggum, 1974; Doll, 1976; 1977; Doll & Køie, 1978; Knudsen & Munck, 1981; Tallberg, 1982; Munck, Bang-Olsen & Stilling; 1986; Tallberg, 1986 and Jensen, 1991). Particularly useful is an extensive contribution by Munck (1972) in which many details about cereal chemistry and screening methods for determining crude protein content, applied in those days, were discussed. We should mention in this respect the established, simple but not very reliable Kjeldahl method which is based on determination of the nitrogen content (crude protein content = $N \times 6.25$), and the DBC-(dye binding capacity) method, which was introduced in the 1950s by Udy (1954; 1956) to estimate the crude protein content in wheat. A good review on selection methods involved in mutation breeding for seed protein improvement was presented by Brock (1979) during a FAO/IAEA conference on this topic, organized in Neuherberg, Germany, in 1978. The contributions made by Tallberg (1982; 1986) are also interesting. This author stresses the need to further increase the knowledge of the underlying biochemical mechanisms and the genetic control of nutritional quality of cereal seeds.

The main purpose of the Scandinavian project was to obtain barley cultivars with an increased lysine content or, more in general, with improved nutritional quality, to improve the use of such cultivars as fodder (in particular for pigs) without the need of supplementing this by additional high lysine proteins (e.g. from imported soybean meal) in order to improve the amino acid balance.

Two mechanisms in high lysine barley are responsible for the lysine increase caused by single recessive genes: either a decrease of the amount of (lysine poor) hordeins (which is the word used in barley for prolamins) is compensated for by increased amounts of free lysine and non-hordein proteins, or the amount of specific lysine rich proteins is increased. The three most promising and best studied 'high lysine' barley genotypes are Hiproly (a 'spontaneous mutant' which is still

the best one and superior to all induced mutants) and the artificially induced mutants 1508 and 7 (Doll, 1976). Increased lysine content (in percentage of grain weight) more or less coincides with reduced (lysine poor) hordeins. In mutant 1508 more free lysine and non-hordein proteins are present. In Hiproly and mutant 7 the contribution of lysine rich proteins of the glutelin fraction has been increased. Mutant 1508, conditioned by gene lys-3a, has a 40% higher lysine content; in Hiproly both lysine and total protein content are 30% higher than in normal barleys. A very modest increase of lysine content (10%) in mutant 7 most probably is also caused by a recessive single-gene mutation (Doll, 1976). Feeding trials for the aforementioned mutants gave different values for specific properties (like digestibility, biological value and net protein utilization).

In 1982 a Final Research Co-ordination Meeting on the subject of cereal grain protein improvement was held at the IAEA in Vienna (for the proceedings see Anon., 1984a). All participants agreed that it was more difficult than expected to convert gains in grain protein content into higher protein yields per unit field area as increased protein is often compensated for by nearly corresponding reduction in grain yield. Continued cross-breeding, however, may result in high protein/high yield combination. For barley it was concluded that it is just a question of time to produce cultivars that are as good as Hiproly for lysine and that can meet total yield of standard cultivars.

So far, it has not been possible to breed new barley cultivars with a substantially improved lysine content as well as an acceptable grain yield. This reduction in yield is mainly caused by reduced grain weight as a result of a disturbed starch production, leading to shrunken seeds in high lysine genotypes. It has therefore been argued by many, that high protein/high lysine could be an artifact caused by disturbed starch deposition in the developing grain. The problem derived probably from the wrong selection method, determining protein and lysine as weight percent of grain dry matter (flour). Many experiments to eliminate this undesired correlation in the aforementioned mutants, by changing the genetic background, have not been successful. It is believed in this respect that mutant 7, despite (or because of?) the low increase in lysine content, appears to be more promising than both other genotypes.

Jensen (1991) reported on 20 low-hordein mutants, selected from 49 000 M_2 seeds, which had arisen from mutagenically treated doubled haploids, obtained from cv. Sultan by the so-called bulbosum technique. Six different mutagenic treatments, 150 or 200 Gy of gamma rays, 1% EMS (for 35–90 min) and 1×10^{-3} mol NaN$_3$, were given. The method of scoring the frequency of chlorophyll mutants (in 400 M_1 spikes per treatment) was used to determine optimal mutagenic treatments. Of 10 mutant lines, selected out of the original 20 low-hordein mutants on the basis of yield performance and hordein content over several years, two numbers had an equal grain yield as cv. Sultan. Moreover, a reduction in hordein content suggests an increased lysine content of about 10%. Jensen (loc.cit.) concludes from these experiments that the combination of a modestly increased lysine content and no reduction in grain yield is possible. He also mentions that still additional mutations may be present in this 'raw' breeding material that may have a negative effect on grain yield. Continued selection for increased yield, therefore, might result in even higher yield, but more recent data showing a breakthrough have not been found in literature.

6.6.2e. Mutants for plant architecture, in particular in grain legumes

In Chapter 1 the subject of mutants for changed plant architecture was briefly introduced and reference was made to a publication by Micke et al. (1990) in which the data collected for induced mutations were grouped, among other things, according to the kind of traits for which mutants had been obtained. The authors reported that at that time in total 336 cases of induced mutations for changed plant architecture had been registered in mutant crop cultivars. It may be noted that the trait that had been improved most often in this category, i.e. 'reduced plant height' (having a positive effect on yield via an improved fertilizer response, high tillering and lodging resistance), mainly was selected in cereals (289 cases), with an almost equal frequency of use of this trait in 'direct' and 'indirect' mutant cultivars. For non-cereal crops, a range of other characters affecting plant architecture, contribute more often (i.e. in 84 cases) to induced changes for plant architecture than mutations for reduced plant height (47 cases). Most examples of changed plant architecture for 'non-cereals' refer to directly released mutant cultivars. Micke (1988a), in a review about the use of induced mutations in grain legumes, mentioned 57 examples of improved plant architecture for this group of crops, including 17 dwarf or bushy-type mutants, 17 with improved lodging resistance, 13 with improved plant architecture, 8 with erect or tall types of growth and 3 cases with a higher harvest index.

In their review, Konzak et al. (1984) briefly treated some specific examples of the successful induction of mutations for growth habit in several crop species, like the early example of induction of an early maturing,

determinate bush habit in cv. Michelite of dry bean – in the USA called Navy pea bean – (*Phaseolus vulgaris*). The original mutant plant, resulting from a mutagenic treatment with X-rays in 1938, still differed too much from, what was considered, the ideal plant type for this dry bean to be useful as a direct mutant cultivar. But, as the observed changes were in the right direction, additional crossing and back-crossing were performed (Down & Anderson, 1956). In these crosses the mutant trait proved to be highly successful as can be concluded from the release of a series of 'indirect' mutant cultivars, like cv. Sanilac, which was released as the first radiation-induced commercial crop cultivar in the USA in 1956; cv. Seaway, released in 1960; cv. Gratiot in 1962; and cv. Seafarer, a cultivar with a wide adaptability, released in 1967. Andersen (1972) reported that by the early 1960s, cv. Sanilac was grown on about 90% of the area planted with dry bean in Michigan State, USA.

Based on these results and starting from cv. Seafarer, Adams (1982) initiated a programme to induce growth habit mutations in dry bean in order to develop a new non-branching so-called 'ideotype' of bean and in 1973 several non-branching plants were obtained from 2.5 ha with M_2 plants. Further crossing work showed that the non-branching habit could be combined with the traits for tall, erect plants and with exceptionally high yields. At the time the review by Konzak et al. (1984) appeared, the new plants did not yet contain all features desired in superior new plant types, but the authors predicted that the newly induced improved adaptation to mechanical harvesting might mean an important break-through for the large scale production of high-yielding, mono-culture-grown field beans and might show the way for breeders of other crops. More recent reports on this topic have not been found in literature.

Other examples that were mentioned by Gottschalk & Wolff (1983), as well as by Konzak et al. (1984), and that will only be briefly discussed here, refer to plant architecture of pea (*Pisum sativum*). Several single gene mutations affect the development of leaves and stipules in pea. (N.B. The nomenclature of the different mutant traits in publications by various authors is not always consistent and sometimes even conflicting.) Blixt (1972) and Snoad & Davies (1972) mention four genes in this respect: *tl*, which results in conversion of all tendrils to leaflets; *st*, leading to a great reduction of the stipule size; *up*, which reduces the number of pairs of leaflets on each leaf to one; and *af*, which converts leaflets to tendrils. In this way the foliage may be reduced or even the architecture completely changed towards plant phenotypes with higher agronomic value.

Gottschalk & Wolff (1983) report on early studies,

mainly in Eastern Europe, on spontaneous and induced mutations for leaf traits in pea and mention that a mutant trait, called *acacia* and known in the literature since 1910, was repeatedly induced by mutagenic treatments of field pea (indicated as *P. arvense* L.s.l.) in Poland (Jaranowski, 1976; Jaranowski & Micke, 1985). Some of these *acacia* genotypes, selected in Bulgaria, were reported to have higher yield than the mother cultivar on the basis of yield of individual plants. Gottschalk & Wolff (1983), on the other hand, point out that because of the presence of the homozygous gene tl^w, tendrils are completely replaced by leaflets; this change (because of lodging) is described as a negative effect and, according to these authors, makes mutant plants carrying this gene unattractive for pea breeding.

The opposite alteration is the development of branched tendrils instead of leaflets, indicated as the *afila* type. Gottschalk & Wolff (1983) give several references to spontaneous as well as induced mutations for this trait but mention that yields were reduced in most cases. Nevertheless, starting from the early 1970s positive reports appeared on the agronomic potential of leafless peas (Snoad & Davies, 1972; Snoad, 1974; Snoad & Hedley, 1981) which indicated that the new phenotypes showed no lower yields than the conventional cultivars under adapted plant conditions. As a result, breeding for 'leafless' pea cultivars started in several countries like the UK, Italy and the Netherlands. Jaranowski & Micke (1985) reported on the 'tendrilness' mutant cv. Wasata, which was released in 1979 and used also in crosses in Poland and two 'indirect' mutants which were released in 1979 and 1981 as cv. Sum and cv. Hamil. (N.B. The term 'tendrilness' apparently was the early term for *afila*, which, in turn, seems to be synonymous to 'semi-leafless'.) Several other induced 'indirect' mutant cultivars of the *afila* type, e.g. cvs. Heiga and Jaran, were released throughout the 1980s in Poland. More data can be found in the mutant variety database presented in *Mutation Breeding Newsletter* 38 (Anon,. 1991a).

Gottschalk & Wolff (1983) refer to an 'entirely new crop' and mention the following main advantages of 'leafless' and 'semileafless' peas: a) a very high standing ability because of the high number of tendrils, b) more uniform ripening, c) a lower susceptibility to pests and diseases, and d) easier drying. In the 'semileafless' types, according to Snoad & Hedley (1981), leaflets are converted to tendrils whereas 'leafless' types, in addition, have very small stipules as the result of the action of another recessive gene. Meanwhile, several cultivars have been introduced in various countries and, although most of them are based on spontaneous mutants, some

cultivars are the result of induced mutations for the aforementioned trait(s) (Micke & Donini, 1982; Micke, 1988a,b). Mutant cultivars, in most cases, are not true 'leafless', but still carry a sufficient photosynthetic area in the form of green tendrils and mostly normal or only slightly reduced stipules.

The quick economic impact of the (semi-)leafless trait can be demonstrated for instance by the observation that in the Netherlands in 1980 no cultivars for pea carrying this trait were officially registered in the Dutch List of Varieties whereas in 1995 all officially released and registered cultivars for common green and yellow dried pea carried this trait. The economic benefit is mainly in the harvesting technology, e.g. green peas harvested by combine.

From the above example of 'leafless' mutants in pea it can be concluded that mutations for this trait, apparently, are easily induced but there is of course no urgent reason to do so since the same gene from spontaneous mutants is already available, which can be easily re-combined with other traits of agronomic importance and, thus, within a short time may lead to new cultivars with a changed architecture. Artificial induction of such mutations would be of practical value only if obstacles would be met in introducing the spontaneous mutant traits into specific cultivars by unfavourable linkages.

The importance of mutagenesis for the improvement of plant architecture would be even considerably higher if a desired trait should be completely lacking in the germplasm of a crop species. A recent example in this respect was provided by van Rheenen, Pundir & Miranda (1994) for chickpea (*Cicer arietinum*).

The authors reported that, despite extensive screening in the ICRISAT germplasm collection of chickpea, which contains about 16 350 accessions, the trait of determinate plant growth had not been found, but that they had been able to induce this desirable trait by mutagenic treatments. Large amounts of seeds (1.5 kg per dose; 100-seed weight = 22 g!) of the widely adapted, wilt resistant kabuli cultivar ICCV 6 were irradiated in 1986 with 150, 300 or 450 Gy of gamma rays. A total of 1.8×10^5 plants were grown in M_2 and in the 150 Gy series one plant was found which showed determinate growth. As this plant was female sterile and did not produce pods, it was crossed with neighbouring M_2 plants from which eleven non-determinate (so-called) F_1 seeds were obtained. In 1988 the F_1 plants were grown, followed by the F_2, F_3 and F_4 generations in 1989/90, 1991/92 and 1993/94 respectively. In the F_2 generation the 'determinate trait' segregated in a postulated digenic epistatic 3 : 13 ratio. Results from F_3 and F_4 confirmed the digenic mode of inheritance for 'determinate growth'. Van

Rheenen *et al.* (1994) further reported that mutation breeding would be continued with the aforedescribed cultivar as well as other plant material in order to obtain fully fertile determinate plants.

An example, showing a successful combined use of induced polyploidy and induced mutagenesis, traces back to Swaminathan (1965; 1969). In the diploid oilcrop *Brassica campestris* var. *toria* an autotetraploid ($2n = 4x = 40$) was produced in 1941, which showed much larger seeds but – as a result of a poorer branching ability and a lack of response to selection for increased numbers of secondary and tertiary branches – a lower total number of siliqua and, hence, a lower total seed yield per plant. Irradiation in 1958 of the most fertile autotetraploids with about 1 and 2 Gy of X-rays (according to Swaminathan, 1969; in Swaminathan, 1965, lower doses are mentioned) resulted in increased variability for branching which enabled further a selection advance for this trait. One of the mutants with the highest degree of branching gave during two consecutive years an average yield of 1449 kg. ha⁻¹ in contrast to 1214 kg. ha⁻¹ for the best diploid. Swaminathan commented that this example illustrates the possibility of changing an allogamous crop for a 'polygenic character' for which little genetic variation exists in the original population.

6.6.2f. Some useful mutants in cross-fertilizing crops (other than maize)

Limitations and draw-backs of mutation breeding in cross-fertilizing crops were mentioned before, for instance in Section 6.4.3, and it was concluded that mutation programmes in such crops in general are not very attractive, except for a number of specific purposes and breeding situations. Mentioned in this respect were the induction of desirable traits that, so far, are not present in a cross-fertilizing crop, the induction of mutations in inbred lines for heterosis breeding and the induction of mutations affecting the flowering system, for instance changes in incompatibility systems or the induction of male sterility (either genic male sterility or cytoplasmic male sterility) in relation to hybrid production. A list of some 35 publications on mutation induction in 'cross pollinators', forming part of an extensive bibliography on the use of induced mutations in cross-breeding, compiled by A. Micke, was added to the Proceedings of an FAO/IAEA Advisory Group Meeting on '*Induced Mutations in Cross-Breeding*' in Vienna, 1975 (Anon., 1976c).

Interestingly, one of the earliest mutant cultivars released refers to cv. Primex (or Svalöf's Primex) of white mustard (*Sinapis alba*), a cross-fertilizing diploid species. Cv. Primex, cultivated as an oil crop, was released in

Sweden in 1950 (see also Chapter 1) and had a higher oil content and a higher seed yield than the original cultivar. Olsson & Persson (1986) reported that in 1952 cv. Svalöf's Primex was the only white mustard cultivar grown in Sweden, but did not refer to its origin as a mutant cultivar. Borg *et al.* (1958), in the discussion of a paper on barley mutant cv. Pallas, reported that through repeated selection in the irradiated population of cv. Primex a further increase in seed yield had been obtained. Olsson & Persson (1986) further discussed this topic and concluded that by recurrent selection in one population over several years the number of seeds per pod, the oil content, the erucic acid content and the level of nematode resistance had been improved without loss in seed yield.

Throughout the years considerable attention has been paid to mutations for incompatibility, but the subject is rather complex and information in the literature sometimes confusing. Therefore, some general principles must be briefly explained here. Incompatibility may be defined as the inability of functional male and female gametes to effect fertilization in particular genotypic combinations. Incompatibility systems can be classified on the basis of different principles, for instance: 1) the site of expression of incompatibility (e.g. on the stigma, in the style or in the ovary), 2) the association with flower morphology (in relatively few cases, like in *Primula* with the 'pin' and 'thrum' type), and 3) the level of gene interaction or the ploidy level of gametes. With respect to the level of gene interaction two principally different systems are distinguished: the gametophytic system and the sporophytic system which both are usually controlled by one locus, the S-locus with several to many alleles.

In the gametophytic system the inhibition of the pollen tube growth occurs in the style; incompatibility occurs when the same S-allele is present in pollen grain and stylar tissue, and the reaction of the pollen grain is completely determined by the haploid genotype. These features make this system well suited for mutational studies. Large-scale screening after mutagenic treatments may lead to the recovery of self-compatible (SC-)mutants.

When sporophytic incompatibility is involved, the behaviour of pollen grain and pistil are not determined by their haploid genotypes but by the dominance relationship between S-alleles in the sporophytic tissue. Dominance relationship in pistil and pollen can be different. Inhibition of the pollen tube may take place either on the stigma or in the ovary. Incompatibility occurs when the same allele comes into expression in pollen and pistil. Changes at the S-locus (or loci) cannot be easily detected. According to de Nettancourt (1977), the sporophytic system is 'not particularly well-suited for the analysis of changes at the S-locus because S-mutations cannot be detected in such a system unless they are established within the entire plant or, at least, in the individual anther, throughout the majority of tapetal cells which determine the incompatibility phenotype of the microspore'. (N.B. The tapetal tissue is maternal tissue.) De Nettancourt continues by saying that 'mutations directly induced in the PMC or in the microspore are not expressed by the pollen grain and, consequently, fail to be selected, upon selfing, by the pistil sieve'. This is caused by the occurrence of chimerism in M$_1$ plants. Therefore, selection can start only during flowering of M$_2$ plants, and temporary use (one time only) is not feasible.

General reviews on incompatibility have been produced by Pandey & de Nettancourt (1976), de Nettancourt (1977), Shivanna & Johri (1985) and Lewis (1979; 1994). Van Gastel (1976) reviewed in particular literature about induced mutations concerning the compatibility system.

Lewis and colleagues (Lewis, 1946; 1948; 1949; Lewis & Crowe, 1954a,b) were successful in inducing self-compatibility (or intraspecific compatibility) in Oenothera and Prunus avium by X-raying flower buds. The author(s) concluded that spontaneous and X-ray induced mutant types did not differ and that self-fertility was induced by permanent loss of either 'stylar activity' or 'pollen activity'.

The first efforts to induce interspecific compatibility by induced mutations trace back to Davies & Wall (1961), who tried to bypass, after radiation of the male and female gametes, the incompatibility barriers. Clear effects were reached only in combinations of Brassica oleracea × Brassica nigra.

Much attention received the work on interspecific compatibility by Pandey (see for instance Pandey, 1969; 1974), who reported that X-irradiation of the anthers of the (self-incompatible) species Nicotiana glauca enabled the pollen of this species to be accepted by the – otherwise incompatible – pistil of Nicotiana forgetiana. Pandey expressed the opinion that this was the result from a mutation in genes that govern self-incompatibility: the so-called S- (or SI-)gene complex. Van Gastel (1976), who worked with Nicotiana alata, reported that various mutagenic treatments may result in the induction of self-compatibility (i.e. loss of S-alleles) but not in the induction of new S-alleles. According to de Nettancourt (1977) the most fundamental problem seems to be to find out 'why such (new S-)alleles cannot be induced by artificial mutagens and how simple inbreeding procedures lead to their generation'. (N.B. The meaning of the word 'generation', in this context, is more or less synonymous to 'appearance' or 'origination'.)

From results published so far it may be concluded that radiation has led to self-compatible mutants in several plant genera like Oenothera, Prunus, Nicotiana, Petunia and Trifolium. The method to irradiate plant material for mutations in the S-complex seems easy but, apparently, little use has been made so far of the ease with which certain self-incompatible crops can be changed into self-pollinating crops. It may be recalled in this respect that radiation, apparently, may lead to the breakdown of the S-function, but not the induction of new S-alleles.

Another category of potentially useful mutations in cross-fertilizing crops refers to plant architecture, in particular when a desired trait is not easily available or (so far) even has not been found at all in germ plasm collections of the specific crop and its near relatives. In Section 6.6.2e the improved branching pattern of the oil crop Brassica campestris var. toria, obtained after mutagenic treatment (Swaminathan, 1965; 1969), was discussed.

There is an increasing interest in mutation breeding for (often cross-fertilizing) crops which produce edible oils as well as in oil crops for industrial use. Several (potentially) useful oil crops have been discovered only recently and quite a few of them are still – from an agronomic point of view – in a semi-wild state. This often implies that, in addition to breeding for increased oil content or improved oil quality, genetic improvement of a range of agronomic traits is required in order to permit commercial cultivation of the crop. Both aspects (increased oil content and quality, and the need of improved agronomic traits) were discussed by Röbbelen (1990) in a general review on mutation breeding for oil crops with 136 references. In this publication also details can be found for the remarkable achievements in altering seed oil quality by mutations in species such as sunflower, safflower, rape seed, linseed, soybean and lupins.

It is often believed that domestication of wild plant species is a promising domain for mutation induction, since many traits favoured for cultivated plants are inherited in a monogenic-recessive way (Anon., 1989c). Worthwhile of being mentioned in this respect are also the reduction of alkaloids in lupins (Micke & Swiecicky, 1988), of glucosides in species of the genus Melilotus which are also cultivated as forage legumes (Scheibe & Micke, 1967) and of glucosinolates as well as erucic acid in species within the genus Brassica (Anon., 1982c; Röbbelen, 1990).

An example not mentioned by Röbbelen, concerns the induction of mutations for non-shattering capsules in the weedy, cross-fertilizing semi-wild oil crop *Euphorbia lagascae*, which was reported by Vogel & Röbbelen (1989), Pascual & Correal (1991; 1992) and Vogel, Pascual-Villalobos & Röbbelen (1993). The seeds of *E. lagascae*, a winter annual herbacious crop which grows in S.E. Spain, contain up to 50% oil. About two thirds of the oil is vernolic acid: a long-chain unsaturated fatty acid for which several applications in the oleochemical industry are known. Vogel & Röbbelen (1989) mention two major problems to be solved before this plant can be taken into cultivation. The most serious problem is that seeds dehisce at an early stage and that – so far – no genetic variation for this trait was found in nature. In addition, *E. lagascae* shows indeterminate growth. According to Vogel *et al.* (1993) treatment of seeds with EMS (0.4% for 4 h at pH 7) resulted in one M_2 plant with indehiscent capsules. This plant, called M24, showed low stability and reduced fertility. Only four plants out of a total of 113 plants derived from this plant showed the desired trait for indehiscence in M_3. Three of these plants produced an M_4 in which on the average 5% of the resulting plants were indehiscent. In addition, in M_3 two other plants with indehiscent capsules were discovered. From these two plants, called M76 and M77, M_4 progenies were obtained which showed the 'indehiscent trait' in 17.5% and 5% of the resulting plants. The authors concluded that the trait for indehiscence must be controlled by more than one recessive gene.

It was further observed that plants with the 'indehiscent trait' had on average two seeds per capsule instead of about three in the normal type with shattering capsules. In addition, four mutants with four carpels ('quatricarpellate') were found in M_2, which produced 10–30% more seeds per plant with the same number of capsules. M_3 quatricarpellate mutants derived from the aforementioned M_2 mutants, transferred this trait to 40–60% of their M_4 progenies. Two additional quatricarpellate mutants were found in M_3.

From the rather limited amount of information presented by the authors (see before), it seems that only 13.2% of the EMS-treated seed (against 32.5% of the control) germinated and that fewer than 5000 M_1 plants formed the base of this mutation-induction programme. This is a very low number for mutation breeding experiments, even when self-fertilizing species (which in general are much easier to handle in mutation breeding programmes than cross fertilizing species) are involved. In a more recent publication, Pascual-Villalobos, Röbbelen & Correal (1994) report on further progress with respect to percentages of indehiscent plants

obtained in later generations (up to M_7) of the project. It was also shown that the various mutants obtained showed different degrees of penetrance (or expression) of the trait for indehiscence as well as different degrees of association with other traits, like the occurrence of partial sterility in mutant M24. Percentages of cross-fertilization and self-fertilization may differ considerably within *E. lagascae*, e.g., from 10 to 50% cross-pollination in the cases described in this publication.

When the facts known so far are put together, the discovery of indehiscent capsules in the mutation programme for *Euphorbia lagascae* may be considered a pleasant surprise, but still much work has to be done before good indehiscent and agronomically acceptable cultivars of *E. lagascae* will become available. Somewhat unexpected results as in this case, again, demonstrate the potential of mutation breeding for obtaining new, agronomically useful traits. Pascual-Villalobos *et al.* (1994) reported that screening of natural stands of the closely related oilseed spurge species *Euphorbia lathyris* had resulted in the discovery of a (spontaneous) indehiscent mutant. The authors further point out that, according to the well-known 'law of homologous series of variation', formulated by Vavilov (see also Chapter 2, Section 2.4), it may be anticipated that such mutants might be found or artificially induced as well in *E. lagascae*.

Finally, in this section reference is made to the report of a Research Co-ordination Meeting on '*Mutation Breeding in Oil Seed Crops*', organized in 1993 by the Joint FAO/IAEA Division in Vienna (Anon., 1994a). In this report details can be found concerning mutagenic treatments with radiation and various chemical mutagens, screening methods, results and recommendations for the *Brassica* group, sesame (*Sesamum indicum*), sunflower (*Helianthus annuus*), opium poppy (*Papaver somniferum*), Castor bean (*Ricinus communis*), *Cuphea* sp., cotton (*Gossypium hirsutum*) and groundnut (*Arachis hypogaea*). For radiation treatments doses were reported to vary between 50 Gy for opium poppy and up to 1500 Gy (!) for *Brassica napus* and *B. juncea*. For chemical mutagens treatments with 0.5–5% EMS for up to 24 h were mentioned for *Brassica* sp., sesame and opium poppy. For *Cuphea* treatments of seed with 1–2 mM Na-azide or NMU for 3 h were reported.

6.6.2g. Mutations for disease and pest resistance
According to many estimations about one third of the total production of agricultural crops is lost due to plant diseases and pests and, therefore, it would be very useful if mutation breeding could contribute to effective, preferably durable resistances in crop plants.

Micke (1980) distinguished four situations when considering whether mutation breeding might be useful to induce resistance.

1. When sources of useful and applicable resistance are present in breeding material of high quality it is most doubtful whether mutation breeding or the use of crosses with resistant wild relatives should be tried.

2. When such sources are not available one could choose between a backcrossing programme with, for instance, a primitive cultivar, which takes much time, or one could decide to start a large-scale mutation programme. If it is known that the resistance is already present in a wild species or primitive cultivar, most breeders will prefer this option, but when the resistance has not been traced yet, mutation breeding may be an effective alternative.

3. When an intensive search for resistance in natural populations has not been successful, mutation breeding or 'genetic manipulation' may be the only way.

4. When, for instance, heterozygous, vegetatively propagated crops with an unique genotype should exist and much of the quality would be lost by crossing (see for instance Chapter 7 on *Mentha*), mutation breeding may be the only way and sometimes even a fast method, when sufficiently large populations are used.

It may be added here that in all cases, irrespective of the source of a resistance, the result depends on an efficient procedure to select resistant genotypes.

The limited number of results of practical significance that have been obtained after mutagenic treatments for the induction of resistances to diseases and pests was discussed already in Chapter 1. Several reasons for the low success were mentioned like insufficient knowledge of phytopathology and of mechanisms of resistance, of epidemiology, of interactions between pathogen and plant populations, use of inadequate screening methods, and in particular the dominant or semi-dominant nature of most of the known effective specific resistances (which indicate active specific gene functions, the creation of which by mutagens is unlikely). Often, newly introduced sources of resistance, irrespective of their origin (e.g. discovered in wild relatives of the crop under study or induced by purposeful mutagenic treatments), at first hand seem very attractive at the laboratory level, in initial greenhouse experiments or even in the field. In many cases, however, such resistances appear not to work effectively or not for long after they have been introduced into

new cultivars. This problem can be observed even more pronounced with respect to many of the 'break-throughs' that are claimed at regular intervals as a result of *in vitro* mutagenesis, 'somaclonal variation' and techniques of genetic engineering (see also Chapter 5). In addition, many claimed 'induced resistances' are suspected to be outcrosses or contaminations.

In more recent years there has been a tendency amongst breeders to concentrate more on durable resistance. Also in these, probably often polygenically inherited types of resistance, it will be difficult to achieve positive results by mutagenesis.

Most of these obstacles were discussed in detail during meetings organized by the Joint FAO/IAEA Division in Vienna and other organizations on mutation breeding for resistance (Anon., 1971; 1974b; 1976a; 1977b; 1983a), on mutation breeding (e.g. Micke, 1983b), or on breeding for resistance (e.g. Jacobs & Parlevliet, eds., 1993), and in various reviews, for instance, in the review on mutation breeding for seed propagated crops by Konzak et al. (1984). Mutations for disease resistance in crop plants (including insect resistance), in particular in barley and rice, were also the main subject of discussion during the 1988 meeting of the *Gamma Field Symposia* in Japan (Anon., 1988b).

Although often doubts have been expressed about claimed induced mutations for resistance, e.g. by Knott (1991) and Jørgensen (1991), Micke (1991b) referred to three cases where mutation breeding for resistance undisputedly had been successful: resistance to *Ascochyta*-blight in chickpea (*Cicer arietinum*), to *Sclerospora graminicola* in pearl millet (*Pennisetum* sp.) and to *Verticillium*-wilt in peppermint (*Mentha piperita*), which is a vegetatively propagated crop. These resistances remained effective for many years (Micke, 1993a). A more detailed history and economic data for peppermint will be discussed in Chapter 7.

Konzak et al. (1984) paid relatively much attention to induced mutations for disease resistance in their review. The first and still much studied example of successfully induced mutations refers to resistance to powdery mildew (*Erysiphe graminis* f.sp. *hordei*) in barley (*Hordeum vulgare*), derived from a recessive allele at the *ml-o* locus. This kind of mutation, which has been mentioned already on several earlier occasions, was first artificially induced by X-rays in the 1940s by Freisleben & Lein (1942; 1943a,b). But apart from this special case it seems that no other new sources of powdery mildew resistance of agronomic significance were obtained by mutagenesis. Mention should be made, however, of the interesting results reported by Röbbelen & Heun (1991) on quantitative ('partial') resistance against powdery mildew, orginally induced by treatments with EMS (0.25% for 6 h at pH 7). The unexpected degree of quantitative variation observed for three consecutive years (M_2 to M_4) in the first experiment, was confirmed in consecutive experiments in which treatments were performed with EMS and sodium azide (NaN_3). The authors concluded that, evidently, single genes may be responsible for quantitative differences in resistance. The quantitative resistances induced by Röbbelen & Heun (1991) have not yet been exploited.

The discussion in Section 3.5.2 of Chapter 3 of the very interesting resistance on the ml-o locus will not be repeated here, but it can be added that it has been proved beyond doubt that this unique, recessive ml-o resistance, which is not of the gene-for-gene type and therefore more durable, and which could not be back-mutated towards susceptibility, indeed has been the result of an induced mutation (Micke, 1992). Konzak et al. (1984) mentioned that the ml-o locus confers a world-wide resistance to all 'races' of the fungus and that in 20 years of study no mildew race virulent on ml-o has been detected. They further pointed out that the ml-o trait is associated with non-pathogenetic necrotic spotting that may be responsible for a decrease in yield by up to 10%. But the level of spotting varies and further crossing was expected to result in cultivars in which the yield remains at the normal level.

The authors further predicted that the induced ml-o resistance will be widely exploited. This has now become true in barley breeding. However, use is made mainly of a spontaneously arisen allelic source of resistance of the ml-o type (named ml-o 11 by Jørgensen) that was detected in Ethiopian barley accessions 30 years after the first artificial induction of ml-o resistance by X-rays in the 1940s (Jørgensen, 1976; 1987; 1991). The first mildew-resistant cultivar carrying the spontaneous ml-o 11 allele, cv. Atem, was released in 1979 in the Netherlands. The cv. Alexis, released in 1986 in Germany, carries the allele ml-o 9 induced by EMS in cv. Diamant.

Jørgensen (1991), in a review on induced mutations for disease resistance in cereals, also confirmed that the high hopes of mutation breeders to be able to induce many useful resistances in different crops have not been fulfilled. He referred in this respect to the very low frequency with which dominant mutants are induced. Jørgensen, to whose view on mutation breeding for resistance we referred earlier in this section, added that in many cases where positive results of mutagenic treatments have been reported, it appears that the observed resistance also can be explained in other ways than by induced mutations, as the screening techniques used do not distinguish between resistances induced by muta-tions and resistances as a result of outcrosses or other contamination. To distinguish between those categories one has to apply special and somewhat costly genetic and/or molecular tests.

Nevertheless, as was said before, a number of cases of induced resistance have been reported in literature throughout the years. Konzak et al. (1984), for instance, already refer to induced resistance to Victoria blight (Helminthosporium victoriae) in oats (Avena sativa), which resulted in two indirect mutant cultivars: cvs Florida 500 and Florida 501. Another example mentioned already concerned resistance to downy mildew (Sclerospora graminicola) in pearl millet (Pennisetum americanum, syn. P. typhoides). Mutants resistant to corn leaf blight (Helminthosporium maydis) race T were isolated in maize (Zea mays) after tissue culture without mutagenic treatment, but the induction of cytoplasmic male sterility which was aimed for, was not achieved.

More examples mentioned by Konzak et al. are gamma ray induced resistance to yellow mosaic virus disease in mung bean (Phaseolus aureus) in Pakistan, a moderate degree of field resistance to rust (Phakopsora pachyrhizi) in soybean (Glycine max) in Thailand and gamma ray induced stem rot resistance in jute (Corchorus capsularis). However, none of these reported cases seem to have yielded spectacular results in terms of important mutant cultivars. Further references can be found in Konzak et al., 1984.

A positive result, reported by Ukai & Yamashita (1987), refers to a recessively conditioned monogenic resistance to the 'soil-borne' barley yellow mosaic virus which causes serious problems in Japan and Europe. The authors reported on the induction of a gamma ray induced, recessive gene ym3, which is not allelic to the two dominant genes Ym1 and Ym2 known to give resistance to the virus. The newly induced ym3 gene produced a mutant plant that is more resistant to the virus than any cultivar known so far. The mutant also showed resistance to another virus (indicated as SBWMV) that often accompanies barley yellow mosaic virus. Jørgensen (1991) comments on this case that the mutant is allelic to another recessive resistance gene, detected later in German cultivars (see Götz et al., 1989).

During the Vienna Symposium in 1990, Nakai (1991) reported on promising mutant lines of rice (Oryza sativa) with resistance to bacterial leaf blight (Xanthomonas campestris pathovar oryzae) and in the poster sessions the results of some other projects on mutation induced resistance were presented, but much of this work had not yet reached the stage of the release of new, officially registered mutant cultivars (see also the earlier mentioned contribution by Röbbelen & Heun, 1991, on induced partial resistance against powdery mildew, *Erysiphe graminis*, f.sp. *hordei* in barley, *Hordeum vulgare*).

Masuda, Yoshioka & Inoue (1994) reported on a new gamma-ray-induced cultivar of pear (*Pyrus communis*), cv. Gold Nijisseiki, with resistance against black spot disease (*Alternaria alternata*). This new cultivar was selected in 1981 after chronic irradiation in a gamma field and registered as a new cultivar in 1992. Some additional details will be presented in Chapter 7.

Apart from an example for rice that was mentioned already before (Wang, 1991), no clear examples of *in vitro* induced mutations for disease resistance which have resulted in economically profitable cultivars, are known to us.

6.6.2h. Some economic figures and other data about successful short-straw mutants and some other mutant types in barley

Short-straw or (semi-)dwarf stature, particularly in cereals, like wheat, barley and rice has become a highly favoured trait because of a lower susceptibility to lodging, higher tillering and a better harvest index (grain yield versus straw).

The best-known and most often described Swedish mutant cultivar, cv. Pallas, which is also known as mutant *ert-k*[32], is a high yielding and highly lodging-resistant two-row cultivar, derived from cv. Bonus. This mutant, officially approved in 1958 and released in 1960, was a direct mutant produced by X-irradiation (75 Gy) of cv. Bonus in 1946. It was grown on large areas in Scandinavia and other European countries till the 1980s. Crosses with cv. Pallas have resulted in many other useful (indirect) mutant cultivars.

The other important primary mutant cultivar of Swedish origin is cv. Mari (*mat-a*[8]), a short-day tolerant and extremely early maturing mutant. This cultivar – that will not be further discussed here – was also used successfully as a crossing parent, from which a range of other indirect mutant cultivars were obtained.

A third high-yielding barley mutant with high lodging resistance, obtained in the first years of the Swedish barley mutation programme after X-treatment of cv. Gull – Sv 44/3 – was not released as a cultivar but only used in crosses.

Many indirect mutant cultivars have been obtained, either as the direct result of a cross between a direct mutant cultivar and another cultivar or by making more complex crosses in which several mutants and/or cultivars or breeding lines have been involved. Mentioned could be for instance: cv. Hellas (cv. Herta × cv. Pallas, approved in 1958), cv. Kristina (cv. Domen × cv. Mari,approved in 1969), cv. Mona (cv. Mari × cv. Monte

Cristo, approved in 1970), cv. Eva and cv. Salve (cv. Birgitta × cv. Mari, approved in 1973 and 1974 respectively), cv. Gunilla (Sv 44/3 × cv. Birgitta, approved in 1970). Among the group of indirect mutant cultivars with a more complex background could be mentioned: cv. Pernilla ((cv. Birgitta × cv. Mari) × cv. Gunilla, approved in 1979), cv. Visir (approved in 1970), cv. Senat (approved in 1974), cv. Troja (year of approval unknown; withdrawn in 1981), cv. Jenny (approved in 1980) and cv. Lina (approved in 1982). Many more details can be found for instance in Gustafsson (1969a,b; 1986), Gustafsson *et al.* (1971), Lundqvist (1991) and in many reports and reviews of meetings organized by the IAEA.

To illustrate the importance and economic value of cv. Pallas and a small number of other European barley mutant cultivars in some selected countries in Western and Central Europe, use is made of figures on areas under cultivation, relative contributions of different cultivars throughout the years and pedigrees of some barley cultivars, compiled by Lupton (1992). In this publication and in the previously mentioned publications by Gustafsson and other authors, many more details about the role of cv. Pallas and other cultivars parents in many other cultivars can be found.

cv. Pallas

Sweden (country of origin of cv. Pallas). Total barley area varying from 0.33×10^6 ha in 1960 up to 0.60×10^6 ha in 1970 and remaining about stable at that level till 1988. Cv.Pallas was grown on a very large scale in the 1960s (exact percentages not known).

Denmark. Total barley area varying from 1.04×10^6 ha in 1965 to 1.58×10^6 ha in 1980 and down again to 1.15×10^6 ha in 1988. In 1963 the relative proportion (%) of cv. Pallas was 10%, in 1965 this was 19% (first ranking cultivar) and in 1967 17%. The cultivar disappeared afterwards because of its susceptibility to powdery mildew.

Spain. Total barley area in 1975 3.18×10^6 ha up to 4.15×10^6 in 1985. The relative proportion of cv. Pallas between 1975 and 1986 went down from 14% to 8%. During most of those years cv. Pallas was the most important of all two-row and six-row cultivars grown. In 1990 cv. Pallas had been almost completely replaced by more modern cultivars.

United Kingdom. Total barley area between 1960 and 1965 varying from 1.55×10^6 ha to 2.18×10^6 ha. The relative proportion of cv. Pallas in 1962 was 18%, which percentage in 1966 had decreased to 3% because of severe attacks by powdery mildew.

cv. Diamant

Cv. Diamant, a Czechoslovakian cultivar derived as a direct mutant from cv. Valticky after gamma irradiation (100 Gy) of seed in 1956, was registered in 1965 (Bouma, 1967; 1976; Bouma & Ohnoutka, 1991). The mutant cultivar shows pleiotropic changes for several traits. Specific characteristics of the mutant (see also Chapter 1, Section 1.3.6) are: a longer juvenile phase, delayed tillering, a considerable (i.e. at least 15 cm) reduction in straw length, more and larger spikes than in cv. Valticky, a longer generative phase and a longer grain-filling period, a higher thousand-grain weight, a better root system and about 11% yield increase as compared to the mother cultivar.

In the early 1970s, mutant cultivar Diamant was the most important barley cultivar, grown in 1971 and 1972 on 37% and 43% (i.e. about 600 000 ha) of the total area for spring barley in Czechoslovakia. Because of its mildew susceptibility cv. Diamant was replaced in the late 1970s by resistant cultivars but the cultivar still remained a most important crossing parent. In 1989 all spring barley cultivars in Czechoslovakia had a cv. Diamant ancestry.

In 1987 already 2.86 million ha with the mutant gene of cv. Diamant were grown all over Europe. This figure refers to 57% of all spring barley cultivars cultivated in the European countries (Bouma & Ohnoutka, 1991, and some more recent additional data). Bouma & Ohnoutka mentioned the names of 113 cultivars, derived after crossing from cv. Diamant in eleven European countries and the – impressive – areas of mutant cultivars of the 'Diamant-type' in those countries. In 1995 more than 120 cultivars of malting barley with the mutant gene from cv. Diamant had been officially released.

cv. Trumpf

Probably the best-known mutant cultivar obtained after further crossing with cv. Diamant is cv. Trumpf, obtained by the Institut für Getreideforschung (Institute for Cereal Research) at Bernburg-Hadmersleben, former German Democratic Republic (DDR), and released in the DDR in 1973. Several sources of resistance genes for, mainly, powdery mildew have been introduced by the crossing parent(s). Some details were mentioned already in Chapter 1.

This very important indirect mutant cultivar is known under two different names; originally in Germany as cv. Trumpf and in the UK as cv. Triumph, reportedly after re-selection within somewhat heterogeneous basic seed. It is also used as a cultivar or as a crossing parent in many other countries.

Comparison of pedigrees as published by Lupton (1992) shows for instance that cv. Trumpf in the German Democratic Republic and in Czechoslovakia is listed as '(Diamant × Hadm.14029/64/6) × Union', whereas for cv. Triumph in Denmark, France, the Netherlands and the UK the pedigree 'Diamant × St.1402964/6' is given. Micke et al. (1990) mention in Mutation Breeding Review 7 that cv. Trumpf resulted from crossing cv. Diamant with 'several sources of disease resistance'. Often, breeders are reluctant to reveal precisely the ancestry of a cultivar but, according to Micke (personal communication), the similarity between the listed ancestry of cv. Trumpf and cv. Triumph makes it certain that one and the same cultivar is meant. Some data about areas occupied by cv. Trumpf in different countries and in different years are given here.

German Democratic Republic (country of origin). The total barley area between 1975 and 1979 was about 0.95 × 10⁶ ha of which in 1975 already 67% were occupied by cv. Trumpf (which was released in 1973!), 39% in 1977, 31% in 1979 and 1% in 1981. Cv. Nadja, the 'sister cultivar' of cv. Trumpf with the same pedigree, was grown on 21% of the spring barley area in the DDR in 1975, on 46% in 1977 and 34% in 1979. During this period cvs. Trumpf and Nadja together occupied 88% (in 1975), 85% (in 1977) and 65% (in 1959). In 1984 both cultivars were not grown anymore in the DDR.

Czechoslovakia. Cv. Trumpf was in 1980 with 8% from the whole area of 0.91 × 10⁶ ha the fifth barley cultivar in this country.

France. The total barley area declined from 2.67 × 10⁶ ha in 1980 to 1.92 × 10⁶ ha in 1988. Top years for cv. Triumph were 1985 with 17% of the whole area, 1986 with 25% (first ranking cultivar grown) and 18% in 1989. Figures for later years are not available.

United Kingdom. The total barley area in UK between 1975 and 1988 went down from 2.3 × 10⁶ ha to 1.9 × 10⁶ ha. Starting from 1979 with 3% of the area for spring barley, cv. Triumph reached 44% in 1981, 54% in 1983, 33% in 1985 and still occupied 14% of the area in 1990. According to Lupton (1992, p. 10) during the period 1970–90 the area for spring barley in the UK went down from about 90% to less than 50% of the total barley area in that country.

cv. Golden Promise

A 'direct' mutant cultivar of spring barley introduced in 1966 by Miln Masters, a UK breeding company, originated after gamma irradiation (exact dose unknown) of cv. Maythorpe in 1956. The cultivar, which was released in 1966, is characterized by short stiff straw, a good yield and good malting quality. The gene for short stiff straw (GP-ert) has pleiotropic effects on yield and quality components. It was recently discovered (Forster, 1994; Forster *et al.*, 1994) that – in comparison with cv. Maythorpe – cv. Golden Promise also has good salt tolerance, which is an important trait for arid regions and, probably, the seashore.

United Kingdom. The total barley area (which at that time consisted of 90% of spring barley), during the years 1969–75 varied around 2.3 × 10^6 ha. Cv. Golden Promise occupied 7% in 1969, 11% in 1972 and 16% in 1975, in which year it was the first cultivar. It remained on the UK National Recommended List of Varieties for over 20 years. The cultivar dominated the Scottish barley area in the 1970s until the mid-1980s. In later years it was replaced because of its mildew susceptibility by several other cultivars.

The cv. Midas, derived from a cross to a sister line of cv. Golden Promise, was released in 1970 and occupied in the period between 1970 and 1973 about 10 000–16 000 ha. Starting from 1981 in particular another mutant cultivar, cv.Triumph (see before), dominated the UK market for spring barleys for several years.

cv. Luther

Barley mutant cultivar Luther was released in the USA in 1966, four years after mutagenic treatment had been performed. It was the first (direct) barley mutant culti-var induced after treatment with a chemical mutagen: DES (conc.: 0.38 mM; duration of treatment: 2 h). This cultivar, induced by R.A. Nilan in cv. Alpine, was characterized by shorter straw, increase grain yield and good lodging resistance, in particular with heavy fertilization. By cross-breeding it gave rise to the induced mutant cvs. Boyer, Hesk, Mal and Eight-Twelve, which were released in the USA in 1974, 1979, 1979 and 1991 respectively, and to cv. Empress, released in 1983 in Canada.

cv. Betina

A direct spring barley mutant cultivar, released in 1968 by Ringot Breeders after EMS treatment of seeds of the Dutch cv. Vada (a cultivar with good medium level mildew resistance, found in particular to be a very good crossing parent).

France. Total area for barley about 2.75 × 10^6 ha between 1970 and 1980, during which period the proportion for spring barley gradually declined from 65% to 35%. Cv. Betina occupied 7% of the spring barley area in 1972 and 12% (second cultivar grown) in 1976, after which it gradually disappeared because of the absence of effective resistance genes for powdery mildew.

7 Mutation breeding in vegetatively propagated crops

7.1. Introduction

7.1.1. Spontaneous somatic mutations and mutant cultivars

Some historical facts

Spontaneous somatic mutants of agronomic value have been utilized since the earliest phases of plant domestication in a range of vegetatively propagated crop species, including fruit trees, currants and tuber crops like potato. In addition, older civilizations had already in an early stage – although most probably somewhat later than for food crops – shown interest in somatic mutants for ornamentals. Reviews on such mutants have appeared in European literature since the second half of the nineteenth century (Carrière, 1865; Darwin, 1868; Cramer, 1907; Chittenden, 1927). In literature spontaneous mutations are often called **bud variations, bud mutations, somatic mutations, bud sports** or, briefly, **sports**.

Darwin (1868) defined bud variations as 'all changes in structure or appearance which occasionally occur in full-grown plants in their flower-buds or leaf-buds' and in many cases ascribed these 'changes' to 'spontaneous variability'. He failed, however, to identify the cause of this variability.

Fruwirth (1929), who described spontaneous somatic variation for potato (*Solanum tuberosum*), mentioned that such variation occurs either as a result of irregular cell divisions, leading to genetically different somatic cells, or after rearrangement of genetically different tissues and layers. Despite the fact that Cramer (1907, p. 430), had already described, also for potato, some earlier cases in which bud variation had led to new cultivars of practical value, several scientists and breeders still expressed their severe doubts during the first decades of the twentieth century about the occurrence of bud mutations, let alone that such mutations could be of practical value to the breeder. This negative opinion, for instance, was expressed for potato by Sutton (1918) and Salaman (1926).

Economic importance of spontaneous somatic mutations

Reliable estimates of the economic importance of spontaneous somatic mutants are scarce. Granhall (1954) mentioned that out of 143 new cultivars of apple (present name: *Malus × domestica*) with known parentage, marketed in the USA and Canada during 1942–52, nearly 25% had originated as bud sports. The percentage of bud sports estimated for the other common American fruits (pears, peaches, nectarines, plums, prunes and cherries) was, on the average, slightly more than 10%.

Wasscher (1956) reported that about 30% of the cultivars of so-called 'florist's flowers' that were grown at that time in the Netherlands, had originated as sports. For carnation (*Dianthus* sp.) the percentage of sports was about 25%, for glass house roses 40% and for winter flowering begonia – with 39 sports of in total 55 cultivars – as high as 71%. Wasscher, in addition, mentioned that within six years of the introduction of the red-coloured cv. William Sim of carnation into the Netherlands, more than 50 bud sport cultivars had been commercialized in that country. Van Tuyl of the Government Breeding Research Institute CPRO-DLO in Wageningen, on the other hand, estimated for hyacinth (*Hyacinthus* sp.), which is one of the most important flower bulb crops in the Netherlands, that in the 1980s only 6% of the total area in this country was occupied by (mainly flower colour) sports for this crop (van Tuyl, personal communication).

Later in this chapter detailed attention will be paid to the highly successful story of mutation breeding in chrysanthemum (official name: *Dendrathema* sp., but still mostly called *Chrysanthemum* sp.). For this second-most important Dutch ornamental – of which 50% is exported – whole 'groups' or 'families' of spontaneous and artificially induced 'sports', mainly for flower colour and flower type, can be produced. One recent example has already been given here. In 1992 and 1993 the 'family' of cv. Reagan, consisting of more than 20 mutants, with sales of more than 400 million flower stalks at an annual base, represented 35–40% of the total Dutch market for chrysanthemum. Of the original cultivar, cv. Reagan itself, only 5.5 million flower stalks were sold at that time. More details and additional examples are presented in Section 7.4.5.

For the contribution of cultivars of potato (*Solanum tuberosum*) which originated by bud sports, various figures have been presented. Dorst (1924) referred to about 3% of the total area planted with two specific cultivars (cvs. Eigenheimer and Rode Star) in the Netherlands; Krantz (1951) to 15% of the production of certified seed potatoes of commercial potato cultivars in the USA in that year, and Turnquist (1960), also for the USA, even to 35%.

Spontaneous sports, nowadays, still often do replace the original clone, not only in ornamentals but also in several fruit crops like apple and in a crop like potato. Lawrence, Slack & Plaisted (1994), for instance, reported that a uniformly russetted sport of the US potato cv. Bake-King, indicated as 'Russet-Bake-King' is going to replace the original clone.

A well-known example for apple (*Malus* × *domestica*) refers to the old apple cultivar Delicious, grown already for more than a century in Europe and the USA. The original cultivar resulted in 1870 from a crossing but nowadays mutant clones – often with specific names – with a better colour, better shape and higher yield have completely replaced the original cultivar.

Another recent example referring to sports in apple was published in 1994 in a local fruit grower's journal in the Netherlands. According to this journal, growers of apples choose their new cultivars very carefully, as those cultivars are intended to be grown for many years because of high costs of starting a new fruit plantation. In 1980 a spontaneous sport of the well-known cultivar Jonagold, characterized by a much more intense, early-red-colouring fruit, was discovered in Belgium. In 1986 this sport became a new, distinct cultivar, named cv. Jonagored. In 1989 already many different mutants of the original cv. Jonagold, belonging to four main groups of mutants (e.g. for colour intensity and the presence of 'colour striping' on the fruit), were distinguished. In 1993, breeder's rights had been given or agreed upon for cv. Jonagored in 28 countries with 15 other countries under negotiation. In Belgium at that time 20 million kilograms of apples were produced on an annual basis from this spontaneous mutant cultivar and it was esti-mated that soon a level of more than 200 million kilo-grams would be reached worldwide.

Of economic importance is also the nectarine or 'hairless peach', a spontaneous mutant of peach (*Prunus persica*). The fruits of the nectarine have a thin, downless skin and firm flesh, and the flavour of the nectarine may be stronger than that of the peach. Fruits of nectarines often are smaller than those of peach, but by selection nectarines with bigger fruits can be obtained. The nectarine trait, most probably, is caused by a monogenic mutation towards recessive with pleiotropic effects. It is also possible to artificially induce mutants for the nectarine trait (Donini, 1976a; Donini & Rosselli, 1977). Such mutants, originally, occurred as mericlinal or periclinal chimeras in which the nectarine trait is limited to (part of) the outer layer (see later this chapter).

Deleterious spontaneous mutations
Mention has been made, so far, of the use of favourable mutations. There is, on the other hand, no doubt that deleterious mutations arise as well at a relatively high rate in most organisms (Peck, 1994). In recent years increased attention has been paid to the relation between deleterious mutations and beneficial mutations in different kinds of plant populations (cross-fertilizing, self-fertilizing, facultative or strictly asexually propagated; large or small populations) and to the relation between different categories of mutations (beneficial, deleterious, evolutionary neutral mutations) and the evolution of sex. Most of these studies are based on mathematical models and it would take us too far to discuss this subject in detail. Some references are given in the following part of this section.

It is sometimes stated that in vegetatively (asexually) propagated populations there is a much bigger chance that beneficial mutations (ultimately) will be lost from the population than in sexually propagated populations, but there seems to be no conclusive evidence for this view.

It is estimated that for *Drosophila* and higher plants with relatively large genomes, given a common spontaneous mutation frequency of for instance 10^{-7} or 10^{-8} per locus per cell per generation, <u>at least</u> one noticeable deleterious mutation occurs per genome per generation (Kondrashov, 1988; Charlesworth, 1990; Charlesworth, Charlesworth & Morgan, 1990). According to Peck (1994) it is likely that when a beneficial mutation arises in an organism with a large genome, this genome is 'contaminated' with many deleterious mutations. It has been calculated that due to the effects of deleterious mutations, in the absence of selection – which may be an unrealistic assumption – approximately 1% reduction in 'fitness' would be expected per generation, and this figure, most probably, is an underestimation of the deleterious mutation pressure that confronts most organisms. In large seed propagated populations natural selection, combined with recombination, is capable of preventing the long-term accumulation of deleterious mutations. On the other hand, in small obligately vegetatively propagated populations in which genetic recombination is lacking, natural selection is not very effective in eliminating deleterious mutations and as such mutations may pile up, under such conditions loss of fitness appears to be inevitable. For natural asexual populations this may result in the reduction of population size. Another consequence is that, because of their limited adaptibility, relatively few obligately asexual populations exist in nature.

One additional point to be made here is, that what may be deleterious in the homozygous condition, could be beneficial – for instance to heterosis – in the heterozygous condition.

The situation for man-made plant populations, because of the human interference, of course, is not fully identical with the situation in nature. Breeders and growers of vegetatively propagated crops who are dealing with large numbers of individuals from one clone, as is the case with potato (*Solanum tuberosum*) or sugarcane (*Saccharum officinarum*), however, should be well aware of the risks of deleterious mutations that may be piling up throughout the years in their plant populations and prevent this by regular mass selection.

7.1.2. Artificially induced mutations and mutant cultivars

In more recent years, mutations useful to man have been artificially induced in various vegetatively propagated crop species. This has been the case in particular in fruit trees like apple (*Malus × domestica*)) and sweet and sour cherry (*Prunus avium* and *Prunus cerasus*, respectively) – where in particular compact or spur types are favoured – but also in various species of citrus (*Citrus* sp.), e.g. for flesh colour, salt tolerance and seedlessness, and in various grasses, sugarcane and many ornamentals.

Mutation breeding for ornamentals has been very successful, in particular in the Netherlands. Micke *et al.* (1990) reported that, on a worldwide scale, 409 mutant cultivars were known to be released for ornamental plant species. In fact, this figure must be a considerable underestimation, mainly due to lack of official registration. For an ornamental like chrysanthemum real figures are certainly much higher than the 169 cultivars that were mentioned by Micke *et al.* Additional economic data and more details about artificially induced mutant

cultivars for these and other crops will be presented later in this chapter.

In accordance with what has been said in the previous section, in particular in vegetatively propagated crops, newly arisen beneficial mutations (in the absence of selection) will be at risk of being lost from a population when in that population also many deleterious mutations occur (Peck, 1994). This statement about newly arisen mutations may also refer to mutations that are favourable to man. In vegetatively propagated plants all mutations occurring in the same cell must be considered as 'linked' due to the absence of recombination. For this reason one may conclude that, in order to keep the number of deleterious mutations low, it is not advisable to apply high doses of radiation in mutagenic treatments. Low doses do <u>not</u> reduce the percentage of deleterious mutations but the number of multiple mutations per chromosome or genome will be reduced in this way, thus avoiding rare beneficial mutations becoming worthless by the presence of several deleterious mutations in the DNA of the same cell.

7.1.3. Main characteristics of vegetatively propagated crops

Many vegetatively propagated crops are grown for harvesting vegetative organs. In such situations flowering and fertility are not major breeding goals whereas in some cases even selection against generative development is performed because flowering could act negatively on yield of the desired vegetative product. In some crops fertility has even vanished almost completely over the years, like in banana (*Musa* sp.) where fruits mostly are seedless (parthenocarp). In some crops like potato, special measures (e.g. grafting on tomato; planting on stone) have been developed to stimulate flowering for breeding purposes.

Other categories of vegetatively propagated crops, however, are grown for their fruits and in those situations flowering is essential for fruit set. In the case of ornamentals, the production of flowers, their timing, number, characteristics, etc. are the ultimate breeding goals.

The present attention to *in vitro* techniques has resulted in a still growing number of new crops that were not originally vegetatively propagated but can in this way be vegetatively propagated and multiplied.

Common characteristics

Most vegetatively propagated crops have a number of characteristics in common.

1) Vegetative propagation produces clones. All individuals belonging to one clone are genotypi-

cally identical, apart from the occurrence of spontaneous mutations.

2) Clonal cultivars – in general – are very heterozygous.

3) The vegetative way of propagation often includes the transmission of certain pathogens (in particular viruses).

4) For breeding, the generative phase is of much importance but often problems are met in this respect.

5) A promising F_1 plant can be used immediately as a new cultivar; the genotype is fixed by clonal propagation.

6) Many vegetatively propagated crops are polyploids (potato, sweet potato, cassava, sugarcane, strawberries).

Obligate and facultative vegetatively propagated crops

For breeding in general but also for mutation breeding in particular, it is useful to make a distinction between obligate vegetatively propagated crops – a small group of species for which (spontaneous or induced) mutants are the only source of genetic variation – and facultative vegetatively reproducing crops in which, for practical reasons, use is made of the possibility of (large scale) vegetative reproduction, but where crossing for breeding purposes is possible. (N.B. As the term 'facultative' in particular fits to the 'natural' situation, it might be preferable to refer to vegetatively reproducing or propagating crops than to – what is common practice – facultative vegetatively propagated crops.)

Asexual reproduction or apomixis has been reported in over 300 species in at least 35 plant families (Hanna & Bashaw, 1987). Facultative apomixis may be very attractive to breeders and growers, both from the view point of easy multiplication as well as because most virus diseases are not transferred by seed. Moreover, hybrid vigour may be fixed by the mechanism of apomixis.

1. Obligate vegetatively propagated crops. As to this group one could, for instance, refer to apomicts like *Poa pratensis* (smooth stalked meadow grass, Kentucky bluegrass) and to man-made triploid hybrids like the ones produced in Bermudagrass (*Cynodon* sp.), sterile and seedless polyploids in banana (*Musa* sp.), and in several ornamentals like *Alstroemeria* sp. Obligate apomicts in fact show reproduction by seed, but in an asexual way.

Triploids usually show highly irregular

meioses, which result in a high level of sterility and in seedlessness. Increased ploidy levels often lead to larger vegetative parts.

Another example concerns the ornamental crop species *Zinnia* × *marylandica*, a disease resistant hybrid that resulted from a cross between *Z. angustifolia* and *Z. elegans*. This hybrid behaves like a segmental allopolyploid with limited or no segregation in subsequent generations because of preferential pairing of homologous chromosomes and elimination of homoeologous pairing. It was observed that as a result of its complicated genetic set-up this hybrid displays no phenotypic variation at all. In such situations artificial induction of mutations may produce the desired genetic variation.

For several crops it was believed in the past that they were obligate (and vegetatively propagated) apomict species, but in a number of cases more detailed studies or expeditions to collect germ plasm in remote areas, have shown that plants with fertile flowers occur as well within the species. Garlic (*Allium sativum*) is an example of such a crop that is, erroneously, still often referred to as an example of an obligate apomictic species

2. Facultative vegetatively reproducing crops. This group contains many economically important crops like potato (*Solanum tuberosum*), cassava (*Manihot esculenta*), sweet potato (*Ipomoea batatas*), sugarcane (*Saccharum officinarum*), pineapple (*Ananas comosus*), many fruit trees, grasses and ornamentals. Many of them are highly heterozygous, polyploid and/or aneuploid and sterile. In a number of cases – in particular in fruit trees and bulb and tuber flowers – they show a prolonged juvenile phase. Such factors may make the application of conventional crossbreeding difficult and often very time and labour consuming. As will be discussed later in this chapter, mutation breeding often may be an attractive addition or alternative to crossbreeding in this group of crops.

Facultative apomixis has been induced by mutagenic treatment in the cereal crop pearl millet (*Pennisetum glaucum*), but for a number of reasons, it has not been used (Hanna, 1995a).

Some bottle-necks of mutation breeding in vegetatively propagated crops

Some particular major bottle-necks of mutation breeding for the group of vegetatively propagated crops should be indicated here already. In Section 7.1.1 a short description was given of the effects of deleterious mutations. The absence of a meiotic sieve in vegetatively propagated crops may result throughout the years in plant populations in which increasing numbers of plants carry different types of natural and, after mutagenic treatment, induced mutations. Gaul (1977) in the 'Manual on Mutation Breeding' (Anon., 1977a, p.87) distinguishes three types of effect which may occur in the first generation after mutagenic treatment: 1) factor or point mutations, 2) chromosomal aberrations, and 3) so-called 'physiological damage' (or primary injury). (N.B. Sequence and phrasing differ slightly from the one used by Gaul.) Factor mutations and chromosomal aberrations, which refer to changes within the DNA, may be transferred to the next generation in seed propagated crops, whereas according to Gaul the so-called 'physiological damage' is restricted to the M_1. Gaul points out that the term 'physiological damage' may encompass effects of chromosomal and non-chromosomal origin and that usually no distinction between both causes can be made. The term is mainly intended as 'descriptive' and does not refer in particular to physiological damage in the biological sense. From a practical point of view – still according to Gaul (1977) – growth retardation and lethality are the two most important kinds of 'physiological damage'. In our opinion the use of this term in the way as suggested by Gaul, is confusing and should be avoided when different types of effect of mutagenesis in M_1 are distinguished. The other term mentioned by Gaul for the same type of effects: 'primary injury', is much more neutral and seems much better suited to indicate effects of mutagenic treatments that are not transferred to the next sexual generation. A final remark in this context is that effects referring to 'primary injury', contrary to the situation for seed propagated plants, may remain present during several generations of vegetative propagation. When a heritable change is defined as a non-lethal genetic change that is passed on to the descendants (Rieger et al., 1991), one could argue that (at least part of) the primary injury refers to genetic damage in somatic cells and is 'heritable' in clones. This part of the primary injury is of the same kind as in cells of the lineage of 'generative' cells (or **germ line**).

A second complication is that mutations may express undesirable pleiotropic effects or be accompanied by close linkage with other mutations. The common method – to use crossing in order to bring the desired mutation in a more favourable genetic background – may not be applicable as easily for vegetatively propagated crops as for seed propagated species. The only alternative then is to use relatively mild mutagenic treatments and to try to induce several mutants with the desired traits and select for a plant in which the expression of negative effects is low.

The risk of clonal transmission of pathogens – in particular viruses – in vegetatively propagated crops was mentioned previously. Symptoms of such diseases sometimes have been mistaken for mutations. All possible precautions (e.g., by discarding suspected plants, hot water or hot air treatments, meristem tip isolation *in vitro*, the use of apomixis or introduction of a seed generation) should be taken to start from plant material that is virus free.

As was mentioned already in Chapter 1 (Section 1.3.8), the induction of mutations in multicellular organisms results in chimerism. A major point of concern is how to deal with chimerism in vegetatively propagated crops and how to manipulate this phenomenon to the benefit of the breeder. Chimerism has been the subject of much research in the past years, and a discussion on which approach to follow when dealing with chimerism in practice will be presented in one of the following sections.

7.1.4. Brief outline of current methods of breeding and maintenance

Breeding procedures

Breeding procedures in vegetatively propagated crops – usually – are relatively simple. As clones often show a high degree of heterozygosity, a selfing or a cross between clones may result in a F_1 generation which displays an enormous heterogeneity and in which selection can be performed immediately. The choice of superior parents with a good combining ability is imperative. Knowledge of the mode of inheritance of important traits and of correlations between different traits should be very useful in this respect. Repeated back-crossing may be necessary, for instance when crosses were made between a modern cultivar and primitive cultivars or wild relatives, for instance to introduce resistance.

One or several generations of selection and testing are required, depending on the crop and on the trait. Selection procedures may be speeded up considerably when early testing is possible. Selection work, on the other hand, may be seriously hampered by a long juvenile phase, as is observed in fruit trees or bulb crops. For tulip (*Tulipa* sp.), a well-known ornamental bulb crop, for instance, it takes four years from seed, obtained from a cross, till a flowering plant, after which the actual testing starts, initially at a small scale but later more extensively and at different locations. Normally, following the first cross, a tulip breeder takes a period of seven years for continuous testing in order to be sure of the potential of his selections. Under normal circumstances it may even take 15–20 years before the breeder has obtained enough material by classical vegetative propagation so that a new cultivar can be offered to the market. For instance, after seven years about 20 bulbs may be available from a selected genotype and about 1500 bulbs twelve years later.

Depending on the crop, an average breeding programme from the first cross to release of a new cultivar may take 10–15 years. If at the end of the whole cycle a clone is obtained that is superior to already existing cultivars, this new breeding product can be released as a new cultivar.

Mutation breeding must be considered a fully fledged breeding method in some vegetatively propagated crops which is of equal standing to the older and more conventional methods. Its merits will be discussed in the following sections.

Maintenance and multiplication

As was mentioned already in Chapter 1, maintenance and multiplication of cultivars in vegetatively propagated crops is difficult and time consuming. The use of appropriate *in vitro* techniques may not only increase the number of different clones but may also speed up breeding cycles. To use tulip again as an example: by making use of *in vitro* vegetative propagation it is possible nowadays, when starting from five bulbs, to obtain 20 000 bulbs within five years. Costs, however, are rather high. The most serious factors that may threaten the marketing of a new cultivar are swopping or mixing up with other cultivars, the occurrence of spontaneous mutations and infection by pathogens, in particular by viruses, but also by bacteria and fungi.

7.2.   Shoot apices, bud types, mutated cells and chimerism

Before paying attention to mutation breeding procedures in vegetatively propagated crops, it is necessary to discuss somewhat more in detail than was done in Chapters 1 and 6, the organization of shoot apices in various kinds of plant shoots and buds, and the fate of a mutated cell in such apices. Additional information will be presented as well on chimerism.

Mutagenic treatments, in particular radiation treatments, not only induce mutations in apical cells, but may also affect the pattern of cellular roles and functions in the predominantly stable organization of such apices. Some understanding of the underlying processes

is also needed in order to make optimal use of such phenomena for mutation breeding purposes.

7.2.1.   Shoot apices

It was reported in 1759 by Kaspar Wolff (see for instance Cutter, 1965) that new plant organs arise from small groups of cells within the apical shoot area (or **shoot apex**) of a plant. These groups of cells are known nowadays as **shoot apical meristems** which, in addition to initiating new organs and new tissues, exercise a certain degree of supervision over other parts of the plant, for instance with respect to apical dominance, and are able to maintain themselves as formative regions. It may be interesting to refer in this respect to a publication by Soma (1973) who, based also on results from his own work, concluded that apical meristems appear to have self-organizing properties. According to Soma this implies that the growth and morphogenesis of the apical meristems may be determined by the apical meristem itself and not by subjacent tissues. In addition, Medford (1992), in a recent review article, points out that modern studies on genetic and molecular aspects of apical shoot meristems may support the opinion expressed already by Wardlaw (1957) that, despite differences in the details, the organization and functioning of all shoot apices may be essentially alike.

The shoot apex occupies the most distal area of the shoot within which various cell types with specialized functions – according to their position – can be distinguished. Depending on the plant species up to some 1000 cells may be present in a shoot apex. Most important are a small number of so-called **initial cells** that divide in two daughter cells, one of which retains the position of the original initial cell whereas the other is added to the meristematic reservoir that, ultimately, gives rise to new tissues and new plant organs.

Within the shoot apex, in addition to the real apical meristem, several other regions are distinguishable like the zone immediately below the apical meristem where the meristems for lateral organs are found, a zone where shoot widening takes place and a region where further cell differentiation becomes apparent. In a later stage of development often a distinction is made between the vegetative apical meristem and floral primordia that arise from this vegetative meristem. The terminology used in this context may be somewhat confusing as some authors apply words like **growing point**, **shoot tip**, **shoot apex**, **(shoot) apical meristem**, **apical dome**, etc. as synonyms whereas other authors may use them for distinct purposes. Further details are outside the scope of this section. For definitions and further reading, reference can be made to a range of publications like Buvat

(1952; 1955), Gifford (1954), Wardlaw (1957; 1968), Dommergues (1964), Cutter (1965), Gifford & Corson (1971), McDaniel (1984), Steeves & Sussex (1989), Medford (1992) and many others.

Initial cells

Hofmeister (1852) postulated the existence of a single initial cell, for which opinion much support was found. Microscopic proof, however, could not be obtained.

If cells exist which are initial cells in the true sense of the word, i.e. 'permanent' cells with a fixed position at the apical tip (Esau, 1965; 1977), mutations induced in such cells would be of particular interest to the plant breeder as each initial cell would produce a lineage of mutated daughter cells. (N.B. The word 'permanent' is put here between quotation marks because one could also argue that a cell that divides in two daughter cells is not a permanent cell anymore.) If, on the other hand, the position of an initial cell carrying a mutation would not be a permanent one, for instance because such a cell would shift to the flanks of the apical area without being replaced by a daughter cell carrying the same mutation as the initial, the production of a lineage of mutated cells will be terminated (unless, of course, in this new position at the flanks the cell continues to divide).

A mutation in an initial cell will be transferred to all plant tissues or structures that ultimately trace back to that cell. This situation might be illustrated by a flower (part) displaying a different colour than all other flowers of the plant or, accordingly, a side branch with hairy leaves in a plant of which the leaves of all other branches are hairless. Later in this chapter we will elaborate on the point that the occurrence of such changes does not necessarily imply that the mutation that was responsible for this phenotypic change, is present in all cells of the 'mutated' plant part and also discuss the consequences of this finding for practical breeding.

It may be clear in this respect that a solidly mutated plant, i.e. a plant with all its somatic cells mutated, does occur only when the whole plant traces back to one (mutated) cell. This situation, of course, is common in the second generation of seed propagated crops, but much more difficult to imagine in vegetatively propagated crops.

Concepts of shoot apex organization

The classical Histogen Theory, presented by Hanstein (1868), implies that some kind of stratification exists in shoot apices of angiospermous plants. Hanstein referred to the presence of a central core which was surrounded by a number of regularly organized layers, covering each other mantle-wise. Each layer was believed to be derived from a few (vertically superimposed) initial cells situated at the ultimate shoot tip. In later years the original, rather rigid, interpretation of Hanstein's ideas on the role of histogenic layers has been abandoned, but the basic concept of stratified shoot apices remained.

As a continuation of Hanstein's concepts, Buder (1928) and his students, in particular Schmidt (1924), developed the Tunica-Corpus Theory. According to this theory – the basic ideas of which are still generally accepted – some tunica (or mantle) layers surround a so-called corpus in which the cells divide in all directions. Tunica layers, of which in dicotyledons normally two are present in the shoot apex, are – to a certain degree – mutually independent because of their plane of cell division which, predominantly, is anticlinal. A mutation in a cell of one tunica layer, therefore, will not be easily transferred to another layer.

Nowadays, two trains of thought are commonly followed to explain shoot apex behaviour. Both views find their origin in the Histogen Theory of Hanstein. According to the first concept, which is mainly based on studies concerning plant chimeras (see for instance Satina, Blakeslee & Avery, 1940; Satina & Blakeslee, 1941; Dermen, 1947; 1951; 1960; Stewart & Dermen, 1970b; 1975; 1979; Stewart, Semeniuk & Dermen, 1974), a number of distinct layers – often called (primary) **histogenic layers** or **germ layers** (see also Rieger et al., 1991) – are distinguished. (N.B. The word 'layer' should not be taken too strictly, as such a layer – except for the original plant epidermis – not necessarily always consists of a single layer of cells.) For dicotyledonous angiosperms usually three distinct layers are indicated, whereas for monocots either two or three layers are assumed to be present. Following Satina et al. (1940), the layers are commonly called L-1, L-2 and L-3 (alternatives: L-I, L-II, L-III; or L1, L2, L3). The indication L-3 normally corresponds to the 'corpus' in the concept of Hanstein. Cell divisions in the outermost L-1 layer are in the anticlinal plane only whereas divisions in L-2 predominantly but not exclusively are anticlinal. In the L-3 area divisions occur in all directions.

The second concept implies the presence of a kind of zonation within the shoot apical meristem, based on cytological studies and patterns of cell division. Often three meristematic areas are distinguished: the central zone at the ultimate tip, the peripheral zone along the flanks (actually a circle or a band) and the so-called 'rib zone' in the central, interior part of the apex area. Such zones, according to Clowes (1961), can only be observed when the meristem is active (for details see for instance Esau, 1977 and Steeves & Sussex, 1989).

Despite the fact that, according to this theory (which,

is sometimes referred to as the 'French School' of Buvat and others), no 'permanent' initial cells exist, the cells in the central zone divide at a lower rate and act as the primary cells for other meristematic cells and tissues, notably for the meristems of the peripheral and rib zone area. It is believed that the initiation of lateral organs like leaves and branches (which include tissue from all histogenic layers!) is the main function of the peripheral zone, whereas the rib zone cells contribute to the formation of the vessels, storage parenchym, etc.

Advanced studies gradually clarified more details of shoot apex behaviour and after it was admitted by Buvat (1955) that the original concept of the 'French School' in which the existence of initial cells was denied, had been too rigid, it appears that – in the end – both concepts are not considered incompatible anymore and can be reconciled (for more details see also Broertjes & van Harten, 1988, and Medford, 1992).

Much more could be said about shoot apices and many more references could be given, but we must limit ourselves here to some comments concerning experimental work to study shoot apex behaviour. It has been tried in many ways to label the (supposed) initial cells, e.g. by isolating shoot tips and growing them in vitro, by puncturing or making small cuttings in the apical area of a shoot or by application of carbon spots and radioisotopes (see for instance Ball, 1974). Plants that are chimeric for a visible trait, like those carrying chlorophyll stripes, have been found to be very useful as well in this respect. Such studies on chimeric plants (for which in particular periclinal chimeras are very well suited; see for instance Bergann, 1985), have proved beyond any doubt that in any case the pattern of mantle-like layers plays an important role in plant histogenesis. The pattern of zonation that was mentioned before may be superimposed on these layers.

In Chapter 6, Section 6.2.8, mention was briefly made of interesting but often neglected results of practical importance for applied mutation breeding by Dermen and Stewart who, on the basis of the behaviour of mutated stripes, concluded that some kind of 'stable' initials must exist. Stewart & Dermen (1970a) demonstrated in this way for *Ligustrum* that two or three initial cells must be present for each histogenic layer. A very low division rate of one division per twelve days was found and the most distal daughter cell continues the role of the original initial cell. This low rate of division suffices to make initial cells the ultimate source of growth. It is often assumed that cells at the circular flanks of the shoot apex divide at much higher rates than in the shoot tip. This, however, is not necessarily true as the exponential growth in that area may be explained

equally well by the presence of many more dividing cells along the flanks.

One additional remark should be made here about leaf initiation. It was originally thought that leaves develop from small groups of initial cells producing local meristems below the top of the shoot apex (Esau, 1977). On later occasions, however, it was suggested (see for instance Poethig & Sussex, 1985) that up to 250 cells, which are not organized in localized meristems, may be involved. Such figures, of course, also depend on the moment of observation, but further details are outside the scope of this section.

The crucial point of what has been explained before about shoot apices and initial cells, was very well phrased by Romberger (1963) in a report that is not often quoted, saying: 'Cells behave as they do because they are where they are'.

The importance of the cell position within apical shoot meristems in relation to cell differentiation to delimit the major tissues, was also emphasized in a brief review by Sussex (1989). This author further expressed the opinion that permanent initial cells (in the strict sense of the word) may not exist.

7.2.2.   Axillary and adventitious buds
In addition to **apical buds**, situated at the end of a stem or shoot, **axillary buds** and **adventitious buds** are also known. Differences between axillary and adventitious buds refer to their origination – either predetermined or not. Adventitious buds may be distinguished according to the place of origin on stem, leaves, etc., the position and number of cells (and cell layers) from which they arise, at least in the initial stage, and to the rate of differentiation of the tissues concerned.

Axillary buds
Axillary buds occur on predetermined places on stems in close association with the leaves at varying distance from the apical meristem. Such buds, or their initials, in principle are found in the axil of each leaf but they do not always develop. It is generally found that axillary buds reproduce the layered pattern (e.g. L-1, L-2, L-3) of the apical buds. Proof for this is often derived from observations on periclinal chimeric plants, which will be discussed in the next section. Some authors, however, claim that the number of independent layers may be lower than in the original shoot apex. Howard, Wainwright & Fuller (1963) for instance report for potato (*Solanum tuberosum*) that only two layers are present instead of the expected three layered structure. This may also depend on the ontogenetic role of the branch.

Axillary buds, in particular those close to the terminal apical bud, may remain (semi-)dormant because of the 'apical dominance' exercised by the apical bud. Meristematic cells in such axillary buds, in that case, when submitted to radiation treatments, may be considerably less radiosensitive than active cells, e.g. in the main bud. This probably can be explained by a much lower rate of division of the (semi-)dormant cells than of the active cells, which implies that considerably fewer cells are in the, very radiosensitive, stage of mitosis, but (other) differences in physiological activity between dormant and active meristems may play a role as well.

Axillary buds are sometimes accompanied by socalled **accessory buds** (see also Section 7.4.6). They may be smaller and less differentiated than the main axillary bud with which they are associated, but as far as is known – once they develop – their structure does not differ from that of the axillary buds.

Adventitious buds

Adventitious buds are initiated in somatic tissues outside the common meristematic areas. They can occur on different plant parts and in various ways: directly or indirectly, 'in vivo' as well as in vitro, naturally or artificially provoked, and may arise from exogenous as well as endogenous tissues. Adventitious buds, after some time, acquire *de novo* the same pattern of organization as normal buds (with their histogenic layers) but they do not constitute tissue deriving from all the histogenic layers of the original plant.

The number of 'initial cells' from which adventitious buds eventually arise is of practical importance and may differ between several cells and one cell. Starting from a 'single cell situation' offers very interesting prospects to the mutation breeder as the resulting plants will not be chimeric. The so-called adventitious bud method, the origin of which traces back to Naylor & Johnson (1937), has been tried out and further perfected in particular for a range of ornamental crops by Broertjes in the 1970s and 1980s. Despite the fact that, as will be discussed later, the scientific foundations and merits of this approach have been challenged on some occasions, the adventitious bud method has been applied very successfully for producing solid mutants after mutagenic treatment of vegetatively propagated crops. Further details will be given in one of the following sections.

7.2.3. The significance of the position of a mutated cell within a plant

Mutations do occur in all kinds of plant tissue and – in principle – are single cell events. Mutations for polyploidy within a tissue may be induced simultaneously in a number of plant cells.

In the past a distinction was often made between mutations within shoot apices and in so-called extra-apical areas. At that time most mutation breeders were already well aware of the fact that potentially valuable mutations for various traits might be present in extra-apical areas, in particular in vegetatively propagated crop plants. If such plants are grown during many (vegetative) generations such mutations may accumulate and, as the action of a meiotic sieve is lacking, deletions and other deleterious mutations may persist as well in a clone.

Mutations in extra-apical plant parts, in most cases, have remained unexploited because it is not easy to detect them, in particular not at an early stage when only one or a few cells carry the mutation. Their way of inheritance, of course, plays a role as well. Moreover, until recently it was difficult to regenerate a single (mutated) cell from outside the germline into a whole mutated plant. In more recent years a range of 'in vivo' and in vitro methods have been developed by which new plants can be grown from a few cells belonging to tissues of practically all areas of the original plant. Various in vitro methods were discussed in Chapter 5 and the adventitious bud technique, that was mentioned already, in particular developed for ('in vivo') mutation breeding in vegetatively propagated crops, will be described in this chapter. As a consequence, it is often possible, nowadays, for the plant breeder to exploit valuable mutations induced in extra-apical plant areas and the original distinction between apical and extra-apical mutations that was made initially, for instance by Bergann (1967), has lost much of its practical relevance.

The fact, however, remains that if a (valuable) mutation should be induced in an apical initial cell, this event normally will have much more impact than an extraapical mutation because, in the first situation, a lineage of mutated daughter cells will be produced without particular efforts by the plant breeder.

In Chapter 6 it was explained already that initial cells present in the different germ layers (or histogenic layers) of a shoot apex do not have the same function. For seed propagated plants it was mentioned in this respect that gametes trace back to the initials of the subepidermal layer (L-2) and, as a consequence, that mutations induced in the initial cells of the other layers (L-1 and L-3) are not inherited by seeds and will become lost, unless special measures are taken. Such a measure could be to try and grow a whole plant from tissue of L-1 or L-3 origin in which the desired mutation is present.

Initials responsible for the outer layer (L-1) mainly contribute to the formation of the epidermis (and other layers or organs that probably could be derived from epidermal tissue) of stem, leaves and flowers. For vegetatively propagated species, predominantly, it is those mutations affecting genes for flower colour in the initial cells of the L-1 layer that are responsible for the changed flower colour in a plant. The flower colour genes in L-2 and L-3 – as well as mutations in those initials – under normal conditions will not – or to a much lesser extent – be expressed. In fact the situation is somewhat more complex and, as will be illustrated for chrysanthemum in Section 7.4.5, flower colour genes in L-1 as well as – at least to a certain extent – flower colour genes in the internal layers (L-1 and L-2) may contribute to the ultimate colour of a flower. Moreover, occasionally histogenic events occur, as a result of which cells from one layer may penetrate into another layer and (partly) replace cells of the other layer (e.g., cells from L-2 penetrating in L-1). Such histogenic events may also affect the flower colour, at least when both layers carried different genes for flower colour. As an aside, it may be mentioned here already that under normal conditions – and contrary to what is often believed – the green colour of leaves is <u>not</u> determined by L-1, as chloroplasts (except for the guard cells of the stomata) are not present in cells of the epidermal layer but in deeper situated cells.

In order to briefly demonstrate the role of mutations in L-3, reference is made to storage organs such as tubers of potato (*Solanum tuberosum*) or roots of cassava (*Manihot esculenta*) which, almost exclusively, consist of cells which trace back to the endogenous (L-3) area of a shoot apex. As a consequence, bud mutations affecting genes responsible for the chemical composition of the storage tissue are effective only if they occur in the L-3 part of apical meristems (unless, of course, it would be possible to regenerate new plants from one or more (mutated) cells of another layer!). The analysis of the genetic nature of supposed periclinal chimeras by stimulating regrowth from endogenous (L-3) tissue is sometimes referred to as the Bateson test, after Bateson who applied this method in 1916 (Bateson, 1916, original not consulted).

7.2.4. Chimerism in vegetatively propagated crops

Types of chimerism

A plant consisting of two or more genetically different somatic tissues is called a **chimera**. The word 'Chimera' (or Chimaera) traces back to Greek mythology and refers to a fire-breathing monster with a lion's head, a goat's body and a serpent's tail which was slain by the hero Bellerophon with the aid of Pegasus (*The Oxford Illustrated Dictionary*, 2nd ed., 1975). Winkler (1907) introduced the word 'chimera' for so-called graft-hybrids which had

resulted from grafting a scion of one plant species on the seedling stock of another species (see Box 7.1) and it was Baur (1910) who suggested applying this word in all situations where the somatic tissue of an organism was found to be genetically heterogeneous. It may not be necessary anymore to recall that a chimeric situation automatically arises when a mutation is induced in a multicellular tissue.

The subject of chimerism in plants is both fascinating and of considerable practical importance. It is impossible to name here all scientists who have contributed significantly to our present fundamental knowledge of this subject, but some names that should be mentioned here in any case are those of E. Baur, F. Bergann, H. Dermen, F. Pohlheim, R.N. Stewart and R. Tilney-Bassett. Names of several other investigators and many references can be found in Tilney-Bassett (1986). A recent review featuring the significance of chimeras for understanding plant development was written by Szymkowiak & Sussex (1996).

In practice a distinction is made between **sectorial chimeras, mericlinal chimeras** and **periclinal chimeras**. The expressions 'sector' and 'sectorial chimera' were used by Baur (1909) for variegated seedlings of *Pelargonium* who carried green-coloured leaves on one half of the plant axis and white leaves on the other half (with sometimes some intermediates on the border area). In the same way leaves may occur which are green for, say, the left part and white for the right part. According to Baur, such plants had a 'sectorially' divided growing point. In early as well as in recent literature, (visibly-)mutated areas (or patches), for instance of leaves, are often called 'mutated <u>sectors</u>' without any proof that the observed area indeed does represent a 'real' sector: a wedge of cake or, more scientifically, a plane figure contained by two radii and the arc of a circle. In order to avoid unnecessary confusion, the expression 'sector' in relation to plants, preferably, should be reserved for an area which is a cross-section including all (three) histogenic layers of a plant and all tissues of a plant from the epidermis to the centre. Consequently, the use of the expressions 'mutated sector' and 'sectorial chimera' should be restricted to such 'true sectors', which are much more rare than mericlinal and periclinal chimeras. Most so-called sectorial chimeras in literature in fact are **mericlinal** chimeras in which, according to the early definition of Jørgensen & Crane (1927), the mutated area is restricted to part of one histogenic layer only.

Mericlinal chimeras, in case of a visible mutation in L-1 – like an aberrant flower or fruit colour, absence of

Box 7.1 Graft-hybrids (some early examples)

A graft-hybrid or graft-chimera is an individual, formed from a graft or scion and a stock, which shows the characteristics of both components. Such graft-hybrids may arise unintentionally when adventitious shoots originate from cells (or callus tissue) from both stock and scion at the graft union. Many interesting facts about graft-hybrids can be found in Baur (1910), in two small books by Neilson-Jones (1934; 1969) and in a chapter of a book on plant chimeras by Tilney-Bassett (1986).

The earliest example that has been recorded became known as the 'Bizarria' orange and refers to the combination of a seedling stock of citron (*Citrus medica*) on which a scion of sour orange (*Citrus aurantium*) was grafted. Tilney-Basset (1986) quotes Strasburger (1907) according to whom the 'Bizarria' orange had been produced in 1644 by a Florentine gardener and was described in 1674 by a physician called Nati. The scion did not take successfully, but callus tissue developed on the area between stock and scion. From this callus tissue a bud arose which developed into a remarkable tree, carrying leaves, flowers and fruit which were either identical with the orange or with the citron and, in addition, fruits were obtained in which the traits of both 'parents' had been combined or mixed in various ways. Strasburger also stated that – at that time – most probably, several, independently 'Bizarria' types were known in different areas (Florence, Naples) of Italy.

Recently, Spena & Salamini (1995), in a review on the formation of chimeras, mutated 'sectors', etc., also included references concerning very early intentional graft hybrids. Mentioned were examples of graft hybrids contain-ing an epidermis of the orange and the internal layers of the lemon. These curiosities were indicated as *Aurantium Virgatum* and A. *Striatum* by Ferrari in 1646. Spena & Salamini in this respect refer to earlier publications on this subject e.g., Pontano in 1505, Columella in 1545 and Della Porta in 1589 and 1592. These original publications have not been consulted.

Another, well-known example of a graft-hybrid traces back to 1825 when a nurseryman, Adam, who lived at Vitry near Paris, grafted a bud of a species of 'broom' (*Cytisus purpureus*) on a stock of common laburnum (*Laburnum anagyroides*). When the bud developed, one of the resulting shoots was more upright and vigorous than the other and produced larger leaves than the common broom. This shoot was selected for further vegetative propagation and became an important garden tree all over Europe under the name + *Laburnocytisus adamii* (The + sign indicates that the plant is a graft-hybrid; sexual hybrids are indicated with the sign x). A remarkable characteristic of this graft-hybrid is that its flowers often recovered the original traits of its asexual parents.

The next-known example refers to 1899 when in a garden in Bronvaux near Metz, France, a hundred-year-old tree was discovered in which a scion of a medlar (*Mespilus germanica*) had been grafted on a hawthorn stock (*Crataegus monogyna*). Two branches had developed from the graft union, one indicated as + *Crataegomespilus asnieresii*, which resembled hawthorn, and + *Crataegomespilus dardari*, which more resembled medlar. This case was also discussed by Strasburger (1907).

At the beginning of the twentieth century artificial graft-hybrids, in particular between members of the genus *Solanum*, became the subject of careful investigations (Winkler, 1907). Many investigators have studied the constitution of these (and other) graft-hybrids and as a general conclusion it has become clear that in the aforementioned examples – when considering a cross-section of a stem of the graft-hybrids – one or more layers of tissue originating from the scion surround a core of tissue that refers back to the stock. For the graft hybrid *Citrus* + 'Bizarria', for instance, a core of *Citrus medica* (citron) is surrounded by – in this case – one single layer (the L-1) from *Citrus aurantium* (sour orange). Accordingly, in + *Crataegomespilus* a core from hawthorn (*Crataegus monogyna*) is surrounded by tissue that originated from medlar (*Mespilus germanica*). The difference observed between + *C. asnieresii* and + *C. dardari* has been explained by the presence of one (L-1 only) or two (L-1 + L-2) layers of medlar surrounding the hawthorn. Finally, in + *Laburnocytisus adamii* a core of common laburnum (*Laburnum anagoroides*) is surrounded by one layer (L-1 only) of broom (*Cytisus purpureus*).

For more details, reference is made to a table presented by Tilney-Bassett (1986), in which all known examples of natural graft-hybrids have been described with the names of their most important investigators. In addition, a similar list has been prepared containing a number of artificially made (or synthetic) graft-chimeras within and between different genera (e.g. *Lycopersicon*, *Nicotiana* and *Solanum*) of the fam. *Solanaceae*.

hairs (like in nectarine) or 'russeting' on a fruit, a different type of hairs on a stem – do not represent a stable situation and in vegetatively propagated crops in general one or two cycles of vegetative propagation suffice to transform them into periclinal chimeras, which is the common (and rather stable) situation for mutated plants (Dermen, 1947). However, as was mentioned for instance by Chevreau, Decourtye & Skirvin (1989) for pear (*Pyrus communis*), depending on the trait and the layer(s) involved, within one plant species periclinal chimeras with different degrees of stability may occur.

In order to explain the presence of a (partly or complete) white leaf margin around a green central area of a leaf blade of *Pelargonium*, Baur (1909) distinguished between two or three peripheral layers of epidermal and sub-epidermal origin on the one side, and 'central cells' on the other side. For leaves with a white margin, Baur concluded that the aforementioned peripheral layers consisted of 'white (albino)' cells whereas the central tissues had remained green. Baur was also the first to use the term 'periclinal chimeras' for such (spontaneous) mutants with a white margin. This subject was also discussed in a lucid way by Tilney-Bassett (1986, p 4–10).

Much progress was made in understanding periclinal chimeras after colchicine was applied and chimeras were produced in which various histogenic layers and their derivatives showed different ploidy levels, e.g. of the 2x-4x-4x or 2x-2x-4x type, etc. Pioneers in this field were, for instance, S. Satina and colleagues in *Datura* (Satina *et al.*, 1940) in the USA, followed by J. Einset, H. Dermen and C. Pratt for various fruit crops, also in the USA and L. Decourtye for apple in France. Many additional data on ploidy-chimeras – including a study of screening methods for the various layers in chimeras – and on the effects of radiation on ploidy-chimeras, accompanied by a list of references on mutations and chimerism in fruit crops, can be found in a nice review by Pratt (1983).

Periclinal chimeras can become (stable) clonal cultivars but of course can be considered as well as raw material for further breeding work (Bergann, 1967). Later in this chapter the merits of breeding work aiming at the production of periclinal mutant cultivars will be discussed and compared to breeding of mutant cultivars which are solid for the mutated trait. Examples of successful cultivars belonging to both categories will be presented.

Bergann (1954) has given a detailed description of the different types of periclinal chimerism that may occur (e.g. L-1 mutated, L-2 + L-3 non-mutated; L-2 mutated, L-1 + L-3 non-mutated; L-1 + L-2 mutated, etc.) and provided names to indicate each type. Most periclinal chimeras,

in nature as well as when grown as mutant cultivars, are of the type: L-1 mutated and L-2 + L-3 non-mutated (the so-called monectochimera or monochlamydious chimera type); L-1 + L-2 mutated and L-3 non-mutated (diectochimera or dichlamydious chimera), or L-1 + L-2 non-mutated and L-3 mutated. The last type may be more difficult to observe in practice, but this does not imply that there should be fewer periclinal chimeras belonging to this category.

All aforementioned types appear to be more stable than a periclinal chimera of the type L-1 normal, L-2 mutated, L-3 normal. In this last case the constitution of the L-2 may easily change, for instance towards L-1 normal and L-2 + L-3 mutant, or L-1 + L-2 mutant and L-3 normal (Howard, 1972). Apparently, the L-2 layer is less stable than the other layers and, hence, the distinction of L-2 is not absolutely safe.

In addition, it may happen that in an existing periclinal chimera a second mutation is induced for the same or a different trait, in the same or in a – so far – non-mutated layer and in this way complicated structures may arise. Bergann & Bergann (1959) for instance reported that cv. Madame Salleron in *Pelargonium* is a so-called trichimera, i.e. a periclinal chimera with L-1, L-2 and L-3 being genetically different (with L-1 genetically normal for chlorophyll (but not expressing this trait!), L-2 having lost the ability to produce chlorophyll and L-3 showing the genetic constitution for dwarf growth). The authors described in detail the nine different types (AAA, BBB, CCC, ABB, AAB, BCC, BBC, ACC and AAC) that were obtained from the original (ABC) type by applying various methods, such as: induction of sprouts on roots, excision of apical and axillary buds, so-called 'parallel incisions' with razor blades in axillary buds, X-irradiation and induction of leaf cuttings. Of course, other combinations could be imagined, e.g. a mutation for flower colour in L-1, a mutation for leaf colour in L-2, and a mutation for starch composition in the L-3 root tissue of tuberous crops. Moreover, by a mutagenic treatment simultaneously different mutations could be induced in the three layers which, eventually, may interact.

The analysis of chimeric structures may be further complicated by the fact that, depending on the species as well as on external factors such as light, some variation may occur with respect to the contribution of the different layers to various tissues. For instance, under different light conditions a relatively smaller or larger central green area and white margin in leaves may be observed or a different number of subepidermal parenchymatic palisade cell layers in the leaf blade may be formed.

Rearrangement of cell layers

Concerning research on the subject of rearrangement of cell layers, mention should be made in particular of the contributions for various plant species by F. and L. Bergann and their students Klopfer, Polheim and Pötsch in Germany; Dermen, Popham and Stewart in the USA, Howard in the UK and Heiken in Sweden.

Despite the fact that the L-1 layer and, to a somewhat lesser extent, the L-2 layer are rather stable because of their (predominantly) anticlinal plane of cell division, irregularities may occur sometimes which may result in rearrangement of layers. Clowes (1957) already reported that plant species may show different degrees of instability. Such instabilities, as a rule, are the result of occasional periclinal divisions in the outer layers and may lead to penetration of cells from one histogenic layer into another (Bergann, 1957a). Irradiation may increase the frequency of such irregularities which may be of interest for breeding work in vegetatively propagated crops. When such irregular divisions occur in (vegetatively propagated) periclinal chimeric plants (e.g. an ornamental plant with the L-1 genetically differing for flower colour from L-2 and L-3), plants may arise which – partly or completely – are phenotypically different from the original periclinal chimera. In a more or less similar way, damaging of the outer tissue of a periclinal chimera may result in a plant in which the damaged layer is substituted by cells from internal layers and, thus, the genetic constitution of L-2 and/or L-3 may become expressed in the outer (L-1) layer. This process is often called 'uncovering', i.e. the genetic constitution of a deeper, genetically different layer can be observed.

Types of tissue rearrangements

Different processes or types of rearrangment of cell layers – in which in particular the L-1 is involved – have been classified by Bergann & Bergann (1959; 1962). Most of the early examples from Bergann & Bergann refer to studies on chimeras of *Pelargonium zonale*. The first two types are so-called vertical processes, i.e. events between layers. They are named perforation (tissue from L-2 and/or L-3 origin penetrates the outer layer(s)) and reduplication (the opposite process). The third type, called transgression, is a so-called horizontal process which refers to lateral replacement of cells within one layer. This, of course, would be of particular interest to the plant breeder when the layer in which the transgression occurs contains cells which are not genetically identical, as is the case in a mericlinal chimera. A combination of two of the aforementioned three types may lead to an inversion of layers within the shoot apex. This phenomenon was called translocation by Bergann & Bergann.

It was mentioned already by van Harten (1978) and Broertjes & van Harten (1988) that Bergann & Bergann's choice of, in particular, the words 'inversion' and 'translocation' for some of the aforementioned processes has been a rather unfortunate one because of the earlier-established meaning of these words in cytogenetics. As a consequence, the use of these words in relation to histogenic events should be avoided.

Dermen (1960) proposed the word replacement for penetration of cells from outer layers into the inner tissue, followed by replacement of the inner cells, and displacement for the opposite process. These terms sufficiently characterize the main processes of tissue rearrangment and should be preferred above the terms used by Bergann & Bergann.

Frequencies of histogenic events

In the work of Bergann & Bergann no figures about the relative frequencies of events have been presented and most attention was paid to displacements (perforations) within the epidermis, which are the easiest events to notice. Stewart & Burk (1970) determined frequencies of dark green spots on the surface of white tobacco leaves in periclinal chimeras with the original constitution L-1 green, L-2 + L-3 white. When dark green patches occurred this indicated that cells from the (genetically) green L-1 had divided in a periclinal direction and had replaced white L-2 cells. Based on these observations it was calculated that (spontaneous) periclinal divisions in L-1 occur in a frequency of 1 per 3100 divisions. Some other studies have produced similar figures. It appears that spontaneous disturbances of the layered pattern in shoot apices occur at rather low frequencies. According to Bergann (1957b; 1967) and others, however, this frequency in nature is high enough to be of practical interest to the plant breeder, in particular to breeders of ornamentals.

Radiation treatments may considerably increase the frequency of histogenic disturbances. Apparently irradiation not only causes an inhibition of cell divisions but also affects the system controlling the direction of mitotic divisions within a meristem. Different radiosensitivities of cells within a meristem may play a role. When, after a radiation treatment, visible changes (e.g. for flower colour) are observed in relatively high frequencies, one has to consider carefully whether this effect was caused by newly induced mutations or by histogenic (non-genetic) effects. One could in this second situation think for instance of destruction of the layered structure of the shoot apex followed by 'penetration' of cells from a genetically different (deeper situated) layer into the epidermis and – ultimately – even complete displacement (see definition by Dermen, 1960, in the previous section) of the original layers. Pötsch (1967; 1969) in this respect reported that irradiation of periclinal chimeras of *Pelargonium zonale* lead to the penetration of cells from deeper layers into the L-1 and added that higher doses (e.g. 20 Gy of X-rays) produced more of such effects.

Detailed experiments on this subject were also performed in the 1970s by van Harten (1978) with a spontaneous bud sport of potato cv. Désirée, called M 52, which was recognized by a yellow/red-splashed tuber skin as a result of a genetic constitution L-1 yellow, L-2 + L-3 red (normal cv. Désirée is genetically red in all three layers). The combined results of a range of experiments involving X-irradiation of so-called tuber eye pieces with various doses led to some tubers which were genetically yellow in all three layers, i.e. representing the genetic constitution of L-1 only, but only after radiation with doses above 20 Gy. At even higher doses the percentage of yellow/red-splashed (periclinal chimeric) tubers declined from 80–90% in the irradiated series to about 50%. The remaining part consisted of tubers with solid and mericlinally red tuber skin. These observations are in good agreement with experiments concerning potato tuber skin colour by Howard (1964) and for potato tuber skin colour and leaf shape by Klopfer (1965a).

Lineberger & Druckenbrod (1985) and Lineberger *et al.* (1993) reported on the vegetative separation ('segregation') of various genotypes that co-exist in periclinal chimeras, by applying tissue culture methods to African violet (*Saintpaulia* sp.) and *Rhododendron*. So-called pinwheel-flowering *Saintpaulia*'s display bicoloured flowers as a result of their periclinal situation. Explants proliferating through adventitious shoots from three different cultivars showed a very high percentage (sometimes up to 100%) vegetative segregation. The predominant phenotype obtained in this way has a unicoloured flower showing the colour of the original corolla margin. The results, according to the authors, suggest that adventitious shoots trace back to cells from L-1 only or from L-1 + L-2, but not from all three histogenic layers together.

Lineberger *et al.* (1993) further reported on rearrangements of cell layers after radiation treatment, in line with the work by Pötsch and van Harten. Different doses of radiation were administered after 4, 14 or 28 days of *in vitro* culture. The highest percentage of rearrangements in *Saintpaulia* cv. Valencia – 1.5% – was obtained after treatment with 50 Gy of thermal neutrons to leaf explants that had been cultured *in vitro* for 28 days. According to the same publication 'spontaneous' stable tissue rearrangements, resulting in an attractive new phenotype with yellow margined leaves, were observed

after a periclinal cultivar of *Rhododendron* (cv. Roosevelt) with a green margined leaf and a yellow central lamina had been brought into *in vitro* culture. The authors concluded that, despite a frequency of such events that may be lower than one per 1000 plantlets brought into tissue culture, this method may be used intentionally to produce new cultivars in ornamentals from periclinal chimeric starting material.

A point to keep in mind when periclinal chimeric plants are subjected to mutagenic treatments is that genetically different cells or tissues may differ in their reaction to the treatment. Moreover, cells in various stages of differentiation and with different specializations will differ in sensitivity to mutagenic treatment of somatic tissue. Effects of competition between cells from genetically different layers also have to be considered, but they will probably be much more important when ploidy chimeras are involved than when layers differ only in one or a few (recessive) genes, e.g. for flower colour, a trait which normally is not expected to affect much the fitness of a mutated cell. If, on the other hand, an aberrant flower colour in one of the histogenic layers of a periclinal chimera would be the result of a large deletion, this certainly might affect fitness.

7.2.5. Chimerism and mutation breeding in vegetatively propagated crops

Mutation breeding in vegetatively propagated crops is impossible without understanding the basic phenomena of chimerism. The different types of chimeras that may occur after a mutation arises in a shoot apex are illustrated by Fig 7.1., derived from Broertjes & van Harten (1988).

Both solid mutants and periclinal chimeras, may have their merits for plant breeding of vegetatively propagated crops. Which system is to be preferred depends strongly on the crop and on the nature of the changes that are desired. This was worked out for instance by Broertjes (1979) for flower colour mutants of chrysanthemum (*Dendrathema morifolium*, previously *Chrysanthemum morifolium*). This subject will be further discussed in Section 7.4.5.

In practice, breeders in most cases do prefer solid mutants, mainly because the risk of 'back-sporting' is much smaller. In fact a clear distinction should be made in this respect between the occurrence of aberrant types as a result of a newly induced mutation – which may occur both in solid mutants and in periclinal chimeras – and the much more common 'back-sporting' as a result of a standard periclinal chimeric situation that did exist already in the cultivar studied. In the latter case the resulting plant will become genotypically identical to one of the components of which the origi-nal periclinal chimera had been made up. Often this may be the genotype of the original (solid) cultivar from which the periclinal chimera had been derived.

A few examples may illustrate the usefulness of some insight in the basic principles of chimerism for plant breeders. An early example refers to efforts by the breeder Vilmorin in France in the nineteenth century to improve the fodder quality of a plant species called com-mon corse (*Ulex europaeus*). This plant species, which has

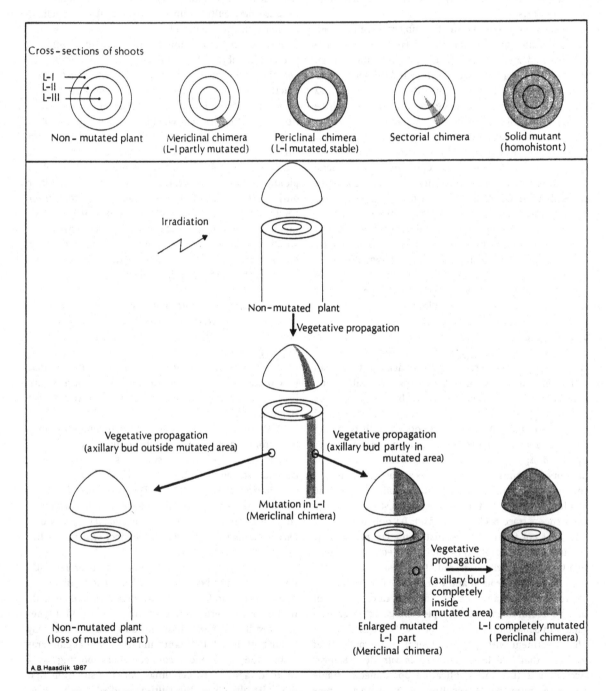

Figure 7.1. Chimerism in shoots. Above: different types of chimeric structures. Below: the fate of a mutated initial cell in L-1 of a vegetatively propagated plant. (From Broertjes & van Harten, 1998, with the permission of Elsevier, Amsterdam.)

only a short history as a domesticated crop, is normally thorned, but some thornless types – which are of course more favourable when the plant is used as a fodder crop – were known. When the thornless type was used in crossing work, this trait was never transmitted. Our present understanding of periclinal chimeras is that the trait 'thornless' must have been restricted in the spontaneous mutants to L-1 only. Further details can be found in Bergann (1954).

Darrow (1931; 1955) reported for so-called Evergreen blackberries (*Rubus laciniatus*) on the first vigorous and productive thornless sport, found about 1931 among many rather small, weak, low-productive and sometimes entirely sterile thornless sports known so far. Darrow investigated whether this vigorous sport, cv. Thornless Evergreen, was (in Darrow's terminology) a 'varietal chimera', that is a chimera with thornless tissue as a thin layer overlying thorny tissue. As the few shoots that were growing from root cuttings were thorny, Darrow concluded that the observed sport, most probably, had a layered chimeric constitution (L-1 thornless; L-2 and L-3 thorned).

Darrow (1955) further referred to a few other examples of (periclinal) chimeric sports for blackberries and dewberries (probably *Rubus hispida*) that had been released throughout the years. The first thornless blackberry sport that had been trademarked, 'Cory Thornless', became prominent already between 1911 and 1918. Darrow further discussed the layered structure (L-1, L-2 and L-3) of the sports and mentioned that in all cases plants propagated by root cuttings as well as seedlings from the described sports were thorny. Later, McPheeters & Skirvin (1983) confirmed the conclusions of Darrow. Both sexual reproduction (relating to the genetic constitution of the L-2) and (*in vitro*) regrowth from root tissue (which relates to L-3) resulted in thorned plants only. Reference is made also to McPheeters & Skirvin (1989) who discussed 'somaclonal variation' as well as variation resulting from dissociation of existing periclinal chimeric structures *in vitro*. This subject has been discussed already in Chapter 5.

An interesting point mentioned before is that when studying thornless sports, solid, non-chimeric thornless plants may be found, which often remain very small and weak when compared to the original thorned type. This finding may be explained by assuming a linkage or a pleiotropic effect of the gene(s) coding for thornless skin and for plant vigour, making the periclinal chimeric thornless plant (L-1 thornless, L-2 + L-3 thorned) superior to solid thornless plants.

One comment to be made here, is that it would be of interest to have more anatomical evidence on the exact way and place of origin (which layer or layers are involved?) of thorns, spines, etc. More information about thornlessness or spinelessness can be found, for instance, in Hall, Cohen & Skirvin (1986) and Nelson *et al.* (1989).

Important methodological improvements in mutation breeding for vegetatively propagated crops have come from Zwintzscher (1955; 1959; 1962), who worked with apple (*Malus pumila*), and Bauer (1957) who performed his studies on black currant (*Ribes nigrum*). These authors showed that repeated back-cutting of (partly mutated) shoots after X-irradiation stimulated basal buds to develop and, in this way, solid mutants could be obtained from plants that, after radiation treatment, must have been highly mericlinal chimeric. This favourable result does not always occur, as has been pointed out by Lacey & Campbell (1987). It is also not always clear whether in such situations the occurrence of solid mutants is the result of regeneration from cells belonging to a single mutated layer in a periclinal chimera, or whether loss of previously induced sectorial or mericlinal chimerism follows logically as a consequence of the normal development of a plant. For this last phenomenon, that was briefly mentioned already in Chapter 1, the expression 'diplontic drift' was proposed by Balkema (1971; 1972).

Several examples are known of repeated, unsuccessful efforts to produce generatively true-breeding cultivars from mutant plants (sports) of a vegetatively propagated ornamental with a different, attractive flower colour. As flower colour in a plant is governed predominantly by the genetic constitution of the L-1, it may be anticipated that the reason for lack of success in such cases mostly can be explained by the occurrence of a periclinal chimeric situation in the mutant plant with the mutation for flower colour being present only in the L-1. As the gametes are known to trace back to the L-2 which, in that case, had remained non-mutated, one should not be surprised that no generatively true-breeding mutant cultivars were obtained. However, by provoking re-arrangement of cells between the different histogenic layers, i.e. transferring the mutant genotype to the L-2 layer, it may be possible for the breeder to reach his goal.

Interesting results have been obtained in this respect from experimental work on periclinal chimeras in potato (*Solanum tuberosum*). This work focussed on a) the contribution of the different histogenic layers to various plant tissues and organs and the independence of such layers, b) on methods to study the genetic constitution of these layers for traits like tuber skin colour and leaf type, and c) the effects of rearrangement of genetically different histogenic layers on the phenotype.

A range of publications on these subjects has been produced by Howard (1958; 1959; 1964; 1967; 1969); Heiken (1960); Heiken & Ewertson (1962); Heiken, Ewertson & Carlström (1963); Klopfer (1965a,b; 1967); van Harten & Bouter (1970) and van Harten (1972), from which some main conclusions are given here. There is ample proof that the potato shoot apex (with only few exceptions) consists of three stable histogenic layers. If spontaneous or induced mutations occur in existing potato cultivars, such mutations – at least initially – are restricted to one layer. Plants in multiplication plots that do not match the official variety description, for instance for tuber skin colour, tuber shape, type of foliage or leaf shape, usually are eliminated during inspection. However, as mutations are only expressed in a 'layer specific' way, there will be a hidden accumulation of spontaneous or induced mutations, which cannot be eliminated by inspection. As a consequence, searching in multiplication fields for plants with interesting aberrant traits is a simple, cheap and attractive option to obtain new cultivars. Off-types (or sports) in vegetatively propagated crops like potato in almost all cases, originally, will be mericlinal and then periclinal chimeric clones. When sufficient plants are inspected, such off-types can be found in all cultivars of potato.

In order to find out whether the off-types indeed are periclinal chimeras (in most cases with only one layer mutated), various experiments can be performed. Such experiments, in addition, may produce new periclinal combinations as well as solid (mutated or non-mutated) potato plants. A common method to check for mutations in the L-3 layer is to grow new plants from tubers of which the eyes had been excised following the so-called Asseyeva-method, known since 1927 (see for instance van Harten, 1978). The plants resulting from adventitious buds only reveal the genetic constitution of the L-3 and, as a consequence, will be always 'solid' and either refer to the original genotype or to a mutant type. As an alternative to the eye-excision method it is also possible to induce adventious buds on roots from which new plants may develop which are exclusively of L-3 origin. When tuber skin colour is the trait studied, in most cases three types (solid 'mutated', solid original phenotype and periclinal chimera – often with a speckled appearance) can be distinguished. Analysis of periclinal tuber skin patterns can be complicated by the fact that the tuber skin does not always relate to the genetic constitution of the original L-1, but may derive from a cork cambium that arose in the L-2 or L-3.

Interpreting chimeric situations may become more

difficult when, for instance, chimeras with only a mutated L-1 (L-2 and L-3 being normal) are compared with the type in which both L-1 and L-2 are mutated (only L-3 normal). By making use of sexual reproduction, as has been discussed already in Chapter 6, only plants are produced which trace back to L-2 and therefore unveil the genetic constitution of L-2. However, making crosses in vegetatively propagated crops may be difficult and genetic analysis will be confronted with much segregation due to heterozygosity of the original clonal cultivar. The tetraploid nature of potato ($2n = 4x = 48$) further complicates genetic analysis.

The application of X-irradiation on tuber eyes or young potato sprouts (e.g. with 30 Gy) may result in all aforementioned kinds of histogenic effects. A periclinal chimera with only L-1 mutated, in this way, may be transformed into all other types (solid original 'wild' type, solid mutant, L-1 + L-2 mutated). Irradiation of a solid mutant (L-1 + L-2 + L-3 mutated), of course, can never induce histogenic effects that result in periclinal chimeras but the treatments, on the other hand, could induce new mutations, including a (back-)mutation for the trait that is studied. Spontaneous mutations may occur as well in all histogenic layers.

To study the genetic constitution of the L-1 for genes that are expressed in that layer, sometimes a study of visible traits may suffice. However, for traits like stem or tuber colour, deeper situated cell layers or tissues may also affect the picture and in that case attempts should be made to grow plants from L-1 origin only. This goal may be achieved by applying the adventitious bud technique that, because of the impact it has had on mutation breeding in vegetatively propagated crops, will be described in detail in Section 7.2.7. Another but less reliable method in this situation was mentioned before, i.e. the use of X-irradiation in order to arrange layer disturbances.

With respect to investigations on the (periclinal) chimeric constitution in potato, and in a few other vegetatively propagated crops as well, it can be concluded that methods are available to study the genetic constitution of each layer separately. This, however, does not imply that all of the aforedescribed methods can be applied easily and for all plant species.

It may be interesting to know that, at least in the past, on several occasions new potato cultivars have been released that were not solid crops but periclinal chimeras, a fact of which the breeder was not aware. Transforming them into solid plant types in some cases altered the characteristic traits of such cultivars to such an extent that the original cultivar went out of production. This, for instance, has been the case with the Dutch mutant cultivar 'Rode Eersteling', registered in 1934 as a spontaneous 'sport' from cv. Eersteling (van Harten & Bouter, 1970). The original cv. Rode Eersteling which entered the variety list, was characterized by red tuber skin as opposed to the yellow-skinned cv. Eersteling from which it was derived. As in this case the red skin colour was (genetically) present only in L-1, now and than 'reversion' to yellow tuber-skin occurred. In later years, solid mutant plants (with L-1 + L-2 + L-3 genetically red for tuber skin colour) were found among the tubers of cv. Rode Eersteling. Such mutants, undoubtedly, resulted from spontaneous histogenic effects, like the aforedescribed process of replacement (Dermen, 1960).

A short inventory on the role of chimerism may be made at this point. Chimerism has not only negative sides. It may be an important phenomenon in relation to the generation of genetic variation, in particular in the earlier stages of mutation breeding programmes. In addition, in some situations, especially when certain ornamental crops are involved, it is possible to release periclinal chimeric structures for favourable traits (e.g. flower colour or shape), as new cultivars within a very short time after their detection. The occurrence of chimerism in breeding programmes, on the other hand, could also be rather troublesome. In any case, some basic understanding of this phenomenon is very useful for a breeder who has to decide on the best way to handle chimeric plant material.

7.2.6. Plant 'variegation' is not necessarily caused by chimerism

A subject that is often – although not always correctly – associated with chimerism, is the occurrence of plants which show stripes and patches of aberrant colours in vegetative parts like flowers, stems and leaves. This phenomenon is commonly indicated as 'variegation'. Plants with, in particular, leaf variegation have become very popular as ornamentals in gardens and at home. Chimerism is by no means the only cause for variegation; in fact a range of different mechanisms and phenomena including the regulation of gene expression in specific tissues, maternal inheritance, sorting out of plastids, (periclinal) chimerism, histogenic effects, the presence of viruses, etc. may be involved. Various patterns of 'variegation' do exist and, as a consequence, an analysis of the actual cause of the observed phenomena is not an easy one. A certain knowledge of the most common mechanisms may be in particular useful for breeders of leafy ornamentals. More details are presented in Box 7.2.

7.2.7. Adventitious bud techniques

As was said before, a mutagenic event in a single cell of a multicellular tissue automatically results in a chimeric situation. In seed propagated plants, after one cycle of propagation by seed, the mutation will be lost when it was not present in one of the gametes from the L-2 layer. For vegetatively propagated crops the situation is different. Plants that – after some vegetative propagation – still carry a mutation that was induced before, will show a periclinal chimeric structure with the mutation in most cases present only in tissues that trace back to a single cell in a particular histogenic layer. When (spontaneous or artificially induced) irregularities in cell divisions occur within the shoot apical area, sometimes tissues belonging to two histogenic layers (e.g. L-1 + L-2 or L-2 + L-3) may carry the mutation, although originally it occurred in only one cell in one of the layers. The occurrence of solid mutants must be considered an exception and generally requires some particular artificial manipulation.

It was mentioned in the previous section that sometimes periclinal chimeras, commercially attractive, e.g. for flower colour, can be directly released and may be officially registered as new and sufficiently stable cultivars. In most situations, however, breeders prefer solid mutants. When resistances are involved, it depends on the particular pathogen or insect pest whether a periclinal structure of resistant tissue may be effective or not.

Depending on the layer(s) in which the mutation is present, different ways must be followed to produce solid mutants. It was pointed out before that propagation by seed results in offspring from L-2 origin. When reproduction takes place by axillary buds, the chimeral situation in the main shoot apex is normally, but not always, reproduced in a rather exact way, which implies that this way of propagation is not particularly suitable to get rid of chimerism. At the same time it can be noticed that shoots grown from axillary buds that developed in later stages of plant development may be less chimeric, i.e. the mutated areas may be larger, but there may also be less mutated areas present. This is mainly due to the transition from mericlinally mutated areas in the main shoot towards periclinally mutated layers in the axillary buds, but is affected as well as by the developmental pattern of the plant species concerned. Later in this chapter we will discuss how use is made of these phenomena in practical mutation breeding, for instance for fruit trees. However, as a method for limiting chimerism the use of axillary buds in most cases is not effective.

Box 7.2 Variegated plants

Variegated plants, following a definition by Kirk & Tilney-Basset (1978), are plants which develop patches of different colours in their vegetative parts. Evenari (1989) mentioned that the phenomenon was observed already in ivy (*Hedera helix*) by Plinius the Elder (23–79 AD) and, nowadays, is observed in hundreds of plant species in all major plant families (Kirk & Tilney-Bassett, 1978). Early studies on leaf variegation in higher plants were performed by Baur (1909) on *Pelargonium* and Correns (1909) on *Mirabilis*. Leaf variegation in those cases is caused by the presence of normal green cells which contain normal plastids, and white or yellowish cells with defective plastids. Interesting details can also be found in Renner (1936a,b).

Several mechanisms may underlie the occurrence of the various types of variegation. Sometimes the variegated appearance is caused by different genetic constitutions of the various histogenic layers (L-1, L-2, L-3) and their derivatives. Certain types of variegation are affected by temperature (Correns, 1909) or light (Chittenden, 1925). Diseases and virus infections may also play a role. To mention just one example in this respect: the yellowish type of variegation in the leaves of the ornamental plant *Abutilon* sp. is caused by virus infection. Certain other categories of variegation will not be further discussed here, like the variegation that results from the action of transposons (see Chapter 3), or from lack of certain mineral nutrients.

According to Tilney-Bassett (1991), two forms of variegation have to be distinguished: those in which contrasting colours do and

those which do not follow cell lineages. It is, indeed, necessary to realize that not all observed variegation represents chimerism (see for instance also Marcotrigiano & Stewart, 1984, and Broertjes & van Harten, 1985). A characteristic of genetically determined 'mosaicism' in plants is that this type of variegation does not break up, even when vegetative regeneration of new plantlets from one single cell of the variegated mother plant would be possible.

Variegation may occur in all vegetative parts, including flowers, stems and leaves. We will concentrate in this box in particular on plants with variegated leaves which are very popular in gardening and as foliage pot plants in homes. Within this group, again, various types of variegation can be distinguished. Well-known variegated pot plants at home are for instance *Dieffenbachia* and *Ficus* (various spp.) with white- and yellow-variegated leaf colours, whereas patterns of reddish-variegated can be often found within the genus *Codiaeum*, as well as in the genera *Caladium* and *Calathea*. Many other genera could be mentioned as well, like *Acanthus*, *Euphorbia*, *Tradescantia*, a few cacteae (*Chamaecereus* and *Opuntia*) and, also for outdoor use, *Acer*, *Agave*, *Aucuba*, *Cornus*, *Fatsia*, *Hedera* (ivy), *Ilex* and conifers like *Chamaecyparis lawsoniana* and some species of *Thuja* and *Taxus*.

Much attention to the phenomenon of variegation in leaves and flowers has been given by Tilney-Bassett (1986; 1991), and in particular in his authoritative book of 1986 many references are given. More recently, Tilney-Bassett (1994) briefly discussed the various

systems of nuclear control of chloroplast inheritance in higher plants (i.e. two types – <u>maternal</u> and <u>paternal</u> – uniparental plastid inheritance and biparental inheritance) in relation to his own investigations on pelargoniums.

The earliest reports on non-Mendelian inheritance were written by Baur (1909) and Correns (1909). Whereas, for instance, plastids in the 'four o'clock plant' (*Mirabilis jalapa*), a famous species studied by Correns in the early 1900s, are maternally inherited, those of *Pelargonium* (see Baur, 1909) are biparentally inherited. Uniparental, maternal inheritance is more common than biparental plastid inheritance. (N.B. Variegation in *Mirabilis* may also be related to transposon effects, but this subject is not further discussed here.)

Other investigators who have contributed significantly to our present knowledge on this subject are, for instance, F. Bergann and L. Bergann, R.J. Chittenden, H. Dermen, H. Dulieu, F. Pohlheim, O. Renner and R.N. Stewart.

Evenari (1989), who reviewed in particular the situation for plants with white-green variegated leaves, categorizes the different ways in which (this) variegation inherits into four groups.

The first way of inheritance mentioned is the non-Mendelian inheritance – a subject that has been briefly discussed already in Chapter 3. Variegation caused by non-Mendelian inheritance may both be due to maternal inheritance as well as to biparental transmission of plastids. Not all details are elucidated yet. The second group concerns Mendelian variegation of which many examples have been given by Kirk & Tilney-Bassett (1978). The nuclear genes involved in such cases are stable and may be either dominant, semi-dominant or recessive. A third type of inheritance mentioned by Evenari relates to the action of mutable nuclear genes. (N.B. Although Evenari clearly refers to effects of transposons, the expression is not used.) The fourth and final group distinguished includes variegation as a result of plastid mutation.

Evenari refers to two types of plastid mutation that may be involved: mutations in plastids caused by recessive mutations in nuclear genes or, as an alternative, autonomous mutations of plastids. However this system of classification does not really elucidate the phenomenon of variegation. For instance, in case of maternal inheritance, all cells should be alike; and for nuclear inheritance likewise. Differentiation of gene expression may be of much more significance in this respect than the mode of inheritance.

Another subject touched upon by Evenari concerns the question whether back mutations from white to green occur. The, prevailing, positive answer to this question has been challenged by Bergann & Bergann (1983) who, for periclinal (green-white-green) chimeras of *Peperomia* sp., showed that the observed green spots could be explained by the fact that different layers contribute to the formation of the leaf-mesophyll.

It may be observed that no two leaves of a variegated plant display exactly the same pattern. All theories which have been developed throughout the years to explain pattern formation of variegated plants have in common that unequal cell divisions are involved, but details are still largely unknown. In some instances, pattern formation may be explained by the way such pattern originate in relation to the way shoot apices are organized.

For instance, when one leaf is found to be white and the other half green, e.g. in ivy (*Hedera*) or privet (*Ligustrum*), or when white stripes run in a regular way from the base to the leaf tip (e.g. in striped chimeral grasses), the pattern of variegation may be explained by the layered organization of the shoot apex or the leaf primordia respectively. In such situations one or more layers, or part of one layer may contain defective chloroplasts, whereas the other part carries normal ones. Checkered patterns may result from the occurrence of normal and mutant plastids which are mixed in the same cells and which, during development, are sorted out. Other possible explanations are the action of the aforementioned mutable genes or transposons and mutations of regulator genes. Among the conifers, some species of *Chamaecyparis* and *Thuja* show the peculiar phenomenon that permanently small white or yellow twigs develop from green ones. Such conifers were called 'eversporting' by Tilney-Bassett (1986).

It has been proven for several species and cultivars of conifers that this situation refers to a two-layered shoot-apex with a white L-1 and a green L-2, in which frequently mutations occur in the L-2 layer. Examples are for instance the cultivars *Chamaecyparis lawsoniana* 'Argente-ovariegata' and 'Fletcher's White'. Twigs that are completely white, of course, have a small 'fitness' and will soon die off. A range of publications on this subject of sporting conifers has been forwarded in the 1970s and early 1980s by F. Pohlheim and it is impossible to mention them all. For an overview see for instance Pohlheim (1980).

Given the many different situations of variegation that do occur, it must be said that all present explanations together cannot fully explain all cases known. As was mentioned before, regulating genes or other factors that control gene-expression, effects exerted by plastids in adjacent tissues, genome/plastome interactions and still other factors unknown so far, may affect patterns of variegation.

From a practical point of view it would be useful for a breeder of ornamental crops to know whether, in a specific situation, leaf variegation is genetically determined (either in a Mendelian or non-Mendelian way), or is caused by periclinal chimerism or still other factors. In the first situation propagation by seed is possible; for periclinal chimerism the variegated pattern will be maintained only by proper vegetative propagation, i.e. by methods of propagation in which the layered structure of the original periclinal chimera is fully maintained. In particular root cuttings, in practically all situations, will not maintain the original periclinal structure and about the same can be said for leaf cuttings. When shoot cuttings are used to propagate periclinal chimeras, the breeder has to reckon with a certain percentage of off-types that will be produced as a result of the instability that is often observed in chimeras as well. Cultivars within one species may differ very much in this respect.

Adventitious shoot formation

Another approach is to induce regrowth from various tissues and on locations that were not predetermined for bud development. A well-known example in this respect is the regeneration by leaf cuttings in African violet (*Saintpaulia ionantha*). In other species this regeneration may be achieved either by natural systems, such as the sympodial growth pattern which automatically leads to reduction or termination of the 'chimeral problem' – as is the case in tomato (Verkerk, 1971) or, as will be discussed later in this section, in rhizomes of *Alstroemeria* – or by deliberate human action, for instance by stimulating shoot formation on roots, or development of sprouts from deeper layers of tubers from which the original eyes have been removed.

Nowadays, various *in vitro* methods suitable for adventitious bud formation are available. Depending also on the starting material, the resulting plants, most often, represent the genetic constitution of L-3 and, commonly, have a multicellular origin. The adventitious buds from which such plants develop (see also Section 7.2.5), are of considerable interest for mutation breeding, as mutations that would be lost or remain undetected otherwise, can be saved in this way. This may be the case in two situations: 1) when mutations occur in somatic plant tissues outside the shoot apical meristem or, 2) when a mutation occurs only in a single cell of one histogenic layer of the shoot apex. The resulting adventitious plants usually trace back to a single histogenic layer and, depending on the plant species and the way of initiation, one or more cells may have been involved in the formation of the adventitious bud. In the case of one cell, of course, the resulting plant traces back to a single histogenic layer as well. If a considerable part of a layer, or even a whole layer, has become mutated for a specific trait, the use of adventitious buds may allow the regeneration of plants with large mutated areas, or even of complete non-chimeric, solid mutants.

Adventitious shoots in mutation breeding

A publication by Sparrow, Sparrow & Schairer (1960) is mostly considered the starting point of studies concerning the application of adventitious buds in mutation breeding of vegetatively propagated plants. These authors reported for African violet (*Saintpaulia ionantha*) that the adventitious plantlets which arose at the base of the petioles of detached leaves, apparently were of single-cell origin. This conclusion was based on the observation that, after irradiation of detached leaflets, all mutated plantlets which appear at the petiole base were non-chimeric. Hence, a single cell origin of such

adventitious plantlets seemed to offer a logical explanation for this result.

The phenomenon itself was described already in botanical studies (on unirradiated plant material) by Naylor & Johnson (1937), but at that time not connected with chimerism as a result of mutagenesis. An inventory by Broertjes, Haccius & Weidlich (1968) showed that various types of formation of adventitious plantlets occurred in nature and had been recorded at that time already for over 350 plant species. Then the idea arose that adventitious plantlets might be very useful in order to produce in a fast and easy way chimera-free mutants in vegetatively propagated plants. In particular Broertjes studied the merits of what became known as 'the adventitious bud method' for a range of ornamental species which mainly belonged to the family of the *Gesneriaceae*, including for instance, *Saintpaulia ionantha* (Broertjes, 1968; 1969a; 1972a,b), *Streptocarpus* (Broertjes, 1969b), *Achimenes* (Broertjes, 1972c) and *Kalanchoë* sp. (Broertjes & Leffring, 1972).

The inventory by Broertjes *et al.* (1968) also made clear that for many important species formation of adventitious plants was not known, or at least not reported in easily accessible publications. Following this review, in quite a few cases efforts to try to develop such methods in various plant species were undertaken but have not been successful. In other cases it has not been tried yet to induce adventitious plantlets.

Because of the potential usefulness of the adventitious bud method, the mutation breeder of vegetatively propagated species should check the literature in this respect but – also when positive results have been reported – always should experiment with the cultivar in which he is specifically interested as considerable differences are found in the ability to produce adventitious shoots.

During recent decades regeneration via adventitious plants has become much more common, because many new *in vitro* methods starting from various types of plant tissue have become available for many plant species (see also Chapter 5). The quickly increasing knowledge of *in vitro* regeneration methods in recent years has led to much more insight into regeneration processes and has also clearly demonstrated the cultivar-dependency of regeneration behaviour. Formation of new plantlets *in vitro* may occur after a longer or shorter callus phase by adventitious buds and in some situations it seems that a callus-phase does not occur at all and regeneration occurs by somatic embryogenesis. To distinguish between both types the latter (somatic embryos) could be identified by the early presence of bipolar structures with a shoot and a root meristem, whereas in the case of the formation of adventitious buds from callus only shoots develop. Rooting of adventitious shoots is to be induced by transferring the separated young shoots to a rooting medium.

Single cell origin of adventitious shoots?
The mutant plants that were obtained for instance in *Saintpaulia* and several other *Gesneriaceae* by combining various mutagenic treatments with adventitious bud techniques, as well as the results from additional cyto-histological investigations, strongly suggested that cells – probably of epidermal origin and situated at the basal part of leaves or leaf petioles – had dedifferentiated, became meristematic and further developed into bud primordia. An important, additional observation was that often more than 90% of the mutant plants obtained via this adventitious bud method, were solid mutants. These results strongly supported the opinion that most of the adventitious shoots were of single cell origin or, to put it more carefully, that the apices of the adventitious shoots traced back to a single cell of, probably, epidermal origin. The low percentage (e.g. 5–10%) of chimeras may be explained by irregularities of various kinds which may occur during bud development. The concept of a single cell origin of adventitious shoots was further substantiated by a mathematical model developed by Broertjes & Keen (1980) that, so far, has not been refuted.

The reason to phrase the single cell hypothesis and the (predominantly) L-1 origin of adventitious shoots in a more careful way than was done in earlier publications, is that the single cell origin of adventitious shoots has been contested on some occasions. Most critical comments came from the Bergann & Bergann (see for instance Bergann & Bergann, 1982) and, on different grounds, from Smith and Norris (Norris, Smith & Vaughn, 1983; Smith & Norris, 1983).

Although details are outside the scope of this book, the way of experimenting and the results presented by the latter group of authors, in our opinion, insufficiently substantiate their objections, and other interpretations to their results could be given as well (for more details and some references, see Broertjes & van Harten, 1985; 1988).

More difficulties, admittedly, are encountered when trying to explain results obtained for *Saintpaulia* sp. by Ando, Akiyama & Yokoi (1986), or to reconcile a – generalized – (L-1) single cell hypothesis with the results obtained after careful experimenting by Bergann & Bergann (1983; 1984) with periclinal chimeras of *Peperomia* sp. It should be noted that *Peperomia* is one of the plant species of which Broertjes (see for instance Broertjes, 1972c) reported that it was proven that single

cell adventitious buds arise at the base of the petiole of separated leaves; an opinion that was strongly contested by Bergann & Bergann! In the 1983 publication of the aforementioned authors, formation of adventitious shoots was reported from the various layers of six different periclinal chimeras of *Peperomia* sp. and the results obtained were related to the histogenesis of leaves for this plant genus. Comparable detailed studies were performed with three (di-)chimeras of the green-white type for *Sedum rubrotinctum*, an angiospermous species with also three distinct apical layers. The results obtained by the Berganns and some other investigators, indeed, do not support the single-cell adventitious bud concept for these species.

Broertjes & van Harten (1988, p.50) also have stated that in many cases 'it is not obvious, or even not to be expected, that the apices of adventitious buds originate from one cell'. In particular buds that arise on stems and roots may originate from inner tissue and it is known that in that case often more cells are involved initially in shoot formation. The authors add that in particular the kind of explants or plant segments that are commonly used *in vitro* and the medium with hormones may often result in the production of adventitious shoots from 'inner' origin in which more than one cell was involved.

In the aforementioned book various methods have been described for a number of crops to reduce chimerism. One interesting example refers to the ornamental crop species *Alstroemeria* where, after irradiation of actively growing young rhizomes, almost exclusively solid-looking (and apparently non-chimeric) mutants were produced from which several were released as commercial cultivars (Broertjes & Verboom, 1974). The initially unexpected solid nature of the mutants most probably is related to the special sympodial growth of the rhizomes which results in dissolving the chimerism that was induced by the mutagenic treatment.

To conclude, it must on one side be admitted that the (L-1) single cell origin hypothesis for adventitious shoots does not apply to all cases. Nevertheless, the adventitious bud method has proved to be a powerful and succesful tool in mutation breeding of vegetatively propagated crops to reduce chimerism. To give one more example, van Harten *et al.* (1981), using the adventitious bud method for explants from rachis and petiole and for leaflet-discs of potato (*Solanum tuberosum*) in *in vitro* mutation experiments (see also Chapter 5), reported for all experiments (based on about 2800 irradiated explants) an average figure for chimeric plants of 1.7% only in combination with an overall mutation frequency (for a range of selected traits) of almost 75%.

When taking all information on adventitious shoot methods together, the general advice when considering mutation breeding in vegetatively propagated crops may be to check whether a suitable adventitious shoot method is available already or whether such a method could be developed. This, certainly, is advisable when dealing with mutation breeding 'in vivo'. More caution is required when advising on the use of such adventitious shoot methods in vitro. The choice of the starting material, the in vitro procedures followed, etc., may largely determine whether one could expect the production of non-chimeric mutant plantlets or not.

7.3.   **Mutation breeding**

7.3.1.   General considerations
The earliest reviews about mutation breeding in vegetatively propagated crops trace back to Cuany (1960; original not consulted), Nybom (1961) and Nybom & Koch (1965). Another early contribution deserving special attention is the Ph.D. thesis on mutagenesis in potato by Heiken (1960).

In the years afterwards, some meetings to discuss the methodology and potential merits of mutation breeding in vegetatively propagated crops were convened by the Joint FAO/IAEA Division of the IAEA in Vienna and reports on these meetings were published by the IAEA (Anon., 1973a; 1975a; 1976b and 1982a). The proceedings of a conference on 'Induced Mutations and Chimera in Woody Plants', organized in 1973 by the Institute of Radiation Breeding NIAS MAF in Ohmiya-machi, Japan, were published in the series Gamma Field Symposia (Anon., 1973b).

The first comprehensive work (with more than 1200 references) regarding applied mutation breeding for vegetatively propagated crops was published by Broertjes & van Harten (1978). At that stage the number of known mutant cultivars for this category was about 150 against about 40 in 1965. In an updated and extended version of the aforementioned book (Broertjes & van Harten, 1988) the corresponding list of induced mutant cultivars already contained more than 300 accessions. In addition, it was mentioned that in many cases induced mutant cultivars – in particular in ornamentals like chrysanthemum – were not identified, reported or officially registered anymore as such.

In a much quoted overview on induced mutant cultivars by Micke et al. (1990), in which all such cultivars known at the FAO/IAEA offices in Vienna until the end of 1989 were included, 40 mutant cultivars for fruit crops which can be vegetatively propagated, 409 for ornamental crops and 29 for other vegetatively propagated crops were mentioned. The authors also add that certainly more mutant cultivars have been produced which remained unnoticed to them. As was mentioned before, this in particular concerns mutant cultivars obtained by private commercial breeders who don't see any benefit in revealing the exact origin of their new cultivars.

In addition to the mentioned publications on mutation breeding for vegetatively propagated crops, reference can be made to some shorter, general chapters on the same subject, for instance by Donini (1975), Broertjes & van Harten (1987) and van Harten & Broertjes (1989). Other publications on mutation breeding for specific (groups of) vegetatively propagated crops were produced, for instance, by Constantin (1984) about the potential of in vitro mutation breeding for vegetatively propagated crops, Donini (1976a,b) and Lapins (1973; 1983) for fruit trees, by van Harten (1978) for potato, and for vegetatively propagated crops in the Netherlands (van Harten, 1982), by Kukimura (1986) and by the IAEA for root and tuber crops (Anon., 1987) and by Lacey & Campbell (1987) for apple. More recently Spiegel-Roy (1990) published in the IAEA series Mutation Breeding Review an overview on mutation breeding in fruit trees with a list of 87 selected most relevant references on this subject. This list, of course, is far from exhaustive and many more contributions have been published on mutation breeding and related subjects in various vegetatively propagated crops, often with special attention to the significant role of chimerism in mutation breeding.

When and how to start mutation breeding in vegetatively propagated crops
As was pointed out already in Chapter 1, a plant breeder, before deciding to try mutation breeding as a possible solution to his breeding problems, should well consider the pros and cons of this method in comparison to conventional breeding approaches, as well as to more recent methods of genetic engineering. Which method, ultimately, should give best prospects, largely depends on the target crop and on the breeding goal(s).

In order to facilitate this decision making and designing a sound, all comprehensive project from the beginning, Broertjes & van Harten (1978; 1988) have formulated a check list of questions to which the breeder should try to find answers. Depending on how far the breeder is already acquainted with his crop and the available literature, such answers may be given more or less easily. For some questions it may not be possible at all to find a proper answer. Answers to some more general questions are needed irrespective of the kind of crop and the preferred breeding method, but for mutation breeding in vegetatively propagated crops some points require special attention. Such points in particular deal with the choice of the starting material – from a genetic point of view as well regarding the plant organ type – and the way to handle the commonly occurring chimerism. It may be useful to emphasize here again that chimerism has negative as well as positive sides for plant breeding.

The most important questions to be raised are as follows.

1. For which traits is variation wished?
2. What is known about the genetics of the trait(s) (dominant/recessive; how many genes; linkages)?
3. What is known of the crop and, more specifically, of the cultivar to be improved (ancestry, degree of heterozygosity; ploidy level; ways of propagation)? Do spontaneous mutations occur? What would be – from a genetic point of view – the best starting material? Differences between cultivars?
4. Why mutation breeding? Is it probably easier, faster or cheaper than other methods?
5. What kind of plant material should be preferred for mutagenic treatment using conventional ways of vegetative propagation or *in vitro* methods (tubers, roots, leaves or leaf parts, single cells, etc.)? What is known in literature about methods to handle chimerism (e.g. adventitious bud methods)?
6. Which mutagenic treatment and under which conditions?
7. How should the material be handled after treatment?
8. Which selection methods are available or should be specifically developed?
9. How should the selected mutants be propagated true-to-type and what further actions are to be taken to reach the stage of an officially released or registered (mutant) cultivar?

It should be stressed once again that the actual breeding of new mutant cultivars, whenever possible, should be performed by trained plant breeders and not by scientists who have their roots in more fundamental sciences. Bio-molecular scientists, phytopathologists, etc., have their specific skills which can be extremely useful for developing new research ideas, new screening methods, etc. but, usually, their knowledge of crop species and growing conditions is very limited and, moreover, they lack the expertise of the experienced plant breeder

to develop new cultivars which indeed do meet the wishes of growers and consumers.

7.3.2. The starting material

Which plant parts?

In Chapter 5 an inventory was made of the vegetative plant parts, either known already in nature (tubers, bulbs, rhizomes, apomictic seeds, etc.) or produced for propagation 'in vivo' or in vitro by man (e.g. stem or leaf cuttings, shoot tip cultures, callus tissue, protoplasts, etc.) which could be used in mutation breeding projects. Recently, Micke & Donini (1993) also presented a list of vegetative plant material used in mutation experiments for a range of vegetatively propagated crops, including for instance tubers; dormant buds, rhizomes and graftwood; dormant shoots; single-budded sets; cuttings; nodal stems; and different kinds of *in vitro* material like shoot tips, leaf rachis and petioles, leaf or leaflet-blades, pedicel segments, ovular callus and pollen mother cells.

The actual choice of the most suitable material for each specific project depends on the different options that are available for a plant species and on the goal of the mutagenic treatment. When, for instance, the main target is to produce a wider range of genetic variation for different traits in a crop or in a specific cultivar, in principle all kinds of propagules could be used by the plant breeder. It is, in this situation, not necessary to pay special attention to the genetic stability of the plant material during the mutation breeding procedure and 'somaclonal variation' would be welcome as well. If, on the other hand, a breeder wishes to improve one single trait (e.g., flower colour or fruit flesh colour) or would like to obtain thornless branches in a particular cultivar of which all other traits should remain unchanged, plant parts like cuttings with existing buds should be used which, even during *in vitro* culture, would assure a stable, true-to-type offspring.

Genetic constitution

Another important point to consider is the genetic constitution of the starting material. Theoretically, the use of genetically heterozygous starting material (*Aa* for diploids and preferably so-called simplex-types: *Aaa* and *Aaaa* for triploids and tetraploids, respectively) would be very attractive, as most mutations go from dominant towards recessive. Such heterozygous genotypes indeed may be frequent in vegetatively propagated crops as a result of accumulated spontaneous mutations and occasional outcrossing – in particular in cultivars of crops that have been cultivated already for many years. However, unless it would be possible to perform test-crosses, it is in most cases not possible to distinguish between homozygous and the different heterozygous genotypes with a dominant expression. When dominant mutations are desired, homozygous recessive starting material (*aa*, *aaa*, etc.) for the trait concerned would be the logical choice. (N.B. One could however also imagine a new dominant mutant allele replacing a dominant 'wild-type' allele.) Dominant mutations are rare, but at least for visible traits can be detected easily in vegetatively propagated plants. Van Harten, Bouter & Schut (1973), for instance, described the analysis of a radiation-induced dominant mutation for leaf shape ('ivy leaf') in the (tetraploid) potato cultivar 'Burmania'.

The ploidy level of the starting material is important as well. In Chapter 4 reference was made to the many studies on radiosensitivity performed in particular by the group of Sparrow at Brookhaven National Laboratory in relation to the Interphase Chromosome Volume (ICV) for a range of plant species. Conger *et al.* (1982) reported that an increased genome size (= higher ploidy level) in plant species within one genus leads to a higher radiosensitivity, but that this effect is more or less compensated by the genetic redundancy as a result of the presence of more than two copies of every chromosome. This subject has been briefly discussed already in Chapters 3 and 6 and we will not repeat this discussion here, but it must be kept in mind that radiosensitivity is not strictly correlated to mutation frequency per locus per dose, let alone to the induction of valuable mutations in vital, fertile plants.

Although the genetic redundancy in polyploids may allow aberrations of chromosomes without loss of viability of the plant and, hence, it may be advantageous to the breeder to induce such mutations in polyploids – provided, of course, that he is interested in such kind of mutations – it is general experience that autopolyploids, but also certain types of allopolyploids, exhibit lower recoverable mutation frequencies than diploids. Studies on this subject were performed in particular for some seed propagated crops and some results and references have been given in Chapter 6.

By contrast, in vegetatively propagated crops, a few examples are known where higher numbers of mutants were observed in cultivars with higher ploidy levels. Broertjes (1976b), for instance, reported for the ornamental crop *Achimenes* sp. – which belongs to the family of the *Gesneriaceae* – very high numbers of mutants in the autotetraploid cv. Tango, a successful cultivar derived by colchicine treatment from the diploid cv. Tarantella. For the induction of mutations in both cultivars use was made of freshly detached leaves which were irradiated with 20–40 Gy of X-rays or 7.5–20 Gy of

fast neutrons. The irradiated leaves were planted in a peaty soil to produce adventitious rhizomes. In total 570 diploid and 961 tetraploid plants from adventitious origin were checked for mutations. The 'overall mutant frequency' (referring to different mutant types for plant habit and foliage, and to different categories of flower mutants) observed for the tetraploid plants of cv. Tango was 20 times (for X-rays) and 40 times (for fast neutrons), respectively, higher than the mutant frequency observed in the corresponding diploid cv. Tarantella. A possible explanation – also based on the higher frequency after neutron treatment – could be that most mutations refer to gross chromosomal aberrations which could not survive intrasomatic competition in diploid adventitious regeneration.

Broertjes (loc.cit.) further pointed out that some ornamentals (e.g. the octo-alloploid (amphidiploid) *Dahlia*, hexaploid *Chrysanthemum*, tetraploid *Gladiolus* and triploid *Begonia*) are known to produce much variation after irradiation, whereas other species (e.g. tetraploid *Freesia*) produce almost no mutants at all. One point that may be of importance in this respect is the way of origin of polyploid plant species, e.g. as auto- or as allo-polyploid. In fact, most natural polyploids may be alloploids.

It seems that the effect of the ploidy level of the mutagenically treated material on radiosensitivity and mutation frequency in vegetatively propagated crops cannot be generalized and one must be careful when drawing conclusions about optimal treatments from previous experiments with other crops, and with cultivars of different origin of the same crop.

The position of the mutated cell

For the present discussion a mutated cell can be described as a cell carrying for instance one (or more?) recessive mutant alleles in a nuclear gene. Because a mutation initially is a single cell event, and as not all plant cells in a multicellular organism act similarly after mutagenic treatment, it is not easy to predict the fate of such a mutated cell (Dommergues, 1964, and many later references). In previous sections it has been pointed out that whether an induced mutation has a chance to become manifest and recognizable (if dominant or homozygous recessive) or not, may be determined by the original 'strategical position' of the mutated cell. In addition, whether a mutated gene will be able to express itself will depend upon the subsequent differentiation and functional specialization of the mutant tissue.

When mutations are induced 'on the wrong place', e.g. when a mutation for the trait 'thornless epidermis' arises in a cell of an inner cell layer of thorny genotypes of blackberry (genus *Rubus*) or gooseberry (genus *Ribes*), this mutation cannot be expressed and the plant remains phenotypically 'thorned'. By vegetative propagation the potentially useful mutation for 'thornless' may be transferred to many consecutive vegetative generations but may remain unobserved until it becomes 'uncovered', either by accident or by applying special methods which have been described already in previous sections. Accordingly, when a mutation in a gene controlling flower colour is induced in a cell of the apical meristem, this mutation could become expressed only when the mutated cell later participates in the development of the flower petals.

From these examples one should deduce that in vegetatively propagated crops, throughout the years, many spontaneous mutations may also have accumulated, which remained unnoticed for many generations when present in `the wrong place'. Such plants, of course, are chimeric. In the past, many publications have appeared in which the induction of useful mutations was reported, whereas in fact only an already present but so far undetected mutant gene had become uncovered and been allowed to express itself. High mutation frequencies for a certain trait should make the breeder suspicious whether the plant material that was treated mutagenically, was not periclinal chimeric already for that particular trait. This problem, of course, is not encountered in seed propagated crops.

When a mutation occurs for a visible trait like leaf or flower colour, it shows itself mostly in vegetative tissue as a stripe (often – but confusingly – called 'mutated sector') representing the (somatic) offspring of the mutated cell. The size of the stripe or patch is related not only to the moment in the ontogeny of the plant at which the mutation was induced, but also to the number and pattern of subsequent cell divisions and the dimension of cell enlargement. Narrow stripes are of course more difficult to observe than wide stripes. Mutations which produce no directly visible effects, like mutations for genes which govern yield, protein content or regeneration ability, can of course not be seen as 'stripes'. Methods are required to detect valuable as well as unwanted hidden genetic variation in vegetatively propagated crops in order to utilize or eliminate it. This subject has been discussed in detail for many crops by Broertjes & van Harten, 1978; 1988).

Mutations occurring in apical or axillary shoot meristems are relatively easy to exploit. Mutations in apical initial cells, which produce a lineage of daughter cells, will result in a lineage of mutated cells and eventually in large mutated areas. A mutated cell lineage may contribute to newly formed shoots, side branches, fruits or leaves. A plant that is completely mutated for a specific trait could occur only if the whole plant originated from a single mutated cell, which is unlikely in nature in this early stage. Only in exceptional cases a whole new side branch may entirely consist of mutated tissue.

When mutations occur in cells outside existing shoot meristems, such cells must be stimulated to develop into shoots in order to get a chance for the mutation to be maintained. Callus in which mutations have been induced has to be stimulated to produce shoots and, consecutively, whole plants must be regenerated before the value of such mutations can be assessed. Attention will be paid as well to this subject later in this chapter.

7.3.3. Mutagenic treatments

General remarks

When mutagenic treatments are to be applied, the general advice is to perform pilot experiments with several genotypes and for the chosen treatment (radiation, chemical mutagen) with a rather wide range of doses and various treatment conditions, in order to determine which material is to be preferred and which treatment(s) and conditions are expected to produce optimal results. In practice, the degree of growth reduction is often taken as a measure for the effectiveness of a treatment. However, it should be remembered that growth reduction or survival are only rough indicators of damage and really do not inform us about the expected mutation rates of particular traits.

In the definitive experiment it is always advisable to look for one or a few specific improvements only (unless the aim is to create a new large 'gene pool' for many traits at the same time), to start from sufficiently large numbers of cuttings, leaves, tubers, etc., from the best available starting material (see Section 7.3.2) and to apply or develop efficient methods to screen large populations for the desired mutation(s). Selection should always imply as well a thorough screening to check that all important crop- and cultivar-specific traits other than the 'target trait(s)' for which mutations are desired have remained basically unchanged. Some small negative additional changes, e.g. when a favourable mutation for flower colour is accompanied by a slightly taller plant size, in many cases do not make the mutant plant useless!

Starting from various kinds of ('in vivo' or *in vitro*) plant material, most mutation work is performed with X-rays or gamma rays. Micke & Donini (1993) presented a table with recommended doses for treatment of a range of vegetatively propagated crops, derived from literature. For vegetative plant parts 'in vivo' in most cases doses from 20–80 Gy are advised; for *in vitro* plant material the advised doses normally range from 8 to 35 Gy. Treatments with chemical mutagens can be performed as well but, as will be discussed further on in this section, are not much in favour and, so far, also have not been very successful for this category of plants.

Irradiation of seeds – in the case of facultatively vegetatively propagated plants – also would be an option, but this approach may lead to other problems, such as a long juvenile phase in fruit trees (Spiegel-Roy, 1990), a low level of seed production and the loss of the cultivar type due to generative segregation of the heterozygous clone. For this group of species irradiation of pollen would be possible as well and would avoid chimerism, but because of the heterozygosity in most vegetatively propagated crops, generative segregation will mask the induced mutations and thus much further breeding work will be required.

The radiation doses that are applied in practice normally are in the range between 5 and 100 Gy (often at dose rates from about 2–20 Gy. min^{-1}) and depend on plant species, cultivar, kind of starting material (plant part, age, dormancy). Although chronic irradiation can be applied for instance to (fruit)trees, in most cases acute irradiation is performed. Most workers prefer the use of relatively low doses. This approach, of course, results in lower mutation frequencies than heavier treatments but, on the other hand, the advantage of administering low doses is that in this way also less chromosomal damage and other negative side effects occur. Such unwanted effects in vegetatively propagated crops are difficult to get rid of because of the absence of a meiotic phase that could sort out such aberrations.

One specific feature of irradiation of vegetatively propagated crops should be mentioned here and will be treated more in detail for fruit trees in Section 7.4.6. In practice, mutants are rather often derived from shoots which originated from buds that were situated somewhat below the apical meristematic zone of the M1 plants. This can be explained as follows. When for instance cuttings (often dormant bud wood) taken from fruit trees are irradiated, usually a main apical bud and primordia for a number of axillary (side) buds are present. Due to the apical dominance exerted by the apical bud, the adjacent lower situated axillary buds may have remained dormant. Irradiation of such shoots with a high enough dose may severely damage or even kill the apical bud tissue and in that situation axillary buds may take over and sometimes new, accessory or adventitious buds may arise. If the lower buds truly follow the organizational pattern of the main bud, a chimeric situation will result from the mutagenic treatment. However, the number of initial cells in axillary buds may be lower

and, in that case, the mutated area may be relatively larger. Moreover, young axillary buds for some plant species may not completely follow the layered structure of the main bud. This should be further investigated for instance by irradiation and other methods applied to mutants with a well-known periclinal chimeric structure (e.g. for leaf colour or for ploidy level) and by studying the resulting plants. Methods have been discussed in Section 7.2 and reference is made for instance to a range of publications on this subject by the Bergann & Bergann and associates.

Another complication that was mentioned before, is that irradiation of buds may induce anomalous histogenic processes, resulting in the loss of integrity of the different histogenic layers in the meristem and in different types of regeneration from various tissues (see also Section 7.2.4). As a consequence, for instance, a mutation for a flower colour gene, present in L-2, can be transferred by irradiation to L-1 and, as such, become visible. Accordingly, chlorophyll-deficient mutations that were unnoticeably present only in L-1, may be transferred by irradiation to L-2 and become expressed as altered green leaf colour or – after seed propagation – become expressed through the germ cells in the sexual offspring.

It should be remembered that a higher radiation dose reduces the number of surviving cells, which increases the chance of recovering solid mutants. The disadvantage of applying higher doses of irradiation, however, is that multiple mutations are induced and that a rapid increase of the level of (non-genetic?) radiation damage can be noted as well which may seriously hamper the normal functioning of cells and the whole plant. Under vegetative propagation, such plants may remain lagging behind for several generations.

Mutagenic treatments 'in vivo'
In order to give a general idea of mutagenic treatments that are performed in practice, some figures are given here for a few vegetatively propagated crops. Mutagenic treatment of *Dendrathema* (better known as *Chrysanthemum*), in particular of the hexaploid *D. morifolium* ($2n = 6x = 54$) as a source of cut flowers, is mostly performed by irradiating cuttings – preferably several hundreds of a selected genotype – with about 15 Gy of X- or gamma rays.

For tulip (*Tulipa*), a mostly diploid ($2n = 2x = 24$) ornamental bulb crop, in the Netherlands series of at least 100 bulbs of a uniform size (e.g. diameter 12 cm) are irradiated with doses between 2.5 and 4.5 Gy during summer time. The question when to irradiate – directly after harvest when the bulblets are in the youngest possible stage of development, or some months later, aiming at the cells in the bulblets which later will form the apices for the secondary bulblets – is not so easy to answer, and also depends on the size of the bulbs. Details have been discussed by Broertjes & van Harten (1988). Selection should be continued for at least four years to allow a mutated cell to grow out and to produce either a solidly mutated bulb or at least a larger, visibly mutated (flower) zone or 'sector' that will be expressed in first or second generation bulblets.

For apples (*Malus pumila*), which are mostly diploid ($2n = 2x = 34$) dormant scions are irradiated with doses of 30–70 Gy. Again at least several hundreds of scions are irradiated at a time. For small fruits (currants, brambles) the same or lower doses are applied to various plant parts. For sugarcane (*Saccharum* sp.), a crop that is highly polyploid (octoploid or higher) with 80 or more chromosomes, often stem cuttings are irradiated acutely with 20–60 Gy X- or gamma rays. More examples of treatments for a wide range of vegetatively propagated crops can be found in Broertjes & van Harten (1988). Some additional comments on optimal doses of radiation, survival rates, etc., will be given in the next section on mutation programmes in which an *in vitro* stage is involved.

Treatments with chemical mutagens, as was said before, are not often performed with larger plant parts like cuttings, tubers and bulbs, mainly because of technical problems with uptake and penetration of the chemical in the plant tissue, the dosimetry and the uniform distribution of the mutagens within the target meristems. It is, as a consequence, difficult to perform experiments that can be fully reproduced. When the treated material is of small size, as is the case with treatments of small explants like excised shoot tips before *in vitro* culture, the problems faced may be smaller.

The chemical mutagens applied, so far, in vegetatively propagated crops in most cases are the commonly used ones like EMS and the nitroso-compounds nitrosoethyl urea and nitrosomethyl urea which have been described in Chapter 4. The use of suitable 'carriers' or agents which improve uptake or penetration of chemical mutagens, like dimethyl sulphoxide (DMSO) may help to increase the effectiveness of chemical mutagens. It may be repeated here that, because of the health hazard, care should always be taken in handling chemical mutagens, e.g. by making use of fumigation hoods, protecting clothing and gloves.

Treatments with chemical mutagens 'in vivo' could be performed for instance by submersion of the petioles of detached leaves in a solution of mutagen, by depositing a wad of cotton with the chemical mutagen on axillary buds, by injecting the mutagen by means of a syringe into nodal areas, etc. These methods were successfully applied for carnation (*Dianthus caryophyllus*), African violet (*Saintpaulia ionantha*), apple (*Malus* sp.), roses and some other crops as well.

As a consequence of the aforementioned technical difficulties, the number of mutant cultivars of vegetatively propagated crops that resulted from chemical treatments has remained very low. Micke *et al.* (1990) as well as Maluszynski *et al.* (1992) in their inventories of induced mutant cultivars, could – out of a total of 523 mutant cultivars in vegetatively propagated crops – mention only 14 cases of chemically induced mutant cultivars. From these cases, 12 referred to ornamentals, one to apple (*Malus* sp., cv. Belrène, with improved earliness, resulting from EMS treatment and released in 1970 in France) and one to mulberry (*Morus alba*, mutant S54, with higher leaf yield – important for raising silk worms – and also resulting from EMS treatment; released in 1974 in India). To give one example for an ornamental crop: for carnation three commercial mutant cultivars were released in 1967 in the former German Democratic Republic after treating leaf axils with 2.5% EMS (see Hentrich & Glawe, 1982).

Colchicine-induced polyploids are often considered as a special category of induced mutants. They are of particular interest in vegetatively propagated crops as the size of vegetative parts (flowers, stems, leaves, storage organs) may be increased whereas the common negative effects on the sexual phase, such as increased sterility, are less harmful in this group of crops. Colchicine is said to induce also gene mutations. The aforementioned IAEA database (Maluszynski *et al.*, 1992) includes six examples of mutant cultivars for ornamentals, obtained after treatment with colchicine with other changes than just polyploidy. In five of these cases, treatments with colchicine were combined with X-rays or gamma rays. Whether colchicine indeed induces gene mutations can not be ascertained from these results as only one mutant cultivar (a chrysanthemum with improved flower colour, released in 1985 in India) resulted from colchicine treatment only. Of course many more cultivars with considerably enlarged vegetative organs have been produced in various crops by treatments with colchicine but genome mutations principally are not included in the inventories by the IAEA.

Mutagenic treatments in vitro

As this subject has been discussed already in Chapter 5, only some general points are summarized and a few additional examples will be presented at this place. Earlier in this chapter mention was made to a specific contribution assessing the potential of *in vitro* mutation breeding in vegetatively propagated crops in a little-quoted IAEA publication of a regional seminar in Latin America by Constantin (1984).

Methods of treatment of *in vitro* material, in particular when radiation is applied, principally, do not differ from '*in vivo*' treatments and the range of doses used is about the same, taking of course into account the kind of tissue used. Often plantlets are irradiated from which afterwards explants are taken. These explants are used for *in vitro* culture, from which – with or without intermediate callus phase – again new plants arise. This procedure may result in a lower level of chimerism of the M_1 plants than when a M_1 generation from seed would be grown, or when for instance tubers would have been produced directly from irradiated plantlets.

Application of chemical mutagens is often claimed to be particularly attractive for *in vitro* material. A short list of suitable chemical mutagens for *in vitro* treatments, including EMS and some nitroso-compounds, was presented by Mathews & Bhatia (1983) in the FAO/IAEA *Mutation Breeding Newsletter* 22 (see also Chapter 4).

It is often stated nowadays that mutagenic treatments of *in vitro* material with radiation or chemical mutagens should not be necessary, as the frequency and spectrum of genetic changes among the 'somaclonal variation' (see Chapter 5) should be high enough. It is not easy to determine whether this view is correct. Most mutagen induced variation as well as 'somaclonal variation' is undirectional, but certain mutagenic agents may favour the induction of specific categories of mutations. For instance, sodium-azide and probably UV-radiation (which may be a good option for some *in vitro* cultures like thin layer cell suspensions) may induce relatively many gene mutations whereas colchicine treatments induce almost exclusively genome mutations.

Because of the unpredictability of 'somaclonal variation' and the instability that is often observed in plantlets derived from *in vitro* cultures, when relatively unorganized systems like cell suspensions and callus cultures are used, it may be better to start *in vitro* cultures from explant material with high embryogenic capacity, thus avoiding the phases of disorganization which favour 'spontaneous' mutations and instead apply a mutagenic treatment. An option for instance would be to irradiate nodal cuttings which contain axillary buds. Regrowth in this case normally occurs from axillary buds which, predominantly, have a stable, layered structure. After some time the resulting shoots can be cut off, transferred to specific rooting media and, at a later stage, be transferred to soil or substrate.

An alternative method would be to take shoot-tips

from plants that are either grown in the field or greenhouse or raised *in vitro*. They could be irradiated before the *in vitro* stage or, for instance, treated with 10–100 Gy in Petri dishes or glass jars, but a callus phase and, thus, the occurrence of 'somaclonal variation' can not be avoided.

An example, not mentioned before in Chapter 5, refers to *in vitro* radiation of detached leaflets from one-month-old shoots of pear (*Pyrus communis*) by Pinet-Leblay, Turpin & Chevreau (1992) with the aim of inducing reduced susceptibility to fire-blight (*Erwinia amylovora*). The leaflets were plated on agar in Petri dishes and, subsequently, treated with up to 70 Gy of gamma rays at a dose rate of 0.04 Gy. min^{-1}, or with up to 1000 J. m^{-2} of UV-light (at 257 nm wavelength emitted by a 30 W germicidal lamp with an energy output of 3 J. m^{-2}. s^{-1}).

In the Annual Report of the National Institute of Agrobiological Resources in Ibaraki, Japan (Anon., 1994b) it was reported for pineapple (*Ananas comosus*), a crop commonly propagated by suckers, that only very few spontaneous mutations occur and that methods had been developed to induce mutations artificially by irradiating leaf cuttings and selecting among regenerants derived from tissue culture. The propagation rate from leaf cuttings ('leaf trimming') and tissue culture was 30x and 500x, respectively, from that by suckers. Effects of acute, semi-acute and chronic irradiation treatments were compared. It was found that suitable doses for mutation induction were below LD 50 and that the dose at which callus induction was blocked approximately corresponded with the dose at which meristems of the irradiated plant material did not survive anymore. It was mentioned that combining tissue culture with an appropriate radiation treatment resulted in efficient mutation induction and good regeneration. Details about suitable doses of irradiation, unfortunately, were not given. Mention was made of the induction of trichomeless mutants, which may be valuable in avoiding respiratory diseases for farm workers.

For quite a few other cases, doses and dose rates for irradiation and concentrations for chemical mutagens were given already in Chapter 5. It was also discussed at that point how the survival rate or the number of regenerated plantlets should be taken into account when determining the most suitable dose. Many researchers, nowadays, prefer a higher survival rate than LD 50, e.g. LD 30, which results in a higher number of regenerated plants but, of course, a lower overall mutation frequency as well (Walther & Sauer, 1985).

It is also possible to apply mutagenic treatments to plant parts or tissues in which the induction of adventitious buds *in vitro*, preferably without or with only a very short callus phase prior to adventitious bud formation, is stimulated. It is common experience that a negative correlation exists between the length of a callus phase *in vitro* and the production of adventitious buds. In other words: adventitious bud formation is favoured by culturing conditions that are not favourable for a (prolonged) callus phase. If adventitious buds give rise to plantlets which trace back to a single mutated cell or to a few cells only, this approach would be an efficient method to prevent or limit chimerism and to obtain solid mutants. The adventitious shoots could be cut off and transferred to a rooting medium and afterwards to soil. Single cell cultures like pollen grains or naked protoplasts should also give rise to non-chimeric mutant plants, if such plantlets would arise directly, i.e. without an intermediate callus phase, but this usually is not the case.

Yet another approach, much used in more fundamental studies on *in vitro* mutagenesis, is to irradiate embryogenic cell suspension cultures consisting of single cells or of small cell clusters. Regeneration and organogenesis from these cells will be completely inhibited above a certain dose, but this upper limit strongly depends on species and genotype. It has been found, for instance, that cell suspensions from carrot (*Daucus carota*) can tolerate up to 200 Gy, which is far more than is the case for 'normal' plants. When cell suspensions are treated, large populations can be handled under controlled conditions. This may even allow the detection of dominant mutations that occur only at very low frequencies. Recessive mutations would not be expressed unless the culture started with haploid material.

The use of cell suspensions and other types of undifferentiated *in vitro* culture material in mutation programmes, on the other hand, is hampered by several drawbacks which may greatly reduce the effectiveness. To mention some of the main complicating factors once again: not all variation observed is genetic in nature, not all genes express under single cell conditions and therefore not all traits can be selected for *in vitro*, not all genotypes can be properly regenerated *in vitro*, etc. The availability of suitable markers, which enable selection *in vitro* for traits which become fully expressed only in adult plants, would give an important stimulus to the application of *in vitro* methods aiming at the genetic improvement of crop plants and a search for such markers has been going on for quite some time. Recently developed molecular methods may become of great importance in this respect.

In conclusion: the answer to the question whether, from a practical point of view *in vitro* mutagenesis of

stable structures, like shoot tips or nodal cuttings with axillary buds, with radiation or chemical mutagens should be preferred to variation arising in unstable cultures, is not always easy to give, although this author, in general, is inclined to prefer the first approach.

7.3.4. Further steps

When a mutation breeding programme is started, the breeder must already have made up his mind about the further steps that, following the mutagenic treatment, should be taken in later stages of the breeding programme, for instance with respect to further handling of the mutagenically treated material and the selection procedures that should be applied. A number of questions to reflect upon in an early stage were mentioned in Section 7.3.1. Not all answers are always easily found and in most cases a profound knowledge of the crop is required to find adequate solutions for various problems. Cooperation between crop specialists and mutation breeding specialists in this respect may be indispensable.

For vegetatively propagated crops an important practical point to consider is always whether – as is most often the case – solid mutants are required or periclinal chimeras, which are often much more easy to obtain, may do as well. In order to illustrate this point once again: for resistance against diseases or pests, solid mutants are to be preferred – unless there is a negative pleiotropy connected with it, in which case, for instance, the central core may be kept free from the resistance gene – whereas for flower colour mutations (chrysanthemum, carnation) a periclinal chimeric situation may be sufficiently stable and, hence, quite acceptable in practice.

As has been pointed out before on several occasions, it has to be anticipated that, after a mutagenic treatment of a multicellular meristematic plant part, the mutated area, most often, will consist of only a narrow stripe or cell-lineage which, at that stage, only in a few cases can be recognized as such. The most important task for the mutation breeder is then to increase the size of the mutated area. How this problem could be tackled has been previously discussed. It is, in any case, not advisable to look for mutations already in the so-called vM_1 (the first vegetative generation after mutagenic treatment of vegetative (v) material in the vM_0) as mutations may remain undetectable because of the small size of the mutated area and also because the desired genetic changes may be masked in vM_1 either by influence from the normal cellular neighbourhood, or by the aberrant plant growth as a result of the general damage caused by the mutagenic treatment.

If useful mutants are found (and confirmed) in consecutive generations, the common procedures for breeding vegetatively propagated crops should be followed. Breeding of vegetatively propagated crops (see also Chapter 1) often starts with the assessment of known cultivars and material collected in centres of origin or from gene banks. For facultative vegetatively propagated crops an alternative approach would be to make – when possible – crosses between promising parent material and to look under different conditions in the F_1 and later generations for attractive phenotypes. The assessment of plants resulting from mutagenic treatments should take place in the same way. In that case, however, it should be kept in mind that part of the general damage present in vM_1 may be also transferred to later vegetative generations.

New mutants, of course, also must be tested for several years and under different conditions in order to assess the stability of the improved traits and the behaviour of the new, putative mutant cultivar before being commercialized. Many so-called 'very promising' mutants are not heard of any more after the first enthusiastic reports. Several reasons may account for this, but often the 'promising mutant' was not checked adequately by crop specialists for negative traits that may be present as well in the new mutant. In addition, mutation breeders, in particular mutation researchers without a background in plant breeding, are not always sufficiently aware of the amount of time required and the complexity of the procedure to get new mutant cultivars tested, registered and commercialized.

Finally, when a new useful mutant trait is obtained, it is always good policy to look for additional mutants in the population or to repeat the process and to try and induce more mutants for that specific trait in the same or similar starting material. Even small differences in the genetic background in which a mutant trait has been induced, may have considerable consequences, for instance for accompanying pleiotropic traits and for modifying pleiotropic effects of the mutated gene and, hence, for the future commercial value of a mutant.

7.3.5. Breeders' rights and mutation breeding

In an increasing number of countries cultivars are protected by plant breeder's rights. In 1968 the International Union for the Protection of New Varieties (UPOV) came into being with the ratification by four countries of a draft dating back to 1961 in which the basis for Plant Breeders' Rights was established. In 1996 there were 31 countries worldwide which adhered to the UPOV convention.

The 1991 act of the UPOV Convention introduced (in article 14, paragraph 15) the principle of <u>essential derivation</u> ('essentially derived varieties' or EDVs). Breeders in this way can establish, if necessary, via court procedures, whether a relation of 'essential derivation' exists between various cultivars. If their case should be accepted this would imply that the breeder of the original cultivar could claim his share of the financial remuneration. For instance, if in an accepted cultivar a gene for resistance to a specific disease has been introduced, followed by the release of the 'improved product' as a new cultivar, the breeder of the new cultivar would have to share profit with the breeder of the original cultivar. The consequences can be very complicated.

This concept may be of particular importance for mutation breeding in vegetatively propagated ornamentals, where a – genetically speaking – small but economically important change (like in flower colour) may result in a new cultivar. So far it could be commercialized without any compensation to the breeder of the original cultivar in which the mutant had been produced or found. The UPOV principle of essential derivation may apply not only to mutants but also to 'somaclonal' variation, backcrossings, transgenic plants, etc., and other products of genetic engineering. It is to be expected that future claims will certainly be made for spontaneous or induced flower colour mutants, but claims for other traits, undoubtedly, will arise as well.

Despite the fact that the intention of the aforementioned UPOV statement is clear, it is still difficult to determine what is 'essential' and, also, what would be the 'minimum distance' between two 'clearly distinguishable' varieties, in particular when several qualitative or quantitative traits are involved. It is also not always clear who would be considered as owner of the original cultivar. It may be expected in this respect that, in an increasing number of cases, claims will be made by farmers or governments of developing countries for the use of ancient cultivars as germplasm.

The UPOV convention will only come into effect after governments of the participating countries have adapted their existing regulations with respect to variety registration and seed production. It appears that the discussion about breeders' rights may continue for quite some time as, in an increasing number of cases, plant breeders as well as companies in the field of biotechnology try to patent the results of their work (new cultivars or 'new' genes) as well as the applied methods (e.g. the use of the 'method of somaclonal variation' in a crop like tomato, or the use of the 'antisense method'). This undoubtedly will considerably affect relations in the 'world of plant breeding', for instance between research workers at research institutes sponsored by governments, private breeders of the original and 'improved' cultivar, owners of a patent, owners of collections, growers, etc. This subject is a minefield for breeders and a potential goldmine for lawyers.

7.4. Achievements of mutation breeding in vegetatively propagated crops

7.4.1. General remarks

As has been done in Chapter 6 for seed propagated crops, a number of examples will be presented in this section to illustrate what has been achieved from a practical point of view by applying mutagenic treatments to vegetatively propagated crops. General figures about numbers of mutant cultivars, groups of crops, mutagenic treatments, etc., have been presented already in Chapter 1 and it may suffice here to repeat that in 1992, some 523 mutant cultivars for vegetatively propagated crops were registered in the FAO/IAEA Mutant Varieties Database in Vienna (Maluszynski et al., 1992). More recently, for the group of ornamental and decorative plants only – which predominantly are vegetatively propagated – 484 mutant cultivars were registered in the aforesaid database (Maluszynski et al., 1995). In particular for ornamentals this number – undoubtedly – is a gross underestimation of the real number of mutant cultivars in vegetatively propagated crops, as was also mentioned already before (see also Broertjes & van Harten, 1988; in particular for chrysanthemum). Therefore figures published since 1985 may not be very realistic anymore. Anyhow, the results obtained show in particular that mutation induction has become an established and successful method of breeding new ornamental plant cultivars.

The most detailed information on methods and results of mutation breeding for vegetatively propagated crops, so far, has been published by Broertjes & van Harten (1978; 1988). Other major contributions on this subject have been produced as well, often under the aegis of activities organized by the IAEA, for instance as proceedings of conferences, panels or consultants' meetings, issues of the FAO/IAEA Mutation Breeding Review, reports on different meetings within the series Gamma Field Symposia in Japan, Ph.D. theses, chapters to more general books, etc. Only a few selected publications can be mentioned here. More references, in particular concerning some of the earliest publications on spontaneous mutations (bud sports), can be found in Chapter 2. To give a few additional examples, mention could be made here of the contributions on spontaneous (bud)

mutations in potato (Solanum tuberosum) by Dorst (1924) and Asseyeva (1927) and on bud mutations in horticultural crops by Shamel & Pomeroy (1936).

Some references to useful reports on induced mutations in (in particular) vegetatively propagated crops are given here in chronological order: Asseyeva & Blagovidova (1935) on induced mutations in potato; Bauer (1957) on the methodology of mutation breeding, with black currant (Ribes nigrum) as an example crop; Heiken (1960) also on potato; Nybom (1961) with an early general review (however without mentioning names of cultivars resulting from mutagenic treatments); Sigurbjörnsson & Micke (1974); van Harten (1978) on induced mutations in potato; Jagathesan (1982) on sugarcane in the report of a FAO/IAEA meeting on 'Induced Mutations in Vegetatively Propagated Plants' in 1980 in Coimbatore, India (Anon, 1982a); van Harten (1982) on results of mutation breeding in vegetatively propagated crops, in particular in ornamentals, in the Netherlands; Lapins (1983) on fruit trees; Donini & Micke (1984) with a list of mutant of mutant cultivars for fruit trees, presented during an IAEA regional seminar in 1982 in Latin America (Anon., 1984b); Langton (1986) with a general account; Anon. (1987) on improvement of root and tuber crops by mutations; Broertjes & van Harten (1987) in a volume on (genetic) improvement of vegetatively propagated crops (Abbot & Atkin, eds, 1987); Lacey & Campbell (1987) on mutant apples in the same volume; Spiegel-Roy (1990) on fruit trees; Datta (1991) on mutation breeding in floriculture; and Donini et al. (1991) on examples of mutation breeding of vegetatively propagated crops in Italy. It would not be difficult to further extend this list with several other publications.

An example that was discussed already in Chapter 5 should be briefly mentioned here. This case refers to the in vitro mutation work by Novak and associates on banana (Omar et al., 1989; Novak et al., 1990). From this programme an early flowering mutant from the Cavendish-type banana 'Grand Nain' (Musa sp.) resulted. Tan et al. (1993) reported about further investigations on this mutant in Malaysia in which material obtained by micropropagation, named 'Fatom-1' was planted on a commercial scale. This mutant was later renamed cv. Novaria. According to information from the IAEA in Vienna (van Zanten, personal communication), in 1997 only a relatively small area (about 150 ha) was planted with this mutant in Malaysia, where the total area for banana at that time could be estimated at about 36 000 ha.

According to a report by Ortiz & Vuylsteke (1996), a much bigger area, namely 67 000 ha, was planted in Cuba with a cultivar of the cooking banana (ABB group), called cv. Burro CEMSA, that should have resulted from mutagenic treatment. However, the contents of the original publication by Rodrígues Nodals *et al.* (1992) – to which Ortiz & Vuylsteke refer – does not substantiate this claim, but merely state that cv. Burro CEMSA had been used as starting material for investigating the potentials of inducing 'somaclonal variation'. Rodrígues Nodals *et al.*, continued by saying that from this work an interesting, small-sized 'somaclone', called 'Burro CEMSA Enano' had been obtained, but further details were not given.

In the following sections some selected examples of mutation breeding in vegetatively propagated crops are discussed. Significant commercial successes have been obtained in particular – but not exclusively – with mutants for ornamental crops. In Section 7.4.5, for instance, successful and large-scale mutation breeding is presented for the important cut flower chrysanthemum in the Netherlands but – be it in most cases on a more modest scale – similar accounts could be presented for other ornamentals like roses, the cut flowers carnation (*Dianthus caryophyllus*), Alstroemeria (Inca lily), the tuber and bulb ornamental crops *Dahlia variabilis* (garden dahlia) and tulip, the flowering pot plants *Begonia* sp., *Achimenes* and *Streptocarpus*.

Much mutation work on vegetatively propagated ornamentals has been performed in the Netherlands, but for some crops like *Begonia* sp., *Streptocarpus*, *Rhododendron simsii* and roses, most mutant cultivars have been produced in other countries like the USA, Germany, France, Belgium and India (for details see Broertjes & van Harten, 1988, or various IAEA reviews, presented in the FAO/IAEA *Mutation Breeding Newsletter* or as FAO/IAEA *Mutation Breeding Review*).

7.4.2. Some mutant cultivars from the early years of mutation breeding

Sigurbjörnsson & Micke (1973), in an appendix to the report of a panel meeting on vegetatively propagated plants in 1972 in Vienna, presented a first list of 39 mutant cultivars for ten different vegetatively propagated crop species. From the 34 mutant cultivars which referred to ornamental crops (rose 1x, carnation 2x, chrysanthemum 9x, dahlia 13x, *Streptocarpus* 5x, *Alstroemeria* 2x and *Azalea* 2x), 20 had been produced in the Netherlands. The five examples for non-ornamentals include three fruit crops (apple 1x, peach 1x and sweet cherry 3x). This list, however, was still far from complete and, for instance, did not yet include several of the earliest cases for this group of crops.

The earliest examples of officially released (or approved) mutant cultivars were later identified by

Broertjes & van Harten (1978; 1988) and had been published in 1949 and 1954 for the ornamental bulb crop tulip (*Tulipa* sp.) in the Netherlands by de Mol (or de Mol van Oud Loosdrecht as he later called himself). The first mutant, cv. Faraday, with an improved white, flushed salmon-pink flower colour, was obtained after mutagenic treatment with X-rays in 1936 of cv. Fantasy. A second tulip mutant, cv. Estella Rijnveld, with a red-flamed white flower colour and induced in cv. Red Champion, was released in 1954. De Mol, who also worked on other ornamental bulb crops like hyacinth (*Hyacinthus* sp.), published hundreds of publications on his radiation work, unfortunately often in Dutch and not in journals that are easily accessible (see for instance de Mol, 1953; 1956). A complete list of De Mol's publications can be found in Thamm (1956).

In the late 1950s and early 1960s several mutation breeding projects, in particular in ornamentals, started (see for instance Broertjes, 1968) and more mutant cultivars for ornamental crops were developed. The mutant cv. UConn White Sim No.1 of carnation (*Dianthus caryophyllus*) was introduced by G.A.L. Mehlquist in 1962 in the USA, after irradiation of rooted cuttings of cv. White Sim. Starting from 1962 several cultivars of the 'Izetka Köpenicker' type were produced in chrysanthemum after irradiation with 10–25 Gy of X-rays by H. Jank in the former German Democratic Republic and, also in the 1960s, several mutant cultivars were obtained in the Netherlands and various other countries in roses (*Rosa* sp.), dahlia (*Dahlia* sp.), azalea (*Azalea* or *Rhododendron*), streptocarpus (*Streptocarpus* sp.) and several other ornamentals.

The first commercial mutant cultivar in a fruit crop, a compact tree type in sweet cherry (*Prunus avium*), named cv. Compact Lambert, was released in 1964 by K.O. Lapins in Canada after irradiation of scions in 1958 with 40 Gy of X-rays (Lapins, 1963; 1965). This mutant was followed by other compact type mutants induced by the same worker. Direct mutant cultivars as well as indirect ones, i.e. after additional crossing work with a mutant, were obtained.

In Germany, Bauer in 1968 produced a mutant cultivar with strong erect habit from black currant (*Ribes nigrum*). The first mutant cultivar of peach (*Prunus persica*), cv. Magnif 135 with bigger and deeper coloured fruits, was released in 1968 in Argentina. The first commercial mutants of apple (*Malus pumila*) were produced in the early 1970s in France and Canada (for references see Broertjes & van Harten, 1988).

Early examples of commercial mutants in sugarcane (*Saccharum* sp.) with improved resistance to red rot (*Physalospora tucumanensis*), obtained after gamma irradiation of buds, were released in 1966 and 1967 by the Sugarcane Breeding Institute in Coimbatore, India. Several 'promising' mutants with various improved traits like glabrous leaf sheath, non-flowering, high yield and improved resistance against various other diseases (e.g., smut, caused by *Ustilago scitaminea*) were reported to be studied in comparative trials for four or five years. More details and references concerning mutants for this crop can be found for instance in Jagathesan (1976; 1979; 1982), Broertjes & van Harten (1988) and Maluszynski *et al.* (1992).

7.4.3. Mutation breeding of grapefruit
During the 1990 FAO/IAEA Symposium on '*Mutation Breeding for Crop Improvement*' (Anon., 1991b) in Vienna, Austria, achievements of practical mutation breeding programmes were assessed and a subject brought to the attention there was the mutation work in grapefruit (*Citrus paradisi*) that has been performed in the USA. In the congress proceedings Hensz (1991) described the origin of grapefruit about 250 years ago in the West Indies, probably as a natural hybrid or as a spontaneous somatic mutation from another citrus crop, the pummelo/pomelo (*Citrus grandis* or, according to others, *C. maxima*). The original grapefruit was white fleshed, had a yellow skin (or 'peel') and carried many seeds. Only after, due to spontaneous mutations, genotypes with seedlessness and pigmentation had been detected, did grapefruit become a commercially important fruit crop.

In 1929 a new cultivar, cv. Ruby Red with red flesh colour and a pigmented (blushed) skin, was discovered in Texas and this became the last successful commercial mutant of spontaneous, somatic origin. Despite its wide acceptance, cv. Ruby Red had some disadvantages such as an increasing loss of flesh colour during the harvest season and a less attractive juice colour. As conventional breeding programmes did not exist for grapefruit, one either had to wait for additional spontaneous mutations – which indeed do occur occasionally – or one could consider the possibilities of trying to artificially induce such mutations.

In grapefruits both seeds (in case of seedy fruits) and vegetative plant parts (cuttings or budsticks) can be used as starting material for mutagenic treatments. Hensz in 1960 and 1977 reported (original publications could not be obtained; for references see Hensz, 1991) that several thousands of seeds as well as budsticks had been sent to Brookhaven National Laboratory, Long Island, New York, for exposure to X-rays or thermal neutrons. Seeds were treated with 50–400 Gy (5–40 kR) of X-rays and with thermal neutrons (flux: 5.8×10^8 thermal neutrons. cm^{-2} for 3–35 h). The budsticks were exposed to 20–100 Gy of

X-rays or 6–15 h of thermal neutrons. Treatment of seeds of cv. Hudson with thermal neutrons resulted in a number of trees that produced fruits that were practically seedless (with for instance four seeds per fruit against 50 in the fruits of the starting material) and showed a much deeper red flesh colour and a more pigmented skin than both cv. Hudson and cv. Ruby Red.

One of the trees from the irradiated material became the starting material for cv. Star Ruby, the release of which was announced by Hensz in 1971, and this mutant became a superior grapefruit cultivar in all major growing areas. Other interesting colour mutants were obtained as well. Hensz (1991) for instance also referred to a strain indicated as 'Hudson Red' with a flesh colour identical to that of cv. Star Ruby, and further reported that the cultivars Star Ruby and Hudson Red were 'three times redder' than the most red natural mutant. In the discussion of the aforementioned paper Hensz added that cv. Star Ruby, despite the fact that it was highly appreciated by consumers, was not as much in favour by the growers because of the relatively low and unstable fruit production of this cultivar.

Continued observation of the plant material obtained from the mutation project revealed the presence of another promising tree, designated 1-48, which was obtained from irradiation with thermal neutrons of budwood of cv. Ruby Red. In the trees that were produced from the 1-48 material, additional genetic instability was observed and in the branches of some of these trees fruits were found that were redder than the parent 1-48 and as red as cv. Star Ruby. From this 'secondary' mutant material, 1200 trees were planted, which all proved to be very vigorous and did not show any further instability in traits of fruit or tree. In 1984 the improved (secondary) material of tree 1-48 was released in Texas as cv. Rio Red. In the next year more than 700 000 propagation buds were distributed to the growers and in his paper, Hensz (1991) further predicted a bright future worldwide for this cultivar. For commercial reasons the Texas growers decided to market cv. Star Ruby and cv. Rio Red under the premium trademark RIO STAR.

Finally, Hensz pointed out that conventional breeding would not have been practical in this specific case and that the success obtained by mutation breeding would not have been achieved. The mutation breeding programme undoubtedly had accelerated the discovery of the two mutants. The method of inducing seedlessness through irradiation has been successfully taken over by other citrus breeders (see for instance Hearn, 1984; 1986).

Some additional data and references about induced seedlessness and improved flesh colour in grapefruit and some other citrus species can be found in a review on mutation breeding in fruit trees by Spiegel-Roy (1990). During the FAO/IAEA Symposium in Vienna in 1990, Spina et al. (1991) reported for Italy on low seed numbers (on average: three) in some gamma induced mutants of cv. Monreal of clementine (*Citrus reticulata*).

More recently, during a FAO/IAEA Symposium in 1995 in Vienna, Tang et al. (1995) reported on successful mutation breeding work for the induction of seedlessness or lower seedy cultivars in sweet orange in China. The authors mentioned that in 1978 and 1987 buds were irradiated with ⁶⁰Co gamma rays at a dose of 80 Gy at 2.5 Gy. min⁻¹. In 1986 and 1988 shoots from previously irradiated trees were re-irradiated at the same dose and about the same dose rate. The treated buds were single-bud grafted on a rootstock. Fruit production started in 1991. Several seedless and less seedy mutants were obtained from which the best were used for production. It was concluded that recurrent irradiation may result in increased mutation frequency for seedlessness. The authors further reported that the plants with shoots showing morphological aberrations in the re-irradiated buds may show more seedless and few-seeded mutants than normal shoots and as such be used as a kind of pre-selection for the occurrence of seedlessness.

7.4.4. Induced durable resistance in peppermint (*Mentha* × *piperita*)

One of the few irrefutable examples of resistance induced by mutation breeding that has remained effective for a considerable number of years and, at the same time, the only case know so far of induced durable resistance in vegetatively propagated crops, refers to resistance to so-called *Verticillium*-wilt, caused by *Verticillium albo-atrum*, in peppermint (*Mentha* × *piperita*), a sterile, allohexaploid (2n = 72) plant that is grown for its aromatic oil and is vegetatively propagated by stolons. The radiation-induced resistance to *Verticillium*-wilt, obtained in 1959 in the USA after irradiation of stolons, was still effective in the 1990s (Todd, 1990; Micke, 1993a).

Since 1890 only one clone of the peppermint plant, the Mitcham variety, with a particularly good yield and high quality of the oil, was grown in the USA. In 1970 the total area in the USA covered about 30 000 ha with a commercial value of 20 million US dollars. The wilt disease, a soil-borne fungus, had been known in 1924 but did not become a serious problem until 1940. The problem was originally tackled by starting a large-scale breeding programme, e.g. by crossing with other species of *Mentha*, but all efforts remained fruitless because it appeared to be impossible in this way to combine the required resistance with the superior quality of the mint oil of the Mitcham clone.

Although it is believed that wilt resistance is based on one or more dominant genes – a situation that commonly does not look very promising for mutation breeding – it was decided in the 1950s, as an ultimate effort, to start a mutation breeding programme. Two positive points for mutation breeding were that it would be easy to irradiate large numbers of stolon cuttings at a time and that, moreover, no complicated screening methods would be required as the large majority of plants, when grown in an area that was heavily infested by wilt, would die off within a few years, thus making it easy to detect amongst the relatively few survivors the resistant plants that were hoped for.

In 1955 smale-scale experiments with irradiating dormant stolons of cv. Mitcham were started by S. Shapiro at Brookhaven National Laboratory in order to determine the optimal doses of thermal neutrons and X-rays. For X-rays the optimal dose was determined at 60 Gy and, according to Murray (1969), the neutron dose 22 × 10¹², without giving further details. This figure, presumably, refers to the 'flux' of thermal neutrons. cm⁻² × duration of treatment (in seconds?).

More than 100 000 irradiated stolon cuttings that were planted in wilt-infested soil, resulted in more than 6 million plants, but after four years of severe wilt attacks this number was reduced to less than 60 000 plants which appeared resistant. From this population, ultimately, only seven highly resistant and five moderately resistant clones were selected, which also had an acceptable oil quality and also were acceptable from an agronomic point of view (Murray, 1969; 1971).

In 1971 a first mutant cultivar, cv. Todd's Mitcham Peppermint, was released by the A.M. Todd Company (director of research: M.J. Murray) in Kalamazoo, Michigan State, USA. This cultivar became officially registered in 1972 by the Crop Science Society of America (Murray & Todd, 1972). The new cultivar, obtained after treatment with thermal neutrons, in addition to its wilt resistance, showed some small morphological differences when compared with cv. Mitcham, as it had a darker green herbage colour, smaller leaves and a more erect and less branched plant habit and, furthermore, matured 5–10 days earlier. Composition and amount of the mint oil, as was mentioned already, did not differ significantly from the established cv. Mitcham. In 1976 a second mutant cultivar, cv. Murray Mitcham, was brought onto the market by the same company and in the same year officially approved by the aforementioned Crop Science Society (Todd, Green & Horner, 1977). This second cultivar, also derived from the plant material that had been treated with neutrons, differed from cv. Todd's Mitcham by a higher yield in the first year. As was mentioned already in Chapter 1, the exact nature of the resistance (monogenic/polygenic) of both mutant cultivars is not known.

Broertjes & van Harten (1988) commented that the simple propagation method of repeatedly using stolon pieces, may have significantly reduced or even eliminated the occurrence of chimerism.

In an issue of the *Mutation Breeding Newsletter*, Todd (1990) reported that in 1989 peppermint was grown in the USA on about 44 000 ha (111 000 acres). Almost 50% of the total area was planted with the original cv. Mitcham, 40% with the mutant cv. Todd's Mitcham and 10% with the mutant cv. Murray Mitcham. The total value of the 1989 crop was estimated to be about 90 million US dollars, of which 40 million could be ascribed to cv. Todd's Mitcham and seven million to cv. Murray Mitcham. In some production districts, for instance the Midwest, the original cultivar cv. Mitcham (40% of the area) as well as both mutant cultivars (with 20% for Todd's Mitcham and 40% for Murray Mitcham) were grown. Some growers in that region find that, in the absence of the disease, the original cv. Mitcham has a better vigour. In some areas like Idaho (8000 ha) no wilt problem exists and only cv. Mitcham is grown. In another district, Willamette with 10 000 ha of peppermint, some 90% of the area is planted with the mutant cv. Todd's Mitcham and 10% with cv. Mitcham. In the Yakima Valley (7000 ha), in fact a resistant cultivar would be required, but neither of the mutant cultivars perform well which, perhaps, could be explained by a high alkali content in the soil.

7.4.5. Mutation breeding of chrysanthemum (with special reference to the Netherlands)

Introduction

Chrysanthemums (present official genus name: *Dendrathema* sp., but still mostly referred to as *Chrysanthemum* sp.), in particular when commercialized as cut flowers, provide the most striking example of highly successful and economically most rewarding mutation breeding work in a vegetatively propagated ornamental. The following, rather detailed, picture in particular describes the situation for the Netherlands where hexaploid chrysanthemums ($2n = 6x = 54$) are grown under short day conditions, predominantly for the production of cut flowers.

The chrysanthemum is, after roses, the second most important Dutch cut flower. In 1993, according to the 1994 'Dutch Descriptive Variety List for Ornamentals and Florist's Crops' (briefly: the variety list), 1.26 billion (!) flower stalks of chrysanthemum, with a commercial value of some dfl 580 million (350–400 million US dollars), were sold at the flower auctions. Approximately 50% of the total production was exported. From the Dutch cultivars grown in the early 1990s about 35% had white flowers, 25% had pink flowers, 20% was yellow, 5% bi-coloured, 5% cream and 4% purple.

Despite the high costs, more than 95% of the Dutch growers do not produce their own starting material – cuttings or young rooted plants – but leave this to 10–15 companies who perform this specialized task. About 25% of the price a grower receives per flower stalk must be paid by him to purchase the starting material.

Spontaneous mutants ('sports') and induced mutants

Sports in chrysanthemum, resulting from spontaneous mutations, have been known for a long time, and it has been common practice in the Netherlands for almost half a century to register good sports as new cultivars. The areas occupied by spontaneous sports as well as by artificially (radiation) induced mutant cultivars and the economic value of such mutant cultivars (in the Netherlands as well as in various other countries) often considerably surpasses that of the original cultivar in which these sports arose. Wasscher (1956) already mentioned that in the Netherlands about 30% of the cultivars had originated as (spontaneously arisen) sports and that these sports occupied an even higher percentage of the total area grown with chrysanthemums than the original cultivars, and this percentage has increased in more recent years (see later). In comparison: for the UK it was reported by Langton (1986) that 40–50% of all applications for plant variety rights for chrysanthemum referred to mutants. This figure, supposedly, includes sports as well as induced mutants.

Some cultivars have an impressive list of successful mutant cultivars that have been derived from them. Broertjes & van Harten (1988) mention several early examples of chrysanthemum cultivars which have given rise to a whole range of 'sports' An account of the way in which within two or three years from a mutant cultivar a whole series of secondary – or tertiary – mutant cultivars (spontaneously arisen as well as induced after recurrent irradiation) can be produced ('A mutant of a mutant of a mutant. . .'), has been given by Broertjes *et al.* (1980) for the so-called 'families' of cv. Horim and cv. Miros. (N.B. The expression 'family' is used by breeders to indicate one original genotype with a whole range of, mainly, flower colour and flower shape mutants.)

A simplified scheme of the most important mutant types which, according to Broertjes *et al.* (1980), arose after recurrent irradiation of a spontaneous pinkish sport of cv. Horim with doses from 15 to 30 Gy of X-rays, derived from *Mutation Breeding Review* 7 (Micke *et al.*, 1990), is presented here as Fig. 7.2. This 'family' of mutant cultivars, for several years, accounted for 30% of the total production of chrysanthemums in the Netherlands.

To mention an example for another country as well, reference can be made to a whole 'family of sports' that have arisen from the well-known cv. Indianapolis in the USA (Stewart & Dermen, 1970b).

Breeding strategies

Chrysanthemum growers switch towards improved cultivars within a remarkably short time after such cultivars have become available. Most new cultivars released in the Netherlands are produced by three major breeding companies. In addition, ten smaller companies are breeding chrysanthemums. Customers and retailers, besides price, are particularly interested in flower colour, flower shape and keepability, whereas the decision by the grower which cultivar to grow is affected as well by a number of other traits; for instance, the number of days to reach flowering under short day conditions, the firmness of the stalk, uniformity, growth vigour, leaf quality and the suitability of a cultivar for year-round production. In recent years breeding for more and smaller flowers and breeding for resistance have become more important.

Each year breeding companies test many new accessions, but only very few of their new products will ever reach a position among the top 25 cultivars grown. The breeding companies are being reimbursed for their work on the base of royalties for their registered cultivars. In 1992 they received 1–1.5 ct per cutting, or about 2–3% of the auction price that a grower receives per flower. In that year, the Dutch breeders, altogether, received dfl 12 million (about 7.5 million US dollars) for their chrysanthemum cultivars.

Taking all Dutch breeding programmes together, selection cycles each year start from about 500 000 seedlings which are obtained from crosses between selected parents. Only some 2–3% of the original seedlings are allowed to enter the next selection cycle. In total four to six selection cycles are performed; two selection cycles may be performed per year. Selection for spontaneous mutants within a cultivar often results in 'strains' which show only small differences from the original cultivar in one or two traits like flower colour and flower size.

It is imperative to the breeder to have at his disposal as quickly as possible a whole range of the common sports, since when such sports are found in the greenhouses of the multiplication companies or growers they are, according to the presently valid breeders' law, not his property any more and can be commercialized by the 'finders' without any financial remuneration for the breeder of the original cultivar (at least, as long as the principle of 'essential derivation', explained in Section 7.3.5 is not generally implemented).

In order to speed up the production of sports and to enlarge the spectrum of colours, radiation treatment has become a routine step in most breeding programmes. The ease with which mutants can be induced has led to a change in the breeding programmes. At present breeders already keep in mind the possibilities offered by mutagenic treatments while evaluating the first results of their cross-breeding work. A striking example in this respect is that, in the past, seedlings with an unattractive pink-like flower colour would have been immediately discarded, even if such seedlings had been outstanding for other traits. Nowadays, such seedlings most probably would be maintained because the breeder knows that it is often possible to produce by routine irradiation treatments within a very short span of time a range of other, more desired or favourable flower colours. The aforementioned mutant family derived from cv. Horim (Broertjes *et al.*, 1980) has become a classical example of the present approach. In general, this method can be successfully applied to amend (small) shortcomings in otherwise promising genotypes obtained from cross-breeding for various other traits.

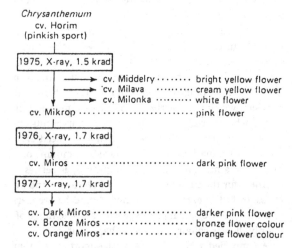

Figure 7.2. Mutant cultivar of chrysanthemum resulting from recurrent irradiation (after Midke *et al.*, 1990).

Mutations for flower colour and other traits

Although many details about the genetics of flower colour in chrysanthemum have not been unravelled yet, it appears (J. de Jong, personal communication) that most phenomena can be explained on the basis of two (major) mutations only: 1. anthocyanin+ (pink colour) → anthocyanin− (white colour); and 2. carotenoids− (white colour) → carotenoids+ (yellow colour). However, it may be expected that a further detailed molecular analysis will prove that still other (modifying?) genes contribute as well to the whole range of mutant cultivars with different flower colours that – as has been demonstrated on several occasions – can be easily produced within a few years time.

It has been shown repeatedly (Jank, 1957; confirmed by Broertjes, 1966) that plants with pink flower colour provide the best starting material for artificially inducing other flower colours. The flower colours that, roughly, follow in a descending line of suitability as starting material, are: white, bronze, red, purple, yellow, salmon, golden orange, yellow-bronze, yellow with red and brown. (N.B. Compare the sequence of colours in Fig. 7.2.)

Within one colour, variation may occur as well and some private breeders express the opinion that, for instance, a dark yellow type refers to plants which are genetically yellow in all three histogenic layers ('solid' or 'homohistont'), that the normal yellow type represents a periclinal situation with two yellow layers (L-1 and L-2), whereas the cream type should have a yellow L-1 only. However, no definite proof has been obtained yet for this opinion and, although the predominant role of the colour genes in the L-1 is generally accepted, opinions still differ as to the relative contribution of the layers derived from L-2 and L-3 in this respect. More insight in the histogenic development of the flower petals would be most welcome.

The aforementioned colour mutations may also arise spontaneously in nature, but it takes more years to obtain the whole colour spectrum, and because of the low frequency of spontaneous mutations (e.g. less than 1×10^{-6}) very large numbers of plants must be inspected. Nevertheless, given the very large numbers of chrysanthemums grown (e.g. in the Netherlands) this low frequency is not considered a major bottle-neck and the method of selecting for spontaneous sports for flower colour and some other traits remains a useful and often rewarding one.

Mutations, induced or spontaneous, are by no means limited to flower colour as, for instance, mutations for traits like flower shape, length and firmness of flower stalks, increased uniformity and low-temperature tolerance may be induced easily as well, although selection for some traits, e.g. for uniformity, may be rather cumbersome. Some chrysanthemum breeders doubt that the spectra of spontaneous and induced mutants are fully identical, but it is not easy to give definite proof for either opinion. An interesting observation is that induced mutant cultivars, in later stages, may be quietly replaced by even better, often spontaneous mutants that are about identical and fit the same botanical description. One could hypothesize in this respect that the spontaneous mutants do not carry unfavourable background mutations that may occur as a result of the irradiation treatment.

Mutation breeding

Most mutation breeding has been performed with *C. morifolium* (synonyms: *C × hortorum* and *C. indica*), a hexaploid, year-round cut flower crop. Most starting material is genetically highly heterozygous.

Due to the multicellular meristems of cuttings mutagenic treatment results in chimerism, and although the production of adventitious plants *in vitro* may result in solid mutants (Broertjes & Lock, 1985; de Jong & Custers, 1986), so far, no reliable 'in vivo' (or *in vitro*) single cell method is available that can be applied in a routine way to all genotypes of chrysanthemum. In general, not too many results should be expected from mutagenic treatments for traits for which selection is not easy and, hence, the general advice is that mutation breeding programmes should focus on easily visible traits like flower colour and flower shape (Broertjes & van Harten, 1988).

In practice, cuttings are submitted to irradiation with X- or gamma rays at doses of about 15 Gy, but sometimes doses of 25 or 30 Gy are applied. When higher doses are applied, in general higher mutation frequencies and larger mutated areas ('sectors') are obtained, but survival rates go down and plant growth may be considerably retarded (Broertjes & van Harten, 1988). After irradiation it is common practice to cut back the growing plants and their outgrowing axillary buds two or three times in order to obtain shoots with larger plant areas mutated (larger 'sectors') before taking cuttings for the flowering trial. Selection is performed on a one-plant basis during flowering, usually under conditions that are identical to those of commercial growing. In this way, throughout the years, hundreds of commercial mutants have been obtained.

Periclinal chimerism

The obtained sports or induced mutants are mainly periclinal chimeric in which it is mostly assumed that only L-1 genetically differs from the original cultivar (Stewart

& Dermen, 1970a,b; Langton, 1980), but rearrangements between the different histogenic layers or tissues derived from them may result in more complicated patterns of chimerism.

Periclinal chimeric mutants in chrysanthemum, in which in most cases a mutated (L-1) layer surrounds the internal, non-mutated layers (L-2 + L-3), often can be directly released as new cultivars, in particular when flower mutations are involved. When during the selection phase sufficient attention has been paid to select, on a clonal base, for stability within the group of periclinal chimeras, it is possible to obtain genotypes that in practice show only a very low percentage of back-sporting. As these periclinal chimeras, in fact, are identical to the original cultivar in almost all traits, it is possible to grow periclinal mutant cultivars as well as the original cultivar together in greenhouses under identical culture conditions. A limitation of periclinal mutants, on the other hand, is that they do not differ from the parent cultivar in cross-breeding if the L-2 layer, from which the gametes arise, has the same genetic constitution as the cultivar in which the (flower) mutations arose. A more detailed overview of the pros and cons of using periclinal chimeras was presented by Broertjes (1979).

Despite the commonly occurring chimerism, present procedures to irradiate (rooted) cuttings of chrysanthemum have been shown to be very effective when traits like flower colour and shape are involved. As this approach is fast, easy and inexpensive, there often is no real need to switch to *in vitro* methods, in particular not as long as such methods often are not generally applicable to a wide range of genotypes and, hence, remain too expensive for private breeding companies. On the other hand, when the chimeric state is not acceptable, for instance when mutation breeding for resistance or for low temperature tolerance is performed, *in vitro* methods in combination with mutagenic treatments can be applied as well. An early report about *in vitro* propagation of chrysanthemum was published by Roest & Bokelmann (1975) and several later references can be found in Broertjes & van Harten (1988). Some recent results of *in vitro* work with chrysanthemum in combination with mutations from outside the Netherlands, are briefly presented in Box 7.3.

Chrysanthemums on the Dutch variety lists

The Dutch variety list for 1994 contains 42 recommended cultivars for chrysanthemum and in 23 cases the cultivar name is accompanied by a number of mutants. In the variety list no distinction is made in this respect between spontaneous 'sports' and induced

mutants. In the discussion of a paper presented during a FAO/IAEA Study Group Meeting in Buenos Aires in 1970, Broertjes (1972b) already pointed out that private plant breeders, e.g. of ornamentals, rather prefer not to indicate the exact origin of their mutants in order to protect their commercial interests, as well as to avoid consumers becoming unnecessarily scared by an association between their products and – what is sometimes believed to be – a 'radioactive treatment'.

We may finish this section with a recent example. One interesting case from the 1994 variety list concerns cv. Reagan, which became a registered cultivar – with breeders' rights protection – only in 1991. In 1990, this cultivar had become the number three cultivar at the Dutch flower auctions with 60 million flower stalks sold in 1990 and 44.5 million in 1991. Together with its mutants (of which in 1994 already twenty were mentioned in the variety list, with names such as 'Bronze Reagan', 'Coral Reagan', 'Nancy Reagan', etc.), the 'Reagan group of cultivars' in 1992 and 1993 represented 35% and 40%, respectively, of the total Dutch market. This implies that from the Reagan 'family' in total more than 400 million flower stalks were sold annually! The economic importance of the original cv. Reagan in 1993 had become small with only 5.5 million flowers offered to the auctions. At that time most mutants of cv. Reagan, although already of considerable practical importance, were still in the stage of applying for breeders rights. In comparison: the two leading cultivars in 1991 as well as in 1994 were cvs. Cassa and Majoor Bosshardt with in 1991 a total of 109 million and 75 million flower stalks, respectively. In 1993, despite the fact that they were still the two most important cultivars, these figures had gone down to 48.4 million and 37.4 million cuttings, respectively, from both original cultivars. (N.B. From these two cultivars several mutants have been released as well. Their production figures are not included.)

7.4.6. Induced mutations in fruit trees

Introduction

Fruit crops, and in particular fruit trees (apple, peach, pear, plum, cherry, citrus, etc.) have been considered attractive material from the view-point of mutation breeding for many years. Main reasons for this are that conventional breeding is time consuming and that results in such crops are rather unpredictable because of their long juvenile phase and their high degree of heterozygosity or generation cycle. Attempts for further improvement of a well-known cultivar by conventional cross-breeding most often result in the loss of the unique genotype without having any guarantee that improve-

Box 7.3 Some remarks on in vitro mutation breeding in chrysanthemum.

Buiatti & Gimelli (1993), in a review paper on in vitro (somaclonal?) variation in ornamentals, described fourteen mutant cultivars derived from in vitro cultures. Three of these cultivars referred to chrysanthemum and in two of them it was reported that the material had been irradiated with 10 and 15 Gy of gamma rays, respectively. (N.B. Author names have been mentioned in the paper but references to the original publications – probably in Russian – are lacking.)

Another example is derived from the Annual report of the National Institute of Agrobiological Resources in Ibaraki, Japan (Anon., 1991c). Plants of cv. Taihei of chrysanthemum were chronically irradiated in a gamma field during 100 days with a total dose of 25–150 Gy. About five months after planting, mutated 'sectors' (= areas) appeared on floral petals of irradiated plants. Petals and buds were dissected and cultured in a modified and enriched MS medium. Induced calli were subcultured for regeneration and the resulting plants were returned to the field. Of the regenerated plants about 40% showed mutations for flower colour, whereas mutations also occurred for shape and size of flowers and leaves. Mutation frequencies obtained in this way, reportedly, were higher than after traditional chronic irradiation. A point of special interest is that, according to this publication, the in vitro method results in non-chimeric mutants.

Recently, Jerzy & Zalewska (1996) reported on the irradiation of leaves of chrysanthemum with 5–25 Gy of X-rays and gamma rays. Adventitious shoots developed in vitro from leaf explants on a MS solid medium with IAA and BA. Changes in colour, size and form of the inflorescence were observed in the regenerated plants. As no undesirable mericlinal or sectorial chimeras were found, it was reported that the adventitious shoots are of single cell origin and, consecutively, that in this way homogeneously mutated tissues can be obtained.

ments for the most important traits would be obtained (Lapins, 1973; 1983). Moreover, fruit growers and breeders are already well aware of the potential economic impact of spontaneous mutants ('sports') from widely grown cultivars. An early, extensive list of 'sports' was produced by Shamel & Pomeroy (1936).

Nowadays growers of fruit trees often are in the position to choose among sports of well-known cultivars. In some countries like the Netherlands, it is possible to submit spontaneous mutants with a different fruit skin colour or striped appearance of the fruits for protection by plant variety rights (or breeders' rights). Such applications are accepted when a) mutants differ sufficiently from an already accepted cultivar, b) the mutant has not been commercialized before the breeders' rights request was made, and c) the sample offered for registration is sufficiently homogenous.

The aforementioned positive experience with spontaneous mutants has led to the idea that improved cultivars for fruit crops might be obtained in even higher frequencies by the use of induced mutagenesis (Spiegel-Roy, 1990). Early experiments to induce mutations by various methods started in the 1930s in the USA, but important results were not obtained.

After the Second World War mutation breeding programmes in various fruit trees, such as in apple (*Malus* sp.), pear (*Pyrus communis*), peach (*Prunus persica*), sweet cherry (*Prunus avium*), apricot (*Prunus armeniaca*) and almond (*Prunus amygdalis*) were initiated in various countries, in particular in Canada, Sweden (during the inital phase of mutation breeding in that country only), the UK, France, Italy, Japan and the Netherlands. In some countries, like Germany, attention was also paid to mutation breeding in bush fruits like black currant (*Ribes nigrum*). Early reports on induced mutations in fruit trees – in particular apple and pear – have been published by Granhall et al. (1949) and Granhall (1953; 1954) in Sweden, and by Bishop (1954) in Canada.

It should be noted here that mutation breeding in some other tree crops (broad-leaved and coniferous trees) and woody plants which are not (or not only) grown for their fruits, often goes more or less along the same lines as for fruit trees. It is impossible to mention them all at this place, but reference could be made for instance to mutation studies on mulberry (*Morus* sp.), a tree grown for its leaves which are used for feeding silk worms (sericulture), e.g. in Japan and China (see for instance Fujita & Takato, 1970; Katagiri, 1973; Nakajima, 1973; and Fujita, 1982); on tea (*Thea sinensis*), which subject was discussed already by Futsuhara (1967); on various forest trees (poplar, spruce, acer); and on sea buckthorn (*Hippophaea rhamnoides*), a small tree grown in Siberia for its edible fruits, juice and medicinal oil (Privalov, 1986). As mutants for this last crop are obtained by treatment of seeds, this subject was briefly discussed earlier in Chapter 6. More details about mutation breeding in the aforementioned crops can be found in Broertjes & van Harten (1988), in some contributions during several, earlier mentioned FAO/IAEA Meetings on vegetatively propagated plants, and in some issues of *Gamma Field Symposia*, e.g. in No.12 on chimerism and mutations in woody plants (Anon., 1973b) and in No.28 (Anon., 1989b), which deals specifically with mutation breeding in tree crops.

We will return now to mutation breeding in 'real' fruit trees. Despite several extensive research projects in this field, Sigurbjörnsson & Micke (1973) during a panel meeting on mutation breeding in vegetatively propagated crops in Vienna in 1972, could only mention five induced mutant cultivars that had been obtained so far for fruit trees, i.e. cv. McIntosh 8F-2-32 for apple (*Malus* sp.), cv. Magnif 135 for peach (*Prunus persica*) and cvs. Compact Lambert, Stella and Early Blenheim for sweet cherry (*Prunus avium*). (N.B. In later years it was corrected that cv. Early Blenheim refers to apricot (*Prunus armeniaca*) instead of sweet cherry.) During the same meeting (Anon., 1973a) a bibliography on mutation breeding in vegetatively propagated crops, including some 60 titles on fruit crops, was presented by Nybom & Micke (1973). Some years later Donini (1977) produced a review on mutation breeding in fruit trees accompanied by 139 references.

Around 1985 the number of mutant cultivars for fruit crops had increased to about 30 mutant cultivars (Broertjes & van Harten, 1988; see also Maluszynski *et al.*, 1992). It appears that since the second half of the 1980s not many new contributions have appeared on pome fruits (apple, pear) and stone fruits (peach, sweet cherry). Most references to mutagenic treatments and mutations in the various species of fruit crops have been given in Broertjes & van Harten (1988).

Much attention has been paid right from the beginning to methodological aspects of mutation breeding in fruit crops (for early reviews see for instance Nybom, 1961, and Nybom & Koch, 1965), in particular with respect to the complications as a result of chimerism that occurs when buds – sometimes with a high number of bud primordia – or cuttings containing a number of (primary and secondary) buds in various stages of development, are treated mutagenically. This methodology will be a main topic of the following sections.

Methodology
Nybom & Koch (1965), Lapins (1983) and, based on this last publication, Spiegel-Roy (1990) have presented a table on physical mutagens (X-rays, gamma rays and thermal neutrons), mentioning doses and dose rates to which various kinds of plant material of different fruit

crops have been submitted. Treatment of various plant parts with chemical mutagens is possible as well, provided that the common problems with penetration and distribution of the chemical can be solved (e.g. by adding DSMO as a carrier). However, it always remains difficult to perform uniform (and reproducible) treatments with chemicals to large somatic plant parts.

Although considerable differences in treatments can be observed, most often (dormant) buds on scions, stem cuttings, etc., are irradiated with X-rays or gamma rays with dosages within the range of 10–60 Gy and dose rates from 0.1 to 15 Gy. min^{-1}. Covering the base of the scions or cuttings with a lead strip during radiation treatment improves results of rooting or grafting. Adventitious buds, induced on explants like root cuttings or leaves, immediately after mutagenic treatment, may be used as well. Lapins (1983) remarked that the chance of discovering mutations is increased by immediate active growth after irradiation and, hence, the use of already grafted scions or already rooted cuttings seems recommendable. According to Bishop (1959) neutrons should be more effective mutagens to induce mutations in apple than X- or gamma rays; this observation is probably based on the usefulness of deletions in highly heterozygous material, but this opinion, later on, has not been much supported.

Lapins, Bailey & Hough (1969) reported that irradiation of winter buds, either dormant or forced directly before irradiation gave higher mutation frequencies than summer buds. The reason, however, is unknown. The LD 50 for dormant buds, depending somewhat on the plant species and the applied dose rate, differs considerably, but mostly varies between 30 and 60 Gy. Lowering the dose rate results in a considerable increase of the LD 50. Moderate exposures, which may result in, for instance, 70% survival, are considered desirable (Visser, 1973) and such doses may produce high mutation frequencies, with about 3–4% of the buds showing mutations (Lapins *et al.*, 1969). Higher doses result in higher mutation frequencies but the probability that they are accompanied by undesirable genetic changes increases as well. On the other hand, higher doses will kill relatively more meristematic cells in apical and axillary buds and, as a result, reduce chimerism and produce relatively more whole-plant (solid) mutants.

In particular Visser and associates (Visser, 1973; Visser, de Vries & Verhaegh, 1969; Visser, Verhaegh & de Vries, 1971) paid much attention to the efficiency of selecting at the nursery stage in order to determine the most effective dose of X-rays. It was found that a direct relationship exists between the survival rate of buds on irradiated scions – which had been irradiated as dormant scions during winter and were kept under water in glass jars until spring – after four weeks, and the percentage of surviving grafts after two seasons in the field.

Chimerism

Much attention has been given to the problem of how to deal with chimerism, for example in fruit crops, and the often assumed loss of mutants by what was called 'intra-individual selection of mutants' by Kaplan (1953; see also Chapter 1). This work in fact started with the methodological investigations performed by Kaplan (loc.cit.) and Bauer (1957) on black currant (*Ribes nigrum*) and by Zwintzscher (1955; 1959; 1962) on apple in Germany.

Kaplan (1953) pointed out that lower buds, developing earlier on a shoot arising from an irradiated bud, carry more mutations than the later formed buds higher on the shoot. These lower buds are already present as primordia in the irradiated buds. Zwintzscher (1959) expressed the opinion that earlier failures of mutation breeding 'without doubt' referred to methodological problems and stressed the importance of developing appropriate procedures to recover mutants. Such procedures were described in 1962 by Zwintzscher (see later this section).

Lapins and associates (see for instance Lapins *et al.*, 1969, and Lapins, 1973; 1983), like several other researchers, mostly irradiated so-called leaf buds of dormant vegetative shoots of various deciduous fruit trees (apple, pears, cherries and peaches). Lapins (1983) mentioned that, on average, a dormant bud in apple contains already seven bud scales, nine leaf primordia and four axillary bud primordia. No axillary buds are present yet at the few highest leaf primordia, although the cells that will form these buds do exist already. The youngest axillary bud primordia can be recognized microscopically at the leaf primordia called P3 and P4 (Lapins, 1973). It was found that the buds 6–10 on the primary shoot – which shoot develops from an irradiated primary bud – carried more recoverable mutations than the buds 1–5 or 11–15. This work puts in more precise terms the earlier mentioned results by Kaplan (1953) and Bauer (1957) with black currants, and Zwintzscher (1955) with apple, which studies showed that more mutations were recovered from secondary buds formed earlier on the primary shoot developing from the original primary bud.

A point to keep in mind is that these secondary buds may show various chimeric patterns. These patterns depend upon their degree of differentiation at the time of treatment, which corresponds to the time of development and the position on the primary shoot. A muta-

tion refers to a single cell event, which automatically implies that originally only one layer is involved. Therefore, after a mutation has been induced, the next step will be the formation of a mericlinal chimera, and when axillary buds arise from the mutated area a periclinal chimeric shoot dose occurs. This periclinal chimeric situation may represent a rather stable situation in vegetatively propagated crops.

In order to avoid at this place further unnecessary details, we may quote here Broertjes & van Harten (1988, pp. 229–30), who summarize the common situation after irradiation of buds on vegetative shoots by saying that

> somewhere on a vM_1 shoot between the parts where the smallest and the largest apices are situated, is the optimum part of the shoot that gives an acceptable mutation frequency and, above all, a 'sector' size which is large enough to ensure detection and recovery after repeated cutting back and vegetative propagation.

Lapins (1973) points out that the main problem in mutation breeding (in fruit trees) is the identification of the mericlinal chimeric areas, carrying the desired mutations, within a multicellular meristem and the further propagation and purification of such mericlinal chimeras. Basically, two ways are available to reach these goals: to subdivide the primary shoot into buds by bud propagation according to the so-called 'isolation method' developed since 1952 by Zwintzscher (1955; 1962; 1977), or to cut back the primary shoot repeatedly, forcing formation of additional axillary buds, following the method that was originally developed by Bauer (1957).

Whereas the lower buds of the primary shoot, apparently, carry relatively more mutations that can be recovered, it was found by Decourtye & Lantin (1971), that larger mutated areas (i.e., areas showing less chimerism) are found in the higher, axillary (secondary) buds that were still undifferentiated in the primary meristem at the time of irradiation. Problems of methodology and chimerism in irradiated fruit crops have also been described and reviewed by various other authors, for instance by Donini (1976b).

In addition to an axillary bud present in a leaf axil there is still another category of buds that should be mentioned in relation to mutation treatments: the accessory buds which can develop next to the main axillary bud in the nodal area. Such accessory buds (see also Section 7.2.2 of this chapter) develop from fewer (irradiated) cells than the main axillary bud, which limits chimerism. However, they most often remain dormant as long as the main bud is intact. In order to limit chimerism one could, therefore, consider damaging axillary buds or remove the axillary shoots in order to stimulate the development of such accessory buds.

Terminology used in vegetatively propagated (fruit)crops

As was mentioned already in Chapter 5 in relation to the starting material used for *in vitro* cultures, the somewhat complicated developmental structures and histological interpretations in buds, scions, rhizomes, explants, etc., and the consecutive generations in a mutation breeding programme (e.g. seed irradiation followed by a number of seed generations or by vegetative propagation *in vitro*, etc.), make it difficult to name the different plant parts and such stages in a uniform and unambiguous way. It is, of course, highly desirable in science and technology to use a generally and internationally accepted terminology.

It may be useful to comment in this respect that there is no disagreement whatsoever about the use of the – established and simple – system of indicating mutated generations for seed propagated crops as M_1, M_2, etc. Accordingly, in order to distinguish between seed propagated crops and vegetatively propagated crops is became accepted at an early stage of mutation breeding to simply add the prefix 'v' in case of vegetatively propagated crops (i.e. vM_1, vM_2, etc.). Nevertheless, sometimes additional information is needed to understand exactly which type of material, and in which (generative or vegetative) generation, was investigated. Such situations, for instance, may occur when an M_1 plant might be self-pollinated or is carried to the M_2 generation by anther culture. In most cases, however, the situation is fully clear and does not require special terminology. In our opinion, and in order to avoid unnecessary confusion by many different, private 'systems of terminology', it would be sufficient to stick to the old terminology (e.g. M_1 and vM_1) and to explain the few 'special situations'. We may comment here briefly on some other terms used in relation to mutations in vegetatively propagated crops.

Lapins (1983) and many other authors use the expression V_1 to indicate the first mutagen treated generation, that consists of primary shoots with primary leaves and their axillary buds. As explained before, we would – parallel to the terms M_1, M_2, etc. – strongly advocate not to omit the 'M' (for 'mutagenic treatment') and to stick to vM_1, vM_2, etc., for the consecutive vegetative generations after mutagenic treatment. Alternatively, Donini *et al.* (1991) indicate the consecutive vegetative generations as M_1V_1 (for the primary shoot developing from an irradiated bud), M_1V_2 (shoots developed from the axillary buds

of an M_1V_1 shoot, etc.). Although one could conceive some situations where Donini's terminology could be useful we still prefer – for the reasons mentioned before – the established and relatively simple system of terminology. Of course, if it would become evident that the 'old' system indeed would not be adequate anymore for many situations, one could switch to a new generally accepted terminology but, in our opinion, this stage has not been reached yet.

It was said before that in very specific situations it may be advisable to describe carefully what exactly is meant rather than to introduce additional systems. This, for instance, could be the case when for facultatively propagated species mutagenic treatments are combined in various ways with vegetative as well as sexual methods of propagation.

One further point still needs some attention. The expression vM_0 should be used for the untreated starting material and therefore the remark by Lapins (1983, p. 85) that 'the irradiated scion with primary buds will produce the first vegetative generation V_1' may be somewhat misleading, because the irradiated scion is already the first vegetative generation. When the apical area of the primary shoot is cut off, the axillary or accessory buds on the primary shoot may develop into secondary shoots which represent already the second vegetative generation (vM_2). All vM_2 trees derived from one vM_1 shoot belong to one '(vegetative) family'.

In any case, it appears to be useful that each author of publications on mutation breeding explains the terms used to describe the experimental procedures.

Handling the plant material after mutagenic treatment

Much attention has been paid by various authors to the question of which buds and how many of them should be taken for further propagation, and various methods have been proposed. The two main considerations in this respect are that lower frequencies of different mutants are obtained from (secondary) buds that are situated closer to the apical area of the primary shoot although (and because), the size of mutated zones near the main bud is bigger.

In the previous section three groups of axillary buds were distinguished. When exposures to radiation remain modest, it appears that buds whose primordia are situated just above the buds that were already differentiated at the moment of irradiation, show the optimal combination of relatively high frequencies of different mutants and sufficiently large mutated zones ('sectors').

Lapins (1983) concluded that mutations for compact

growth can be selected in primary shoots of intact vM_1 plants (V_1 according to Lapins) which shoots, following repeated pruning, carry only the buds of group 3 (vM_2) and therefore, propagation to obtain vM_2 (V_2) plants would be unnecessary. This method, however, is useful only for breeders who are interested in mutants for an easily recognizable trait like compact growth. Previously, Visser (1973, and in earlier publications by Visser and associates) already pointed out that if the desired mutant traits appear to be lacking in these primary shoots and their buds, shoots from group 1 and 2 origin should be forced to develop as vM_2 by pruning, which procedure may be repeated in order to produce an vM_3 for further inspection for useful mutations.

Lapins (1983) continued by saying that, on the other hand, propagation towards vM_2 and vM_3 plants on their own roots or even grafting on appropriate rootstocks would provide better plant material for identification and selection of mutants that are not easily recognizable. In several publications, e.g. by Donini (1975), schematic illustrations are provided to further elucidate the procedures that are advocated.

The conclusion must be that breeders of fruit crops, interested in applying mutation breeding techniques, should familiarize themselves very well with this subject before starting extensive programmes.

Different mutations and useful mutant types

As has been mentioned already in Section 7.4.2, the first induced mutant cultivar for fruit trees was cv. Compact Lambert of sweet cherry (*Prunus avium*), that was released in 1964 by Lapins in Canada (Lapins, 1963; 1965; 1983). Scions of cv. Lambert were irradiated in 1958 with 40 Gy of X-rays. The obtained mutant combines compact growth and dwarf growth with earliness and a good cropping quality. Lapins (1983) added that the mutant failed to transfer the mutant trait(s) to its progeny.

The first mutant cultivar released for small fruits is cv. Westra in black currant (*Ribes nigrum*), which was released in 1968. This mutant, differing from the mother cultivar by its strong erect habit, was obtained in Germany after irradiation in 1949 of cv. Westwick Choice with 15 Gy of X-rays (Bauer, 1974).

Spiegel-Roy (1990) distinguishes the following categories of traits for which it has been tried to induce mutations: 1) compact types (shorter stems or short internodes), 2) fruit colour (anthocyanins, lycopenes), 3) self fertility, 4) seedlessness, 5) time of fruit maturity, 6) disease resistances and, in addition, the induction of polyploids is mentioned. This list more or less corresponds to the categories of mutations distinguished by Lapins (1983).

Broertjes & van Harten (1988) have presented for the different fruit crops lists of all mutant cultivars that were known to them at that time and, as can be also concluded from the FAO/IAEA Data Base published in *Mutation Breeding Newsletter* 39 (Maluszynski et al., 1992; closing date: January 1991) by the IAEA in Vienna, and during the following years only a few new cases have been reported. This may be partly due to the fact that at that time, basically, the intensive worldwide search for mutant cultivars by the IAEA staff had been abandoned (Micke, personal communication), probably because sufficient convincing evidence about valuable mutant cultivars had been brought together.

For fruit crops the IAEA list includes, for instance, nine mutant cultivars for apple (*Malus pumila*), seven for sweet cherry (*Prunus avium*), four for sour cherry (*Prunus cerasus*), three for orange/mandarin (*Citrus* sp.), two for grapefruit (*Citrus grandis*), two for peach (*Prunus persica*) and one for apricot (*Prunus armeniaca*), plum (*Prunus domestica*), olive (*Olea europaea*), grape (*Vitis vinifera*), black currant (*Ribes nigrum*) and a few less known crops.

This limited list indicates that mutation breeding in this category of crops, so far, has been considerably less successful than for instance in several vegetatively propagated ornamental crops. Broertjes & van Harten (1988), Spiegel-Roy (1990) and others as well have discussed the reasons why certain mutation breeding programmes have been successful and other projects failed. The use of well adapted, universally grown cultivars, the improvement of one or two – preferably monogenically inherited and easily recognizable – traits only, the scale of the trait, careful testing and selection, etc., are important factors in this respect.

It was mentioned before that in practically all cases mutations have been induced by radiation treatments. The only exception (see also Section 7.3.3): the apple mutant cv. Belréne with improved earliness and more even ripening, obtained by L. Decourtye and B. Lantin of the Institut National de Recherches Agronomiques, Angers, France, released in 1970, was obtained after treatment of growing shoots of cv. Reine des Reinettes in 1961 with 1% EMS. Because of its somewhat reduced yield, cv. Belrène has not become a widely grown cultivar. For more details and references concerning the release of this mutant see Broertjes & van Harten (1988).

In order to illustrate the mutation work that has been performed on fruit crops, some examples concerning mutant traits of (potential) economic importance will be briefly discussed in the following sections. Successful mutations for seedlessness and improved flesh colour in citrus have been discussed already in Section 7.4.3.

Disease resistance. Despite the importance of improved resistance against various diseases in fruit crops, only very few successful examples of induced mutations are known. In 1991 a mutant resistant to black spot disease (*Alternaria alternata* pv. Japanese pear) in Japanese pear (*Pyrus serotina*), discovered after chronic irradiation of grafted trees of the susceptible cv. Nijisseiki in the Gamma Field of the Institute of Radiation Breeding in Ohmiya-machi, Japan in 1981, was registered as a new mutant cultivar 'Gold Nijisseiki'. Black spot disease is one of the most serious diseases in Japanese pears. Resistant cultivars are homozygous recessive for this trait. Similar experiments with different cultivars and also with acute instead of chronic irradiation have also yielded positive results and local adaptability tests are under way (Sanada, 1986; Sanada, Nishida & Ikeda, 1986; Sanada *et al.*, 1993; Masuda *et al.*, 1994).

Mutants with still higher levels of resistance to black spot disease, obtained after acute or chronic radiation of cv. 'Gold Nijisseiki', were reported in the Annual Report 1995 of the National Institute of Agrobiological Resources (Anon., 1996). One mutant was tentatively called 'Super Gold'.

Not many other unequivocal examples of induced resistances against diseases and pests in fruit crops are known. In two cases a difference in susceptibility to apple powdery mildew (*Podosphaera leucotricha*) was observed in induced mutants of cv. McIntosh (McIntosh & Lapins, 1966) and cv. Cox (Campbell & Wilson, 1977). The induction of resistance against downy mildew (*Plasmopora viticola*) in grapevine (*Vitis vinifera*) was reported by Coutinho (1975).

Colour and structure of fruit skin. Hundreds of spontaneous sports for improved fruit colour, e.g. by increased reddening of the skin, are known in particular in several well-known cultivars of apples. For other cultivars such mutations may show up as rather exceptional cases only. According to Spiegel-Roy (1990) no useful red-skinned mutations have been found in apples that are entirely yellow or greenskinned.

Whereas Lapins (1983) expresses the opinion that skin colour of apple and pear fruits largely depends on the genetic constitution of L-2, Spiegel-Roy (1990) mentions that visible colour of the skin depends mainly on anthocyanins in the vacuoles of the cells in the epidermal layer (L-1) and in the three outer 'layers' of the hypodermis, which are of L-2 origin. The 'brightness' of the red colour is negatively correlated to the amount of chlorophyll in the L-2 layer of the skin. Production of anthocyanins, apparently, is based on the action of dominant genes. Irradiation treatments have resulted in mutations towards more or less anthocyanin in the skin and towards a darker or more bright red colour. A genetic explanation for these phenomena has not been found in literature. As an example: Ikeda (1974) reported on the induction of improved (increased) skin colour mutants of the outstanding apple cultivar Fuji in Japan. The mutants resulted from a (semi-chronic) mutagenic treatment for 10 days in the gamma field from the Institute of Radiation Breeding in Ohmiya, Japan. Skin colour mutants in apple may be periclinal chimeras. In citrus the rind colour is said to be determined by the genetic constitution of the L-2.

Mutagenic treatment may also affect russeting. One of the most successful radiation induced mutants is a 'russet free' mutant obtained in France from the well-known apple cv. Golden Delicious (the most important cultivar in that country at that time), called cv. Lysgolden (also called Goldenir). This cultivar, with slightly reduced yield, was released in 1970 by L.Decourtye and B.Lantin of the Fruit Research Institute INRA in Angers, France, after mutagenic treatment with 50 Gy of gamma rays in 1963. According to unconfirmed information from a source that could not be traced again, during the period 1985–90 one third of all the apple trees sold in France should have referred to cv. Lysgolden. However, from official data obtained from the Dutch Research Station for Fruit Growing in Wilhelminadorp, The Netherlands (H.Kemp, personal communication), it could only be retrieved that the area on which cv. Lysgolden was grown in France increased from 197 ha in 1985 to 587 ha in 1995. According to the same source, the total area in France on which cv. Golden Delicious and related mutant cultivars were grown during that period was 15 000–20 000 ha.

Compact tree growth and rootstocks. Compact trees (with reduced size) are much favoured in modern plantations because they allow, for instance, dense planting, easier pruning and mechanical harvesting. Spontaneous mutations are known for this trait, in particular in apple and many breeding efforts are directed towards the production of compact types. As the only alternative, mutation breeding may be used to induce compact growth. Lapins (1983) points out that the compact growth habit is mainly determined by genetic constitution of the L-3 layer. Sexual transmittance of this trait would be possible only when the mutant gene(s) for compact growth are present as well in the L-2 and, as a consequence, failure to sexually transmit this trait may be explained by the presence of the compact mutation in L-3 only.

In sweet cherry (*Prunus avium*) compact types are very much desired because of the high picking costs.

Reversion to the original type may occur and in some cases growth vigour may be lower, which meant that induced mutants for compact growth are not widely used. Nevertheless, several economically interesting compact mutants have been induced by irradiation in Canada, starting already from 1964 and, in later years, in the USA and in Italy. Mutagenic treatments in most cases were performed with gamma rays in the range of 30–60 Gy. Lapins (1983; see also earlier references, mentioned in this publication) reported that the self-compatible, compact growing mutant cv. Compact Stella found ready commercial acceptance in N.America. Spiegel-Roy (1990) comments that it is claimed that the Italian compact mutants in sweet cherry, in contrast to the compact mutants that were produced in Canada and the USA, do not show reversion. For more details and many references, see Broertjes & van Harten (1988) and Spiegel-Roy (1990).

Interesting work on apples has been performed in the 1970s in Poland by Zagaja and associates (Zagaja, 1976; Zagaja & Przybyla, 1973; 1976). The original aim was to develop compact type mutants in apple. For this purpose one-year-old dormant shoots of various cultivars were irradiated with 25 Gy of gamma rays. So-called three-bud scions were grafted on 10-year-old trees. Procedures were according to Zwintzscher (1962). Starting from 1974 the mutants were grown in a trial orchard and Zagaja & Przybyla (1976) reported that several compact types showed the favourable trait of a much lower growth vigour than the known spontaneous mutant compact type. The authors explained that at least three vegetative propagations were required to assess the stability of new clones. Other authors, however (e.g. Lacey, see later this section) report that two vegetative propagations may suffice to reach stability. Experiments similar to those for apple were also performed by the Polish group in sour cherry (*Prunus cerasus*). Starting from 1972, Zagaja and colleagues irradiated dormant rooted 'yearlings' of apple with gamma rays in order to induce less vigorous rootstocks. Although the authors reported that promising results had been obtained (Zagaja & Przybyla, 1976), and referred to further experiments to assess the results obtained so far, we have not come across results from later years in the literature.

Much work has been performed also with apple in various other countries. Ikeda (1977) also reported interesting results concerning dwarf growth mutants in apple in Japan after treatments with up to 50 Gy chronic gamma radiation at a dose rate of 0.2 Gy per day. Mention was made before to the work by Visser and associates in the Netherlands (Visser, 1973; Visser et al., 1969; 1971). Compact types, obtained after irradiation of dormant scions with 45–70 Gy were grafted on rootstocks and after three successive years of selection, 4 out of every 5 clones were found to be stable both in apple and pear.

The most extensively studied example of inducing compactness by mutation breeding refers to the work by Lacey & Campbell in the UK with the apple cv. Bramley's Seedling. This more than 100 years old, overvigorous triploid cultivar, for which no natural compact types were known, was one of the two most planted cultivars in the UK in the 1980s. The well-organized work of the aforementioned workers is documented in various papers (Campbell & Lacey, 1973; Lacey, 1982; Lacey & Campbell, 1979a, b, c; 1987). Starting from 1970, dormant scions were irradiated with 60 Gy of gamma rays (from ^{60}Co at a dose rate of 10 Gy. h^{-1}), followed by selection in vM_2 and vegetative propagation (ten cuttings for each selected type). Twelve compact types have been selected among 100 compact vM_3 clones. Evaluation was continued for more than seven years with seven clones that looked uniform without detrimental side-effects, e.g. did show a yield at or near to that of the control. Both periclinal mutants – some of which were sufficiently stable to be commercially acceptable – and (fully stable) homohistonts were obtained. However, as Spiegel-Roy (1990) reported, apple trees of the Bramley's type – for economic reasons – are not much replanted any more in the UK. As a consequence, it is most unlikely that the induced improved compact mutants of this type will ever be used on a large scale.

For apple, two gamma ray-induced mutant cultivars with drastically reduced tree size: cvs. Courtagold and Courtavel, produced by Decourtye and associates in Angers, France, were released in 1972. Spiegel-Roy (1990) commented that chances that compact types should be used in practice look rather small as many natural mutants for these and other traits occur as well and also because nowadays the use of rootstocks in apple enables growers to manipulate tree size in this way.

Conclusion

A general conclusion about the merits of mutation breeding in fruit crops, drawn already by Lapins (1983) and completely in line with the prevailing opinions expressed by many other authors and on various occasions (Broertjes & van Harten, 1978; 1988), is that induced mutations may supplement, rather than replace cross breeding and, at the same time, that spontaneous or induced mutants, changed in one or two traits, constitute valuable parent material for further crossing work.

7.4.7. **Induced mutations in vegetatively propagated grasses, with special reference to bermuda turfgrass**

Grasses, in particular turfgrasses which are used for lawns and recreational areas, are economically important and the turfgrass industry, in several countries, is one of the most rapidly growing plant-related industries. In the USA alone, for instance, it is estimated that in the 1990s over 10 million hectares (lawns, playing grounds, etc.) were grown with various turfgrasses.

A description of the prospects of mutation breeding in grasses is limited here to some grasses that can be vegetatively propagated, either by apomixis (asexual reproduction through seed, see also Hanna, 1995a) or by stolons, runners or other vegetative plant parts. Two categories of apomictic plants can be distinguished: facultative apomicts like *Poa pratensis* (meadow grass, smooth stalked meadow grass, blue grass or Kentucky blue grass), and the (sub)tropical *Panicum maximum*, and obligate apomicts to which group, for instance, belongs (see Burton, 1979) the pentaploid form ($2n = 5x = 50$) of the tropical grass *Paspalum dilatatum* (Dallis grass).

For obligate apomicts, vegetative propagation is the only possible way of propagation and the same applies to sterile interspecific cultivars, which for instance may refer to triploid or tetraploid F_1 hybrids. In such situations spontaneous or induced mutations would be indispensable as the only way to obtain genetic variation. Mutations – if favourable – could be directly used as new cultivars. Reference should be made in this context to a review on apomixis by Vielle Calzada, Crane & Stelly (1996). According to these authors, inheritance studies do suggest that recurrent apomixis is controlled by a single dominant locus. They further mention that, so far, no recurrently apomictic mutants have been recovered which, in their opinion, suggests that recurrent apomixis requires 'gains in function'.

In Sweden, Julen (1954, 1958) and Gustafsson & Gadd (1965) already showed that X-irradiation of cv. Falking of *Poa pratensis* (meadow grass) increased the frequency of plantlets showing morphological aberrations from 2 to 14%. Most aberrant types were weaker, but some more vigorous mutant plants were observed as well. Much attention was paid in these studies to cytogenetic constitution and mode of reproduction. Hanson & Juska (1959; 1962) irradiated cv. Merion of *Poa pratensis* (Kentucky bluegrass) with thermal neutrons and obtained a higher resistance to stem rust (*Puccinia graminis*). However, as was discovered later on, the mutant, unfortunately, lost resistance to another disease (*Drechslera poae*).

Very few cases of officially registered or released induced mutant cultivars for grass species are known and before discussing some examples more in detail, these mutants will be briefly mentioned here. Broertjes & van Harten (1988) refer to four reported cases of commercialized mutants for grasses. Best known became cv. Tifway-2, an interspecific hybrid (*Cynodon dactylon* × *C. transvaalensis*), officially registered in 1981. This cultivar belongs to the important group of bermudagrasses (*Cynodon* sp.) and will be discussed later in this section. Cultivars of bermudagrass are mainly used as turfgrasses, in particular on lawns and golf courses in the USA. Taliaferro (1995) pointed out that, until recently, practically all turf cultivars of *Cynodon* were single plant, clonally propagated selections.

The other mutant cultivars mentioned by Broertjes & van Harten were cv. AU Centennial for the turf grass *Eremochloa ophiuroides* (Centipedegrass) and cvs. TXSA 8202 and 8212 for *Stenotaphrum secundatum* (St. Augustine grass). Centipedegrass is a perennial grass that can be propagated vegetatively or by seed. The Centipedegrass mutant cultivar 'AU Centennial', which was registered in the USA by Pedersen & Dickens (1985), was obtained after irradiation in 1976 of a population of seeds (caryopses) from *E. ophiuroides* with 300–400 Gy of gamma rays. Starting from approximately 8000 seedlings, 95 plants with good turf potential, plant vigour and morphological aberrations were selected after four months. Continued selection led to 44 clones which were multiplied and studied in plots of about 0.9 × 1.5 m (with three replicates). From these 44 mutant clones the aforementioned mutant was selected in 1983 as the most desirable turf type. Mutant cultivar AU Centennial is a vegetatively propagated dwarf clone with higher leaf density, shorter internodes, shorter 'seed heads' and a darker green colour than the 'common' Centipedegrass that was used as starting material. Hanna (1995b), in addition, refers to his own unpublished data about three to five cycles of recurrent gamma irradiation (dose 100 Gy) to which seed of 'common' Centipedegrass had been submitted in the 1980s in order to increase winterhardiness. Plant material (code TC312) with better cold tolerance as well as tolerance to low soil pH was obtained, but whether this material has been officially registered as a new cultivar was not mentioned.

The two mutants for the vegetatively propagated, triploid St. Augustine grass were registered for the USA as 'germplasm' carrying resistance to a specific strain of *Panicum* mosaic virus by Toler *et al.* (1985). Also in this case it is not known whether they ever reached the stage of cultivar. The mutants resulted from irradiation of single node stolon cuttings of cv. Floratam with 45–70 Gy of gamma rays.

The aforementioned mutant cultivars were also included in the FAO/IAEA mutant varieties database (see Maluszynski *et al.*, 1992, in *Mutation Breeding Newsletter* 39). This list also contained a second mutant cultivar for bermudagrass (that had remained unnoticed by Broertjes & van Harten, 1988): cv. Tifgreen-2, which was officially registered in 1983 (see also *Mutation Breeding Newsletter* 33, Anon., 1989a). Some additional remarks about the bermudagrasses will be made further in this section.

Probably the most important stimulus to mutation breeding in grasses was given in the late 1960s by the large-scale pioneering work of Burton and associates at the Coastal Plains Experiment Station at Tifton, Georgia, USA. The work of these authors has been documented in an extensive range of publications, dealing mainly with bermudagrasses for turf but also as a forage grass, some other grasses and some apomictic crop plants, as well as sorghum and millets, all belonging to the family of the *Graminaea* (Burton, 1974; 1975; 1976; 1979; 1981; 1985; Burton *et al.*, 1980; Burton & Hanna, 1977; Burton, Hanna & Powell, 1982; Burton & Jackson, 1962; Hanna, 1990; and Powell, Burton & Young, 1974).

Burton and colleagues developed the predominant Tifton series of turf bermudagrasses, including the cvs. Tifgreen, Tifway and Tifdwarf, which are triploid F_1 hybrids, obtained from crosses between tetraploid *Cynodon dactylon* and diploid *C. transvaalensis*. These cultivars, which resulted from planned breeding programmes and largely replaced earlier selections within naturally occurring variation, are vegetatively propagated and sterile and, hence, cannot be further improved by conventional breeding methods. Cv. Tifdwarf, a turfgrass widely used on golf courses, is a presumed natural mutant of cv. Tifgreen.

Despite their many favourable qualities, the bermudagrasses of the 'Tif' series, for instance, lack resistance to nematodes and several diseases and, in addition, an improved level of winterhardiness would be desirable for certain regions. Because additional interspecific crosses did not result in the production of improved bermudagrasses, and stimulated by the success of the spontaneous mutant cv. Tifdwarf, Burton and Powell started in 1969 to produce mutants of cvs. Tifdwarf, Tifway and Tifgreen by treating cuttings with one or two nodes, derived from dormant stolons, with EMS and gamma rays. The intention was to produce cultivars which should retain most of the superior traits of the 'Tif' bermudagrasses but add variation for traits such as plant colour, size and resistance to pests and diseases.

The EMS series failed (see Burton, 1979), but gamma treatments of these cuttings in the winter of 1970–1 resulted in 71 mutants for various traits. Two years later irradiation of cuttings of cvs. Tifgreen, Tifdwarf and Tifway resulted in an additional 81 mutants. The applied doses of gamma ray varied from 70 to 120 Gy but best results were obtained when doses of 70–90 Gy of gamma rays from ^{60}Co were applied. The irradiated 'sprigs' consecutively were planted in sterile soil in the greenhouse and, after being well established, space planted in the field.

Burton (1976) and Burton & Hanna (977) reported that up to 6% of the M1 plants produced mutant clones in subsequent reproduction of which about 70% were found to be uniform after several cycles of vegetative propagation. A daily search in the field quickly revealed mutants with improved colour and such mutants were removed immediately and grown in the greenhouse in order to prevent them being overgrown by normal plants and being lost. Careful evaluation of the mutants for prolonged periods demonstrated that at least three years were required before the really good mutants could be identified with a reasonable certainty. Mutants smaller than the aforementioned (spontaneous mutant) cv. Tifdwarf proved not to be attractive in practice. Evaluation for nematode resistance, in particular to root knot nematode (*Meloidogyne graminis*), was found to be difficult because reliable laboratory screening methods were not available and uniform infestation in the field is difficult to achieve.

In 1972 and 1973 one million sprigs of cv. Coastcross-1 – a completely sterile tetraploid, highly digestible but not winterhardy forage (pasture) F_1 hybrid cultivar of bermudagrass (*Cynodon dactylon*), obtained from a cross between 'Coastal' bermudagrass and a bermudagrass from Kenya and released in 1967 – had been irradiated with 70 Gy of gamma rays at Oak Ridge, Tennessee, USA. Its susceptibility to winter killing meant that the cv. Coastcrosss-1 could not be grown in most areas of the USA. The main objective was therefore to induce mutations for winterhardiness; this goal, because of the complete sterility, cannot be reached by conventional breeding methods. The possibility to propagate useful mutants vegetatively, on the other hand, makes this approach very attractive. The irradiated sprigs were disked into the soil in an area with, in most cases, heavy winter frost. However, conditions with sufficient frost, necessary for testing for winterhardiness, did not occur before the winter of 1976–7. Burton (1976) mentioned that among the few plantlets from the irradiated material that survived the moderate winter conditions of that year, no plants were found that had spread by rhizomes. Irradiation of again one million sprigs in 1977 produced some chlorophyll-deficient mutants, but mutants with good rhizomes were not found (Burton, 1979). In a publication by Burton *et al.* (1980) and in a conference review presented by Burton in 1980 in Coimbatore, India (but printed two years later as Burton *et al.*, 1982), it was mentioned that one surviving mutant plant from the material that had been irradiated in 1971, called Coastcross-1-M3, developed a rhizome about 30 cm long, a characteristic that was common in 'Coastal' bermudagrass (one of the parents of Coastcross-1), but that was not observed in cv. Coastcross-1 itself. By vegetative propagation enough clones were produced to make larger replicated experimental plots of clones from cv. Coastcross-1, Coastcross-1-M3 and 'Coastal' bermudagrasses. During the severe winter of 1977–8, 98% of the grass in the clipped plots with cv. Coastcross-1 and with clonal material obtained from 'Coastcross-1-M3' was destroyed whereas plots with 'Coastal' bermudagrasses suffered very little loss of stand. It was concluded that the material derived from Coastcross-1-M3 was only slightly – but significantly – more winterhardy than Coastcross-1, but not enough to very much extend the zone in which material of cv. Coastcross-1 can be grown in the USA.

The authors further reported that in 1979 some 700 000 sprigs of Coastcross-1-M3 were irradiated and planted for further testing. As no later reports on this subject are known, it must be concluded that further experiments either yielded negative results or have been discontinued for some reason.

The previously mentioned mutant cv. Tifway-2 of (turf) bermudagrass was officially registered in 1981 and the main trait improved according *Mutation Breeding Newsletter* 18 (Anon., 1981b, pp. 8–10) was an increased nematode resistance, whereas some other traits also slightly differed from those for the original cv. Tifway. A second gamma-induced mutant, officially registered in 1983 as Tifgreen-2, was in particular characterized by an improved vigour under minimal management but also by much better nematode resistance than cv.Tifgreen (*Mutation Breeding Newsletter* 33, Anon., 1989a).

Recently, Taliaferro (1995) referred to a report by Hanna (1990) about the use of gamma rays for the production of dwarfed mutations of cv. Midiron, a cold tolerant interspecific triploid hybrid of *C. dactylon* × *C. transvaalensis*. In a short summary it was mentioned that fine-textured mutations had been induced after irradiating dormant rhizomes of cv. Midiron with 80 Gy of ^{60}Co gamma radiation. After 6 years of clipping at one-half inch height, about 65% of the 69 mutants that were originally selected had maintained the improved leaf structure.

Burton (1979) reported that mutagenic treatment of the (obligate apomictic) prostrate dallisgrass (*Paspalum dilatatum* var. *pauciciliatum*) for 20 h with thermal neutrons (flux 6.48 × 104 cm^{-2}. s^{-1}) resulted in a more than four-fold increase of the frequency of vegetative and floral mutants, but no increased forage yield, seed yield or resistances were observed. A broad-leaved mutant, which therefore looked attractive as having better forage quality, unfortunately, was less vigorous than the original type and it was impossible to transfer this 'leafy character' because of the obligate apomictic character of mutant as well as normal type.

Earlier, Burton (1974) mentioned irradiation experiments that had been started in 1971 with the outstanding apomictic hybrid 'Tifton 54' of bahiagrass (*Paspalum notatum*) with 40–120 Gy of gamma rays from a ^{60}Co source with the objectives to improve seed yield and the percentage of seed set, to break apomixis and to produce 'useful mutants'. No later reports on this topic have been found.

7.4.8. Amylose-free starch mutants of potato (*Solanum tuberosum*)

A publication by de Nettancourt & Dijstra (1969) showed that spontaneous mutations leading to so-called waxy-like microspores do occur at very low frequencies (i.e. two mutants among 2.5 million microspores analyzed) in the species *Solanum verrucosum*, a diploid, tuber-bearing wild relative of the common potato (*S. tuberosum*). Earlier work by other scientists on microspores of, for instance, maize (*Zea mays*) and tomato (*Lycopersicon esculentum*) had shown that normal (starchy) microspores contain two components: amylopectine and amylose , whereas in the starch of waxy-like microspores of the aforementioned crops the amylose component was absent. De Nettancourt & Dijstra (loc.cit) concluded that the same explanation might hold for the waxy-type microspores observed in *S. verrucosum*.

Amylose-less or amylose-free (amf-)microspores can be detected in a relatively simple way by staining with a solution of potassium-iodine (I_2-KI), or Lugol's solution. After treatment of the anthers with Lugol's solution a dark blue colour is observed in the microspores when both amylopectine and amylose are present, and a reddish-brown colour when the starch is amylose-free.

It was suggested by de Nettancourt & Dijstra that the observed waxy-like microspores of *S. verrucosum* may represent a spontaneous mutation in a locus controlling starch composition and that the observed low frequency of this mutation probably could be increased by mutagenic treatment of the flower buds. Their next suggestion was to produce (diploid) lines of *S. verrucosum* which breed true for the waxy character in the tubers, followed by crossing these lines with the common tetraploid potato.

Based on the aforementioned publication a group of scientists at the University of Groningen, the Netherlands, decided to investigate whether plants of the common potato (Solanum tuberosum) with amylose-free tubers could be produced. Such mutants would be highly interesting for the (Dutch) potato starch industry as they could considerably extend the range of applications of potato starch. To give an impression of present applications: about two thirds of the total starch production in Western Europe is used in food and beverages, as a thickener, sweetener, fat replacer or confectionary. The remaining part is used in non-food products like paper, textiles, biodegradable plastics, cosmetics, pharmaceuticals, etc. The Netherlands is the largest producer and exporter of industrial starch derived from potato. Starch, of course, can be obtained from various other crops as well.

As is the case for the composition of starch present in common microspores, the reserve starch of most plant species, accumulated in storage organs (endosperm, tubers) has two components: amylose and amylopectin. Amylose is a linear polymer of 100–10 000 glucose units linked by $\alpha(1,4)$ linkages. In potato about 20% of the tuber starch consists of amylose. The other component, amylopectin, represents about 80% of the total starch in potato and is characterized by short linear $\alpha(1,4)$ chains which are connected by $\alpha(1,6)$ linkages.

Mutants which affect the size of the contribution by one of the components have been known for many years in a number of seed propagated crops like maize, barley and rice. The best-known example refers to the amylose-free (waxy) mutant for the endosperm of maize (Zea mays). At the time that the experiment in Groningen started, no mutants were known in which the composition of the reserve starch in a vegetative storage organ like the potato tuber had been altered.

As the common potato is a tetraploid crop ($2n = 4x = 48$), selection of recessive single gene mutants in the M_2 generation is hardly possible. Experiments, therefore were started with a monoploid (or monohaploid, i.e. $2n = x = 12$) potato clone in which recessive mutations are not masked by dominant alleles for the same gene and, thus, can be immediately observed. As, occasionally, the aforementioned reddish-brown colour (which reveals the absence of amylose) had been observed in some cells of this monoploid clone after exposure to a diluted Lugol's solution, it was decided to start an in vitro mutation programme in order to obtain solid amylose-free mutant plants from this clone.

Leaves were irradiated with 8.5. Gy of X-rays (dose rate 1.9 Gy. min⁻¹) just before transferring them to the in vitro medium (for details see Hovenkamp-Hermelink et al., 1987), at which dose still moderate regeneration took place. From the adventitious shoots that regenerated in vitro on the irradiated leaf explants, stem segments were taken which, subsequently, produced so-called minitubers (Hovenkamp-Hermelink et al., 1988). Starch composition was analyzed in tubers. For characterization of starch in tubers the Lugol test was somewhat adapted. After screening of about 12 000 minitubers an amylose-free potato mutant was obtained. Biochemical and molecular methods further proved that the character amylose-free (amf) is determined by the absence of the GBSS gene, a gene that controls the synthesis of the enzyme granule bound starch synthase (or GBSS). The amf-character appears to be a monogenic Mendelian recessive gene. The phenotype of the amf-mutant can be detected in various plant parts, such as the columella cells of root tips, plastids of the guard cells in stomata, microspores and tubers (Jacobsen et al., 1989; 1991).

The amf-mutant which originated from a monohaploid clone, not unexpectedly, showed many drawbacks, such as lack of vigour and sterility. Hovenkamp-Hermelink et al. (1988) and Jacobsen et al. (1989) reported that the obtained mutant after the in vitro phase, rather unexpectedly, still had the monoploid constitution ($2n = x = 12$), but that a second round of adventitious shoot formation yielded a diploid type. The method has been described already by Hermsen et al. (1981). As the 'doubled' amf-clones of the original amf-monoploid (clone 86.040) did show good flowering but poor pollen stainability of only 8.2%, they were used as female parents in crosses with diploid amylose-containing plants. From over 1400 pollinations only 13 berries were obtained which dropped off prematurely. Nevertheless, by embryo rescue 20 F_1 plants could still be obtained from the 78 ovules present in those berries. The resulting F_1 plants all contained amylose starch and were used for further studies on the genetic nature of the amf-character (see before).

Subsequently, the amf-character had to be incorporated into suitable diploid clones that produce high frequencies of so-called 2n-gametes. Such clones might be used to return to the tetraploid ($2n = 4x = 48$) level of the common potato, which is an absolute 'must' to produce amf-potato cultivars which can be used in the starch industry. Suitable diploid clones must be highly vigorous, fertile, devoid of lethal genes, have a broad genetic base and combine good agronomic characters. To obtain such clones and to develop tetraploid cultivars from the original recessive mutant by conventional breeding methods requires much time but may be highly rewarding. In the late 1980s research on amf-mutants was transferred from Groningen University to the Department of Plant Breeding of Wageningen Agricultural University where the original radiation-induced amf-mutant and its derivatives are still used extensively for fundamental studies concerning, for instance, formation of 2n-gametes and 2n-eggs by E. Jacobsen, M.S. Ramanna and associates. The amf-mutant is also still used as basic plant for producing amf-cultivars of potato by combining methods of plant breeding and plant biotechnology.

In the meantime, the Dutch starch industry and the Department of Plant Breeding in Wageningen together have developed tetraploid amf-clones with acceptable agronomic characters. These clones, however, did not result from the aforementioned radiation induced amf-mutant, but by applying the so-called antisense technique (see also Chapter 1). Two clones obtained in this way were submitted in 1996 for official registration in the Netherlands as new amylose-free potato cultivars and were grown even at that time (with official permission) on several hundred hectares. The somewhat lower tuber yield of the amf-clones when compared with the common yield of potatoes for the starch industry, is compensated by a higher kilogramme price received by the farmer for amf-cultivars. This, however, is common practice, for instance for sugarbeet (*Beta vulgaris*) where farmers are paid a premium for higher sugar content.

Finally, it may be expected that further (conventional) breeding work, starting either from the mutant resulting from the irradiated material or from the 'antisense' material, in due course will lead to potato cultivars for starch production with yields similar to those of the potato cultivars that are commonly grown for starch production.

References

Abbott, A.J. & Atkin, R.K. (eds). (1987). *Improving Vegetatively Propagated Crops.* London: Academic Press.

Adams, M.W. (1982). Plant architecture and yield breeding. *Iowa State Journal of Research*, **56**, 225–54.

Ahloowalia, B.S. (1986). Limitations to the use of somaclonal variation in crop improvement. In *Somaclonal Variation and Crop Improvement*, ed. J. Semal, pp.14–27. Dordrecht: Martinus Nijhoff.

Ahloowalia, B.S. (1995). In vitro mutagenesis for the improvement of vegetatively propagated crops. In *Induced Mutations and Molecular Techniques for Crop Improvement* (Proceedings FAO/IAEA Symposium, Vienna), pp.531–41. Vienna: IAEA.

Ahmad, H. & Khawaja, H.I.T. (1992). Radiation-induced nullisomy in *Hordeum vulgare* L. *Plant Breeding*, **108**, 332–4.

Ahnström, G. (1989). Mechanism of mutation induction. In *Science for Plant Breeding* (Proceedings 12th Congress EUCARPIA, Göttingen, Germany F.R.), pp.153–60. Berlin: Paul Parey.

Alpen, E.L. (1990). *Radiation Biophysics.* London: Prentice-Hall Int.

Altenburg, E. (1934). The artificial production of mutants by ultraviolet light. *The American Naturalist*, **68**, 491–507.

Ames, B.N. (1989). Endogenous DNA damage as related to cancer and aging. *Mutation Research*, **214**, 41–6.

Andersen, A.L. (1972). The Sanilac story. In *Mutation Breeding Workshop* (Proceedings Mutation Breeding Workshop, University of Tennessee, Knoxville, USA). Knoxville, Tennessee, USA: Agricultural Experiment Station, University of Tennessee.

Andersson, G. & Olsson, G. (1954). Svalöfs *Primex* white mustard. A market variety selected in X-ray treated material. *Acta Agriculturae Scandinavica*, **4**, 574–7.

Ando, T., Akiyama, Y. & Yokoi, M. (1986). Flower colour sports in *Saintpaulia* cultivars. *Scientia Horticulturae*, **29**, 191–7.

Anonymous (1857). *Gardeners Chronicle*, p.613 & p.629 (original not consulted).

Anonymous (1951). *Genes and Mutations.* Cold Spring Harbor Symposia on Quantitative Biology, vol.16. Cold Spring Harbor, Long Island, New York: The Biological Laboratory.

Anonymous (1955). *Mutation* (Brookhaven Symposia in Biology), Number 8. Upton, New York: Biology Department, Brookhaven National Laboratory.

Anonymous (1956a). *Radioactive Isotopes and Ionizing Radiations in Agriculture, Physiology and Biochemistry* (Proceedings of the International Conference on the Peaceful Uses of Atomic Energy, Geneva, 1955), vol.12. New York: United Nations.

Anonymous (1956b). *Genetics in Plant Breeding* (Brookhaven Symposia in Biology), Number 9. Upton, New York: Biology Department, Brookhaven National Laboratory.

Anonymous (1958). *Isotopes in Agriculture* (Proceedings of the Second United Nations International Conference on the Peaceful Uses of Atomic Energy, Geneva), vol.27. New York, USA: United Nations.

Anonymous (1959). *'EUCARPIA' Second Congress of the European Association for Plant Breeding.* Köln, Germany.

Anonymous (1961a). *Mutation and Plant Breeding* (Proceedings Symposium Cornell University, Ithaca, N.Y., 1960). National Academy of Sciences-National Research Council, Publ. 891, Washington D.C.

Anonymous (1961b). *Effects of Ionizing Radiations on Seeds* (Proceedings FAO/IAEA Symposium, Karlsruhe, 1960). Vienna: IAEA.

Anonymous (1961c). *Fundamental Aspects of Radiosensitivity* (Brookhaven Symposia in Biology), Number 14. Upton, New York: Biology Department, Brookhaven National Laboratory.

Anonymous (1965). *The Use of Induced Mutations in Plant Breeding* (Report of the FAO/IAEA Technical Meeting, Rome, 1964). Oxford: Pergamon Press.

Anonymous (1967). *Use of Chronic Irradiation in Mutation Breeding.* Gamma Field Symposia 6, Ohmiya-machi, Ibaraki-ken, Japan: Institute of Radiation Breeding MAF.

Anonymous (1969). *Induced Mutations in Plants* (Proceedings FAO/IAEA Symposium on the Nature, Induction and Utilization of Mutations in Plants, Pullman, Washington). Vienna: IAEA.

Anonymous (1970). *Manual on Mutation Breeding.* Technical Reports Series No.119. Vienna: IAEA.

Anonymous (1971). *Mutation Breeding for Disease Resistance* (Proceedings FAO/IAEA Panel, Vienna, 1970). Vienna: IAEA.

Anonymous (1973a). *Induced Mutations in Vegetatively Propagated Plants* (Proceedings FAO/IAEA Panel, Vienna, 1972). Vienna: IAEA.

Anonymous (1973b). *Induced Mutation and Chimera in Woody Plants.* Gamma Field Symposia 12. Ohmiya-machi, Ibaraki-ken, Japan: Institute of Radiation Breeding NIAS MAF.

Anonymous (1974a). *Mutation Breeding Newsletter 4.* Vienna: IAEA.

Anonymous (1974b). *Induced Mutations for Disease Resistance in Crop Plants* (Proceedings FAO/IAEA/SIDA Research Co-ordination Meeting, Novi Sad, 1973). Vienna: IAEA.

Anonymous (1974c). *Polyploidy and Induced Mutations in Plant Breeding* (Proceedings of two meetings, Joint FAO/IAEA Division and EUCARPIA, Bari, Italy, 1972). Vienna: IAEA.

Anonymous (1975a). *Improvement of Vegetatively Propagated Plants through Induced Mutations* (Proceedings FAO/IAEA Research Co-ordination Meeting, Tokai, Japan, 1974). IAEA-TECDOC-173. Vienna: IAEA.

Anonymous (1975b). *Efficiency of Breeding Method for Screening of Mutants.* Gamma Field Symposia 14. Ohmiya-machi, Naka-gun, Ibaraki-ken, Japan: Institute of Radiation Breeding NIAS MAF.

Anonymous (1976a). *Induced Mutations for Disease Resistance in Crop Plants* (Proceedings FAO/IAEA/SIDA Research Co-ordination Meeting, Ames, Iowa, USA, 1975). IAEA-TECDOC-181. Vienna: IAEA.

Anonymous (1976b). *Improvement of Vegetatively Propagated Plants and Tree Crops through Induced Mutations* (Proceedings Second FAO/IAEA Research Coordination Meeting, Wageningen, The Netherlands). IAEA-TECDOC-194. Vienna: IAEA.

Anonymous (1976c). *Induced Mutations in Cross-Breeding* (Proceedings FAO/IAEA Advisory Group Meeting, Vienna, 1975). Vienna: IAEA.

Anonymous (1976d). *Mutants in Physiological Research of Crop Plants.* Gamma Field Symposia 15. Ohmiya-machi, Ibaraki-ken, Japan: Institute of Radiation Breeding NIAS MAFF.

Anonymous (1977a). *Manual on Mutation Breeding* (2nd edn). Vienna: IAEA.

Anonymous (1977b). *Induced Mutations Against Plant Diseases* (Proceedings FAO/IAEA/SIDA Symposium, Vienna). Vienna: IAEA.

Anonymous (1978). *Plant Breeding for Resistance to Insect Pests: Considerations about the Use of Induced Mutations* (Proceedings FAO/IAEA Advisory Group Meeting, Dakar, Senegal, 1977). IAEA-TECDOC-215. Vienna: IAEA.

Anonymous (1979a). *Seed Protein Improvements in Cereals and Grain Legumes* (Proceedings FAO/IAEA/GSF Symposium, Neuherberg, Germany, 1978), vol.1 & 2. Vienna: IAEA.

Anonymous (1979b). *Induced Mutations for Crop Improvement in Africa* (Proceedings Regional FAO/IAEA/IITA Seminar, Ibadan, Nigeria, 1978). Vienna: IAEA.

Anonymous (1981a). *Induced Mutations – A Tool in Plant Research* (Proceedings FAO/IAEA Symposium, Vienna). Vienna: IAEA.

Anonymous (1981b). *Mutation Breeding Newsletter 18.* Vienna: IAEA.

Anonymous (1982a). *Induced Mutations in Vegetatively Propagated Plants. II. (Proceedings Final FAO/IAEA Research Co-ordination Meeting, Coimbatore, India, 1980).* Vienna: IAEA.

Anonymous (1982b). *Mutation Breeding Newsletter 19.* Vienna: IAEA.

Anonymous (1982c). *Improvement of Oil Seed and Industrial Crops by Induced Mutations* (Proceedings FAO/IAEA Advisory Group Meeting, Vienna, 1980). Vienna: IAEA.

Anonymous (1983a). *Induced Mutations for Disease Resistance in Crop Plants II* (Proceedings FAO/IAEA/SIDA Research Co-ordination Meeting, Risø, Denmark, 1981). Vienna: IAEA.

Anonymous (1983b). *Chimerism in Irradiated Dicotyledonous Plants* (Reports FAO/IAEA Consultants' Meeting, Vienna, 1981). IAEA-TECDOC-289. Vienna: IAEA.

Anonymous (1983c). *Mutation Breeding Newsletter 22.* Vienna: IAEA.

Anonymous (1983d). *Mutation Breeding Newsletter 23,* p.20. Vienna: IAEA.

Anonymous (1984a). *Cereal Grain Protein Improvement* (Proceedings Final FAO/IAEA/GSF/SIDA Research Co-ordination Meeting, Vienna, 1982). Vienna: IAEA.

Anonymous (1984b). *Induced Mutations for Crop Improvement in Latin America* (Proceedings Regional FAO/IAEA Seminar, Lima, Peru, 1982). IAEA-TECDOC-305. Vienna: IAEA.

Anonymous (1985). *Mutation Breeding for Disease Resistance Using In-Vitro Culture Techniques* (Report FAO/IAEA Advisory Group Meeting, Vienna, 1984). IAEA-TECDOC-342. Vienna: IAEA.

Anonymous (1986a). *In Vitro Technology for Mutation Breeding* (Report of two FAO/IAEA Co-ordination Meetings, Vienna 1983 and 1985). IAEA-TECDOC-392. Vienna: IAEA.

Anonymous (1986b). *Nuclear Techniques and In Vitro Culture for Plant Improvement* (Proceedings FAO/IAEA Symposium, Vienna, 1985). Vienna: IAEA.

Anonymous (1987). *Improvement of Root and Tuber Crops by Induced Mutations* (Conclusions and Recommendations of 2 FAO/IAEA Research Co-ordination Meetings, Pattaya, 1984/Vienna, 1986). IAEA-TECDOC-411. Vienna: IAEA.

Anonymous (1988a). *Mutation Breeding Newsletter 31.* Vienna: IAEA.

Anonymous (1988b). *Evaluation and Availability of the Mutations for Disease Resistance.* Gamma Field Symposia 27. Ohmiya-machi, Naka-gun, Ibaraki-ken, Japan: Institute of Radiation Breeding.

Anonymous (1989a). *Mutation Breeding Newsletter 33.* Vienna: IAEA.

Anonymous (1989b). *Production of Mutants in Tree Crops.* Gamma Field Symposia 28.Ohmiya-machi, Naka-gun, Ibaraki-ken, Japan: Institute of Radiation Breeding NIAR MAFF.

Anonymous (1989c). *Plant Domestication by Induced Mutation* (Proceedings FAO/IAEA Advisory Group Meeting, Vienna, 1986). Vienna: IAEA.

Anonymous (1990a). *Mutation Breeding Newsletter 35.* Vienna: IAEA.

Anonymous (1990b). *Plant Tissue Culture Techniques for Mutation Breeding: A Training Manual.* Plant Breeding Unit, Joint FAO/IAEA Programme, IAEA Laboratories-Seibersdorf. Seibersdorf, Austria: IAEA.

Anonymous (1991a). *Mutation Breeding Newsletter 38.* Vienna: IAEA.

Anonymous (1991b). *Mutation Breeding for Crop Improvement* (Proceedings FAO/IAEA Symposium, Vienna, 1990), vols 1 and 2. Vienna: IAEA.

Anonymous (1991c). *Annual Report National Institute of Agrobiological Resources* (NIAR), Kannondai, Tsukuba, Ibaraki, Japan.

Anonymous (1992a). *Utilization of Novel Mutants for Development of Biological Functions in Crops.* Gamma Field Symposia 31. Ohmiya-machi, Naka-gun, Ibaraki-ken, Japan: Institute of Radiation Breeding NIAR MAFF.

Anonymous (1992b). *Mutation Breeding Newsletter 39.* Vienna: IAEA.

Anonymous (1993). *Mutation Breeding Newsletter 40.* Vienna: IAEA.

Anonymous (1994a). *Mutation Breeding of Oil Seed Crops* (Proceedings FAO/IAEA Research Co-ordination Meeting, Vienna, 1993). IAEA-TECDOC-781. Vienna: IAEA.

Anonymous (1994b). *Annual Report 1994.* National Institute of Agrobiological Resources. Kannondai, Tsukuba, Ibaraki 305, Japan.

*Anonymous (1994c). *Mutation Breeding Newsletter 41.* Vienna: IAEA.

Anonymous (1995). *Induced Mutations and Molecular Techniques for Crop Improvement* (Proceedings FAO/IAEA Symposium, Vienna). Vienna: IAEA.

Anonymous (1996). *Annual Report 1995.* National Institute of Agrobiological Resources. Tsukuba, Japan, p.53.

Arihara, A., Kita, T., Igarashi, S., Goto, M. & Irikura, Y. (1995). White Baron: A non-browning somaclonal variant of Danshakuimo (Irish Cobbler). *American Potato Journal,* **72,** 701–5.

Ashri, A. (1968). Genic-cytoplasmic interactions affecting growth habit in peanuts (*A. hypogaea*), II. A revised model.*Genetics,* **60,** 807–10.

Ashri, A. & Levy, A. (1974). Sensitivity of developmental stages of peanut (*A. hypogaea*) embryos and ovaries to several chemical mutagen treatments. *Radiation Botany,* **14,** 223–8.

Ashri, A., Offenbach, R., Cahaner, A. & Levy, A. (1977). Transmission of acriflavine-induced trisomic mutants affecting branching pattern in peanuts, *Arachis hypogaea* L. *Zeitschrift für Pflanzenzüchtung,* **79,** 210–8.

Asseyeva, T. (1927). Bud mutations in the potato and their chimerical nature. *Journal of Genetics,* **19,** 1–26.

Asseyeva, T. & Blagovidova, M. (1935). (Artificial mutations in the potato). *Bulletin for Applied Botany, Genetics and Plant Breeding.* (Leningrad) Ser.A (15), 81–5 (in Russian).

Atkinson, G.F. (1897). Report upon some preliminary experiments with the Röntgen rays on plants. *Nature,* **56,** 600.

Auerbach, C. (1943). Chemical induced mutations and re-arrangements. *Drosophila Information Service,* **17,** 48–50.

Auerbach, C. (1947). The induction by mustard gas of chromosomal instabilities in *Drosophila melanogaster.* *Proceedings of the Royal Society, Edinburgh, UK,* Section B, Vol.62, 307–20.

Auerbach, C. (1951). Induction of changes in genes and chromosomes. Problems in chemical mutagenesis. In *Genes and Mutations. Cold Spring Harbor Symposia on Quantitative Biology,* vol.16, pp.199–213. Cold Spring Harbor, Long Island, New York: The Biological Laboratory.

Auerbach, C. (1960). Chemical mutagenesis in animals. In *Chemische Mutagenese* (Proceedings Erwin-Baur-Gedächtnisvorlesungen I, Gatersleben, 1959), ed. H.Stubbe, pp.1–3. Berlin: Akademie-Verlag.

Auerbach, C. (1961). Chemicals and their effects. In *Mutation and Plant Breeding* (Proceedings Symposium Cornell University, Ithaca, N.Y., USA, 1990), Publ.891, 120–44. Washington D.C., USA: National Academy of Sciences-National Research Council.

Auerbach, C. (1976). *Mutation Research (Problems, results and perspectives).* London: Chapman and Hall.

Auerbach, C. (1978). Forty years of mutation research: a pilgrim's progress. *Heredity,* **40,** 177–87.

Auerbach, C. & Kilbey, B.J. (1971). Mutation in Eukaryotes. *Annual Review of Genetics,* **5,** 163–218.

Auerbach, C. & Robson, J.M. (1944). Production of mutations by allyl isothiocyanate. *Nature,* **154,** 81.

Auerbach, C. & Robson, J.M. (1946). Chemical production of mutations. *Nature,* **157,** 302.

Auerbach, C. & Robson, J.M. (1947). The production of mutations by chemical substances. *Proceedings of the Royal Society, Edinburgh, UK,* Section B, Vol.62, 271–83.

Auerbach, C., Robson, J.M. & Carr, J.G. (1947). The chemical production of mutations. *Science,* **105,** 243–7.

Auerbach, C. & Westergaard, M. (1960). A discussion of mutagenic specificity. In *Chemische Mutagenese* (Proceedings Erwin-Baur-Gedächtnisvorlesungen I, Gatersleben, 1959), ed. H.Stubbe, pp.116–23. Berlin: Akademie-Verlag.

Aung, T. & Thomas, H. (1976). Transfer of mildew resistance from the wild oat *Avena barbata* into the cultivated oat. *Nature,* **260,** 603–4.

Auni, S., Daskalov, S. & Filev, K. (1978). Radiogenetic effect of gamma irradiation under different ontogenetic states of sweet pepper. *Comptes Rendus de L'Academie Bulgare des Sciences,* **31,** 1357–60 (original not consulted).

Avery, A.G., Satina, S. & Rietsema, J. (1959). Extra chromosomal types. In: Blakeslee: *The Genus Datura,* pp.86–109. New York: The Ronald Press Company.

Awoleye, F., van Duren, M., Dolezel, J. & Novak, F.J. (1994). Nuclear DNA content and *in vitro* induced somatic polyploidization cassava (*Manihot esculenta* Crantz) breeding. *Euphytica,* **76,** 195–202.

Bacq, Z.M. & Alexander, P. (1958). *Grundlagen der Strahlenbiologie,* (2 Aufl.). Stuttgart: Thieme jr.

Bacq, Z.M. & Alexander, P. (1961). *Fundamentals of Radiobiology.* Oxford: Pergamon Press.

Bain, H.F. & Dermen, H. (1944). Sectorial polyploidy and phyllotaxy in the cranberry (*Vaccinium macrocarpon* Ait.). *American Journal of Botany,* **31,** 581–7.

Bajaj, Y.P.S. (1990a). In vitro production of haploids and their use in cell genetics and plant breeding. In *Biotechnology in Agriculture and Forestry, Vol.12. Haploids in Crop Improvement,* ed. Y.P.S.Bajaj, pp.1–44. Berlin: Springer Verlag.

Bajaj, Y.P.S. (1990b). Somaclonal variation. Origin, induction, cryopreservation, and implications in plant breeding. In *Biotechnology in Agriculture and Forestry, Vol.11. Somaclonal Variation in Crop Improvement I,* ed. Y.P.S.Bajaj, pp.3–48. Berlin: Springer-Verlag.

Balkema, G.H. (1971). *Chimerism and Diplontic Selection.* Thesis, Landbouwhogeschool Wageningen, Rotterdam: A.A. Balkema.

Balkema, G.H. (1972). Diplontic drift in chimeric plants. *Radiation Botany,* **12,** 51–5.

Ball, E.A. (1974). Experimental observations on the living shoot apex. *Revue de Cytologie et de Biologie Végétales,* **37,** 353–70.

Barabas, Z. (1962). Observation of sex differentiation in *Sorgum* by use of induced male-sterile mutants. *Nature,* **195,** 257–9.

Bates, G.W., Hasenkampf, C.A., Contolini, C.L. & Piastuch, W.C. (1987). Assymmetric hybridization in *Nicotiana* by fusion of irradiated protoplasts. *Theoretical and Applied Genetics,* **74,** 718–26.

Bateson, W. (1916). Root-cuttings, chimaeras and "sports". *Journal of Genetics* (Bangalore, Hyderabad), **6,** 75–80 (original not consulted).

Bateson, W. & Pellow, C. (1915). On the origin of "rogues" among culinary peas (*Pisum sativum*). *Journal of Genetics* (Bangalore, Hyderabad), **5,** 15–36 (original not consulted).

Bauer, R. (1957). The induction of vegetative mutations in *Ribes nigrum*. *Hereditas*, **43**, 323–37.

Bauer, R. (1974). Westra, an X-ray-induced erect-growing black-currant variety and its use in breeding. In *Polyploidy and Induced Mutations in Plant Breeding* (Proceedings of two Meetings, Joint FAO/IAEA Division and EUCARPIA, Bari, 1972), pp.13–20. Vienna: IAEA.

Baur, E. (1909). Das Wesen und die Erblichkeitsverhältnisse der "Varietates albomarginatae hort." von *Pelargonium zonale*. *Zeitschrift für induktive Abstammungs- und Vererbungslehre*, **1**, 330–51.

Baur, E. (1910). Propfbastarde. *Biologisches Zentralblatt*, **30**, 497–514.

Baur, E. (1924). Untersuchungen über das Wesen, die Entstehung und die Vererbung von Rassenunterschieden bei *Antirrhinum majus*. *Bibliotheca Genetica*, **4**, 1–170.

Baur, E. (1930). Mutations-Auslösung bei *Antirrhinum majus*. *Zeitschrift für Botanik*, **23**, 676–702.

Baur, E. (1932). Der Einfluß von chemischen und physikalischen Reizungen auf die Mutationsrate von *Antirrhinum majus*. *Zeitschrift für induktive Abstammungs- und Vererbungslehre*, **60**, 467–73.

Beale, G. (1993). The discovery of mustard gas mutagenesis by Auerbach and Robson in 1941. *Genetics*, **134**, 393–9.

Beale, G. & Knowles, J. (1978). *Extranuclear Genetics*. London: Edward Arnold.

Beard, B.H. (1970). Estimating the number of meristem initials after seed irradiation: A method, applied to flax stems. *Radiation Botany*, **10**, 47–57.

Becquerel, H. (1901). Sur quelques effects chimiques produits par le rayonnement du radium. *Comptes rendus*, **133**, 712 (original not consulted).

Behnke, M. (1979). Selection of potato callus for resistance to culture filtrates of *Phytophthora infestans* and regeneration of resistant plants. *Theoretical and Applied Genetics*, **55**, 69–71.

Behnke, M. (1980a). General resistance to late blight of *Solanum tuberosum* plants regenerated from callus resistant to culture filtrates of *Phytophthora infestans*. *Theoretical and Applied Genetics*, **56**, 151–2.

Behnke, M. (1980b). Selection of dihaploid potato callus for resistance to the culture filtrate of *Fusarium oxysporum*.

Zeitschrift für Pflanzenzüchtung, **85**, 254–8.

Bekendam, J. (1961). X-ray induced mutations in rice. In *Effects of Ionizing Radiations on Seeds* (Proceedings FAO/IAEA Symposium, Karlsruhe, 1960), pp.609–29. Vienna: IAEA.

Benzer, S. (1961). On the topography of the genetic fine structure. *Proceedings of the National Academy of Sciences, USA*, **47**, 403–15.

Bergann, F. (1954). Praktische Konsequenzen der Chimärenforschung für die Pflanzenzüchtung. *Wissenschaftliche Zeitschrift der Karl-Marx-Universität, Leipzig (Mathematisch-Naturwissenschaftliche Reihe)*, **4**, 281–91.

Bergann, F. (1957a). Gelungene experimentelle Entmischungen und Umlagerungen bei bekannten oder vermuteten Periklinalchimären. *Berichte der Deutschen Botanischen Gesellschaft*, **70**, 355–60.

Bergann, F. (1957b). Die züchterische Auswertung der intraindividuellen (somatischen) Variabilität von Kulturpflanzen durch bewußte Auslösung von Regenerationsvorgängen. *Wissenschaftliche Zeitschrift der Pädagogischen Hochschule Potsdam (Mathematisch-Naturwissenschaftliche Reihe)*, **3**, 105–9.

Bergann, F. (1967). The relative instability of chimerical clones-the basis for further breeding. In *Induzierte Mutationen und ihre Nutzung* (Proceedings Erwin-Baur-Gedächtnisvorlesungen IV, Gatersleben, 1966), ed. H.Stubbe, pp.287–300. Berlin: Akademie-Verlag.

Bergann, F. (1985). Das Studium von Periklinalchimären – eine legitime Methode der Histogeneseforschung. *Biologisches Zentralblatt*, **104**, 735–8.

Bergann, F. & Bergann, L. (1959). Über experimentell ausgelöste vegetative Spaltungen und Umlagerungen an chimärischen Klonen, zugleich als Beispiele erfolgreicher Staudenauslese. I. *Pelargonium zonale* Ait.' Madame Salleron'. *Der Züchter*, **29**, 361–74.

Bergann, F. & Bergann, L. (1962). Über Umschichtungen (Translokationen) an den Sproßscheiteln periklinaler Chimären. *Der Züchter*, **32**, 110–19.

Bergann, F. & Bergann.L. (1982). Zur Entwicklungsgeschichte des Angiospermenblattes 1. Über Periklinalchimären bei *Peperomia* und ihre experimentelle Entmischung und

Umlagerung. *Biologisches Zentralblatt*, **101**, 485–502.

Bergann, F & Bergann, L. (1983). Zur Entwicklungsgeschichte des Angiospermenblattes 2.Über die Blattmusterbildung bei meso- und diektochimärischen Formen von *Peperomia*-Arten, insbesondere über die Beteiligung des Dermatogens an der Mesophyllbildung. *Biologisches Zentralblatt*, **102**, 403–29.

Bergann, F. & Bergann, L. (1984). Zur Entwicklungsgeschichte des Angiospermenblattes 4. Über Periklinalchimären bei *Sedum rubrotinctum* R.T. Clausen. *Biologisches Zentralblatt*, **103**, 147–71.

Beversdorf, W.D. & Kott, L.S. (1987). An in vitro mutagenesis/selection system for *Brassica napus*. *Iowa State Journal of Research*, **61**, 435–43.

Bhatia, C.R. & Abraham, V. (1983). Handling of the first and second generations following mutagenic treatment of seeds in dicotyledonous plants. In *Chimerism in Irradiated Dicotyledonous Plants* (Report FAO/IAEA Consultants' Meeting, Vienna, 1981), pp. 25–9. IAEA-TECDOC-289. Vienna: IAEA.

Bianchi, A., Mariani, G. & Uberti, P. (1961). Mutations induced in endosperm and seedlings of maize following X-irradiation and diepoxybutane treatments of mature pollen. In *Effects of Ionizing Radiation on Seeds* (Proceedings FAO/IAEA Symposium, Karlsruhe, 1960), pp.419–30. Vienna: IAEA.

Bidney, D.L. & Shepard, J.F. (1981). Phenotypic variation in plants regenerated from protoplasts: The potato system. *Biotechnology and Bioengineering*, **23**, 2691–701.

Binding, H., Binding, K. & Straub, J. (1970). Selektion in Gewebekulturen mit haploiden Zellen. *Naturwissenschaften*, **3**, 138–9.

Bird, R.McK. & Neuffer, M.G. (1987). Induced mutations in maize. *Plant Breeding Reviews*, **5**, 139–80.

Bishop, C.J. (1954). Mutations in apples induced by X-irradiation. *Journal of Heredity*, **45**, 99–104.

Bishop, C.J. (1959). Radiation-induced fruit colour mutations in apples. *Canadian Journal of Genetics and Cytology*, **1**, 118–23.

Blakeslee, A.F. (1935). Hugo de Vries 1848–1935. *Science*, Vol **81**, No.2111, 581–2.

Blakeslee, A.F. (1936). Twenty-five years of genetics, 1910–1935. *Brooklyn Botanic Garden Memoirs*, **4**, 29–40.

Blakeslee, A.F. (1937). Dédoublement du nombre de chromosomes chez les plantes par traitement chimique. *Comptes rendus hebdomadaires des seances de l'Académie des Sciences*, **205**, 476–9.

Blakeslee, A.F. & Avery, A.G. (1937a). Methods of inducing chromosome doubling in plants by treatment with colchicine. *Science*, **86**, 408.

Blakeslee, A.F. & Avery, A.G. (1937b). Methods of inducing doubling of chromosomes in plants, by treatment with colchicine. *Journal of Heredity*, **28**, 392–411.

Blau, M. & Altenburger, K. (1922). Über einige Wirkungen von Strahlen. II. *Zeitschrift für Physik*, **12**, 315–29.

Blixt, S. (1961). Quantitative studies of induced mutations in peas. V. Chlorophyll mutations. *Agri Hortique Genetica*, **19**, 402–47.

Blixt, S. (1972). Mutation genetics in *Pisum. Agri Hortique Genetica*, **30**, 1–293.

Blixt, S. (1975). The pea. In *Handbook of Genetics*, vol.2, ed. R.C. King, pp.181–222. New York: Plenum Press.

Blixt, S., Ehrenberg, L. & Gelin, O. (1958). Quantitative studies of induced mutations in peas. I. Methodological investigations. *Agri Hortique Genetica*, **16**, 238–50.

Blixt, S., Ehrenberg, L. & Gelin, O. (1960). Quantitative studies of induced mutations in peas. III. Mutagenic effect of ethylene imine. *Agri Hortique Genetica*, **18**, 109–23.

Blixt, S., Ehrenberg, L. & Gelin, O. (1963). Studies on induced mutations in peas. VII. Mutation spectrum and mutation rate of different mutagenic agents. *Agri Hortique Genetica*, **21**, 178–216.

Blixt, S., Gelin, O., Mossberg, R., Ahnström, G., Ehrenberg, L. & Löfgren, R.A. (1964). Studies of induced mutations in peas. IX. Induction of leafspots in peas. *Agri Hortique Genetica*, **22**, 186–94.

Blixt, S. & Gottschalk, W. (1975). Mutations in the Leguminosae. *Agri Hortique Genetica*, **33**, 33–85.

Blonstein, A.D., Stirnberg, P. & King, P.J. (1991). Mutants of *Nicotiana plumbaginifolia* with specific resistance to auxin. *Molecular & General Genetics*, **228**, 361–71.

Bogyo, T.P. (1991). Numerical aspects of mutation breeding programmes. In *Plant Mutation Breeding for Crop Improvement* (Proceedings FAO/IAEA Symposium, Vienna, 1990), vol.2, pp.273–98. Vienna: IAEA.

Borg, G., Fröier, H. & Gustafsson, Å. (1958). Pallas barley, a variety produced by ionizing radiation: its significance for plant breeding and evolution. In *Isotopes in Agriculture* (Proceedings of the Second United Nations International Conference on the Peaceful Uses of Atomic Energy, Geneva), vol.27, 341–9. New York: United Nations.

Börner, T. & Sears, B.B. (1986). Plastome mutants. *Plant Molecular Biology Reporter*, **4**, 69–92.

Borojevic, K. (1969). Genetic changes in quantitative characters of irradiated population. *Japanese Journal of Genetics*, **44**, suppl.1, 404–16.

Bouharmont, J. & Dabin, P. (1986). Application des cultures in vitro a l'amélioration du Fuchsia par mutation. In *Nuclear Techniques and In Vitro Culture for Plant Breeding* (Proceedings FAO/IAEA Symposium, Vienna, 1985), pp.339–47. Vienna: IAEA.

Bouma, J. (1967). A new variety of spring barley 'Diamant' in Czechoslovakia. In *Induzierte Mutationen und Ihre Nutzung* (Proceedings Erwin-Baur-Gedächtnisvorlesungen IV, Gatersleben, 1966), ed. H. Stubbe, pp.177–82. Berlin: Akademie-Verlag.

Bouma, J. (1976). The spring barley mutant cultivar 'Diamant', its economic importance and breeding value. *Mutation Breeding Newsletter* **7**, 2–3.

Bouma, J. & Ohnoutka, Z. (1991). Importance and application of the mutant 'Diamant' in spring barley breeding. In *Plant Mutation Breeding for Crop Improvement* (Proceedings IAEA/FAO Symposium, Vienna, 1990), vol.1, pp.127–33. Vienna: IAEA.

Bozzini, A. (1991). The role of plant breeding for the future of mankind and the need for genetic resources and opportunities for mutagenesis or gene engineering. Introduction. In *Plant Mutation Breeding for Crop Improvement* (Proceedings FAO/IAEA Symposium, Vienna, 1990), vol.2, p.466. Vienna, IAEA.

Brand, A.J. & Bridgen, M.P. (1989). 'UConn White': A white flowered *Torenia fournieri.HortScience*, **24**, 714–15.

Brettell, R.I.S. & Ingram, D.S. (1979). Tissue culture in the production of novel disease-resistant crop plants. *Biological Review*, **54**, 329–45.

Brettell, R.I.S., Thomas, E. & Ingram, D.S. (1980). Reversion of Texas male-sterile cytoplasm in maize culture to give fertile T-toxin resistant plants. *Theoretical and Applied Genetics*, **58**, 55–8.

Brewbaker, J.L. & Emery, G.C. (1962). Pollen radiobotany. *Radiation Botany*, **1**, 101–54.

Bright, S.W.J., Ooms, G., Foulger, D., Karp, A. & Evans, N. (1986). Mutation and tissue culture. In *Plant Tissue Culture and its Agricultural Applications*, ed. L.A. Whithers & P.G. Alderson, pp.431–49. Stoneham, Mass., USA: Butterworth.

Briggs, R.W. (1974). Cytoplasmic male sterility research: M_2 generation from streptomycin treatments. *Maize Genetics Cooperation Newsletter*, **48**, 32–5.

Briggs, R.W. (1975). Cytoplasmic male sterility research: M_3 generation from streptomycin treatments. *Maize Genetics Cooperation Newsletter*, **49**, 35.

Brink, R.A. (1956). A genetic change associated with the R locus in maize which is directed and potentially reversible. *Genetics*, **41**, 872–89.

Brink, R.A. (1958). Paramutation at the R locus in maize. *Cold Spring Harbor Symposia on Quantitative Biology*, **23**, 379–91.

Brink, R.A. (1960). Paramutation and chromosome organisation. *Quarterly Review of Biology*, **35**, 120–37.

Brink, R.A. (1964). Genetic repression in multicellular organisms. *American Naturalist*, **98**, 193–211.

Brink, R.A. (1973). Paramutation. *Annual Review of Genetics*, **7**, 129–52.

Brink, R.A., Styles, E.D. & Axtell, J.D. (1968). Paramutation: directed genetic change. *Science*, **159**, 161–70.

Britt, A.B. (1996). DNA damage and repair in plants. *Annual Review of Plant Physiology and Plant Molecular Biology*, **47**, 75–100.

Brock, R.D. (1965). Induced mutations affecting quantitative characters. In *The Use of Induced Mutations in Plant Breeding* (Report FAO/IAEA Technical Meeting, Rome, 1964), pp.451–64. Oxford: Pergamon Press.

Brock, R.D. (1967). Quantitative variation in *Arabidopsis thaliana* induced by ionizing radiation. *Radiation Botany*, **7**, 193–203.

Brock, R.D. (1970). Mutations in quantitatively inherited traits induced by neutron irradiation. *Radiation Botany*, **10**, 209–23.

Brock, R.D. (1971). The role of induced mutations in plant improvement. *Radiation Botany*, **11**, 181–96.

Brock, R.D. (1976). Quantitatively inherited variation in *Arabidopsis thaliana* induced by chemical mutagens. *Environmental and Experimental Botany*, **16**, 241–53.

Brock, R.D. (1977). When to use mutations in plant breeding. In *Manual on Mutation Breeding*, 2nd ed., pp.213–19. Vienna: IAEA.

Brock, R.D. (1979). Mutation plant breeding for seed protein improvement. In *Seed Protein Improvement in Cereals and Grain Legumes* (Proceedings FAO/IAEA Symposium, Neuherberg, Germany, 1978), vol.1, pp.43–55. Vienna: IAEA.

Brock, R.D. (1980). Mutagenesis and crop improvement. In *The Biology of Crop Productivity* (Chapter 10), ed. P.S. Carlsson, pp.383–409. New York: Academic Press.

Brock, R.D. & Micke, A. (1979). Economic aspects of using induced mutations in plant breeding. In *Induced Mutations for Crop Improvement in Africa* (Proceedings Regional FAO/IAEA/IITA Seminar, Ibadan, Nigeria, 1978). IAEA-TECDOC-222, pp.19–32. Vienna: IAEA.

Broertjes, C. (1966). Mutation breeding in *Chrysanthemum. Euphytica*, **15**, 156–62.

Broertjes, C. (1968). Mutation breeding in vegetatively propagated crops. In *Mutations in Plant Breeding. II* (Proceedings FAO/IAEA Panel, Vienna, 1967), pp.59–62.

Broertjes, C. (1969a). Induced mutations and breeding methods in vegetatively propagated species. In *Induced Mutations in Plants* (Proceedings FAO/IAEA Symposium, Pullman, USA), pp.325–9. Vienna: IAEA.

Broertjes, C. (1969b). Mutation breeding of *Streptocarpus. Euphytica*, **18**, 333–9.

Broertjes, C. (1972a). *Use in Plant Breeding of Acute, Chronic or Fractionated Doses of X-rays or Fast Neutrons as Illustrated with Leaves of Saintpaulia*. Thesis, Wageningen Agricultural University. Agricultural Research Reports 776. Wageningen: PUDOC.

Broertjes, C. (1972b). Improvement of vegetatively propagated plants by ionizing radiation. In *Induced Mutations and Plant Improvement* (Proceedings FAO/IAEA Study Group Meeting, Buenos Aires, 1970), pp.293–9. Vienna: IAEA.

Broertjes, C. (1972c). Mutation breeding of *Achimenes. Euphytica*, **21**, 48–63.

Broertjes, C. (1976a). Is DTT a means to improve the mutation spectrum of vegetatively propagated plants?. In *Improvement of Vegetatively Propagated Plants and Tree Crops through Induced Mutations* (Proceedings Second FAO/IAEA Research Coordination Meeting, Wageningen, The Netherlands), pp.13–23. IAEA-TECDOC-194. Vienna: IAEA.

Broertjes, C. (1976b). Mutation breeding of autotetraploid Achimenes cultivars. In *Improvement of Vegetatively Propagated Plants and Tree Crops through Induced Mutations* (Proceedings Second FAO/IAEA Research Coordination Meeting, Wageningen, The Netherlands), pp.1–12. IAEA-TECDOC-194. Vienna: IAEA.

Broertjes, C. (1979). The improvement of *Chrysanthemum morifolium* Ram. by induced mutations. In *Proceedings EUCARPIA Meeting on Chrysanthemum (1978)*, pp.93–102. Littlehampton, UK.

Broertjes, C. (1982). Significance of in vitro adventitious bud techniques for mutation breeding of vegetatively propagated crops. In *Induced Mutations in Vegetatively Propagated Plants.II. (Proceedings* Final FAO/IAEA Research Co-ordination Meeting, Coimbatore, India, 1980), pp.1–10. Vienna: IAEA.

Broertjes, C., Haccius, B. & Weidlich, S. (1968). Adventitious bud formation on isolated leaves and its significance for mutation breeding. *Euphytica*, **17**, 321–44.

Broertjes, C. & Keen, A. (1980). Adventitious buds: do they develop from one cell? *Euphytica*, **29**, 171–6.

Broertjes, C., Koene, P. & van Veen, J.W.H. (1980). A mutant of a mutant of a...: Irradiation of progressive radiation-induced mutants in a mutation breeding programme with *Chrysanthemum morifolium* Ram. *Euphytica* **29**, 525–30.

Broertjes, C. & Leffring, L. (1972). Mutation breeding of *Kalanchoë. Euphytica*, **21**, 415–23.

Broertjes, C. & Lock, C.A.M. (1985). Radiation-induced low-temperature tolerant solid mutants of *Chrysanthemum morifolium* Ram. *Euphytica*, **34**, 97–103.

Broertjes, C. & van Harten, A.M. (1978). *Application of Mutation Breeding Methods in the Improvement of Vegetatively Propagated Crops*. Amsterdam: Elsevier Scientific Publishing Company.

Broertjes, C. & van Harten, A.M. (1985). Single cell origin of adventitious buds. *Euphytica*, **34**, 93–5.

Broertjes, C. & van Harten, A.M. (1987). Application of mutation breeding methods. In *Improving Vegetatively Propagated Crops*, ed. A.J. Abbott & R.K. Atkin, pp.335–48. London: Academic Press.

Broertjes, C. & van Harten, A.M. (1988). *Applied Mutation Breeding for Vegetatively Propagated Crops*. Amsterdam: Elsevier.

Broertjes, C. & Verboom, H. (1974). Mutation breeding in *Alstroemeria. Euphytica*, **23**, 39–44.

Brunner, H. (1992). Radiation treatment services for mutation induction in crop plants. *Mutation Breeding Newsletter*, **39**, 4–5

Buder, J. (1928). Der Bau des phanerogamen Sprossvegetationspunktes und seine Bedeutung für die Chimärentheorie. *Berichte der Deutschen Botanischen Gesellschaft*, **46**, 20–1.

Buiatti, M. (1989). Use of cell and tissue cultures for mutation breeding. In *Science for Plant Breeding* (Proceedings of the 12th Congress of EUCARPIA, Göttingen, Germany), pp.179–200. Berlin: Paul Parey.

Buiatti, M. & Gimelli, F. (1993). Somaclonal variation in ornamentals. In *Creating Genetic Variation in Ornamentals* (Proceedings of the XVIIth EUCARPIA Symposium, San Remo), ed. T. Schiva & A. Mercuri, pp.5–24. San Remo: Istituto Sperimentale per la Floricoltura.

Burk, L.G., Stewart, R.N. & Dermen, H. (1964). Histogenesis and genetics of a plastid-controlled chlorophyll variegation in tobacco. *American Journal of Botany*, **51**, 713–24.

Burr, B. & Burr, F.A. (1989). Transposable element-induced mutations. In *Science for Plant Breeding* (Proceedings of the 12th Congress of EUCARPIA, Göttingen, Germany), pp.161–4. Berlin: Paul Parey.

Burton, G.W. (1974). Radiation breeding of warm season forage and turf grasses. In *Polyploidy and Induced Mutations in Plant Breeding* (Proceedings of two

Meetings, Joint FAO/IAEA Division and EUCARPIA, 1972, Bari, Italy), pp.35–9. Vienna: IAEA.

Burton, G.W. (1975). Improving sterile turf and forage bermudagrass hybrids by gamma radiation. In *Improvement of Vegetatively Propagated Plants through Induced Mutations* (Proceedings FAO/IAEA Resaearch Co-ordination Meeting, Tokai, Japan), pp.32. IAEATECDOC-173. Vienna: IAEA.

Burton, G.W. (1976). Using gamma irradiation to improve sterile turf and forage bermudagrasses. In *Improvement of Vegetatively Propagated Plants and Tree–Crops through Induced Mutations* (Proceedings Second FAO/IAEA Research Coordination Meeting, Wageningen, The Netherlands), pp.25–32. IAEA-TECDOC-194. Vienna: IAEA.

Burton, G.W. (1979). Induced mutations for improving millets, apomictic crop plants and vegetatively propagated grasses. In *Induced Mutations for Crop Improvement in Africa* (Proceedings Regional FAO/IAEA/IITA Seminar, 1978, Ibadan, Nigeria), pp.33–40. IAEATECDOC-222. Vienna: IAEA.

Burton, G.W. (1981). Tifway-2 bermudagrass. *Mutation Breeding Newsletter*, **18**, 8–10.

Burton, G.W. (1985). Registration of Tifway-2 bermudagrass. *Crop Science*, **25**, 364.

Burton, G.W., Constantin, M.J., Dobson, J.W.Jr., Hanna, W.W. & Powell, J.B. (1980). An induced mutant of Coastcross 1 bermudagrass with improved winterhardiness. *Environmental and Experimental Botany*, **20**, 115–17.

Burton, G.W. & Hanna, W.W. (1976). Ethidium bromide induced cytoplasmic male sterility in pearl millet. *Crop Science*, **16**, 731–2.

Burton, G.W. & Hanna, W.W. (1977). Performance of mutants induced in sterile triploid turf bermudagrass. *Mutation Breeding Newsletter*, **9**, 4.

Burton, G.W. & Hanna, W.W. (1982). Stable cytoplasmic male-sterile mutants induced in Tift 23DB1 pearl millet with mitomycin and streptomycin. *Crop Science*, **22**, 651–2.

Burton, G.W., Hanna, W.W. & Powell, J.B. (1982). Mutation breeding of vegetatively propagated turf and forage bermudagrass. In *Induced Mutations in Vegetatively Propagated Plants. II.* (Proceedings FAO/IAEA Research Co-ordination Meeting, Coimbatoire, India, 1980), pp.167–74. Vienna: IAEA.

Burton, G.W. & Jackson, J.E. (1962). Radiation breeding of apomictic prostrate dallis grass, *Paspalum dilatatum* var. *pauciciliatum*. *Crop Science*, **2**, 491–4.

Buvat, R. (1952). Structure, évolution et fonctionnement du méristème de quelques dicotylèdones. *Annales des Sciences Naturelles Botanique* (11 Sér.), **13**, 202–3.

Buvat, R. (1955). Le méristème apical de la tige. *l'Année Biologique*, 3(31), 595–656.

Cairns, J., Overbaugh, J. & Miller, S. (1988). The origin of mutants. *Nature*, **335**, 142–5.

Caldecott, R.S. (1956). Ionizing radiations as a tool for plant breeders. In *Radioactive Isotopes and Ionizing Radiations in Agriculture, Physiology, and Biochemistry* (Proceedings of the International Conference on the Peaceful Uses of Atomic Energy, Geneva, 1955), vol.12, 40–5. New York: United Nations.

Caldecott, R.S. (1958). Post-irradiation modification of injury in barley-its basic and applied significance. In *Isotopes in Agriculture* (Proceedings of the Second United Nations International Conference on the Peaceful Uses of Atomic Energy, Geneva), vol. 27, 260–9. New York: United Nations.

Caligari, P.D.S., Ingram, N.R. & Jinks, J.L. (1981). Gene transfer in *Nicotiana rustica* by means of irradiated pollen. I. Unselected progenies. *Heredity*, **47**, 17–26.

Campbell, A. (1993). Barbara McClintock. *Annual Review of Genetics*, **27**, 1–6.

Campbell, A.I. & Lacey, C.N.D. (1973). Compact mutants of Bramley's Seedling apple induced by gamma radiation. *Journal of Horticultural Science*, **48**, 397–402.

Campbell, A.I. & Wilson, D. (1977). Prospects for the development of disease-resistant temperate fruit plants by mutation breeding. In *Induced Mutations against Plant Diseases* (Proceedings FAO/IAEA/SIDA Symposium, Vienna), pp.215–26. Vienna: IAEA.

Campbell, T.A. (1987). *Cuphea tolucana*, a promising new oil crop. *Plant Sciences*, **67**, 909–17.

Carlson, P.S. (1970). Induction and isolation of auxotrophic mutants in somatic cell cultures of *Nicotiana tabacum*. *Science*, **168**, 1366–8.

Carlson, P.S. (1973). Methionine sulfoximine-resistant mutants of tobacco. *Science*, **180**, 487–9.

Carrière, E.A. (1865). *Production et fixation des variétés dans les végétaux*. Paris.

Cartledge, J.L. & Blakeslee, A.E. (1934). Mutation rate increased by aging seeds as shown by pollen abortion. *Proceedings of the National Academy of Sciences of the USA*, **20**, 103–10.

Cassells, A.C. & Jones, P.W., ed. (1995). *The Methodology of Plant Genetic Manipulation: Criteria for Decision Making* (Proceedings EUCARPIA Plant Genetic Manipulation Section Meeting, Cork, Ireland, 1994). Dordrecht, The Netherlands: Kluwer Academic Publishers.

Cassells, A.C. & Periappuram, C. (1993). Diplontic drift during serial subculture as a positive factor influencing the fitness of mutants derived from the irradiation of *in vitro* nodes of *Dianthus* 'Mystere'. In *Creating Genetic Variation in Ornamentals* (Proceedings of the XVIIth EUCARPIA Symposium, San Remo), ed. T. Schiva & A. Mercuri, pp.71–81. San Remo: Istituto Sperimentale per la Floricoltura.

Cassells, A.C., Walsh, C. & Periappuram, C. (1993). Diplontic selection as a positive factor in determining the fitness of mutants of *Dianthus* 'Mystère' derived from X-irradiation of nodes in *in vitro* culture. *Euphytica*, **70**, 167–74.

Catcheside, D.G. (1948). Genetic effects of radiations. *Advances in Genetics*, **2**, 271–358.

Catcheside, D.G. & Lea, D.E. (1943). The effect of ionization distribution on chromosome breakage by X-rays. *Journal of Genetics*, **45**, 186–96.

Chadwick, K.H. & Leenhouts, H.P. (1981). *The Molecular Theory of Radiation Biology*. Berlin: Springer-Verlag.

Chaleff, R.S. (1981). *Genetics of Higher Plants (Applications of Cell Culture)*. Cambridge: Cambridge University Press.

Chaleff, R.S. (1983). Isolation of agronomically useful mutants from plant cell cultures. *Science*, **219**, 676–82.

Chaleff, R.S. & Carlson, P.S. (1974). Somatic cell genetics of higher plants. *Annual Review of Genetics*, **8**, 267–78.

Chang, T.T. (1995). Rice. In *Evolution of Crop Plants* (2nd ed.), ed. J. Smartt & N.W. Simmonds, pp.147–55. Harlow, Essex, UK: Longman Scientific & Technical.

Chaplin, J.F & Mann, T.J. (1978). Evaluation of tobacco mosaic resistance factor transferred from burley to flue-cured tobacco. *Journal of Heredity*, **69**, 175–8.

Charlesworth, B. (1990). Mutation-selection balance and the evolutionary advantage of sex and recombination. *Genetic Research*, **55**, 199–221.

Charlesworth, B., Charlesworth, D. & Morgan, M.T. (1990). Genetic loads and estimates of mutation rates in highly inbred plant populations. *Nature*, **347**, 380–2.

Chauhan, S.P. & Patra, N.K. (1993). Mutagenic effects of combined and single doses of gamma rays and EMS in opium poppy. *Plant Breeding*, **110**, 342–5.

Cheng, X.Y., Gao, M.W., Liang, Z.Q. & Liu, K.Z. (1990). Effect of mutagenic treatments on somaclonal variation in wheat (*Triticum aestivum* L.). *Plant Breeding*, **105**, 47–52.

Cherney, J.H., Cherney, D.J.R., Akin, D.E. & Axtell, J.D. (1991). Potential of brown-midrib, low-lignin mutants for improving forage quality. *Advances in Agronomy*, **46**, 157–98.

Chevreau, E., Decourtye, L. & Skirvin, R.M. (1989). A review of pear chimeras: Their identification and separation into pure types. *HortScience*, **24**, 32–4.

Chittenden, R.J. (1925). Studies in variegation.II. *Hydrangea* and *Pelargonium*. *Journal of Genetics*, **16**, 43–61.

Chittenden, R.J. (1927). Vegetative segregation. *Bibliographia Genetica*, **3**, 355–442.

Clement, G. & Poisson, C. (1988). Genetic determinants of semi-dwarf characters induced in *Oryza sativa* for upland rice cultivation. *Mutation Breeding Newsletter*, **31**, 5–6.

Cline, M.G. & Salisbury, F.B. (1966). Effects of ultraviolet radiation on the leaves of higher plants. *Radiation Botany*, **6**, 151–63.

Clowes, F.A.L. (1957). Chimeras and meristems. *Heredity* (London), **11**, 141–8.

Clowes, F.A.L. (1961). *Apical Meristems*. Oxford: Blackwell Scientific Publications.

Coe, E.H. Jr. (1966). The properties, origin and mechanism of conversion-type inheritance at the *B* locus in maize. *Genetics*, **53**, 1035–63.

Coe, E. (1993). Gene list and working maps. *Maize Genetics Cooperation Newsletter*, **67**, 133–66.

Coe, E.H. Jr. & Neuffer, M.G. (1977). The genetics of corn. In *Corn and Corn Improvement* (2nd ed.), ed. G.F. Sprague, pp.111–223. Madison, Wisconsin: American Society of Agronomy.

Coe, E.H. Jr., Thompson, D. & Walbot, V. (1988). Phenotypes mediated by the *iojap* genotype in maize. *American Journal of Botany*, **75**, 634–44.

Cohen, S.N. & Shapiro, J.A. (1980). Transposable genetic elements. *Scientific American*, **242**(2), 36–45.

Colijn, C.M., Kool, A.J. & Nijkamp, H.J.J. (1979). An effective chemical mutagenesis procedure for *Petunia hybrida* cell suspensions. *Theoretical and Applied Genetics*, **55**, 101–6.

Conger, A.D., Sparrow, A.H., Schwemmer, S.S. & Klug, E.E. (1982). Relation of nuclear volume and radiosensitivity to ploidy level (haploid to 22-ploid) in higher plants and a yeast. *Environmental and Experimental Botany*, **22**, 55–74.

Conger, B.V. & Carabia, J.V. (1977). Mutagenic effectiveness and efficiency of sodium azide versus ethyl methanesulfonate in maize: induction of somatic mutations at the *yg*$_2$ locus by treatment of seeds differing in metabolic state and cell population. *Mutation Research*, **46**, 285–95.

Conger, B.V., Trigiano, R.N. & Gray, D.J. (1986). *Dactylis glomerata*: A potential system for in vitro mutagenesis and mass propagation in cereals and grasses. In *Nuclear Techniques and In Vitro Culture for Plant Improvement* (Proceedings FAO/IAEA Symposium, Vienna, 1985), pp.371–83. Vienna: IAEA.

Constantin, M.J. (1976). Mutations for chlorophyll-deficiency in barley: Comparative effects of physical and chemical mutagens. *Barley Genetics III* (Proceedings of the Third International Barley Genetics Symposium, 1975, Garching, Germany), ed. H. Gaul, pp.96–112. München: Verlag Karl Thiemig.

Constantin, M.J. (1983). Mutagen-induced chimerism in dicotyledonous plants. In *Chimerism in Irradiated Dicotyledonous Plants* (Report FAO/IAEA Consultants Meeting, Vienna, 1981), pp.19–21. IAEA-TECDOC-289. Vienna: IAEA.

Constantin, M.J. (1984). Potential of in vitro mutation breeding for the improvement of vegetatively propagated crop plants. In *Induced Mutations for Crop Improvement in Latin America* (Proceedings FAO/IAEA Regional Seminar, Lima, Peru, 1982). IAEA-TECDOC-305, pp.59–78. Vienna: IAEA.

Contant, R.B., Devreux, M., Ecochard, R.M., Monti, L.M., de Nettancourt, D., Scarascia Mugnozza, G.T. & Verkerk, K. (1971). Radiogenetic effects of gamma- and fast neutron irradiation on different ontogenetic stages of the tomato. *Radiation Botany*, **11**, 119–36.

Cooper, W.E. & Gregory, W.C. (1960). Radiation-induced leaf spot resistant mutants in the peanut (*Arachis hypogaea* L.). *Agronomy Journal*, **52**, 1–4.

Corley, R.H.V., Lee, C.H., Law, I.H. & Wong, C.Y. (1986). Abnormal flower development in oil palm clones. *The Planter*, **62**, 233–40.

Correns, C. (1909). Vererbungsversuche mit blass(gelb)grünen und buntblättrigen Sippen bei *Mirabilis jalapa*, *Urtica pilulifera* und *Lunaria annua*. *Zeitschrift für induktive Abstammungs- und Vererbungslehre*, **1**, 291–329.

Coutinho, M.P. (1975). L'application des radiations pour l'obtention de vignes résistantes au *Plasmopora*. *Vitis*, 281–6.

Cramer, P.J.S. (1907). *Kritische Übersicht der bekannten Fälle von Knospenvariation*.Natuurkundige Verhandelingen der Hollandsche Maatschappij van Wetenschappen, Haarlem 3.6.

Creissen, G.P. & Karp, A. (1985). Karyotypic changes in potato plants regenerated from protoplasts. *Plant Cell, Tissue and Organ Culture*, **4**, 171–82.

Crinò, P., Lai, A., Di Bonito, R & Saccardo, F. (1990). Tomato mutants induced by in vitro cultured cotyledons. *Acta Horticulturae*, **280**, 435–8.

Crispi, M.L., Ullrich, S.E. & Nilan, R.A. (1987). Investigation of partial sterility in advanced generation, sodium azide-induced lines of spring barley. *Theoretical and Applied Genetics*, **74**, 402–8.

Crowther, J.A. (1924). Some considerations relative to the action of X-rays on tissue cells. *Proceedings of the Royal Society London*, B **96**, 207–11.

Cuany, R.L. (1960). Nature of somatic mutations induced by radiation in flowering plants. In *Proceedings 2nd Inter-American Conference on Peaceful Applications of Nuclear Energy*. Buenos Aires, 1959, 29–37.

Cuany, R.L., Sparrow, A.H. & Pond, V. (1958). Genetic response of *Antirrhinum majus* to acute and chronic plant irradiation. *Zeitschrift für induktive Abstammungs-und Vererbungslehre*, **89**, 7–13.

Cullis, C.A. (1973). DNA differences between flax genotrophs. *Nature*, **243**, 515–6.

Cullis, C.A. (1977). Molecular aspects of the environmental induction of heritable changes in flax. *Heredity*, **38**, 129–54.

Cullis, C.A. (1979). Quantitative variation of ribosomal RNA genes in flax genotrophs. *Heredity*, **42**, 237–46.

Cullis, C.A. (1986). Phenotypic consequences of environmentally induced changes in plant DNA. *Trends in Genetics*, **2**, 307–9.

Cutter, E.G. (1965). Recent experimental studies of the shoot apex and shoot morphogenesis. *The Botanical Review*, **31**, 7–113.

Daly, M.J. & Minton, K.W. (1995). Resistance to radiation. *Science*, **270**, 1318.

D'Amato, F. (1950). Mutazioni clorofilliane nell'orzo indotte da derivati acridinici. *Caryologia*, **3**, 211–20.

D'Amato, F. (1951). Nuovi dati sull'attivitá mutagena dei derivati dell'acridina. *Caryologia*, **3**, 311–26.

D'Amato, F. (1952). Further investigations on the mutagenic activity of acridines (XXXIII-LI). *Caryologia*, **4**, 388–414.

D'Amato, F. (1965). Chimera formation in mutagen-treated seeds and diplontic selection. In *The Use of Induced Mutations in Plant Breeding* (Report FAO/IAEA Technical Meeting, Rome, 1964), pp.303–16. Oxford: Pergamon Press.

D'Amato, F. (1975). The problem of genetic stability in plant tissue and cell culture. In *Crop Genetic Resources for Today and Tomorrow*, ed. Frankel, O.H. & Hawkes, J.G., pp.333–48. Cambridge: Cambridge University Press.

D'Amato, F. (1986). Spontaneous mutations and somaclonal variation. In *Nuclear Techniques and In Vitro Culture for Plant Improvement* (Proceedings FAO/IAEA Symposium, Vienna, 1985), pp.3–10. Vienna: IAEA.

D'Amato, F. (1992). Induced mutations in crop improvement: basic and applied aspects. In *Perspectives for Agriculture and Society in the Third Millenium* (Proceedings Scientific Meeting Faculty of Agriculture, University of Pisa, Italy, 1991), ed. L. Iacoponi, R. Fiorentini & M. Giovannetti. *Agricoltura Mediterranea*, vol.122 (special issue).

D'Amato, F. & Hoffmann-Ostenhof, O. (1956). Metabolism and spontaneous mutations in plants, *Advances in Genetics*, VIII, 1–28.

Darlington, C.D. (1973). *Chromosome Botany and the Origins of Cultivated Plants*, 3rd edn, London: George Allen & Unwin Ltd.

Darrow, G.M. (1931). A productive thornless sport of the Evergreen blackberry. *Journal of Heredity*, **22**, 405–6.

Darrow, G.M. (1955). Nature of thornless blackberry sports. *Fruit Varieties and Horticultural Digest*, **10**(1), 14–15. East Lansing Mich., USA: Dept. of Hort., Mich. State College.

Darwin, C. (1859). *The Origin of Species by Means of Natural Selection*. London: Murray.

Darwin, C. (1868). *The Variation of Animals and Plants under Domestication*. 10th impr. of the 2nd ed. Vol.I (1921). London: Murray.

Daskalov, S. (1986). Mutation breeding in pepper. *Mutation Breeding Review*, **4**, 1–26.

Datta, S. (1991). Role of mutation breeding in floriculture. In *Plant Mutation Breeding for Crop Improvement* (Proceedings FAO/IAEA Symposium, Vienna, 1990), vol.1, pp.273–81. Vienna: IAEA.

Daub, M.E. (1986). Tissue culture and the selection of resistance to pathogens. *Annual Review of Phytopathology*, **24**, 159–86.

Davidson, D., Pertens, E. & Armstrong, S.W. (1987). Changes in frequencies of variegated leaves in NMU treated tobacco: evidence for a differential response to NMU. *Theoretical and Applied Genetics*, **73**, 915–19.

Davies, D.R. (1983). Chimerism in irradiated dicotyledonous plants. In *Chimerism in Irradiated Dicotyledonous Plants* (Proceedings FAO/IAEA Consultants Meeting, Vienna, 1981), pp.13–21. IAEA-TECDOC-289. Vienna: IAEA.

Davies, D.R. & Wall, E.T. (1961). Gamma radiation and interspecific incompatibility in plants. In *Effects of Ionizing Radiations on Seeds* (Proceedings FAO/IAEA Symposium, Karlsruhe, 1960), pp.83–101. Vienna: IAEA.

Day, P. (1993). Integrating plant breeding and molecular biology: accomplishments and future promise. In *International Crop Science I*, ed. D.R. Buxton *et al.*, pp.517–23. Madison, Wisconsin, USA: Crop Science Society of America, Inc.

Deane, C.R., Fuller, M.P. & Dix, P.J. (1995). Selection of hydroxyproline-resistant proline-accumulating mutants of cauliflower (*Brassica oleracea* var. *botrytis*). *Euphytica*, **85**, 329–34.

Decourtye, L. & Lantin, B. (1971). Considérations méthodologiques sur l'isolement de mutantes provoqués chez le pommier et le poirier. *Annales de l'Amélioration des Plantes*, **21**, 29–44.

de Fossard, R.A. (1976). *Tissue Culture for Plant Propagators*. Armidale, N.S.W., Australia: University of New England Printery (original not consulted).

de Fossard, R.A. (1977). The horizons of tissue culture propagation. In *Seminar NSW ASS. Nurserymen*, pp.1–147. University of Sydney (original not consulted).

Degenhart, N.R., Werner, B.K. & Burton, G.W. (1995). Forage yield and quality of a brown mid-rib mutant in pearl millet. *Crop Science*, **35**, 986–8.

de Jong, J. & Custers, J.B.M. (1986). Induced changes in growth and flowering of *Chrysanthemum* after irradiation and *in vitro* culture of pedicels and petal epidermis. *Euphytica*, **35**, 137–48.

de Jong, J., Huitema, J.B.M & Preil, W. (1991). Use of in vitro techniques for the selection of stress tolerant mutants of *Chrysanthemum morifolium*. In *Plant Mutation for Crop Improvement* (Proceedings FAO/IAEA Symposium, Vienna, 1990), vol.2, pp. 149–55. Vienna: IAEA.

de Kleijn, E.H.J.M., Boers, G.J. & Heijbroek, A.M.A. (1992). *Visie op de Internationale Concurrentiekracht in het Uitgangsmateriaal*. Den Haag, The Netherlands: LEIDLO and Rabobank Nederland.

de Kleijn, E.H.J.M. & Heijbroek, A.M.A. (1992). *A View of International Competitiveness in the Flower Bulb Industry*. Den Haag, The Netherlands: LEI-DLO and Rabobank Nederland.

de Klerk, G.-J. (1990). How to measure somaclonal variation. *Acta Botanica Neerlandica*, **39**, 129–44.

Delaunay, L.N. (1930). Die Chromosomenaberranten in der Nachkommenschaft von röntgenisierten Ähren einer reinen Linie von *Triticum vulgare albidum* All. *Zeitschrift für induktive Abstammungs- und Vererbungslehre*, **55**, 352–5.

Delaunay, L.N. (1931). Resultate eines dreijährigen Röntgenversuchs mit Weizen. *Der Züchter*, **3**, 129–37.

Dellaert, L.M.W. (1979). Comparison of selection methods for specified mutants in self-fertilizing crops. Theoretical approach. In *Seed Protein Improvement in Cereals and Grain Legumes* (Proceedings FAO/IAEA Symposium, Neuherberg, Germany, 1978), vol.1, pp.57–75. Vienna: IAEA.

Dellaert, L.M.W. (1980). Segregation frequencies of radiation-induced viable mutants in *Arabidopsis thaliana* (L.)Heynh. *Theoretical and Applied Genetics*, **57**, 137–43.

Dellaert, L.M.W. (1982). The use of induced mutations in improvement of seed propagated crops: selection methods. In *Induced Variability in Plant Breeding* (Proceedings international symposium of the section Mutation and Polyploidy of EUCARPIA, Wageningen, The Netherlands, 1981), pp.14–18. Wageningen: Centre for Agricultural Publishing and Documentation (PUDOC).

Dellaert, L.M.W. (1983). Efficiency of mutation programmes. In *Chimerism in Irradiated Dicotyledonous Plants* (Report FAO/IAEA Consultants Meeting, Vienna, 1981), pp.33–4. IAEA-TECDOC-289. Vienna: IAEA.

de Mol, W.E. (1931). Somatische Variation der Blumenfarbe der Hyazinthe durch Röntgenbestrahlung und andere äussere Umstände. *Zeitschrift für induktive Abstammungs- und Vererbungslehre*, **59**, 280–3.

de Mol, W.E. (1933). Mutation sowohl als Modifikation durch Röntgenbestrahlung und die 'Teilungshypothese'. *La Cellule*, **42**, 149–62.

de Mol, W.E. (1953). X-raying of hyacinths and tulips from the beginning, before thirty years (1922) till today (1952). *Japanese Journal of Breeding*, **3**, 1–8.

de Mol van Oud Loosdrecht, W.E. (1956). Der Einfluss der Röntgenstrahlen auf die Entwicklung des Pollens und der Sprosse bei Tulpen, pp.31–128. Bonn, München, Wien: Bayerischer Landwirtschafstverlag.

Derks, F.H.M., Hall, R.D. & Colijn-Hooymans, C.M. (1992). Effect of gamma-irradiation on protoplast viability and chloroplast DNA damage in *Lycopersicon peruvianum* with respect to donor-recipient protoplast fusion. *Environmental & Experimental Botany*, **32**, 255–64.

Dermen, H. (1940). Colchicine polyploidy and technique. *The Botanical Review*, **6**, 599–635.

Dermen, H. (1945). The mechanism of colchicine-induced cytohistological changes in cranberry. *American Journal of Botany*, **32**, 387–94.

Dermen, H. (1947). Histogenesis of some bud sports and variegations. *Proceedings of the American Society for Horticultural Science*, **50**, 51–73.

Dermen, H. (1951). Ontogeny of tissues in stem and leaf of cytochimeral apples. *American Journal of Botany*, **38**, 753–60.

Dermen, H. (1960). Nature of plant sports. *American Horticultural Magazine*, **39**, 123–73.

Dertinger, H. & Jung, H. (1970). *Molecular Radiation Biology*. New York: Springer-Verlag.

Dessauer, F. (1922). Über einige Wirkungen von Strahlen.I. *Zeitschrift für Physik*, **12**, 38–47.

Devreux, M., Donini, B. & Scarascia-Mugnozza, G.T. (1968). Chimeric situation avoided by gametophyte irradiation in barley and *durum* wheat. In *Proceedings 12th International Congress of Genetics*, Tokyo, Japan. vol.I, 252.

Devreux, M., Magnien, E. & Dalschaert, X. (1986). Cellules vegetales in vitro et radiations ionisantes. In *Nuclear Techniques and In Vitro Culture for Plant Improvement* (Proceedings FAO/IAEA Symposium, Vienna, 1985), pp.93–101. Vienna: IAEA.

de Vries, H. (1900). Das Spaltungsgesetz der Bastarde. *Berichte der Deutschen Botanischen Gesellschaft*, **18**, 83–90. Opera VI, 208–80 (original not consulted).

de Vries, H. (1901). *Die Mutationstheorie I*. Leipzig: Veit & Co.

de Vries, H. (1903). *Die Mutationstheorie II*. Leipzig: Veit & Co.

de Vries, H. (1905). *Species and Varieties: Their Origin by Mutation*. Chicago: The Open Court Publishing Company.

Dillon, D. & Stadler, D. (1994). Spontaneous mutation at the *mtr* locus in Neurospora: The molecular spectrum in wild-type and a mutator strain. *Genetics*, **138**, 61–74.

Dolezel, J. & Novak, F.J. (1984). Effect of plant tissue culture media on the frequency of somatic mutations in *Tradescantia* stamen hairs. *Zeitschrift für Pflanzenphysiologie*, **114**, 51–8.

Dolezel, J., Novak, F.J. & Havel, L. (1986). Cytogenetics of garlic (*Allium sativum* L.) in vitro culture. In *Nuclear Techniques and In Vitro Culture for Plant Improvement* (Proceedings FAO/IAEA Symposium, Vienna, 1985), pp.11–19. Vienna: IAEA.

Doll, H. (1976). Genetic studies of high-lysine barley mutants. In *Barley Genetics III* (Proceedings 3rd International Barley Genetics Symposium Garching, Germany, 1975), ed. H.M. Gaul, pp.542–6. München: Karl Thiemig.

Doll, H. (1977). Storage proteins in cereals. In *Genetic Diversity in Plants*, ed. Amir Muhammed, Rustem Aksel & R.C. von Borstel, pp.337–47. New York: Plenum Press.

Doll, H. & Køie, B. (1978). Influence of the high-lysine gene from barley mutant 1508 on grain, carbohydrate and protein yield. In *Seed Protein Improvement by Nuclear Techniques* (Proceedings two FAO/IAEA Research Co-ordination Meetings, Baden, Austria, 1977; Vienna, 1977), pp.107–14. Vienna: IAEA.

Doll, H, Køie, B. & Eggum, B.O. (1974). Induced high lysine mutants in barley. *Radiation Botany*, **14**, 73–80.

Dommergues, P. (1964). La destinée de la cellule mutée: conséquences dans le cas des plantes à multiplication végétative et dans les plantes à reproduction sexuée. In *EUCARPIA* (Proceedings Third Congress European Association for Research on Plant Breeding, Paris, 1962), pp.115–39. Paris: EUCARPIA.

Donini, B. (1975). Induction and isolation of somatic mutations in vegetatively propagated plants. In *Improvement of Vegetatively Propagated Plants through Induced Mutations* (Proceedings FAO/IAEA Research Coordination Meeting, Tokai, Japan), pp.35–51. IAEATECDOC-173. Vienna: IAEA.

Donini, B. (1976a). Use of radiations to induce useful mutations in fruit trees. *Mutation Breeding Newsletter*, **8**, 7–8.

Donini, B. (1976b). The use of radiations to induce useful mutations in fruit trees. In *Improvement of Vegetatively Propagated Plants and Tree Crops through Induced Mutations* (Proceedings Second FAO/IAEA Research Coordination Meeting, Wageningen, The Netherlands), pp.55–67. IAEA-TECDOC-194. Vienna: IAEA.

Donini, B. (1977). Breeding methods and applied mutagenesis in fruit plants. In *The Use of Ionizing Radiation in Agriculture* (Proceedings Workshop Association Euratom-ITAL, Wageningen, 1976), pp.453–87.

Donini, B. (1982). Mutagenesis applied to improve fruit trees: Techniques, methods and evaluation of radiation-induced mutations. In *Induced Mutations in Vegetatively Propagated Plants. II.* (Proceedings Final FAO/IAEA Research Coordination Meeting, Coimbatore, India, 1980), pp.29–35. Vienna: IAEA.

Donini, B., Kawai, T. & Micke, A. (1984). Spectrum of mutant characters utilized in developing improved cultivars. In *Selection in Mutation Breeding* (Proceedings FAO/IAEA Consultants Meeting), pp.7–31. Vienna: IAEA.

Donini, B., Mannino, P., Ancora, G., Sonnino, A., Fideghelli, C., Della Strada, G., Monasta, F., Quarta, R., Faedi, W., Albertini, A., Rivalta, L., Pennone, F., Rosati, P., Caló, A., Costacurta, A., Cersosima, A., Cancellier, S., Petruccioli, G., Filipucci, B., Panelli, G., Russo, F., Starrantino, A., Roselli, G., Romisondo, P., Me, G. & Radicati, L. (1991). Mutation breeding programmes for the genetic improvement of vegetatively propagated plants in Italy. In *Plant Mutation Breeding for Crop Improvement* (Proceedings FAO/IAEA Symposium, Vienna, 1990), vol.1, pp.237–55. Vienna: IAEA.

Donini, B. & Micke, A. (1984). Use of induced mutations in improvement of vegetatively propagated crops. In *Induced Mutations for Crop Improvement in Latin America* (Proceedings FAO/IAEA Regional Seminar, Lima, Peru, 1982). IAEA-TECDOC-305, pp.79–98. Vienna: IAEA.

Donini, B. & Rosselli, G. (1977). The use of mutation breeding to obtain nectarines. *Rivista della Ortoflorofrutticoltura Italiana*, **63**, 175–81 (in Italian with English summary).

Dooner, H.K. & Robbins, T.P. (1991). Genetic and developmental control of anthocyanin biosynthesis. *Annual Review of Genetics*, **25**, 173–99.

Döring, H.P. (1989). Tagging genes with maize transposable elements. An overview. *Maydica*, **34**, 73–88.

Döring, H.P. & Starlinger, P. (1986). Molecular genetics of transposable elements in plants. *Annual Review of Genetics*, **20**, 175–200.

Dorst, J.C. (1924). Knopmutatie bij den aardappel en hare betekenis voor den landbouw. *Genetica*, **6**, 1–123.

Douglas, G.C. (1986). Effects of gamma radiation on morphogenesis and mutagenesis in cultured stem explants of poplar. In *Nuclear Techniques and In Vitro Culture for Plant Improvement* (Proceedings FAO/IAEA Symposium, Vienna, 1985), pp.121–8. Vienna: IAEA.

Down, E.E. & Anderson, A.L. (1956). Agronomic use of an X-ray induced mutant. *Science*, **124**, 223–4.

Drake, J.W. (1969). Mutagenic mechanisms. *Annual Review of Genetics*, **3**, 247–68.

Drake, J.W. (1970). *The Molecular Basis of Mutation*. San Francisco: Holden-Day.

Drake, J.W. (1991). Spontaneous mutation. *Annual Review of Genetics*, **25**, 125–46.

Driscoll, C.J. & Jensen, N.F. (1963). A genetic method for detecting induced intergeneric translocations. *Genetics*, **48**, 459–68.

Duc, G. (1995). Mutagenesis of faba bean (*Vicia faba* L.) and the identification of five different genes controlling no nodulation, ineffective nodulation or supernodulation. *Euphytica*, **83**, 147–52.

Dulieu, H. (1967). Sur les différents types de mutations extranucléaires induites par le méthane sulfonate d'éthyle chez *Nicotiana tabacum* L. *Mutation Research*, **4**, 177–89.

Duncan, R.R., Waskom, R.M. & Nabors, M.W. (1995). *In vitro* screening and field evaluation of tissue-culture-regenerated sorghum (*Sorghum bicolor* (L.) Moench) for soil stress tolerance. In *The Methodology of Plant Genetic Manipulation: Criteria for Decision Making* (Proceedings EUCARPIA Plant Genetic Manipulation Section Meeting, Cork, Ireland, 1994), ed. A.C. Cassells & P.W. Jones, pp.373–80. Dordrecht, The Netherlands: Kluwer Academic Publishers.

Duncan, D.R. & Widholm, J.M. (1986). Cell selection for crop improvement. *Plant Breeding Reviews*, **4**, 153–73.

Duron, M. & Decourtye, L. (1986). Effets biologiques des rayons gamma appliqués à des plantes de *Weigela* cv. 'Bristol Ruby' cultivées in vitro. In *Nuclear Techniques and In Vitro Culture for Plant Improvement* (Proceedings FAO/IAEA Symposium, Vienna, 1985), pp.103–11. Vienna: IAEA.

Duron, M. & Decourtye, L. (1990). In vitro variation in *Weigela*. In *Somaclonal Variation in Crop Improvement I*, ed. Y.P.S. Bajaj, pp.606–23. Berlin: Springer-Verlag.

Durrant, A. (1958). Environmental conditioning of flax. *Nature*, **181**, 928–9.

Durrant, A. (1959). A new facet of the chromosome theory? *New Scientist*, **6**, 293–5.

Durrant, A. (1962). The environmental induction of heritable change in *Linum*. *Heredity*, **17**, 27–61.

Durrant, A. (1971). Induction and growth of flax genotrophs. *Heredity*, **27**, 277–98.

Durrant, A. (1981). Unstable genotypes. *Philosophical Transactions of the Royal Society of London*, series B, vol.**292** (no.1062), 467–74.

Dustin, A.P., Havas, L. & Lits, F. (1937). Action de la colchicine sur les divisions cellulaires chez les végétaux. *Comptes Rendus de l'Association des Anatomistes*, **32**, 170–6

Ehrenberg, L. (1960). Chemical mutagenesis: Biochemical and chemical points of view on mechanisms of action. In *Chemische Mutagenese* (Proceedings Erwin-Baur-Gedächtnisvorlesungen I, Gatersleben, 1959), ed. H. Stubbe, pp.124–36. Berlin: Akademie-Verlag.

Ehrenberg, L., Ekman, G., Gustafsson, Å, Jansson, G. & Lundqvist, U. (1965). Variation in quantitative and biochemical characters in barley after mutagenic treatment. In *The Use of Induced Mutations in Plant Breeding* (Report FAO/IAEA Technical Meeting, Rome, 1964), pp.477–90. Oxford: Pergamon Press.

Ehrenberg, L., Fedorcsák, I & Näslund, M. (1973). Possible biochemical mechanisms of "radiostimulation" in living cells. *Stimulation Newsletter*, **5**, 1–14.

Ehrenberg, L., Gustafsson, Å. & Lundqvist, U. (1956). Chemically induced mutation and sterility in barley. *Acta Chemica Scandinavica*, **10**, 492–4.

Ehrenberg, L. & Gustafsson, Å. (1957). On the mutagenic action of ethylene oxide and di-epoxybutane in barley. *Hereditas*, **43**, 595–602.

Ehrenberg, L., Gustafsson, Å. & Lundqvist, U. (1959). The mutagenic effects of ionizing radiations and reactive ethylene derivatives in barley. *Hereditas*, **45**, 351–68.

Ehrenberg, L., Gustafsson, Å. & Lundqvist, U. (1961). Viable mutants induced in barley by ionizing radiations and chemical mutagens. *Hereditas*, **47**, 243–82.

Ehrenberg, L., Lundqvist, U & Ström, G. (1958). The mutagenic action of ethylene imine in barley. *Hereditas*, **44**, 330–6.

Eigsti, O.J. & Dustin, P.Jr. (1955). *Colchicine in Agriculture, Medicine, Biology and Chemistry*. Ames, Iowa: The Iowa State College Press.

Ekberg, I. (1969). Different types of sterility induced in barley by ionizing radiations and chemical mutagens. *Hereditas*, **63**, 257–78.

Emerson, R.A. (1914). The inheritance of recurring somatic variation in variegated ears of maize. *American Naturalist*, **48**, 87–115 (original not consulted).

Emerson, R.A. (1917). Genetical studies of variegated pericarp in maize. *Genetics*, **2**, 1–35.

Emerson, R.A. (1929). The frequency of somatic mutation in variegated pericarp of maize. *Genetics*, **14**, 488–511.

Emmerling, M.H. (1955). A comparison of X-ray and ultraviolet effects on chromosomes of *Zea mays*. *Genetics*, **40**, 697–714.

Engler, D.E. & Grogan, R.G. (1984). Variation in lettuce plants regenerated from protoplasts. *Journal of Heredity*, **75**, 426–30.

Epp, M.D. (1973). Nuclear gene-induced plastome mutations in *Oenothera hookeri* I. Genetic analysis. *Genetics*, **75**, 465–83.

Esau, K. (1965). *Plant Anatomy* (2nd ed.). New York: John Wiley & Sons.

Esau, K. (1977). *Anatomy of Seed Plants* (2nd ed.). New York: John Wiley and Sons.

Evans, D.A. (1986). Practical use of genetic variation derived from *in vitro* culture. In *Nuclear Techniques and In Vitro Culture for Plant Improvement* (Proceedings FAO/IAEA Symposium, Vienna, 1985), pp.331–9. Vienna: IAEA.

Evans, D.A. (1987). Somaclonal variation. In *Tomato Biotechnology*, ed. D.J. Nevins & R.A. Jones, pp.59–69. New York: Alan R. Liss Inc.

Evans, D.A. (1988). Applications of somaclonal variation. In *Biotechnology in Agriculture*, ed. A. Mizrahi, pp.203–23. New York: Alan R.Liss Inc.

Evans, D.A. (1989). Somaclonal variation – Genetic basis and breeding applications. *Trends in Genetics*, **5**, 46–50.

Evans, D.A. & Sharp, W.R. (1983). Single gene mutations in tomato plants regenerated from tissue culture. *Science*, **221**, 949–51.

Evans, D.A. & Sharp, W.R. (1986). Applications of somaclonal variation. *Biotechnology*, **4**, 528–32.

Evans, D.A. & Sharp, W.R. (1988). Somaclonal variation and its application in plant breeding. *International Association for Plant Tissue Culture (IAPTC) Newsletter*, **54**, 2–10.

Evans, D.A., Sharp, W.R. & Medina-Filho, H.P. (1984). Somaclonal and gameto-clonal variation. *American Journal of Botany*, **71**, 759–74.

Evans, H.J. & Sparrow, A.H. (1961). Nuclear factors affecting radiosensitivity. II. Dependence on nuclear and chromosome structure and organization. In *Fundamental Aspects of Radiosensitivity* Brookhaven Symposia in Biology), Number 14, pp.101–27. Upton, New York: Biology Department, Brookhaven National Laboratory.

Evans, H.J. (1962). Chromosome aberrations induced by ionizing radiation. *International Review of Cytology*, **13**, 221–321.

Evenari, M. (1989). The history of research on white-green variegated plants. *The Botanical Review*, **55**, 106–39.

Fahmy, O.G. & Fahmy, M.J. (1956a). Mutagenicity of 2-chloroethyl methanesulphonate in *Drosophila melanogaster*. *Nature*, **177**, 996–7.

Fahmy, O.G. & Fahmy, M.J. (1956b). Differential genetic response to the alkylating mutagens and X-radiation. *Journal of Genetics*, **54**, 146–64.

Fahmy, O.G. & Fahmy, M.J. (1957). Mutagenic response to the alkyl-methanesulphonates during spermatogenesis in *Drosophila melanogaster*. *Nature*, **180**, 31–4.

Falconer, D.S. (1981). *Introduction to Quantitative Genetics*, 2nd edn. London: Longman.

Farmelo, G. (1995). The Discovery of X-rays. *Scientific American*, **273**, 68–73.

Favret, E.A. (1964). Genetic effects of single and combined treatment of ionizing radiations and ethyl methanesulphonate on barley seeds. In *Barley Genetics I* (Proceedings First International Barley Genetics Symposium, Wageningen, 1963), ed. S. Broekhuizen, G. Dantuma, H. Lamberts & W. Lange, pp.68–81. Wageningen: PUDOC.

Favret, E.A. & Ryan, W.W. (1964). Two cytoplasmic male-sterile mutants induced by X-rays and EMS. *Barley Newsletter*, **8**, 42.

Fedoroff, N.V. (1983). Controlling elements in maize. In *Mobile Genetic Elements*, ed. J.A. Shapiro, pp.1–63. New York: Academic Press.

Fedoroff, N.V. (1984). Transposable genetic elements in maize. *Scientific American*, **250** (6), 65–74.

Fedoroff, N.V. (1989). Maize transposable elements. In *Mobile DNA*, ed. D.E. Berg & M.M. Howe, pp.375–411. Washington, D.C.: American Society for Microbiology.

Fedoroff, N.V. (1994). Barbara McClintock (Obituary). *Genetics*, **136**, 1–10.

Fedoroff, N. & Bottstein, D. (ed.). (1992). *The Dynamic Genome*. New York: Cold Spring Harbor Laboratory Press.

Feenstra, W.J. & Jacobsen, E. (1980). Isolation of a nitrate reductase deficient mutant of *Pisum sativum* by means of selection for chlorate resistance. *Theoretical and Applied Genetics*, **58**, 39–42.

Fendrik, I. & Bors, J. (1991). *Strahlenschäden an Pflanzen*. Berlin: Paul Parey.

Fieldes, M.A. (1994). Heritable effects of 5-azacytidine treatments on the growth and development of flax (*Linum usitatissimum*) genotrophs and genotypes. *Genome*, **37**, 1–11.

Fincham, J.R.S. & Sastry, G.R.K. (1974). Controlling elements in maize. *Annual Review of Genetics*, **8**, 15–50.

Fluhr, R., Aviv, D., Galun, E, & Edelman, M. (1985). Efficient induction and selection of chloroplast-encoded antibiotic-resistant mutants in *Nicotiana*. *Proceedings of the National Academy of Sciences of the USA*, **82**, 1485–9.

Forster, B.P. (1994). Salt tolerance of the barley mutant 'Golden Promise'. *Mutation Breeding Newsletter*, **41**, 4.

Forster, B.P., Pakniyat, H., Macaulay, M., Matheson, W., Phillips, M.S., Thomas, W.T.B. & Powell, W. (1994). Variation in the leaf sodium content of the *Hordeum vulgare* (barley) cultivar Maythorpe and its derived mutant cv. Golden Promise. *Heredity*, **73**, 249–53.

Franzke, C.J. & Ross, J.G. (1952). Colchicine induced variants in *Sorghum*. *Journal of Heredity*, **43**, 107–15.

Freisleben, R. & Lein, A. (1942). Über die Auffindung einer mehltauresistenten Mutante nach Röntgenbestrahlung einer anfälligen reinen Linie von Sommergerste. *Naturwissenschaften*, **30**, 608.

Freisleben, R. & Lein, A. (1943a). Vorarbeiten zur züchterischen Auswertung röntgeninduzierter Mutationen. I. Die in der Behandlungsgeneration (X_1) sichtbare Wirkung der Bestrahlung ruhender Gerstenkörner. *Zeitschrift für Pflanzenzüchtung*, **25**, 235–54.

Freisleben, R. & Lein, A. (1943b). Vorarbeiten zur züchterischen Auswertung röntgeninduzierter Mutationen. II. Mutationen des Chlorophyllapparates als Testmutationen für die mutationsauslösende Wirkung der Bestrahlung der Gerste. *Zeitschrift für Pflanzenzüchtung*, **25**, 255–83.

Freisleben, R.A. & Lein, A. (1944a). Möglichkeiten und praktische Durchführung der Mutationszüchtung. *Kühn-Archiv*, **60**, 211–25.

Freisleben, R. & Lein, A. (1944b). Röntgeninduzierte Mutationen bei Gerste. *Der Züchter*, **16**, 49–64.

Frey, K.J. (1965). Mutation breeding for quantitative attributes. In *The Use of Induced Mutations in Plant Breeding* (Report FAO/IAEA Technical Meeting, Rome, 1964), pp.465–75. Oxford: Pergamon Press.

Friedberg, E.C. (1985). *DNA Repair*. New York: Freeman & Company.

Fruwirth, C. (1929). Über eine durch spontane Variabilität entstandene Kartoffelform und über spontane Variabilität der Kartoffel überhaupt. *Zeitschrift für Pflanzenzüchtung*, **14**, 35–79.

Frydenberg, O., Doll, H. & Sandfaer, J. (1964). The mutation frequency in different spike categories in barley. *Radiation Botany*, **4**, 13–25.

Fujita, H. (1982). Studies on mutation breeding in mulberry (*Morus* sp.). In *Induced Mutations in Vegetatively Propagated Plants II* (Proceedings Final FAO/IAEA Research Co-ordination Meeting, Coimbatore, India, 1980), pp.249–80. Vienna: IAEA.

Fujita, H. & Takato, S. (1970). An entire leaf mutant in mulberry. *Technical News No.5*. Ohmiya, Ibaraki, Japan: Institute of Radiation Breeding.

Futsuhara, Y. (1967). Studies of radiation breeding in the tea plants. In *Use of Chronic Irradiation in Mutation Breeding*. Gamma Field Symposia 6, 107–20. Ohmiya-machi, Ibaraki-ken, Japan: Institute of Radiation Breeding MAF.

Futsuhara, Y. (1968). Breeding of a new rice variety "Reimei" by gamma-ray irradiation. In *The Present State of Mutation Breeding*. Gamma Field Symposia 7, 87–109. Ohmiya-machi, Ibaraki-ken: Japan: Agriculture, Forestry and Fisheries Research Council and Institute of Radiation Breeding MAF.

Futsuhara, Y., Toriyama, K. & Tsunoda, K. (1967). On the breeding of rice variety Reimei by irradiation of gamma rays. *Japanese Journal of Breeding*, **17**, 85–90.

Gager, C.S. (1908). *Effects of the Rays of Radium on Plants*. New York: Memoirs of the New York Botanical Garden, Vol.IV.

Gager, C.S. & Blakeslee, A.F. (1927). Chromosome and gene mutations in *Datura* following exposure to radium rays. *Proceedings of the National Academy of Sciences of the USA*, **13**, 75–9.

Gao, M.W., Cheng, X.Y. & Liang, Z.Q. (1991). Effect of in vitro mutagenesis on the frequency of somaclonal variation in wheat. In *Plant Mutation Breeding for Crop Improvement* (Proceedings FAO/IAEA Symposium, Vienna, 1990), vol.2, pp.423–34. Vienna: IAEA.

Gardner, C.O. (1961). An evaluation of effects of mass selection and seed irradiation with thermal neutrons on yield of corn. *Crop Science*, **1**, 241–5.

Gardner, C.O. (1969). Genetic variation in irradiated and control populations of corn after ten cycles of mass selection for high grain yield. In *Induced Mutations in Plants* (Proceedings FAO/IAEA Symposium, Pullman, USA), pp.469–77. Vienna: IAEA.

Gardner, C.O. (1972). Utilization of genetic variability induced by mutation breeding of cross-fertilized plants. In *Mutation Breeding Workshop* (Proceedings Mutation Breeding Workshop, University of Tennessee, Knoxville, USA). Knoxville, Tennessee, USA: Agricultural Experiment Station, University of Tennessee.

Gaul, H. (1957). Die verschiedenen Bezugssysteme der Mutationshäufigkeit bei Pflanzen angewendet auf Dosis-Effektkurven. *Zeitschrift für Pflanzenzüchtung*, **38**, 63–76.

Gaul, H. (1958). Present aspects of induced mutations in plant breeding. *Euphytica*, **7**, 275–89.

Gaul, H. (1959). Über die Chimärenbildung in Gerstenpflanzen nach Röntgenbestrahlung von Samen. *Flora* (Jena) Abt.B, **47**, 209–41.

Gaul, H. (1960). Critical analysis of the methods for determining the mutation frequency after seed treatment with mutagens. *Genetica Agraria*, **12**, 297–318.

Gaul, H. (1961a). Studies on diplontic selection after X-irradiation of barley seeds. In *Effects of Ionizing Radiations on Seeds* (Proceedings FAO/IAEA Symposium, Karlsruhe, 1960), pp.117–36. Vienna: IAEA.

Gaul, H. (1961b). Use of induced mutants in seed-propagated species. In *Mutation and Plant Breeding* (Proceedings Symposium Cornell University, Ithaca, N.Y., USA, 1990), Publ. 891, 206–51. Washington D.C., USA: National Academy of Sciences-National Research Council.

Gaul, H. (1963). Mutationen in der Pflanzenzüchtung. *Zeitschrift für Pflanzenzüchtung*, **50**, 194–307.

Gaul, H. (1964a). Mutations in plant breeding. *Radiation Botany*, **4**, 155–232.

Gaul, H. (1964b). Induced mutations in plant breeding. In *Genetics Today* (Proceedings XI International Congress of Genetics, The Hague, The Netherlands, 1963), vol.3, ed. S.J. Geerts, pp.689–709. Oxford: Pergamon Press.

Gaul, H. (1965a). The concept of macro- and micro-mutations and results on induced micro-mutations in barley. In *The Use of Induced Mutations in Plant Breeding* (Report FAO/IAEA Technical Meeting, Rome, 1964), pp.407–28. Oxford: Pergamon Press.

Gaul, H. (1965b). Selection in M$_1$ generation after mutagenic treatment of barley seeds. In *Induction of Mutations and the Mutation Process* (Proceedings of a Symposium of the Czechoslovak Academy of Sciences, Prague, 1963), ed. J. Véleminský & T. Gichner, pp.62–72. Prague: Publ. House of the Czechoslovak Academy of Sciences.

Gaul, H. (1967). Studies on populations of micro-mutants in barley and wheat without and with selection. In *Induzierte Mutationen und ihre Nutzung* (Proceedings Erwin-Baur-Gedächtnisvorlesungen IV, Gatersleben, 1966), ed. H. Stubbe, pp.269–81. Berlin: Akademie-Verlag.

Gaul, H. (1977). Plant injury and lethality. In *Manual on Mutation Breeding*, 2nd edn, pp.87–8. Vienna: IAEA.

Gaul, H., Frimmel, C., Gichner, T. & Ulonska, E. (1972). Efficiency of mutagenesis. In *Induced Mutations and Plant Improvement* (Proceedings FAO/IAEA Study Group Meeting, Buenos Aires, 1970), pp.121–39. Vienna: IAEA.

Gaul, H. & Mittelstenscheid, L. (1960). Hinweise zur Herstellung von Mutationen durch ionisierende Strahlen in der Pflanzenzüchtung. *Zeitschrift für Pflanzenzüchtung, 43*, 404–22.

Gaul, H. & Mittelstenscheid, L. (1961). Untersuchungen zur Selektion von Kleinmutationen bei Gerste. *Zeitschrift für Pflanzenzüchtung, 45*, 300–14.

Gavazzi, G., Tonelli, C., Todesco, G., Arreghini, E., Raffaldi, F., Vecchio, F., Barbuzzi, G., Biasini, M.G. & Sala, F. (1987). Somaclonal variation versus chemically induced mutagenesis in tomato (*Lycopersicon esculentum* L.). *Theoretical and Applied Genetics, 74*, 733–8.

Gecheva, A.M.P. (1983). New mutant soybean variety "Boriana". *Mutation Breeding Newsletter, 23*, 3–4.

Geier, T. (1983). Induction and selection of mutants in tissue cultures of Gesneriaceae. *Acta Horticulturae, 131*, 329–37.

Geier, T. (1989). An improved type of 'Kohleria' obtained through in vitro chemical mutagenesis. *Mutation Breeding Newsletter, 34*, 7–8.

Gengenbach, B.G., Green, C.E. & Donovan, C.M. (1977). Inheritance of selected pathotoxin resistance in maize plants regenerated from cell cultures.

Proceedings of the National Academy of Sciences of the USA, 74, 5113–17.

Gierl, A. & Saedler, H. (1992). Plant-transposable elements and gene tagging. *Plant Molecular Biology, 19*, 39–49.

Gierl, A., Saedler, H. & Peterson, P.A. (1989). Maize transposable elements. *Annual Review of Genetics, 23*, 71–85.

Gifford, E.M. Jr. (1954). The shoot apex in angiosperms. *The Botanical Review, 20*, 477–529.

Gifford, E.M. Jr. & Corson, G.E. Jr. (1971). The shoot apex in seed plants. *Botanical Review, 37*, 143–229.

Gill, B.S., Kam-Morgan, L.N.W. & Shepard, J.F. (1986). Origin of chromosomal and phenotypic variation in potato protoclones. *Journal of Heredity, 77*, 13–16.

Gleba, Y.Y., Butenko, R.G. & Sytnik, K.M. (1975). Fusion of protoplasts and parasexual hybridization in *Nicotiana tabacum* L.. *Doklady Akademii Nauk SSSR, 221*, 1196–8 (in Russian).

Gleba, Y.Y. & Shlumukov, L.R. (1990). Somatic hybridization and cybridization. In *Plant Tissue Culture: Applications and Limitations*, ed. S.S. Bhojwani, pp.316–45. Amsterdam: Elsevier.

Gonzáles, A.J. (1994a). Radiation safety: New international standards. *Bulletin International Atomic Energy Agency, 1994(2)*, 2–11.

Gonzáles, A.J. (1994b). Biological effects of low doses of ionizing radiation: A fuller picture. *Bulletin International Atomic Energy Agency, 1994(4)*, 37–45.

Goodspeed, T.H. (1929). The effects of X-rays and radium on species of the genus *Nicotiana*. *Journal of Heredity, 20*, 243–59.

Gottschalk, W. (1987). Mutation: Higher Plants. *Progress in Botany, 49*, 216–30.

Gottschalk, W. & Kaul, M.L.H. (1975). Gene-ecological investigations in *Pisum* mutants. I. The influence of climatic factors upon quantitative and qualitative characters. *Zeitschrift für Pflanzenzüchtung, 75*, 182–91.

Gottschalk, W. & Wolff, G. (1983). *Induced Mutations in Plant Breeding*. Monographs on Theoretical and Applied Genetics No.7. Berlin: Springer Verlag.

Götz, R., Friedt, W., Kaiser, R. & Foroughi-Wehr, B. (1989). Genetic diversity for breeding of durable barley yellow mosaic virus resistance. In *Science for Plant Breeding/Book of Poster Abstracts*

(XII.EUCARPIA Congress, Göttingen, Germany), part 1, poster 3, 14. Gelsenkirchen-Buer, Germany: Th. Mann Publishers.

Goulden, C.H. (1939). Problems in plant selection. In *Proceedings 7th International Genetics Congress* (Edinburgh, UK), ed. R.C. Burnett, pp.132–3. Cambridge: Cambridge University Press (original not consulted).

Grabau, E.A., Hanlon, R. & Pesce, A. (1995). Mutagenesis and selection for oligomycin resistance in soybean (*Glycine max* L. Merr) suspension culture cells. *Plant Cell, Tissue and Organ Culture, 42*, 121–7.

Granhall, I. (1953). X-ray mutations in apples and pears. *Hereditas, 39*, 149–55.

Granhall, I. (1954). Spontaneous and induced bud-mutations in fruit trees. *Acta Agriculturae Scandinavica, 4*, 594–600.

Granhall, I., Gustafsson, Å., Nilsson, Fr. & Oldén, E.J. (1949). X-ray effects in fruit trees. *Hereditas, 35*, 269–79.

Gregory, W.C. (1955). X-ray breeding of peanuts (*Arachis hypogaea* L.). *Agronomy Journal, 47*, 396–9.

Gregory, W.C. (1956). Induction of useful mutations in the peanut. In *Genetics in Plant Breeding* (Brookhaven Symposia in Biology, Number 9), pp.177–90. Upton, New York: Biology Department, Brookhaven National Laboratory.

Gregory, W.C. (1961). The efficacy of mutation breeding. In *Mutation and Plant Breeding* (Proceedings Symposium Cornell University, Ithaca, N.Y. USA, 1990), Publ. 891, 461–86. Washington D.C., USA: National Academy of Sciences/National Research Council.

Gregory, W.C. (1965). Mutation frequency, magnitude of change and the probability of improvement in adaptation. In *The Use of Induced Mutations in Plant Breeding* (Report FAO/IAEA Technical Meeting, Rome, 1964), pp.430–41. Oxford: Pergamon Press.

Gregory, W.C. (1966). Mutation breeding. In *Plant Breeding*, ed. K.J. Frey, pp. 189–218. Ames, Iowa, USA: Iowa University Press.

Griesbach, R.J. (1989). Selection of a dwarf *Hemerocallis* through tissue culture. *HortScience, 24*, 1027–8.

Griesbach, R.J. & Semeniuk, P. (1987). Use of somaclonal variation in the improvement of *Eustoma grandiflorum*. *Journal of Heredity, 78*, 114–16.

Grun, P. (1976). *Cytoplasmic Genetics and Evolution*. New York: Columbia University Press.

Grunewaldt, J. (1988). In vitro selection and propagation of temperature tolerant *Saintpaulia*. In *Proceedings International Symposium on Propagation of Ornamental Plants* (Geisenheim, Germany, 1987). *Acta Horticulturae*, **226**, 271–6.

Guha, S. & Maheshwari, S.C. (1964). *In vitro* production of embryos from anthers of *Datura*. *Nature*, **204**, 497.

Guilleminot, H. (1908). Effets des rayons X et des rayons du radium sur la cellule vegetale. *Journal de Physiologie et de Pathologie Générale*, **10**, 1–16.

Gupta, P.P., Schieder, O. & Gupta, M. (1984). Intergeneric nuclear gene transfer between somatically and sexually incompatible plants through asymmetric protoplast fusion. *Molecular & General Genetics*, **197**, 30–5.

Gustafsson, Å. (1938). Studies on the genetic basis of chlorophyll formation and the mechanism of induced mutating. *Hereditas*, **24**, 33–93.

Gustafsson, Å. (1940). *The Mutation System of the Chlorophyll Apparatus*. Lunds Universitets Årsskrift. N.F. Avd.2. Bd 36. Nr 11. Lund: C.W.K. Gleerup.

Gustafsson, Å. (1941). Mutation experiments in barley. *Hereditas*, **27**, 225–42.

Gustafsson, Å. (1942). The plastid development in various types of chlorophyll mutations. *Hereditas*, **28**, 483–92.

Gustafsson, Å. (1947). Mutations in agricultural plants. *Hereditas*, **33**, 1–100.

Gustafsson, Å. (1954). Mutations, viability, and population structure. *Acta Agriculturae Scandinavica*, **4**, 601–32.

Gustafsson, Å. (1960). Chemical mutagenesis in higher plants. In *Chemische Mutagenese* (Proceedings Erwin-Baur-Gedächtnisvorlesungen I, Gatersleben, 1959), ed. H. Stubbe, pp.14–29. Berlin: Akademie-Verlag.

Gustafsson, Å. (1963). Productive mutations induced in barley by ionizing radiations and chemical mutagens. *Hereditas*, **50**, 211–63.

Gustafsson, Å. (1969a). A study on induce mutations in plants. In *Induced Mutations in Plants* (Proceedings FAO/IAEA Symposium, Pullman, USA), pp.7–31. Vienna: IAEA.

Gustafsson, Å. (1969b). Positive Mutationen und ihre Verwendung in der Züchtung hochleistender Gersten-Sorten. *Bericht Arbeitstagung 1969, Gumpenstein, Österreich*, pp.63–88. Gumpenstein, Österreich: Bundesversuchsanstalt für alpenländische Landwirtschaft.

Gustafsson, Å. (1979). The genetic analysis of phenotype patterns in barley. In *Induced Mutations for Crop Improvement in Africa* (Proceedings Regional FAO/IAEA/IITA Seminar, Ibadan, Nigeria, 1978). IAEA-TECDOC-222, pp.41–53. Vienna: IAEA.

Gustafsson, Å. (1986). Mutation and gene recombination-principal tools in plant breeding. In *Svalöf 1886–1986. Research and Results in Plant Breeding*, ed. G. Olsson, pp.76–84. Stockholm: LTs förlag.

Gustafsson, Å. & Ehrenberg, L. (1959). Ethylene imine: a new tool for plant breeders. *The New Scientist* (19-3-1959), **5**, 624–5.

Gustafsson, Å. & Gadd, J. (1965). Mutations and crop improvement. IV. *Poa pratensis* J. (*Gramineae*). *Hereditas*, **53**, 90–103.

Gustafsson, Å. & Gadd, I. (1966). Mutations and crop improvement. VII. The genus *Oryza* L. (*Gramineae*). *Hereditas*, **55**, 273–357.

Gustafsson, Å., Hagberg, A. & Lundqvist, U. (1960). The induction of early mutants in Bonus barley. *Hereditas*, **46**, 675–99.

Gustafsson, Å., Hagberg, A., Lundqvist, U. & Persson, G. (1969). A proposed system of symbols for the collection of barley mutants at Svalöv. *Hereditas*, **62**, 409–14.

Gustafsson, Å., Hagberg, A., Persson, G. & Wiklund, K. (1971). Induced mutations and barley improvement. *Theoretical and Applied Genetics*, **41**, 239–48.

Gustafsson, Å, Lundqvist, U. & Ekberg, I. (1966). The viability reaction of gene mutations and chromosome translocations in comparison. In *Mutations in Plant Breeding* (Proceedings of a FAO/IAEA Panel, Vienna, 1966), 103–7. Vienna: IAEA.

Gustafsson, Å & Lundqvist, U. (1980). Hexastichon and intermedium mutants in barley. *Hereditas*, **92**, 229–36.

Gustafsson, Å. & MacKey, J. (1948). The genetical effects of mustard gas substances and neutrons. *Hereditas*, **34**, 371–86.

Hagberg, A. (1986). Induced structural rearrangments. In *Genetic Manipulation in Plant Breeding* (Proceedings International Symposium EUCARPIA, Berlin, 1985). ed. W. Horn, C.J. Jensen, W. Odenbach & O. Schieder, pp.17–36. Berlin: Walter de Gruyter & Co.

Hagberg, A. & Åkerberg, E. (1962). *Mutations and Polyploidy in Plant Breeding*. Stockholm: Svenska Bokförlaget Bonniers.

Hagemann, R. (1958). Somatische Konversion bei *Lycopersicon esculentum* Mill. *Zeitschrift für Vererbungslehre*, **89**, 587–613.

Hagemann, R. (1964). *Plasmatische Vererbung*. Jena: VEB Gustav Fischer Verlag.

Hagemann, R. (1969). Somatische Konversion (Paramutation) am *sulfurea* locus von *Lycopersicon esculentum* Mill. IV. Die genotypische Bestimmung der Konversionshäufigkeit. *Theoretical and Applied Genetics*, **39**, 295–305.

Hagemann, R. (1976). Plastid distribution and plastid competition in higher plants and the induction of plastom mutations by nitroso-urea compounds. In *Genetics and Biogenesis of Chloroplasts and Mitochondria*, ed. Th. Bücher, W. Neupert, W. Sebald & S. Werner, pp.331–8. Amsterdam: Elsevier/North Holland.

Hagemann, R. (1979). Genetics and molecular biology of plastids of higher plants. *Stadler Symposia*, **11**, 91–115.

Hagemann, R. (1982). Induction of plastome mutations by nitroso-urea-compounds. In *Methods in Chloroplast Molecular Biology*, ed. M. Edelman, R.B. Hallick & N-H. Chua, pp.119–27. Amsterdam: Elsevier Biomedical Press.

Hagemann, R. & Berg, W. (1977). Vergleichende Analyse der Paramutationssysteme bei höheren Pflanzen. *Biologisches Zentralblatt*, **96**, 257–301.

Hagemann, R. & Snoad, B. (1971). Paramutation(somatic conversion) at the *sulfurea* locus of *Lycopersicon esculentum*. *Heredity*, **27**, 409–18.

Hake, S. (1992). Unraveling the knots in plant development. *Trends in Genetics*, **8**, 109–14.

Hake, S., Vollbrecht, E. & Freeling, M. (1989). Cloning *Knotted*, the dominant morphological mutant in maize using *Ds2* as a transposon tag. *The EMBO Journal*, **8**, 15–22.

Hall, B.G. (1993). Selection-induced mutations. In: *Volume of Abstracts* (17th International Congress of Genetics, Birmingham, UK), 8.

Hall, H.K., Cohen, D. & Skirvin, R.M. (1986). The inheritance of thornlessness from tissue culture-derived 'Thornless Evergreen' blackberry. *Euphytica*, **35**, 891–8.

Hall, H.K., Quazi, H. & Skirvin, R.M. (1986). Isolation of a pure thornless Loganberry by meristem tip culture. *Euphytica*, **35**, 1039–44.

Hall, H.K., Skirvin, R.M. & Braam, W.F. (1986). Germplasm release of 'Lincoln Logan', a tissue-derived genetic thornless 'Loganberry'. *Fruit Varieties Journal*, **40**, 134–5 (original not consulted).

Handro, W. (1981). Mutagenesis and *in vitro* selection. In *Plant Tissue Culture: Methods and Applications in Agriculture*, ed. T.A. Thorpe, pp.155–80. New York: Academic Press.

Hanna, W.W. (1990). Induced mutations in Midiron and Tifway bermuda grasses. *Agronomy Abstracts*, p.175. San Antonio, Texas: American Society of Agronomy.

Hanna, W.W. (1995a). Use of apomixis in cultivar development. *Advances in Agronomy*, **54**, 333–50.

Hanna, W.W. (1995b). Centipedegrass-diversity and vulnerability. *Crop Science*, **35**, 332–4.

Hanna, W.W. & Bashaw, E.C. (1987). Apomixis: its use and identification in plant breeding. *Crop Science*, **27**, 1136–9.

Hanna, W.W. & Burton, G.W. (1985). Morphological characteristics and genetics of two mutations for early maturity in pearl millet. *Crop Science*, **25**, 79–81.

Hanstein, J. (1868). Die Scheitelzellgruppe im Vegetationspunkt der Phanerogamen. *Festschrift Niederrheinische Gesellschaft für Natur- und Heilkunde*. Bonn, pp.1–26 (original not consulted).

Hanson, A.A. & Juska, F.V. (1959). A 'progressive'mutation in *Poa pratensis* L. by ionizing radiation. *Nature*, **184**, 1000–1.

Hanson, A.A. & Juska, F.V. (1962). Induced mutations in Kentucky bluegrass. *Crop Science*, **2**, 369–71.

Harle, J.R. (1972). A revision of mutation breeding procedures in *Arabidopsis* based on a fresh analysis of the mutant sector problem. *Canadian Journal of Genetics and Cytology*, **14**, 559–72.

Harle, J.R. (1974). Mutation breeding and the mutant sector problem in *Arabidopsis*. *Canadian Journal of Genetics and Cytology*, **16**, 476–80.

Harris, M. (1964). *Cell Culture and Somatic Variation*. New York: Holt, Rinehart and Winston.

Harte, C. (1994). *Oenothera (Contributions of a Plant to Biology)*. Berlin: Springer-Verlag.

Haynes, R.H. & Kunz, B.A. (1988). Metaphysics of regulated deoxyribonucleotide biosynthesis. *Mutation Research*, **200**, 5–10.

Hearn, C.J. (1984). Development of seedless orange and grapefruit cultivars through seed irradiation. *Journal of the American Society for Horticultural Science*, **109**, 270–3.

Hearn, C.J. (1986). Development of seedless grapefruit cultivars through budwood irradiation. *Journal of the American Society of Horticultural Science*, **111**, 304–6.

Heath-Pagliuso, S., Pullman, J. & Rappaport, L. (1989). 'UC-T3 Somaclone': Celery germplasm resistant to *Fusarium oxysporum* f.sp.*apii.*, race 2. *HortScience*, **24**, 711–12.

Heddle, J.A., Whissell, D. & Bodycote, D.J. (1969). Changes in chromosome structure induced by radiations: A test of the two chief hypotheses. *Nature*, **221**, 1158–60.

Heiken, A. (1960). *Spontaneous and X-ray Induced Somatic Aberrations in Solanum tuberosum L.* Stockholm: Almqvist and Wiksell.

Heiken, A. & Ewertson, G. (1962). The chimaerical structure of a somatic *Solanum* mutant revealed by ionizing irradiation. *Genetica*, **33**, 88–94.

Heiken, A., Ewertson, G. & Carlström, L. (1963). Studies on a somatic subdivided-leaf mutant in *Solanum tuberosum*. *Radiation Botany*, **3**, 145–53.

Heinz, D.J. (1973). Sugar-cane improvement through induced mutations using vegetative propagules and cell culture techniques. In *Induced Mutations in Vegetatively Propagated Plants* (Proceedings FAO/IAEA Panel, Vienna, 1972), pp.53–9. Vienna: IAEA.

Heinz, D.J., Krishnamurthi, M., Nickell, L.G. & Maretzki, A. (1977). Cell, tissue, and organ culture in sugarcane improvement. In *Applied and Fundamental Aspects of Plant Cell, Tissue and Organ Culture*, ed. J. Reinert & Y.P.S. Bajaj, pp.3–17. Berlin: Springer-Verlag.

Heinz, D.J. & Mee, G.W.P. (1969). Plant differentiation from callus tissue of *Saccharum* species. *Crop Science*, **9**, 346–8.

Heinz, D.J. & Mee, G.W.P. (1971). Morphologic, cytogenetic, and enzymatic variation in *Saccharum* species hybrid clones derived from callus tissue. *American Journal of Botany*, **58**, 257–62.

Hensz, R.A. (1991). Mutation breeding of grapefruit (*Citrus paradisi* Macf.). In *Plant Mutation Breeding for Crop Improvement* (Proceedings FAO/IAEA Symposium, Vienna, 1990), vol.1, pp. 533–6. Vienna: IAEA.

Hentrich, W. & Glawe, M. (1982). Züchtung von Edelnelken (*Dianthus caryophyllus*) durch ÄMS-Applikation in die Blattachseln von Jungpflanzen. *Archiv für Züchtungsforschung* (Berlin), **12**, 197–207.

Hermelin, T., Brunner, H., Daskalov, S. & Nakai, H. (1983). Chimerism in M₁ plants of *Vicia faba*, *Capsicum annuum* and *Linum usitatissimum*. In *Chimerism in Irradiated Dicotyledonous Plants* FAO/IAEA Consultants Meeting, Vienna, 1981), pp.43–9. IAEA-TECDOC-289. Vienna: IAEA.

Hermsen, J.G.Th., Ramanna, M.S., Roest, S. & Bokelmann, G.S. (1981). Chromosome doubling through adventitious shoot formation on in vitro cultivated leaf explants from diploid interspecific potato hybrids. *Euphytica*, **30**, 239–46.

Hertwig, P. (1927). *Partielle Keimesschädigungen durch Radium und Röntgenstrahlen*. Handbuch der Vererbungswissenschaften (ed. E. Baur & M. Hartmann), Vol. III, part 1. Berlin: Verlag von Gebr. Borntraeger.

Heslot, H. (1960). Action d'agents chimiques mutagènes sur quelques plants cultivées. In *Chemische Mutagenese* (Proceedings Erwin-Baur-Gedächtnisvorlesungen I, Gatersleben, 1959), ed. H. Stubbe, pp.106–8. Berlin: Akademie-Verlag.

Heslot, H. (1965). The nature of mutations. In *The Use of Induced Mutations in Plant Breeding* (Report of the FAO/IAEA Technical Meeting, Rome, 1964), pp.3–45. Oxford: Pergamon Press.

Heslot, H. & Ferrary, R. (1958). Action génétique comparée des radiations et de quelques mutagènes sur l'orge. *Annales de l'Institute National Agronomique*, **44**, 133–52.

Heslot, H., Ferrary, R., Lévy, R.& Monard, C. (1959). Recherches sur les substances mutagènes (halogéno-2-éthyl): amines, dérivés oxygénés du sulfure de bis-(chloro-2 éthyle), esters sulfoniques et sulfuriques. *Comptes Rendus des Séances de l'Académie des Sciences*, **248**D, 729–32.

Heslot, H., Ferrary, R., Lévy, R. & Monard, C. (1961). Induction de mutations chez l'orge.Efficacité relative des rayons gamma, du sulfate d'éthyle, du méthane sulfonate d'éthyle et de quelques autres substances. In *Effects of Ionizing Radiations on Seeds* (Proceedings FAO/IAEAO Symposium, Karlsruhe, 1960), pp.243–50. Vienna: IAEA.

Heslot, H., Ferrary, R. & Tempé, J. (1966). The relative mutagenic effects of some nitrosamines on barley seeds. *Mutation Research*, **3**, 354–5.

Hibberd, K.A. & Green, C.E. (1982). Inheritance and expression of lysine plus threonine resistance selected in maize tissue culture. *Proceedings of the National Academy of Sciences of the USA*, **79**, 559–63.

Highkin, H.R. (1958). Temperature-induced variability in peas. *American Journal of Botany*, **45**, 626–31.

Hildering, G.J. & Verkerk, K. (1965). Chimeric structure of the tomato plant after seed treatment with EMS and X-rays. In *The Use of Induced Mutations in Plant Breeding* (Report FAO/IAEA Technical Meeting, Rome, 1964), pp.317–20. Oxford: Pergamon Press.

Hill, J. (1967). The environmental induction of heritable changes in *Nicotiana rustica* parental and selection lines. *Genetics*, **55**, 735–54.

Hill, J. & Perkins, J.M. (1969). The environmental induction of heritable changes in *Nicotiana rustica*. Effects of genotype-environment interactions. *Genetics*, **61**, 661–75.

Hodgdon, A.L., Nilan, R.A. & Kleinhofs, A. (1979). Azide mutagenesis-varietal response, pregermination conditions and concentration. *Barley Genetics Newsletter*, **9**, 29–33.

Hofmeister, W. (1852). Zur Entwickelungsgeschichte der *Zostera*.

Botanische Zeitung, **10**, 121–31; 137–49; 157–8.

Holliday, R. (1987). The inheritance of epigenetic defects. *Science*, **238**, 163–70.

Horn, W., Jensen, C.J., Odenbach, W. & Schieder, O. (eds). (1986). *Genetic Manipulation in Plant Breeding* (Proceedings International Symposium EUCARPIA, Berlin, 1985). Berlin: Walter de Gruyter.

Horsch, R.B. (1995). Transgenic crops. Processes, products and problems. In *Induced Mutations and Molecular Techniques for Crop Improvement* (Proceedings FAO/IAEA Symposium, Vienna), pp.357–69. Vienna: IAEA.

Hosticka, L.P. & Hanson, M.R. (1984). Induction of plastid mutations in tomatoes by nitrosomethylurea. *Journal of Heredity*, **75**, 242–6.

Hovenkamp-Hermelink, J.H.M., Jacobsen, E., Ponstein, A.S., Visser, R.G.F., Vos-Scheperkeuter.G.H., Bijmolt, E.W., de Vries, J.N. & Feenstra, W.J. (1987). Isolation of an amylose-free starch mutant of the potato (*Solanum tuberosum* L.). *Theoretical and Applied Genetics*, **75**, 217–21.

Hovenkamp-Hermelink, J.H.M., Jacobsen, E., Pijacker, L.O., de Vries, J.N., Witholt, B. & Feenstra, W.J. (1988). Cytological studies on adventitious shoots and minitubers of a monoploid potato clone. *Euphytica*, **39**, 213–19.

Howard, H.W. (1958). Transformation of a monochlamydius into a dichlamydius chimaera by X-ray treatment. *Nature*, **182**, 1620.

Howard, H.W. (1959). Experiments with a potato periclinal chimera. *Genetica*, **30**, 278–91.

Howard, H.W. (1964). The use of X-rays in investigating potato chimeras. *Radiation Botany*, **4**, 361–71.

Howard, H.W. (1967). Further experiments on the use of X-rays and other methods in investigating potato chimeras. *Radiation Botany*, **7**, 389–99.

Howard, H.W. (1969). A full analysis of a potato chimera. *Genetica*, **40**, 233–41.

Howard, H.W. (1972). The stability of an L3 mutant potato chimera. *Potato Research*, **15**, 374–7.

Howard, H.W., Wainwright, J. & Fuller, J.M. (1963). The number of independent layers at the stem apex in potatoes. *Genetica*, **34**, 113–20.

Howland, G.P. & Hart, R.W. (1977). Radiation biology of cultured plant cells. In *Applied and Fundamental Aspects of Plant Cell, Tissue and Organ Culture*, ed. J. Reinert & Y.P.S. Bajaj, pp.731–56. Berlin: Springer-Verlag.

Hsu, C-M, Yang, W-P., Chen, C-C., Lai, Y-K. & Lin, T-Y. (1993). A point mutation in the chloroplast *rps12* gene from *Nicotiana plumbaginifolia* confers streptomycin resistance. *Plant Molecular Biology*, **23**, 179–83.

Hu, J. & Rutger, J.N. (1991). A streptomycin induced no-pollen male sterile mutant in rice (*Oryza sativa* L.). *Journal of Genetics & Breeding*, **45**, 349–52.

Hu, J. & Rutger, J.N. (1992). Pollen characteristics and genetics of induced and spontaneous genetic male-sterile mutants in rice. *Plant Breeding*, **109**, 97–107.

Huang, C. & Liang, J. (1980). Plant breeding achievements in ancient China. *Agronomic History Research*, **1**, 1–10. Beijing: Agronomy Press (in Chinese, original not consulted).

Huitema, J.B.M., Gussenhoven, G., Dons, J.J.M. & Broertjes, C. (1986). Induction and selection of low-temperature-tolerant mutants of *Chrysanthemum morifolium* Ramat. In *Nuclear Techniques and in Vitro Culture for Plant Improvement* (Proceedings FAO/IAEA Symposium, Vienna, 1985), pp.321–7. Vienna: IAEA.

Huitema, J.B.M., Preil, W., Gussenhoven, G.C. & Schneidereit, M. (1989). Methods for the selection of low-temperature tolerant mutants of *Chrysanthemum morifolium* Ram. by using irradiated cell suspension cultures. I.Selection of regenerants *in vivo* under suboptimal temperature conditions. *Plant Breeding*, **102**, 140–7.

Ichijima, K. (1934). On the artificially induced mutations and polyploid plants of rice occurring in subsequent generations. *Proceedings of the Imperial Academy of Japan, Tokyo*, **10**, 388–91.

Ichikawa, S. (1972). Radiosensitivity of a triploid clone of *Tradescantia* determined in its stamen hairs. *Radiation Botany*, **12**, 179–89.

Ichikawa, S. (1992). *Tradescantia* stamen-hair system as an excellent botanical tester of mutagenicity: Its responses to ionizing radiations and chemical mutagens and some synergistic effects found. *Mutation Research*, **270**, 3–22.

Ichikawa, S. (1994). Sectoring patterns of spontaneous and radiation-induced somatic pink mutations in the stamen hairs of a temperature-sensitive mutable clone of *Tradescantia*. *Japanese Journal of Genetics*, **69**, 577–91.

Ichikawa, S., Imai, T. & Nakano, A. (1991). Comparison of somatic mutation frequencies in the stamen hairs of one mutable and two stable clones of *Tradescantia* treated with small doses of gamma rays. *Japanese Journal of Genetics*, **66**, 513–25.

Ichikawa, S., Shima, N., Ishii, C., Kanai, H., Sanda-Kamigawara, M.S. & Matsuura-Endo, C. (1996). Variation of spontaneous somatic mutation frequency in the stamen hairs of *Tradescantia* clone BNL 02. *Genes & Genetic Systems*, **71**, 159–65.

Ichikawa, S., Yamaguchi, A. & Okumura, M. (1993). Synergistic effects of methyl methanesulfonate and X rays in inducing somatic mutations in the stamen hairs of *Tradescantia* clones, KU 27 and BNL 4430. *Japanese Journal of Genetics*, **68**, 277–92.

Iida, S. & Amano, E. (1987). A method to obtain mutants in outcrossing crops. Induction of seedling mutants in cucumber using pollen irradiation. *Technical News No.32*. Ohmiya-machi, Ibaraki, Japan: Institute of Radiation Breeding, NIAR, MAFF.

Ikeda, F. (1974). Radiation-induced fruit colour mutation in the apple var.Fuji. *Technical News No.15*, Ohmiya-machi, Ibaraki-ken, Japan: Institute of Radiation Breeding MAF.

Ikeda, F. (1977). Induced mutation in dwarf growth habits of apple trees by gamma-rays and its evaluation in practical uses. In *Use of Dwarf Mutation*. Gamma Field Symposia **16**, 63–81. Ohmiya-machi, Ibaraki-ken: Institute of Radiation Breeding NIAS MAFF.

Ikenaga, M. & Mabuchi, T. (1966). Photoreactivation of endosperm mutations induced by ultraviolet light in maize. *Radiation Botany*, **6**, 165–9.

Ingram, D.S. & MacDonald, M.V. (1986). In vitro selection of mutants. In *Nuclear Techniques and in Vitro Culture for Plant Improvement* (Proceedings FAO/IAEA Symposium, Vienna, 1985), pp.241–58. Vienna; IAEA.

Ipser, J. (1993). Effect of dimethylsulfoxide on methylmethanesulfonate-induced chromosomal aberrations in *Crepis capillaris* cultivated in vitro. *Biologia Plantarum*, **35**, 137–9.

Iqbal, R.M.S., Chaudhry, M.B., Aslam, M. & Bandesha, A.A. (1991). Economic and agricultural impact of mutation breeding in cotton in Pakistan. In *Plant Mutation Breeding for Crop Improvement* (Proceedings FAO/IAEA Symposium, Vienna, 1990), vol.1, pp.187–201. Vienna: IAEA.

Iqbal, R.M.S., Chaudhry, M.B., Aslam, M. & Bandesha, A.A. (1994). Development of a high yielding cotton mutant, NIAB-92 through the use of induced mutations. *Pakistan Journal of Botany*, **26**, 99–104.

Itoh, T. & Kondo, S. (1992). Tracing development of soybean leaf primordia by marking progenitor cells with mutant markers. *Japanese Journal of Genetics*, **67**, 61–70.

Jackson, J.A. & Dale, P.J. (1989). Somaclonal variation in *Lolium multiflorum* L. and *L. temulentum* L. *Plant Cell Reports*, **8**, 161–4.

Jackson, J.F. (1987). DNA repair in pollen. A review. *Mutation Research*, **181**, 17–29.

Jacobs, Th. & Parlevliet, J.E., eds. (1993). *Durability of Disease Resistance*. Dordrecht: Kluwer Academic Press.

Jacobsen, E. (1984). Modification of symbiotic interaction of pea (*Pisum sativum* L.) and *Rhizobium leguminosarum* by induced mutations. *Plant and Soil*, **82**, 427–38.

Jacobsen, E. & Feenstra W.J. (1984). A new pea mutant with efficient nodulation in the presence of nitrate. *Plant Science Letters*, **33**, 337–44.

Jacobsen, E., Hovenkamp-Hermelink, J.H.M., Krijgsheld, H.T., Nijdam, H., Pijnacker, L.P., Witholt, B. & Feenstra, W.J. (1989). Phenotypic and genotypic characterization of an amylose-free starch mutant of potato. *Euphytica*, **44**, 43–8.

Jacobsen, E., Ramanna, M.S., Huigen, D.J. & Sawor, Z. (1991). Introduction of an amylose-free (*amf*) mutant into breeding of cultivated potato, *Solanum tuberosum*, L. *Euphytica*, **53**, 247–53.

Jacobsen, P. (1966). Demarcation of mutant-carrying regions in barley plants after ethylmethane-sulfonate seed treatment. *Radiation Botany*, **6**, 313–28.

Jacquot, M. (1986). Mutagenesis of upland rice at IRAT. *Mutation Breeding Newsletter*, **28**, 10–12.

Jagathesan, D. (1976). Induction and isolation of mutants in sugarcane. In *Improvement of Vegetatively Propagated Plants and Tree Crops through Induced Mutations* (Proceedings Second FAO/IAEA Research Coordination Meeting, Wageningen, The Netherlands), pp.69–82. IAEA-TECDOC-194. Vienna: IAEA.

Jagathesan, D. (1979). Mutation breeding in sugarcane. In *Induced Mutations for Crop Improvement in Africa* (Proceedings FAO/IAEA/IITA Seminar, Ibadan, Nigeria, 1978), pp.67–82. IAEA-TECDOC-222. Vienna: IAEA.

Jagathesan, D. (1982). Mutation breeding in sugarcane. In *Induced Mutations in Vegetatively Propagated Plants II* (Proceedings FAO/IAEA Co-ordination Meeting, Coimbatore, India, 1980), pp.139–52. Vienna: IAEA.

James, J. (1961). Some observations on radiation and roses. *American Rose Magazine*, **16**, p.4 and p.23.

Jan, C.C. & Rutger, J.N. (1988). Mitomycin C- and streptomycin-induced male sterility in cultivated sunflower. *Crop Science*, **28**, 792–5.

Jank, H. (1957). Experimentelle Mutationsauslösung durch Röntgenstrahlen bei *Chrysanthemum indicum*. *Der Züchter*, **27**, 223–31.

Jaranowski, J.K. (1976). Gamma-induced mutations in *Pisum arvense* (L.s.l.). *Genetica Polonica*, **17**, 479–95 (original not consulted).

Jaranowski, J.K. & Micke, A. (1985). Mutation breeding in peas. *Mutation Breeding Review*, **2**, 1–23.

Jende-Strid, B. (1988). Coordinator's Report: Anthocyanin genes. Stocklist of *ant* mutants kept at the Carlsberg Laboratory. *Barley Genetics Newsletter*, **18**, 74–9.

Jende-Strid, B. (1990). Co-ordinator's Report: Anthocyanin genes. *Barley Genetics Newsletter*, **20**, 87–8.

Jende-Strid, B. (1991). Gene-enzyme relations in the pathway of flavonoid biosynthesis in barley. *Theoretical and Applied Genetics*, **81**, 668–74.

Jende-Strid, B. (1994). Coordinator's Report: Anthocyanin genes. *Barley Genetics Newsletter*, **24**, 162–5.

Jende-Strid, B. & Lundqvist, U. (1978). Diallelic tests of anthocyanin-deficient mutants. *Barley Genetics Newsletter*, **8**, 57–9.

Jenkins, M.T. (1924). Heritable characters of maize. XX. Iojap-striping, a chlorophyll defect. *Journal of Heredity*, **15**, 467–72.

Jensen, C.J. (1974). Chromosome doubling techniques in haploids. In *Haploids in Higher Plants: Advances and Potential*

(Proceedings of the First International Symposium, Guelph, Ontario, Canada), ed. K.J. Kasha, pp.153-90. Guelph: The University of Guelph.

Jensen, J. (1991). New high yielding, high lysine mutants in barley. In *Plant Mutation Breeding for Crop Improvement* (Proceedings FAO/IAEA Symposium, Vienna, 1990), vol.2, pp.31-41. Vienna: IAEA.

Jensen, K.A., Kirk, I., Kølmark, G. & Westergaard, M. (1951). Chemically induced mutations in *Neurospora*. In *Genes and Mutations. Cold Spring Harbor Symposia on Quantitative Biology*, vol.16, pp.245-61. Cold Spring Harbor, Long Island, New York: The Biological Laboratory.

Jerzy, M. & Zalewska, M. (1996). Polish cultivars of *Dendrathema grandiflora* Tzelev and *Gerbera jamesonii* bred *in vitro* by induced mutations. *Mutation Breeding Newsletter*, **42**, 19.

Jinks, J.L., Caligari, P.D.S. & Ingram, N.R. (1981). Gene transfer in *Nicotiana rustica* using irradiated pollen. *Nature*, **291**, 586-8.

Johannsen, W. (1903). *Über Erblichkeit in Populationen und reinen Linien*. Jena: Fischer (original not consulted).

Johannsen, W. (1909). *Elemente der Exacter Erblichkeitslehre*. Jena: Gustav Fischer Verlag.

Johnson, E.L. (1936). The effects of X-rays upon green plants. In *The Biological Effects of Radiation*, ed. B.M. Duggar, pp.961-1013. New York and London: McGraw Hill Book Co.

Johnston, M.O. & Schoen, D.J. (1995). Mutation rates and dominance levels of genes affecting total fitness in two angiosperm species. *Science*, **267**, 226-9.

Joosten, M.H.A.J., Cozijnsen, T.J. & de Wit, P.J.G.M. (1994). Host resistance to a fungal tomato pathogen lost by a single base-pair change in an avirulence gene. *Nature*, **367**, 384-6.

Jørgensen, C.A. & Crane, M.B. (1927). Formation and morphology of *Solanum* chimeras. *Journal of Genetics*, **18**, 247-73.

Jørgensen, J.H. (1976). Identification of powdery mildew resistant barley mutants and their allelic relationship. In *Barley Genetics III* (Proceedings 3rd International Barley Genetics Symposium, Garching, 1975), ed. H. Gaul, pp.446-55. München, Germany: Karl Thiemig.

Jørgensen, J.H. (1987). Three kinds of powdery mildew resistance in barley. In *Barley Genetics V* (Proceedings 5th International Barley Genetics Symposium, Okayama, Japan, 1986), ed. S. Yasuda & T. Konishi, pp.583-92. Nakasange, Okayama: Sanyo Press.

Jørgensen, J.H. (1991). Experience from mutation studies on cereal disease resistance. In *Plant Mutation Breeding for Crop Improvement* (Proceedings FAO/IAEA Symposium, Vienna, 1990), vol.2, pp.81-91. Vienna: IAEA.

Jørgensen, J.H. (1992). Discovery, characterization and exploitation of Mlo powdery mildew resistance in barley. In *Breeding for Disease Resistance*, ed. R. Johnson & G.J. Jellis. *Euphytica*, **6-3**, 141-52.

Jørgensen, J.H. (1993). Durability of resistance in the pathosystem: barley-powdery mildew. In *Durability of Disease Resistance*, ed. Th. Jacobs & J.E. Parlevliet, pp.159-76. Dordrecht, The Netherlands: Kluwer Academic Publishers.

Jorgensen, R. (1993). The germinal inheritance of epigenetic information in plants. *Philosophical Transactions of the Royal Society of London*, ser.B, vol.**338**, 173-81.

Joshi, S.N. & Frey, K.J. (1969). Mutagen induced variability for oat seed weight in selected and unselected populations. *Radiation Botany*, **9**, 501-7.

Julen, G. (1954). Observations on X-rayed *Poa pratensis*. *Acta Agriculturae Scandinavica*, **4**, 585-93.

Julen, G. (1958). Über die Effekte der Röntgenbestrahlung bei *Poa pratensis*. *Der Züchter*, **28**, 37-40.

Kaplan, R.W. (1953). Über Möglichkeiten der Mutationsauslösung in der Pflanzenzüchtung. *Zeitschrift für Pflanzenzüchtung*, **32**, 121-31.

Karp, A. (1991). On the current understanding of somaclonal variation. In *Oxford Surveys of Plant Molecular and Cell Biology*, vol.7, ed. B.J. Miflin, pp.1-58. Oxford: Oxford University Press.

Karp, A. (1995). Somaclonal variation as a tool for crop improvement. *Euphytica*, **85**, 295-302.

Karp, A. & Bright, S.W.J. (1985). On the causes and origins of somaclonal variation. In *Oxford Surveys of Plant Molecular and Cell Biology*, vol.2, ed. B.J. Miflin, pp.199-234. Oxford: Oxford University Press.

Kastenbaum, M.A. & Bowman, K.O. (1970). Tables for determining the statistical significance of mutation frequencies. *Mutation Research*, **9**, 527-49.

Katagiri, K. (1973). Radiation damage in winter buds and relation of shoot cutting-back to mutation frequencies and spectra in acutely gamma-irradiated mulberry. In *Induced Mutation and Chimera in Woody Plants*. Gamma Field Symposia 12, 63-79. Ohmiya-machi, Ibaraki-ken, Japan: Institute of Radiation Breeding NIAS MAF.

Kaul, M.L.H. (1988). *Male Sterility in Higher Plants*. Berlin: Springer-Verlag.

Kaul, M.L.H. & Bhan, A.K. (1977). Mutagenic effectiveness and efficiency of EMS, DES and gamma-rays in rice. *Theoretical and Applied Genetics*, **50**, 241-6.

Kawai, T. (1983). M_1 chimerism following mutagen treatment of seeds in rice and some other cereals. In *Chimerism in Irradiated Dicotyledonous Plants* (Report FAO/IAEA Consultants Meeting, Vienna, 1981), pp.7-11. IAEA-TECDOC-289. Vienna: IAEA.

Kawai, T. & Amano, E. (1991). Mutation breeding in Japan. In *Plant Mutation Breeding for Crop Improvement* (Proceedings FAO/IAEA Symposium, Vienna, 1990), vol.1, pp.47-66.

Kermicle, J.L., Eggleston, W.B. & Alleman, M. (1995). Organization of paramutagenicity in *R-stippled* maize. *Genetics*, **141**, 361-72.

Khotyljova, L.V., Khokhlova, S.A. & Khokhlov, I.V. (1988). Modification of genetic effects of gamma radiation by laser radiation. *Mutation Breeding Newsletter*, **32**, 15.

Khush, G.S. & Rick, C.M. (1968). Cytogenetic analysis of the tomato genome by means of induced deficiencies. *Chromosoma*, **23**, 452-84.

Kihlman, B.A. (1950). 8-Ethoxycaffeine, an ideal inducer of structural chromosome changes in the root tips of *Allium cepa*. *Experimental Cell Research*, **1**, 135-8.

Kihlman, B.A. (1966). *Actions of chemicals on dividing cells*, p.169-71. Englewood Cliffs, N.J., USA: Prentice-Hall.

Kihlman, B.A. (1977). *Caffeine and Chromosomes*. Amsterdam: Elsevier Scientific Publishing Company.

Kihlman, B. & Levan, A. (1951). Localized chromosome breakage in *Vicia faba*. *Hereditas*, **37**, 382-8.

Kilbey, B.J. (1995). Charlotte Auerbach (1899–1994). *Genetics*, **141**, 1–5.

Kimball, , R.F. (1987). The development of ideas about the effects of DNA repair on the induction of gene mutations and chromosomal aberrations by radiation and by chemicals. *Mutation Research*, **186**, 1–34.

Kinoshita, T. & Takahashi, M. (1969). Induction of cytoplasmic male sterility by gamma irradiation in sugar beets. *Japanese Journal of Breeding*, **19**, 445–57.

Kinoshita, T., Takahashi, M. & Mikami, T. (1979). Induction of cytoplasmic male sterility by sugar beets (a preliminary report). *Seiken Zihô*, **27/28**, 66–71.

Kinoshita, T., Takahashi, M. & Mikami, T. (1982). Cytoplasmic male sterility induced by chemical mutagens in sugar beets. *Proceedings Japanese Academy*, **58**, Series B, 319–22.

Kirk, J.T.O. & Tilney-Bassett, R.A.E. (1967). *The Plastids*. San Francisco: Freeman and Comp.

Kirk, J.T.O. & Tilney-Bassett, R.A.E. (1978). *The Plastids, Their Chemistry, Structure, Growth and Inheritance* (2nd ed.). Amsterdam: Elsevier/North-Holland Biomedical Press.

Kivi, E.I. (1981). Earliness mutants from sodium azide treated six-row barley. In *Barley Genetics, IV* (Proceedings of the Fourth International Barley Genetics Symposium, Edinburgh, UK), pp.855–7. Edinburgh: Edinburgh University Press.

Kivi, E.I. (1991). Realization of mutation programmes for cereal breeding. In *Plant Mutation Breeding for Crop Improvement* (Proceedings FAO/IAEA Symposium, Vienna, 1990), vol.1, pp.119–26. Vienna: IAEA.

Kleese, R.A. (1993). Use of new technologies in selection for biochemical traits. In *International Crop Science I*, ed. D.R. Buxton *et al.* Madison, Wisconsin, USA: Crop Science Society of America, Inc.

Kleffel, B., Walther, F. & Preil, W. (1986). X-ray-induced mutability in embryogenic suspension cultures of *Euphorbia pulcherrima*. In *Nuclear Techniques and In Vitro Culture for Plant Improvement* (Proceedings FAO/IAEA Symposium, Vienna, 1985), pp.113–20. Vienna: IAEA.

Kleinhofs, A., Hodgdon, A.L., Owais, W.M. & Nilan, R.A. (1984). Effectiveness and safety of sodium azide mutagenesis. In *Induced Mutations for Crop Improvement in Latin America* (Proceedings Regional FAO/IAEA Seminar, Lima, Peru, 1982), IAEA-TECDOC-305, pp.53–8. Vienna: IAEA.

Kleinhofs, A., Owais, W.M. & Nilan, R.A. (1978). Azide. *Mutation Research*, **55**, 165–95.

Kleinhofs, A., Sander, C., Nilan, R.A. & Konzak, C.F. (1974). Azide mutagenicity – mechanism and nature of mutants produced. In *Polyploidy and Induced Mutations in Plant Breeding* (Proceedings of two Meetings, Joint FAO/IAEA Division and EUCARPIA, Bari, Italy, 1972), pp.195–9. Vienna: IAEA.

Kleinhofs, A., Warner, R.L., Muehlbauer, F.J. & Nilan, R.A. (1978). Induction and selection of specific gene mutations in *Hordeum* and *Pisum*. *Mutation Research*, **51**, 29–35.

Klopfer, K. (1965a). Erfolgreiche experimentelle Entmischungen und Umlagerungen periklinalchimärischer Kartoffelklone. *Der Züchter*, **35**, 201–14.

Klopfer, K. (1965b). Über den Nachweis von drei selbständigen Schichten im Sproßscheitel der Kartoffel. *Zeitschrift für Pflanzenzüchtung*, **53**, 67–87.

Klopfer, K. (1967). Methods of demonstrating and breeding chimerical potato clones. In *Induzierte Mutationen und ihre Nutzung* (Erwin Baur-Gedächtnisvorlesungen IV, Gatersleben, 1966), ed. H. Stubbe, pp.305–9. Berlin: Akademie-Verlag.

Knott, D.R. (1991). What determines the success of mutation breeding?. In *Plant Mutation Breeding for Crop Improvement* (Proceedings FAO/IAEA Symposium, Vienna, 1990), vol.1, pp.111–18. Vienna: IAEA.

Knudsen, K.E.B. & Munck, L. (1981). The feasibility of breeding barley for feed quality exemplified by the nutritional analysis of the botanical components of Bomi and the 1508 high lysine barley. In *Barley Genetics IV* (Proceedings 4th International Barley Genetics Symposium, Edinburgh, UK), pp.320–9. Edinburgh: Edinburgh Press.

Koernicke, M. (1904a). Über die Wirkung von Röntgenstrahlen auf die Keimung und das Wachstum. *Berichte der Deutschen Botanischen Gesellschaft*, **22**, 148–55.

Koernicke, M. (1904b). Die Wirkung der Radiumstrahlen auf die Keimung und das Wachstum. *Berichte der Deutschen Botanischen Gesellschaft*, **22**, 155–66.

Koernicke, M. (1905). Über die Wirkung von Röntgen und Radiumstrahlen auf pflanzliche Gewebe und Zellen. *Berichte der Deutschen Botanischen Gesellschaft*, **23**, 404–14.

Kohalmi, S.E. & Kunz, B.A. (1988). Role of neighbouring bases and assessment of strand specificity in ethylmethanesulphonate and N-methyl-N'-nitro-N-nitrosoguanidine mutagenesis in the *SUP4-o* gene of *Saccharomyces cerevisiae*. *Journal of Molecular Biology*, **204**, 561–8.

Kølmark, G. (1953). Differential response to mutagens as studied by the *Neurospora* reverse mutation test. *Hereditas*, **39**, 270–6.

Kølmark, G. (1956). Mutagenic properties of certain esters of inorganic acids investigated by the *Neurospora* back-mutation test. *Comptes rendus des travaux du Laboratoire Carlsberg*, Série physiologique, **26**, 205–20.

Kølmark, G. & Westergaard, M. (1953). Further studies on chemically induced reversions at the adenine locus of *Neurospora*. *Hereditas*, **39**, 209–24.

Kondrashov, A.S. (1988). Deleterious mutations and the evolution of sexual reproduction. *Nature*, **336**, 435–40.

Konzak, C.F. (1956). Induction of mutations for disease resistance in cereals. In *Genetics in Plant Breeding* (Brookhaven Symposia in Biology), Number 9, pp.157–76. Upton, New York: Biology Department, Brookhaven National Laboratory.

Konzak, C.F. (1984). Role of induced mutations. In *Crop Breeding, A Contemporary Basis*, ed. P.B. Vose & S.G. Blixt, chapter 9, pp.216–92. Oxford: Pergamon Press.

Konzak, C.F. (1993). Induction of mutations in oats. *Mutation Breeding Newsletter*, **40**, 6.

Konzak, C.F., Kleinhofs, A. & Ullrich, S.E. (1984). Induced mutations in seed-propagated crops. In *Plant Breeding Reviews*, vol.2, ed. J. Janick, pp.13–72. Westport, Connecticut: AVI Publishing Company.

Konzak, C.F., Nilan, R.A., Wagner, J. & Foster, R.J. (1965). Efficient chemical mutagenesis. In *The Use of Induced Mutations in Plant Breeding* (Report FAO/IAEA Technical Meeting, Rome, 1964), pp.49–70. Oxford: Pergamon Press.

Korschinsky, K. (1901). Heterogenesis und Evolution. *Flora*, **89**, 240–363.

Kostoff, D. (1935a). Mutations and the ageing of seeds. *Nature*, **135**, 107.

Kostoff, D. (1935b). Chromosome alterations by centrifuging. *Zeitschrift für induktive Abstammungs- und Vererbungslehre*, **69**, 301–2.

Kostoff, D. (1938). The effect of centrifuging upon the germinated seeds from various plants. *Cytologia*, **8**, 420–42.

Krantz, F.A. (1951). Potato breeding in the United States. *Zeitschrift für Pflanzenzüchtung*, **29**, 388–93.

Kranz, A.R. (1984). Experience with in-vitro mutant selection of intact plants. In *Selection in Mutation Breeding* (Proceedings FAO/IAEA Consultants Meeting, Vienna, 1982), pp.145–53. Vienna: IAEA.

Krausse, G.W. (1982). Induktion, Selektion und Nutzung von Mutanten bei Soja 1. Mitt.Chlorophyllmutationsraten nach Samenbehandlung mit N-Nitroso-N-Methylharnstoff in Abhängigkeit von der Nachkommenschaftsgröße und der Chimärennatur der M₁-Pflanzen. *Archiv für Züchtungsforschung*, **12**, 11–22.

Krebbers, E., Hehl, R., Piotrowiak, R., Lönnig, W.E., Sommer, H. & Saedler, H. (1987). Molecular analysis of paramutant plants of *Antirrhinum majus* and the involvement of transposable elements. *Molecular & General Genetics*, **209**, 499–507.

Krishnamurthi, M. & Tlaskal, J. (1974). Fiji disease resistant *Saccharum officinarum* var. Pindar subclones from tissue cultures. *Proceedings Congress International Society of Sugar Cane Technologists*, **15**, 130–7.

Krumbiegel, G. (1979). Response of haploid and diploid protoplasts from *Datura innoxia* Mill. and *Petunia hybrida* L. to treatment with X-rays and a chemical mutagen. *Environmental & Experimental Botany*, **19**, 99–103.

Krythe, J.M. & Wellensiek, S.J. (1942). Five years of colchicine research. *Bibliographia Genetica*, **14**, 1–132.

Kucera, J., Lundqvist, U. & Gustafsson, Å. (1975). Induction of breviaristatum mutants in barley. *Hereditas*, **80**, 263–78.

Kukimura, H. (1986). Mutation breeding on root and tuber crops – A review. In *Research around the Institute of Radiation Breeding – 25 Years of the Symposia and the Next*. Gamma Field Symposia 25, 109–30. Ohmiya-machi, Naka-gun, Ibaraki-ken, Japan: Institute of Radiation Breeding NIAR MAFF.

Künzel, G. & Scholz, F. (1971). The frequency of chlorophyll mutations in relation to M₁ sterility. *Barley Genetics Newsletter*, **1**, 26–8.

Lacadena, J.R. (1974). Spontaneous and induced parthenogenesis and androgenesis. In *Haploids in Higher Plants: Advances and Potential* (Proceedings of the First International Symposium, Guelph, Ontario, Canada), ed. K.J. Kasha, pp.13–32. Guelph: The University of Guelph.

Lacey, C.N.D. (1982). The stability of induced compact mutant clones of Bramley's seedling apple. *Euphytica*, **31**, 452–9.

Lacey, C.N.D. & Campbell, A.I. (1979a). The characteristics and stability of a range of Cox's Orange Pippin apple mutants showing different growth habits. *Euphytica*, **28**, 119–26.

Lacey, C.N.D. & Campbell, A.I. (1979b). The positions of mutated sectors in shoots from irradiated buds of Bramley's Seedling apple. *Environmental and Experimental Botany*, **19**, 145–52.

Lacey, C.N.D. & Campbell, A.I. (1979c). The characters of some selected mutant clones of Bramley's Seedling apple, and their stability during propagation. *Proceedings, EUCARPIA Fruit Section Symposium on Tree Fruit Breeding* (Angers, France), pp.301–6.

Lacey, C.N.D. & Campbell, A.I. (1987). Selection, stability and propagation of mutant apples. In *Improving Vegetatively Propagated Crops*, ed. A.J. Abbott & R.K. Atkin, pp.349–62. London: Academic Press.

LaChance, L.E. (1958). Ingestion of ethylenediaminetetraacetic acid and the effect on life span of irradiated and control *Habrobracon* females. *Nature*, **182**, 870–1.

Lal, R. & Lal, S. (1990). Somaclonal variation in crop improvement. In *Crop Improvement Utilizing Biotechnology*, pp.1–71. Boca Raton, Florida, USA: CRC Press.

Langton, F.A. (1980). Chimerical structure and carotenoid inheritance in *Chrysanthemum morifolium* (Ramat.). *Euphytica*, **29**, 807–12.

Langton, F.A. (1986). Mutation breeding and its role in the improvement and commercialization of vegetatively propagated crops. In *Intraspecific Classification of Wild and Cultivated Plants* (Proceedings Symposium Oxford, U.K., 1984), ed. B.T. Styles, pp.263–76. Oxford: Claredon Press.

Lapins.K.O. (1963). Note on compact mutants of Lambert cherry produced by ionizing radiation. *Canadian Journal of Plant Science*, **43**, 424–5.

Lapins, K.O. (1965). The Lambert Compact cherry. *Fruit Varieties and Horticultural Digest*, **19**, 23 (original not consulted).

Lapins, K.O. (1973). Induced mutations in fruit trees. In *Induced Mutations in Vegetatively Propagated Plants* (Proceedings FAO/IAEA Panel, Vienna, 1972), pp.1–19. Vienna: IAEA.

Lapins, K.O. (1974). Compact Stella sweet cherry introduced. *Mutation Breeding Newsletter*, **3**, 18.

Lapins, K.O. (1983). Mutation breeding. In *Methods in Fruit Breeding*, ed. J.N. Moore & J. Janick, pp.74–99. W. Lafayette, Ind.: Purdue University Press.

Lapins, K.O., Bailey, C.H. & Hough, L.F. (1969). Effects of gamma rays on apple and peach leaf buds at different stages of development. I. Survival, growth and mutation frequencies. *Radiation Botany*, **9**, 379–89.

Larkin, P.J. (1987). Somaclonal variation: History, method and meaning. *Iowa State Journal of Research*, **61**, 393–434.

Larkin, P.J., Banks, P.M., Bhati, R., Brettell, R.I.S., Davies, P.A., Ryan, S.A., Scowcroft, W.R., Spindler, L.H. & Tanner, G.J. (1989). From somatic variation to variant plants: mechanisms and applications. *Genome*, **31**, 705–11.

Larkin, P.J., Ryan, S.A., Brettell, R.I.S. & Scowcroft, W.R. (1984). Heritable somaclonal variation in wheat. *Theoretical and Applied Genetics*, **67**, 443–55.

Larkin, P.J. & Scowcroft, W.R. (1981). Somaclonal variation – a novel source of variability from cell cultures for plant improvement. *Theoretical and Applied Genetics*, **60**, 197–214.

Larkin, P.J. & Scowcroft, W.R. (1983). Somaclonal variation and eyespot toxin tolerance in sugarcane. *Plant Cell, Tissue and Organ Culture*, **2**, 111–21.

Larsson, H.E.B. (1985a). Morphological analysis of laxatum barley mutants. *Hereditas*, **103**, 239–53.

Larsson, H.E.B. (1985b). Genetic analysis of laxatum barley mutants. *Hereditas*, **103**, 255–67.

Larsson, H.E.B. (1985c). Linkage studies with genetic markers and some laxatum barley mutants. *Hereditas*, **103**, 269–79.

Lawley, P.D. (1974). Some chemical aspects of dose-response relationships in alkylation mutagenesis. *Mutation Research*, **23**, 283–95.

Lawrence, C.W. (1965). Radiation-induced polygenic mutation. In *The Use of Induced mutations in Plant Breeding* (Report FAO/IAEA Technical Meeting, Rome, 1964), pp.491–6. Oxford: Pergamon Press.

Lawrence, C.W. (1971). *Cellular Radiobiology*. Studies in Biology no. 30. London: Edward Arnold (Publishers) Ltd.

Lawrence, C.W. (1991). Classical mutagenesis techniques. *Methods in Enzymology*, vol.194(18), pp.273–80. Academic Press.

Lawrence, D.F., Slack, S.A. & Plaisted, R.L. (1994). Russet-Bake-King: A uniform russeted sport of Bake-King. *American Potato Journal*, **71**, 127–9.

Lea, D.E. (1946). *Actions of Radiations on Living Cells*. Cambridge: Cambridge University Press.

Lea, D.E. & Catcheside, D.G. (1942). The mechanism of the induction by radiation of chromosome aberrations in *Tradescantia.Journal of Genetics*, **44**, 216–45.

Lea, D.E. & Coulson, C.A. (1949). The distribution of the number of mutants in bacterial populations. *Journal of Genetics*, **49**, 264–85.

Lee, M. & Phillips, R.L. (1987). Genomic rearrangements in maize induced by tissue culture. *Genome*, **29**, 122–8.

Lee, M. & Phillips, R.L. (1988). The chromosomal basis of somaclonal variation. *Annual Review of Plant Physiology and Molecular Biology*, **39**, 413–37.

Levan, A. (1945). A haploid sugar beet after colchicine treatment. *Hereditas*, **31**, 399–410.

Levitt, J. (1972). *Responses of Plants to Environmental Stresses*. New York: Academic Press.

Levy, A. & Ashri, A. (1973). Differential physiological sensitivity of peanut varieties to seed treatments with the mutagens ethidium bromide, MNNG and sodium azide. *Radiation Botany*, **13**, 369–73.

Levy, A. & Ashri, A. (1975). Ethidium bromide – An efficient mutagen in higher plants. *Mutation Research*, **28**, 397–404.

Levy, A. & Ashri, A. (1978). Induced plasmon mutations affecting the growth habit of peanuts, *A. hypogaea* L. *Mutation Research*, **51**, 347–60.

Levy, A., Ashri, A. & Rubin, B. (1979). Ethidium bromide uptake by peanut seeds and its relationship to varietal sensitivity and mutagenic efficiency. *Experimental and Environmental Botany*, **19**, 49–57.

Lewin, B. (1987). *Genes III*. New York: John Wiley & Sons.

Lewin, B. (1994). *Genes V*. Oxford: Oxford University Press.

Lewis, D. (1946). Useful X-ray mutations in plants. *Nature*, **158**, 519–20.

Lewis, D. (1948). Structure of the incompatibility gene I. Spontaneous mutation rate. *Heredity*, **2**, 219–36.

Lewis, D. (1949). Structure of the incompatibility gene II. Induced mutation rate. *Heredity*, **3**, 339–55.

Lewis, D. (1979). Sexual incompatibility in plants. *Studies in Biology no 110*, London: Edward Arnold.

Lewis, D. (1994). Gametophytic-sporophytic incompatibility. In *Genetic Control of Self-incompatibility and Reproductive Development in Flowering Plants*, ed. E.G. Williams, A.E. Clarke & R.B. Knox, pp.88–101. Dordrecht: Kluwer Academic Publishers.

Lewis, D. & Crowe, L.K. (1954a). Structure of the incompatibility gene. IV. Types of mutations in *Prunus avium* L. *Heredity*, **8**, 357–63.

Lewis, D. & Crowe, L.K. (1954b). The induction of self-fertility in tree fruits. *Journal of Horticultural Science*, **29**, 220–5.

Li, L.-C. & Chu, E.H.Y. (1987). Evaluation of methods for the estimation of mutation rates in cultured mammalian cell populations. *Mutation Research*, **190**, 281–7.

Li, H.W., Hu, C.H., Chang, W.T. & Weng, T.S. (1961). The utilization of X-radiation for rice improvement. In *Effects of Ionizing Radiations on Seeds* (Proceedings FAO/IAEA Symposium, Karlsruhe, 1960), pp.485–93. Vienna: IAEA.

Li, S.L. & Rédei, G.P. (1969). Estimation of mutation rate in autogamous diploids. *Radiation Botany*, **9**, 125–31.

Lindgren, D. (1972). The temperature influence on the spontaneous mutation rate I. Literature review. *Hereditas*, **70**, 165–78.

Lindgren, D., Erikson, G. & Šulovská, K. (1970). The size and appearance of the mutated sector in barley spikes. *Hereditas*, **65**, 107–32.

Lindstrom, E.W. (1933). Hereditary radium-induced variation in the tomato. *Journal of Heredity*, **24**, 128–37.

Lineberger, R.D. & Druckenbrod, M. (1985). Chimeral nature of the pinwheel flowering African violets (*Saintpaulia*, *Gesneriaceae*). *American Journal of Botany*, **72**, 1204–12.

Lineberger, R.D., Pogany, M., Malinich, T., Druckenbrod, M. & Warner, A. (1993). Genotypic segregation and chimeral rearrangments in tissue culture. A potential source of new ornamental cultivars. In *Creating Genetic Variation in Ornamentals* (Proceedings of the XVIIth Symposium of EUCARPIA, Section Ornamentals, San Remo), ed. T. Schiva & A. Mercuri, pp.83–92. San Remo: Istituto Sperimentale per la Floricoltura.

Liu, M.-C. & Chen, W.-H. (1976). Tissue and cell culture as aids to sugarcane breeding.1. Creation of genetic variation through callus culture. *Euphytica*, **25**, 393–403.

Lönneborg, A. & Jansson, C. (1993). Isolation of plant genes by T-DNA and transposon mutagenesis-gene tagging. *Progress in Botany*, **54**, 295–305.

Lonsdale, D.M. (1987). Cytoplasmic male sterility: a molecular perspective. *Plant Physiology and Biochemistry*, **25**, 265–71.

Lopriore, G. (1897). Azione del raggi X sul protoplasma della cellula vegetale vivente. *Estr. dall Nuova Rassegua*, Catania (original not consulted).

Lörz, H. & Brown, P.T.H. (1986). Variability in tissue culture derived plants – possible origins; advantages and drawbacks. In *Genetic Manipulation in Plant Breeding* (Proceedings EUCARPIA Symposium, Berlin, 1985), ed. W. Horn, C.J. Jensen, W. Odenbach & O. Schieder, pp.513–35. Berlin: Walter de Gruyter.

Loveless, A. (1958). Increased rate of plaque-type and host-range mutation following treatment of bacteriophage *in vitro* with ethyl methane sulphonate. *Nature*, **181**, 1212–13.

Lundqvist, U. (1964). Induction of mutations in barley pollen by ultra-violet and X-rays. In *Barley Genetics I* (Proceedings First International Barley Genetics Symposium, Wageningen, 1963), ed. S. Broekhuizen, G. Dantuma, H. Lamberts & W. Lange, pp.92–5. Wageningen: PUDOC.

Lundqvist, U. (1986). Barley mutants-diversity and genetics. In *Svalöf 1886-1986. - Research and Results in Plant Breeding.* ed. G. Olsson, pp.85-8. Stockholm: LTs förlag.

Lundqvist, U. (1991). Swedish mutation research in barley with plant breeding aspects (A historical review). In *Plant Mutation Breeding for Crop Improvement* (Proceedings FAO/IAEA Symposium, Vienna, 1990), vol.1, pp.135-48. Vienna: IAEA.

Lundqvist, U. (1992). *Mutation Research in Barley* (Dissertation). Svalöv, Sweden: Dept. of Plant Breeding Research, The Swedish University of Agricultural Sciences.

Lundqvist, U., Akebe, B. & Lundqvist, A. (1989). Gene interaction of induced *intermedium* mutations of two-row barley. *Hereditas,* **111,** 37-47.

Lundqvist, U. & Lundqvist, A. (1987). An *intermedium* gene present in a commercial six-row variety of barley. *Hereditas,* **107,** 131-5.

Lundqvist, U. & Lundqvist, A. (1988a). Mutagen specificity in barley for 1580 *eceriferum* mutants localized to 79 loci. *Hereditas,* **108,** 1-12.

Lundqvist, U. & Lundqvist, A. (1988b). Induced *intermedium* mutants in barley: origin, morphology and inheritance. *Hereditas,* **108,** 13-26.

Lundqvist, U. & Lundqvist, A. (1988c). Gene interaction of induced *intermedium* mutations of two-row barley. *Hereditas,* **108,** 133-40.

Lundqvist, U. & Lundqvist, A. (1989). The co-operation between *intermedium* genes and the six-row gene *hex-v* in a six-row variety of barley. *Hereditas,* **110,** 227-33.

Lundqvist, U. & Lundqvist, A. (1990). Progressive promotion of lateral floret development in three- and four-gene combinations of *intermedium* genes in barley. *Hereditas,* **113,** 237-42.

Lundqvist, U. & Lundqvist, A. (1991). Dominant resistance to barley powdery mildew race D1, isolated after mutagen treatments in four highbred barley varieties. *Hereditas,* **115,** 241-53.

Lundqvist, U., Meyer, J. & Lundqvist, A. (1991). Mutagen specificity for 71 lines resistant to barley powdery mildew race D1 and isolated in four highbred barley varieties. *Hereditas,* **115,** 227-39.

Lundqvist, U & von Wettstein, D. (1962). Induction of *eceriferum* mutants in barley by ionizing radiations and

chemical mutagens. *Hereditas,* **48,** 342-62.

Lundqvist, U., von Wettstein-Knowles, P. & von Wettstein, D. (1968). Induction of *eceriferum* mutants in barley by ionizing radiations and chemical mutagens, II. *Hereditas,* **59,** 473-504.

Lupton, F.G.H. (1992). Changes in varietal distribution of cereals in central and western Europe (Agro-ecological Atlas of Cereal Growing in Europe, vol.IV). Wageningen: PUDOC.

Luria, S.E. & Delbrück, M. (1943). Mutations of bacteria from virus sensitivity to virus resistance. *Genetics,* **28,** 491-511.

MacArthur, J.W. (1934). X-ray mutations in the tomato. *Journal of Heredity,* **25,** 75-8.

MacDonald, M.V., Ahmad, I., Menten, J.O.M. & Ingram, D.S. (1991). Haploid culture and *in vitro* mutagenesis (UV light, X-rays and gamma rays) of rapid cycling *Brassica napus* for improved resistance to disease. In *Plant Mutation Breeding for Crop Improvement* (Proceedings FAO/IAEA Symposium, Vienna, 1990), vol.2, pp.129-38. Vienna: IAEA.

MacKey, J. (1951). Neutron and X-ray experiments in barley. *Hereditas,* **37,** 421-64.

MacKey, J. (1954a). Mutation breeding in polyploid cereals. *Acta Agriculturae Scandinavica,* **4,** 549-57.

MacKey, J. (1954b). The biological action of mustards on dormant seeds of barley and wheats. *Acta Agriculturae Scandinavica,* **4,** 419-29.

MacKey, J. (1956). Mutation breeding in Europe. In *Genetics in Plant Breeding* (Report Brookhaven Symposia in Biology), Number 9, pp.141-56. Upton, New York: Biology Department. Brookhaven National Laboratory.

MacKey, J. (1981). Value of induced mutation research for improving genetic knowledge (Invited paper). In *Induced Mutations - A Tool in Plant Research* (Proceedings FAO/IAEA Symposium, Vienna, 1981), pp.3-22. Vienna: IAEA.

MacKey, J. (1984). Selection problems and objectives in mutation breeding. In *Selection in Mutation Breeding* (Proceedings FAO/IAEA Consultants Meeting, Vienna, 1982), pp.35-48. Vienna: IAEA.

Maddox, J. (1995). The sensational discovery of X-rays. *Nature,* **375,** 183.

Mak, C., Ho, Y.W., Tan, Y.P. & Rusli Ibrahim (1996). Novaria - un mutant de bananier induit par irradiation gamma. *InfoMusa,* **5,** 35-6.

Makarova, S.I. (1967). Experimental mutations in *Pisum sativum* L. In *Mechanism of Mutation and Inducing Factors* (Proceedings of a Symposium on the Mutational Process, Prague, 1965), ed. Z. Landa, pp.321-2. Prague: Academia (Publishing House of the Czechoslovak Academy of Sciences).

Malepszy, S., Eberhardt, J. & Maluszynski, M. (1973). Mutagenic effects of NMH, NEH and HM in barley. *Genetica Polonica,* **14,** 47-59.

Maliga, P. (1980). Isolation, characterization and utilization of mutant cell lines in higher plants. *International Review of Cytology,* Suppl. 11A, pp.225-50.

Maliga, P. (1983). Protoplasts in mutant selection and characterization. *International Review of Cytology,* Suppl.16, pp.161-7.

Maliga, P. (1984). Isolation and characterization of mutants in plant cell culture. *Annual Review of Plant Physiology,* **35,** 519-42.

Malinovskii, B.N., Zoz, N.N. & Kitaev, A.I. (1973). Induction of cytoplasmic male sterility in sorghum by chemical mutagens. *Soviet Genetics,* **9,** 682-8 (original not consulted).

Maluszynska, J. & Maluszynski, M. (1983). MNUA and MH mutagenic effect after double treatment of barley seeds in different germination periods. *Acta Biologica* (Katowice, Poland), **11,** 238-48.

Maluszynski, M. (1982). The high mutagenic effectiveness of MNUA in inducing a diversity of dwarf and semi-dwarf forms of spring barley. *Acta Societatis Botanicorum Poloniae,* **51,** 429-40.

Maluszynski, M., Ahloowalia, B.S. & Sigurbjörnsson, B. (1995). Application of *in vivo* and *in vitro* mutation techniques for crop improvement. *Euphytica* (Selected Papers, Meeting EUCARPIA, Section Genetic Manipulation in Plant Breeding, Cork, Ireland), vol.85, 303-15.

Maluszynski, M., Eberhardt, J. & Cudny, H. (1974). Effect of NMH on the extent of somatic damages, on the number of cells subject to division and on DNA synthesis in barley seedlings. In *Bulletin de l'Académie Polonaise des Scien-*

ces (Série des sciences biologiques), Cl.2, vol.**22**, 10, pp.667–73.

Maluszynski, M., Fuglewicz, A., Szarejko, I. & Micke, A. (1989). Barley mutant heterosis. In *Current Options for Cereal Improvement* (Proceedings FAO/IAEA Research Co-ordination Meeting, 1986, Guelph, Canada), ed. M. Maluszynski, pp.129–46. Dordrecht: Kluwer Academic Publishers.

Maluszynski, M, & Maluszynska, J. (1977). Increasing of the mutagenic efficiency of MNUA and MH on treatment of barley seeds at different stages of germination. *Acta Biologica* (Katowice, Poland), **4**, 37–50.

Maluszynski, M., Micke, A & Donini, B. (1986). Genes for semidwarfism in rice induced by mutagenesis. In *Rice Genetics* (Proceedings International Rice Genetics Symposium, Manila, 1985), pp.729–37. Manila: International Rice Research Institute.

Maluszynski, M., Micke, A., Sigurbjörnsson, B., Szarejko, I. & Fuglewicz, A. (1987). The use of mutants for breeding and for hybrid barley. In *Barley Genetics V* (Proceedings of the Fifth International Barley Genetics Symposium, 1986, Okayama, Japan), ed. S. Yasida & T. Konishi, pp.969–77. Okayama, Japan: Sanyo Press.

Maluszynski, M., Sigurbjörnsson, B., Amano, E., Sitch, L. & Kamra, O. (1991). Mutant varieties-data bank, FAO/IAEA database. *Mutation Breeding Newsletter*, **38**, 16–49.

Maluszynski, M., Sigurbjörnsson, B., Amano, E., Sitch, L. & Kamra, O. (1992). Mutant varieties-databank, FAO/IAEA data base, part II. *Mutation Breeding Newsletter*, **39**, 14–33.

Maluszynski, M., van Zanten, L., Ashri, A., Brunner, H., Ahloowalia, B., Zapata, F.P. & Weck, E. (1995). Mutation techniques in plant breeding. In *Induced Mutations and Molecular Techniques for Crop Improvement* (Proceedings FAO/IAEA Symposium, Vienna), pp.489–504. Vienna: IAEA.

Marcotrigiano, M. & Jagannathan, K. (1988). *Paulownia tomentosa* 'Somaclonal Snowstorm'. HortScience, **23**, 226–7.

Marcotrigiano, M. & Stewart, R.N. (1984). All variegated plants are not chimeras. *Science*, **223**, 505.

Marx, J. (1994). DNA repair comes into its own. *Science*, **266**, 728–30.

Masrizal, Simonson, R.L. & Baenzinger, P.S. (1991). Response of different wheat tissues to increasing doses of ethyl methanesulfonate. *Plant Cell, Tissue and Organ Culture*, **26**, 141–6.

Masuda, T., Yoshioka, T. & Inoue, K. (1994). Selection of mutants resistant to black spot disease using the AK-toxin in japanese pears irradiated with gamma-rays. In *Mutation Breeding with Novel Selection Techniques*.Gamma Field Symposia 33, 91–102. Ohmiya-machi, Naka-gun, Ibaraki-ken, Japan: Institute of Radiation Breeding NIAR MAFF.

Mather, K. (1941). Variation and selection of polygenic characters. *Journal of Genetics*, **41**, 159–93.

Mathews, H. & Bhatia, C.R. (1983). Experimental mutagenesis of in-vitro cultured plant cells and protoplasts. *Mutation Breeding Newsletter*, **22**, 12–17.

Matsumoto, K. & Yamaguchi, H. (1991). Induction and selection of aluminium tolerance in the banana. In *Plant Mutation Breeding for Crop Improvement* (Proceedings FAO/IAEA Symposium, Vienna, 1990), vol.2, pp. 249–56. Vienna: IAEA.

Matsuo, T. & Onozawa, Y. (1961). Mutations induced in rice by ionizing radiations and chemicals. In *Effects of Ionizing Radiations on Seeds* (Proceedings FAO/IAEA Symposium, Karlsruhe, 1960), pp.495–500. Vienna: IAEA.

Matzke, M., Matzke, A.J.M. & Mittelstenscheid, O. (1994). Inactivation of repeated genes – DNA-DNA Interaction? In *Homologous Recombination and Gene Silencing in Plants*, ed. J. Paszkowski, pp.271–307. Dordrecht: Kluwer Academic Press.

Mavor, J.W. (1925). The attack on the gene. *Scientific Monthly*, **21**, 355–63 (original not consulted).

Mayr, E. (1963). *Animal Species and Evolution*. Cambridge, MA: Harvard University Press.

Mc Cabe, P.F., Timmons, A.M. & Dix, P.J. (1989). A simple procedure for the isolation of streptomycin resistant plants in Solanaceae. *Molecular & General Genetics*, **216**, 132–7.

McClintock, B. (1948). Mutable loci in maize. *Carnegie Institution of Washington Yearbook*, **47**, 155–69.

McClintock, B. (1949). Mutable loci in maize. *Carnegie Institution of Washington Yearbook*, **48**, 142–54.

McClintock, B. (1950a). Mutable loci in maize. *Carnegie Institution of Washington Yearbook*, **49**, 157–67.

McClintock, B. (1950b). The origin and behavior of mutable loci in maize. *Proceedings of the National Academy of Sciences of the USA*, **36**, 344–55.

McClintock, B. (1951). Chromosome organization and genic expression. In *Genes and Mutations. Cold Spring Harbor Symposia on Quantitative Biology*, vol.16, pp.13–47. Cold Spring Harbor, Long Island, New York: The Biological Laboratory.

McClintock, B. (1954). Mutations in maize and chromosomal aberrations in *Neurospora. Carnegie Institution of Washington Yearbook*, **53**, 254–61.

McClintock, B. (1956). Controlling elements and the gene. *Cold Spring Harbor Symposia on Quantitative Biology*, vol.21, 197–216. Cold Spring Harbor, Long Island, New York: The Biological Laboratory.

McClintock, B. (1984). The significance of responses of the genome to challenge. *Science*, **226**, 792–801.

McDaniel, C.N. (1984). Shoot meristem development. In *Positional Controls in Plant Development*, ed. P.W. Barlow & D.J. Carr, pp.319–47. Cambridge: Cambridge University Press.

McIntosh, D.L. & Lapins, K. (1966). Differences in susceptibility to apple powdery mildew observed in McIntosh clones after exposure to ionizing radiation. *Canadian Journal of Plant Science*, **46**, 619–23.

McKelvie, A.D. (1963). Studies in the induction of mutations in *Arabidopsis thaliana* (L.) Heynh. *Radiation Botany*, **3**, 105–23.

McPheeters, K. & Skirvin, R.M. (1983). Histogenic layer manipulation in chimeral 'Thornless Evergreen' trailing blackberry. *Euphytica*, **32**, 351–60.

McPheeters, K. & Skirvin, R.M. (1989). Somaclonal variation among ex vitro 'Thornless Evergreen' trailing blackberries. *Euphytica*, **42**, 155–62.

McPherson, M.J. (ed.), (1991). *Directed Mutagenesis: A Practical Approach*. Oxford: Oxford University Press.

Medford, J.I. (1992). Vegetative apical meristems. *The Plant Cell*, **4**, 1029–39.

Mehandjiev, A.D. (1991). Application of experimental mutagenesis in soybean. In *Plant Mutation Breeding for Crop Improvement* (Proceedings FAO/IAEA Symposium, Vienna, 1990), vol.1, pp.407–12. Vienna: IAEA.

Meins Jr., F. (1983). Heritable variation in plant cell culture. *Annual Review of Plant Physiology*, **34**, 327–46.

Melchers, G. (1974). Haploids for breeding by mutation and recombination. In *Polyploidy and Induced Mutations in Plant Breeding* (Proceedings of two Meetings, Joint FAO/IAEA Division and EUCARPIA, Bari, Italy, 1972), pp.221–31. IAEA: Vienna.

Melchers, G. & Bergmann, L. (1959). Untersuchungen an Kulturen von haploiden Geweben von *Antirrhinum majus. Berichte der Deutschen Botanischen Gesellschaft*, **78**, 21–9.

Melchers, G., Sacristan, M.D. & Holder, A.A. (1978). Somatic hybrid plant of potato and tomato regenerated from fused protoplasts. *Carlsberg Research Communications*, **43**, 203–18.

Mericle, L.W. & Mericle, R.P. (1965). Biological discrimination of differences in natural background radiation level. *Radiation Botany*, **5**, 475–92.

Mericle, L.W. & Mericle, R.P. (1967). Genetic nature of somatic mutations for flower color in *Tradescantia*, clone 02. *Radiation Botany*, **7**, 449–64.

Mericle, L.W. & Mericle, R.P. (1971). Somatic mutations in clone 02 *Tradescantia. The Journal of Heredity*, **62**, 323–8.

Mertz, E.T. (1992). Discovery of high lysine, high tryptophan cereals. In *Quality Protein Maize*, ed. E.T. Mertz, pp.1–8. St. Paul MN, USA: American Association of Cereal Chemists.

Mertz, E.T., Bates, L.S., Nelson, O.E. (1964). Mutant gene that changes protein composition and increases lysine contents of maize endosperm. *Science*, **145**, 279–80.

Meyer, P., Heidmann, I., Forkmann, G.& Saedler, H. (1987). A new petunia flower colour generated by transformation of a mutant with a maize gene. *Nature*, **330**, 677–8.

Meyer, P., Linn, F., Heidmann, I., Meyer z.A., H., Niedenhof, I. & Saedler, H. (1992). Endogenous and environmental factors influence 35S promoter methylation of a maize *A1* gene construct in transgenic petunia and its colour phenotype. *Molecular & General Genetics*, **231**, 345–52.

Michaelis, P. (1969). *Über Plastiden-Restitutionen (Rückmutationen). Cytologia* (Tokyo), **34**, Supplement, pp.1–115.

Micke, A. (1970). Genetic aspects of selection for protein after mutation induction. In *Improving Plant Protein by Nuclear Techniques* (Proceedings FAO/IAEA Symposium), pp.229–36. Vienna: IAEA.

Micke, A. (1980). Möglichkeiten der Resistenzzüchtung mit Hilfe induzierter Mutationen. In *Bericht Arbeitstagung 1980 der Arbeitsgemeinschaft der Saatzuchtleiter*, pp.69–84. Irdning, Austria: Bundesversuchsanstalt für alpenländische Landwirtschaft Gumpenstein.

Micke, A. (1983a). International research programmes for the genetic improvement of grain proteins. In *Seed Proteins (Biochemistry, Genetics, Nutritive Value)*, ed. W. Gottschalk & H.P. Müller, pp.25–44. The Hague: Martinus Nijhoff.

Micke, A. (1983b). Some considerations on the use of induced mutations for improving disease resistance of crop plants. In *Induced Mutations for Disease Resistance in Crop Plants II* (Proceedings FAO/IAEA/SIDA Research Co-ordination Meeting, Risø, Denmark, 1981), pp.3–19. Vienna: IAEA.

Micke, A. (1984). Mutation breeding of grain legumes. *Plant and Soil*, **82**, 337–57.

Micke, A. (1988a). Genetic improvement of grain legumes using induced mutations. An overview. In *Improvement of Grain Legume Production Using Induced Mutations* (Proceedings FAO/IAEA Workshop, Pullman, USA, 1986), pp.1–51. Vienna: IAEA.

Micke, A. (1988b). Genetic improvement of food legumes in developing countries by mutation induction. In *World Crops: Cool Season Food Legumes* (Proceedings International Food Legume Research Conference, Spokane, Washington, USA, 1986), ed. R.J. Summerfield, pp.1031–47. Dordrecht: Kluwer Academic Publishers.

Micke, A. (1991a). Induced mutations for crop improvement. In *Biotechnology and Mutation Breeding*. Gamma Field Symposia 30, 1–21. Ohmiya-machi, Naka-gun, Ibaraki-ken, Japan: Institute of Radiation Breeding, NIAR MAFF.

Micke, A. (1991b). Plant mutation breeding: its future role, the methodology needed, training and the research priorities. In *Plant Mutation Breeding for Crop Improvement* (Proceedings FAO/IAEA Symposium, Vienna, 1990), Vol.2, 473–5. Vienna: IAEA.

Micke, A. (1992). 50 years induction of mutations for improving disease resistance of crop plants. *Mutation Breeding Newsletter*, **39**, 2–4.

Micke, A. (1993a). Durability of resistance in induced mutants. In *Durability of Disease Resistance*, ed. Th. Jacobs & J.E. Parlevliet, p.336. Dordrecht: Kluwer Academic Press.

Micke, A. (1993b). Mutation breeding of grain legumes. *Plant and Soil*, **152**, 81–5.

Micke, A. & Donini, B. (1982). Use of induced mutations in improvement of seed propagated crops. In *Induced Variability in Plant Breeding* (Proceedings International Symposium of the Section Mutation and Polyploidy of EUCARPIA, Wageningen, The Netherlands, 1981), pp.2–9. Wageningen: Centre for Agricultural Publishing and Documentation (PUDOC).

Micke, A. & Donini, B. (1993). Induced mutations. In *Plant Breeding. Principles and Prospects*, ed. M.D. Hayward, N.O. Bosemark, & I. Romagosa, p.52–62. London: Chapman & Hall.

Micke, A., Donini, B. & Maluszynski, M. (1987). Induced mutations for crop improvement – a review. *Tropical Agriculture (Trinidad)*, **64**, 259–78.

Micke, A., Donini, B. & Maluszynski, M. (1990). Induced mutations for crop improvement. *Mutation Breeding Review*, **7**, 1–41.

Micke, A., Maluszynski, M & Donini, B. (1985). Plant cultivars derived from mutation induction or the use of induced mutants in cross breeding. *Mutation Breeding Review*, **3**, 1–92.

Micke, A. & Swiecicky, W. (1988). Induced mutations in lupin. In *Proceedings 5th International Lupin Conference, Poznan, Poland*. pp.110–27. Poznan: PWRIL.

Mikami, T., Kinoshita, T. & Takahashi, H. (1980). Induction of cytoplasmic mutations on male sterility by acridine dyes and streptomycin in sugar beets. *Proceedings Sugar Beet Association of Japan*, **22**, 48–54.

Minocha, J.L. & Gupta, R.K. (1988). Induction of male sterility in rice using chemical mutagens. *Mutation Breeding Newsletter*, **32**, 5–6.

Moès, A. (1958). Les mutations induites par les rayons X chez l'orge distique (*Hordeum vulgare* Jess. subsp. *distichum*).

I. *Bulletin de l'Institut Agronomique et des Stations de Recherches de Gembloux*, **26**, 335–61.

Moès, A. (1959). Les mutations induites par les rayons X chez l'orge distique (*Hordeum vulgare* Jess. subsp. *distichum*). II. *Bulletin de l'Institut Agronomique et des Stations de Recherches de Gembloux*, **27**, 167–221.

Morejohn, L.C., Bureau, T.E., Bajer, M., Bajer, A.S. & Fosket, D.E. (1987). Oryzalin, a dinitroaniline herbicide, binds to plant tubulin and inhibits microtubule polymerization *in vitro*. *Planta*, **172**, 252–64.

Morel, G. (1972). The impact of plant tissue culture on plant breeding. In *The Way Ahead in Plant Breeding* (Proceedings 6th Congress of EUCARPIA, Cambridge, 1971), ed. F.G.H. Lupton, G. Jenkins & R. Johnson, pp.185–94. Cambridge.

Morgan, T.H. (1926). Genetics and the physiology of development. *The American Naturalist*, **60**, 48–515.

Moyer, J.W. & Collins, W.W. (1983). 'Scarlet' sweet potato. *HortScience*, **18**, 111–12.

Müller, A.J. & Gichner, T. (1964). Mutagenic activity of 1-methyl-3-nitro-1-nitrosoguanidine on *Arabidopsis*. *Nature*, **201**, 1149–50.

Muller, H.J. (1927). Artificial trans-mutation of the gene. *Science*, **66**, 84–7.

Muller, H.J. (1928). The production of mutations by X-rays. *Proceedings National Academy of Sciences of the USA*, **14**, 714–26 (original not consulted).

Munck, L. (1972). Improvement of nutritional value in cereals. *Hereditas*, **72**, 1–128.

Munck, L., Bang-Olsen, K. & Stilling, B. (1986). Genome adjustment by breeding to balance yield defects in high-lysine mutants of barley. In *Genetic Manipulation in Plant Breeding* (Proceedings International Symposium EUCARPIA, Berlin, 1985), ed. W. Horn, C.J. Jensen, W. Odenbach & O. Schieder), pp.49–60. Berlin: Walter de Gruyter & Co.

Munck, L., Karlsson, K.E. & Hagberg, A. (1971). Selection and characterization of a high-protein, high-lysine variety from the world barley collection. In *Barley Genetics II* (Proceedings of the Second International Barley Genetics Symposium, 1969, Pullman, Washington), ed. R.A. Nilan, pp.544–58. Pullman: Washington State University Press.

Munck, L., Karlsson.K.E., Hagberg, A. & Eggum, B.O. (1970). Gene for improved nutritional value in barley seed protein. *Science*, **168**, 985–7.

Müntzing, A. (1942). Frequency of induced chlorophyll mutations in diploid and tetraploid barley. *Hereditas*, **28**, 217–21.

Murashige, T. & Nakano, R. (1967). Chromosome complement as a determinant of the morphogenic potential of tobacco cells. *American Journal of Botany*, **54**, 963–70.

Murfet, I.C. & Reid, J.B. (1993). Developmental mutants. In *Peas: Genetics, Molecular Biology and Biotechnology*, ed. R. Casey & D.R. Davies, pp.165–216. Wallingford, UK: CAB International.

Murray, M.J. (1969). Succesful use of irradiation breeding to obtain *Verticillium*-resistant strains of pepper-mint. In *Induced Mutations in Plants* (Proceedings FAO/IAEA Symposium, Pullman, USA), pp.345–71. Vienna: IAEA.

Murray, M.J. (1971). Additional observations on mutation breeding to obtain *Verticillium*-resistant strains of peppermint, *Mentha piperita* L. In *Mutation Breeding for Disease Resistance* (Proceedings FAO/IAEA Panel, Vienna, 1970), pp.171–95. Vienna: IAEA.

Murray, M.J. & Todd, W.A. (1972). Registration of Todd's Mitcham peppermint. *Crop Science*, **12**, 128.

Mutschler, M. (1990). Use of biotechnology to create or transfer novel traits in tomato. *HortScience*, **25**, 1521–2.

Nagatomi, S., Kamijyo, M., Narusawa, H., Iwasaki, T., Okzaki, T. & Maruta, Y. (1996). Three mutant varieties in *Eustoma grandiflorum* induced through in vitro culture of chronic irradiated plants. *Technical News No. 53*. Ohmiya-machi, Ibaraki, Japan: Institute of Radiation Breeding NIAR MAFF.

Nakai, H. (1991). Practical value of induced mutants of rice resistant to bacterial leaf blight. In *Plant Mutation Breeding for Crop Improvement* (Proceedings FAO/IAEA Symposium, Vienna, 1990), vol.2, pp.113–27. Vienna: IAEA.

Nakajima, K. (1973). Induction of useful mutations of mulberry and roses by gamma rays. In *Induced Mutations in Vegetatively Propagated Plants* (Proceedings FAO/IAEA Panel, Vienna, 1972), pp.195–216. Vienna: IAEA.

Naumann, C.H., Underbrink, A.G. & Sparrow, A.H. (1975). Influence of radiation dose rate on somatic mutation induction in *Tradescantia* stamen hairs. *Radiation Research*, **62**, 79–96.

Navashin, M. (1933a). Altern der Samen als Ursache von Chromosomen-mutationen. *Planta*, **20**, 233–43.

Navashin, M. (1933b). Origin of spontaneous mutations. *Nature*, **131**, 436.

Navashin, M. & Shkvarnikov, P. (1933). Process of mutation in resting seeds accelerated by increased temperature. *Nature*, **132**, 482.

Naylor, E.E. & Johnson, B. (1937). A histological study of vegetative reproduction in *Saintpaulia ionantha*. *American Journal of Botany*, **24**, 673–8.

Negrutiu, I. (1990). *In vitro* mutagenesis. In *Plant Cell Line Selection: Procedures and Applications*, ed. P.J. Dix, pp.19–38. New York: VCH, Weinheim.

Neilson-Jones, W. (1934). *Plant Chimaeras and Graft Hybrids*. London: Methuen & Co.

Neilson-Jones, W. (1969). *Plant Chimeras*. London: Methuen & Co.

Nelson, E.H., Francesco, J.T.D., Martin, L. & Skirvin, R.L. (1989). Evergreen black-berry screening for genetic thornless-ness. *Acta Horticulturae*, **262**, 113–18.

Nelson, O.E. (1981). The mutants opaque-9 through opaque-13. *Maize Genetics Cooperation Newsletter*, **55**, 68–4.

Nelson, O.E. (ed.) (1988). *Plant Transposable Elements*. New York: Plenum Press.

Nelson, O.E., Mertz, E.T. & Bates, L.S. (1965). Second mutant gene affecting the amino acid pattern of maize endosperm proteins. *Science*, **150**, 1469–70.

Nettancourt, D. de (1977). *Incompatibility in Angiosperms*. Berlin: Springer Verlag.

Nettancourt, D. de & Dijstra, M. (1969). Starch accumulation in the microspores of a *Solanum* species and possible implications in mutation breeding. *American Potato Journal*, **46**, 239–42.

Neuffer, M.G. & Chang, M.T. (1989). Induced mutations in biological and agronomic research. In *Science for Plant Breeding* (Proceedings 12th Congress of EUCARPIA, Göttingen, Germany F.R.), pp.165–78. Berlin: Paul Parey.

Neuffer, M.G. & Coe, E.H. (1978). Paraffin oil technique for treating mature corn pollen with chemical mutagens. *Maydica*, **23**, 21–8.

Neuffer, M.G. & Ficsor, G. (1963). Mutagenic action of ethyl methanesulfonate in maize. *Science*, **139**, 1296–7.

Nevers, P., Shepherd, N.S. & Saedler, H. (1986). Plant transposable elements. In *Advances in Botanical Research*, **12**, 103–203.

Newton, K.J. & Coe, E.H. Jr. (1986). Mitochondrial DNA changes in abnormal growth (nonchromosomal stripe) mutants of maize. *Proceedings of the National Academy of Sciences of the USA*, **83**, 7363–6.

Nickell, L.G. & Heinz, D.J. (1973). Potential of cell and tissue culture techniques as aids in economic plant improvement. In *Genes, Enzymes and Populations*, ed. A.M. Srb, pp.109–28. New York: Plenum Press.

Nilan, R.A. (1981a). Recent advances in barley mutagenesis. In *Barley Genetics IV* (Proceedings of the Fourth International Barley Genetics Symposium, Edinburgh), pp.823–31. Edinburgh: Edinburgh University Press.

Nilan, R.A. (1981b). Induced gene and chromosome mutants. *Philosophical Transactions of the Royal Society of London*, series B, vol.**292** (no.1062), 457–66.

Nilan, R.A. (1987). Trends in barley mutagenesis. In *Barley Genetics V.* (Proceedings of the Fifth International Barley Genetics Symposium, Okayama, Japan, 1986), ed. S. Yasuda & T. Konishi, pp.241–9. Okayama, Japan: Sanyo Press.

Nilan, R.A., Kleinhofs, A. & Sander, C. (1976). Azide mutagenesis in barley. In *Barley Genetics III.* (Proceedings of the Third International Barley Genetics Symposium, 1975, Garching, Germany), ed. H. Gaul, pp.113–22. München: Verlag Karl Thiemig.

Nilan, R.A., Konzak, C.F.Heiner, R.E. & Froese-Gertzen, E.E. (1964). Chemical mutagenesis in barley. *Barley Genetics I.* (Proceedings First International Barley Genetics Symposium, Wageningen, 1963), ed. S. Broekhuizen, G. Dantuma, H. Lamberts & W. Lange, pp.35–54. Wageningen: PUDOC.

Nilan, R.A., Sideris, E.G., Kleinhofs, A., Sander, C. & Konzak, C.F. (1973). Azide – a potent mutagen. *Mutation Research*, **17**, 142–4.

Nilsson, N.H. (1931). Sind die induzierten Mutanten nur selektive Erscheinungen? *Hereditas*, **15**, 320–8.

Nitsch, J.P., Nitsch, C. & Péreau-Leroy, P. (1969). Obtention de mutants à partir de *Nicotiana* haploïdes issus de grains de pollen. *Comptes rendus hebdomadaires des séances de l'Academie des Sciences* (Paris, France), Série D, **269**, 1650–2.

Norris, R., Smith, R.H. & Vaughn, K.C. (1983). Plant chimeras used to establish de novo origin of shoots. *Science*, **220**, 75–6.

Novak, F.J. (1991a). *In vitro* mutation system for crop improvement. In *Plant Mutation Breeding for Crop Improvement* (Proceedings FAO/IAEA Symposium, Vienna, 1990), vol.2, pp.327–42. Vienna: IAEA.

Novak, F.J. (1991b). Mutation breeding using tissue culture techniques. In *Biotechnology and Mutation Breeding*. Gamma Field Symposia **30**, 23–32. Ohmiya-machi, Naka-gun, Ibaraki-ken, Japan: Institute of Radiation Breeding NIAR MAFF.

Novak, F.J., Afza, R., Daskalov, S., Hermelin, T. & Lucretti, T. (1986a). Assessment of somaclonal and radiation-induced variability in maize. In *Nuclear Techniques and In Vitro Culture for Plant Improvement* (Proceedings FAO/IAEA Symposium, Vienna, 1985), pp.29–33. Vienna: IAEA.

Novak, F.J., Afza, R., van Duren, M. & Omar, M.S. (1990). Mutation induction by gamma irradiation of *in vitro* cultured shoot-tips of banana and plantain (*Musa* cvs.). *Tropical Agriculture* (Trinidad), **67**, 21–8.

Novak, F.J., Brunner, H., Afza, R. & van Duren, M. (1993). Mutation breeding of *Musa* sp. (banana, plantain). *Mutation Breeding Newsletter*, **40**, 2–4.

Novak, F.J., Daskalov, S., Brunner, H., Nesticky, M., Afza, R., Dólezelova, M., Lucretti, S., Herichova, A. & Hermelin, T. (1988). Somatic embryogenesis in maize and comparison of genetic variability induced by gamma radiation and tissue culture techniques. *Plant Breeding*, **101**, 66–79.

Novak, F.J., Hermelin, T., Daskalov, S. & Nesticky, M. (1986b). *In vitro* mutagenesis in maize. In *Genetic Manipulation in Plant Breeding* (Proceedings International Symposium EUCARPIA, Berlin, 1985), ed. W. Horn, C.J. Jensen, W. Odenbach & O. Schieder, pp.563–76. Berlin: Walter de Gruyter & Co.

Novak, F.J. & Micke, A. (1988). Induced mutations and *in vitro* techniques for plant breeding. In *Plant Breeding and Genetic Engineering* (Proceedings SABRAO Int. Symposium and Workshop, Kuala Lumpur, Malaysia, 1987), ed. A.H. Zakri, pp.63–86. Kuala Lumpur, Malaysia: City Reprographic Services.

Nuffer, M.G. (1957). Additional evidence on the effect of X-ray and ultraviolet radiation on mutation in maize. *Genetics*, **42**, 273–82.

Nybom, N. (1954). Mutation types in barley. *Acta Agriculturae Scandinavica*, **4**, 430–56.

Nybom, N. (1956). On the differential action of mutagenic agents. *Hereditas*, **42**, 211–17.

Nybom, N. (1961). The use of induced mutations for the improvement of vegetatively propagated plants. In *Mutation and Plant Breeding* (Proceedings Symposium Cornell University, Ithaca, N.Y., 1960), Publ. 891, pp.252–94. Washington D.C., USA: National Academy of Sciences/National Research Council.

Nybom, N., Gustafsson, Å., Granhall, I. & Ehrenberg, L. (1956). The genetic effects of chronic gamma irradiation in barley. *Hereditas*, **42**, 74–84.

Nybom, N. & Koch, A. (1965). Induced mutations and breeding methods in vegetatively propagated plants. In *The Use of Induced Mutations in Plant Breeding* (Report FAO/IAEA Technical Meeting, Rome, 1964), pp.661–78. Oxford: Pergamon Press.

Nybom, N. & Micke, A. (1973). Bibliography: Mutation breeding of vegetatively propagated plants and woody perennials (Annex 2). In *Induced Mutations in Vegetatively Propagated Plants* (Proceedings FAO/IAEA Panel, Vienna, 1972), pp.203–20. Vienna: IAEA.

Oehlkers, F. (1943). Die Auslösung von Chromosomenmutationen in der Meiosis durch Einwirkung von Chemikalien. *Zeitschrift für induktive Abstammungs- und Vererbungslehre*, **81**, 313–41.

Oehlkers, F. (1946). Weitere Versuche zur Mutationsauslösung durch Chemikalien. *Biologisches Zentralblatt*, **65**, 176–86.

Oehlkers, F. (1949). Mutationsauslösung durch Chemikalien. *Sitzungsbericht der Heidelberger Akademie der Wissenschaften, Mathematisch-naturwissenschaftliche Klasse*, 149, Abhand. 1–40.

Oehlkers, F. (1953). Chromosome breaks influenced by chemicals. In *Symposium* on *Chromosome Breakage* (held at the John Innes Horticultural Institution, 1952). *Supplement to Heredity*, Vol.6, 95–105. London.

Oehlkers, F. (1956). Die Auslösung von Mutationen durch Chemikalien bei *Antirrhinum majus*. *Zeitschrift für induktive Abstammungs- und Vererbungslehre*, **87**, 584–9.

Oh, M.-H., Lee, H.S., Song, J.Y., Choi, D.-W, Kwon, Y.M., Lee, J.S. & Kim, S.-G. (1995). Origin of tetraploidization in protoplast cultures of petunia (*Petunia hybrida*). *Journal of Heredity*, **86**, 461–6.

Oka, H.I., Hayashi, J. & Shiojiri, I. (1958). Induced mutation of polygenes for quantitative characters in rice. *Journal of Heredity*, **49**, 11–14.

Old, R.W. & Primrose, S.B. (1989). *Principles of Gene Manipulation* (fourth edition), pp.87–98. Oxford: Blackwell Scientific Publications.

Olsson, G. & Persson, C. (1986). Recurrent selection in white mustard (*Sinapis alba* L.). In *Svalöf 1886–1986. Research and Results in Plant Breeding*, ed. G. Olsson, pp.53–6. Stockholm: LTs förlag.

Omar, M.S., Novak, F.J. & Brunner, H. (1989). *In vitro* action of ethylmethanesulphonate on banana shoot tips. *Scientia Horticulturae*, **40**, 283–95.

Ortiz, R. & Vuylstek D. (1996). Recent advances in *Musa* genetics, breeding and biotechnology. *Plant Breeding Abstracts*, **66**, 355–63.

Orton, T.J. (1983). Experimental approaches to the study of somaclonal variation. *Plant Molecular Biology Reporter*, **1**, 67–76.

Orton, T.J. (1984). Somaclonal variation: Theoretical and practical considerations. In *Gene Manipulation in Plant Improvement* (16th Stadler Genetics Symposium, Univ. of Missouri, Columbia, USA), ed. J.P. Gustafson, pp.427–68. New York: Plenum Press.

Orton, T.J. (1987). Genetic instability in celery tissue and cell cultures. *Iowa State Journal of Research*, **61**, 481–98.

Osone, K.I. (1963). Studies on the developmental mechanism of mutated cells induced in irradiated rice seeds. *Japanese Journal of Breeding*, **13**, 1–13.

Pandey, K.K. (1969). X-irradiation-induced S-gene mutations, accumulated centric chromosome fragments and evolution of B-chromosomes. In *Induced Mutations in Plants* (Proceedings FAO/IAEA Symposium, Pullman, USA), pp.792–5. Vienna: IAEA.

Pandey, K.K. (1974). Overcoming interspecific pollen incompatibility through the use of ionising radiation. *Heredity*, **33**, 270–84.

Pandey, K.K. (1975). Sexual transfer of specific genes without gametic fusion. *Nature*, **256**, 310–13.

Pandey, K.K. (1978). Gametic gene transfer in *Nicotiana* by means of irradiated pollen. *Genetica*, **49**, 53–69.

Pandey, K.K. & de Nettancourt, D. (1976). Radiation-induced mutation and sexual incompatibility in flowering plants. In *Induced Mutations in Cross-Breeding* (Proceedings FAO/IAEA Advisory Group Meeting, Vienna, 1975), pp.157–77. Vienna: IAEA.

Pandey, K.K., Przywara, L. & Sanders, P.M. (1990). Induced parthenogenesis in kiwifruit (*Actinidia deliciosa*) through the use of lethally irradiated pollen. *Euphytica*, **51**, 1–9.

Parlevliet, J.E. & Zadoks, J.C. (1977). The integrated concept of disease resistance: a new view including horizontal and vertical resistance in plants. *Euphytica*, **26**, 5–21.

Pascual, M.J. & Correal, E. (1991). Mejora de *Euphorbia lagascae* por mutagenesis. *Actas de Horticultura*, **8**, 309–12.

Pascual, M.J. & Correal, E. (1992). Mutation studies of an oilseed spurge rich in vernolic acid. *Crop Science*, **32**, 95–8.

Pascual-Villalobos, M.J., Röbbelen, G. & Correal, E. (1994). Production and evaluation of indehiscent mutant genotypes in *Euphorbia lagascae*. *Industrial Crops and Products*, **3**, 129–43.

Pastink, A., Vreeken, C., Nivard, M.J.M., Searles, L.L. & Vogel, E.W. (1989). Sequence analysis of N-ethyl-N-nitrosourea-induced *vermilion* mutations in *Drosophila melanogaster*. *Genetics*, pp.123–9.

Pastink, A., Heemskerk, E., Nivard, M.J.M., van Vliet, C.J. & Vogel, E.W. (1991). Mutational specificity of ethyl methanesulfonate in excision-repair-proficient and -deficient strains of *Drosophila melanogaster*. *Molecular & General Genetics*, **229**, 213–18.

Pate, J.B. & Duncan, E.N. (1963). Mutations in cotton induced by gamma-irradiation of pollen. *Crop Science*, **3**, 136–8.

Paul, D.B. & Krimbas, C.B. (1992). Nikolai V. Timoféeff-Ressovsky. *Scientific American*, **266**(2), 64–70.

Peck, J.R. (1994). A ruby in the rubbish: Beneficial mutations, deleterious mutations and the evolution of sex. *Genetics*, **137**, 597–606.

Pedersen, J.F. & Dickens, R. (1985). Registration of AU Centennial centipedegrass. *Crop Science*, **25**, 364.

Perkins, J.M., Eglington, E.G. & Jinks, J.L. (1971). The nature of inheritance of permanently induced changes in *Nicotiana rustica*. *Heredity*, **27**, 441–57.

Persson, G. & Hagberg, A. (1969). Induced variation in a quantitative character in barley. Morphology and cytogenetics of *erectoides* mutants. *Hereditas*, **61**, 115–78.

Perthes, G. (1904). Versuche über den Einfluss der Röntgenstrahlung und Radiumstrahlen auf die Zellteilung. *Deutsche medizinische Wochenschrift*, 17–18.

Peschke, V.M. & Phillips, R.L. (1992). Genetic implications of somaclonal variation in plants. *Advances in Genetics*, **30**, 41–75.

Peterson, P. (1987). Mobile elements in plants. *CRC Critical Reviews of Plant Science*, **6**, 105–208.

Peterson, P. (1993). Transposable elements in maize: their role in creating plant genetic variability. *Advances in Agronomy*, **51**, 79–124.

Peto, F.H. (1933). The effect of ageing and heat on the chromosomal mutation rates in maize and barley. *Canadian Journal of Research*, **9**, 261–4.

Petrov, D.F., Fokina, E.S. & Zhelagnova, N.B. (1971). Method of producing cytoplasmic male sterility in maize. *U.S.Patent* 3 594 152.

Phillips, R.L. (1989). Somaclonal and gametoclonal variation. *Genome*, **31**(2), 1119–20.

Phillips, R.L., Kaeppler, S.M. & Olhoft, P. (1994). Genetic instability of plant tissue cultures: breakdown of normal controls. *Proceedings of the National Academy of Sciences of the USA*, **91**, 5222–6.

Phillips, R.L., Kaeppler, S.M. & Peschke, V.M. (1990). Do we understand somaclonal variation? In *Progress in Plant Cellular and Molecular Biology*, ed. H.J.J. Nijkamp, L.H.W. van der Plas & J. van Aartrijk, pp.131–41. Dordrecht: Kluwer Academic Publishers.

Phillips, R.L., Plunkett, D.J. & Kaeppler, S.M. (1992). Novel approaches to the induction of genetic variation and plant breeding implications. In *Plant Breeding in the 1990s* (Proceedings of the Symposium on Plant Breeding in the 1990s, North Carolina State University, Raleigh, NC, USA, 1991). ed. H.T. Stalker & J.P. Murphy, pp.389–408. Wallingford, UK: C.A.B. International.

Pickersgill, B. & Heiser Jr., C.B. (1976). Cytogenetics and evolutionary change under domestication. *Philosophical Transactions of the Royal Society of London*. Series B.275(936), 55–69.

Pierik, R.L.M. (1987). *In Vitro Culture of Higher Plants*. Dordrecht: Martinus Nijhoff.

Pinet-Leblay, C., Turpin, F.X. & Chevreau, E. (1992). Effect of gamma and ultra-violet irradiation on adventitious regeneration from *in vitro* cultured pear leaves. *Euphytica*, 62, 225–33.

Pirovano, A. (1957). *Elettrogenetica – Esperimenti su Vegetali*. Roma: Istituto di Frutticultura e di Elettrogenetica.

Plewa, M.J. & Wagner, E.D. (1993). Activation of promutagens by green plants. *Annual Review of Genetics*, 27, 93–113.

Poethig, R.S. & Sussex, I.M. (1985). The cellular parameters of leaf development in tobacco: a clonal analysis. *Planta*, 165, 170–84.

Pohlheim, F. (1974). Nachweis von Mischzellen in variegaten Adventivsprossen von *Saintpaulia*, enstanden nach Behandlung isolierter Blätter mit N-Nitroso-N-Methylharnstoff. *Biologisches Zentralblatt*, 93, 141–8.

Pohlheim, F. (1980). Zur Sproßvariation bei den *Cupressaceae*. *Wissenschaftliche Zeitschrift der Humboldt-Universität Berlin* (Mathematisch-Naturwissenschaftliche Reihe), 29, 295–306.

Pohlheim, F. (1981). Genetischer Nachweis einer NMH-induzierten Plastommutation bei *Saintpaulia ionantha* H.Wendl. *Biologische Rundschau*, 19, 47–50.

Pohlheim, F. & Beger, B. (1974). Erhöhung der Mutationsrate im Plastom bei *Saintpaulia* durch N-Nitroso-N-Methylharnstoff. *Biologische Rundschau*, 12, 204–6.

Polsoni, L., Kott, L.S. & Beversdorf, W.D. (1988). Large-scale microspore culture technique for mutation-selection studies in *Brassica napus*. *Canadian Journal of Botany*, 66, 1681–5.

Porter, K.S., Axtell, J.D., Lechtenberg, V.L. & Colenbrander, V.F. (1978). Phenotype, fiber composition, and in vitro dry matter disappearance of chemically induced brown midrib (bmr) mutants of sorghum. *Crop Science*, 18, 205–8.

Postma, J.G. (1990). Mutants of *Pisum sativum* (L.) altered in the symbiosis with *Rhizobium leguminosarum*. Ph.D. thesis, University of Groningen, The Netherlands.

Potrykus, I. (1970). Mutation und Rückmutation extrachromosomal vererbter Plastidmerkmale von *Petunia*. *Zeitschrift für Pflanzenzüchtung*, 63, 24–39.

Pötsch, J. (1967). On the dissociation of chimerical shoot-variants by the use of X-rays. In *Induzierte Mutationen und ihre Nutzung* (Proceedings Erwin-Baur-Gedächtnisvorlesungen IV, Gatersleben, 1966), ed. H. Stubbe, pp.301–4. Berlin: Akademieverlag.

Pötsch, J. (1969). Die Abhängigkeit röntgeninduzierter Histogeneseanomalien von der Höhe der Bestrahlungsdosis bei *Pelargonium zonale* Ait. 'Madame Salleron'. *Wissenschaftliche Zeitschrift der Pädagogischen Hochschule Potsdam*, 13, 129–37.

Powell, J.B., Burton, G.W. & Young, J.R. (1974). Mutations induced in vegetatively propagated turf bermudagrasses by gamma irradiation. *Crop Science*, 14, 327–32.

Powell, W., Caligari, P.D.S. & Hayter, A.M. (1983). The use of pollen irradiation in barley breeding. *Theoretical and Applied Genetics*, 65, 73–6.

Prakken, R. (1959). Induced mutation. *Euphytica*, 8, 270–322.

Prasad, G. & Tripathi, D.K. (1987). Study of chemical mutagenesis in barley. In *Barley Genetics V* (Proceedings of the Fifth International Barley Genetics Symposium, Okayama, Japan), ed. S. Yasuda & T. Konishi, pp.271–7. Okayama, Japan: Sanyo Press.

Pratt, C. (1983). Somatic selection and chimeras. In *Methods in Fruit Breeding*, ed. J.N. Moore & J. Janick, pp.172–85. West Lafayette, Indiana, USA: Purdue University Press.

Pratt, D. (1983). Genetic variability induced in *Nicotiana sylvestris* by protoplast culture. *Theoretical and Applied Genetics*, 64, 223–30.

Prina, A.R. (1992). A mutator nuclear gene inducing a wide spectrum of cytoplasmically inherited chlorophyll deficiencies in barley. *Theoretical and Applied Genetics*, 85, 245–51.

Prina, A.R., Hagberg, A. & Favret, E.A.

(1986). Inheritable sterility induced by X-rays and sodium azide in barley. *Genetica Agraria*, 40, 309–19.

Privalov, G.F. (1986). "Zyrianka" a mutant variety of sea buckthorn. *Mutation Breeding Newsletter*, 28, 4–5.

Prosser, J. (1993). Detecting single-base mutations. *Trends in Biotechnology*, 11, 6(113), 238–46.

Rapoport, J.A. (1946). (Carbonyl compounds and the chemical mechanism of mutation.) *C.R. (Doklady) Acad. Sci. U.R.S.S.*, 54, 65–7 (in Russian with English summary; original not consulted).

Rapoport, J.A. (1948). (The action of ethylene oxide, glycine- and glycoloxides on gene mutation.) *C.R. (Doklady) Acad. Sci. U.R.S.S.*, 60, 469–72 (in Russian; original not consulted)

Rapoport, I.A., Zoz, N.N., Makarova, S.I. & Salnikova, T.V. (1966). *Supermutagenes*. Moscow: Publishing House 'Nauka' (in Russian with English summaries)

Raquin, C. (1985). Induction of haploid plants by *in vitro* culture of *Petunia* ovaries, pollinated with irradiated pollen. *Zeitschrift für Pflanzenzüchtung*, 94, 166–9.

Raquin, C. (1986). Etude des conditions d'obtention de pétunias haploïdes gynogénétiques par culture in vitro d'ovaires de plantes pollinisées par du pollen irradié. In *Nuclear Techniques and In Vitro Culture for Plant Improvement* (Proceedings FAO/IAEA Symposium, Vienna, 1985), pp.207–11. Vienna: IAEA.

Raquin, C., Cornu, A., Farcy, E., Maizonnier, D., Pelletier, G. & Vedel, F. (1989). Nucleus substitution between *Petunia* species using gamma ray-induced androgenesis. *Theoretical and Applied Genetics*, 78, 337–41.

Rasmusson, D.C. & Phillips R.L. (1997). Plant breeding progress and genetic diversity from *de novo* variation and elevated epistasis. Review and interpretation. *Crop Science*, 37, 303–10.

Razoriteleva, E.K., Beletskii, Y.D. & Zhdanov, Y.A. (1970). The genetical nature of mutation induced by N-nitroso-N-methylurea.I. The variegated plants. Genetika, USSR), 6, 1072–6 (in Russian with English summary).

Reboud, X. & Zeyl, C. (1994). Organelle inheritance in plants. *Heredity*, 72, 132–40.

Reddy, V.R.K. (1992). Mutagenic parameters in single and combined treatments of gamma rays, EMS and sodium azide in triticale, barley and wheat. *Advances in Plant Sciences*, **5**, 542-53.

Rédei, G.P. (1970). *Arabidopsis thaliana* (L.) Heynh. A review of the genetics and biology. *Bibliographia Genetica*, **XX**, 1-151.

Rédei, G.P. (1971). A portrait of Lewis John Stadler 1896-1954. *Stadler Genetics Symposia*, vol.1 and 2, eds G. Kimber & G.P. Rédei, pp.5-20. Columbia, Missouri.

Rédei, G.P. (1973). Extra-chromosomal mutability determined by a nuclear gene locus in *Arabidopsis*. *Mutation Research*, **18**, 149-62.

Rédei, G.P. (1974a). Economy in mutation experiments. *Zeitschrift für Pflanzenzüchtung*, **73**, 87-96.

Rédei, G.P. (1974b). Analysis of the diploid germline of plants by mutational techniques. *Canadian Journal of Genetics and Cytology*, **16**, 473-6.

Rédei, G.P. (1975). *Arabidopsis* as a genetic tool. *Annual Review of Genetics*, **9**, 111-27.

Rédei, G.P. (1982a). *Genetics*. New York: MacMillan Publishing Company.

Rédei, G.P. (1982b). Mutagen assay with *Arabidopsis*. A report of the U.S. Environmental Protection Agency Gene-Tox Program. *Mutation Research*, **99**, 243-55.

Rédei, G.P. & Acedo, G. (1976). Biochemical mutants in higher plants. In *Cell Genetics in Higher Plants* (Proceedings International Training Course, Szeged, Hungary), ed. D. Dudits, G.L. Farkas & P. Maliga, pp.39-58. Budapest: Akadémiai Kiadó.

Reisch, B.I. (1983). Genetic variability in regenerated plants. In *Handbook of Plant Cell Culture*, vol.1 (Techniques for Propagation and Breeding), ed. D.A. Evans, W.R. Sharp, P.V. Ammirato & Y. Yamada, pp.748-69. New York: MacMillan Publishing Company.

Renner, O. (1936a). Zur Kenntnis der nichtmendelnden Buntheit der Laubblätter. *Flora*, **130**, 218-90.

Renner, O. (1936b). Zur Entwicklungsgeschichte randpanaschierter und reingrüner Blätter von *Sambucus, Veronica, Pelargonium, Spiraea, Chlorophytum*. *Flora*, **30**, 454-66.

Revell, S.H. (1959). The accurate estimation of chromatid breakage, and its relevance to a new interpretation of chromatid aberrations induced by ionizing radiations. *Proceedings of the Royal Society of London, Series B*, **150**, 563-89.

Revell, S.H. (1974). The breakage-and-reunion theory and the exchange theory for chromosomal aberrations induced by ionizing radiations: a short history. *Advances in Radiation Biology*, **4**, 367-416.

Rhoades, M.M. (1936). The effect of varying gene dosage on aleurone colour in maize. *Journal of Genetics*, **33**, 347-54.

Rhoades, M.M. (1938). Effect of the Dt gene on the mutability of the a_1 allele in maize. *Genetics*, **23**, 377-97.

Rhoades, M.M. (1943). Genic induction of an induced cytoplasmic difference. *Proceedings of the National Academy of Sciences of the USA*, **29**, 327-9.

Rhoades, M.M. (1956). Lewis John Stadler. *Genetics*, **41**, 1-3.

Rick, C.M. (1986). Tomato mutants: freaks, anomalies and breeders resources. *HortScience*, **21**, 918 and 987.

Rieger, R. (1963). *Die Genommutationen (Ploidiemutationen)*. Jena: VEB Gustav Fischer Verlag.

Rieger, R., Michaelis, A. & Green, M.M. (1976). *Glossary of Genetics and Cytogenetics*, 4th ed., Berlin: Springer-Verlag.

Rieger, R., Michaelis, A. & Green, M.M. (1991). *Glossary of Genetics*, 5th ed., Berlin: Springer-Verlag.

Röbbelen, G. (1959). 15 Jahre Mutationsauslösung durch Chemikalien. *Der Züchter*, **29**, 92-5.

Röbbelen, G. (1966). Chloroplastdifferenzierung nach geninduzierter Plastommutation bei *Arabidopsis thaliana* (L.) Heynh. *Zeitschrift für Pflanzenphysiologie*, **55**, 387-403.

Röbbelen, G. (1972). Untersuchungen zur genetischen Charakterisierung von induzierten phänotypischen Reversionen bei *Arabidopsis*-Mutanten. *Zeitschrift für Pflanzenzüchtung*, **67**, 177-96.

Röbbelen, G. (1990). Mutation breeding for quality improvement, a case study for oilseed crops. *Mutation Breeding Review*, no.6. Vienna: IAEA.

Röbbelen, G. & Heun, M. (1991). Genetic analysis of partial resistance against powdery mildew in induced mutants of barley. In *Plant Mutation Breeding for Crop Improvement* (Proceedings FAO/IAEA Symposium, Vienna, 1990), vol.2, pp.93-111. Vienna: IAEA.

Robertson, D.S. (1978). Characterization of a mutator system in maize. *Mutation Research*, **51**, 21-8.

Rodríguez Nodals, A., de la Ventura Martín, J.C., Rodríuez-Rodriguez, R., Lopez-Torres, J., Pino-Algoprogram, J. & Román-Martínez, M.I (1992). Banana genetic improvement program at INVIT in Cuba: A progress report. *InfoMusa*, **1**, 3-5.

Roest, S. & Bokelmann, G.S. (1975). Vegetative propagation of *Chrysanthemum morifolium* Ram. in vitro. *Scientia Horticulturae*, **3**, 317-30.

Roest, S. & Bokelmann, G.S. (1980). *In vitro* adventitious bud techniques for vegetative propagation and mutation breeding of potato (*Solanum tuberosum* L.) 1. Vegetative propagation *in vitro* through adventitious shoot formation. *Potato Research*, **23**, 167-81.

Romberger, J.A. (1963). Meristems, growth and development in woody plants. *U.S. Department of Agriculture Technical Bulletin 1293*. Washington D.C., USA.

Rosielle, A.A. & Frey, K.J. (1975). Estimates of selection parameters associated with harvest index in oat lines derived from a bulk population. *Euphytica*, **24**, 121-31.

Rossi, L. (1979). Mutation breeding of durum wheat. In *Induced Mutations for Crop Improvement in Africa* (Proceedings Regional FAO/IAEA/IITA Seminar, Ibadan, Nigeria, 1978), IAEA-TECDOC-222, pp.97-114. Vienna: IAEA.

Rowland, G.G. (1991). An EMS-induced low-linolenic-acid mutant in McGregor flax. *Canadian Journal of Plant Science*, **71**, 393-6.

Rowland, G.G., McHughen, A. & Bhatty, R.S. (1989). Andro flax. *Canadian Journal of Plant Science*, **69**, 911-13.

Rowland, G.G., McHughen, A., Gusta, L.V., Bhatty, R.S., MacKenzie, S.L. & Taylor, D.C. (1995). The application of chemical mutagenesis and biotechnology to the modification of linseed (*Linum usitatissimum* L.). *Euphytica*, **85**, 317-21.

Russell, G.E. (1978). *Plant Breeding for Pest and Disease Resistance* (Studies in the Agricultural and Food Sciences). London: Butterworths.

Russell, P.J. (1992). *Genetics* (Third Edition). New York: HarperCollins Publishers.

Rutger, J.N. (1991). Mutation breeding of rice in California and the United States of America. In *Plant Mutation Breeding for Crop Improvement*

(Proceedings FAO/IAEA Symposium, Vienna, 1990), vol.1, pp.155–65. Vienna: IAEA.

Rutger, J.N. (1992). Impact of mutation breeding in rice-a review. *Mutation Breeding Review*, **8**, 1–23.

Rutger, J.N., Peterson, M.L. & Hu, C.H. (1977). Registration of Calrose 76 rice. *Crop Science*, **17**, 978.

Rybínsky, W., Patyna, H. & Przewózny, T. (1993). Mutagenic effect of laser and chemical mutagens in barley (*Hordeum vulgare* L.). *Genetica Polonica*, **34**, 337–43.

Saccardo, F. (1983a). Chimera formation in M₁ pea plants raised from mutagen-treated seeds. In *Chimerism in Irradiated Dicotyledonous Plants* (Report FAO/IAEA Consultants Meeting, Vienna, 1981), IAEA-TECDOC-289, pp.23–4. Vienna: IAEA.

Saccardo, F. (1983b). Gametophyte irradiations in some dicotyledonous crop species: Prospects and results. In *Chimerism in Irradiated Dicotyledonous Plants* (Report FAO/IAEA Consultants Meeting, Vienna, 1981). IAEA-TECDOC-289, pp.31–2. Vienna: IAEA.

Saccardo, F. & Monti, L.M. (1984). Mutation breeding in higher plants by gamete irradiation technique. In *Induced Mutations for Crop Improvement in Latin America* (Proceedings FAO/IAEA Regional Seminar, Lima, Peru, 1982). IAEA-TECDOC-305, pp.225–34. Vienna: IAEA.

Saccardo, F., Monti, L.M., Frusciante, L., Crinò, P., Vitale, P., Chiaretti, D., Lai, A., Soressi, G.P., Alllavena, A. & Fala-vigna, A. (1991). Mutation breeding programmes for the genetic improvement of grain legumes and vegetable crops in Italy. In *Plant Mutation Breeding for Crop Improvement* (Proceedings FAO/IAEA Symposium, Vienna, 1990), vol.1, pp.537–46. Vienna: IAEA.

Sacristán, M.D. (1982). Resistance responses to *Phoma lingam* of plants regenerated from selected cell and embryogenic cultures of haploid *Brassica napus*. *Theoretical and Applied Genetics*, **61**, 193–200.

Sagan, M., Duc, G., Cornu, A. & Messager, A. (1991). Mutagenesis in pea (*Pisum sativum* L.) as a tool for studying plant *Rhizobium* symbiosis. In *Plant Mutation Breeding for Crop Improvement* (Proceedings FAO/IAEA Symposium, Vienna, 1990), vol.1, pp.469–77. Vienna: IAEA.

Sager, R. (1960). Genetic systems in *Chlamydomonas*. *Science*, **132**, 1459–65.

Sager, R. (1972). *Cytoplasmic Genes and Organelles*. New York: Academic Press.

Sakuramoto, F. & Ichikawa, S. (1996). Effects of X-ray dose fractionations with various intervals in inducing somatic mutations in the stamen hairs of *Tradescantia* clone KU 9. *Genes & Genetic Systems*, **71**, 355–61.

Salaman, R. (1926). *Potato Varieties*. Cambridge: Cambridge University Press.

Salnikova, T.V. (1993a). Chemical mutagenesis for crop breeding – Achievements in the former USSR. *Mutation Breeding Newsletter* **40**, 11–12.

Salnikova, T.V. (1993b). Iosif Abramovich Rapoport (1912–1990). *Mutation Breeding Newsletter*, **40**, 1–2.

Sanada, T. (1986). Induced mutation breeding in fruit trees: resistant mutant to black spot disease of Japanese pear. In *Research around the Institute of Radiation Breeding – 25 Years of the Symposia and the Next*. Gamma Field Symposia 25, 87–106. Ohmiya-machi, Naka-gun, Ibaraki-ken: Institute. Radiation Breeding NIAR MAFF.

Sanada, T., Kotobuki, K., Nishida, T., Fujita, H. & Ikeda, F. (1993). A new Japanese pear cultivar "Gold Nijiseiki" resistant mutant to black spot disease of Japanese pear. *Japanese Journal of Breeding*, **43**, 455–61.

Sanada, T., Nishida, T. & Ikeda, F. (1986). Resistant Mutant to Black Spot Disease of Japanese Pear. *Technical News No. 29*. Ohmiya-machi, Ibaraki, Japan: Institute of Radiation Breeding NIAR MAFF.

Sanford, J.C., Chyi, Y.S. & Reisch, B.I. (1984a). An attempt to induce 'egg transformation' in *Lycopersicon esculentum* Mill. using irradiated pollen. *Theoretical and Applied Genetics*, **67**, 553–8.

Sanford, J.C., Chyi, Y.S. & Reisch, B.I. (1984b). Attempted 'egg transformation' in *Zea mays* L., using irradiated pollen. *Theoretical and Applied Genetics*, **68**, 260–75.

Sanford, J.C., Weeden, N.F. & Chyi, Y.S. (1984). Regarding the novelty and breeding value of protoplast-derived variants of Russett Burbank (*Solanum tuberosum* L.). *Euphytica*, **33**, 709–15.

Sangwan, R.S. & Sangwan, B.S. (1986). Effets des rayons gamma sur l'embry-ogénèse somatique et l'androgénèse chez divers tissus vegetaux cultivés in vitro. In *Nuclear Techniques and In Vitro Culture for Plant Improvement* (Proceedings FAO/IAEA Symposium, Vienna, 1985), pp.181–5. Vienna: IAEA.

Sankaranarayanan, K. (1993). Ionizing radiation, genetic risk estimation and molecular biology: impact and inferences. *Trends in Genetics*, **9**, 79–84.

Sapehin, A.A. (1930). Röntgen-Mutationen beim Weizen (*Triticum vulgare*). *Der Züchter*, **2**, 257–9.

Sapehin, A.A. (1936). X-ray mutants in soft wheat. *Trudy Prikl.Bot.Genet.i.Selektsii*, Ser.9, 3–47 (in Russian with English summary).

Satina, S. & Blakeslee, A.F. (1941). Periclinal chimeras in *Datura stramonium* in relation to development of leaf and flower. *American Journal of Botany*, **28**, 862–71.

Satina, S., Blakeslee, A.F. & Avery, A.G. (1940). Demonstration of the three germ layers in the shoot apex of *Datura* by means of induced polyploidy in periclinal chimeras. *American Journal of Botany*, **27**, 895–905.

Satoh, H. & Omura, T. (1981). New endosperm mutations induced by chemical mutagens in rice, *Oryza sativa* L. *Japanese Journal of Breeding*, **31**, 316–26.

Savitzky, H. (1975). Hybridization between *Beta vulgaris* and *B. procumbens* and transmission of nematode (*Heterodera schachtii*) resistance to sugarbeet. *Canadian Journal of Genetics and Cytology*, **17**, 197–209.

Savitzky, H. (1978). Nematode (*Heterodera schachtii*) resistance and meiosis in diploid plants from interspecific *Beta vulgaris* x *B. procumbens* hybrids. *Canadian Journal of Genetics and Cytology*, **20**, 177–86.

Sax, K. (1938). Chromosome aberrations induced by X-rays. *Genetic*, **23**, 494–516.

Sax, K. (1940). An analysis of X-ray induced chromosomal aberrations in *Tradescantia*. *Genetics*, **25**, 41–68.

Sax, K. (1941). Types and frequencies of chromosomal aberrations induced by X-rays. *Cold Spring Harbor Symposia on Quantitative Biology*, **9**, 93–103.

Scarascia-Mugnozza, G.T., D'Amato, F., Avanzi, S., Bagnara, D., Belli, M.L., Bozzini, A., Brunori, A., Cervigny, T., Devreux, M., Donini, B., Giorgi, B., Martini, G., Monti, L.M., Moschini, E., Mosconi, C., Porreca, G. & Rossi, L.

(1991). Mutation breeding programme for *durum* wheat (*Triticum turgidum* ssp. *durum* Desf.) improvement in Italy. In *Plant Mutation Breeding for Crop Improvement* (Proceedings FAO/IAEA Symposium, Vienna, 1990), vol.1, pp.95-109. Vienna: IAEA.

Scarascia-Mugnozza, G.T., D'Amato, F., Avanzi, S., Bagnara, D., Belli, M.L., Bozzini, A., Brunori, A., Cervigni, T., Devreux, M., Donini, B., Giorgi, B., Martini, G., Monti, L.M., Moschini, E., Mosconi, C., Porreca, G., & Rossi, L. (1993). Mutation breeding for *durum* wheat (*Triticum turgidum* ssp. *durum* Desf.) improvement in Italy. *Mutation Breeding Review*, **10**, 1-28.

Schaeffer, G.W. & Sharpe Jr, F.T. (1990). Modification of amino acid composition of endosperm proteins from *in-vitro* selected high lysine mutants in rice. *Theoretical and Applied Genetics*, **80**, 841-6.

Scheibe, A. & Micke, A. (1967). Experimentally induced mutations in leguminous forage plants and their agronomic value. In *Induzierte Mutationen und Ihre Nutzung* (Proceedings Erwin-Baur-Gedächtnisvorlesungen IV, Gatersleben, 1966), ed. H. Stubbe, pp.231-6. Berlin: Akademie-Verlag.

Schieder, O. (1976). Isolation of mutants with altered pigments after irradiating haploid protoplasts from *Datura innoxia* Mill. with X-rays. *Molecular & General Genetics*, **149**, 251-4.

Schieder, O. (1977). Hybridisation experiments with protoplasts from chlorophyll-deficient mutants of some *Solanaceous* species. *Planta*, **137**, 253-7.

Schiemann, E. (1912). Mutationen bei *Aspergillus niger* van Tieghem. *Zeitschrift für induktive Abstammungs- und Vererbungslehre*, **8**, 1-35.

Schmidt, A. (1924). Histologische Studien an phanerogamen Vegetationspunkten. *Botanisches Archiv*, **7/8**, 345-404.

Schoenmakers, H.C.H. (1993). *Somatic hybridization between Lycopersicon esculentum and Solanum tuberosum*, Ph.D thesis, Agricultural University Wageningen. Wageningen, The Netherlands.

Schubert, I. & Rieger, R. (1977). On the expressivity of aberration hot spots after treatment with mutagens showing delayed or non-delayed effects. *Mutation Research*, **44**, 337-44.

Schwarz, A.G., Cook, P.R. & Harris, H. (1971). Correction of a genetic defect in a mammalian cell. *Nature New Biology*, **230**, 5-8.

Schy, W.E. & Plewa, M.J. (1985). Induction of forward mutation at the yg_2 locus in maize by ethylnitrosourea. *Environmental Mutagenesis*, **7**, 155-62.

Scossiroli, R.E. (1965). Value of induced mutations for quantitative characters in plant breeding. In *The Use of Induced Mutations in Plant Breeding* (Report FAO/IAEA Technical Meeting, Rome, 1964), pp.443-50. Oxford: Pergamon Press.

Scossiroli, R.E. (1977). Mutations in characters with continuous variation. In *Manual on Mutation Breeding*, 2nd edn, pp.118-23. Vienna: IAEA.

Scowcroft, W.R. (1985). Somaclonal variation: The myth of clonal uniformity. In *Genetic Flux in Plants*, ed. B. Hohn & E.S. Dennis, pp.217-45.

Scowcroft, W.R., Brettell, R.I.S., Ryan, S.A., Davies, P.A. & Palotta, M.A. (1987). Somaclonal variation and genetic flux. In *Plant Tissue and Cell Culture*, ed. C.E. Green et al., pp.275-86. New York: Allen R.Liss.

Scowcroft, W.R. & Larkin, P.J. (1982). Somaclonal variation: A new option for plant improvement. In *Plant Improvement and Somatic Cell Genetics*, ed. I.K. Vasil, W.R. Scowcroft & K.J. Frey, pp.159-78. New York: Academic Press.

Scowcroft, W.R., Larkin, P.J. & Brettell, R.I.S. (1983). Genetic variation from tissue culture. In *Use of Tissue Culture and Protoplasts in Plant Pathology*, ed. J.P. Helgeson & B.J. Deverall, pp.139-62. New York: Academic Press.

Scowcroft, W.R. & Larkin, P.J. (1988). Somaclonal variation. In *Applications of Plant Cell and Tissue Culture* (Ciba Foundation Symposia 137), ed. G. Bock & J. Marsh, pp.21-35. Chichester, U.K.: John Wiley & Sons.

Sears, B.B. & Sokalski, M.B. (1991). The *Oenothera* plastome mutator: effect of UV irradiation and nitroso-methyl urea on mutation frequencies. *Molecular & General Genetics*, **229**, 245-52.

Sears, E.R. (1944). Cytogenetic studies with polyploid species of wheat. II. Additional chromosomal aberrations in *Triticum vulgare*. *Genetics*, **29**, 232-46.

Sears, E.R. (1954). *The Aneuploids of Common Wheat*. Missouri Agricultural

Experiment Research Bulletin 572. Miss. USA.

Sears, E.R. (1956). The transfer of leaf-rust resistance from *Aegilops umbellulata* to wheat. In *Genetics in Plant Breeding* (Brookhaven Symposia in Biology, Number 9), pp.1-22. Upton, New York: Biology Department, Brookhaven National Laboratory.

Secor, G.A. & Shepard, J.F. (1981). Variability of protoplast-derived potato clones. *Crop Science*, **21**, 102-5.

Seetharami Reddi, T.V.V. & Prabhakar, G. (1983). Azide induced chlorophyll mutants in grain *Sorghum* varieties. *Theoretical & Applied Genetics*, **64**, 147-9.

Shama, R.H.K. & Sears, E.R. (1964). Chemical mutagenesis in *Triticum aestivum*. *Mutation Research*, **1**, 387-99.

Shamel, A.D. & Pomeroy, C.S. (1936). Bud mutations in horticultural plants. *Journal of Heredity*, **27**, 487-94.

Shen Y., Cai Q., Gao, M. & Liang, Z. (1995). Isolation and genetic characterization of somaclonal mutants with large-sized grain in rice. *Cereal Research Communications*, **23**, 235-41.

Shepard, J.F. (1981). Protoplasts as sources of disease resistance in plants. *Annual Review of Phytopathology*, **19**, 145-66.

Shepard, J.F., Bidney, D. & Shahin, E. (1980). Potato protoplasts in crop improvement. *Science*, **208**, 17-24.

Shigematsu, K. & Matsubara, H. (1972). The isolation and propagation of the mutant plant from sectorial chimera induced by irradiation in *Begonia rex*. *Japanese Journal of the Society of Horticultural Science*, **41**, 196-200 (in Japanese with English summary).

Shivanna, K.R. & Johri, B.M. (1985). *The Angiosperm Pollen. Structure and Function*. New Delhi: Wiley Eastern Limited.

Siddiq, E.S. (1968). Effect of mutagen and dose on the size of the mutated sector in rice. *Indian Journal of Genetics and Plant Breeding*, **28**, 301-4.

Sideris, E.G., Nilan, R.A. & Bogyo, T.P. (1973). Differential effect of sodium azide on the frequency of radiation-induced chromosome aberrations vs. the frequency of radiation-induced chlorophyll mutations in *Hordeum vulgare*. *Radiation Botany*, **13**, 315-22.

Sideris, E.G., Nilan, R.A. & Konzak, C.F. (1969). Relationship of radiation-induced damage in barley seeds to the inhibition of certain oxidoreductases by sodium azide. In *Induced Mutations in Plants* (Proceedings FAO/IAEA Symposium, Pullman, USA), pp. 313–22. Vienna: IAEA.

Sigurbjörnsson, B. (1976). The improvement of barley through induced mutation. In *Barley Genetics III* (Proceedings of the Third International Barley Genetics Symposium, Garching, Germany, 1975), ed. H. Gaul, pp.84–95. München: Verlag Karl Thiemig.

Sigurbjörnsson, B. (1991). Opening address. In *Plant Mutation Breeding for Crop Improvement* (Proceedings FAO/IAEA Symposium, Vienna, 1990), vol.1, pp.3–5. Vienna: IAEA.

Sigurbjörnsson, B. & Micke, A. (1969). Progress in mutation breeding. In *Induced Mutations in Plants* (Proceedings FAO/IAEA Symposium, Pullman, USA), pp.673–98. Vienna: IAEA.

Sigurbjörnsson.B. & Micke, A. (1973). List of varieties of vegetatively propagated plants developed by utilizing induced mutations (Annex 1). In *Induced Mutations in Vegetatively Propagated Plants* (Proceedings FAO/IAEA Panel, Vienna, 1972), pp.195–202. Vienna: IAEA.

Sigurbjörnsson, B. & Micke, A. (1974). Philosophy and accomplishments of mutation breeding. In *Polyploidy and Induced Mutations in Plant Breeding* (Proceedings of two Meetings Joint FAO/IAEA Division and EUCARPIA, Bari, Italy, 1972), pp.303–43. Vienna: IAEA.

Simpson, D.J. & von Wettstein, D. (1991). Coordinator's report: Nuclear genes affecting the chloroplast. Stock list of mutants kept at the Carlsberg Laboratory. *Barley Genetics Newsletter*, **21**, 102–8.

Singh, R.J. (1993). *Plant Cytogenetics*. Boca Raton, USA: CRC Press.

Singleton, W.R. (1954). The effect of chronic gamma radiation on endosperm mutations in maize. *Genetics*, **39**, 587–603.

Singleton, W.R. (1955). The contribution of radiation genetics to agriculture. *Agronomy Journal*, **47**, 113–17.

Singleton, W.R. (1962). *Elementary Genetics*. Princeton, N.J.D. Van Nostrand Company.

Singleton, W.R. (1964). Induction of mutations and plant breeding. In *Eucarpia, Paris, 1962* (Proceedings Third Congress European Association for Research on Plant Breeding), pp.143–55. Paris: Eucarpia.

Skirvin, R.M. (1978). Natural and induced variation in tissue culture. *Euphytica*, **27**, 241–66.

Skirvin, R.M. & Janick, J. (1976a). 'Velvet Rose' *Pelargonium*, a scented geranium. *HortScience*, **11**, 61–2.

Skirvin, R.M. & Janick, J. (1976b). Tissue culture – induced variation in scented *Pelargonium* spp. *Journal of the American Society of Horticultural Science*, **101**, 281–90.

Skirvin, R.M., McPheeters, K.D. & Norton, M. (1994). Sources and frequency of somaclonal variation. *HortScience*, **29**, 1232–7.

Smith, E.L., Schlehuber, A.M., Young Jr, H.C. & Edwards, L.H. (1968). Registration of Agent Wheat. *Crop Science*, **8**, 511–12.

Smith, J.E., ed. (1958). *A Selection of the Correspondence of Linnaeus and Other Naturalists from the Original Manuscripts*, Vol.1. London: Longman (original not consulted).

Smith, M. (1985). In vitro mutagenesis. *Annual Review of Genetics*, **19**, 423–62.

Smith, R.H., Duncan, R.R. & Bhaskaran, S. (1993). In vitro selection and somaclonal variation. In *International Crop Science I.*, ed. D.R. Buxton *et al.*, pp.629–32. Madison, Wisconsin, USA: Crop Science Society of America, Inc.

Smith, R.H. & Norris, R.E. (1983). In vitro propagation of African violet chimeras. *HortScience*, **18**, 436–7.

Smith, S.E. (1989). Biparental inheritance of organelles and its implications in crop improvement. *Plant Breeding Reviews*, **6**, 361–93.

Snape, J.W., Parker, B.B., Simpson, E., Ainsworth, C.C., Payne, P.I. & Law, C.N. (1983). The use of irradiated pollen for differential gene transfer in wheat (*Triticum aestivum*). *Theoretical and Applied Genetics*, **65**, 103–11.

Snedecor, G.W. & Cochran, W.G. (1976). *Statistical Methods*, 6th edn. Ames, Iowa, U.S.A.: The Iowa State University Press.

Snoad, B. (1974). A preliminary assessment of 'leafless peas'. *Euphytica*, **23**, 257–65.

Snoad, B. & Davies, D.R. (1972). Breeding peas without leaves. *Span*, **15**, 87–9.

Snoad, B. & Hedley, C.L. (1981). Potential for redesigning the pea crop using spontaneous and induced mutations. In *Induced Mutations – A Tool in Plant Research* (Proceedings FAO/IAEA Symposium, Vienna), pp.111–26.

Vienna: IAEA.

Søgaard, B. & von Wettstein-Knowles, P. (1987). Barley: genes and chromosomes. *Carlsberg Research Communications*, **52**, 123–96.

Soh, A.C. (1987). Abnormal oil palm clones. Possible causes and implications: further discussions. *The Planter*, **63**, 59–65.

Soma, K. (1973). Experimental studies on the morphogenesis in the vegetative shoot apex. In *Induced Mutation and Chimera in Woody Plants*. Gamma Field Symposia **12**, pp.83–94. Ohmiya-machi, Ibaraki-ken, Japan: Institute of Radiation Breeding NIAS MAF.

Song, H.S., Lim, S.M. & Widholm, J.M. (1994). Selection and regeneration of soybeans resistant to the pathotoxic culture filtrates of *Septoria glycines*. *Phytopathology*, **84**, 948–51.

Sonnino, A., Ancora, G. & Locardi, C. (1986). In vitro mutation breeding of potato. In *Nuclear Techniques and In Vitro Culture for Plant Improvement* (Proceedings FAO/IAEA Symposium, Vienna, 1985), pp.385–94. Vienna: IAEA.

Sparrow, A.H. (1954). Somatic mutations induced in plants by treatment with X and γ radiation. In *Proceedings of the 9th International Congress of Genetics* (Florence, Italy, 1953), ed. G. Montalenti & A. Chiaruga, p.1105–6. Florence: "Caryologia", vol.6, suppl.

Sparrow, A.H., Binnington, J.P.& Pond, V. (1958). Bibliography on the *Effects of Ionizing Radiations on Plants, 1896–1955* (Brookhaven National Laboratory Report 504/Biology and Medicine). Upton, N.Y.: Brookhaven National Laboratory.

Sparrow, A.H., Cuany, R.L., Miksche, J.P. & Schairer, L.A. (1961). Some factors affecting the responses of plants to acute and chronic radiation exposures. *Radiation Botany*, **1**, 10–34.

Sparrow, A.H. & Evans, H.J. (1961). Nuclear factors effecting radiosensitivity. I. The influence of nuclear size and structure, chromosome complement, and DNA content. In *Fundamental Aspects of Radiosensitivity* (Brookhaven Symposia in Biology, No.14), pp.76–100. Upton, New York: Biology Department, Brookhaven National Laboratory.

Sparrow, A.H., Price, H.J. & Underbrink, A.G. (1972). A survey of DNA content per cell and per chromosome of prokaryotic and eukaryotic organisms: Some evolutionary considerations. In *Evolution of Genetic Systems* (Brookhaven Symposia in Biology, No.23), p.451–94. Biology Department, Brookhaven National Laboratory, Upton, New York. New York: Gordon and Breach.

Sparrow, A.H., Schairer, L.A. & Sparrow, R.C. (1963). Relationship between nuclear volumes, chromosome numbers, and relative radiosensitivities. *Science*, **141**, 163–6.

Sparrow, A.H. & Singleton, W.R. (1953). The use of radiocobalt as a source of gamma rays and some effects of chronic irradiation on growing plants. *The American Naturalist*, **87** (832), 29–48.

Sparrow, A.H., Sparrow, R.C. & Schairer, L.A. (1960). The use of X-rays to induce somatic mutations in *Saintpaulia*. *African Violet Magazine*, **13**, 32–7.

Sparrow, A.H., Sparrow, R.C., Thompson, K.H. & Schairer, L.A. (1965). The use of nuclear and chromosomal variables in determining and predicting radiosensitivities. In *The Use of Induced Mutations in Plant Breeding* (Report of the FAO/IAEA Technical Meeting, Rome, 1964), p.101–32. Oxford: Pergamon Press.

Sparrow, A.H. & Sparrow, R.C. (1976). Spontaneous somatic mutation frequencies for flower color in several *Tradescantia* species and hybrids. *Environmental and Experimental Botany*, **16**, 23–43.

Sparrow, A.H., Underbrink, A.G. & Rossi, H.H. (1972). Mutations induced in *Tradescantia* by small doses of X-rays and neutrons: Analysis of dose-response curves. *Science*, **176**, 916–18.

Spena, A. & Salamini, F. (1995). Genetic tagging of cells and cell layers for studies of plant development. In *Methods in Plant Cell Biology, Part A* (Methods in Cell Biology, vol.49), ed. D.W. Galbraith, H.J. Bohmert & D.P. Bourque, pp.331–354. San Diego: Academic Press.

Spiegel-Roy, P. (1990). Economic and agricultural impact of mutation breeding in fruit trees. *Mutation Breeding Review*, 5. Vienna: IAEA.

Spina, P., Mannino, P., Reforgiato Recupero, G. & Starrantino, A. (1991). Use of mutagenesis at the Istituto Sperimentale per l'Agrumicoltura, Acireale (Results and prospects for the future). In *Plant Mutation Breeding for Crop Improvement* (Proceedings FAO/IAEA Symposium, Vienna, 1990),

vol.1, pp.257–61. Vienna: IAEA.

Springen, K. (1987). Improving on mother nature. *Newsweek*, **109**, 3.

Sree Ramulu, K., Dijkhuis, P. & Roest, S. (1983). Phenotypic variation and ploidy level of plants regenerated from protoplasts of tetraploid potato (*Solanum tuberosum* L. cv.'Bintje'). *Theoretical and Applied Genetics*, **65**, 329–38.

Sree Ramulu, K., Dijkhuis, P., Roest, S., Bokelmann, G.S. & de Groot, B. (1986). Variation in phenotype and chromosome number of plants regenerated from protoplasts of dihaploid and tetraploid potato. *Plant Breeding*, **97**, 119–28.

Sree Ramulu, K. & van der Veen, J.H. (1973). Sulphhydryl protection against ionizing radiations. *Arabidopsis Information Service*, **10**, 30–1.

Stadler, D., MacLeod, H. & Dillon, D. (1991). Spontaneous mutation at the mtr locus of *Neurospora*: the spectrum of mutant types. *Genetics*, **129**, 39–45.

Stadler, J., Phillips, R. & Leonard, M. (1989). Mitotic blocking agents for suspension cultures of maize 'Black Mexican Sweet' cell lines. *Genome*, **32**, 475–8.

Stadler, L.J. (1928a). Genetic effects of X rays in maize. *Proceedings of the National Academy of Sciences of the USA*, **14**, 69–75.

Stadler, L.J. (1928b). Mutations in barley induced by X-rays and radium. *Science*, **68**, 186–7.

Stadler, L.J. (1929). Chromosome number and the mutation rate in *Avena* and *Triticum*. *Proceedings of the National Academy of Sciences of the USA*, **15**, 876–81.

Stadler, L.J. (1930). Some genetic effects of X-rays in plants. *Journal of Heredity* **21**, 3–19.

Stadler, L.J. (1932). On the genetic nature of induced mutations in plants. In *Proceedings of the Sixth International Congress of Genetics* (Ithaca, New York), vol.1, 274–94. Brooklyn, New York: Brooklyn Botanic Garden.

Stadler, L.J. (1939). Genetic studies with ultra-violet radiation. In *Proceedings of the Seventh International Congress of Genetics*, Edinburgh, Scotland, pp.260–76.

Stadler, L.J. (1942). *Some observations on gene variability and spontaneous mutation*.The Spragg Memorial Lectures on Plant Breeding (Third

Series), East Lansing, USA: Michigan State College.

Stadler, L.J. (1944). The effect of X-rays upon dominant mutation in maize.*Proceedings of the National Academy of Sciences of the USA*, **30**, 123–8.

Stadler, L.J. (1946). Spontaneous mutation at the R locus in maize. I. The aleurone-color and plant-color effects. *Genetics*, **31**, 377–94.

Stadler, L.J. (1951). Spontaneous mutation in maize. In *Genes and Mutations. Cold Spring Harbor Symposia in Quantitative Biology*, vol.16, pp.49–63. Cold Spring Harbor, Long Island, New York: The Biological Laboratory.

Stadler, L.J. (1954). The gene. *Science*, **120**, 811–19.

Stadler, L.J. & Roman, H. (1943). The genetic nature of X-ray and ultraviolet induced mutations affecting the gene A in maize (abstract). *Genetics*, **28**, 91.

Stadler, L.J. & Sprague, G.F. (1936a). Genetic effects of ultra-violet radiation in maize. I. Unfiltered radiation. *Proceedings of the National Academy of Sciences of the USA*, **22**, 572–8.

Stadler, L.J. & Sprague, G.F. (1936b). Genetic effects of ultraviolet radiation in maize. II. Filtered radiations. *Proceedings of the National Academy of Sciences of the USA*, **22**, 579–91.

Stadler, L.J. & Sprague, G.F. (1937). Contrasts in the genetic effects of ultra-violet radiation and X-rays. *Science*, **85**, 57–8.

Starzycki, S. (1990). Ionophoren – a new method of mutation induction in plants. *Mutation Breeding Newsletter*, **36**, 13–14.

Stebbins Jr, G.L. (1950). *Variation and Evolution in Plants*. New York: Columbia University Press.

Steeves, T.A. & Sussex, I.M. (1989). *Patterns in Plant Development* (2nd ed.). New York: Cambridge University Press.

Steffensen, D.M. (1968). A reconstruction of cell development in the shoot apex of maize. *American Journal of Botany*, **55**, 354–69.

Stein, E. (1922). Über den Einfluß von Radiumbestrahlung auf *Antirrhinum*. *Zeitschrift für induktive Abstammungs- und Vererbungslehre*, **29**, 1–15.

Stein, E. (1926). Untersuchungen über die Radiomorphosen von *Antirrhinum*. *Zeitschrift für induktive Abstammungs- und Vererbungslehre*, **43**, 1–87.

Stein, E. (1929). Über Gewebe-Entartung in Pflanzen als Folge von Radium-bestrahlung (zur Radiomorphose von *Antirrhinum*). *Biologischen Zentralblat*, **49**, 112–26.

Stein, E. (1930). Weitere Mitteilung über die durch Radiumbestrahlung induzierten Gewebe-Entartungen in *Antirrhinum* (Phytocarcinome) und ihr erbliches Verhalten (Somatische Induktion und Erblichkeit). *Biologischen Zentralblatt*, **50**, 129–58.

Stettler, R.F. (1968). Irradiated mentor pollen: its use in remote hybridization of black cottonwood. *Nature*, **219**, 746–7.

Stewart, F.M. (1994). Fluctuation tests: How reliable are the estimates of mutation rates? *Genetics*, **137**, 1139–46.

Stewart, R.N. & Burk, L.G. (1970). Independence of tissues derived from apical layers in ontogeny of the tobacco leaf and ovary. *American Journal of Botany*, **57**, 1010–16.

Stewart, R.N. & Dermen, H. (1970a). Determination of number and mitotic activity of shoot apical initial cells by analysis of mericlinal chimeras. *American Journal of Botany*, **57**, 816–26.

Stewart, R.N. & Dermen, H. (1970b). Somatic genetic analysis of the apical layers of chimeral sports in chrysanthemum by experimental production of adventitious shoots. *American Journal of Botany*, **57**, 1061–71.

Stewart, R.N. & Dermen, H. (1975). Flexibility in ontogeny as shown by the contribution of the shoot apical layers to leaves of periclinal chimeras. *American Journal of Botany*, **62**, 935–47.

Stewart, R.N. & Dermen, H. (1979). Ontogeny in monocotyledons as revealed by studies of developmental anatomy of periclinal chloroplast chimeras. *American Journal of Botany*, **66**, 47–58.

Stewart, R.N., Semeniuk, P. & Dermen, H. (1974). Competition and accommodation between apical layers and their derivatives in the ontogeny of chimeral shoots of *Pelargonium* × *hortorum*. *American Journal of Botany*, **61**, 54–67.

Strasburger, E. (1907). Über die Individualität der Chromosomen und die Propfhybriden-Frage. *Jahrbücher für wissenschaftliche Botanik*, **44**, 482–555.

Stubbe, H. (1929). Über die Möglichkeiten der experimentellen Erzeugung neuer Pflanzenrassen durch künstliche Auslösung von Mutationen. *Der Züchter*, **1**, 6–11.

Stubbe, H. (1930a). Untersuchungen über experimentelle Auslösung von Mutationen bei *Antirrhinum majus*. I. *Zeitschrift für induktive Abstammungs-und Vererbungslehre*, **56**, 1–38.

Stubbe, H. (1930b). Untersuchungen über experimentelle Auslösung von Mutationen bei *Antirrhinum majus*. II. *Zeitschrift für induktive Abstammungs-und Vererbungslehre*, **56**, 202–32.

Stubbe, H. (1932). Untersuchungen über experimentelle Auslösung von Mutationen bei *Antirrhinum majus*. III. *Zeitschrift für induktive Abstammungs-und Vererbungslehre*, **60**, 474–513.

Stubbe, H. (1934). Einige Kleinmutationen von *Antirrhinum majus* L.*Der Züchter*, **6**, 299–303.

Stubbe, H. (1935). Samenalter und Genmutabilität bei *Antirrhinum majus* L. (Nebst einigen Beobachtungen über den Zeitpunkt des Mutierens während der Entwicklung). *Biologisches Zentralblatt*, **55**, 209–14.

Stubbe, H. (1936). Die Erhöhung der Genmutationsrate in alternden Gonen von *Antirrhinum majus*. *Biologisches Zentralblatt*, **56**, 562–7.

Stubbe, H. (1937a). *Spontane und strahleninduzierte Mutabilität*. Leipzig: Georg Thieme.

Stubbe, H. (1937b). Der gegenwärtige Stand der experimentellen Erzeugung von Mutationen durch Einwirkung von Chemikalien. *Angewandte Chemie*, **13**, 241–6.

Stubbe, H. (1938). *Genmutation I. Allgemeiner Teil*. Handbuch der Vererbungswissenschaft, Band II F. Berlin: Verlag von Gebrüder Bornträger.

Stubbe, H. (ed.) (1960). *Chemische Mutagenese* (Proceedings Erwin-Baur-Gedächtnisvorlesungen I, Gatersleben, 1959). Berlin: Akademie-Verlag.

Stubbe, H. (ed.). (1962). *Strahleninduzierte Mutagenese* (Proceedings Erwin-Baur-Gedächtnisvorlesungen II, Gatersleben, 1961). Berlin: Akademie-Verlag.

Stubbe, H. (1963). *Kurze Geschichte der Genetik bis zur Wiederentdeckung der Vererbungsregeln Gregor Mendels*. Jena: VEB Gustav Fischer Verlag.

Stubbe, H. (1966). *Genetik und Zytologie von Antirrhinum L. sect. Antirrhinum*. Jena: VEB Gustav Fischer Verlag.

Stubbe, H. (ed.) (1967). *Induzierte Mutationen und Ihre Nutzung* (Proceedings Erwin-Baur-Gedächt-nisvorlesungen IV, Gatersleben, 1966). Berlin: Akademie-Verlag.

Stubbe, H. (1972). *History of Genetics*. Cambridge, Mass.: The MIT Press.

Stubbe, H. & von Wettstein, D. (1941). Über die Bedeutung von Klein- und Grossmutationen in der Evolution. *Biologisches Zentralblatt*, **61**, 265–97.

Subrahmanyam, N.C. & Kasha, K.J. (1975). Chromosome doubling of barley haploids by nitrous oxide and colchicine treatments. *Canadian Journal of Genetics and Cytology*, **17**, 573–83.

Suda, H., Matsubara, H. & Kudo, T, (1982). A method for production of a dwarf type variant plant of *Begonia masonia* Irmscher. *Japan Patent 57-22298* (original not consulted).

Sussex, I.M. (1989). Developmental programming of the shoot meristem. *Cell*, **56**, 225–9.

Sutton, A. (1918). Do potatoes give rise to new and distinct varieties by bud variation? *Bulletin No.9, Messrs Sutton and Sons (UK)*.

Suzuki, D.T., Griffiths, A.J.F., Miller, J.H. & Lewontin, R.C. (1989). *An Introduction to Genetic Analysis*, 4th edn, New York: W.H. Freeman and Company.

Swaminathan, M.S. (1965). A comparison of mutation induction in diploids and polyploids. In *The Use of Induced Mutations in Plant Breeding* (Report of the FAO/IAEA Technical Meeting, Rome, 1964), p.619–41. Oxford: Pergamon Press.

Swaminathan, M.S. (1969). Role of mutation breeding in a changing agriculture. In *Induced Mutations in Plants* (Proceedings FAO/IAEA Symposium, Pullman, USA), pp. 719–34. Vienna: IAEA.

Swaminathan, M.S., Chopra, V.L. & Bhaskaran, S. (1962). Chromosome aberrations and the frequency and spectrum of mutations induced by ethylmethane sulphonate in barley and wheat. *Indian Journal of Genetics and Plant Breeding*, **22**, 192–207.

Swanson, E.B., Herrgesell, M.J., Arnoldo, M., Sippell, D.W. & Wong, R.S.C. (1989). Microspore mutagenesis and selection: Canola plants with field tolerance to the imidazolinones. *Theoretical and Applied Genetics*, **78**, 525–30.

Sybenga, J. (1983). Genetic manipulation in plant breeding: somatic versus generative. *Theoretical and Applied Genetics*, **66**, 179–201.

Syukur, S., Jacobs, M. & Negrutiu, I. (1991). Analysis of mutant plants resistant to salt or water stress and to proline analogues obtained from the protoplasts of *Nicotiana plumbaginifolia viviani*. In *Plant Mutation Breeding for Crop Improvement* (Proceedings FAO/IAEA Symposium, Vienna, 1990), vol.2, pp.265-9. Vienna: IAEA.

Szarejko, I., Maluszynski, M., Polok, K. & Kilian, A. (1991). Doubled haploids in the mutation breeding of selected crops. In *Plant Mutation Breeding for Crop Improvement* (Proceedings FAO/IAEAO Symposium, Vienna, 1990), vol.2, pp.355-78. Vienna: IAEA.

Szymkowiak, E.J. & Sussex, I.M. (1996). What chimeras can tell us about plant development. *Annual Review of Plant Physiology and Plant Molecular Biology*, 4, 351-76.

Taliaferro, C.M. (1995). Diversity and vulnerability of bermuda turfgrass species. *Crop Science*, 35, 327-32.

Tallberg, A. (1982). Characterization of high-lysine barley genotypes. *Hereditas*, 96, 229-45.

Tallberg, A. (1986). Improvement of protein quality in cereal grain. In *Svalöf 1886-1986, Research and Results in Plant Breeding*, ed. G.Olsson, pp.165-72. Stockholm: LTs förlag.

Tan, Y.P., Ho, Y.W., Mak, C. & Ibrahim, Rusli(1993). "Fatom-1" – An early-flowering mutant derived from mutation induction of Grand Nain, a cavendish banana. *Mutation Breeding Newsletter*, 40, 5-6.

Tang, X., Wu S., Peng C., Li Z., Yi G., Luo M., Wu C. & Huang H. (1995). Development of a seedless citrus cultivar through bud re-irradiation. In *Induced Mutations and Molecular Techniques for Crop Improvement* (Proceedings FAO/IAEA Symposium, Vienna), pp.646-8. Vienna: IAEA.

Tegenkamp, T.R. (1969). Mutagenic effects of magnetic fields on *Drosophila melanogaster*. In *Biological Effects of Magnetic Fields*, vol.2, ed. M.F. Barnothy, pp.189-206. New York: Plenum Press.

Thamm, K.J.J. (1956). Rückblick auf die bisherige wissenschaftliche Arbeit des Amsterdamer Biologen Willem Eduard de Mol van Oud Loosdrecht, pp.7-30. Bonn, München, Wien: Bayerischer Landwirtschaftsverlag.

Thomas, E., Bright, S.W.J., Franklin, J., Lancaster, V.A., Miflin, B.J. & Gibsson, R. (1982). Variation amongst protoplast-derived potato plants (*Solanum tuberosum* cv.'Maris Bard'). *Theoretical and Applied Genetics*, 62, 65-8.

Thomas, H. & Grierson, D. (ed.) (1987). *Developmental Mutants in Higher Plants*. Society for Experimental Biology, Seminar Series 32. Cambridge: Cambridge University Press.

Thompson, D., Walbot, V. & Coe Jr, E.H. (1983). Plastid development in iojap- and chloroplast mutator-affected maize plants. *American Journal of Botany*, 70, 940-50.

Tilney-Bassett, R.A.E. (1970). The control of plastid inheritance in *Pelargonium*. *Genetic Research*, 16, 49-61.

Tilney-Bassett, R.A.E. (1975). Genetics of variegated plants. In *Genetics and Biogenesis of Mitochondria and Chloroplasts* (First Colloquium of Biological Sciences of the Ohio State University), ed. C.W. Birky Jr, P.S. Perlman & T.J. Byers, pp.268-308. Columbus, USA: Ohio State University Press.

Tilney-Bassett, R.A.E. (1986). *Plant Chimeras*. London: Edward Arnold.

Tilney-Bassett, R.A.E. (1987). The chimeral problem. In *Improving Vegetatively Propagated Crops*, ed. A.J. Abbott & R.K. Atkin, pp.271-84. London: Academic Press.

Tilney-Bassett, R.A.E. (1991). Genetics of variegation and maternal inheritance in ornamentals. In *Genetics and Breeding of Ornamental Species*, ed. J. Harding, F. Singh & J.N.M. Mol, pp.225-49. Dordrecht: Kluwer Academic Publishers.

Tilney-Bassett, R.A.E. (1994). Nuclear controls of chloroplast inheritance in higher plant. *Journal of Heredity*, 85, 347-54.

Timmons, A.M. & Dix, P.J. (1991). Influence of ploidy on plastome mutagenesis in *Nicotiana*. *Molecular & General Genetics*, 227, 330-3.

Timoféeff-Ressovsky, N.W. (1937). *Experimentelle Mutationsforschung in der Vererbungslehre. Beeinflüssung der Erbanlagen durch Strahlen und andere Faktoren*. Dresden, Leipzig: Theodor Steinkopf Verlag.

Timoféeff-Ressovsky, N.W., Zimmer, K.G. & Delbrück, M. (1935). Über die Natur der Genmutation und der Genstruktur. *Nachrichten von der Gesellschaft der Wissenschaften zu Göttingen* (Math.Phys.Klasse, Neue Folge, Fachgruppe II Biologie), Bd.1, Nr.13, 189-245.

To, K.Y., Chen, C.C. & Lai, Y.K. (1989). Isolation and characterization of streptomycin-resistant mutants in *Nicotiana plumbaginifolia*. *Theoretical and Applied Genetics*, 78, 81-6.

Todd, A.J. (1990). Performance of disease resistant peppermint mutants in the USA. *Mutation Breeding Newsletter*, 36, 14-15.

Todd, W.A., Green, R.J. & Horner, C.E. (1977). Registration of Murray Mitcham peppermint. *Crop Science*, 17, 188.

Toler, R.W., Beard, J.B., Grisham, M.P. & Crocker, R.L. (1985). Registration of TXSA 8202 and TXSA 8218 St. Augustinegrass germplasm resistant to *Panicum* mosaic virus St. Augustine decline strain. *Crop Science*, 25, 371.

Tollenaar, D. (1934). Untersuchungen über Mutation bei Tabak. I. Entstehungsweise und Wesen künstlich erzeugter Gen-Mutanten. *Genetica*, 16, 111-52.

Tollenaar, D. (1938). Untersuchungen über Mutation bei Tabak.II. Einige künstlich erzeugte Chromosom-Mutanten. *Genetica*, 20, 285-94.

Travis, D.M., Stewart, K.D. & Wilson, K.G. (1975). Nuclear and cytoplasmic chloroplast mutants induced by chemical mutagens in *Mimulus cardinalis*: Genetics and ultrastructure. *Theoretical and Applied Genetics*, 46, 67-77.

Truhaut, R. & Deysson, G. (1972). Action du bromure d'éthidium sur la division des cellules méristématiques radiculaires d'*Allium sativum* L.*Comptes rendus hebdomadaires des séances de l'Academie des Sciences* (Paris, France), Série D, 274D(9), 1286-8.

Tschermak-Seysenegg, E. von(1951). Rediscovery of Gregor Mendel's work. An historical retrospect. *Journal of Heredity*, 42, 163-71.

Turnquist, O.C. (1960). Production of certified seed potatoes by varieties 1959. In *Potato Handbook 1960*, ed. J.C. Campbell, pp.55-9. New Brunswick, N.Y.: Potato Association of America.

Udy, D.C. (1954). Dye-binding capacities of wheat flour protein fractions. *Cereal Chemistry*, 31, 389-95.

Udy, D.C. (1956). Estimation of protein in wheat and flour by ion-binding. *Cereal Chemistry*, 33, 190-7.

Ukai, Y. (1983). "Crossing-within-spike-progeny-method". An Effective Method for Selection of Mutants in Cross-fertilizing Plants. *Technical News No. 25*, Ohmiya-machi, Ibaraki-ken, Japan: Institute of Radiation Breeding NIAS.

Ukai, Y. (1990). Application of a new method for selection of mutants in cross-fertilizing species to recurrently mutagen-treated populations of Italian ryegrass. In *Reproduction System and Mutation*. Gamma Field Symposia 29, 55–67. Ohmiya-machi, Naka-gun, Ibaraki-ken, Japan: Institute of Radiation Breeding NIAR MAFF.

Ukai, Y. & Yamashita, A. (1974). Theoretical considerations on the problems of screening of mutants. I. Methods for selection of a mutant in the presence of chimera in M_1 spikes. *Acta Radiobotanica et Genetica (Bulletin of the Institute of Radiation Breeding)*, 3, 1–44. Ohmiya, Ibaraki, Japan: Institute of Radiation Breeding NIAS.

Ukai, Y. & Yamashita, A. (1981). Early maturing mutants induced by ionizing radiations and chemicals in barley. In *Barley Genetics IV* (Proceedings of the Fourth International Barley Genetics Symposium, Edinburgh), pp.846–54. Edinburgh: Edinburgh University Press.

Ukai, Y. & Yamashita, A. (1987). Induced mutants highly resistant to barley yellow mosaic virus. In *Barley Genetics V* (Proceedings 5th International Barley Genetics Symposium, Okayama, Japan, 1986), ed. S, Yasuda & T. Konishi, pp.279–86. Nakasange, Okayama: Sanyo Press.

Underbrink, A.G., Schairer, L.A. & Sparrow, A.H. (1973). *Tradescantia* stamen hairs: a radiobiological test system applicable to chemical mutagenesis. In *Chemical Mutagens: Principles and Methods for their Detection*, vol.3, ed. A. Hollaender, pp.171–207. New York: Plenum Press.

Underbrink, A.G., Sparrow, A.H. & Pond, V. (1968). Chromosomes and cellular radiosensitivity-II. Use of interrelationships among chromosome volume, nucleotide content and D_0 of 120 diverse organisms in predicting radiosensitivity. *Radiation Botany*, 8, 205–37.

Underbrink, A.G., Sparrow, R.C., Sparrow, A.H. & Rossi, H.H. (1970). Relative biological effectiveness of x-rays and 0.43-MeV monoenergetic neutrons on

somatic mutations and loss of reproductive integrity in *Tradescantia* stamen hairs. *Radiation Research*, 44, 187–203.

Upton, A.C. (1982). The biological effects of low-level ionizing radiation. *Scientific American*, 246, 29–37.

van den Bulk, R.W. (1991). Application of cell and tissue culture and *in vitro* selection for disease resistance breeding – a review. *Euphytica*, 56, 269–85.

van den Bulk, R.W. & Dons, J.J.M. (1993). Somaclonal variation as a tool for breeding tomato for resistance to bacterial canker. *Acta Horticulturae*, 336, 347–55.

van den Bulk, R.W., Löffler, H.J.M., Lindhout, W.H. & Koornneef, M. (1990). Somaclonal variation in tomato: effect of explant source and a comparison with chemical mutagenesis. *Theoretical and Applied Genetics*, 880, 817–25.

van der Krol, A.R., Mur, L.A., Beld, M., Mol, J.N.M. & Stuitje, A.R. (1990). Flavonoid genes in *Petunia*: Addition of a limited number of gene copies may lead to a suppression of gene expression. *The Plant Cell*, 2, 291–9.

van der Leij, F.R., Visser, R.G.F., Ponstein, A.S., Jacobsen, E. & Feenstra, W.J. (1991). Sequence of the structural gene for granule-bound starch synthase of potato (*Solanum tuberosum* L.) and evidence for a single point deletion in the amf allele. *Molecular & General Genetics*, 228, 240–8.

van der Veen, J.H.G., van Brederode, G.H.M. & Vis, N.F. (1969). Sulfhydryl protection against X-rays in *Arabidopsis*. *Arabidopsis Information Service*, 6, 23.

van Duren, M., Morpurgo, R., Dolezel, J. & Afza, R. (1996). Induction and verification of autotetraploids in diploid banana (*Musa acuminata*) by *in vitro* techniques. *Euphytica*, 88, 25–34.

van Gaasbeek, A.F., Heijbroek, A.M.A., Vaandrager, P. & Boers, G.J. (1994). The World Seed Market: Developments and Strategy. Den Haag, the Netherlands: LEI-DLO, Rabobank Nederland and Ministry of Agriculture, Nature Management and Fisheries.

van Gastel, A.J.G. (1976). *Mutability of the self-incompatibility locus and identification of the S-bearing chromosome in Nicotiana alata*. Ph.D.thesis. Wageningen: Wageningen Agricultural University.

van Harten, A.M. (1972). A suggested

method for investigating L1 constitution in periclinal potato chimeras. *Potato Research*, 15, 73–5.

van Harten, A.M. (1978). *Mutation Breeding Techniques and Behaviour of Irradiated Shoot Apices of Potato*. Ph.D. thesis. Agricultural Research Reports 873. Wageningen: Centre for Agricultural Publishing and Documentation (PUDOC).

van Harten, A.M. (1982). Mutation breeding in vegetatively propagated crops with emphasis on contributions from the Netherlands. In *Induced Variability in Plant Breeding* (Proceedings International Symposium of the Section Mutation and Polyploidy of EUCARPIA, Wageningen, the Netherlands, 1981), pp.22–30. Wageningen: Centre for Agricultural Publishing and Documentation (PUDOC).

van Harten, A.M. & Bouter, H. (1970). Rode Eersteling: a false name for a well-known tuber colour mutant. *Potato Research*, 13, 353–5.

van Harten, A.M., Bouter, H. & Broertjes, C. (1981). In vitro adventitious bud techniques for vegetative propagation and mutation breeding of potato (*Solanum tuberosum* L.). II. Significance for mutation breeding. *Euphytica*, 30, 1–8.

van Harten, A.M., Bouter, H. & Schut, B. (1973). Ivy leaf of potato (*Solanum tuberosum*), a radiation-induced dominant mutation for leaf shape. *Radiation Botany*, 13, 287–92.

van Harten, A.M. & Broertjes, C. (1989). Induced mutations in vegetatively propagated crops. *Plant Breeding Reviews*, 6, 55–91. Portland, Oregon: Timber Press.

van Raalte, D. (1969). Het Handboek voor de Bloemisterij. 4th edn, vol.IV. Assen, Amsterdam, The Netherlands: Born Periodieken NV.

van Rheenen, H.A., Pundir, R.P.S. & Miranda, J.H. (1994). Induction and inheritance of determinate growth habit in chickpea (*Cicer arietinum* L.). *Euphytica*, 78, 137–41.

van Tuyl, J.M., Meijer, B. & Diën, M.P. van (1992). The use of oryzalin as an alternative for colchicine in *in-vitro* chromosome doubling of *Lilium* and *Nerine*. *Acta Horticulturea*, 325, 625–30.

van Voorst, A. & Arends, J.C. (1982). The origin and chromosome numbers of cultivars of *Kalanchoe blossfeldiana* von

Poelln.: Their history and evolution. *Euphytica*, **31**, 573–84.

Vasil, I.K. (1990). The realities and challenges of plant biotechnology. *Biotechnology*, **8**, 296–301.

Vasileva, M.Stefanov, V., Naidenova, N., Peicheva, S., Ancheva, M. & Milanova, G. (1991). (Cytogenetic effect of helium-neon and argon lasers in *Pisum sativum*. *Genetika i Selektsiya*, **24**, 90–8). (In Russian; original not consulted. See *Plant Breeding Abstracts* (1992), **62**, 3255).

Vavilov, N.I. (1922). The law of the homologous series in variation. *Journal of Genetics*, **12**, 47–89.

Véleminský, J. & Gichner, T., eds. (1965). *Induction of Mutations and the Mutation Process* (Proceedings of a Symposium of the Czechoslovak Academy of Sciences, Prague, 1963). Prague: Publishing House of the Czechoslovak Academy of Sciences.

Véleminský, J. & Gichner, T. (1968). The mutagenic activity of nitrosoamines in *Arabidopsis thaliana*. *Mutation Research*, **5**, 429–31.

Véleminský, J. & Gichner, T. (1978). DNA repair in mutagen-injured higher plants. *Mutation Research*, **55**, 71–84.

Véleminský, J. & Gichner, T., eds(1987). *DNA Damage and Repair in Higher Plants and Their Relation to Genetic Damage*. *Mutation Research*, 181 (special issue).

Véleminský, J. & Gichner, T. (1988). Mutagenic activity of promutagens in plants. Indirect action. *Mutation Research*, **197**, 221–42.

Verhoeven, H.A., Sree Ramulu, K. & Dijkhuis, P. (1990). A comparison of the effects of various spindle toxins on metaphase arrest and formation of micronuclei in cell-suspension culture of *Nicotiana plumbaginifolia*. *Planta*, **182**, 408–14.

Verkerk, K. (1971). Chimerism of the tomato plant after seed irradiation with fast neutrons. *Netherlands Journal of Agricultural Science*, **19**, 197–203.

Vielle Calzada, J.-P., Crane, C.F. & Stelly, D.M. (1996). Apomixis: The Asexual Revolution. *Science*, **274**, 1322–3.

Vig, B.K. (1973). Somatic crossing over in *Glycine max* (L.) Merrill: effect of some inhibitors of DNA synthesis on the induction of somatic crossing over and point mutations. *Genetics*, **73**, 583–96.

Vig, B.K. (1975). Soybean (*Glycine max*): a new test system for study of genetic parameters as affected by environmental mutagens. *Mutation Research*, **31**, 49–56.

Villegas, E., Vasal, S.K. & Bjarnason, M. (1992). Quality protein maize: What it is and how it was developed. In *Quality Protein Maize*, ed. E. Mertz, pp.27–48. St.Paul MN, USA: American Association of Cereal Chemists.

Visser, T. (1973). Methods and results of mutation breeding in deciduous fruits, with special reference to the induction of compact and fruit mutations in apple. In *Induced Mutations in Vegetatively Propagated Plants* (Proceedings FAO/IAEA Panel, Vienna, 1972), pp.21–33. Vienna: IAEA.

Visser, T., de Vries, D.P. & Verhaegh, J.J. (1969). Pre-determination of survival rate of apple and pear scions after X-ray treatment. *Euphytica*, **18**, 352–4.

Visser, T., Verhaegh, J.J. & de Vries, D.P. (1971). Pre-selection of compact mutants induced by X-ray treatment in apple and pear. *Euphytica*, **20**, 195–207.

Vizir, I.Y., Anderson, M.L., Wilson, Z.A. & Mulligan, B.J. (1994). Isolation of deficiencies in the *Arabidopsis* genome by γ-irradiation of pollen. *Genetics*, **137**, 1111–19.

Vogel, R. & Röbbelen, G. (1989). Breeding of *Euphorbia lagascae* Spreng., a possible new oil crop for industrial uses. In *Science for Plant Breeding/Book of Poster Abstracts* (XII. EUCARPIA Congress, Göttingen, Germany), part 1, poster 15, 4. Gelsenkirchen-Buer, Germany: Th. Mann Publishers.

Vogel, R., Pascual-Villalobos, M.J. & Röbbelen, G. (1993). Seed oils for new chemical applications. 1.Vernolic acid produced by *Euphorbia lagascae*. *Angewandte Botanik*, **67**, 31–4.

Vollbrecht, E., Veit, B., Sinha, N. & Hake, S. (1991). The developmental gene *Knotted-1* is a member of a maize homoeobox gene family. *Nature*, **350**, 241–3.

von Sengbusch, R. (1942). Süsslupinen and Öllupinen. *Landwirtschaftliches Jahrbuch*, **91**, 723–880.

von Wettstein, D. (1995). Breeding of value added barley by mutation and protein engineering. In *Induced Mutations and Molecular Techniques for Crop Improvement* (Proceedings FAO/IAEA Symposium, Vienna), pp.67–76. Vienna: IAEA.

von Wettstein, D., Gustafsson, Å. & Ehrenberg, L. (1959). Mutationsforschung und Züchtung (Meeting Arbeitsgemeinschaft für Forschung des Landes Nordrhein-Westfalen, 1957), pp.7–50. Köln and Opladen, Germany: Westdeutscher Verlag.

von Wettstein, D., Henningsen, K.W., Boynton, J.E., Kannangara, G.C. & Nielsen, O.F. (1971). The genic control of chloroplast development in barley. In *Autonomy and biogenesis of mitochondria and chloroplasts*, ed. N.K. Boardman, A.W. Linnane & R.M. Smillie, pp.205–23. Amsterdam: North-Holland Publishing Company.

von Wettstein, D., Jende-Strid, B., Ahrenst-Larsen, B. & Sorensen, J.A. (1977). Biochemical mutant in barley renders chemical stabilization of beer superfluous. *Carlsberg Research Communication*, **42**, 341–51.

von Wettstein, D., Jende-Strid, B., Ahrenst-Larsen, B. & Erdal, K. (1980). Proanthocyanidin-free barley prevents the formation of beer haze. *Technical Quarterly of the Master Brewery Association of America*, **17**, 16–23.

von Wettstein, D., Kahn, A., Nielsen, O.F. & Gough, S. (1974). Genetic regulation of chlorophyll synthesis analyzed with mutants in barley. *Science*, **184**, 800–2.

von Wettstein, D., Nilan, R.A., Ahrenst-Larsen, B., Erdal, K., Ingversen, J., Jende-Strid, B., Kristiansen, K.N., Larsen, J., Outtrup, H. & Ullrich, S.E. (1985). Proanthocyanidin-free barley for brewing: Progress in breeding for high yield and research tool in polyphenol chemistry. *Technical Quarterly of the Master Brewery Association of America*, **22**, 41–52.

Vose, P.B. & Blixt, S.G. (1984). *Crop Breeding: A Contemporary Basis*. Oxford: Pergamon Press.

Vuylsteke, D., Swennen, R. & De Langhe, E. (1991). Somaclonal variation in plantains (*Musa* spp., ABB group) derived from shoot-tip culture. *Fruits*, **46**, 429–39.

Vuylsteke, D., Swennen, R.L. & De Langhe, E.A. (1996). Field performance of somaclonal variants of plantain (*Musa* spp., AAB group). *Journal of the American Society for Horticultural Science*, **121**, 42–6.

Walbot, V. (1983). Suggestion for efficient recovery of visible mutants. *Plant Molecular Biology Reporter*, 1(4), 30–1.

Walbot, V. (1992). Strategies for mutagenesis and gene cloning using transposon tagging and T-DNA insertional mutagenesis. *Annual Review of Plant Physiology and Plant Molecular Biology*, **43**, 49–82.

Walker, G.C. (1985). Inducible DNA repair systems. *Annual Review of Biochemistry*, **54**, 425–57.

Walles, H. (1971). Plastid inheritance and mutations. In *Structure and Functions of Chloroplasts*, ed. M. Gibbs, pp.51–88. New York: Springer.

Walters, T., Moynihan, M.R., Mutschler, M.A. & Earle, E.D. (1990). Cytoplasmic mutants from seed mutagenesis of *Brassica campestris* with NMU. *Journal of Heredity*, **81**, 214–16.

Walther, F. & Sauer, A. (1985). Analysis of radiosensitivity – A basic requirement for *in vitro* somatic mutagenesis. I. *Prunus avium* L. *Acta Horticulturae*, **169**, 97–104.

Walther, F. & Sauer, A. (1986a). Analysis of radiosensitivity: A basic requirement for *in vitro* somatic mutagenesis. II. *Gerbera jamesonii*. In *Nuclear Techniques and In Vitro Culture for Plant Improvement* (Proceedings FAO/IAEA Symposium, Vienna, 1985), pp.155–9. Vienna: IAEA.

Walther, F.W. & Sauer, A. (1986b). *In vitro* mutagenesis in *Gerbera jamesonii*. In *Genetic Manipulation in Plant Breeding* (Proceedings EUCARPIA Symposium, Berlin, 1985), ed. W. Horn, C.J. Jensen, W. Odenbach & O. Schieder, pp.555–62. Berlin: Walter de Gruyter.

Walther, F.W.& Sauer, A. (1991). Split dose irradiation of *in vitro* derived microshoots. In *Plant Mutation Breeding for Crop Improvement* (Proceedings FAO/IAEA Symposium, Vienna, 1990), vol.2, pp.343–53. Vienna: IAEA.

Wan, S.Y., Deng, Z.A., Deng, X.X., Ye, X.R. & Zhang, W.C. (1991). Advances made in *in vitro* mutation breeding in citrus. In *Plant Mutation Breeding for Crop Improvement* (Proceedings FAO/IAEA Symposium, Vienna, 1990), vol.1. pp.263–9. Vienna: IAEA.

Wang, L.Q. (1991). Induced mutation for crop improvement in China, a review. In *Plant Mutation Breeding for Crop Improvement* (Proceedings FAO/IAEA Symposium, Vienna, 1990), vol.1, pp. 9–32. Vienna: IAEA.

Wardlaw, C.W. (1957). On the organization and reactivity of the shoot apex in vascular plants. *American Journal of Botany*, **44**, 176–85.

Wardlaw, C.W. (1968). *Morphogenesis in Plants*. London: Methuen & Co Ltd.

Wasscher, J. (1956). The importance of sports in some florist's flowers. *Euphytica*, **5**, 163–70.

Weiling, F. (1960). Über die Häufigkeit der nach Röntgenbestrahlung von Samen überlebenden Corpusinitialen. *Biometrische Zeitschrift*, **2**, 145–63.

Weiling, F. & Gottschalk, W. (1961). Die genetische Konstitution der X_1-Pflanzen nach Röntgenbestrahlung ruhender Samen. *Biologisches Zentralblatt*, **80**, 570–612.

Wen, X. & Qu, L. (1996). Crop improvement through mutation techniques in Chinese agriculture. *Mutation Breeding Newsletter*, **42**, 3–6.

Wenzel, G. & Foroughi-Wehr, B. (1990). Progeny tests of barley, wheat, and potato regenerated from cell cultures after *in vitro* selection for disease resistance. *Theoretical and Applied Genetics*, **80**, 359–65.

Wersuhn, G. (1989). Obtaining mutants from cell culture. *Plant Breeding*, **102**, 1–9.

Wessler, S.R. & Varagona, M.J. (1985). Molecular basis of mutations at the waxy locus of maize: Correlation with the fine structure map. *Proceedings of the National Academy of Sciences of the USA*, **82**, 4177–81.

Westergaard, M. (1957). Chemical mutagenesis in relation to the concept of the gene. *Experientia*, **XIII**, 224–34.

Westergaard, M. (1960). Chemical mutagenesis as a tool in macromolecular genetics. In *Chemische Mutagenese* (Proceedings Erwin-Baur-Gedächtnisvorlesungen I, Gatersleben, 1959), ed. H. Stubbe, pp.30–44. Berlin: Akademie-Verlag.

Widholm, J.M. (1988). In vitro selection with plant cell and tissue cultures: An overview. *Iowa State Journal of Research*, **62**, 587–97.

Williams, N. (1995). Closing in on the complete yeast genome sequence. *Science*, **268**, 1560–1.

Wilson, G.B. (1950). Cytological effects of some antibiotics. *Journal of Heredity*, **41**, 227–31.

Winkler, H. (1907). Über Propfbastarde und pflanzliche Chimären. *Berichte der Deutschen Botanischen Gesellschaft*, **25**, 568–76.

Winkler, H. (1930). *Die Konversion der Gene*. Jena: Gustav Fischer Verlag.

Winkler, H. (1935). Chimären und Burdonen. Die Lösung des Propfbastardproblems. *Der Biologe*, **4**(9), 279–90.

Wisman, E. (1993). *Genetic (in)stability in tomato*. Ph.D. Thesis Wageningen Agricultural University.

Wisman, E., Ramanna, M.S. & Koornneef, M. (1993). Isolation of a new paramutagenic allele of the *sulfurea* locus in the tomato cultivar Moneymaker following *in vitro* culture. *Theoretical and Applied Genetics*, **87**, 289–94.

Wisman, E. & Ramanna, M.S. (1994). A reinvestigation of the instability at the *yv* locus in tomato. *Heredity*, **72**, 536–46.

Witkin, E.M. (1976). Ultraviolet mutagenesis and inducible DNA repair in *Escherichia coli*. *Bacteriological Reviews*, **40**, 869–907.

Witherspoon, W.D., Wernsman, E.A., Gooding, G.V. & Rufty, R.C. (1991). Characterization of a gametoclonal variant controlling virus resistance in tobacco. *Theoretical and Applied Genetics*, **81**, 1–5.

Wolff, F. (1909). Über Modifikationen und experimentell ausgelöste Mutationen von *Bacillus prodigiosus* und anderen Schizophyten. *Zeitschrift für induktive Abstammungs- und Vererbungslehre*, **2**, 90–132.

Wolff, S. & Luippold, H.E. (1956). The production of two chemically different types of chromosomal breaks by ionizing radiations. *Proceedings of the National Academy of Sciences of the USA*, **42**, 510–14.

Xu, Y.S., Murto, M., Dunckley, R., Jones, M.G.K. & Pehu, E. (1993). Production of asymmetric hybrids between *Solanum tuberosum* and irradiated *S. brevidens*. *Theoretical and Applied Genetics*, **85**, 729–34.

Yamaguchi, H., Tano, S. & Tatara, A. (1974). Mutations induced by 5-bromodeoxyuridine in presoaked seeds of barley. *Mutation Breeding Newsletter*, **4**, 5.

Yatou, O. & Amano, E. (1991). DNA structure of mutant genes in the waxy locus in rice. In *Plant Mutation Breeding for Crop Improvement* (Proceedings FAO/IAEA Symposium, Vienna, 1990), vol.1, pp.385–9. Vienna: IAEA.

Yonezawa, K. (1975). Method and efficiency of mutation breeding for quantitative characters. In *Efficiency of Breeding Methods for Screening of Mutants*. Gamma Field Symposia **14**, pp.39–56. Ohmiya-machi, Ibaraki-ken, Japan: Institute of Radiation Breeding NIAS MAF.

Yonezawa, K. & Yamagata, Y. (1975a). Practical merit of delayed selection after single and recurrent mutagenic treatments-II. Optimum generation for selection. *Radiation Botany*, **15**, 169–84.

Yonezawa, K. & Yamagata, H. (1975b). Comparison of the scoring methods for mutation frequency in self-pollinating disomic plants. *Radiation Botany*, **15**, 241–56.

Yonezawa, K. & Yamagata, Y. (1977). On the optimum mutation rate and optimum dose for practical mutation breeding. *Euphytica*, **26**, 413–26.

Yoshida, Y. (1962). Theoretical studies on the methodological procedures of radiation breeding. 1. New methods in autogamous plants following seed irradiation. *Euphytica*, **11**, 95–111.

Yu, C.K. & Dodson, E.O. (1961). Effects of centrifugal force upon chromosomal mutation in barley. *Genetics*, **46**, 1411–23.

Zagaja, S.W. (1976). Compact type mutants in apples and sour cherries. *Mutation Breeding Newsletter*, **8**, 8–9.

Zagaja, S.W. & Przybyla, A. (1973). Gamma-ray mutations in apples. In *Induced Mutations in Vegetatively Propagated Plants* (Proceedings FAO/IAEA Panel, Vienna, 1972), pp.35–40. Vienna: IAEA.

Zagaja, S.W. & Przybyla, A. (1976). Compact type mutants in apple and sour cherries. In *Improvement of Vegetatively Propagated Plants and Tree Crops through Induced Mutations* (Proceedings 2nd FAO/IAEA Research Coordination Meeting, Wageningen, Netherlands), IAEA-TECDOC-194, pp.171–84. Vienna: IAEA.

Zakri, A.H. (ed) (1988). *Plant Breeding and Genetic Engineering* (Proceedings SABRAO Int.Symposium and Workshop, 1987, Kuala Lumpur, Malaysia). Kuala Lumpur, Malaysia: City Reprographic Services.

Zamecnik, M.V. & Zamecnik, P.C. (1967). Mutation of chloroplasts in *Ageratum* following treatment with 5-bromo-deoxyuridine. *Experimental Cell Research*, **45**, 218–29.

Zanone, L. (1965). Effect of mutagenic agents in *Vicia faba* L. Comparison between effects of ethyl methane sulphonate, ethylene imine and X-rays on the induction of chlorophyll mutations. In *The Use of Induced Mutations in Plant Breeding* (Report FAO/IAEA Technical Meeting, Rome, 1964), pp.205–13. Oxford: Pergamon Press.

Zehr, B.E., Williams, M.E., Duncan, D.R. & Widholm, J.M. (1987). Somaclonal variation in the progeny of plants regenerated from callus cultures of seven inbred lines of maize. *Canadian Journal of Botany*, **65**, 491–9.

Zeven, A.C. & de Wet, J.M.J. (1982). *Dictionary of Cultivated Plants and Their Regions of Diversity*, p.23. Wageningen, The Netherlands: Centre for Agricultural Publishing and Documentation (PUDOC).

Zhang, Y.X. & Lespinasse, Y. (1991). Pollination with gamma-irradiated pollen and development of fruits, seeds and parthenogenetic plants in apple. *Euphytica*, **54**, 101–9.

Zimmer, K.G. (1961). *Studies on Quantitative Radiation Biology*. Edinburgh-London: Oliver & Boyd.

Zoz, N.N. (1966). Chemical mutagenesis in higher plants. In *Supermutagenes*, ed. I.A. Rapoport et al., pp.93–104. Moscow: Publishing House 'Nauka' (in Russian with English summary).

Zoz, N.N. (1967). Chemical mutagenesis in higher plants and new, highly active mutagens. In *Mechanism of Mutation and Inducing Factors* (Proceedings of G. Mendel Memorial Symposium on the Mutational Process, Prague, 1965), ed. Z. Landa, pp.317–20. Prague: Publishing House of the Czechoslovak Academy of Sciences.

Zwintzscher, M. (1955). Die Auslösung von Mutationen als Methode der Obstzüchtung.1. Die Isolierung von Mutanten in Anlehnung an primäre Veränderungen. *Der Züchter*, **25**, 290–302.

Zwintzscher, M. (1959). Die Auslösung von Mutationen als Methode der Obstzüchtung. In "*Eucarpia*" (Proceedings, Second Congres of the European Association for Research on Plant Breeding, Köln, Germany), pp.202–11.

Zwintzscher, M. (1962). Methoden zur Isolierung induzierter Mutanten der Baumobstgehölze.*Mitteilungen Obst und Garten, Klosterneuburg, Österreich*, Ser. B, 12B, 125–34.

Zwintzscher, M. (1977). Examples for induced variation by X- and γ-rays in Cox Orange Pippin and other apple varieties. *Acta Horticulturae*, **75**, 35–41.

Index

Abelmoschus esculentus (okra), 231
Abutilon, 270
accelerators *see* radiation types (beta radiation)
accessory buds, 260, 293
Acanthus, 270
Acer, 270, 291
Achimenes, 272, 276
acridines *see* chemical mutagens
acute irradiation *see* mutagenic treatments (irradiation)
additions *see* chromosome mutations
adenine *see* DNA structure
adventitious buds, 118, 260, 281, 289
  single-cell origin, 272–3, 289
  bud techniques, 198, 260, 269–74, 281, 301
adventitious meristems *see* adventitious buds
adventitious shoot formation *see* adventitious buds
Aegilops umbellulata, 60, 80, 207
Agave, 270
ageing of seed *see* seed ageing
Ageratum corymbosum, 155
African violet *see* *Saintpaulia ionantha*
Agrobacterium tumefaciens, 36–7, 169
Agropyron elongatum, 80
alkaloid content *see* mutant traits
alkylating agents *see* chemical mutagens
Allium ampeloprasum, 220
Allium cepa (onion), 88, 155
Allium sativum (garlic), 156, 171, 255
allo(poly)ploidy *see* polyploidy
almond *see* *Prunus amygdalis*
alpha (α-)radiation *see* radiation types
Alstroemeria, 21, 255, 272, 273
Alternaria alternata (black spot disease) resistance in pear, 249, 296
Alternaria resistance in tomato, 86
amphidiploid *see* polyploidy
amylose-free (*amf-*) potatoes, 300–2

see also mutant traits (amylose-free)
Ananas comosus (ananas), 280
androgenesis, 82
  see also anther culture, haploids
aneuploidy *see* mutation types
anther culture, 82
Anthirrinum majus (snapdragon), 46, 47, 52, 74–5, 87, 89, 90, 96, 153, 193
antibiotics *see* chemical mutagens
antisense (RNA) method *see* biotechnology
apical buds *see* apical meristems
apical dominance, 260
apical meristems, 212–17, 257, 259
  GECN (number of genetically effective cells), 208–10
  germ layers (germ line, histogenic layers), 208, 209, 258, 288
  initial cells, 27, 208–9, 212–17, 222, 257–8, 260, 266
  see also shoot apices
Apium graveolens (celery), 174, 179
apomictic crops, apomixis, 255, 299
apple *see* *Malus × domestica*
apricot *see* *Prunus armeniaca*
Arabidopsis thaliana (mouse ear cress), 11, 30, 66, 88, 97, 98, 100, 209, 211, 227
Arachis hypogaea (groundnut, peanut), 96, 97, 156
Ascochyta-blight resistance in chickpea, 24, 247
Aspergillus niger, 76
'Asseyeva method', 268
Aucuba, 270
Auerbach, C., 54, 57
autonomous elements *see* transposons
autopolyploidy *see* polyploidy
autumn crocus *see* *Colchicum autumnale*
Avena barbata, 80
Avena sativa (oats), 20, 80, 97, 161, 248
axillary buds, 216, 259–60, 280, 289, 292

see also apical meristems
Azalea, 284
azides *see* chemical mutagens (sodium azide)

Bacillus prodigiosus, 46
back-mutations *see* mutation types
back-sporting, 265
  see also chimera formation
bacterial leaf blight in rice *see* *Xanthomonas campestris*
Bahiagrass *see* *Paspalum notatum*
banana, plantain *see* *Musa* spp.
barley *see* *Hordeum vulgare*
barley yellow mosaic virus (BYMV) resistance in barley, 248
base analogues *see* chemical mutagens
base pair *see* DNA structure
base pair substitution *see* mutation types
Bateson test, 261
beans *see* *Phaseolus vulgaris*
Begonia masoniana, 45, 196, 197, 276
Begonia rex, 196, 197, 252, 284
Bermudagrass (Bermuda turfgrass) *see* *Cynodon*
Beta patellaris, 80
Beta procumbens, 80
Beta vulgaris (sugarbeet), 80, 155, 156
beta (β-) rardiation *see* radiation types
biomagnetism *see* radiation types ((electro)magnetic radiation),
biotechnology, 34–9, 164–5
  antisense (RNA) method, 35, 37, 77, 185, 302
  chromosome fragmentation, 135–6
  chromosome transplantation, 35–6
  cybrids (cytoplasmic hybrids), 36, 135–6, 164
  DNA fragmentation, 135, 165
  electrofusion, 36, 135–6
  electroporation, 36
  gene tagging, 73